The Routledge Handbook of Place

The handbook presents a compendium of the diverse and growing approaches to place from leading authors as well as less widely known scholars, providing a comprehensive yet cutting-edge overview of theories, concepts and creative engagements with place that resonate with contemporary concerns and debates.

The volume moves away from purely western-based conceptions and discussions about place to include perspectives from across the world. It includes an introductory chapter, which outlines key definitions, draws out influential historical and contemporary approaches to the theorisation of place and sketches out the structure of the book, explaining the logic of the seven clearly themed sections. Each section begins with a short introductory essay that provides identifying key ideas and contextualises the essays that follow. The original and distinctive contributions from both new and leading authorities from across the discipline provide a wide, rich and comprehensive collection that chimes with current critical thinking in geography. The book captures the dynamism and multiplicity of current geographical thinking about place by including both state-of-the-art, in-depth, critical overviews of theoretical approaches to place and new explorations and cases that chart a framework for future research. It charts the multiple ways in which place might be conceived, situated and practised.

This unique, comprehensive and rich collection will be an essential resource for undergraduate and graduate teaching, for experienced academics across a wide range of disciplines and for policymakers and place-marketers. It will provide an invaluable and up-to-date guide to current thinking across the range of disciplines, such as Geography, Sociology and Politics, and interdisciplinary fields such as Urban Studies, Environmental Studies and Planning.

Tim Edensor is Professor of Human Geography, Manchester Metropolitan University, UK, and a Principal Research Fellow, School of Geography, University of Melbourne, Australia.

Ares Kalandides is a practicing urban and regional planner based in Berlin, Germany, and Athens, Greece. He is Professor of Place Management at Manchester Metropolitan University, UK, and Director of the Institute of Place Management.

Uma Kothari is Professor of Migration and Postcolonial Studies, University of Manchester, UK, and Professor of Human Geography, University of Melbourne, Australia.

The Routledge Handbook of Place

Edited by Tim Edensor, Ares Kalandides and Uma Kothari

LONDON AND NEW YORK

First published 2020
by Routledge
4 Park Square, Milton Park, Abingdon, Oxon OX14 4RN
605 Third Avenue, New York, NY 10017

First issued in paperback 2023

Routledge is an imprint of the Taylor & Francis Group, an informa business

British Library Cataloguing-in-Publication Data
A catalogue record for this book is available from the British Library

Library of Congress Cataloging-in-Publication Data
Names: Edensor, Tim, 1957- editor. | Kalandides, Ares, editor. | Kothari, Uma, editor.
Title: The Routledge handbook of place / edited by Tim Edensor, Ares Kalandides and Uma Kothari.
Other titles: Handbook of place
Description: Abingdon, Oxon ; New York, NY : Routledge, 2020. |
Includes bibliographical references and index.
Identifiers: LCCN 2019055753 | ISBN 9781138320499 (hardback) |
ISBN 9780429453267 (ebook)
Subjects: LCSH: Place (Philosophy)
Classification: LCC B105.P53 R68 2020 |
DDC 304.2/3–dc23
LC record available at https://lccn.loc.gov/2019055753

ISBN: 978-1-03-257047-1 (pbk)
ISBN: 978-1-138-32049-9 (hbk)
ISBN: 978-0-429-45326-7 (ebk)

DOI: 10.4324/9780429453267

Typeset in Bembo
by Swales & Willis, Exeter, Devon, UK

Publisher's Note
The publisher has gone to great lengths to ensure the quality of this reprint but points out that some imperfections in the original copies may be apparent.

Contents

Figures

Maps

Tables

Contributors

Justin Armstrong is a writer and anthropologist who studies ghost towns and remote islands. He teaches writing and anthropology at Wellesley College in Massachusetts, USA. He has done fieldwork in the rural High Plains of the US, northern Saskatchewan, Newfoundland, the Faroe Islands and Iceland. He holds a PhD in Cultural Anthropology from McMaster University in Hamilton, Canada. Born and raised in northern Canada, he now lives in Boston with his wife and their dogs.

Bradley Austin is an Assistant Professor of Geography at Anne Arundel Community College in Arnold, MD, USA. His research revolves around applying critical and intersectional approaches to understanding human–environment interactions and how these perspectives can inform an equitable praxis of geographic education.

David Beel is Senior Lecturer in Political Economy at Manchester Metropolitan University, Manchester, UK, based within the Future Economies team. His primary research focus is on economic geography and the processes of city-region building. He is currently carrying out research on 'Spaces of New Localism' as part of the WISERD Civil Society Research Centre. Previously, he was a Senior Researcher at the Centre for City Region Dynamics and the Department of Geography, Staffordshire University, Stoke-on-Trent, UK.

Beau B. Beza is a Registered Landscape Architect, with a Bachelor of Science in Landscape Architecture (UC, Davis), a Master of Urban Planning (University of Melbourne) and a PhD (Royal Melbourne Institute of Technology). He is Associate Head of School, Teaching & Learning in the School of Architecture and Built Environment – Deakin University, Geelong, Australia. He has more than 15 years' experience in teaching/research and has widely published works in urban space production.

David Bissell is Associate Professor and Australian Research Council Future Fellow in the School of Geography at the University of Melbourne, Melbourne, Australia. He combines qualitative research on embodied practices with social theory to explore mobile lives. He is author of *Transit Life: How Commuting is Transforming Our Cities*, co-editor of *Stillness in a Mobile World* and the *Routledge Handbook of Mobilities*, and managing editor of *Social & Cultural Geography*.

Quintin Bradley is a Senior Lecturer in Planning and Housing, Leeds Beckett University, Leeds, UK, and leads a research programme into neighbourhood and community planning with a particular focus on public engagement in questions of housing supply and housing allocations. His research interests include popular contention in planning and housing and the role of social movements in urban policy. He is co-editor of *Localism and Neighbourhood Planning: Power to the People?* (Policy Press, 2017) and author of *The Tenants' Movement* (Routledge, 2015).

Cathrine Brun is a human geographer, professor and Director of the Centre for Development and Emergency Practice (CENDEP) at Oxford Brookes University, Oxford, UK. Her research interests, some of which are introduced in a recent paper with Anita Fábos 'Mobilising home for long term displacement: a reflection on the durable solutions' (*Journal of Human Rights Practice*, 2017), concern forced migration and conflict, housing and home; theory, ethics and practice of humanitarianism.

Jon Coaffee is Professor of Urban Geography based in the School of Politics and International Studies at the University of Warwick, Coventry, UK, and an exchange professor at New York University's Center for Urban Science and Progress. He is a recognised international expert in counter-terrorism, urban security and resilience. His research focusses upon the interplay of physical, technical and socio-political aspects of urban vulnerability.

Mike Collier, Professor of Visual Art at the University of Sunderland, is a lecturer, writer, curator and artist. He co-founded WALK (Walking, Art, Landscape and Knowledge), a research centre at the University of Sunderland, Sunderland, UK, which looks at the way we creatively engage with the world as we walk through it. Much of his work is place-specific, exploring our relationship to a 'more than human' world, paying close attention to the environment through which he walks. For more information see www.mike.collier.eu and www.walk.uk.net.

David Cooper is a Senior Lecturer in English at Manchester Metropolitan University, Manchester, UK, where he co-directs (with Rachel Lichtenstein) the Centre for Place Writing. A founding co-editor of the interdisciplinary journal *Literary Geographies*, he has published widely on contemporary British place writing. He has also published extensively on digital literary mapping and is currently Co-Investigator on 'Chronotopic Cartographies' (Arts and Humanities Research Council). In addition, David is increasingly experimenting with creative-critical approaches to place.

Cara Courage is a placemaking, and arts, activism and museums academic and practitioner, and Head of Tate Exchange (Tate, London, UK), Tate's platform dedicated to socially engaged art. Cara is author of *Arts in Place: The Arts, the Urban and Social Practice* (Routledge, 2017), the co-editor of *Creative Placemaking and Beyond* (Routledge, 2018) and editor of the forthcoming *Placemaking Handbook* (Routledge, 2020).

Patrick Devine-Wright is an environmental social scientist with expertise spanning human geography and environmental psychology. His primary interests are place attachment and community engagement with energy transitions. He sits on the editorial boards of journals including *Global Environmental Change*, *Energy Research and Social Science* and the *Journal of Environment Psychology* and is Lead Author for the IPCC Working Group 3 on Climate Mitigation.

Kim Dovey is Professor of Architecture and Urban Design at the University of Melbourne, Melbourne, Australia, where he is also Director of InfUr- the Informal Urbanism Research Hub. Kim has published and broadcast widely on social issues in architecture, urban design and planning. Books include *Framing Places*, *Fluid City*, *Becoming Places* and *Urban Design Thinking*. He leads research projects on urban morphology and informal settlement.

Hélène B. Ducros holds a Juris Doctor and PhD in Geography from the University of North Carolina. Her research has examined placemaking and attachment to place, landscape perception, the rural–urban nexus, and modes of rural heritage preservation. She is an editor for *EuropeNow* at the Council for European Studies (Columbia University, NY, USA), where she leads the Research and Pedagogy Committees and is founding Chair of the Critical European Studies Research Network.

Michelle Duffy is Associate Professor in Human Geography, Centre for Urban and Regional Studies, University of Newcastle, Australia. Her research explores the role of art and sound practices in creating and/or challenging notions of identity and belonging in public spaces and public events; the significance of emotion and affect in creating notions of belonging and exclusion; and explores the body as a means of embodied, emotional and affective communication.

Nick Dunn is Executive Director of Imagination, an open and exploratory design research lab at Lancaster University, Lancaster, UK, where he is Professor of Urban Design. He is Senior Fellow of the Institute for Social Futures, leading research on future cities. A keen nightwalker, his book, *Dark Matters: A Manifesto for the Nocturnal City* (Zero, 2016), is an exploration of the importance of nocturnal urban environments in the face of late capitalism.

Marita Dyson is a multidisciplinary artist working in songwriting, performance and visual art, and is co-founder of The Orbweavers with Stuart Flanagan. The Orbweavers have released three studio albums, created soundtracks for film, television and radio, and were 2016 State Library of Victoria Creative Fellows researching Melbourne waterways. Marita has exhibited at National Gallery of Victoria and Immigration Museum. Her song maps are held in the State Library of Victoria collection.

Tim Edensor is Professor of Human Geography, Manchester Metropolitan University, Manchester, UK, and a Principal Research Fellow, School of Geography, University of Melbourne, Melbourne, Australia. Tim is the author of *Tourists at the Taj* (1998), *National Identity, Popular Culture and Everyday Life* (2002), *Industrial Ruins: Space, Aesthetics and Materiality* (2005), *From Light to Dark: Daylight, Illumination and Gloom* (2017) and *Stone: Stories of Urban Materiality* (2020). He is also the editor of *Geographies of Rhythm* (2010) and co-editor of *Into the Dark: Reappraising the Meaning and Experience of Gloom* (2020).

Cristina Garduño Freeman is a Research Fellow at the University of Melbourne, Melbourne, Australia. Her focus is on people's connections with places within architectural history, critical heritage and digital humanities. Her first book, *Participatory Culture and the Social Value of an Architectural Icon: Sydney Opera House*, was published by Routledge in 2018.

Costis Hadjimichalis is Professor Emeritus and teaches graduate courses at the Geography Department, Harokopio University, Athens, Greece. Among his recent books are: *Space in*

Left Thought (co-author Dina Vaiou, 2012, Nisos, in Greek), *Schuldenkrise und Landraub in Griechenland*, Westfälishes Dampfboot, 2016, *Crisis Spaces: Structures, Struggles and Solidarity in Southern Europe*, Routledge, 2017 and *Landscapes of Crisis: Paths of Uneven Geographical Development*, Alexandria, 2018 (in Greek).

Jaime Hernández-Garcia is a Colombian Architect, with a PhD in Architecture and Planning from the UK. He is Professor at the Aesthetics Department of School of Architecture and Design, Pontificia Universidad Javeriana in Bogotá, Colombia. He has more than 20 years' experience in teaching and research, particularly on informal settlements. He has widely published in English and Spanish, focussing on informal settlements/housing, open/public space, placemaking/place branding, environment/sustainability and community participation.

Ysanne Holt is Professor of Art History and Visual Culture, University of Northumbria, Newcastle upon Tyne, UK. Her current research is on the experience and representations of the UK north, focussing on identities of marginal or 'at edge' sites such as borders and island locations. She recently co-edited *Visual Culture in the Northern UK Archipelago: Imagining Islands* (Routledge, 2018) and published a chapter on interwar art, crafts and design in *Rural Modernity in Britain: A Critical Intervention* (Edinburgh University Press, 2018).

Ray Hudson is Professor Emeritus at Durham University, Durham, UK, and as a radical geographer is a specialist in development, economic and political geography. He has published extensively on these issues, as author or editor of over 20 books and the author or co-author of over 100 journal articles and book chapters. He was head of the Department of Geography from 1992 to 1997. From 2003 to 2007, he was Director of the Wolfson Research Institute and in 2012 he was appointed Deputy Vice-Chancellor of the University, becoming the second most senior academic. He is currently a visiting professor at University College Dublin and an honorary professor at Cardiff University.

Mark Jayne is Professor of Human Geography at Sun Yat-sen University, P.R. China. Mark is a social and cultural geographer who has published 80 journal articles, book chapters and official reports, undertaking research in the UK, Ireland, Slovakia, The Netherlands, New Zealand, Australia, USA and China. Amongst his books are *Cities and Consumption*, *Urban Theory Beyond the West: A World of Cities*, *Urban Theory: New Critical Perspectives* and *Chinese Urbanism: Critical Perspectives*.

Cuttaleeya Jiraprasertkun is an Assistant Professor at the Faculty of Architecture, Kasetsart University, Thailand. Her research interests encompass both theoretical and methodological concerns of how to read and interpret space and place. She authored several book chapters published by Routledge, Springer VS and NUS Press and articles in a number of local and international journals. Her current works focus on community public space in various urban contexts of Thailand.

Martin Jones is Deputy Vice-Chancellor and Professor of Human Geography, Staffordshire University, Stoke-on-Trent, UK. He works on the interface between economic and political geography. He is the co-editor of the journal *Territory, Politics, Governance*. Martin has conducted empirical work with local community groups and public policy officials to seek to improve the life chances of disadvantaged people. Current ESRC-funded research examines the geographies of state intervention and civil society through localism and city-region building.

Rhys Jones is a Professor of Political Geography at Aberystwyth University and a former Head of the Department of Geography and Earth Sciences. He has written extensively about the geographies of the state and nationalism and has written a book – *Placing the Nation* (University of Wales Press, 2008) – which examines the significance of the university town of Aberystwyth for the reproduction of Welsh nationalism.

Maarja Kaaristo is a Research Associate in Tourism Mobilities at Manchester Metropolitan University, Manchester, UK. Her research interests include inland waterways, rural tourism, mobilities, materialities, placemaking, ethnographical methods and the history of European ethnology.

Ares Kalandides is a practicing urban and regional planner based in Berlin, Germany, and Athens, Greece. He is Professor of Place Management at Manchester Metropolitan University, Manchester, UK, and Director of the Institute of Place Management. He is currently working on issues of citizen participation in urban development.

Evgenia (Jenny) Kanellopoulou is a Senior Lecturer in Law at Manchester Law School, Manchester Metropolitan University, Manchester, UK, and holds a PhD in Law from the University of Edinburgh. She has been teaching and researching in law since 2012 and is currently leading a British Academy-funded project on the regulation of urban squats, influenced by theories developed for the digital environment. Her main interest is legal interdisciplinarity from a variety of perspectives.

Uma Kothari is Professor of Migration and Postcolonial Studies, University of Manchester, Manchester, UK, and Professor of Human Geography, University of Melbourne, Melbourne, Australia. Her research interests include colonialism and humanitarianism, mobilities and borders and cultural geographies of seafarers. She is Vice President of the European Association of Development Institutes, on the advisory board of In Place of War, a Fellow of the UK Academy of Social Sciences and a recipient of the Royal Geographical Society's Busk Medal.

Bastian Lange teaches at the University of Leipzig, Germany. He taught as guest professor (2011–2012) at Humboldt University Berlin (2011–2012) and University of Vechta, Germany (2018–2019). In 2008 he founded the independent research and consultancy office Multiplicities in Berlin, which supports politics, business and creative ways to sustainable urban regions in the European context. Multiplicities advises municipalities, cities, federal states and EU programmes on the development of innovative places with collaborative participation processes.

James Lesh is a Research Fellow at the University of Melbourne. He researches heritage conservation and urban history. He has published widely on urban heritage and is preparing a monograph on evolutions in urban conservation in twentieth-century Australia.

Michele Lobo explores race, encounter and co-belonging in the Anthropocene. She serves as Editor of *Social & Cultural Geography*, Book Review Editor of *Postcolonial Studies Journal* and Convenor of the Institute of Australian Geographers (Cultural Geography Study Group). She has recently published in *Economic and Political Weekly* (2019), *Geoforum* (2019), *Area*

(2019), *Geohumanities* (2019) and *Urban Studies* (2018). She is a Lecturer in Geography, School of Humanities and Social Sciences, Deakin University, Melbourne, Australia.

Dawn Lyon is Reader in Sociology in the School of Social Policy, Sociology and Social Research at the University of Kent, Canterbury, UK. She has published in the fields of the sociology of work, time, gender, migration, community, and visual and sensory sociology. She is particularly interested in the rhythms of work and working lives. Her book, *What is Rhythmanalysis?*, was published by Bloomsbury Academic in 2018.

Ali Madanipour is Professor of Urban Design and a founding member of the Global Urban Research Unit (GURU), School of Architecture, Planning and Landscape, Newcastle University, Newcastle upon Tyne, UK. His research has focussed on the social and theoretical dimensions of the planning, design, development and management of cities. His latest books include *Cities in Time: Temporary Urbanism and the Future of the City* (Bloomsbury, 2017) and the *Handbook of Planning Theory* (Routledge, 2018).

Judith Mair is Associate Professor and Leader of the Tourism Discipline Group, Business School, University of Queensland, Australia. Her research interests include the impacts of tourism and events on community and society; consumer behaviour in tourism and events; the relationship between events and climate change; and business and major events. She is the author and/or editor of four books, *The Routledge Handbook of Events, Conferences and Conventions: A Research Perspective*; *Events and Sustainability*; and *Festival Encounters*, as well as over 40 academic papers.

Deborah Martin is Professor at the Graduate School of Geography at Clark University in Worcester, Massachusetts, USA. Her research examines place identity, urban politics, governance and legal geographies. She has published in journals such as *Annals of the Association of American Geographers, Antipode, Environment and Planning A, International Journal of Urban and Regional Research, Progress in Human Geography, Transactions of the Institute of British Geographers, Urban Geography* and *Urban Studies*.

Dominic Medway is Professor of Marketing in the Institute of Place Management at Manchester Metropolitan University, Manchester, UK. Dominic's work is primarily concerned with the complex interactions between places, spaces and those who manage and consume them, reflecting his academic training as a geographer. He has published extensively in a variety of leading academic journals, including *Environment and Planning A, Tourism Management, Journal of Environmental Psychology, Urban Geography, Cities, European Journal of Marketing* and *Marketing Theory*.

Glenda Mejía is a Senior Lecturer in Global and Languages Studies with the School of Global, Urban and Social Studies at RMIT University, Melbourne, Australia. Some of her research interests cover migration (belonging, placemaking, language and identity). Her latest article in migration studies is (2018) 'A sense of belonging: social media use of Latin American migrants in Australia' in *Transnational Migrations in the Asia-Pacific. Transformative Experiences in the Age of Digital Media* (Rowman and Littlefield International, 2018).

Helen Mort is five-times winner of the Foyle Young Poets of the Year Competition and her work has been shortlisted for the Costa Prize and the T.S. Eliot Prize. Her poetry collections *Division Street* and *No Map Could Show Them* are published by Chatto & Windus. Helen also writes creative non-fiction and has been published in the *Independent* and in

Mount London, an anthology of ascents in the vertical city. She has also edited *Waymaking*, a book of adventure-inspired art and literature by women. She is a Lecturer in Creative Writing at Manchester Metropolitan University, Manchester, UK, and is a Fellow of the Royal Society of Literature.

Nikos Ntounis is a senior research associate at the Institute of Place Management, Manchester Metropolitan University, Manchester, UK. His PhD thesis explored the theoretical development of place management and its establishment as an interdisciplinary field. His current research is focussed on advancing the theory and practice of various aspects of place management and urban geography, including: place branding, place governance, urban autonomy and alternative spaces. Results of his work have been published in academic journals, including *Environment and Planning A: Economy and Space* and the *Journal of Place Management and Development*.

Paul O'Hare is a Senior Lecturer in Geography and Development at Manchester Metropolitan University, Manchester, UK. His research interest revolves around efforts to engage 'the public' in governance and planning decision-making processes, and in the contribution that civil society organisations may make to policy formulation and implementation.

Catherine Parker is Professor of Retail and Marketing Enterprise at Manchester Metropolitan University, Manchester, UK, and Co-Chair of the Institute of Place Management, the professional institute for all those involved in making better places. Cathy is regarded as an international expert and leader in place management and is one of the co-authors of the recent UK Government-commissioned report 'High Street 2030: Achieving Change'. She is frequently asked to commentate on place- and retail-related topics and has recently appeared on *BBC Breakfast*, *BBC Sunday Politics* (NW), BBC Radio 5 live and *ITV Tonight*.

David Paton is a lecturer, artist-researcher and craftsperson with a specialism in Cornish granite. Since competing his PhD in cultural geography (2015) he has carried out a series of research projects under the title 'Tracing Granite'. His practice is a synthesis of sculpture and quarrying, performance, film, auto-ethnographic and collaborative field-work that jointly examine the relationship between place, making, material and body.

Mike Pearson is Emeritus Professor of Performance Studies, Aberystwyth University, Aberystwyth, Wales, UK. He is co-author with Michael Shanks of *Theatre/Archaeology* (2001) and author of *In Comes I: Performance, Memory and Landscape* (2006); *Site-Specific Performance* (2010); *Mickery Theater: An Imperfect Archaeology* (2011); and *Marking Time: Performance, Archaeology and the City* (2013). He creates theatre as a solo artist; with Good News From The Future; and for National Theatre Wales including *The Persians* (2010), *Coriolan/us* (2012) and *Iliad* (2015).

Luis Eduardo Perez Murcia holds a PhD in Development Policy and Management from the University of Manchester and is currently a post-doctoral researcher at the HOMInG Project, University of Trento, Trento, Italy. His research interests include displacement and home, refugees and transnational migration. Recent publications include '"The sweet memories of home have gone": displaced people searching for home in a liminal space' (*Journal of Ethnic and Migration Studies*, 2018) and 'Where the heart is and where it hurts: conceptions of home for people fleeing conflict' (*Refugee Survey Quarterly*, 2019).

Chris Perkins is Reader in Geography at the University of Manchester, Manchester, UK. His interests lie at the interface between mapping technologies and social and cultural practices, with on-going research into performative aspects of contemporary mapping behaviour and play, alongside an emerging interest in island studies. His most recent book is *Time for Mapping: Cartographic Temporalities* (2018).

James Petty is an Honorary Fellow in Criminology at the University of Melbourne, Melbourne, Australia. His academic work focusses on homelessness and housing, spatial regulation, urban environments and welfare. James works in the community sector in the area of alcohol and other drugs, focussing on evidence-based policy and reform.

Catherine Phillips is Senior Lecturer in the School of Geography at the University of Melbourne, Melbourne, Australia. She combines research on everyday practices with social theory to explore human–environment relations and their implications for environmental governance. Her recent work focusses on food and agriculture, waste and discard politics, and urban natures.

Tracey Potts is Assistant Professor in Critical Theory and Cultural Studies at the University of Nottingham, Nottingham, UK. Her research is concentrated in the areas of material culture, aesthetics and everyday life with an especial focus on the conjunction of taste, class, space and affect. She is the co-author of *Kitsch! Cultural Politics and Taste* (with Ruth Holliday) and is currently working on a monograph on clutter and procrastination entitled 'Neither Use Nor Ornament'.

Pravin S. Rana is Assistant Professor in Tourism Management at Banaras Hindu University, Varanasi, India, and has published two dozen papers on the issues of cultural tourism and pilgrimages, and has published books including *Tourism Geography* (2006) and *Pilgrimage Tourism* (2014). He is the Secretary of ACLAI – the Asian Cultural Landscape Association of India (Varanasi). He has presented papers in Norway, Canada and Singapore.

Les Roberts is a Reader in Cultural and Media Studies at the University of Liverpool, Liverpool, UK. The core focus of his work is situated within the interdisciplinary fields of spatial anthropology and spatial humanities. His publications include the monographs *Film, Mobility and Urban Space* (2012) and *Spatial Anthropology: Excursions in Liminal Space* (2018), and the edited collections *Deep Mapping* (2016) and *Locating the Moving Image: New Approaches to Film and Place* (2014).

Sarah Robertson is a Research Fellow in the Centre for Urban Research at RMIT University in Melbourne, Australia. Her interdisciplinary research emerges from cultural and urban geographies and examines human–nonhuman relations and sustainable interventions at the residential scale.

Ben Rogaly is Professor of Human Geography at the University of Sussex, Brighton, UK. His latest book *Stories From a Migrant City: Living and Working Together in the Shadow of Brexit* was published by Manchester University Press in 2020.

Henrik Schultze is a post-doctoral scholar at the department of Urban and Regional Studies at Humboldt-University Berlin, Germany. He is currently working in the Collaborative Research Centre 'Re-Figuration of Spaces', Subproject 'The World Down My Street:

Resources and Networks Used by City Dwellers' funded by the DFG. He completed his PhD on the role of place in respect to social constructions of belonging in 2017.

Rosemary Shirley is Associate Professor in Museum Studies at the University of Leicester, Leicester, UK. She has published widely on aspects of rural modernity including her monograph *Rural Modernity, Everyday Life and Visual Culture* (Routledge, 2015) and in the edited collection *Rural Modernity: A Critical Intervention* (EUP, 2018). She co-curated the major critical landscape exhibition 'Creating the Countryside' (2017) at Compton Verney Gallery and edited the accompanying publication of the same name.

Rana P.B. Singh has been Professor of Cultural Landscapes & Heritage Studies at Banaras Hindu University, Varanasi, India, and is the President of ACLA – the Asian Cultural Landscape Association (SNU Seoul, Korea). He has published extensively on the issues of sacred landscapes, heritage and pilgrimages based on his studies in north India and has also carried out field studies in Japan, Sweden, Italy and Korea.

Rob St. John is an artist and writer. Working primarily with sound, film and photography, his practice is attentive to entanglements of nature and culture in contemporary landscapes and is often based on slow, sustained periods of fieldwork.

Shanti Sumartojo is Associate Professor of Design Research at Monash University's Faculty of Art, Design and Architecture in Melbourne, Australia. She is also a member of the Emerging Technologies Research Lab at Monash University, where she leads the 'Future Shared Environments' research theme. Using ethnographic methodologies, she investigates how people experience their spatial surroundings, including both material and immaterial aspects, with a particular focus on the built environment, design and technology.

Dina Vaiou is Professor Emeritus of Urban Analysis and Gender Studies in the Department of Urban and Regional Planning of the National Technical University of Athens (NTUA), Greece. Her research interests include: the feminist critique of urban analysis; the changing features of local labour markets, with special emphasis on women's work and informalisation processes; the impact of mass migration on southern European cities and women's migration in particular; the gendered effects of the crisis and local solidarity responses. She has coordinated several research projects and has published extensively in Greece and abroad.

Ellen van Holstein is Research Fellow in Urban Geography at the University of Melbourne, Melbourne, Australia. Her research focusses on opportunities for participation and inclusion in urban community life and how these might be promoted or hampered by the technologies, policies and everyday practices that shape community groups. She is currently working on the study 'The Disability Inclusive City' where she analyses these questions in relation to people with intellectual disability.

Phillip Vannini is a Professor in the School of Communication & Culture and Canada Research in Public Ethnography at Royal Roads University in Victoria, BC, Canada. He is the author/editor of over a dozen books including his weather-related ethnography of people living with renewable resources in Canada, *Off the Grid* (published by Routledge in 2013).

Alexander Vasudevan is Associate Professor in Human Geography at the School of Geography and the Environment at University of Oxford, Oxford, UK. Alex's work explores the city as a site of political contestation. He is the author of *The Autonomous City: A History of Urban Squatting* (Verso, 2017), *Metropolitan Preoccupations: The Spatial Politics of Squatting in Berlin* (Wiley-Blackwell, 2015) and co-editor of *Geographies of Forced Evictions: Dispossession, Violence, Insecurity* (Palgrave, 2017).

Veronica Vickery is an artist, cultural geographer and an Honorary Research Associate at the University of Exeter, Exeter, UK. Her practice-based research interrogates the experience of alienation in the face of the often-violent geo-politics of landscaping processes and the socio-politics of contemporary exclusions. Worked through immersed and often long-term sited enquiry, these interests are materialised in solo performance, painting/installation, text, online projects and social interventions.

Gary Warnaby is Professor of Retailing and Marketing based in the Institute of Place Management at Manchester Metropolitan University, Manchester, UK. His research interests focus on the marketing of places (particularly in an urban context) and retailing. Results of this research have been published in various academic journals in both the management and geography disciplines.

Saskia Warren is a Cultural and Social Geographer and Senior Lecturer in Human Geography at the University of Manchester, Manchester, UK. She currently holds an Arts and Humanities Research Council Leadership Fellowship, 2017–2020. The programme of research investigates the interplay of religious faith and gender in the sub-sectors of visual arts, fashion and digital media. Saskia has published widely, including in leading international journals in Geography and cognate disciplines.

Paul Watt is Professor of Urban Studies at the Department of Geography, Birkbeck, University of London, London, UK. He has published widely on neighbourhoods, housing and urban regeneration. Books include *Social Housing and Urban Renewal: A Cross-National Perspective* (Emerald, 2017) and *London 2012 and the Post-Olympics City: A Hollow Legacy?* (Palgrave Macmillan, 2017). The following monograph will be published in 2020: 'Estate Regeneration and its Discontents: Public Housing, Place and Inequality in London' (Policy Press).

Iain White is Professor of Environmental Planning at the University of Waikato, Hamilton, New Zealand. Iain has authored or co-authored *Water and the City* (Routledge, 2010), *Environmental Planning in Context* (Palgrave, 2015), *The Routledge Companion to Environmental Planning* (Routledge, 2019) and *Why Plan? Theory for Practitioners* (Lund Humphries, 2019).

Ilan Wiesel is Senior Lecturer in Geography at the University of Melbourne, Melbourne, Australia. His research focusses on social and economic inequalities in cities that are caused or exacerbated by disability, income and wealth, or by the unequal distribution of public urban infrastructures such as affordable housing. He currently leads the ARC Discovery-funded study titled 'The Disability Inclusive City', which examines the inclusivity of mainstream urban services.

Catherine Wilkinson is a Senior Lecturer in Education at Liverpool John Moores University, Liverpool, UK. Prior to this, Catherine worked as a Lecturer in Children, Young

People and Families in the Faculty of Health and Social Care, Edge Hill University, Ormskirk, UK. Catherine completed her PhD in Environmental Sciences at the University of Liverpool, funded by an ESRC CASE award. Catherine's research interests include qualitative research, mixed methods, ethnography, participatory research, young people, youth voice and community radio.

Samantha Wilkinson is a Senior Lecturer in Childhood and Youth Studies at Manchester Metropolitan University, Manchester, UK. Prior to this, she was a Lecturer in Human Geography at the same institution. She has a PhD in Human Geography, an MSc in Environmental Governance and a BA (Hons) in Human Geography, all from the University of Manchester. Samantha's diverse research interests include young people and alcohol, home care and dementia, hair and identity, the sharing economy, animal geographies and innovative qualitative methods.

Rachel Woodward is Professor of Human Geography in the School of Geography, Politics and Sociology at Newcastle University, Newcastle upon Tyne, UK. Her research interests include military geographies and landscapes, and the sociology of military forces. Key publications include *Military Geographies* (Blackwell, 2004), *Sexing the Soldier* (with Trish Winter, Routledge, 2007), *Bringing War to Book* (with K. Neil Jenkings, Palgrave, 2018) and the edited *A Research Agenda for Military Geographies* (Edward Elgar, 2019).

Sarah Wright is a geographer of English/Welsh descent living on unceded Gumbaynggirr Jagun in Australia. She is part of the Yandaarra Collective, led by Gumbaynggirr Elder Aunty Shaa Smith, which aims to shift camp together towards Gumbaynggirr-led decolonising ways of caring for and as Country. She is also part of the Bawaka Collective and has had the privilege of working/living/loving with Yolngu grandmothers and other family for over 12 years. She is Professor of Geography and Critical Development Studies, and Future Fellow, at the University of Newcastle, Callaghan, Australia.

Alison Young is Francine V. McNiff Professor of Criminology, School of Social and Political Sciences, University of Melbourne, Melbourne, Australia, and Professor in the Law School at City University, London, UK. She has authored numerous works on the intersection of crime, law, culture and place, including *Street Art World* (2016), *Street Art, Public City* (2014) and *Street/Studio* (2010). She is the founder of the Urban Environments Research Network, an interdisciplinary and international group of academics, artists and architects.

Introduction

Thinking about place – themes and emergent approaches

Tim Edensor, Ares Kalandides and Uma Kothari

The power of place

The Routledge Handbook of Place seeks to provide a state-of-the-art snapshot of the rich and varied thinking about place across disciplines and theoretical currents. In inviting contributions from a diverse range of authors from different disciplines, geographical contexts and career stages, we offer a single volume that includes a plethora of fresh, often innovative accounts of place. New voices and those from established thinkers offer accounts that emerge from various strands of geographical thinking, sociology, anthropology, architecture and planning, history of art, business and place management, and creative practice, amongst others. Needless to say, although we believe this is a comprehensive collection of accounts about place, such an endeavour cannot claim to be exhaustive: no single account or approach can hope to account for the complexity of place as concept and entity. Yet together, these chapters constitute an alluring assemblage of perspectives, approaches, themes and practices.

It is often argued that globalisation increasingly threatens place, that it dissolves its difference and its protective boundaries. We are warned that places are losing their identity and specificity to a uniform global space, becoming homogeneous and interchangeable. Though by no means a new debate – with notions of local communities menaced by uncontrollable 'others' having circulated for centuries – what is pressingly pertinent in current times is the paradoxical realisation that we all live in one place, namely a planet that we may be about to render uninhabitable, while simultaneously living in different places on this planet. This constant tension between place as the simultaneous existence of 'the here and there' lies at the heart of contemporary conceptions. A look at current anti-globalisation movements is instructive: while some defend place in a protectionist, introverted sense based on 'othering' and exclusion, others acknowledge both the interdependence and unequal relations between place and interrogate global processes that subordinate one place to another. Yet at the core of all approaches lies a reignited preoccupation with place. While we observe a new rise in localisms, regionalisms and nationalisms, there is also an increased sense that we need to acknowledge the 'responsibilities of place' (to use Doreen Massey's expression) within global interconnections. In editing this collection, we are strongly motivated by the belief that there is a need for a *Handbook of Place* at this particular moment, one that will address both academic and professional enquiries, but also the heightened existential *angst* which currently links humans to the fate of their place or places.

The hold of place as an epistemological and ontological force, as a common-sense way of understanding our location in the world, persists, enabling us to make sense of the distinctiveness of the sites through which we move and in which we live. Along with a central focus on ideas about space and landscape, as Tim Cresswell (1992) notes, the notion of place is central to geographical enquiry. Place is thus subject to a compendium of diverse approaches and expanding definitions, being variously conceived as a geographical location, a physical entity with boundaries, a demarcated space of governance and regulation, a series of shared cultural meanings that engender a sense of belonging, and a potentially exclusive and contested container of those possessing distinctive cultural identities. More recently, other notions have accompanied such considerations: place is conceived as a relational entity shaped by its relations with others within larger networks and spaces of flows; as irrevocably shaped by fluid local and global forces – and, relatedly, as a site configured by the movement of capital and economic strategies; as a multi-temporal setting rather than a space frozen in time; as conditioned by human as well as non-human agencies; and as an affectively and sensorially inhabited environment. These recent accounts especially overcome the perils of sedentarism that characterised earlier narratives that tended to suggest that place is bounded, classifiable and static. Moreover, the emergence of these innovative supplementary approaches also underpin why it is a timely moment to publish a collection such as this. Indeed, the structure of the book has been informed by what we regard as enduringly salient themes through which place has been analysed and which are given a contemporary inflection in new chapters. We also include organising parts that take account of more recent thinking about place to acknowledge the vitality of new approaches and deepen place thinking. First, however, we briefly summarise the evolution of academic engagements with place.

Early preoccupations with place can be traced back to the colonial era where newly occupied territories had to be described in detail to defend the colonisers' interests (Blais et al., 2011). Comprehensive accounts of each place's features were thus produced: its geology, its fauna and flora, but also its human systems and settlements. Geography itself was born as the study of places, an interdisciplinary endeavour that brought together in one place elements from diverse established and emerging sciences. This was the world as a mosaic: one that was made up of a patchwork of places, each with its individual features and distinct characteristics.

This initiated a broader engagement, for, with the development of the natural sciences, place too became the object of an emerging 'spatial science'. New 'objective' delineations identified social customs, religious and class constitution, and typical geology, natural history or historical lineage, amongst other characteristics, as definitive (Kitchin, 2014). The distanced observer was provided with a range of new measuring techniques to turn formerly descriptive accounts, primarily based on observation, into scientific facts. In such accounts only the quantifiable was authoritative.

It was against this scientistic, quantitative backdrop that a new humanistic geography developed, critiquing depictions of the identifiable characteristics of place. Its adherents insisted that geographers needed to pay attention to the deeper connection of people with place within a world of places (Tuan, 1977). In such accounts, people subjectively bestow meaning onto place and they derive meaning from place. A 'sense of place' is thus conceived as part of the human condition. With links to existentialism and phenomenology this approach moves away from the distanced observer of the spatial science who measures the world. In humanistic studies, people are immersed in place; indeed, for some thinkers, place only exists because humans give it meaning. Place is thus conceptualised as a *meaningful* location.

Yet, subsequently, a focus on the experiential was accompanied by understandings that place was invariably a social construct and could only be understood as the result of a set of meanings that were likely to be informed by ideologies. In some influential accounts, places are understood to be predominantly structured by capitalist relations (Harvey, 1989). Capital, some argued, moves freely across space, while places are fixed. Accordingly, preoccupation with place would be an endeavour bound to fail, or would be regarded as inherently conservative, a sideshow to the global spaces of capital that are properly deserving of attention. Doreen Massey (1993) significantly overcame such deterministic assumptions by arguing that places matter: places are not merely locations at which capital manifests itself, realises values and fixes its crises; equally, places and the social relations that constitute them shape the way capital functions. Furthermore, she insisted that it was not solely class relations that influenced the formation and constitution of place but that other power relations were also pertinent, particularly gendered relations.

This book consciously engages with many of the above understandings of place, acknowledging the creative tension that is produced by bringing so many different approaches together: places are both distinctive and similar; they are unique and interconnected; they have ontological existence and are the products of the human mind; they are both things and processes; they are physical, social and imaginary; they are constituted by and constitute class, gender, racial or sexual relations; they are produced by representations, practices, norms and materiality. Yet, as we discuss shortly, we augment these earlier approaches with a range of more recent preoccupations and themes through which place has been examined.

The parts of the book

The parts of this book are thus organised to account for perspectives that are heterodox but which all go beyond earlier reductive investigations of place, following the spirit of a contemporary catholic approach, wide in scope and varied in conceptualisation.

Its ontological lure as a site in which to locate ourselves in the world foregrounds how place is constituted at a range of spatial and temporal scales. By looking at some of these scalar formations in **Part I**, we highlight the complexities and fluid contexts within which place is situated and acknowledge its conceptual richness. We explore how place is constituted and at what scales, how place is connected to other places and how place is enfolded by larger scalar entities. **Part II** emphasises the qualities of place, those attributes not so easily amenable to scientific classification but which are integral to the ways in which place is inhabited and experienced in unreflexive, habitual ways. This part is closely aligned with the discussion about non-representational theories discussed below and on emergent discussions about affect and sensation. **Part III** investigates how multiple identities are invariably entangled with place and, similarly, that place is shaped by the different ways in which identities are accommodated within it. This focus also considers how place is understood according to notions of who does and does not belong, and with relational understandings about who is other to place and who can be located beyond place, but, ultimately, in order to persist, places must accommodate and negotiate across forms of difference based on class, age and gender for instance.

Part IV is concerned to examine how place is managed, regulated and ordered and the challenges that emerge to counter such organising processes. Certain powerful groups seek to fix the meanings and functions of place through a medley of strategies that range from inscribing presence on space through imposing particular forms of architecture, commemoration and heritage, to installing processes of policing, law, managing risk and militarising place. Yet such processes require constant vigilance and there are always those who contest authoritative practices by fleetingly occupying place or establishing sites in which to carry out alternative practices.

Part V focusses on processes of displacement and replacement to take account of the contemporary processes through which place is made unstable, volatile and insecure, and is also subject to processes of relocation, settlement and abandonment. The unprecedented numbers of people seeking refuge in another country since the Second World War underpins the contemporary importance of considering people's relationship with displacement. In acknowledging the recent emergence of mobilities thinking, such a focus foregrounds the unequal ways in which some move fluidly and frequently whereas others stay in place, unable to access the riches and connections that would allow them to move.

In referring back to themes explored in Part I, **Part VI** takes note of the increasingly complex ways in which place is incorporated into larger networks of supply, manufacture and finance. It explores how place managers and promoters strive to become drawn into such frameworks, as part of an intensified competition between places that has eventuated from the need to mark out distinctiveness within larger spatial contexts, through place branding, place marketing, tourist promotion, retail provision and lifestyle lures for potential high-income inhabitants. Such economic strategies broadcast certain attributes and ignore others in seeking to gain competitive advantage.

Finally, we take account of the recent surge in inventive, creative engagements with place in **Part VII**. This resonates with the rise of geohumanities, as David Cooper details in greater depth in his introduction. As Michael Dear (2015: 22) remarks, these transdisciplinary approaches have 'required a nonexclusionary openness to all forms of knowing' and a willingness to suspend specific disciplinary modes of writing, practising research and learning. Innovative creative engagements blend geographic analysis with literary study, laboratory work, political activism, creative writing and poetry, historical research, and creative practices in photography, music, art and film. These considerations of place, then, have moved well beyond the provenance of geographers but have also augmented and embellished geographical understandings of place (Boyd and Edwardes, 2019).

Emergent themes

Besides these dedicated parts, we also want to draw attention to some other pertinent theoretical considerations that have been applied to thinking about place, as motifs that run through the book.

In extending Doreen Massey's key concept of a progressive sense of place (1993), an insistence that place is always relational is signified by the rise of **network** thinking, which conceives places as continuously remade through their relationships with multiple elsewheres. In recent times, academic accounts from many disciplinary perspectives have identified how, for instance, particular places might be situated within tourist itineraries, production networks and flows of commodities, information, technology and money (Appadurai, 1990). These networked understandings have had the effect of foregrounding place-consciousness in response to the perceived threats of globalisation and such anxieties have promoted reifications that posit a pre-global era in which places were stable and consistent across time, occupied by homogeneous communities and reproduced by enduring traditions and hierarchies. Such notions resonate with earlier, equally nostalgic accounts that depicted emergent cities as replete with anomie, detachment and instrumentality, conceived as inimical to the tradition-bound, pre-modern, more 'authentic' communities with their communal obligations, as portrayed in Ferdinand Tönnies' (1955) conception of *gemeinschaft* – loosely meaning 'community' – as 'the lasting and genuine form of living together'.

And yet, as Doreen Massey (1993) points out, the multiple flows that swirl through, into and from place also mobilise understandings that obviate essentialist assumptions and attempts to draw boundaries around a place. For so diverse are these multiple connections in the ongoing, dynamic reconfiguring of place that any attempt to fix place is invariably and absurdly partial. For instance, over the past 500 years, quarries from various locations have supplied the building stone of Manchester (Edensor, 2011). Initially, a local quarry close to the city centre provided all the stone for important buildings but it was of poor quality, and as soon as the industrial revolution inspired the creation of canal and railway networks into which the city was incorporated, other sources of stone supply became available. Consequently, this local quarry was abandoned and stone was imported from sites in Derbyshire, Lancashire, Cheshire and Staffordshire; reliable sources of supply that lasted for a hundred years or so. The advent of architectural modernism and building technologies diminished the importing of building stone into Manchester, yet stone was still imported from regional sources but in the form of the aggregate used to produce concrete for these 20th-century structures. In recent years stone veneer has once more been imported into the city for high prestige urban renewal projects. However, this highly polished material derives from the global market, coming from countries as diverse as China, South Africa, Brazil and Italy. Building stone thus indicates how place is continuously made and remade by connections that are typically constituted at diverse scales.

This example also foregrounds how considering cities as composed by the flows that move from, to and through them can alert us to spatial hierarchies that highlight how certain places – typically large cities – are densely connected to global flows in comparison with others, and also identify those who may move between places and those who have to stay in place. These inequalities constitute what Doreen Massey (1993) terms power-geometry, the uneven distribution of financial, political, class and economic power across space.

These questions of power-geometry, and the highly specific, selective decisions that ascertain which aspects of a place's heritage, cultural identity and design should be privileged, highlight another key concern of this book: to explore how places are always crosscut with shifting patterns of inequality and power. This also foregrounds a call for a broader grasp of the ways in which place is **contested**. As the chapters in Part III detail, places are shaped by class, ethnicity, age, (dis)ability and gender, and the strategies to shape place by the powerful are apt to be challenged by others, a process exemplified by the rise of heritage from below (Atkinson, 2008). For instance, the centre of Manchester, the city in which we all reside at times, has been populated by officially sanctioned forms of commemoration, bronze and stone statues raised high on plinths. Queen Victoria lingers in Piccadilly Gardens, a series of war memorials adorn the refashioned St Peters Square. In the nearby Albert Square are five 19th-century figurative monuments: Prince Albert, national politician W. E. Gladstone, local statesmen John Bright, philanthropist Oliver Heywood and a long-dead Bishop of Manchester. Yet, until recently, two other renowned local residents and a notorious event were not commemorated. There were no officially recognised memorials to leading suffragette Emmeline Pankhurst or to the co-author of the Communist Manifesto, Friedrich Engels, and no material commemoration of the slaughter of 18 protestors who had gathered to demand the reform of parliamentary recognition in what became known as the Peterloo Massacre in 1819 and were killed by the 15th Hussars and the local yeomanry. Yet, in 2017, a large stone social-realist sculpture of Engels from Ukraine was installed in Tony Wilson Place, in late 2018 a bronze likeness of Emmeline Pankhurst was situated in St Peters Square and in August 2019 a circular stepped memorial to the Peterloo Massacre was unveiled outside the Manchester Convention Centre. Here, then, the memorial landscape has been augmented by three memorials that have put key events and figures from Manchester's history back into place and, in the case of the Pankhurst and Peterloo memorials, following long campaigns to redress this forgetting. The commemorations preferred by an urban elite have now been complemented by these suffragist and socialist memorials, representing identities formerly denied commemorative representation.

This contestation of place is not only limited to that of symbolic meaning and interpretation. Social movements occupy places in the most physical sense, by **activists** who repurpose them as the turf for political claims. Thus squatters may challenge the concept of private property itself (Vasudevan, 2015); countercultural enclavic places, such as Christiania in Copenhagen, may propose a different urban organisation in the heart of capitalist cities (Ntounis and Kanellopoulou, 2017); and LGBT movements constitute and are constituted by 'gay villages' around the world (Nash, 2006). More recently, the Occupy activists in 2011–2012 set up their humble tents in the heart of the City of London and Wall Street, visibly occupying the nodes of financial globalisation (Nolan and Featherstone, 2015). Place-based activism may be explicitly local in scope while also having ambitions to change the world order.

Another strand of enquiry that has emerged from assertions about the diminution of place in the face of global flows has contended that a **placelessness** has emerged that blurs place distinctiveness, echoing earlier alarms raised by Joseph Meyrowitz and Edward Relph. According to contemporary expressions of placelessness, the effects of global flows have eventuated in serial monotony and clone towns. They draw on Marc Augé's distinction between anthropological place, which is lived, familiar, localised, organic, social and characterised by fixity or repeated movements and associations, and non-places which are highly mediated, ubiquitous and marked by experiences of solitude, detachment and alienation. These festival marketplaces, airports, malls and similar sites are typified by Augé (1996: 178) as 'spaces of circulation, communication and consumption, where solitudes coexist without creating any social bond or even a social emotion'. Peter Merriman (2004: 162) argues that, on the contrary, places are 'more contingent, open, dynamic and heterogeneous' than Augé proposes. He notes that a more relational and emergent approach reveals that place is shaped by the 'folding of particular spaces, times and materials into these places' as well as the flows of people and vehicles through them. As Merriman further contends, different people 'dwell in, move through and inhabit these spaces of travel and exchange in different ways' (Ibid: 151–152), entering as workers, managers, planners, tourists or shoppers, foregrounding the relational and situational qualities of place and disavowing any universal experiences and perspectives of them. In addition, as Degen et al. submit, such sites are not merely passively consumed as empty and spectacular environments but are sensed in multimodal, complex ways that are shaped by material affordances, memory and personal attachments, and detached assessment. They also emphasise that they are 'always open to reversals, destabilisations, and inversions of the intended model of behaviour and experience' (2008: 1908).

In many countries, the politics of decentralisation and devolution have rendered the small scale, the town or even the neighbourhood, as the privileged place of policy implementation (Rodríguez-Pose and Gill, 2003). In this context, **place management** has emerged as a governance model that is based on the principle of fostering local partnerships among different players to tackle place-specific issues (Warnaby and Medway, 2013). Here, small scale is paramount: it is argued that spatial proximity, trust, face-to-face interaction and the sharing of common resources creates opportunities for governance that do not exist at larger scales. On the contrary, critics of the concept argue that it is not possible to solve systemic problems (such as homelessness or unemployment) locally, particularly in times of reduced funding (Tomaney, 2016).

Nonetheless, place promotion strategies have included culture-led attempts to construct art galleries, signature buildings, heritage districts and design-led renovation. Also prominent has been the trend towards eventification and festivalisation (Jakob, 2012: 448). Yet though such programmes aim to demarcate one place from another, they all too often reproduce somewhat serial, homogeneous designs, events and cultural offerings, resulting in artistic compromise and the production of pervasive, generic designs. The cultural strategies of an international 'creative class' (Julier, 2005) thus may 'fail to connect with the specificities of the places within which they are located' (Quinn, 2010, 271–272). Yet elsewhere, other regeneration strategies are more attuned to history and culture in devising designs and events that accord with the identity of place. For instance, the redesign of the seafront promenade at the north-west English traditional coastal resort of Blackpool takes account of the popular cultural practices and performances, the carnivalesque aesthetics and the sensory experience of the town, thereby avoiding generic design formulae and articulating place-specificity (Edensor and Millington, 2018).

Perhaps most persuasively, Doreen Massey (1993) maintains that the specificity of place derives from how each place is the focus of a distinct mixture of wider and more local social relations, and these are accompanied by the accumulated histories of place and its previous connections. And such relationalities should be borne in mind to challenge defensive, essentialist and exclusionary ideas that seek to banish placelessness by restoring place to an authentic, original condition, notions that can be scaled up to support limited, circumscribed ideas about who belongs to the nation.

A key theoretical development across geography and the social sciences that is highly pertinent for thinking about place has been the emergence of **non-representational theory** (Thrift, 2008). In asserting that there is much more to social life than that which can be represented or discursively imparted, non-representational ideas focus on the emergent, fluid qualities of life, on the affective, sensory and emotional qualities that are rarely articulated – indeed may not be amenable to representation but are nevertheless integral to the ways in which space is experienced. Accordingly, we move away from the supposedly objective, rational criteria through which place is identified – for instance, according to its demographic constituency, physical environment, or traditions and customs and ethnicity, for instance – and towards the more evanescent but no less powerful qualities through which place is experienced. Such accounts emphasise the transpersonal, the potency of affect and atmosphere, and the agencies of other, non-human forces. This has further promoted a wider consideration of the *qualities* of place, the theme of Part II, which focusses on those elements that are less objectively identifiable, but also recur throughout the book, especially in Part VII.

We regard the discussion of these more evanescent attributes as critical in challenging reified and essentialist conceptions of place, which usually overlook the more grounded ways in which place is experienced. This neglect seems surprising when it seems obvious that though official accounts or classificatory approaches prevail across discourse, for all people, the most common encounter with the places in which we live and move through is utterly mundane. Inhabitants of place are clearly not continually making meaning or acting according to preconditioned meanings but simply living, moving, sensing, resting and socialising in place, making and consolidating affective and emotional attachments, and performing unreflexive routines. Yet place is not an inert backdrop to these experiences but is itself a dynamic realm in which multiple forces and agencies surge to attention or carry on unnoticed. Atmospheres swirl through place, affective intensities gather at particular locations and at certain times, attentive attunements to occurrences wax and wane. In such a context, place, like the humans and non-humans that occupy it, is continuously emergent as life 'takes place' through movements, intensities and encounters (Lorimer, 2005).

Accordingly, besides learning from non-representational conceptions, conceptions of place have also been bedevilled by an anthropocentrism which disregards the many ways in which place might be **created by non-humans** in alliance with humans but also irrespective of human concerns. How can a more-than-human, even posthuman understanding of place be considered? In echoing the relational ontologies of Actor Network Theory, Jones and Cloke (2008) demonstrate how non-human agencies co-constitute and reconstitute the fabrics of everyday places. They explore how trees are 'active in the creation and folding fields of relations' that constitute place by growing in distinctive ways according to qualities of light, soil and other material elements of place, and responding to human interventions. Moreover, they solicit human interventions in a continuous ongoing relationship that confounds anthropocentric notions that place is a purely social human construct and foregrounds a processual, dynamic emergence perpetrated by a multitude of agencies. The relational agency of trees has recently been conceived as far more communally distributed between trees, chemicals and mycorrhizal fungi than has previously been imagined, through the emergence of the concept of the 'wood-wide web', further interrogating ideas about place and its interconnectivities beyond the human (Helgason et al., 1998). In this collection, these considerations are echoed in the chapter by Deborah Martin and extended in Catherine Phillips and Sarah Roberston's account of the place-making capacities of bees, as well as resonating through David Paton's account of the agencies of stone in place.

A great problem in considering place in a global frame, as with many areas of enquiry, are the continued **ethnocentric assumptions** perpetrated by much western theory, where ideas formulated in distinctive scholarly and cultural settings are generalised as if they apply universally. Conceptions of place are no exception to such generalisations and explain why we fail to comprehend the sheer diversity of places that exist and the multiple processes through which they evolve, are inhabited and are practiced. As Jenny Robinson (2003: 275) remarks with regard to the practice of urban theory, a 'very parochial scholarship has paraded the world in the clothes of universalism for some time'. Generalised concepts that originate in specific western urban settings theory are uncritically applied to cities in which they may lack salience and non-western cities are often contextualised within a developmental or comparative lens. For instance, Mbembe and Nuttall (2004: 348) claim that African cities all too often end up epitomising the 'intractable, the mute, the abject, or the other-worldly'; the same often applies to academic portrayals of a wide range of non-western places, in addition to the colonial and tourist representations that circulate through popular culture (Kothari, 2015).

Recent work has challenged these dominant Eurocentric urban theories, as Asian, Latin American and African scholars' conceptions of urban lives, processes and structures have bypassed influential western concepts in favour of situated forms of understanding and contextual analyses (Edensor and Jayne, 2012). Through his work on a range of cities and places outside the west, AbdouMaliq Simone (for instance, see Simone, 2005) draws attention to the demise of coherent and enduring locales constituted through stable social connections and forms of regulation and the propensity of inhabitants to maximise potential opportunities by establishing connections with as many people, networks and scenes as possible, fluid affiliations through which place is continuously remade (for similar analyses see Mbembe, 2004; Mehrotra, 2008).

Though similarly substantive explorations have not been widely disseminated with regard to non-Eurocentric conceptions or empirical research about place, there are some pertinent examples. The broad arguments of Arturo Escobar (2010) emphasise how policies of interculturality, a mix of indigenous and modern knowledge and *buen vivir* (good living for all) decentre current neo-liberal preoccupations with individuality, nature–culture dichotomies, capitalised abstractions of space and the primacy of the market that so often shape places in western settings. More specifically, Waleed Hazbun (2010) explores how the distinctive place of the Moroccan beach combines distinctive religious practices, familial gatherings and camping, while Hooshmand Alizadeh unpacks distinctively Kurdish understanding of private and public space in Kurdish settings. We particularly want to pick out chapters in this volume where such alternative imaginings and conceptions of place are explored. Cuttaleeya Jiraprasertkun outlines a distinctively Thai conception of place, Rana Singh and Pravin Rana reveal the complex ways in which different kinds of Indian places are encompassed within sacred geographies and Sarah Wright details the deeply intimate and sophisticated Aboriginal engagements with place in Gumbaynggirr Country. The chapter by Freeman et al. investigates how the Spanglish words used to describe key aspects of Latin American places capture affects, sensations and meanings that cannot be similarly conjured by English words, revealing how language distinctively shapes how place is understood.

This volume does not pretend to formulate definitive answers to all of the salient issues present here, along with many of the other perspectives that seek to conceptualise place. Rather, our intention is to revisit and broaden existing discussions about place to embrace the varied landscape of diverging viewpoints and disciplines. Accordingly, the book should be read as part of the long process through which place has been explored in multiple and contesting ways. The richness of place as concept and lived reality is that it can bring together a vast array of diverse trajectories, stories, objects, humans, non-humans and processes: place is where geography may meet poetry or where biology encounters the visual arts. In this book we have sought to honour the abundance of such fertile notions about place and the multiplicities that centre upon it.

References

Alizadeh, H. (2012) 'The concept of privacy and space in Kurdish cities', in T. Edensor and M. Jayne (eds) *Urban Theory Beyond the West: A World of Cities*, pp. 137–155, London: Routledge.

Appadurai, A. (1990) 'Disjuncture and difference in the global cultural economy', *Theory, Culture and Society*, 7(2–3), 295–310.

Atkinson, D. (2008) 'The heritage of mundane places', in B. Graham and P. Howard (eds) *The Ashgate Research Companion to Heritage and Identity*, pp. 381–396.

Augé, M. (1996) *Non-Places: Introduction to an Anthropology of Supermodernity*, trans. John Howe, London: Verso.

Blais, H., Deprest, F. and Singaravelou, P. (2011) 'French geography, cartography and colonialism: introduction'. *Journal of Historical Geography*, 37(2), 146–148.
Boyd, C. and Edwardes, C. (eds) (2019) *Non-Representational Theory and the Creative Arts*, London: Palgrave Macmillan.
Cresswell, T. (1992) *In Place/Out of Place: Geography, Ideology and Transgression*, Minneapolis: University of Minnesota Press.
Dear, M. (2015) 'Practicing geohumanities'. *GeoHumanities*, 1(1), 20–35.
Degen, M., DeSilvey, C. and Rose, G. (2008) 'Experiencing visualities in designed urban environments: learning from Milton Keynes'. *Environment and Planning A*, 40(8), 1901–1920.
Edensor, T. (2011) 'Entangled agencies, material networks and repair in a building assemblage: the mutable stone of St Ann's Church, Manchester'. *Transactions of the Institute of British Geographers*, 36(2), 238–252.
Edensor, T. and Jayne, M. (eds) (2012) *Urban Theory Beyond the West: A World of Cities*, London: Routledge.
Edensor, T. and Millington, S. (2018) 'Learning from Blackpool promenade: re-enchanting sterile streets'. *The Sociological Review*, 66(5), 1017–1035.
Escobar, A. (2010) 'Latin America at a crossroads'. *Cultural Studies*, 24(1), 1–65.
Harvey, D. (1989) *The Condition of Postmodernity*. Oxford: Blackwell.
Hazbun, W. (2010) 'Modernity on the beach: a postcolonial reading from Southern shores'. *Tourist Studies*, 9(3), 225–244.
Helgason, T., Daniell, T., Husband, R., Fitter, A. and Young, J. (1998) 'Ploughing up the wood-wide web?' *Nature*, 394(6692), 431.
Jakob, D. (2012) 'The eventification of place: urban development and experience consumption in Berlin and New York City'. *European Urban and Regional Studies*, 20(4), 447–459.
Jones, O. and Cloke, P. (2008) 'Non-human agencies: trees in place and time', in C. Knappett and L. Malafouris (eds) *Material Agency*, pp. 79–96, Boston, MA: Springer.
Julier, G. (2005) 'Urban designscapes and the production of aesthetic consent'. *Urban Studies*, 42(5/6), 869–887.
Kothari, U. (2015) 'Reworking colonial imaginaries in post-colonial tourist enclaves'. *Tourist Studies*, 15(3), 248–266.
Lorimer, H. (2005) 'Cultural geography: the busyness of being 'more than representational'. *Progress in Human Geography*, 29(1), 83–94.
Massey, D. (1993) 'Power-geometry and a progressive sense of place', in J. Bird, B. Curtis, T. Putnam and L. Tickner (eds) *Mapping the Futures: Local Cultures, Global Change*, pp. 59–69, London: Routledge.
Mbembe, A. (2004) 'Aesthetics of superfluity'. *Public Culture*, 16(3), 373–405.
Mbembe, A. and Nuttall, S. (2004) 'Writing the world from an African metropolis'. *Public Culture*, 16(3), 347–372.
Mehrotra, R. (2008) 'Negotiating the static and kinetic cities: the emergent urbanism of Mumbai', in A. Huyssen (ed.) *Other Cities, Other Worlds: Urban Imaginaries in a Globalizing Age*, pp. 113–140, Durham, NC: Duke.
Merriman, P. (2004) 'Driving places: Marc Augé, non-places, and the geographies of England's M1 motorway'. *Theory, Culture & Society*, 21(4–5), 145–167.
Nash, C. J. (2006) 'Toronto's gay village (1969–1982): plotting the politics of gay identity'. *Canadian Geographer/Le Géographe Canadien*, 50(1), 1–16.
Nolan, L. and Featherstone, D. (2015) 'Contentious politics in austere times'. *Geography Compass*, 9(6), 351–361.
Ntounis, N., and Kanellopoulou, E. (2017) 'Normalising jurisdictional heterotopias through place branding: the cases of Christiania and Metelkova'. *Environment and Planning A*, 49(10), 2223–2240.
Quinn, B. (2010) 'Arts festivals and urban tourism and cultural policy'. *Journal of Policy Research in Tourism, Leisure and Events*, 2(3), 264–279.
Robinson, J. (2003) 'Postcolonialising geography: tactics and pitfalls'. *Singapore Journal of Tropical Geography*, 24(3), 273–289.
Rodríguez-Pose, A., and Gill, N. (2003) 'The global trend towards devolution and its implications'. *Environment and Planning C: Government and Policy*, 21(3), 333–351.

Simone, A. (2005) 'Urban circulation and the everyday politics of African urban youth: the case of Douala, Cameroon'. *International Journal of Urban and Regional Research*, 29(3), 516–532.
Thrift, N. (2008) *Non-Representational Theory: Space, Politics, Affect*, London: Routledge.
Tomaney, J. (2016) 'Limits of devolution: localism, economics and post-democracy'. *The Political Quarterly*, 87(4), 546–552.
Tönnies, F. (1955) *Community and Association*, London: Routledge (first published 1887).
Tuan, Y. F. (1977) *Space and Place: The Perspective of Experience*. Minneapolis: University of Minnesota Press.
Vasudevan, A. (2015) *Metropolitan Preoccupations: The Spatial Politics of Squatting in Berlin*, Hoboken, NJ: John Wiley and Sons.
Warnaby, G. and Medway, D. (2013) 'What about the "place" in place marketing?' *Marketing Theory*, 13 (3), 345–363.

Part I
Situating place

Tim Edensor

Introduction: situating place

The mayor of the small Addu Atoll, Maldives' southernmost atoll, declared that he was proud to belong to an island in the Southern Hemisphere. It made the people of the island special, he said. They see themselves as geographically distinct but they also have connections to places that the rest of the country do not have and are outward-looking folk, not parochial like other Maldivians. In the past, they sailed to Ceylon with the southwest wind and subsequently had to wait in Ceylon until the northeast monsoon would stir up the wind so that it was behind them for the return voyage. Moreover, in colonial times, the British wanted to install a military base in the Southern Hemisphere and many in Addu feel that they were specially chosen and benefited enormously from British influences, always having the best schools and hospitals. The distinctive location of the atoll is underlined by the presentation to visiting tourists of a certificate to verify that they have crossed the equator

The mayor is articulating the key theme of this section; that of scale and the location of place. While the islands may be considered somewhat remote, their special location within the Maldives, in the Southern Hemisphere and as part of the colonial British sphere of influence, is further augmented by the romantic allure of Addu for visiting tourists who wish to mark passage into the Southern Hemisphere. Other places similarly broadcast their location within variously scaled scalar contexts. Some announce their presence within the Arctic Circle or on the Equator, while others claim to occupy the furthest point west, east, north or south of a continent or country. More contentiously, Dunsop Bridge in the northern English county of Lancashire asserts that it is located at the exact centre of the British Isles, an accolade marked in the village's only public telephone box.

Spatial scales vary enormously, from the human body to the household, from the neighbourhood to the city or larger metropolitan area, from the state or county to the nation, and from the continent to the globe. The location of place within these spatial contexts is also supplemented by other supra-national political and sacred spaces, and to networks constituted at diverse scales. Place is also always temporally situated and a vast array of temporal frames may be adopted by inhabitants, place-managers and visitors, from deep geological time to the recent past, through a pre-historical or colonial lens, or from a perspective that primarily focusses on the present or signifies the future. It also seems that these spatial and temporal contexts of place are becoming more multiple and multi-layered.

Many narrative accounts for a pre-modern or traditional place consider that social, cultural and economic life was overwhelmingly organised around the local and, in an era before massive urbanisation, this was primarily a rural place. Thus, food would likely be supplied by local farmers, shoes made by neighbourhood cobblers and clothes stitched together by local tailors. Extensive travel was infrequent for most people and squires, landowners, aristocrats and priests tended to wield power. People worked for local farmers, merchants and traders, routines were somewhat fixed and based on church and local festivities and customs, and most social relations were forged in the local arena. While such depictions are perhaps slightly stereotypical and over-general, there is no doubt that the changes wrought by modernity transformed these overpoweringly local places into places that were shaped according to a plethora of connections to external authorities, businesses and media. According to Anthony Giddens, most places were subsequently subject to disembedding processes, in that the primary rupture between traditional and modern places was that they became 'thoroughly penetrated by and shaped in terms of social influences quite distant from them' (1990: 19). Since the advent of modernity, people are far more likely to obtain clothes, shoes, food, work and information from connections that penetrate place from further afield, whereas previously there were limits on co-ordinating activities between places. In addition, in terms of temporal scales, instead of being tethered to the time of place insofar as it related to agricultural work and customs, for instance, time was largely discerned to an abstract, extended temporality.

The disembedding of the local from place has in recent times been primarily attributed to the workings of an increasingly global, footloose capital that is able to choose where to invest, speculate and set up industrial operations that are invariably part of a large network of production networks. Places have accordingly been brought into the orbit of these decisions and typically compete to attract corporate businesses and retail outlets. Consequently, geographers have emphasised how places are entangled in the making and re-making of social, political and economic scales of organisation, processes that invariably produce what Doreen Massey (1993) refers to as power geometries. For instance, Neil Brenner (1999) contends that subnational metropolitan regions are often more important in the geography of economic change and that the national scale is becoming less important (Brenner, 1999). As a consequence, certain places are seen to be more connected than others; with their myriad connections to global flows of money, technology and information, cities such as London and Tokyo are evidently more economically, culturally and politically influential than, say, a small village in Burkina Faso. This dispersal of economic power across places also has consequences for the labour movement. Where once it was possible to build national networks that enrolled place(s) into a territorial alliance, at a global scale such potentialities diminish, whereas global capital may adopt strategies that enfold places within production networks at larger scales as part of a process of uneven development (Smith, 1990). The consequence is that certain places in which competitive advantage no longer pertains are apt to be cut off from industrial production and investment, becoming cut adrift from the networks to which they once belonged. This also engenders a certain volatility where the character of place cannot be guaranteed to remain stable for very long.

Besides the locating of place within economic networks, places also became disembedded from local political processes with the rise of the nation-state and the division of that political entity into regional states and provinces. Political power now tends to be concentrated at regional, national and supra-national scales. Moreover, since the 16th and 17th century, the expanse of colonial power also brought places that were formerly locally or regionally governed under colonial rule through British, French and Belgian political expansion. Such effects linger though colonialism has faded, replaced by other imperialist strategies, proxy wars, coercion and the making of alliances that similarly detach place from political power. Thus, place has become entangled in a range of political spaces of ascending expanse, regional, national, supra-national, continental and global.

The incorporation of place within these different scalar contexts has been described as 'nested' relations, akin to a Russian doll, a conception that perpetrates a spatial hierarchy that has been critiqued as too rigid and neglectful of the inter-penetration and melding of these different scalar realms, not to mention the fluidity through which certain scales become salient, only to be subsequently pre-empted by others.

Above all, however, it is commonly argued that the most pertinent contemporary enfolding of place is its incorporation within global flows, so that a 'global sense of place' devolves. Of course, all places are subject to the looming environmental crisis, though some more critically than others, and are at threat from sea level rise and climate change. Increasing flows of refugees, migrants and asylum seekers enter places hitherto inured to such processes, while most settings receive rolling international news coverage and can access televised global sporting events.

Though there is much salience to the notion of a global sense of place, certain assertions about its overarching salience can be hyperbolic. As we discuss in the next part, in focussing on the economic and the meta-political, they neglect the ways in which people inhabit places in grounded, sensory and pragmatic ways through their everyday routines and habits. Moreover, even though their built environments and economies may be primarily shaped by forces from outside the local sphere, such processes are always situated in a specific spatial context, are domesticated in settings where they supplement already existing features and practices. Since places are layered with traces of past social relations and events, and have been shaped by situated human and non-human processes, they continuously (re)emerge out of durable and resonant historical formations.

Furthermore, as many critics of accounts of global homogeneity point out, imported architectural styles, fashions and popular cultural forms and practices are typically hybridised or 'glocalized' (Robertson, 1994) as they are absorbed and transformed in local spheres. Hence, despite elements of 'serial monotony' (Harvey, 1989), contemporary places may be better conceived as sites of cultural, social and economic negotiation and adaptation. Moreover, as Andrew Herod (2010: 2) points out: 'whereas many have argued that globalization heralds the evisceration and colonization of scales such as the local, some have suggested that the global's increasing power is bringing with it not the undermining of the local but, perhaps paradoxically, its reassertion'.

Importantly, Herod's contention underlines how places are not settings in which there is passive acceptance of the external processes that penetrate them; instead, we must take account of the agency of place and its inhabitants. For there is a sense in which certain mooted singularities of place are marketed and championed in response to the forces within which they become embroiled. Local pubs and restaurants offer food to gastrotourists sourced from local produce, local festivals are initiated or revived and there has been a surge in amateur enthusiasts who research local histories. In a different vein, bioregionalists attend to the subtle specificities that mark out places as distinct from others (Berthold-Bond, 2000). Such responses can also be profoundly defensive, as with 'NIMBY' protests against perceived malign intrusions from elsewhere and reactionary actions prompted by anxieties about the settlement of migrants in place, not to mention the rise in populist Trumpian narcissism that exclusively aligns self and place that we are currently witnessing. Yet the well-known slogan 'Think Globally, Act Locally' also demonstrates how global processes and concerns can become sedimented in place and reveal that global political networks can be organised in and across places.

Finally, it is also significant to consider how those places assumed to be marginal to these global flows, often alleged to be backward or traditional, may possess the potential to inspire independent cultural and political practices. For marginal or liminal places may offer an escape from normative modes of organisation and greater scope for the subaltern, subcultural and alternative practices, as well as accommodating oppositional groups who can devise political strategies that challenge mainstream authority and convention (Edensor and Smith, forthcoming).

Overall, it seems as if place is continuously enfolded into a range of differently scaled spatial entities ranging from small locales to neighbourhoods, cities, regions and nations that serve as spheres of belonging at different times (Kusenbach, 2008). Notions about the hierarchical constitution of these scales diminishes the fluid ways in which people continuously move between these contexts, reworking and reproducing them through everyday, routine practices. Inhabitants of place connect to local, regional, national and global networks at different times depending on circumstance and a shifting sense of allegiance, denying assertions that there is any overarching scalar context in which place is essentially embedded.

To commence this part, Kim Dovey presents a conception of place based on assemblage thinking and the work of Deleuze and Guattari, highlighting how this connects multi-disciplinary perspectives by foregrounding the relational, multi-scalar and anti-reductionist attributes of place, as well as critiques of power and territory. In considering it within an assemblage framework, place is also connected to urban morphology, but also disavows notions that places are closed, purified and static towards those that are open, multiple and dynamic. Dovey thus champions places of becoming rather than those in which identity formation is fixed, those that are immanent rather than transcendent, and he emphasises that place as assemblage is grounded in the particularities and practices of everyday life.

Ares Kalindides revisits Massey's 1991 seminal text, 'A global sense of place', to assess its contemporary significance. He makes five main arguments: first, and critically in the context of this part, that the concept of 'power-geometries' introduced in the text provides us with an understanding of the unequal positions of places in a global system of power relations. Second, that by asking spatial questions, Massey's paper enriched Marxist geography's focus on capital with debates on gender and race. Third, that Massey shows how a preoccupation with place can indeed be a progressive endeavour, raising questions of geographical responsibility. Fourth, that our conceptualisations of spatial relations, far from being just an intellectual exercise, have implications for our politics in place and across places. Finally, Kalindides contends that Massey's choice of medium and language are conscious choices to undermine the distanced authority of the academic intellectual and find a more direct way of talking to broader audiences.

In exploring the spatial scales through which place is constituted, the national incorporation of place seems particularly salient in contemporary times, symbolising the incorporation of difference into the larger political entity of the nation. Yet, as Rhys Jones maintains, common-sense understandings tend to view places and nations as geographical entities that operate at two distinct scales. In his chapter, Jones challenges these assumptions by demonstrating the inter-connectedness of place and nation. First, he draws on examples based in Wales, a region and nation located on the western seaboard of the UK, showing how the discourse of nationalism portrays nations and national territories as places that should possess meaning for members of the nation. The second kind of inter-connectedness he discusses relates to how nations come to inhabit 'local' places in various ways. He concludes by highlighting the potential for place to act as a source of national reconciliation.

Drawing on the work of Anssi Paasi's work on region building, Beel and Jones explore the salience of the regional as a context in which place can be accommodated. The authors emphasise the contingency and relationality of place, as well as the less evanescent social, economic and political structures and practices that are configured around a place. They also highlight how the material coherence of place, together with an imaginary coherence, provide vehicles for collective action and shared identity, and consolidate affinity to place. In developing new ways of thinking about place vis-à-vis region, they consider how the region too might provide a further spatial context within which place can be productively aligned.

Places that are located adjacent to a border may be regarded as threatened by an all-too-close otherness or, alternatively, might afford opportunities for diversity and hybrid identities to emerge, along with creative practices borne out of the trans-border movements of people and ideas. Ysanne Holt focusses on the progressive potentialities that inhere in this second eventuality by examining how the extraordinary LYC Museum and Art Gallery at the village of Banks in rural north Cumbria, close to the Northumberland and Scottish borders, was the site of a longstanding intermingling of social and cultural relations, skills and resources. Holt argues that the practices, networks and notion of place with which the artist and curator Li-Yuan Chia was associated throughout the 1970s exemplifies the expanded geographical and conceptual process of 'cross-bordering'. Through its activities and interconnections, this hybrid borderland site was a meeting place of different cultures of time and place, mingling ideals of community through national and international conceptual, environmental and land art practices and local skills and craft traditions. This analysis also disavows any sense of its rural location as remote or marginal.

Rana Singh and Pravin Rana's chapter explores how places are enfolded into variously scaled and constituted spatial frameworks shaped by their sacred symbolic constitution. By investigating the multiple ways in which places are integrated into a Hindu faithscape in India, they show how the distinctive sacrality of such sites are reproduced at different scales. The authors first acknowledge a sense of sacred place at a local scale, which they align with the notion of *genus loci*. In a larger spatial framework, seven sacred cities are supplemented by the twelve most important Shiva abodes and the four abodes of Vishnu that are located across India. Both small and pan-Indian pilgrimage places are reproduced as sacred sites by repeated visits and rituals.

In identifying a supra-national spatial context into which places were absorbed, Uma Kothari's chapter examines the production of colonial imaginaries and the ways in which they were manifest and contested in colonised places. Kothari considers the specificities of place in accounts created by colonialists in order to appreciate the situatedness of colonial rule and the entanglements of its material and discursive forces. The chapter begins with an examination of how power was inscribed onto the fabric of colonised places through the creation of dual cities in which separate European cantonments and native quarters were established. Drawing on research with former colonial administrators, the chapter goes on to reveal the contingent and heterogeneous character of these colonial places. It further explores collaborations between anti-colonial leaders who were exiled by the British to the Seychelles to demonstrate how colonial imaginaries of colonised places and the practices they informed were not seamlessly transferred but resisted. The chapter concludes with a discussion of contemporary postcolonial legacies.

Scholarly debates have long grappled with the question of what mobility might be doing to place. Although some thinkers have been concerned that mobility is antithetical to place, David Bissell's chapter explains how mobility and place are inextricably linked. The chapter summarises some of the most significant academic literature that has explored the relationship between place and mobility, and it appraises these ideas through fieldwork from a research project that explores the lives of mobile workers in Australia whose lives routinely take them away from their places of home for days or weeks at a time. Arguing that place is ultimately contingent on all kinds of different mobilities, Bissell explains how the field of mobility studies encourages a nuanced consideration of place that is attentive to relative and contingent fluidities and fixities.

To conclude this part, Deborah Martin considers place as more-than-human in extending a relational approach to understanding place production. In exploring notions of place and by investigating environmental policies as integral to urban place-making, the chapter focusses on a formal governance program to plant trees in a number of cities in Massachusetts, USA, the *Greening the Gateway Cities* program. Martin considers a number of human and non-human actors that may shape a place through tree planting. A state government agency and its forester-actors plant trees and foster tree stewardship on primarily private property. Community non-profit partners help to promote the program and individual residents and property owners steward the newly planted trees. In mobilising this energy and climate-adaptation formal governance policy, multi-scalar actors are brought together around a shared caring for trees. The resulting relationships and landscape demonstrates the ongoing, emerging and multi-actor making of places as human-environments.

References

Berthold-Bond, D. (2000) 'The Ethics of "Place": Reflections on Bioregionalism', *Environmental Ethics*, 22 (1): 5–24.

Brenner, N. (1999) 'Beyond State-centrism? Space, Territoriality, and Geographical Scale in Globalization Studies', *Theory and Society*, 28 (1): 39–78.

Edensor, T. and Smith, T. (forthcoming) 'Commemorating Economic Crisis at a Liminal Site: Memory, Creativity and Dissent at Achill Henge, Ireland', *Environment and Planning D: Society and Space*.

Giddens. A. (1990) *The Consequences of Modernity*, Cambridge: Polity.

Harvey, D. (1989) *The Condition of Postmodernity*, Oxford: Blackwell.

Herod, A. (2010) *Scale*, London: Routledge.

Kusenbach, M. (2008) 'A Hierarchy of Urban Communities: Observations on the Nested Character of place', *City and Community*, 7 (3): 225–249.
Massey, D. (1993) 'Power-geometry and a Progressive Sense of Place', in: J. Bird, B. Curtis, T. Putnam and L. Tickner (eds), *Mapping the Futures: Local Cultures, Global Change*, London: Routledge, 60–70.
Robertson, R. (1994) 'Globalisation or Glocalisation?', *Journal of International Communication*, 1 (1): 33–52.
Smith, N. (1990) *Uneven Development: Nature, Capital and the Production of Space*, Oxford: Blackwell Publishers.

1

Place as assemblage

Kim Dovey

Introduction

'Place' is a peculiar kind of concept in that we all know what it means in everyday life yet in academic discourse its definitions are deeply contested. In his excellent introduction to geographic theories of place Cresswell (2004) traces a history from the development of a phenomenology of place by Relph (1976) and others to the idea of a 'progressive' sense of place developed by Massey (1993). What emerges here is a central conundrum of place theory – a tension between ontological and phenomenological approaches on the one hand and social constructionist approaches on the other. The ontology of place has been traced by Casey (1997) to its emergence (as 'topos') in early Greek philosophy (most notably Aristotle) where it was seen as a form of ontological ground. In this sense place is inseparable from being – to exist is to exist in a place. This approach was revived and extended by Heidegger (1962) for whom there is no 'being' without 'being-in-the-world'. Existence is inextricably spatial as well as temporal and social. Heidegger's work represents not only a spatial turn but also an essentialist turn wherein place is pre-given and deep-rooted. This was the foil for Massey's avowedly anti-Heideggerian theory of place as 'open', 'progressive' and 'global', a valorization of 'routes' over 'roots' and the idea that places are the products of social and spatial inter-connections (1993, 2000). This open sense of place is outward-looking, defined by multiple and fluid identities; its character comes from connections and interactions rather than origins and boundaries. For Massey the Heideggerian sense of place is driven by the desire for singular identities rooted in authentic histories and protected by boundaries that exclude the other.

There have long been suggestions that these seemingly opposed views might be reconciled in a theory of place that is at once ontological and socially constructed (Casey, 1997; Cresswell, 2004; Thrift, 1999). Massey (2000) has also explored some of the contradictions of place and the difficulty of any simple categorizations of open/closed and global/local; she acknowledges the ambiguities of 'home' as a place of safety and yearning, but also of dark secrets and the unhomely (Massey, 1992). I suggest that assemblage thinking, based in the work of Deleuze and Guattari (1987), offers the prospect for a more encompassing approach that cuts across any binary between place as routes or roots, open or closed. In this conception Heidegger's 'being-in-the-world' might be replaced with a more Deleuzian 'becoming-in-the-world' – a more dynamic and open sense of place as a multiplicitous assemblage.

What is at stake here is much more than a theoretical point. Place is a concept with a deep potency both in everyday experience and in the modes of thought that are transforming our planet at multiple scales. The concept of 'place' spans disciplines and scales, from architecture through urban design and planning to global politics, from a place at the table to the nation state. We have seen the emergence of new fields of education and practice such as 'placemaking', 'place marketing' and 'place management'. New developments are often advertized as creating a 'sense of place' and political action to stop them is often framed as a defence of something similar. These everyday practices of placemaking, marketing, management and political action are infused with tacit theories of place; the tensions outlined above play out in everyday practice, even where there is a presumed shared understanding of what 'place' means.

Assemblage

The key theoretical base for assemblage thinking lies in Deleuzian philosophy and particularly in the conceptual toolkit outlined in Deleuze and Guattari's (1987) book *A Thousand Plateaus*; my interpretation also owes a considerable debt to DeLanda (2006). While 'place' is not a concept that is deployed here, the concept of the 'plateau' from the title of the book gives a first clue in a shared etymology (L: *platea*) to which I will return.

The term 'assemblage' is a translation of the French 'agencement' which is akin to a 'layout', 'arrangement' or 'alignment' and to some degree 'agency' – it suggests at once a dynamic process and a diagrammatic spatiality, both verb and noun. An assemblage is a whole that is formed from the interconnectivity and flows between constituent parts wherein the identities and functions of both parts and wholes emerge from the flows, alliances and synergies between them. For Deleuze 'It is never filiations which are important, but alliances, alloys' (Deleuze and Parnet, 1987: 69) and 'Don't ask what it means, show how it works'. Assemblage is a concept that resists any singular definition and is understood in multiple ways; my interpretation is based on those aspects of the theory that I find useful in urban research.

Assemblage thinking embodies what is often called a 'flat' ontology that can be understood as a set of reciprocal relations between the particular and the general. It is 'flat' in the sense that it does not recognize a transcendent order to which the immanent world can be reduced, particulars cannot be reduced to the general, the local cannot be reduced to the global. The particular is not just an instance of a general rule; rather, what becomes the rule emerges in part from the interactions of particulars. Thus, assemblage is multi-scalar thinking without any presumed hegemony of scale. Incremental change at the local scale can transform the larger assemblage. While sometimes accused of having a fetish for the particular (Scott and Storper, 2015), assemblage thinking is best understood as a search for the general within the particular.

Assemblage thinking is fundamentally relational – interconnections are prior to the things being connected. From this view a street or neighbourhood is not a thing nor a collection of discrete things, it is the multi-scale relations of buildings-sidewalk-road-city; the flows of traffic, people and goods; the interconnections of public to private space, and of this street to the city, that make it a 'street' and distinguish it from other places such as parks, plazas, freeways and shopping malls. The assemblage is dynamic – trees and people grow and die, buildings are constructed and demolished. It is the flows of life, traffic, goods and money that give the street its intensity and its sense of place.

Assemblage thinking offers an approach to theories of place without the reductionism and essentialism that has weighed down such discourse for so long. It is empirical without the reductionism of empirical science; it seeks to understand the social construction of place without reduction to discourse; and it seeks to understand the experience of place without the essentialism that can afflict phenomenology. While place is often seen as a primary mode of stabilizing being, assemblage is a philosophy of becoming.

Power/desire

Assemblage is a theory of power with its roots in the Foucault (1977) critique of power as embodied in micropractices, as distributed and capillary rather than simply held. The Foucaultian apparatus of the panopticon is a key source and model of an assemblage, yet Deleuze and Guattari exploit this revolution in thinking about power for its emancipatory potential as well as a critique of discipline. For Deleuze power is based in flows of desire which are the immanent productive forces of life, not limited to the human world. From this view the desire of a plant for light and water is not fundamentally different from the desire for decent housing. People and places are not subject to practices of power so much as they are produced by desires. Desire is a form of becoming that precedes being and identity; as desires become coded and organized they become identities, thing, places. All of the places we inhabit can be seen as products of desire. Streets, doors, corridors and freeways are products of desires to connect between places; a corner office with a commanding view emerges from desires for status, light and prospect; suburbs reflect desires for individual household identities. To see desire as the basis of power is to see it as positive, productive and as operating at a micropolitical level. It is axiomatic to assemblage thinking that power is distributed and embodied in material spatial arrangements – that agency is embodied in the materiality of places. This is not some kind of environmental determinism but a recognition that power is not simply held by human agents. Power is slippery and hidden within the material world and it is a key task of assemblage thinking to show how this works.

Territory

One of the key dimensions of assemblage thinking for an understanding of place is the concept of territory – more specifically the ways social and spatial boundaries are both inscribed and erased through a twofold movement of territorialization/deterritorialization. Territorialization emerges from the desire for order and a closed sense of place or home, the desire to keep chaos and difference at bay (Deleuze and Guattari, 1987: 310–2). A territory is a stabilised assemblage that works through a mix of both materiality and meaning – walls, fences, doors and gates on the one hand together with representation and expressive discourses on the other. Territorialization is multi-scalar, from Brecht's famous inscription of a chalk circle isolating the self from society to a national border, from leaving your coat on a chair in a restaurant to designing a gated community. Deterritorialization is the counter-movement through which territories are eroded (buildings are demolished, nations are invaded); deterritorialized elements are then recombined into new assemblages through a process of reterritorialization.

Territorialization produces more or less spatial segmentarity – a socio-spatial production of identity by social class, ethnicity, gender and age – toilets, schools, retirement villages, men's clubs, gated communities, squatter settlements and nation states are all segments of various kinds and scales. Public places such as streets, parks and plazas are relatively deterritorialized compared with the more private bedrooms, kitchens and offices which are relatively segmented and ideologically controlled. Segmentarity often works to eliminate space for hybridity, to fix identities through binary spatial categories and open/closed forms of placemaking that limit social mobility. Deleuze and Guattari (1987: 212, 224) make a distinction between 'supple' and 'rigid' segmentarities where rigid segmentarities are organized in a hierarchic structure of concentric segments (such as nation-city-neighbourhood-house) that resonate together. Large corporate placemakers (such as hotel chains or fast food restaurants) often work in a similar way where each branch is made to resonate with global formulae. Supple segmentarities by contrast involve a fluidity of lateral connections with potential for old segments to dissolve and new segments to form. This supple segmentarity is based on the power of networks and a fundamental distinction between tree-like and rhizomic structures and practices.

Rhizome/tree

A Thousand Plateaus begins with a chapter called 'Rhizome' which sets up a contrast between tree-like systems organized hierarchically with roots, stem and branches, and the rhizome (grass, potato, bamboo) which is characterized by horizontal lines of movement, networks and connectivity. Tree-like thinking organizes our world hierarchically under the branches of a transcendent idea (state, corporation, family, church) while rhizomic thought is identified by lateral movement of networks. Spatial structures are nearly always a mix of rhizome and tree – public life is fundamentally linked to networks of encounter and exchange in permeable spatial structures while privacy is more linked to enclosure and *cul de sacs*. There are however many assemblages that embody enclosed networks: permeable and open places of supple encounter that are rigidly enclosed on their boundaries. Open-plan offices, shopping malls and gated enclaves often embody such a structure where a supple space of flows is captured within a rigid territory.

The conceptual contrast between rhizome and tree which begins *A Thousand Plateaus* finds a parallel in the penultimate chapter on striated and smooth space (Deleuze and Guattari, 1987). The term 'striated' captures the etymological links to the Latin *stringere* 'to draw tight', linked to 'strict' and 'stringent'. This is contrasted with the 'smooth', which is intended to be understood as an absence of boundaries or joints rather than homogeneous. Smoothness implies a slipperiness and movement where one slides seamlessly from one identity, meaning or image to another. These are not different types of space so much as spatial properties. Striated space is where identities and spatial practices have become stabilized in strictly bounded territories with scripted spatial practices and socially controlled identities. Smooth space is identified with movement and instability through which stable territories are erased and new identities and spatial practices become possible. The smooth and the striated are not types of space since every real place is a mixture of the two in a reciprocal relation where they are constantly 'enfolded' into each other (Deleuze, 1993). A good example of such principles in placemaking practice can be found in current school design where architects and educators seek to design more effective places for learning – moving away from the disciplinary school towards one that enables student-based learning; this is based on the realization that while a strict tree-like segmentarity serves the discipline of teaching, learning is a multi-modal practice suited to a more flexible placemaking (Dovey and Fisher, 2014).

Twofolds

The set of conceptual oppositions I have introduced here – deterritorialization/territorialization, rhizome/tree, smooth/striated – are not to be understood as binary oppositions but as twofold concepts where the focus is on the interconnections and dynamism between them. These twofold concepts are part of a much larger cluster of conceptual oppositions that resonate with each other and with the broader ontological twofolds of difference/identity and becoming/being (Deleuze and Guattari, 1987). For Deleuze the privileging of identity over difference is a central tenet of western metaphysics that needs to be overturned – identities emerge from a field of differences; being is not pre-given but is the outcome of becoming. What makes this cluster of twofold concepts resonate together is the connection of the rhizomic, smooth and deterritorializing movement with an ontology of 'becoming'. Tree-like thinking, striation and territory are the means by which identity is constructed out of difference. Yet these categories are never separate: 'smooth space allows itself to be striated, and striated space reimparts a smooth space. Perhaps we must say that all progress is made by and in striated space, but all becoming occurs in smooth space' (Deleuze and Guattari, 1987: 486). This passage cuts across any simple interpretation of 'smooth' or open as good and 'striated' or closed as bad. One of the most common mistakes I see is in architecture students who believe they are designing places of emancipation simply by removing enclosure. Design is striation in the sense that it fixes a formal outcome – the task is to understand how different morphologies enable different kinds of becoming.

This conceptual opposition between change and stability, between 'wings and roots' – to add another metaphor – makes it tempting to add the conceptual opposition of 'space' versus 'place' and to identify space with freedom and movement in contrast to the stability and rootedness of place. I think this is a mistake and that place is best conceived as the assemblage rather than one part or kind of assemblage. The concept of place has been widely misrecognized as a tree-like concept that organizes spatial meanings around an essentialized stem. This view of place is understandable since it meets a primary human desire for a stabilized identity. Yet the identification of place with singular modes of rooted sedentary dwelling and stabilized identity is a narrow and insular view. Place is an assemblage that stabilizes dwelling but also encompasses lines of movement and processes of becoming.

Sense of place

Thus far I have addressed the open/closed question but what of the experiential dimension, the sense of place? In *The Logic of Sense*, Deleuze (1990) analyzes the encounter with the sensory world and argues that while we operate everyday on a basis of the 'common sense' of a taken for granted world, the logic of sense is infused with paradox. Sensation operates at a pre-reflective level, prior to cognition and meaning. The encounter with a place is experienced first and analyzed second. Deleuze is a realist philosopher inasmuch as sensation is seen as based in a material world yet is not reducible to materiality – sensation is an event that connects the material and expressive poles of an assemblage. The quest to 'make sense' of this experience is an impossible task because language can name this sense but is powerless to define it (1990: 19–20). Deleuze likens this quest to Lewis Carroll's *Hunting of the Snark*:

> They sought it with thimbles, they sought it with care;
> They pursued it with forks and hope;
> They threatened its life with a railway share;
> They charmed it with smiles and soap.
>
> (Carroll, 1876)

The poem is a nonsense that connects material things (forks, thimbles and soap) with ethereal expressions (care, smiles and hope). When we try to extract the 'sense' from the assemblage in which it is lodged we neutralize it; we are left with the 'smile' without the 'cat' as it were (Deleuze, 1990: 32). We say that an event 'takes place', but the event also creates place. Thus, we all understand 'place' in everyday life through this encounter, yet we are powerless to define it. When we experience a 'sense of place' it is a short step to presuppose an origin or essence as the foundation of the 'sense'. The concept of a 'sense of place' can become what Deleuze calls a 'despotic signifier' that seeks to stabilize identity in the service of a transcendent sense of power (Colebrook, 2002: 120). For Deleuze signs are aspects of the production of desire, we should ask not what they mean but what they do and how they work. What the discourse of 'place' so often does is to fix place identity and thereby close down processes of becoming.

When we speak of the 'sense' of place we also imply something more than location or spatial extension, that places have 'intensity' – a word we use to describe qualities that are indivisible such as temperature, density, colour, pleasure and encounter. It is the intensity that is most strongly linked to the sense and affect of place – the intensity of sunlight; the beauty of form; the buzz of conversation; the whiteness of the walls; the vastness of the sea; the sound of birds; the smell of coffee. Intensities are directly desired affects or qualities rather than meanings (Colebrook, 2002: 43–5); however, where desires become coded as everyday experience, they are reduced to signified identities like in a tourist brochure. This is what we mean when we say a place has become commodified – the sense of place becomes a cliché, a pre-packaged meaning for consumption. Thus 'white walls' are re-packaged as style: the Greek islands, modern architecture and so on.

Morphogenesis

An assemblage approach to place seeks to abandon any quest for a transcendent account of place identity and instead seeks an understanding of how places emerge through morphogenic processes. For Deleuze, morphogenesis is based in difference, not in a lack of resemblance between pre-given things but in a multiplicity of relations and flows of desire as the drivers of becoming (DeLanda, 2002). The Deleuzian ontology replaces essences with morphogenic processes. Assemblage is a realist philosophy – there is no sense of 'holding the world in brackets' as phenomenology often does, nor of reducing it to discourse as representational theory often does. For Deleuze, however, the real is not limited to the actual world we encounter but encompasses a world of possibility, the real embodies a set of capacities for change. What most matters about the material world – the morphologies of place – are the range of capacities and possibilities that are embedded within it. This concept of capacity is close to what Gibson (1979) calls 'affordance'. Instead of seeing the morphology of place as a set of fixed forms that stabilize identity, we need to understand the ways in which it enables new kinds of becoming.

My account of place as assemblage is inspired in part by the urban theory of Jacobs (1961) who wrote about the ways that everyday sidewalk life settles into certain patterns of social and economic life. Her analysis shows how a city works as a concentrated mix of people and buildings, intensive public/private interfaces and richly interconnected street networks; the place in this sense is the emergent outcome of a complex set of relations and processes that produce urban vitality. With various colleagues I have sought to apply assemblage thinking to a range of morphological dimensions of urban life including public/private interfaces, creative clusters, urban density, functional mix and access networks (Dovey et al., 2018). Jacobs' work is presented in a disarmingly simple language, and it is widely applied in a superficial and formularized manner. The complex synergies of 'density', 'mix' and 'access' that she explores can be understood as an urban 'DMA' – a set of morphological conditions that does not determine urban outcomes any more than human DNA determines who we are, but is crucial to the ways in which any city works and the ways place identity emerges at a range of scales (Dovey, 2016). Density concentrates more people, buildings and practices within a given area; shortening distances

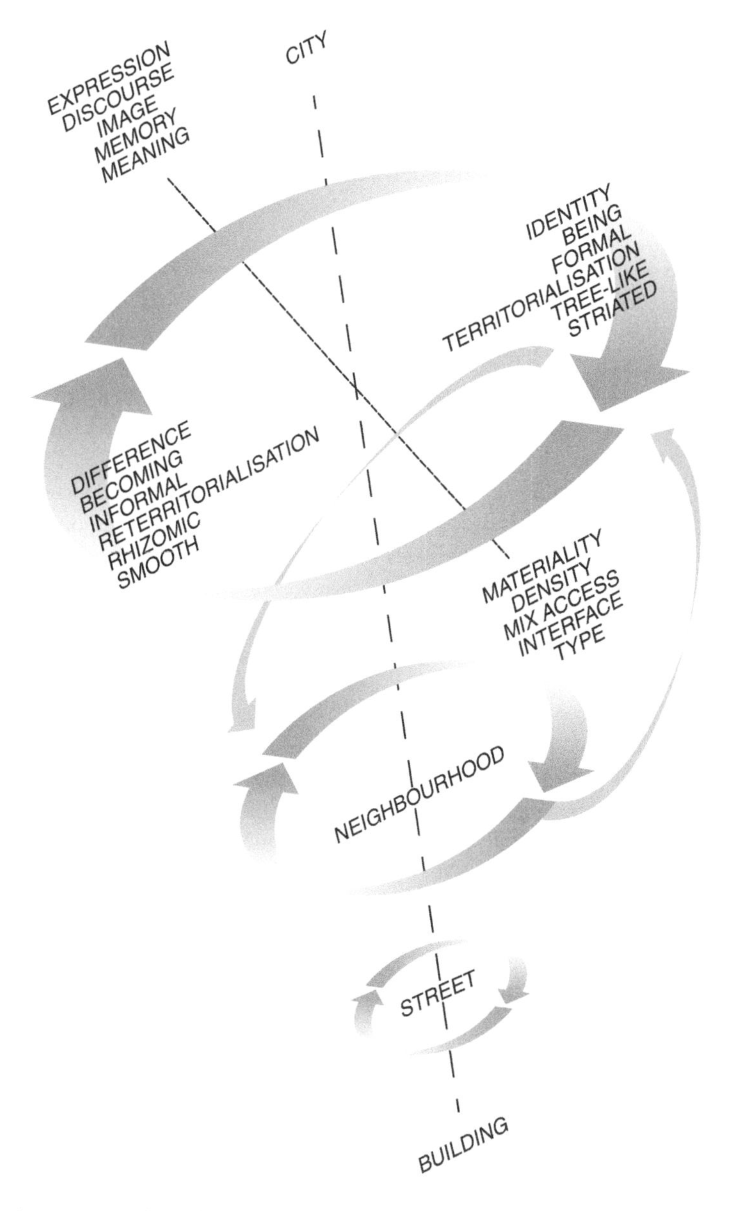

Figure 1.1 Place as complex adaptive assemblage.

between places and constructing those places through these connections. Mix is fundamentally about differences and juxtapositions between activities, attractions and people; at once a social, functional and formal mix – a mix of mixes between home, work and play. Much more than 'diversity', mix is a means of generating random encounters and of enabling flows between different kinds of people, buildings and functions. Access is about how we get around via access networks at multiple scales and with different modes of transport. If we want to understand 'walkability' as a capacity of any place then the urban DMA is the key (Dovey and Pafka, 2019).

Complex adaptive assemblage

Assemblage thinking has connections to the science of emergence, complex adaptive systems and resilience thinking (Holling and Gunderson, 2002; Johnson, 2001; Walker and Salt, 2006). In this sense places can be understood as complex adaptive systems where over time a regime or identity with certain characteristics emerges, settles down and becomes more or less resilient. Resilience is defined as the capacity of the system to adapt to change without crossing a threshold into a new 'regime' or 'identity' (Walker and Salt, 2006: 32). Resilience in this sense is not the capacity to maintain or return to a single stable state but rather a dynamic capacity to move between a range of adaptive states without crossing a threshold of no return. The forms of stability that we recognize as an enduring place identity can be understood in this way as an emergent regime, sustained by complex sets of relations and adaptive capacities. In this sense places are constantly subject to pressures for regime change – new buildings, new people and new roads all change the density, mix and access networks. Resilience thinking involves the study of thresholds of change in 'key slow variables' where incremental change can push the system across a threshold into a new regime, for better or worse (Walker and Salt, 2006: 113). Such variables may include land value, gentrification, traffic, density, parking, social mix, street life, climate, crime and public transport. As any of these variables changes, other variables adapt. Taller buildings lead to greater density and intensity of street life, greater viability for public transport and less sunshine. The assemblage may escalate and it may plateau. Resilience thinking suggests that places undergo adaptive cycles of expansion, conservation, collapse and re-organization within a multi-scalar model where every cycle is interconnected to cycles at higher and lower scales. Smaller and faster cycles of change at the level of a room or a building (from furniture to renovations) are geared to larger and slower cycles at the level of a neighbourhood or city (from gentrification to public transport investment).

Complex adaptive systems theory derives from the physical sciences and assemblage thinking would resist any reduction of place to a science. However, there are many interconnections that suggest a 'complex adaptive assemblage' may be a useful label for an understanding of place (Dovey, 2012). The movement between territorialization/deterritorialization resonates strongly with the adaptive cycle of resilience thinking and Figure 1.1 shows these cycles, axes and scales in diagrammatic form. The series of twofold concepts of striated/smooth, identity/difference, being/becoming, tree/rhizome can be similarly identified. This axis with its cluster of twofolds is, however, only one of two axes of the assemblage – the other is the distinction between the material and the expressive. In this regard resilience thinking is largely limited to the material pole of this axis with little to say about discourse, image, symbolism or art. This is the traditional division between the sciences on the one hand and the arts and humanities on the other; any understanding of place must bring them together. The assemblage is at once both material and representational, it resists reduction to either textual analysis or material structure.

Place as plateau

The title of *A Thousand Plateaus* is adapted from a concept in Bateson's (2000 [1972]) book, *Steps to an Ecology of Mind*, where he describes the ways in which dynamic cultural systems can stabilize on a plane or 'plateau' between polarities (note the shared etymology of these *pla* words). For Bateson, a 'plateau' emerges where the tendency for a system to escalate out of control (an arms race, escalating crime, environmental degradation) is countered by a plateau or plane of stability that co-exists with constant change (2000: 113). This tendency towards escalation can be linked to gentrification, to what Jacobs (1961) called the 'self-destruction of diversity' and to cycles of creative destruction under neoliberal capitalism (Harvey, 1985). What we sense as the stability of place is often a plateau of development produced by locally sustainable limits. Yet to perceive place as static is to misrecognize it as a thing rather than an assemblage of differences:

> an assemblage is first and foremost what keeps very heterogeneous elements together: ... How do things take on consistency? How do they cohere? Even among very different things, an intensive continuity can be found. We have borrowed the word 'plateau' from Bateson precisely to designate these zones of intensive continuity.
>
> *(Deleuze, in Deleuze and Parnet, 1987: 179)*

The sense of place always involves consistency and coherence, often misrecognized as uniformity and regularity. Place intensity is a dynamic intensity where tensions are sustained and sustainable. To understand place as assemblage is to conceive it as a socio-spatial plateau, a plane of consistency that is open to change but is resilient to unrestrained escalation.

Places of becoming

Assemblage is not a theory where all the parts fall into line with some orthodoxy; rather, it is one where concepts might be put to work to connect different modes of thought. Most theories are territories where intellectual capital is bounded and defended along with the reputations of those who contribute or sign up. Boundaries between theories are constructed, partly through jargon but also through more direct discourses of exclusion. Assemblage is a theory with multiplicity and difference at its core, where connections between territories are the primary source of new ideas. It is a philosophy where the process of becoming takes priority over any stabilized sense of being, where the Heideggerian 'being-in-the-world' is replaced by a Deleuzian 'becoming-in-the-world'. It enables an understanding of place in terms of both actual and possible worlds – at once a set of actual arrangements and a range of capacities for what it might become. This is what DeLanda (2011) calls a 'space of possibility' – a capacity to become transformed, designed and planned in a range of different ways. Assemblage therefore offers a theory of place that is closely geared to the creative and critical practices of architecture, urban design, urban arts and planning. Boundaries between research and practice erode as we experiment on the real city as the only possible laboratory of urban studies (Thrift, 2011). An assemblage is at once multi-scalar yet flattens any hierarchies of scale – the particulars of everyday life become as crucial as policy frameworks. Assemblage thinking cuts across divisions between formal and informal urbanism, opening up understandings of incremental adaptation and emergence. An assemblage approach is productive of alliances between the sciences, humanities and arts; between spatiality and sociality; between materiality and meaning. The academic monopolies held for so long by numbers and words in the sciences and humanities respectively are eroded as maps, diagrams and images also take on key roles as forms of spatial knowledge (Dovey et al., 2018).

I want to finish here by returning to the idea of an ontology of difference – that difference comes first and identity emerges from difference in itself. Difference is intrinsic to the everyday conception of place where even formularized places are rendered different from each other through everyday practice. Such differences *between* places largely construct the identities of places in everyday life; distinctions that are well-embedded in practices of power and often driven by the quest for social distinction. What Massey identifies as open or progressive are places *of* difference – where productive difference is embodied within a place; places of becoming that open new possibilities. However, this distinction is not about literal openness or closure – a bedroom or seminar room can be a place of becoming that relies on closure for this capacity. There are many different kinds of places and they embody difference in multiple ways.

We need to understand the concept of place as being ontological without becoming essentialist, deep-seated in everyday life without being deep-rooted in fixed origins. Place is immanent rather than transcendent; grounded in the particularities and practices of everyday life. Place as assemblage is a conceptual framework that connects the 'sense' of a place to the built form, the social to the material. The transformation we need most is from places that are closed, purified and static towards those that are open, multiple and dynamic. The difference that makes a difference is that between places of difference and places of purity; between places of becoming and those where identity formation is fixed and finished.

Acknowledgement

This chapter draws upon arguments first published in *Becoming Places* (Dovey, 2010: Ch. 2) and further applied in *Urban Design Thinking* (Dovey, 2016: Chs 12 and 30).

References

Bateson, G. (1972) *Steps to an Ecology of Mind*, New York: Ballantyne.
Carroll, L. (1876) *The Hunting of the Snark*, London: Macmillan.
Casey, E. (1997) *The Fate of Place*, Berkeley, CA: University of California Press.
Colebrook, C. (2002) *Understanding Deleuze*, Sydney: Allen and Unwin.
Cresswell, T. (2004) *Place: A Short Introduction*, Oxford: Blackwell.
DeLanda, M. (2002) *Intensive Science and Virtual Philosophy*, London: Bloomsbury.
DeLanda, M. (2006) *A New Philosophy of Society*, New York: Continuum.
DeLanda, M. (2011). *Philosophy and Simulation*, New York: Continuum.
Deleuze, G. (1990) *The Logic of Sense*, New York: Columbia University Press.
Deleuze, G. (1993) *The Fold*, Minneapolis, MI: U. of Minnesota Press.
Deleuze, G. and Guattari, F. (1987) *A Thousand Plateaus*, London: Athlone Press.
Deleuze, G. and Parnet, C. (1987) *Dialogues II*, New York: Columbia UP.
Dovey, K. (2010) *Becoming Places*, London: Routledge.
Dovey, K. (2012) 'Informal Settlement and Complex Adaptive Assemblage', *International Development Planning Review*, 34 (3): 371–390.
Dovey, K. (2016) *Urban Design Thinking*, London: Bloomsbury.
Dovey, K. and Fisher, K. (2014) 'Designing for Adaptation: The School as Socio-Spatial Assemblage', *Journal of Architecture*, 19 (10): 43–63.
Dovey, K. and Pafka, E. (2019) 'What is Walkability?: The Urban DMA', *Urban Studies*. Advance online publication: DOI: 10.1177/0042098018819727
Dovey, K., Pafka, E. and Ristic, M. (eds) (2018) *Mapping Urbanities*, New York: Routledge.
Foucault, M. (1977) *Discipline and Punish*, New York: Vintage.
Gibson, J. (1979) *The Ecological Approach to Visual Perception*, Boston, MA: Houghton Mifflin Harcourt.
Harvey, D. (1985) *The Urbanization of Capital*, Baltimore, MD: Johns Hopkins UP.
Heidegger, M. (1962) *Being and Time*, New York: Harper Row.

Holling, C. and Gunderson, L. (2002) 'Resilience and Adaptive Cycles', in: L. Gunderson and C. Holling (eds), *Panarchy*, Washington, DC: Island Press, 25–62.
Jacobs, J. (1961) *The Death and Life of Great American Cities*, New York: Random House.
Johnson, S. (2001) *Emergence*, London: Penguin.
Massey, D. (1992) 'A Place Called Home?' *New Formations*. 3–15.
Massey, D. (1993) 'Power-geometry and a Progressive Sense of Place', in J. Bird (ed), *Mapping the Futures*, London: Routledge, 59–69.
Massey, D. (2000) 'Space-time and the Politics of Location', in: A. Read (ed), *Architecturally Speaking*, London: Routledge, 49–61.
Relph, E. (1976) *Place and Placelessness*, London: Pion.
Scott, A. and Storper, M. (2015) 'The Nature of Cities', *International Journal of Urban and Regional Research*, 39 (1): 1–15.
Thrift, N. (1999) 'Steps to an Ecology of Place', in: D. Massey, Allen, J. and Sarre, P. (eds), *Human Geography Today*, Cambridge: Polity, 295–322.
Thrift, N. (2011) 'Lifeworld Inc – and What To Do About It', *Environment and Planning D*, 29, 5–26.
Walker, B. and Salt, D. (2006) *Resilience Thinking*, Washington, DC: Island Press.

2

Doreen Massey's 'a global sense of place' revisited

Ares Kalandides

Introduction

'A global sense of place', Doreen Massey's seminal essay of 1991, is one of the most frequently referenced articles in geography (Cresswell, 2006; Vaiou, 2017), although its popularity waxes and wanes over time. It appeared in June 1991, the last year of publication of *Marxism Today*, the theoretical and discussion journal of the Communist Party of Great Britain, and as such does not constitute an academic article in the narrow sense. It can be understood as a contribution to a broader discussion inside Marxism and a retort to other colleagues' formulations on place. The article is one of Massey's many attempts to reach a broader audience, both through her choice of the medium and the language of communication. It only features five references, most of which are not academic. However, the one pointing to Massey's own *Spatial Divisions of Labour* (1984) is of greater significance, as it positions the piece within her research projects of the late 1970s and early 1980s. Local specificity, and a preoccupation with *place*, are already addressed in *Spatial Divisions of Labour*. Research programmes inspired by this, labelled 'locality research', examined economic change in selected localities (localities here are conceptualized as places), triggering a controversy around the study of *place* in radical geography. *Place* became a main object of scrutiny for Massey at the beginning of the 1990s, with a series of articles written in its defence, that more or less develop the same arguments: 'A global sense of place' (1991a), 'The political place of locality studies' (1991b) and 'Questions of locality' (1993a), as well as a public lecture in 1990 published three years later, 'Power-geometries and a progressive sense of place' (1993b).

In this chapter I proceed as follows: in the next section I point out the main arguments in 'A global sense of place' and demonstrate their significance; in the second section I place the article in the context of Massey's own work and follow the development of concepts introduced here; in the final section I consider the relevance of the article today and why it clearly positions Massey not only as an influential geographer, but also as an important public intellectual.

Rereading 'A global sense of place'

'A global sense of place' makes a series of arguments and introduces concepts that Massey will develop further over subsequent decades. Its point of departure is the Marxian 'annihilation of space by time' and the so-called 'time–space-compression', namely that the internationalization of capital, the speeding up of global connections, movement and communications have caused spatial barriers to become (almost) obsolete, as well as our experience of that. This, it is argued, creates dislocation and uncertainty for local communities who lose their connections to place, retracting into competitive localisms, reactionary nationalisms and other particularisms. Sense of place becomes an introspective, exclusionary reaction to the threat of accelerated globalization. Communities seek refuge in sanitized versions of heritage, looking for the 'real' meaning of places. Massey's article questions both assumptions and shows on the one hand that time–space-compression is highly heterogeneous and on the other that there is a way of conceptualizing an alternative, *progressive*, sense of place.

She observes that the concept of 'time–space-compression' as a new feature may represent a Western viewpoint, whereas the history of colonialism with the sudden disruption of places by the violent presence of colonising others has been part of the experience of the colonised world for centuries. Also, limiting our experience of space as determined by money and capitalism obscures the ways that both gender and race enable or hinder mobility: women's experience of space, with the limitations of social norms and the threat of physical violence, has been more determined by men than it has by capital. Whereas the speed-up of global connections is indeed codetermined by the free movement of capital, the way we experience space and place is tied to much more than this. Furthermore, 'time–space-compression' does not happen equally for everyone in all spheres of activity: at the same time as the growth of air travel has increased the mobility of businessmen, the concurrent and related decline of shipping may have caused the further isolation of islanders.

In brief, 'different social groups and different individuals, are placed in very distinct ways in relation to these flows and interconnections' – something one might call the 'power-geometry of it all' (Massey, 1991a: 25). Such differentiation not only concerns who gets to move and who doesn't, as well as our experience of it, but also who gets to control such movement: the power over 'time–space-compression'. We thus find those who can both turn global flows to advantage and control them and, at the very bottom, those who can neither profit from nor have power over them. In between there is a 'highly complex social differentiation […] in the degree of movement and communication, but also in the degree of control and of initiation' (Massey, 1991a: 26).

Building upon the arguments on the social differentiation of 'time–space-compression' and the power-geometry in relation to this global net of connections, Massey sets out to reconceptualize 'sense of place' and 'place' itself. Across the article, 'sense of place' is associated both with 'rootedness' (p. 26), i.e., the particular feelings of belonging, and with 'identity' (p. 26), 'character' (p. 27), 'specificity' (p. 27) or 'uniqueness' (p. 29) of place. Rejecting the equation of 'time' with movement and progress on the one hand and 'space/place' with stasis and reaction on the other, she develops a series of propositions that may help us think of a 'progressive sense of place', 'which is extroverted, which includes a consciousness of its links with the wider world, which integrates in a positive way the global and the local' (p. 28).

Taking us for a stroll through a 'real place, and certainly one not defined primarily by administrative or political boundaries' (p. 28), Massey takes us through her own neighbourhood, Kilburn High Road north west of London, where we encounter the manifold presences of global connections: news, signs, produce, people and vehicles. There are two observations she makes from this virtual walk: first, that although 'Kilburn may have a character of its own, it is absolutely not a seamless, coherent identity, a single sense of place that everyone shares' (p. 28). If people have multiple identities, then so can places. Second, it would be wrong to conflate 'community' and 'place'. There are communities that can span the world – 'from networks of friends with like interests, to major religious, ethnic or political communities' (p. 28) – and also different communities with their own 'senses of place' in one place. Furthermore, communities are internally differentiated; as observed above, a woman's sense of place may be very different to that of a man's. Not only does Kilburn have many identities but, with parts of the world connected to it, these are certainly not introverted but rather provoke in us a 'global sense of place'. Among other things this observation defies the need to define place by drawing its enclosing boundaries.

With the geography of social relations changing, where some may stretch over long distances and others remain closer,

> it is possible to envisage an alternative interpretation of place [in which] what gives place its specificity is not some long internalised history but the fact that it is constructed out of a particular constellation of social relations, meeting and weaving together at a particular locus.
>
> *(p. 28)*

Each 'place' can be seen as a particular point of the intersection of networks, social relations and communications – a meeting place. Instead of places being understood as 'areas with boundaries around, they can be imagined as articulated moments in networks of social relations and understandings' (p. 28). There are very real relations between places and the world – economic, political, cultural and others.

These arguments formulate a 'progressive sense of place' which includes the following: first, place is not static nor frozen in time. It is produced by the interweaving of social relations and, since social relations evolve and mutate, then so will place. Places are processes. Second, although boundaries may be necessarily political or administrative, these are not necessary for the conceptualization of place, at least not understood as simple enclosures. It is through linkages to the outside that places acquire their uniqueness. Third, places do not have single identities, but are full of internal conflict. And finally, none of the above denies the specificity of space, but this specificity is created by the way that the differentiated space of global connections, the power-geometries, interlink in the particular locus and interact with the accumulated history of place. Understanding place through its connections to the local allows for a 'global sense of the local, a global sense of place' (p. 29).

Although the contribution of the article to academic debate and its importance in Massey's own work will be made clearer in the next section, we can consider the following.

First, by asking spatial questions 'A global sense of place' enriches the Marxist focus on capital with debates on gender and race – a turn that is not necessarily always greeted positively, especially at a moment in which Marxism faces an existential crisis. The fact that Massey decides to include this article in the 1994 collection of essays which she entitles *Space, Place and Gender* points to its importance for her.

Second, in this article she introduces the concept of power-geometries, which she will further develop. This allows her to rethink both space and place as unequal, relational formations that are carved out of the interweaving of social relations.

Third, more broadly, she expands the relational understanding of space – which by 1991 is well accepted in more radical strands of geography (Vaiou and Hadjimichalis, 2012) – to place, showing not only that a preoccupation with place can be a progressive endeavour too, but also that seeing place in this way allows us to raise questions of geographical responsibility.

Fourth, Massey does not just raise theoretical questions but shows how our conceptualization of the world has implications for our politics. For example, understanding the interrelatedness of power-geometries, how the movement of one can actively reduce the mobility of others and deepen the 'spatial imprisonment of other groups' (1991a: 26), points to the need to develop a politics of mobility and access. Or, the realization that the multiple identities of place, 'can either be a source of richness or a source of conflict, or both' (p. 28), the disjuncture of community and place leads to basic questions of a place-based politics.

Finally, Massey's choice of medium and language of communication is a conscious decision to talk, in the most direct sense, to a non-academic audience, including activists and anybody interested in the role of place in politics. Her avoidance of academic jargon, her choice of narrative and description – of Kilburn High Road, the use of *I* and *we*, and even the relative lack of academic reference – create a directness that reduce the distance between reader and author.

Place and space in the work of Doreen Massey

Relational place/space

Massey's relational conceptualization of place and space starts well before the 'A global sense of place' and finds first comprehensive expression in the 1984 *Spatial Divisions of Labour*. It is subsequently developed further and provisionally condensed in the 2005 *For Space*.

Places, argues Massey, are not just about buildings, nor merely about 'capital momentarily imprisoned' (1991b: 275). Instead, they are the intersections of dynamic social activities and social relations. And the identity of place, like that of individuals, is neither single nor pre-given. It has its own internal contradictions, tensions and conflicts that develop and change over time. The fact that we can conceptualize place as a process does not mean it ceases to exist (Featherstone, Bond, and Painter, 2013). An examination of the tension between the two – place as a *thing* and as a *process* – can be a promising avenue for further theoretical exploration (Dell'Agnese, 2013).

Over the two decades following 'A global sense of place', Massey develops a conceptualization of place/space, culminating in the 2005 volume *For Space* (2005). *Space*, argues Massey, is the sphere of the possibility of multiplicity. It is the dimension of simultaneous co-existence of social relations at all geographical scales, from the local to the global. We need to think of space and *time* together: one as the dimension that allows multiplicity and the other as the dimension of the interaction and transformation of this multiplicity.

If we understand space as a four-dimensional notion of 'space–time', i.e., not only as an abstract dimension but also as the simultaneous coexistence of social relations at all geographical levels – from the personal space of the home to global interconnections – we can re-think place in the same way:

> [A] 'place' is formed out of the particular set of social relations which interact at a particular location. And the singularity of any particular place is formed in part out of the specificity of the interactions which occur at that location (nowhere else does this precise mixture occur) and in part out of the fact that the meeting of those social relations at that location (their partly happenstance juxtaposition) will in turn produce new social effects.
>
> *(1994a: 168)*

Place is constituted by the interaction of trajectories at a particular location. Like the social relations that constitute it, place is always in flux and imbued with power. Although places may be heterogeneous and their boundaries porous, that does not mean they are not unique. Rather, we should think of that specificity as dynamic and relationally constructed – as a 'global sense of place'. However, place and space are not only constituted by social relations but can also (re)produce them. The 'throwntogetherness' (Massey's word) of people and things in place can always give rise to something new. Thus, place is more than an outcome – it also generates new social phenomena.

As Massey is neither the first not the only one to develop a relational understanding of space, I argue that one of her main contributions to the debate is that she avoids conceptualizing time, space and place *against* each other, but rather understands them as interrelated, albeit distinct concepts. In particular, she urges us to avoid creating a false competition between *space* and *place*: 'Don't let's counterpose place and space', as she will call her 2002 commentary (Massey, 2002). Her second major contribution is her understanding of space (and place) not simply as relational but as generative: the 'throwntogetherness' in place is not just the outcome of spatial relations, but can itself produce new ones.

Space, place and gender

Social relations and power-geometries, says Massey already in *Spatial Divisions of Labour* (1984), the way that inequalities are spatially organized, are not just of an economic nature. Spaces and places – as well as the sense that we have of them – are gendered. This not only reflects the ways in which gender is socially understood, but also has consequences for the way gender itself is constituted (Massey, 1994b). Also, '[w]ays of thinking, of apprehending the world, have to be produced and maintained. They are products of real histories' (Massey, 1997: 27) that have 'involved constant struggles: over the meanings of genders, the articulation of spaces and the construction of knowledge and the rights to its possession' (Ibid: 33).

Gendered relations are part of the socioeconomic change that produces a new spatial division of labour (Massey, 1983a). For example, in mining areas, the profession of the miner, as well as participation in unions, formed part of the constitution of a male working-class identity, which was accompanied by a simultaneous absence of female paid work. Whilst a man's place is in the mine, at the union or in the pub, a woman's place is at home, taking care of the man who earns the family's livelihood. As jobs for older men become scarce, an increasing number of women enter paid work, in particular in the low-paid service sector. The reorganization of capital leads to an overall decline in income, creating tensions *inside* the working class. As it is constituted through work in the mine, male identity is thus under attack both from unemployment and the rise of female paid work. But, says Massey, something else is happening: capital not only influences gender relations, but also uses them. Women, who did not have a tradition of paid work or the same organization in unions and their unions' support, accept lower wages, thus posing fewer obstacles to the needs of capital. This is the dual relationship between the reorganization of the production process and gendered relations: capital uses local gender relations for its own profit but at the same time interferes in re-forming them. Thus, local gender relations – as they were historically formed through the multiple layers of production relations – influence and are influenced by the particular form of the spatial division of labour. Class and patriarchal relations have created a reserve of female labour. Such reserves are a social construct, not the outcome of biological differences.

In order to understand the form that recession took here, we need to recognize that the regions were not only marked by working-class identity and the strong presence of the unions, but also that local identity was particularly *male* (Massey, 1983b).

Together with Linda McDowell, Massey shows the ways in which the conditions that sustain and reproduce male hegemony are created by examining the economic development of four English regions from the 19th to the late 20th century (McDowell and Massey, 1984). Further research in one of the more recent areas of rapid economic growth, Cambridge, showed that men's total devotion to work (90% of all those in the workforce are men) meant that there must be somebody else to take care of the household (Massey, 1995). This kind of work, in other words, is based on the absence of men from reproductive work, as somebody else is responsible for care (Massey, 1994b).

Massey's irritation with her (male) colleagues' writings takes a surprising turn in 'Flexible sexism' (Massey, 1991c), where she takes on Harvey's *The Condition of Postmodernity* (1989) and Soja's *Postmodern Geographies* (1989). Her objections include issues of style – the opaque language and the single (male) authoritarian voice of the author – and the choice of illustrations. However, she mostly takes issue with the authors' approach to modernity, with no mention of patriarchy and sexism outside production relations or even about the ways that places – in particular the city and public space – are always gendered.

We can trace Massey's feminism back to the political movements of the 1960s and 1970s and, in her own words, to an Althusserian re-reading of Marxism (Featherstone, Bond, and Painter, 2013; Saldanha, 2013). With the possibilities that the postmodern 'deconstruction of the whole notion of identity' opened up (Freytag and Hoyler, 1999: 88), Massey develops concepts that link the mutual constitution of 'space, place and gender'. In Massey's own words: 'doing feminist geography is about a lot more than studying gender specifically. It is an outlook on life and a way of doing things which is a lot more subversive than that' (Featherstone, Bond, and Painter, 2013: 259). It is about the ways in which we think and speak/write about space, it is about challenging hierarchies and performing the 'real two-way movement between the conceptual and the political' (Ibid). Above all it is 'about how we "live" our theoretical positions' out there in real life, in academia and beyond.

Power-geometries and the politics of place

A major argument in 'A global sense of place' is that a cohesive theoretical approach to *place* will permit progressive place-based political practice (Massey, 1991a). It should also give us the tools to understand globalization as a product of social relations, where the local and global intersect, thus constituting unequal 'power-geometries' (Massey, 1999b). Space–time compression does not annihilate space for everybody in the same way. One person's mobility can mean immobility for somebody else. The actual form of time–space compression is defined by power-geometries between individuals, groups and places (Massey, 1991a).

Globalization is produced in places that include nodes which control relations spanning the planet. London is such a node, from which not only a large portion of financial flows are controlled, but also a particular narrative about globalization is exported (Massey, 2004a). Thus, certain places – and associated local politics – carry with them a 'geographical responsibility'. In order to understand this, we need a 'geographical imagination' which will allow us to recognize the unequal ways in which places are interconnected and the specific role – and hence responsibility – of some places in this 'power-geometry' (Massey, 2004b).Massey's conceptualization of space leads us to think of globalization as a particular form of intensified, unequal global interconnections in a neoliberal context. In globalization, individuals, groups and places hold different positions that interact through 'power-geometries'. The nodes of globalized social relations, the control centres of globalization, are also the privileged spaces of political struggle:

> When we were thinking about this kind of analysis at the beginning of the Thatcher period, our analysis remained largely national – that's how we thought about hegemony, and there were real reasons for doing so. But we cannot do that now. And the change is not just one of empirical focus: globalisation means that the whole concept of hegemony no longer operates in quite the same way.
>
> *(Hall and Massey, 2010: 69)*

The consequences of the above conceptualization are manifold for the ways in which we understand the political economy of places and the relations among them: if places are constituted relationally, then it is not possible to look at each place's development in isolation. Instead, a place must also be seen as the outcome of inter-place relations. These may be relations of competition (as the neoliberal mantra goes), but can also be relations of co-determination, dependency, subordination and cooperation. As both *Spatial Divisions of Labour* (1984) and *World City* (2007) show, a place's 'economic success' may depend on another place's decline, contingent on how they are both positioned within global 'power-geometries'. If we follow this perspective, our reading of the winners and losers of the 2008 financial crisis is going to be very different from mainstream interpretation (see Hall and Massey, 2010).

If globalization is produced in particular places, then progressive local politics in these places will need to recognize the responsibility this entails and question these places' global dominance. The ways in which we practice politics in a city such as London need to include all viewpoints that allow ways to challenge globalization. Massey herself participated in the Occupy London movement, an experience she would later describe in spatial terms, as the humble tents at the feet of the grand altars of 'God and Mammon' (Hall, Massey, and Rustin, 2013).

This should by no means reduce the importance of local politics, but simply reformulate it in different terms. As Massey herself argued, there is nothing inherently progressive or reactionary in place and space (Featherstone, Bond, and Painter, 2013), but there is in how we think about and act in relation to place/space. People come together in places and through this '"throwntogetherness" that occurs in place, we are forced to construct a public realm to find ways to live with each other' (Wills, 2013: 143). In this reading, places are privileged – but not the only – arenas of political engagement around collective interests.

If we want to envisage 'a politics of place beyond place' (Massey, Bond, and Featherstone, 2009), it is important to imagine globalization in spatial terms through power-geometries in space–time (Massey, 1999b).

The language of place

For Massey, it is not just the concepts that matter for a place-based politics – it is also the words and language we use. We can compare 'A global sense of place' (in *Marxism Today*) and 'The political place of locality studies' (in *Environment and Planning A*), both from 1991, to see how she changes register to talk differently about similar things to different readerships. The same basic notion with all its complexity – place as the open-ended interweaving of distinct trajectories at one location – is conveyed with the lightness of everyday storytelling in the former and the intricacies of academic language in the latter:

> The language we use has effects in moulding identities and characterising social relationships. It is crucial to the formation of the ideological scaffolding of the hegemonic common sense. Discourse matters. Moreover, it changes, and it can – through political work – be changed.
>
> *(Massey, 2013: 9)*

If space is the sphere of the simultaneous coexistence of difference, then there are myriad possible stories on the planet. However, the hegemonic narrative of globalization is told through the eyes of the colonising West (Massey, 1999b). Every time we speak of 'developed' or 'developing' countries our vocabulary turns space into time. Language puts places in a temporal succession, where developed ones are at the front and developing ones are at the back of an imaginary queue; it is a question of time until the developing arrive at the position of the developed. From simultaneous co-existence we create temporal succession. However, by ascertaining spatiality both conceptually and in our language, we recognize that different trajectories are both relatively autonomous and interconnected.

Challenging authoritative language is raised to a political position in Massey's writing and you will look for dichotomous expressions or thinking patterns in vain: it is not place *versus* space or space *versus* time or male *versus* female (Massey, 1991c, 2005). It is place *and* space and time and *gender* together – us *and* them. Her thoughts are full of small question marks, slight fissures in a complex – yet not rigid – system that invites us to consider the premise and express doubts. She allows herself to revisit her former arguments, complete them, partly correct them, leaving open formulations or expressing doubts. Her texts – always in the first person – are full of 'I think', 'I believe' and 'I argue', even when they become philosophical. Her language uses everyday words and avoids eccentric neologisms. The goal is communication – not authority. The 'Understanding Cities' series for the Open University and the publication of *Soundings* (with Stuart Hall and Michael Rustin) from 1995 on are further testimonies of her engagement with communication. In her texts, words of the everyday are invested with new meaning. Terms that she has coined – 'global sense of place', 'power-geometries', 'spatial divisions of labour' – draw their power from the unexpected juxtaposition of ordinary words. She shows us ways to understand terms (space, time, place, identity) because that understanding makes sense in the particular place-time and because it serves a particular progressive politics. In 'Vocabularies of the economy' (2013) she reminds us of the power of words, the force of language, and invites us to take back words hijacked by neoliberalism.

Doreen Massey as a public intellectual

As Massey herself observes, theories do not stem from some eternal and objective reality, but rather address needs in a specific spatial-temporal moment and from a particular political viewpoint (Massey, 1999a). So, what do the ideas developed in 'A global sense of place' mean today and what is their relevance for our world? I argue that they matter in several ways.

First of all, the article matters in terms of academic method. Set in the broader context of the ways Massey has conceptualized place throughout her work, it provides us with some important clues on how to study places. If place is constituted through intersections of social relations in a particular location, then any study of place would need to examine both the make-up of those social relations and the location at which they intersect. Examining the spatiality of social relations explores the ways in which they may or may not span the globe, whether they're closer or farther away, connected or disconnected, the ways in which they are inserted in a complex power-geometry and their development over time. A study of location needs to investigate the smaller scale, the nitty-gritty of the everyday, the creation of meaning and the drawing of boundaries, as well as the materialities of place. The interplay between the physical and the social, the representational and the performative, the ontological and the processual, the fragmented and the integral, the dismantling and reassembling of scale – all only give us some glimpses into the complexity of the study of place.

'A global sense of place' allows us to conceptualize place identity in an outward-looking way at a time when localisms and nationalisms seem to be on the rise. Whilst recognizing both place specificity and people's attachment to place, this concept questions the existence of unique and stable place identities, founded on some long-internalized history. It makes clear that places are not internally homogeneous nor are they interrelated with other places in some abstract way. Rather, they are both internally structured around power relations and part of an uneven global power-geometry.

Furthermore, understanding places as processes means that we should not interpret change as loss – and, in particular, not as a loss of identity. What matters for its outcome and consequences is the way in which individuals and social groups have power over that change. This also creates a productive tension between place as process and place as a 'thing out there'.

What is already distinguishable in 'A global sense of place' – Massey's bridging of the divide between academia and politics – subsequently becomes more pronounced: the Open University book series, Massey's involvement in publications such as *Marxism Today*, *Red Pepper* and *Soundings*, as well as her engagement in social movements and policy-making around the world (Nicaragua, Venezuela, Greece, GLC, Occupy London). Massey was simultaneously an academic, educator, activist and policy advisor, an exemplary public intellectual, and 'A global sense of place' has become the paradigmatic text of this engagement.

A careful rereading set in the context of Massey's life and oeuvre can reveal the power of a text that was originally published in a party journal and can throw light onto why it has become such a milestone in contemporary geographical thought. Although many of the ideas developed in it have become part of the academic core, the article retains its radical potential, as this chapter has tried to demonstrate. It provides us with a set of tools, not only to understand and question, but to change the way we *do* academia and politics.

References

Cresswell, T. 2006. *Place. A Short Introduction*. Malden, MA: Blackwell Publishing.

Dell'Agnese, E. 2013. 'The Political Challenge of Relational Territory.' In *Spatial Politics: Essays for Doreen Massey*, edited by D. Featherstone and J. Painter. Chichester: Wiley-Blackwell, 115–124.

Featherstone, D., S. Bond, and J. Painter. 2013. '"Stories so far": A Conversation with Doreen Massey.' In *Spatial Politics: Essays for Doreen Massey*, edited by D. Featherstone, and J. Painter. Chichester: Wiley-Blackwell, 253–266.

Freytag, T., and M. Hoyler. 1999. "I Feel as if I've Been Able to Reinvent Myself' – A Biographical Interview with Doreen Massey.' In *Power-geometries and the Politics of Space-time*, edited by D. Massey. Heidelberg: Department of Geography, University of Heidelberg, 83–90.

Hall, S., and D. Massey. 2010. 'Interpreting the Crisis.' *Soundings: A Journal of Politics and Culture*, 44: 57–71.

Hall, S., D. Massey, and M. Rustin. 2013. 'After Neoliberalism: Analysing the Present.' *Soundings: A Journal of Politics and Culture*, 54: 3–19.

Harvey, D. 1989. *The Condition of Postmodernity*. Oxford: Blackwell.

Massey, D. 1983a. 'Industrial Restructuring as Class Restructuring: Production Decentralization and Local Uniqueness.' *Regional Studies* 17(2): 73–89.

Massey, D. 1983b. 'The Shape of Things to Come.' *Marxism Today*, April: 18–27.

Massey, D. 1984. *Spatial Divisions of Labour: Social Structures and the Geography of Production*. London: McMillan.

Massey, D. 1991a. 'A Global Sense of Place.' *Marxism Today*, June: 24–29.

Massey, D. 1991b. 'The Political Place of Locality Studies.' *Environment and Planning A: Economy and Space* 23(2): 267–281.

Massey, D. 1991c. 'Flexible Sexism.' *Environment and Planning D: Society and Space* 9(1): 31–57.

Massey, D. 1993a. 'Questions of Locality.' *Geography* 78(2): 142–149.

Massey, D. 1993b. 'Power-geometry and a Progressive Sense of Place'. In *Mapping the Futures: Local Cultures, Global Change*, edited by J. Bird, B. Curtis, and T. Putnam, G. Robertson and L. Tickner. London and New York: Routledge, 60–70.

Massey, D. 1994a. 'A Place Called Home?' In *Space, Place and Gender*, edited by D. Massey. Cambridge: Polity Press, 157–173.

Massey, D. 1994b. 'Space, Place and Gender.' In *Space, Place and Gender*, edited by D. Massey. Cambridge: Polity Press, 185–190.

Massey, D. 1995. 'Masculinity, Dualisms and High Technology.' *Transactions of the Institute of British Geographers* 20(4): 487–499.

Massey, D. 1997. 'Economic/Non-economic.' In *Geographies of Economies*, edited by R. Lee and J. Wills. London: Arnold, 27–36.

Massey, D. 1999a. 'Philosophy and Politics of Spatiality: Some Considerations. The Hettner-Lecture in Human Geography.' *Geographische Zeitschrift* 87(1): 1–12.

Massey, D. 1999b. 'Imagining Globalization: Power-geometries of Time–space.' In *Global Futures: Migration, Environment and Globalization*, edited by A. Brah, M. Hickman, M.M. Mic, and M. Ghaill. Basingstoke: Palgrave Macmillan, 27–44.

Massey, D. 2002. 'Don't Let's Counterpose Place and Space.' *Development* 45(1): 24–25.

Massey, D. 2004a. 'Geographies of Responsibility.' *Geografiska Annaler: Series B, Human Geography* 86(1): 5–18.

Massey, D. 2004b. 'The Responsibilities of Place.' *Local Economy* 19(2): 97–101.

Massey, D. 2005. *For Space*. Thousand Oaks, CA: Sage.

Massey, D. 2007. *World City*. Cambridge: Polity Press.

Massey, D. 2013. 'Vocabularies of the Economy.' *Soundings: A Journal of Politics and Culture* 54: 9–22.

Massey, D., S. Bond, and D. Featherstone. 2009. 'The Possibilities of a Politics of Place Beyond Place? A Conversation with Doreen Massey.' *Scottish Geographical Journal* 125(3–4): 401–420.

McDowell, L., and D. Massey. 1984. 'A Woman's Place?' In *Geography Matters!* edited by D. Massey and J. Allen. Cambridge: Cambridge University Press, 128–147.

Saldanha, A. 2013. 'Power-Geometry as Philosophy of Space.' In *Spatial Politics: Essays for Doreen Massey*, edited by D. Featherstone and J. Painter. Chichester: Wiley-Blackwell, 44–55.

Soja, E. 1989. *Postmodern Geographies: The Reassertion of Space in Critical Social Theory*. New York: Verso.

Vaiou, D. ed. 2017. 'Αφιέρωμα στην Doreen Massey [A tribute to Doreen Massey].' *ΓΕΩΓΡΑΦΙΕΣ [GEOGRAPHIES]* 29: 3–59.

Vaiou, D. and C. Hadjimichalis. 2012. *Ο χώρος στην αριστερή σκέψη [Space in Left Thought]*. Athens: Nisos.

Wills, J. 2013. 'Place and Politics.' In *Spatial Politics: Essays for Doreen Massey*, edited by D. Featherstone and J. Painter. Chichester: Wiley-Blackwell, 135–145.

3

Place and nation

Rhys Jones

Introduction

Common-sense understandings of geographic scale tend to view places and nations as geographical entities that operate at two distinct scales. Places tend to be considered as geographical entities that exist at the local scale. Some academic conceptualisations of place reinforce such a perception. John Agnew's (1987) three-fold definition of place tends, at least implicitly, to convey an understanding of place that is local in character. After all, how can one meaningfully describe a place's location (in terms of its latitude and longitude) and its locale (its material basis for human existence) at scales other than the local? Similarly, our common-sense understandings of nations tends to draw our attention to a distinct spatial scale; the national scale. A key aspect of any nation, of course, is said to be its strong association with a particular territory (Wiebe, 2002: 5). Such an emphasis reinforces the notion that discourses of nationalism are ultimately concerned with one particular kind of spatial imagination; one centred on the national territory and scale.

My main aim in this chapter is to challenge these geographical imaginations by demonstrating the inter-connectedness of place and nation. I do so in two ways. I begin by showing how the discourse of nationalism portrays nations and national territories as places that should have some meaning for members of the nation; a degree of meaning that leads to a situation in which members of the nation should be willing to die for it, if the need arises. The second kind of inter-connectedness I discuss relates to how nations come to inhabit 'local' places in various ways. Nations take on meaning and are reproduced in different places and a detailed examination of this process provides us with an insight into the character of national discourses in general.

I have been a little guilty up until this point of taking some things for granted; things of which the reader might not be aware or with which they might not agree. Let me clarify two issues. First, I consider nations to be primarily the contingent product of discourses produced and performances undertaken by a range of actors. Nations do not possess agency as such but, rather, must be reproduced through a 'group-making project' (Brubaker, 2004). Second, national discourses are inherently geographical in their character. While much attention has been directed traditionally towards understanding the histories and times of nations and nationalism, for several years geographers and others have attempted to shine a light on the geographical themes that are folded into nationalist discourses and performances (Gruffudd, 1994). The current chapter is part of this recent tradition in that it seeks to demonstrate the many ways in which the notion of place is implicated in the reproduction of the nation.

The chapter proceeds as follows. In section 2, I discuss how successful nationalist 'group-making projects' promote an ideal that the national territory is a place with which members of the nation can positively identify. I then proceed to discuss various ways in which nations and nationalist discourses inhabit more localised places. In section 3, I elaborate on how such places take on a key role in allowing nations to be represented through various media and reproduced through a range of practices. In section 4, I consider the attempts that have been made to study places as geographical venues within which nations and nationalist discourses can take on material form and elicit affective responses. In section 5, I change tack somewhat by examining the way in which places can allow one to understand the conflict that lies at the heart of all nationalist discourses. I also provide a more upbeat discussion of how places can act as fora within which nationalist reconciliation can emerge. The examples I discuss are largely based on the specific case study of Wales, a region and nation located on the western seaboard of the UK, although I also refer to other locations and nations where appropriate.

National territory as place

The first way in which we can consider the close interrelationship between place and nation is in relation to the attempts made within nationalist discourses to view national territories as places. At heart, nationalist discourses are predicated on the need to protect and enhance national territories or homelands; some argue that therein lies a large part of their discursive power (Paasi, 1996). The ultimate goal of nationalist discourses, when viewed from a geographical perspective, is to inculcate a sense of belonging amongst the members of a nation towards the nation itself and, by extension, towards the national territory. Likewise, the existence of a national territory becomes one key way of promoting a sense of national distinctiveness. Nations are differentiated from others through the promotion of a discourse of difference, while also being simultaneously subject to a discourse of integration, which highlights their internal homogeneity (Van Houtum and Van Naerssen, 2002). The national territory becomes one key vehicle for enabling this process to occur, with one national place – characterised by an alleged internal homogeneity and eliciting a group sense of belonging – being contrasted with other national places.

The border, boundary or frontier between different national territories becomes a key site where one can witness this place-making process operating at the national scale. Anssi Paasi (1996) examines the significance of the border region of Karelia for the construction of a Finnish national territory. State agents and individuals in civil society are involved in the production and performance of a Finnish national identity, especially pertinent in a border region like Karelia, given its disputed political status vis-á-vis the Soviet Union and, latterly, the Russian Federation. Place-making in this region, therefore, takes on a broader national and geopolitical significance. A similar connection between national territories and identity and belonging comes to the fore in Wales. As part of a research project on the campaign for bilingual road signs in Wales – road signs that contained place names and instructions for road users in the Welsh and English languages – some individuals interviewed described the significance of these road signs as markers of a Welsh national territory and, by extension, of Wales as a distinctive place. One individual (original emphasis) said as follows:

> I feel a certain exhilaration as I cross the Severn Bridge near to Cas-Gwent [Chepstow], and it's 'Cas-gwent' that the signs say. And 'Casnewydd' [Newport] and 'Dim Parcio' [No Parking] and so on. From *the very boundary* [of Wales]. To compare that with what existed fifty years ago, it's nigh-on miraculous.

We witness here the distinctiveness of bilingual road signs as markers of a Welsh national territory (Jones and Merriman, 2012). The feeling of exhilaration described by the interviewee was linked to a feeling of pride and a sense that these road signs were helping to create a more bilingual place with which this individual could identify and feel a sense of belonging (see Figure 3.1).

The above quote highlights how the border or boundary acts as a particularly powerful location within which national territories take on the status of place. References to road signs, however, also draw our attention to the way in which the existence of national territories – viewed as national places – extend well beyond the border into the 'heartlands' of the nation. It is instructive to return to the campaign in favour of bilingual road signs in Wales. During the 1960s and 1970s, many politicians described campaigners as vandals because of their tendency to either deface or destroy monolingual English road signs. For many Welsh nationalist campaigners, the English road signs themselves, rather than the defaced signs, were deemed 'eyesores' and acts of cultural vandalism:

> If our road-signs fulfil the demands of aesthetic standards, they also destroy completely the standards of Welshness. In the eyes of the Welshman [sic], they are ugly, unbearably ugly. And the only way … to convince everyone of this, is by offending other standards, the standards of superficial aesthetics, i.e. by painting English road-signs and leaving them – for all to see – untidy and illegible.
>
> *(Iwan, 1968: no page)*

Figure 3.1 An example of a bilingual road sign.

Members of the Welsh nationalist movement possessed a totally different set of understandings of the meaning of monolingual road signs (Jones and Merriman, 2009), which were viewed as materials and objects that undermined the ability of Welsh speakers to belong to Wales as a national territory. The act of defacing road signs was necessary, therefore, as a way of drawing attention to the linguistic and cultural defilement that was associated with their presence within the Welsh national territory. Bilingual road signs were viewed differently. They played an important material role in creating a Welsh national territory with which Welsh speakers could identify. In all this, we witness how national territories can take on a significance that extends well beyond being a spatial container for nationalist discourses. If nationalist discourses are to be effective, national territories – at both the border and within the 'heartland' – take on a particular significance as sources of cultural meaning for members of the nation.

The process of place-making can also exist in more practised and performative contexts. Tim Edensor (2004: 109) shows how the performance of driving within different nations, 'in which we unreflexively carry out quotidian manoeuvres and modes of dwelling as habituated body subjects', helps to reinforce a 'national habitus' (2002: 89–93) or a shared identity that acts as a common frame of reference for members of the nation. Similarly, the communal practices of members of a nation on a national day of remembrance or celebration can help to mark out the national territory in performative terms (Edensor, 2002: 69–70). In a very real sense, therefore, a national territory 'becomes materialized through … sets of social practices' (Kingsbury, 2008: 53). And in both these cases – driving and acts of commemoration – national territories become places through shared practices. I develop this theme in more detail in the following section.

Place and nation 1: representation and reproduction

As I noted earlier, one can also study the inter-connectedness of place and nation by examining the way in which national discourses land or make use of certain local places. One key set of associations exists in relation to the way in which nationalist discourses are: 1) represented through certain places; 2) reproduced within particular places.

Certain places play a key role in *representing* nations. These places are often highly localised and yet contribute to the representation of nations in far-reaching ways. A number of theoretical and empirical contributions in Geography and beyond have sought to explore the key role played by specific places in our comprehension of given nations. Johnson (1995) examines the importance of key places within the Dublin cityscape as coming to symbolise the national struggle within Ireland. She explains how understandings of Irish nationalism were played out in the context of the statuary that was erected in the city, so that in many ways, the statues themselves came to reflect the wider currents affecting the Irish nation and its struggle for political independence. Appleton (2002), too, examines how geographical scales other than the national scale are implicated in the representation of nations. Drawing on an in-depth study of the *Saturday Evening Post*, the long-running and immensely popular weekly gazetteer published in the United States during much of the twentieth century, she argues that the magazine represented American nationalist discourse in different ways, using various spatial scales. An important set of scalar narratives centred on the scale of the home and the local scale. Certain places were seen as representing American ideals and their use in the pages of the *Saturday Evening Post* helped the magazine's readers to make sense of what American identity was supposed to mean.

And yet, localities and places possess a significance that extends well beyond merely representing the nation. We also need to consider the way in which localities and places are key sites for the *reproduction* of nationalist discourse in various ways. First, the places where individuals live, work and socialise are key sites within which they make sense of their relationship with the nation (MacLaughlin, 2001). Indeed, one can question whether individuals can ever make sense of their own position within the nation without considering how those understandings are conditioned by local circumstances. In this sense, nationalist discourse is always amended and contested within particular places, even when it is ostensibly generated elsewhere. In these everyday acts within different places, one witnesses how nationalist discourse is reproduced in small-scale ways. For example, Fevre et al. (1999) demonstrate how understandings of nationalism are re-worked within Welsh local settings. They discuss the way in which many people in north Wales use the processes that operate within the local housing market, in which Welsh-speakers cannot afford to compete with English newcomers, to help them reaffirm their sense of Welsh nationalism. A lack of housing in rural areas, therefore, is recast as a national issue (see also Thomson and Day 1999).

Second, it is evident that nationalist discourse is always generated within certain places and, as such, local politics and cultures always have the potential to play an important role in influencing the character of that discourse. Much sociological and historical literature focusses on the role of the intellectual and intelligentsia in producing and transmitting nationalist discourse (Kornprobst, 2005: 403). Focussing our attention on the role of such agents also behoves us to examine the material and spatial contexts within which these individuals operate. There is a need to examine, therefore, the embedded and place-based relations of these individuals and how these come to inflect the generation of nationalist discourses.

Aberystwyth is one such place that has played a key role in the generation of Welsh nationalist discourse. Political activities in Aberystwyth were instrumental in the

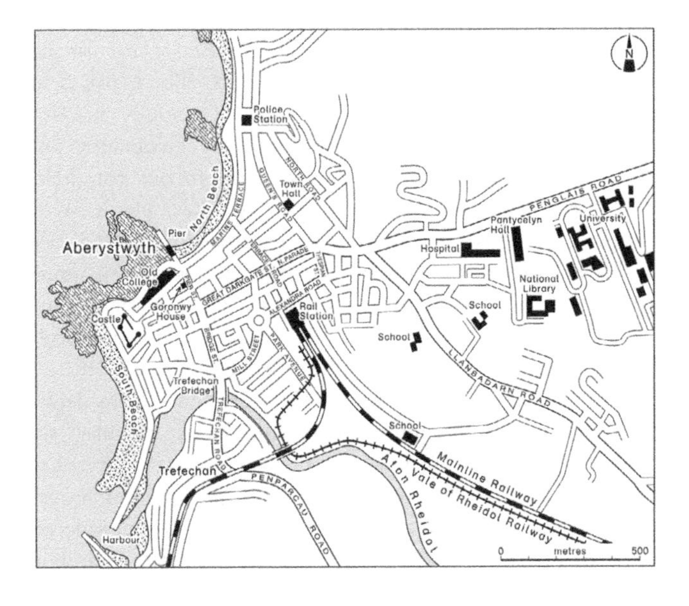

Map 3.1 Aberystwyth's key locations for nationalist debate.

development of a broader Welsh linguistic nationalism during the 1960s. Certain locations within Aberystwyth acted as a pivotal location in the early stages of the development of Cymdeithas yr Iaith Gymraeg, the campaign group formed in 1962 to promote greater legal status for the Welsh language. Almost all of the formative meetings and protests took place in various locations within the town; university halls of residence, cafes and rooms in a of private houses and public houses, such as the Coopers Arms and the Black Lion (see Map 3.1). In addition, these political activities were centred on a fairly small group of core individuals comprising Aberystwyth's Welsh nationalist movement.

Aberystwyth's key locations for nationalist debate

It is significant, too, that Cymdeithas yr Iaith Gymraeg's first major protest occurred on Trefechan Bridge, on the main approach into the town from the south. The events that took place on 2 February 1963 focus our attention, first, on Aberystwyth as a key place within the nationalist politics of the time and, second, on locations within Aberystwyth that served as loci for these seminal moments in the evolution of Welsh nationalism: the upstairs room of the Home Café, which became an informal headquarters for the protest, the Post Office, which was the original target of the student protests, and Trefechan Bridge, which witnessed a sit-in and some violence instigated by disgruntled local residents. This one day of protest, furthermore, cemented the importance of Aberystwyth as a focal point for future Welsh language campaigning and as a key place in the production of Welsh nationalism. The words of EG Millward, one of the leaders of the protest, emphasised this point: 'we are making Aberystwyth a focal point for this campaign' (*The Times*, 1963). We witness, here, how particular places become identified as key locations for the generation of nationalist discourse, becoming nodes of nationalist debate for intellectuals and intelligentsia.

Place and nation 2: materiality and affect

The discussion in the previous section points to the need to understand the materiality of place (Hetherington, 1998) and the impact that this has on the representation and reproduction of nations and nationalist discourse. We witnessed the key role played by statues and monuments of different kinds and how nationalist practices became entwined with the materiality of certain places. But to understand the impact of the materiality of place on the representation and reproduction of the nation, we also need to pay heed to how these material markers of the nation elicit particular kinds of responses among the members of the nation. What kinds of emotions do members of a nation feel when they view a national monument? What affective responses are engendered when an actor encounters an object imbued with certain nationalist significance? Is it a feeling of joy, pride, indifference or, as in Figure 3.2, anger?

Recent work in Geography and beyond has begun to examine these affective experiences of material and place-based nations. Materials, according to this literature, are 'lively, elemental, excessive, forceful, interrogative, distributed, more-than-solid, more-than-earthly, emergent, and in process' (Merriman and Jones, 2017: 602) and, as such, play a constitutive, varied and unpredictable role in shaping affective nationalisms. Each of the materials we might associate with nationalism – documents, signs, monuments, indeed all kinds of objects that are enrolled into our everyday experience of nationalism – generate affective forces. Some affective responses may be intended and anticipated while others may be unforeseen. Such sentiments echo Brubaker's (2004) comments about the contingent character of the group-making project associated with

Figure 3.2 Affective nationalism: destroying monolingual road signs.

nationalism. Some objects and materials may well generate the kind of affective response designed by intellectuals and the intelligentsia, while others may fail miserably. Other objects, which might be considered to be ostensibly apolitical, lying beyond the realm of nationalist discourse and practice, can, under particular circumstances, be reframed as a result of the unintended emotions that they elicit. The political protests concerning road signs in Wales in the 1960s and 1970s provides an excellent illustration of this point. Seemingly mundane and apolitical issues, such as the colour and font of place names, took on distinctly political and affective qualities as a result of the unintended ways in which they were perceived by nationalist actors (Jones and Merriman, 2009).

Angharad Closs Stephens (2016: 181) describes this embodied, affective and material aspect as akin to national atmospheres; ones which 'congeal around particular objects and bodies and echo as part of an assemblage'. She uses the example of the London Olympics of 2012 to illustrate the significance of these national atmospheres, as a particular place – London in this case – was transformed into a generator of object/body/emotion relations. The Olympic Stadium, in particular, became a significant node in the generation of emotion. Viewers' experiences of being in an enclosed stadium and watching inspiring sporting endeavours played a key role in generating positive emotions, such as happiness, pride and togetherness. But, of course, there is no guarantee that the materiality of the place will always have the desired effect. The booing that accompanied George Osborne's (the UK's Chancellor of the Exchequer at the time) presentation of medals for the Men's 400 metres T38 Paralympic event testifies to the unpredictability of affective responses to staged nationalist celebrations. In Closs Stephen's (2016: 186) words, in this specific instance 'something unpredictable … disrupted the otherwise carefully choreographed atmospheres'.

A focus on place and affect can also help us approach another key material aspect of nations and nationalism, namely their reliance on infrastructures of different kinds. Williams and Smith (1983: 511) describe how the process of nation-building involves the transformation of a territory 'by new cities, by a network of roads and railways, by dams and power stations, by making deserts bloom and tundra yield their riches, by multiplying factories and plants'. It is as a result of such infrastructures that a territory is transformed into a national territory. Yet, Williams and Smith do not consider the important affective qualities of these national infrastructures. What are the emotions and identities that are elicited when one enters a large railway station in any state and sees various destinations – dotted throughout the national territory – being listed on the departure boards? For those living in the UK, does entering a hospital or GP surgery elicit a more embodied feeling of pride in the National Health Service than does merely thinking about that organisation in the abstract?

Recent work by Merriman and Jones (2017) examines such issues. They focus on the affective qualities associated with road infrastructures in Wales. The case of the A470 trunk road is particularly instructive in this respect. Ever since it was designated as a new trunk road in 1972, it has played a significant role in, first, enabling embodied and mobile connections between North and South Wales and, second, instilling a series of emotional responses among its many travellers, including excitement and anticipation as individuals travel to see friends and relatives in other parts of Wales, and boredom and frustration as a result of the tediousness associated with travelling along its whole length. Third, it also becomes a linear place in its own right, as a result of the many public proclamations and private reflections on its importance as a physical infrastructure with which individuals feel a sense of belonging. In short, it has become 'a road that ... provide[s] an affective and relational glue' for the Welsh nation (Ibid: 611). It has, moreover, become an accepted part of Welsh popular culture. Musician Cerys Matthews, for instance, travelled along the A470 in 2014 to 'find out what it can tell me about Welsh identity and the essence of Welshness' (BBC, 2014).

Such themes illustrate the need for any study of the significance of the concept of place for nationalism – and of nationalism for place – to examine the links between materiality and affect. Even if they are sometimes fleeting, 'intermittent' and 'flickering' (Merriman and Jones, 2017: 600), these associations are important and help to practically demonstrate how 'national feelings touch us, take hold and become infectious' (Closs Stephens, 2016: 183).

Place and nation 3: conflict and reconciliation

One undercurrent in the preceding discussion is the way in which places can become sites of nationalist conflict. Inevitably, nationalist conflicts occur in certain locations and can play a constitutive role within place-making processes. At the same time, some argue that it is possible to view places as sites and scales that provide an opportunity for national reconciliation. I now discuss these more conflictual and peaceable associations between place and nation.

Nationalist conflict within place can arise for many reasons but one key area of enquiry has been in relation to place names. Place and street names represent key ways in which nationalism becomes embedded within particular places. Azaryahu and Kook (2002: 199) maintain that part of the significance of place and street names is that they 'introduce historical memory into a sphere of human activity that seems to be separated from the realm of ideology'. They are significant, therefore, because of their banal qualities but can also be contested features of place that may well reflect and further more extreme versions of nationalism. Street names in Israel represent one example of the contested process of naming places. Azarayahu and Kook's (Ibid: 206) research shows how the naming of street names in the town of Umm el Fahm during the 1990s 'conflated urban planning and ideological self-assertion'. The re-naming of streets by the leaders of a municipal council, controlled by an Islamic movement since 1988, shows how hot forms of nationalism may be reflected in, and contribute to, banal landscape features. Other work shows that it is not solely the histories commemorated within such names that are at stake but also the language used. Research by Gade (2003) examines the divisive politics of the language associated with road signs in Québec and Catalonia (see also Raento, 1997).

Another notable example of the way in which place and street names can generate nationalist conflict within place arises in the work of Nash (1999). She studies the link between language, political conflict and place names in Northern Ireland and shows how the act of naming places has been a long-standing colonial practice. These practices have been witnessed most clearly in the replacing of indigenous place names with those associated with a metropolitan English or British culture. This practice has elicited a strong response from Irish nationalists, with calls for allegedly 'inauthentic' English place names to be replaced by original and indigenous Irish names. Place – and the place names inscribed in the landscape – become key issues of nationalist struggle and contestation.

One of the most striking aspects of Nash's (1999) work is the attempt that has been made to view place names as cultural resources that can help to reconcile Northern Irish Protestants and Catholics. Far from being a source of discord and conflict, place names have been viewed as one way of enabling all groups in Northern Ireland to celebrate cultural diversity. Nash describes how the Ulster Federation of Local Studies, for instance, received money from the Cultural Traditions Programme, whose aim has been to promote an appreciation of cultural diversity in Northern Ireland. Similarly, place names featured in school projects that attempted to emphasise the varied origins of place names (Ibid: 471). Nash's work importantly points the way towards viewing places not as sites within which nationalist conflict is played out but as locations that can allow for some form of co-habitation and reconciliation to emerge.

A similar emphasis on the positive potentiality of place can be found in some of the fundamental principles that underpin the education system in Wales. In broad terms, it has long been recognised as a fundamental tenet of Welsh politics that there are many, equally valid, ways of being Welsh and, significantly, that these can be practised differently in many different parts of Wales (Cloke et al., 1998). And certainly, this kind of approach has been adopted within the education system. A Welsh Curriculum guidance document, published in 2003, for instance, states that '[b]ecause Welsh society is very diverse, there can be no single view of what it is to be Welsh' and that '[b]ecause of the variety and diversity within Wales, the Curriculum Cymreig will take different forms in different schools' (ACCAC, 2003: 5; see also Welsh Government, 2015: 14).

While such statements might be viewed as signifying a desire to devolve understandings of Welshness to the local scale – for practical and political reasons – they also reveal attempts to promote an understanding of Welsh nationalism that is actively negotiated within schools and particular places (Erickson, 1995). This can be further displayed in the ways in which these ideals are implemented by teachers and received by pupils. A teacher based in a large institution located near the border in north-east Wales referred to how teachers who taught Welsh as a subject – along with those who taught other subjects through the medium of Welsh – had to become 'diplomats' and 'negotiators' within the classroom (cf. Benwell, 2014). Teachers had to become skilled at working out how the Welsh language and culture could be introduced to audiences that possessed mixed identities. We see here how the place of the school became a zone of contact between different individuals and groups, and one in which an open and accommodating form of Welshness could be – indeed had to be – developed. Place thus acted as a site within which different individuals and groups could be reconciled with each other.

Conclusions

The aim of this chapter has been to examine the interconnectedness of place and nation in conceptual and empirical contexts. These two spatial and scalar categories are intimately entwined. National territories – if they are to have any meaningful connection with the members of the nation – take on a status as places in their own right. Conversely, nations and national discourses become embedded in particular places. I outlined different ways of approaching this second kind of interconnection, focussing on ideas of representation and reproduction, materiality and affect, and conflict and reconciliation. This threefold division is an heuristic device since there are clear overlaps between these different themes. There is a need to examine the manifold connections between these different ways of embedding nations within places; connections that were, perhaps, underplayed as a result of the structure used here.

One of the most significant themes discussed is that it is possible for place to act as a source of national reconciliation. Without overstating the significance of such themes, it seems to me that a focus on the local manifestations of, and variations in, nations and nationalist discourse can potentially act to counter some of the exclusionary and essentialist versions of nationalism being peddled at present across the world. If this is so, then a study of the connections between place and nation becomes something of more than academic significance. It should be of interest to active and concerned citizens, along with all progressive politicians and policy-makers.

Acknowledgements

The themes discussed in this chapter draw on approximately twenty years of research on nationalism and its connection to place, for which I am grateful to various organisations, most notably the UK's Arts and Humanities Research Council (grants AH/E503586/1 and AH/J011436/1) and the Economic and Social Research Council (grant ES/L009099/1). Thanks to the editors for their useful comments. All errors are mine.

References

ACCAC (2003) Developing the Curriculum Cymreig (Qualifications, Curriculum and Assessment Authority for Wales (ACCAC), Cardiff).

Agnew J (1987) *Place and Politics: the Geographical Mediation of State and Society* (Allen and Unwin, London).

Appleton L (2002) 'Distillations of something larger: the local scale and American national identity', *Cultural Geographies* 9: 421–447.

Azaryahu M and Kook R (2002) 'Mapping the nation: street names and Arab-Palestinian identity: three case studies', *Nations and Nationalism* 8: 195–213.

BBC (2014) 'The Welsh M1', available at: www.bbc.co.uk/programmes/b03nt8g7 (accessed 30 April 2016).

Benwell M (2014) 'From the banal to the blatant: expressions of nationalism in secondary schools in Argentina and the Falkland Islands', *Geoforum* 52: 51–60.

Brubaker R (2004) *Ethnicity Without Groups* (Harvard University Press, Cambridge MA).

Cloke P, Goodwin M and Milbourne P (1998) 'Cultural change and conflict in rural Wales: competing constructs of identity', *Environment and Planning A* 30: 463–480.

Closs Stephens A (2016) 'The affective atmospheres of nationalism', *Cultural Geographies* 23: 181–198.

Edensor T (2002) *National Identity, Popular Culture and Everyday Life* (Berg, Oxford).

Edensor T (2004) 'Automobility and national identity: representation, geography and driving practice', *Theory, Culture and Society* 21: 101–120.

Erickson RJ (1995) 'The importance of authenticity for self and society', *Symbolic Interaction* 18: 121–144.

Fevre R, Borland J and Denney D (1999) 'Nation, community and conflict: housing policy and immigration in North Wales', in R Fevre and A Thompson (eds) *Nation, Identity and Social Theory: Perspectives from Wales* (University of Wales Press, Cardiff) pp. 129–148.

Gade DW (2003) 'Language, identity and the scriptorial landscape in Québec and Catalonia', *The Geographical Review* 93: 429–448.

Gruffudd P (1994) 'Back to the land: historiography, rurality and the nation in interwar Wales', *Transactions of the Institute of British Geographers* 19: 61–77.

Hetherington K (1998) 'In place of geometry: the materiality of place', *The Sociology Review* 43: 185–199.

Iwan D (1968) 'We are not asking for the moon: a letter to *Y Cymro*', April 30, 1968. Welsh Office translation, The National Archives BD 43/139.

Johnson NC (1995) 'Cast in stone: monuments, geography and nationalism', *Environment and Planning D: Society and Space* 13: 51–65.

Jones R and Merriman P (2009) 'Hot, banal and everyday nationalism: bilingual road signs in Wales', *Political Geography* 28: 164–173.

Jones R and Merriman P (2012) 'Network nation', *Environment and Planning A* 45: 937–953.

Kingsbury P (2008) 'Did somebody say juoissance? On Slavoj Zizek, consumption and nationalism', *Emotion, Space and Society* 1: 48–55.

Kornprobst M (2005) 'Episteme, nation-builders and national identity: the re-construction of Irishness', *Nations and Nationalism* 11: 403–421.

MacLaughlin J (2001) *Reimagining the Nation-State: the Contested Terrains of Nation-Building* (Pluto Press, London).

Merriman P and Jones R (2017) 'Nations, materialities and affects', *Progress in Human Geography* 41: 600–617.

Nash C (1999) 'Irish place names: postcolonial locations', *Transactions of the Institute of British Geographers* 24: 457–480.

Paasi A (1996) *Territories, Boundaries and Consciousness: The Changing Geographies of the Finnish-Russian Border* (John Wiley, Chichester).

Raento P (1997) 'Political mobilization and place-specificity: radical nationalist street campaigning in the Spanish Basque Country', *Space and Polity* 1: 191–204.

Thomson A and Day G (1999) 'Situating welshness: "local" experience and national identity', In R Fevre and A Thompson (eds) *Nation, Identity and Social Theory: Perspectives from Wales* (University of Wales Press, Cardiff) pp. 27–47.

The Times (1963) 'Summonses in Welsh: group's campaign in Aberystwyth', 4 February, p. 6.

Van Houtum H and Van Naerssen T (2002) 'Bordering, ordering and othering', *Tijdschrift Voor Economische En Sociale Geografie* 93: 125–136.

Welsh Government (2015) *Successful Futures: Independent Review of Curriculum and Assessment Arrangements in Wales* (Welsh Government, Cardiff).

Wiebe R (2002) *Who We Are: A History of Popular Nationalism* (Princeton University Press, Princeton NJ).

Williams CH and Smith AD (1983) 'The national construction of social space', *Progress in Human Geography* 7: 502–518.

4

Region, place, devolution

Geohistory still matters

David Beel and Martin Jones

Introduction

In an article published nearly 15 years ago, MacLeod and Jones (2001) carefully reviewed, situated, extended and above all celebrated the enormous intellectual contributions of Anssi Paasi to the scholarly project of doing 'regions in geography'. Situated within, and going beyond, the 'new regional geography' movement in human geography and the social sciences more broadly, they looked at Paasi's thinking on regionalization processes, abstracted in four stages, which collectively allowed them to advance (as they claimed) a meaningful understanding of regional change. Rolling forward the research clock to the likes of the Northern Powerhouse and other 'devolution deals' and events across the UK, we maintain that Passi's framework *remains* a cutting-edge theoretical framework in and through which to examine region-building processes and practices – particularly the relationship between region and place in a 'foregrounded regional studies' (Paasi and Metzger, 2017). This chapter accordingly looks at the 'new new localism' and suggests the need to think about the dawn of a 'new new regional geography'. In doing so, the chapter suggests that city-regions involve a new politics of place-making, which opens up new ways of thinking about place vis-à-vis region. The implications of this are outlined.

New new localism

City-Region-based agglomerations are currently riding high on the political and policy agenda across the world. Their emergence is not accidental; they are being built in direct response to the deep ideological thinking exposed in key documents such as the World Bank's *Word Development Report 2009: Reshaping Economic Geography*. This set in train a series of 'new economic geography' influenced arguments closely following the work of policy advisors such as Krugman and Glaeser (Peck, 2016). These collectively claim that, first, urbanization is a global phenomenon to be embraced at all costs and, within this, city-regions are the principal scale at which this happens and people experience lived reality. Second, somewhat provocatively, the economic basis of city-regions rests on concentration and specialization, which allows spatial agglomeration to take place. Third, cosmopolitan policy management is required with a bold and confident voice, working with the grain of market logistics and new 'spatial orderings' (such as governance frameworks) to lubricate agglomeration and provide efficiency by lowering transaction costs and promoting proximity, and thereby liberating growth and allowing it to spread geographically (for an overview, see Storper, 2013).

In the UK, this motif is clearly evident in interventions over the last few years in the wake of RSA's City Growth Commission, which argued for the unleashing of metro growth through a series of city-regions or 'metros' – defined as the 'larger constellation of cities and towns that constitute a functional economy within build up areas' – as the main drivers of economic growth in an increasingly knowledge-driven global economy (RSA, 2014). The UK Conservative Government, through policy discourses and narratives of devolution, localism, rebalancing and the Northern Powerhouse, is taking these agendas forward as a response to hold-down the global and also for finding a way around the messy nature of austerity and local state restructuring (see Conservative Party, 2015; Jones, 2019).

The authors have been involved in a three-year research project that is probing on the missing socially and spatially disembedded sphere of these competitive relationships, equilibrating tendencies and, critically, the vacuum around the policies and politics of assembling city-regions. In short, there is little research being undertaken on *City-Region Building*, i.e., which civil society stakeholders are involved and what the motives are for engagement or a lack of engagement. Added to this, there is no critical assessment of whether and how marginalization (by interest groups and by geographical location) and uneven development (the relationship between regions, cities and places) operates and, in turn, whether this fuels, sustains or destroys economic agglomeration, development and growth. The project is, therefore, addressing this gap within the research field of human geography and the social sciences more broadly.

The authors have deployed case study research based on three sites in Wales (Cardiff Capital Region, Swansea Bay City Region and the North East of Wales) and two sites in England (Sheffield City Region and Greater Manchester City Region). This involves interviews with around 20–25 stakeholders in each location – and we are currently undertaking a comparative study of stakeholder and civil society organizational involvement in the *City-Region Building* agenda. By focussing on the institutions of economic governance, the project is specifically looking at those involved in Local Enterprise Partnerships, various City Deals, Enterprise Zones and city-region development in general. The following research questions are being asked: what policy, strategy and institutional changes have taken place, and are currently taking place, in the landscape of economic development since 2010 in England and Wales? How do these changes affect and involve civil society organizations? What are the narratives of devolution and community engagement in the LEPs, EZs, City Deals and City-Regions? How are these being worked into policies and procedures for stakeholder engagement? Who is involved in the new localism and how does this relate to forms of associational life and political engagement? In turn, what are the compositions of LEP, EZ, City Deal and City-Region boards, and their sub-groups and other structures of engagement? And, how successful are the *City-Region Builders* and the new localism in realizing the objectives of agglomeration, economic development and growth, and social empowerment?

Geographers have positioned the above as part of a 'new localist' political and policy discourse, given the arguments around the reanimation of place-based civil society as a means of stimulating localist economic development (see Clark, 2014; Clark and Cochrane, 2013; Jones, 2019). The localism is not new though: it is a reworked policy narrative (see Peck, 1995), and one that will doubtless recur again, and I prefer to note this as an instance of 'new new localism' (Jones and Jessop, 2010). This is because the latest variant of localist thinking draws extensively on some key antecedents. According to the

'Big Society' guru Norman, localism 'is a coherent and logistical expression of a conservative tradition which goes back to the 18th century' (Norman, 2011: 201). Edmund Burke's 'little platoons' pepper this literature and are presented as progressive enablers for a democratic form of civil society-centred economic and social policy. The Conservatives' new localism, then, stresses a

> three-way relationship between individuals, institutions and the state. It is when this relationship is functioning well that societies flourish. This requires each element in the triad to be active and energised in its own right ... Societies should be thought of as ecosystems.
>
> *(Ibid: 201)*

We would like to suggest in this chapter that Paasi's treatise on regions and places increasingly allows a window into the study of such 'new new localist' ecosystems, thereafter raising questions on how we construct and deploy notions of place and region as spatial concepts and constructs.

New new regional geography

If the new regional geography (Gilbert, 1988) was launched to capture a coalescing concern with local responses to capitalist processes, cultural identifications and identifying the region as a medium for social interaction, then Anssi Paasi's has clearly gone well beyond this; hence the suggested label of a 'new new regional geography'. As noted previously (MacLeod and Jones, 2001), Paasi (1986: 110) sought to transcend the dualism between Marxism and humanism by seeing regions 'not as static frameworks for social relations but as concrete, dynamic manifestations of the development of a society'. Areal extent though is a misnomer, as regions are to be analysed reflexively within the context of their very cultural, political and academic conception (Paasi, 1991, 1996, 2010). Notions of *institutionalization* come into play here, which is not a short-hand with the study of institutions; instead, attention is paid to *geohistorical* socio-spatial processes during which territorial units emerge as part of the spatial structure of a society and become established and clearly identified in different spheres of social action and social consciousness. They are at once lines on the map and also geographical reference points in popular and political culture. This is operationalized through a methodology of abstraction: abstract to concrete and simple to complex in the identification of phenomenon (cf. Brenner et al., 2004; Sayer, 1992).

Stage 1

Paasi has deconstructed the regionalization process by abstracting four stages, which rather than implying a linear sequence, of course, are to be understood as mutually constituting, reciprocal and recursive processes of structuration only distinguishable from each other analytically for the purposes of grounded research, hence why they are abstractions. The first of these concerns the assumption of territorial awareness and shape, where a territory assumes some bounded configuration in individual and collective consciousness and becomes identified as a distinct unit in the spatial structure of society. At the heart of this stage one can point to a series of struggles relating to cognitive mapping and the hegemony of one geographical imagination over others, the politics of scale, difference, identity and subjectivity, and the stretching and bounding of power relations (MacLeod and Jones, 2001).

Rolling things forward, this clearly connects with the drawing of, and designation of, the city-region boundaries of Sheffield, Manchester, Swansea and Cardiff noted above, where power-holding actors in a territory (or outside it even) have defined and symbolized the spatial and social limits of membership and create the discourses and practices for inclusion and exclusion, to the extent that territorial shaping refers not only to the creation of boundaries but also to their representation, to their roles both as social institutions and symbols of territory. Relatedly, territorial awareness and shape can be used to shine light on the ongoing and somewhat cul-de-sac debate in English-speaking human geography on territorial (seemingly bounded) versus relational conceptions (networked and mosaic) notions of space and statehood (see Jones and MacLeod, 2011). The illuminated perspective is that these processes are co-constituted: not either/or, but and/both, and the balance between them depends on institutionalization practices and the balance and roles of those actors involved and their geographical dependency (see Jones and Paasi, 2013, 2015; Paasi, 2010, 2013).

In following Paasi's first stage, Map 4.1 highlights well both the bounded, mapped nature of producing city regions but also the ways in which this can be contested. The city region as a whole is largely based upon what is termed the 'functional economic area' surrounding

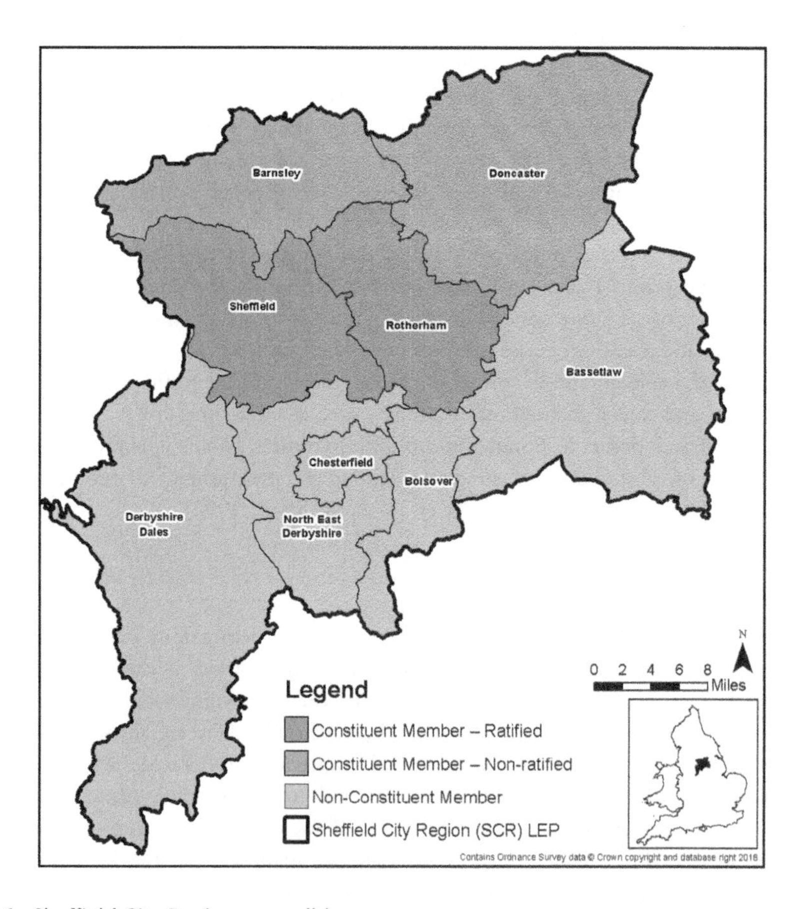

Map 4.1 Sheffield City Region geopolitics.

Sheffield, which in turn focusses upon commute work patterns and employment opportunities (see Etherington and Jones, 2016). This represents the city region as whole but, as Map 4.1 suggests, this is far from simple when trying to create a contiguous city region. Here, the economic geography of the region has been contested by the cultural and historical identities of the different local authorities. This means that only Barnsley, Doncaster, Rotherham and Sheffield chose to constitute themselves within the SCR and be part of the 2018 Mayoral elections (Figure 4.1), whereas the remaining local authorities (with interests in Derbyshire and Nottinghamshire) chose to wait for a possible future devolution deal with their historic county regions. In Chesterfield, these overlapping senses of 'regionality' (Painter, 2008) became set against the regionalization of the city region within the context of austerity. The following letter represents the deep geohistory of Derbyshire in Chesterfield:

> What on earth is our council thinking? Chesterfield is in Derbyshire. What a plan, live in Derbyshire and be controlled by Sheffield ... I do hope, fingers and toes crossed, they choose the Derbyshire and Nottinghamshire path.
>
> (Derbyshire Times, *Letters, 3 March 2016)*

Whereas the response from the Chesterfield Borough Council took a different approach, choosing to follow the potential offer of the city region, emphasizing the economic necessity of the decision:

> Full membership would align with Chesterfield's economic geography and provide opportunities for new and existing businesses on a footprint that makes sense to them ... Chesterfield would be able to benefit fully from the two existing SCR devolution deals, including the £30m p.a. additional funding over 30 years, and continuing negotiations with government for further devolution.
>
> *(Chief Exec. Unit, 25 February 2016)*

Therefore, the cognitive mapping and the economic hegemony of the city region is disrupted by the pre-existing geohistory, this in part contests the processes of regionalization in play with SCR. It causes the SCR to split, as Map 4.1 highlights, creating an uneven geography moving forward.

Stage 2

For Paasi, of course, this leads on to the second stage, the formation of the conceptual and symbolic shape of regions, which is neither pure nor uncontested but is instead subject to continuous negotiation, translation and a hybridity of cultural expression. That said, power-holding elites will endeavour to press that such negotiations and translation manifest in a hegemonic territorial grid of meaning whereby only a selection of invented traditions, histories and remembrances are established and creatively implicated in the constitution of a territory's social relations. Paasi's work mentions the importance here of power-laden symbols such as cartographies, flags, memorabilia, histories, etc. (Paasi, 1996, 2013), but in relation to my research on city-regions, attention is also drawn to the very naming of a region, which helps to connect its image and place consciousness both of insiders and of outsiders. The case of city-region building in South Wales is important in this regard.

In 2011, the Welsh Government established a task and finish group to consider the potential role of city-regions in future economic development. The task was to decide, on

Figure 4.1 Post-democracy City Region Building.

Figure 4.2 Cardiff Capital Region symbolism.

the basis of objective evidence, whether a spatially focussed city-region approach to economic development, as opposed to the (national) Wales Spatial Plan, could deliver an increase in jobs and prosperity for Wales. Drawing on evidence mainly from Europe and North America, three arguments for adopting a city-region approach were made: improving the planning system; improving connectivity; and driving investment through a stronger and more visible offering from an agglomerated wider region (see Jones et al., 2015). Two distinctive city-regions were proposed – the South East Wales City Region and the Swansea Bay City Region – with the proviso being that all this has to be about creating urban engines and power-houses of growth by harnessing the beneficiaries of transport, housing, inward investment and funding opportunities. In following Paasi's tract, the South East Wales City Region naming history is illuminating, as over time it has morphed to being called the Cardiff Capital Region (see Figure 4.2).

This was initially to purposefully distinguish Cardiff from neighbouring Newport for external promotion purposes, then later to the full renaming of the city region in order to acknowledge both capital city power status and the stretched-out variegated geography of city-region building. This points to a metro-centric focus upon Cardiff, as the agglomerative centre to the region, and highlights the way in which city region governance is funnelled with regards to the future growth of the city region.

Stage 3

These processes are constituted in particular structures of expectation, themselves critical in facilitating the third stage, the emergence of institutions, where Paasi sought and still seeks to capture the identity-framing vehicles of education, law, local politics and organizations rooted in civil society (local media, working clubs, arts and literature organizations), as well as informal conventions such as economic ties or proximity and social mores. The entrenchment of these processes into the spatial matrix of society can also foster symbolic shape. For example, as more city-regional scale organizations are instituted into an activity such as economic development, the very consciousness of some place-based agendas may be intensified (MacLeod and Jones, 2001). All of which helps in providing an effective means of reproducing the material and mental existence of territories in question.

This, again, closely connects to the city-region research agenda noted above, particularly the roles played by key activists (either those involved or outside the representational institutional governance structures of the four city-regions) in colouring the territorial consciousness and at the same time reproducing the very power assigned to such institutional roles. Indeed, for Paasi (1986, 1996, 2013), it is the institutions of a territory (and associated infrastructural power as state theorists would put it) that eventually become the most important factors in the macro-reproduction of the region. Within the context of England, there have been attempts to create new 'soft spaces' of governance (Haughton et al., 2013) for city regions through the creation of Local Enterprise Partnerships (LEPs). These have sought to bring business into the processes of producing a growth coalition for the city region. It strategically places economic interest at the centre of the regionalization process, framing the way in which sub-national devolution and growth will take place. Within the context of city regions, this can make LEPs powerful institutions that enable business elites to have a stronger voice whilst at the same time pushing other voices to the periphery:

> Trickle down doesn't work for the most vulnerable and disadvantaged and you have to have strategies around social regeneration (for want of a better word) alongside economic regeneration. Those two things should come together and I don't think they do because the LEP is very purely focussed on the economic policy ... Feels like I'm in a rowing boat and my colleagues are in a rowing boat and we're trying to turn round this big tanker.
>
> *(Interview 1, Sheffield, 2015)*

The local civil society actor in the quote above highlights how the governance structures and actors involved in the Sheffield City Region shape the processes in a purely economic direction. This means two things for the respondent: one, it fails to address the social problems existing in the city region and, two, the civil society actor has little voice in addressing this through the current structures of governance put in place. Hence, a place-based agenda for growth is intensified, which territorializes the strategic interest of the LEP towards an agglomerative growth model, as it reshapes the representational regime of the city region (Jessop, 1990, 2016).

Stage 4

Every theory has its limitations and previously MacLeod and Jones (2001) noted that it was only fair to acknowledge that Paasi's key research objective has been to uncover the more localized or bottom-up articulations involved in the reproduction of sociospatial consciousness and regional shaping of society (though see emerging research on spatial planning, Paasi, 2013; Paasi and Zimmerbauer, 2016; Zimmerbauer et al., 2017). The final stage in this latter process concerns the establishment of a region in the spatial structure and popular consciousness, where it assumes the form of an institutionalized 'territorial unit' and as an identifiable constituent in the regional division of society. In practical terms, the region is ready to be mobilized for such purposes as place marketing or as a weapon in an ideological struggle over resources and power. Further, if provided with administrative status, it comes to assume the material expression of the end to which state power is applied (Paasi, 1991).

The last few years in England demonstrates the relevance of Paasi's thinking. The full map of Local Enterprise Partnerships (LEPs) is becoming embedded and is now being superimposed by further voluntary arrangements of local authorities through City Deals and Local Growth Deals with government and proposals for devolution to five initial (indirectly elected but legally recognised as strategic coordinating bodies) Combined Authorities (see Sandford, 2019). Whilst the result is complex, these point towards the endurance of a *de facto* city region scalar and institutional fix. Indeed, each of the three main political parties appears to be wedded to such a fix, subject to proposing modifications. The South Yorkshire Sheffield City Region (SCR), which straddles the 'traditional' administrative geography of counties and regions and internalizes a new scale of policymaking, is becoming an established region. As part of the 'Northern Powerhouse' movement, which has been establishing in the past five years as a means of addressing austerity and rebalanced development, Sheffield has secured a deal with the UK Government to transfer more powers over transport, housing and economic growth to the city region. The Sheffield city devolution deal, the second agreed in England (after Manchester), although not involving additional money, is being presented as a shift in power from Whitehall to the Combined Authorities in the region. This includes responsibility for the majority of the adult skills budget, greater control over transport schemes and greater power to decide which assets to sell for development (compare Etherington and Jones, 2016; HM Government, 2014).

These processes hence go against the grain of Paasi's fourth stage as a form of central government localism is created not a bottom-up flourishing of local and regional identity. This has significant impact to the processes of city regional regionalization because the scalar jumps in governance from the local authority to the combined authority circumvents existing institutions, coalitions, popular identities and civil society actors working at the local authority scale:

> At one point they talk about localism but if you look at regionalization, it's huge, it's huge and actually the local voluntary community sector can't even hope to engage with, let alone deliver against that agenda. Therefore, civil society is finding itself squeezed behind/between a rhetoric that emphasizes its importance but a reality which mitigates against its ability to capture the resources to deliver against that agenda.
>
> *(Interview 12, Bolsover, 2016)*

The Bolsover respondent above highlights how the scalar change in governance marginalizes both their ability to work at the local level (to address the needs of the communities they serve) and their ability to have influence upon processes of governance at a combined authority level. This is further reflected below in Cardiff also:

> All of a sudden we become completely insignificant so whereas at the moment locally we can lobby quite hard and push the direction on certain things, all of that power would go away and how to influence rather than power. So that for us would cause quite a significant problem. If we start working more collaboratively with other similar organizations then great, we can form a nice little consortium and then we can retain the same level of perceived power and all will be well with the world. But it doesn't fit well with how any of us work really; we work with quite defined communities, we do quite tailored things for them.
>
> *(Interview 2, Cardiff, 2015)*

Whereby the respondent identifies how changes in the scalar relationship to governmental structures deeply weakens their position as an actor in the local community and state. This means that, despite the language of localism being threaded through the centralized processes of devolution, there is in fact a further distancing of the local from the structures of governance created by the introduction of city regions and their combined authority governmental structures.

Conclusion: rethinking place and region

In following Paasi's treatise on regions in geography, we feel this can provide (once again) fresh thinking for today, even 30 years after the original argument was put down in *Fennia*, and in doing so it still offers powerful methodological means and conceptual tools with which to advance an imaginative and progressive understanding of regional change. In particular, as MacLeod and Jones (2001) argued previously, Paasi's geohistorial approach still provides much scope with which to unravel the political, economic and cultural process that enables individual and institutional place-based biographies to coalesce in the form of a distinctive territorial unit with the overall regionalization of society (MacLeod and Jones, 2001). Moreover, by placing the institutionalization process, its multiple and overlapping 'stage', and the critical role played by discourse and symbolic orderings of space at the centre of his treatise, Paasi still enables us to locate many of the complex forces at work in constructing the regionalization of society.

Further, and in the context of city-region building research, Paasi's framework permits us to problematize the reciprocal relationships that can exist between the whole gamut of institutional forms relating to economic behaviour (LEPs, EZs, City-Region Boards, Combined Authorities, etc.), the politics of representation, political power geometries, scale, and identity, and the sedimentation of these practices into regions. In most accounts of city-regions, questions pertaining to the social construction of boundaries, territorial shape and the very becoming of region and their associated institutional fixes remain hidden from view (compare HM Government, 2014; RSA, 2014; Storper, 2013). In contrast, Paasi's stress on region building as an active and ongoing process, rich in political strategy and cultural expression, still sanctions useful insights for researchers and regional strategies alike to uncover the very formation of economic and political life. Perhaps, then, it is time to think about a 'new new regional geography' where the interrelationships between region and place can be considered once more.

Based on the discussions here, this has three initial implications for spatial thinking. First, it provides a revised model for understanding place that does not take places and regions as given bounded spatial units, but instead emphasizes the contingency and relationality of space. Second, approaches, therefore, require identification and description of the place(s) to be incorporated as an intrinsic part of the research process, rather than treating place and region as taken-for-granted backdrops. This approach further recognizes that the shape, reach and orientation of place might differ according to the research questions being examined. Third, new city-region-making notions consequently demand a new body of research concerned with establishing the material and imagined coherences of place (see Jones and Woods, 2013), employing mixed-method strategies.

Material coherence here refers to the particular social, economic and political structures and practices that are configured around a place. Thus, material coherence may be provided by the territorial ambit of a local authority, by the geographical coverage of an economic development initiative, by the catchment area of a school or hospital, by a travel-to-work area, by the reach of a supermarket or shopping centre, or by any combination of the above and other similar structures and practices. Material coherence hence alludes to the institutional structures that hold places together and provide vehicles for collective action.

Imagined coherence here relates to collective resident consciousness and the sense of shared identity and affinity with a place, resulting in a perceived community with shared patterns of behaviour and common geographical reference points. Imagined coherence, therefore, makes place meaningful as a space of collective action. There are territorial units that exhibit material coherence but lack a strong imagined coherence (such as artificially amalgamated local authority areas) and there are territories with an imagined coherence but only a weak material coherence (for example, where institutional boundaries bisect contiguous urban areas or where areas with strongly developed popular consciousness exist within much larger institutional units).

But, *both* material coherence and imagined coherence are also important in fixing (through multiple intersections) the scale at which place and regions can be identified. Imagined coherence is framed around perceived shared forms of behaviour, whether linked to common patterns of collective consumption, shared affinity with sporting or cultural institutions, or common geographical/historical reference points. However, this imagined coherence is not founded on direct inter-personal connection between residents. In this sense it differs from the social coherence of a neighbourhood – which may share some of the above attributes but is framed around the probability of direct interaction between members. It also differs from the imagined coherence of a region, which is a looser affiliation that draws more on perceived cultural and political identities and economic interests. Similarly, material coherence should be denser and more complex than that found at place or regional scale. The material coherence of a neighbourhood will be restricted by its situation within a larger geographical area for employment, administrative and many service provision functions, while the material coherence of a region could be fragmented by the inclusion of several different labour markets, local authority areas, sub-regional shopping centres and so on. These attributes do not easily translate into discrete territorial units with fixed boundaries. Labour market areas overlap, as do shopping catchment areas; residents may consider themselves to be part of multiple places for different purposes and at different times; the reach of a town as an education centre may be different to its reach as an employment centre; and so on. The boundaries that might be ascribed will vary depending on the issue in question (Orford and Webb, 2017). Savage's (2009) work on 'granular space' is illustrative of these concerns:

> People do not usually see places in terms of their nested or relational qualities: town against country: region against nation, etc. but compare different places with each other without a strong sense of any hierarchical ordering. I further argue that the culturally privileged groups are highly 'vested' in place, able to articulate intense feelings of belonging to specific fixed locations, in ways where abstract and specific renderings of place co-mingle. Less powerful groups, by contrast, have a different cultural geography, which hives off fantasy spaces from mundane spaces.
>
> *(Savage, 2009: 3)*

The application of the approach discussed logically leads us to start by identifying places by their cores – whether these be towns or cities or geographical areas – rather than as bounded territories, and working outwards to establish an understanding of their material and imagined coherence. This process will necessarily require mixed methods, combining cartographic and quantitative data on material geographies with qualitative evidence of imagined coherence and performed patterns and relations. This is more than just an exercise in boundary-drawing. Whilst it may be possible to identify fixed territorial limits for the reach of a locality with respect to certain governmental competences or policy fields, applying proxy boundaries to imagined places must necessarily assume a degree of permeable, and that places may be configured differently depending on the object of inquiry. Through these mechanisms, then, whilst research on place and region can be spatially focussed, it should not be spatially constrained, and needs to be prepared to follow networks and relations across scales and spaces in order to reveal the full panoply of forces and actors engaged in the constitution of city-region making.

References

Brenner N, Jessop B, Jones M and MacLeod G (2004) Introduction: state space in question, in N Brenner, B Jessop, M Jones and G MacLeod (eds) *State/Space: A Reader*, pp. 1–26. Blackwell: Oxford.

Clark J and Cochrane A (2013) Geographies and politics of localism: the localism of the United Kingdom's coalition government. *Political Geography* 34: 10–23.

Clark N (2014) Locality and localism: a view from British human geography. *Policy Studies* 34: 492–507.

Conservative Party (2015) *The Conservative Party Manifesto 2015: Strong Leaderships, A Clear Economic Plan, A Brighter, More Secure Future*. London: Conservative Party.

Etherington D and Jones M (2016) The city-region chimera: the political economy of metagovernance failure in Britain. *Cambridge Journal of Regions, Economy and Society* 9: 371–389.

Gilbert A (1988) The new regional geography in English and French-speaking countries. *Progress in Human Geography* 12: 208–228.

Haughton G, Allmendinger P and Oosterlynck S (2013) Spaces of neoliberal experimentation: Soft spaces, postpolitics, and neoliberal governmentality. *Environment and Planning A* 45(1): 217–234. doi:10.1068/a45121

HM Government (2014) *Sheffield City Region Agreement on Devolution*. London: HM Government.

Jessop B (1990). *State Theory: Putting the Capitalist State in Its Place*. Cambridge: Polity Press.

Jessop B (2016). *The State: Past, Present, Future*. Cambridge: Polity Press.

Jones M (2013) Polymorphic spatial politics: tales from a grassroots regional movement, in W Nicholls, B Miller and J Beaumont (eds) *Spaces of Contention: Spatialities and Social Movements*, pp. 103–120. Aldershot: Ashgate.

Jones M (2019) *Cities and Regions in Crisis: The Political Economy of Subnational Economic Development*. Cheltenham: Elgar.

Jones M and Jessop B (2010) Thinking state/space incompossibly. *Antipode* 42: 1119–1149.

Jones M and MacLeod G (2011) Territorial/relational: conceptualising spatial economic governance, in A Pike, A Rodrigues-Pose and J Tomaney (eds) *Handbook of Local and Regional Development*, pp. 259–270. London: Routledge.

Jones M, Orford S and Macfarlane V (eds) (2015) *People, Places and Policy: knowing Contemporary Wales through New Localities*. London: Routledge.

Jones M and Paasi A (2013) Guest editorial: regional world(s): advancing the geography of regions. *Regional Studies* 47: 1–5.

Jones M and Paasi A (eds) (2015) *Regional Worlds: Advancing the Geography of Regions*. London: Routledge.

Jones M and Woods M (2013) New localities. *Regional Studies* 47: 29–42.

MacLeod G and Jones M (2001) Renewing the geography of regions. *Environment and Planning D: Society and Space* 19: 669–695.

Norman J (2011) *The Big Society*. Buckingham: Buckingham University Press.

Orford S and Webb B (2017) Mapping the interview transcript: identifying spatial policy areas from daily working practices. *Area* 50: 529–541.

Paasi A (1986) The institutionalisation of regions: a theoretical framework for understanding the emergence of regions and regional identity. *Fennia* 164: 105–146.

Paasi A (1991) Deconstructing regions: notes on the scale of spatial life. *Environment and Planning A* 23: 239–256.

Paasi A (1996) *Territories, Boundaries and Consciousness: The Changing Geographies of the Finnish-Russian Border*. Chichester: Wiley.

Paasi A (2010) Regions are social constructs, but who or what 'constructs' them? Agency in question. *Environment and Planning A* 42: 2296–2301.

Paasi A (2013) Regional planning and the mobilisation of 'regional identity: from bounded spaces to relational complexity. *Regional Studies* 47: 1206–1219.

Paasi A and Metzger J (2017) Foregrounding the region. *Regional Studies* 51: 19–30.

Paasi A and Zimmerbauer K (2016) Penumbral borders and planning paradoxes: relational thinking and the questions of borders in spatial planning. *Environment and Planning A* 48(1): 75–93.

Painter J (2008) Cartographic anxiety and the search for regionality. *Environment and Planning A* 40: 342–361.

Peck J (1995) Moving and shaking: business elites, state localism and urban privatism. *Progress in Human Geography* 19: 16–46.

Peck J (2016) Economic rationality meets celebrity urbanology: exploring Edward Glaeser's city. *International Journal of Urban and Regional Research* 40: 1–30.
RSA (2014) *Unleashing Metro Growth: Final Recommendations of the City Growth Commission*. London: Royal Society of Arts.
Sandford M (2019) Giving power away? The 'de-words' and the downward transfer of power in mid-2010s England. *Regional & Federal Studies*, doi: 10.1080/13597566.2019.1640682.
Savage M (2009) *Townscapes and landscapes*. New York, England: Mimeograph, Department of Sociology, University of New York.
Sayer A (1992) *Method in Social Science: A Realist Approach*. London: Routledge.
Storper M (2013) *Keys to the City: How Economics, Institutions, Social Interaction, and Politics Shape Development*. Princeton, NJ: Princeton University Press.
Zimmerbauer K, Riukulehto S and Suutari T (2017) Killing the regional Leviathan? Deinstitutionalization and stickiness of regions. *International Journal of Urban and Regional Research* 41: 676–693.

5

Rethinking place at the border through the LYC Museum and Art Gallery

Ysanne Holt

Considerations of place at the border have primarily focussed on the conflictual relations that occur at hard geo-political borders as, for example, between nation states. Studies of visual culture in these contexts have generally examined how diverse types of representation underline the character of these border zones through, for instance, the documentation of repeated acts and forms of behaviour that appear to typify experience and contribute to the perceived identity of such places. One example in this context includes artist Katie Davies' short 2008 film *38th Parallel*, documenting the ritualized actions of border guards between North and South Korea. Repeated behaviour in these terms can be understood to effectively 'perform' place (Butler, 1997). There is a tendency here to perceive a hard border as a line marking two distinct territories with distinctively different ways of life, unique social, political and cultural identities and so on. In contrast to these dominant assumptions, however, are studies of soft border regions, where the perception is not of a border-line, but of a borderland. This borderland might usefully be considered a hybrid, or even a third space where place and identities are seen to be multiple, emergent and overlapping (Massey, 1995; Licona, 2012; Holt, 2018a). With this conceptualization in mind it is possible to discern generative processes of 'cross' or 'de-bordering' and an emphasis on place as characterized by fluidity and exchange, encouraging activities which in themselves deny both boundaries and conventional hierarchies. This chapter is concerned with one such soft-border region and with notions of place and identity so defined. It considers certain artistic and curatorial practices within specific locations and notions of the 'artist-in-place' which have instilled both conceptual and practical acts of 'de-bordering', with specific value, so I argue, in terms of very recent social and political circumstances.

In 1974 the art critic Paul Overy described a visit to a museum and art gallery at a location in rural north-east Cumbria, close by the borders with both Northumberland and Scotland. The 'LYC' had recently been established by the Chinese conceptual and kinetic artist and poet Li Yuan-chia. Having left China for Taiwan in 1949, Li arrived in London in *c.*1966, via Bologna and Milan, showing his Cosmic Point artworks, which drew together Eastern Buddhist and Western abstract influences, at the then avant-garde Lisson and Signals galleries. Encouraged by a friend with family connections in this part of (then) Cumberland, in 1968 Li moved north, acquiring a dilapidated row of stone buildings at the village of Banks from the painter Winifred

Nicholson. Here he gradually developed the museum which, for the following ten years, was to be perhaps the artist's most significant art work. For Paul Overy, the nearby 'Hadrian's Wall was an outpost of a far-flung empire, a barrier to prevent the invasion of one culture by another [whereas] The LYC Museum is a meeting point of different cultures, of time and place' (Brett and Sawyer, 1999: 131).

In reality, however, Hadrian's Wall too was always a meeting point for exchange between peoples, an intermingling of races and a permeable border rather than a solid barrier. As has been observed, the Roman army itself was comprised of numerous ethnicities and the Northern Frontier was a site of multi-cultural flows throughout the period of its construction (Tolia Kelly and Nesbitt, 2009). The following then thinks about time and place in a rural 'soft-border' location focussing on issues generated by Li's museum, and with reflection too on the value of his activities for progressive notions of place, culture and creativity in the present-day Anglo-Scottish border region. This is a location which has received much attention following the 2014 Scottish independence referendum and the current post-Brexit context, inducing widespread reflection on questions of place and identity at a moment which threatens increased levels of social and economic division.

Place in this border region is marginal or peripheral, as in 'at edge', and often of 'solitary topographies' – Peter Davidson's phrase to encapsulate northern landscapes (2005: 21). This is a place with a long past violent history associated with the Border Reivers, or moss troopers, who with expert knowledge navigated the surrounding bogs on their nightly raids between England and Scotland, ultimately intermingling across and along the border. This then is a place of once mutual hardships and conflict, but of longer-standing interactions, shared identities, traditions, heritage and common material resources too. And it is certainly not 'remote' in any static or 'timeless' sense. Rather this is an evolving, a lived and relational place.

With this relational understanding as opposed to any settled or overly rooted notion, a place, like a landscape, in this context is understood not as a space or terrain to be perceived, but as an evolving site of habitation. This is to think of landscape, for example, as 'seen with' rather than 'looked at', as 'lived with', not 'lived off' (Wylie, 2007). And this particular conception frames the importance of the LYC Museum and Art Gallery. Place then as mutually constituted, over-layered with human and non-human interactions with its common material resources – wood, soil, stone, wool – its objects and commodities. In this sense we might speak of the 'material communities' of the borderland.

This is borderland place as palimpsest, overwritten through time and with a legacy of meanings and accumulated values. Like Li's museum in fact, with its gathering of objects and artefacts, its displays of Roman remains loaned from nearby Vindolanda; its drawings, locally woven rag rugs, found or repurposed objects, pottery and contemporary artworks from near and far. The LYC was full of traces of connections and interactions and, like the Wall and its surroundings, with evidence of deep rootedness and of mobility and transience; a particular form of performed, or curated place.

To think further of place as over-layered, in this sense, we can usefully reflect back on the crucially important interwar period and the distinct appeal of the farmhouse 'Bankshead' at Banks and its surrounding environment for its owner the painter Winifred Nicholson, with her then husband Ben Nicholson and the modernist artists who visited, Christopher Wood and Ivon Hitchens amongst them. That interwar moment on the heels of the Great War is more broadly characterized by a 'simple life' spirit and by an elemental aesthetic. Such a context fostered a cultural flight to distant places, far-flung coasts and edges, including Cornwall and the Scottish islands, sites within which to evade modernity or else to imagine ideal alternatives (Martin-Jones, Holt and Jones, 2018). Something of this impulse, combined with distaste for sophisticated metropolitan modernity, underlay the importance of Bankshead for these artists. Their renovated farmhouse was in fact termed 'a painters' place', a place to draw like-minded others together (Collins, 1991).

The interior of Bankshead at the time presented a mixture of vernacular furniture and stone flags but with a Mondrian painting on the wall which combined to produce a simple, uncluttered and harmonious atmosphere that was at once local and international. As Chris Stephens noted, there is something of a revivified Arts and Crafts aesthetic here (with pertinent historical connections in this area) and with a wider context in that interwar crafts revival, a modernizing of past traditions, of weaving, studio pottery and so on (Stephens, 2002). And in terms of my focus here on evolving engagements with material resources as producing material communities, an attitude to regional crafts displays the complex knots and loops typical of the period. In this context we might place Winifred Nicholson's interest in the tradition of hooky and proggy mats made from saved-up home-dyed scraps of blankets and old clothes – or from remnants brought from the nearby mill at Otterburn in Northumberland – a craft born out of necessity and limited means. All of which allies to the interest in interwar craftsmanship at the heart of British modernism which more broadly underpinned bodies such as the 1920s Rural Industries Bureau, aiming to boost struggling local and national economies while at the same time maintaining craft traditions passed down through generations but in danger of decline (Holt, 2018b). So there is both a utopian idealism and a practical spirit at play here – a connecting of past and present with concerns for a sustainable future for this place and, in the emphasis on the continuation of regional skills and the use of local materials, a rejection of conventional boundaries, or borders, between hierarchical categories of fine art and craft, of international modernism and vernacular tradition.

We can trace this sensibility, transformed too in many ways, forward into the 1970s and to the values that the LYC Museum and Art Gallery – just along the road from Bankshead – represented. The 1970s, of course, was a period of increased environmental and ecological consciousness as witnessed in the emergence of contemporary forms of Land art and revived interests more widely in nature, the countryside, natural resources and materials; and all of these were amply registered in the predominantly rural border regions of North East England, Cumbria and Southern Scotland at this time.

A key touring exhibition of this period organized by the Arts Council and curated by art historian Andrew Causey was 'Nature as Material' (1980) which spoke to many of these cultural, social and environmental concerns. Causey had visited Banks, where the artist Paul Nash – an important interest of his – had also visited in the 1920s. Indeed Causey wrote an article in *Studio International* that drew parallels between Paul Nash and recent British Land art. For Causey, an observable tendency in their art had a 'pragmatic', 'non-hierarchic' attitude to materials, a 'predilection for the wastelands of the Pennines and the Celtic fringe' (so edges and border regions) and unlike the work of many contemporary American Land artists, something of a pastoral society's respect for nature, 'a co-operation rather than a competition with nature' (1977: 122, 126). In other words, a concern for 'living with' not 'living off' the land.

Included in 'Nature as Material' and having already by that date shown at the LYC, was sculptor David Nash, then resident in the 'Celtic fringe' of Wales. Through a practice involving the specificity of local, natural materials and living trees, Nash produced or assembled minimalist structures. His works are process and time-based and connected to a particular vision relating art and life. In his own words, he seeks 'a simple approach to living and doing ... a life and work that reflects the balance and continuity of nature' (Alfrey, Sleeman, and Tufnell, 2013: 56). His ambition, rather as the artist Hamish Fulton (2005: 242) was and is to be 'woven into nature'. Nash designed and made a window for the museum featuring Li's interlocking initials. His was a spirit fundamentally attuned to Li's own in its response to the unique material characteristics of place and locality.

Another key LYC artist in its earliest years was sculptor and photographer Andy Goldsworthy, who performed and documented works there such as *Hazel Stick Throw, Banks* (1980) – a form of drawing in space in which the sticks are thrown up from the earth and with the resulting pattern produced by the elements of wind and air. His works were also included in 'Nature as Material' and his focus was also with process and with materials specifically related to place. His concern with the temporal and the ephemeral was common to much visual arts practice of the 1970s and early 1980s, beyond simply Land or Environmental art. Works such as *Hazel Stick* convey a sense of embodied space; of bodily presence, of a physical and sensory engagement in the world. Embodied space here is understood in Miles Richardson's terms as the location where human experience and consciousness take on material and spatial form, 'embodied space is "being-in-the-world" and suggests the existential and phenomenological reality of place, its smell, its feel and colour' (Hanson, 1982: 3). The concept of being-in-the-world resonates with Li's own sense of and relation to his museum – as artwork; his environment, his idea of an artist, his presence in the landscape.

Goldsworthy, soon to settle in Dumfriesshire in south-west Scotland, was, like Nash, also shown in the 1982 exhibition at Carlisle art gallery: 'Presences of Nature: Words and Images of the Lake District'. Despite the title, the diverse works by contemporary painters, sculptors, writers, photographers and craftspeople were made of, in and in response to the county beyond the lakes too: 'the lush farmland of the Eden valley, the bleak rolling fellsides, Hadrian's Wall' and so on (Hanson, ed. 1982: 9). The exhibition catalogue contained poetry by Frances Horovitz (including 'A Dream', dedicated to Winifred Nicholson), Roger Garfitt and Rodney Pybis, who had all previously contributed to an LYC press publication, *Wall*. A multiplicity of forms and practices were combining to evoke a sense of place.

Before this exhibition the LYC regularly showed the work of local and visiting crafts people – weavers, wood workers, ceramicists. We can say therefore that Li's programming was effectively always 'cross-bordering' between diverse forms of cultural production – writing, craft, different kinds of fine art practice from the innovative and experimental to the more popular and populist – and consciously extending to draw local children into its activities. It was inclusive and boundary-defying in nature, it was 'non-hierarchic' in attitude. And in the broader context of a 1970s 'craft revival', the writer and artist Andy Christian, who lived locally for a period and was very connected to LYC, has recalled the 'drift to the area in the 70s and early 80s of artists and craftspeople of an anti-urban impulse, in despair at the gradual defeat of socialism' (Christian, 2008). Echoing that interwar flight to the margins and edges of the land, like Li, they rented or restored inexpensive stone houses, with several taking up teaching in the surrounding art colleges in Carlisle, Newcastle or Sunderland in fine art, ceramics and textiles. Andy Christian remembered 'a real sense of idealism, of rural hope and political will in the countryside'. We can determine from the many artist-craftspeople drawn to this borders area who worked with local materials and natural forms – many of whom showed regularly at the LYC – a purposefully progressive sense of place, one that was collaborative, boundary-defying and far from nostalgic or escapist.

An influx of talent became ever more entangled with a local culture of making, as witnessed in continual LYC workshops (puppets, rugs, tapestry and felt-making, etc.) and in the surrounding area. In this context the artist Audrey Barker came north to rural Cumbria following productive encounters with the experimental artist Joseph Cornell in New York and an exhibition of her own assemblages at the Lisson Gallery in 1967. In the 1970s she established with her husband Denis 'The Barkers of Lanercost' at Abbey Mill, a few miles from Li's village of Banks. There they supported the reproduction of Roman artefacts, as well as screen-prints, soft toys and hooked rugs made with tweed from the Otterburn mill in Cumbrian designs and using the skills of local people. Denis Barker (n.d.) spoke of an 'untapped source of rural energy needing a domestic industry' and of a local community with a will to create. Something of that 1920s Rural Industries spirit here, re-energised within a particular 1970s ecological sensibility. What we see is the continual evolution of material communities – formed through those dynamic interactions with natural resources – wool, wood, stone – fed through cultural traditions, local skills and forms of tacit knowledge and all underlining the natural and cultural assets of place.

There is an entanglement of resources at play here. This is to follow social anthropologist Tim Ingold who exchanges Bruno Latour's concept of a network (of objects, ideas, processes, matter) for a 'meshwork'. For Ingold it is in the 'entanglement of lines, not the connection of points that the mesh is constituted' – 'every relation is one line in a meshwork of interwoven trails' (2008: 81, 90). In this particular place, these 'textile-based metaphors' are very apt for understanding the density and 'texture' of social connections, the 'fabric' and 'web' of social life, the 'interweaving' and 'interlocking' 'pattern' of relationships, the close or loose ties through which social actions are organized (Scott, 2013: 2; Holt, 2018b). All of this engenders an endlessly shifting set of interrelations between people, objects, nature and materials, and takes us back to the 'tangled mesh of interwoven and complexly knotted strands'. The LYC can be understood to be at the heart of this mesh. A material community is actively constituted through an on-going layering and weaving together of 'place-forming' matter and material, the human and the non-human, history and heritage, habitual forms of response and interaction. There is a deep-rooted culture of making, or production, here and, strikingly, a conceptualization of the artist-within-place as a skilled worker or artisan engaging with local materials, and not remotely an 'artist-outsider' in any more conventional framing of artistic identity.

This particular sense of an artistic persona appears in the local Cumbrian press in 1975 with an article entitled 'Why Li is looking for friends' and headed 'Li – the deserving artist': 'the plucky Chinese artist who has … laboured long and hard over the years to transform what was a tumble-down old property into a modern Museum and Art Gallery, where he maintains a high standard of exhibits'. Li, the journalist goes on, is looking for people to become Friends of his Museum, to

> back up their membership with a modest sum of cash, they'll receive discount on items, free access to catalogues, use of the Arts Theatre and the LYC library. The ultimate ambition to have an art school running at the Banks, where people can come and express themselves on canvas.
>
> (Evening News and Star, *1975, 6)*

Li was 'looking for friends' so the museum might survive, but this was underpinned by strong desire to be useful to the local community – to be relevant – always border-crossing between people and place, and between forms and practices.

To cross, at this point, the nearby geographical border to Scotland, an early contact of Li, the Edinburgh-based artist and gallery director Richard Demarco, has described the LYC museum 'as a place maker' (British Library, Artists' Lives interview). Demarco records that he introduced Li to the influential German Fluxus and Conceptual artist Joseph Beuys at the Edinburgh Festival in 1970. Beuys' practice in the 1970s and particularly his Social Sculpture actions are especially important for this discussion, based as they were on the belief that art can make a difference to society by enabling participation and interaction. For Beuys, every aspect of life could be approached creatively and, as a result, everyone has the potential to be an artist. That thinking and the model of the Free University underscored Demarco's programme of experimental summer schools, Edinburgh Arts, between 1972 and 1980 – initiated with the University of Edinburgh and the North American Students' Association as part of a trans-Atlantic cultural dialogue; many of the students who participated were from universities in the USA but students, artists, writers, performers, teachers and many others from Britain and Europe were also involved. As described, this was a liberal, experimental approach to arts education – Demarco's stated ambition was 'to open his participants to new pathways for self-determined creative action, based on encounters and exchanges with diverse places, people, artefacts and events' (Richard Demarco, on-line archive).

Edinburgh Arts summer schools evolved into a series of journeys, with the journey itself central to the learning experiences. Demarco spoke of 'weaving together' geography and cultures, the past and the present, prompting unexpected intellectual vistas and the chance of self-discovery and creative growth. Visits were made to the artist and concrete poet Ian Hamilton Finlay's garden, Little Sparta, in the Pentland Hills and over the border to the LYC – where attentive travellers were made welcome by Li. The archive contains a letter from Li to Demarco, thanking him for his visit and underlining this notion of a socially relevant artist-in-place with the lines:

> the arts are life plus beauty. The arts always have a good relationship with our daily lives. If we exist the arts exist. So we know the arts and we know our lives. We will know what our life is all about yesterday, today and where we will go tomorrow.
>
> *(Li-Yuan Chia. Cumbria. Richard Demarco Archive, www.demarco-archive.ac.uk)*

Despite yearly visitor numbers of over 30,000, financial support for Li and the LYC's daily life was limited. Its survival, like those other non-commercial galleries and art centres listed by Paul Overy as from 'the mouth of the Tyne to the Solway Firth across the neck of Britain' was largely due to the support of the Northern Arts Association, 'one of the first and the liveliest of the regional arts associations which so often act more humanly and intelligently than the central bureaucracy of the Arts Council' (Overy, 1974: 130). Northern Arts, begun in 1961, was indeed very enterprising at this point with dedicated arts, crafts, theatre and literature officers who were deeply invested in the region and lobbied on behalf of the most significant issues. Changes to a national model of support and declining resources for the Arts in general in the early 1980s, however, accounted for Li's struggles. Fund-raising ventures, such as his 'buy a brick' scheme, were sadly not sufficient and alongside personal circumstances Li's decision was to close the LYC in 1982.

The 'extraordinary and exhilarating atmosphere' of the museum that Overy reported on in 1974 was gone and an exceptional episode in cross- and de-bordering both conceptually and practically came to a close. There are, though, current and surviving examples of arts ventures deeply invested in place, other models within the 'remoter' parts of rural Cumbria and Northumbria with a similar ethos, if different in terms, for example, of the types of arts practices they might support or encourage. All are fundamentally perceived as 'arts organizations located in rural places' rather than 'rural arts organizations'. All have a clear concern to dismantle the tired but enduring polarities between the rural and the urban; to engage with the global as much as the local, and all understand 'place' itself as shaped and reshaped by the continuous interactions between the near-at-hand and the larger contexts.

A similar ethos underpins Allenheads Contemporary Arts (ACA) some 15 miles from Banks in the North Pennines AONB in what was at one time a remote but very active lead-mining region. Allenheads Contemporary Arts was established in 1995 by two 'incomer' (via Wales and New York) artists and curators Helen Ratcliffe and Alan Smith. Allenheads Contemporary Arts is an 'integration of life, work and contemporary art in a rural location' begun through renovation of derelict buildings in what had been deemed 'England's Dying Village' in the local press in 1985 (Ratcliffe and Smith, 2009: 7–9). Far from an 'idyllic bubble' this is a 'place to operate through dialogue', in an area 'shaped by successive human generations and their manipulation of matter as well as by natural forces'. 'It's not a community arts venue; rather we are part of the community.'

From initiating local children's workshops and engaging with local annual shows to developing collaborations with national and international artists, ACA, as, I would argue, the LYC, 'facilitates a crossing of cultures and participates in networks that are about a mutual relevance, meaningful contact and a synergy between people' (Ibid). As a result, an evolving programme of events and residencies has produced a series of diverse and overlayered interpretations of the local environment. Departing from the sculptural practices of artists such as David Nash and Andy Goldsworthy however, ACA has supported contemporary art and science collaborations in projects such as the 2018 'As Above, So Below'. Allenheads Contemporary Arts has therefore encouraged the exploration of the vast Dark Skies of the north Pennines in the development of a community observatory, and has brought visiting artists to engage with the subterranean environment too – the hollow earth – through experiential walks and periods of immersion in the disused lead mines around the village resulting in, for example, unique sound-recordings or soundscapes and 'border-crossing' collaborations between visual artists, writers and musicians. The ongoing residency programme is perceived as a 'testing ground and an incubator' and an opportunity for artists to 'develop work without boundaries'. At time of writing ACA is showing an exhibition of work by Newcastle-based artist James Davoll entitled 'Bound', which provides a timely exploration of the significance of what has and may again function as a hard geo-political border in Ireland by way of a sound, video and photographic installation documenting the many Irish border-crossing roads that bear the physical traces of changes in jurisdiction over time. The viewing of this exhibition on what had been intended as 'Brexit Day', 29 March, in the gallery in Allenheads village, underlines ACA's continual focus on uniting local, national and international engagements with place and their wider social, political or environmental significance. Place here is constituted through combined interconnections, through transient interactions as much as long-established relations.

While first showing at the LYC in the 1970s, the artists David Nash and Andy Goldsworthy also made work in the early days of Grizedale Sculpture, the residency-based public sculpture park in Grizedale Forest in the Lake District. The more recent organization, Grizedale Arts based at the Coniston Institute and Lawson Park Farm, also resonates interestingly however with the initial values of the LYC in relation to the artist-in-place. The current director, Adam Sutherland, has stated in interview that 'rural exile' has run its course, that 'rural culture needs to validate itself … express ideas, some values through creative activity, demonstrate ways of living creatively, consciously' (Sutherland, 2010: n.p.). He speaks of a

> new model for artists/creative people, where there is a connection to place and taking part in community, being part of a local community. This doesn't mean you have to be on good terms, but it does mean you have the same rights as everyone else and that you "belong". For artists being part of a local community means serving a useful purpose ....
>
> *(Ibid)*

For Alistair Hudson, the focus of Grizedale Arts in its varied activities is to connect rural places to 'the current global conditions, and to see rural places such as its own, as part of a contemporary complex that is shaping the way we all live and work' (Byrne, 2012: 103). This is a view of community and of the artist and place that is at once both rooted and mobile, fostering encounter and collaboration, that is boundary-dissolving and alert to the possibilities of new ways of being. In this context a socially engaged model of arts practice is articulated as a continuation of the legacy of John Ruskin, an honorary member of the Coniston Mechanics Institute in the later 1870s and constant advocate of the necessary value of artists to wider society. Like LYC and ACA, Grizedale Arts clearly understands the equal significance of collaboration with local school children and with community crafts groups as well as with national and international arts practitioners. In terms of that departure from the specific sculptural aesthetic of Nash and Goldsworthy, as Catherine Wood (2013: 23–24) writes, 'In recent years, Grizedale has invited a new generation of artists for whom site specific, found material can mean a car-boot sale or a jubilee celebration, as much as sticks, stones and leaves'. The focus is on a non-hierarchic, boundary crossing 'mobile, conceptual art drawing on existing community happenings and subcultures' as witnessed in the diverse involvements of artists such as Liam Gillick, Marcus Coates and Jeremy Deller.

As noted above, the nature and evolution of these 'arts in rural place' venues and organizations as responsive to a mobile and progressive concept of place is especially significant at the present historical moment and in this particular region, underlining as they do a need for working together across borders of all kinds, not just geographically but across skills, disciplines and practices; for thinking in fact not of borders, or of boundaries, but of connecting bridges. Through relational concepts of place and through consideration of models such as these there is a sense in which 'rural peripheries' or 'borderlands', however they may be framed, might function as deliberative spaces for developing new ecologies of living. This then is to turn from that still insistent focus on centralized organization and activity and away from outworn notions of the rural as traditional, backward and disconnected from urgent global conditions, and instead to view rural places as sites of innovation, to forge more positive notions of rurality. All of this speaks precisely on one level to the lived experience of those inhabiting this soft border or borderland region. It suggests the real possibilities of a generative reframing of place, in part by learning from the past and recent creative methodologies and collaborative actions of artists in the northern borderland – beginning, for example, with Li Yuan-chia's Museum and Art Gallery on Hadrian's Wall.

Acknowledgments

An early version of this chapter was delivered at the conference 'The LYC Museum and Art Gallery and the Museum as Practice' at Manchester Art Gallery and University of Manchester, March 2019. My thanks to Hammad Nasar and Sarah Turner of the Paul Mellon Centre for Studies in British Art for the invitation to participate. My thanks also to Alan Smith and Helen Ratcliffe for valuable conversation about the genesis and ethos of Allenheads Contemporary Arts.

References

Alfrey, N. Sleeman, J. and Tufnell, B. eds (2013). *Uncommon Ground: Land Art in Britain, 1966–1979*, London: Hayward Publishing, n.p.

Artists' Lives C466, National Life Stories, the British Library, interview with Richard Demarco, https://sounds.bl.uk/Oral-history/Art/021M-C0466X0242XX-0024V0 (last accessed 15 April, 2019).

Barker, D. (n.d.) 'The Barkers of Lanercost', in *The Makers: Craftsmen from Four Northern Counties, Northumberland, Co.Durham, Cumberland, Westmorland*, exhibition catalogue, Sunderland: Ceolfrith Press.

Butler, J. (1997). *Excitable Speech: A Politics of the Performative*, New York: Routledge.

Byrne, J. (2012). 'Grizedale Arts: Use Value and the Little Society', *Afterall: A Journal of art, Context and Enquiry*, 30 (Summer), pp. 101–107.

Causey, A. (1977). 'Space and Time in British Land Art', *Studio International*, 193, March-April, pp. 122–130.

Christian, A. (2008). Personal correspondence with author.

Collins, J. (1991) *A Painters' Place: Banks Head, Cumberland, 1924–31*, Abbott Hall Art Gallery, Bristol: Redcliffe Press.

Davidson, P. (2005). *The Idea of North*, London: Reaktion.

Demarco Archive, www.richarddemarco.org/index.html, (last accessed 15 April 2019).

Demarco, R. National Life Stories: Artists' Lives', https://sounds.bl.uk

Fulton, H. in Kastner, J. and Wallis, B. eds. (2005). *Land and Environmental Art*, London: Phaidon.

Hanson, N. ed. (1982). *Presences of Nature: Words and Images of the Lake District*, Carlisle: Carlisle Museum and Art Gallery.

Holt, Y. (2018a). 'Performing the Anglo-Scottish Border: Cultural Landscapes, Heritage and Borderland Identities', *Journal of Borderlands Studies*, 33, no. 1, pp. 53–69.

Holt, Y. (2018b). 'Borderlands: Visual and Material Culture in the Interwar Anglo-Scottish Borders', in Bluemel, K. and McCluskey, M. eds., *Rural Modernity in Britain: A Critical Intervention*, Edinburgh: Edinburgh University Press, pp. 167–187.

Ingold, T. (2008). *Lines: A Brief History*, London: Routledge.

Licona, A. (2012). *Zines in Third Space: Radical Cooperation and Borderlands Rhetoric*, New York: Suny Press.

Li-Yuan C. Cumbria. Richard Demarco Archive, www.demarco-archive.ac.uk

Low, S.M. and Lawrence-Zúñiga, D. (2003). 'Introduction: locating culture,' in S.M. Low and D. Lawrence-Zúñiga, eds., *The Anthropology of Space and Place: Locating Culture*, pp. 1–48, London: Wiley Blackwell.

Martin-Jones, D., Holt, Y., and Jones, O. eds. (2018) *Visual Culture in the Northern British Archipelago: Imagining Islands*, London & New York: Routledge.

Massey, D. (1995) 'The Conceptualisation of Place', in Massey, D. and Jess, P. eds., *A Place in the World?: Places, Cultures, and Globalization*, pp. 45–77, Oxford: Oxford University Press.

Masters, M. (1975). 'Why Li is Looking for Friends', *Evening News and Star*, 8 July.

Overy, P. (1974) 'Museum on Hadrian's Wall', *The Times*, 4 September, reprinted in Brett, G. and Sawyer, N (1999). *Li Yuan-chia: Tell Me What Is Not Yet Said*, London: Institute of International Visual Arts.

Ratcliffe, H. and Smith, A. (2009). Introduction to Warr, T. ed. *Setting the Fell on Fire*, Manchester: Art Editions North.

Scott, J. (2013). 'The History of Social Network Analysis' in *Social Network Analysis*, 1st ed., p. 2, 1991, London: Sage.

Stephens, C. (2002). 'Ben Nicholson: Modernism, Craft and the English Vernacular', in Peters Corbett, D., Holt Y. and Russell, F. eds, *The Geographies of Englishness: Landscape and the National Past, 1880–1940*, pp. 199–225, London and New Haven: Yale University Press.

Sutherland, A. (2010) 'Grizedale, the Lake District and Rural art', e- interview with White, R. *artcornwall.org*. www.artcornwall.org/interviews/Adam_Sutherland_Grizedale2.htm, (last accessed 15 April 2019).

Tolia-Kelly, D., and Nesbitt, C., eds. (2009). *An Archaeology of 'Race': Exploring the Northern Frontier in Roman Britain*, exhibition catalogue, Segedunum and Tullie House, Durham University: Durham.

Wood, C. (2013). 'Roadshow: Grizedale Arts', *Art Monthly*, May, pp. 23–25.

Wylie, J. (2007). *Landscape*, London and New York: Routledge.

6

Faith and place

Hindu sacred landscapes of India

Rana P.B. Singh and Pravin S. Rana

Introducing sacrality, spatiality and faith system

Sacred landscape combines the absoluteness of space, the relativeness of places and the comprehensiveness of landscape. Altogether, this constitutes a 'wholeness' that conveys the inherent spirit of 'holiness', which here we call '*sacredscapes*'; these are regulated and reproduced by those of faith and in their sacred rituals. Accordingly, as adherents of faith within sacred space, we form a sense of ourselves and the sense of our-place at varying scales of space–time. We begin from the local scale, and here we may first experience the sacred message through the spirit of place, its *genus loci*, and the power of place: place speaks, place communicates. In Hindu cosmology, the *Matsya Purāṇa* (*c.*CE 400) enumerates a large number of sacred places with descriptions of associated schedules, gestures, dreams and auspicious signs and symbols. The seven sacred cities within this schema (*Sapta-purīs*) are Mathura, Dvaraka, Ayodhya, Haridvar, Varanasi, Ujjain and Kanchipuram. Rather differently, the twelve most important Shiva abodes are scattered all over India and are known as *Jyotir lingas tīrthas*, with the four abodes of Vishnu in the four corners of India serving as another group of popular pan-Indian pilgrimage places. This chapter will focus on particularly vivid examples, illustrating Hindu reciprocal relationships between sacred places and the faith system. These are illustrated within the taxonomic frame of sacred places, ritualscapes, festivities, sacred water and aspects of spatial transposition that link locality and universality.

At one level, geographical enquiry can investigate the power of sacred places by searching for the cosmic designs embedded in ritual landscapes and the ways in which spatial orientations are directed towards astronomical phenomena. Particular sacred cities where such arrangements exist can be considered as a *mesocosm*, geometrically linking the celestial realm of the *macrocosm* with the *microcosmic* realm of human consciousness and cultural traditions of text, tradition and rituals. At the largest scale, both classical and modern Hindu literature is full of reverence for 'Mother India' (*Bhārat Mātā*) and 'Mother Earth' (*Bhū Devī*), with the 'land (and the Earth)' conceived as personified goddess. This notion is conceptualized by relating all geographical features – mountains, hills, rivers, caves and unique sites – as lived and imagined landscapes within Mother Earth; in this sense, these sites and places automatically become embedded in the sacred geography of ancient India (Eck, 2012: 11). In Hindu tradition this is called '*divya kshetra*' (a pious/divine territory) or *tīrtha* (holy place).

The Hindu *tīrthas* of India provide examples of a deep interrelatedness between faith-generated cultural continuity and the landscape. *Tīrthas* are 'crossing-over' places (from mundane to sacred) with many levels of meaning. The word '*tīrtha*' means a 'ford' or river-crossing and, by extension, these are places that allow passage between mundane and spiritual realms. Each Hindu pilgrimage is a '*tīrthayātrā*' (*tīrtha* journey) and the geographical manifestation of each '*tīrthayātrā*' evokes a new kind of landscape that for the devotee overlays sacred and symbolic meaning upon a physical and material base. If touring is an outer journey in geographical space then pilgrimage is the geographical expression of an inner journey. If touring is something largely pleasure- or curiosity-oriented then pilgrimage combines spiritual and worldly aspirations in places where the immanent and the transcendent mesh. Today, most Hindu sacred places are hybrid spaces that blend the religious and the mundane in complex, often contradictory, forms. Each '*sacredscape*' of sacred places, religious ritual performances and religious functions is embedded within the socio-economic-environmental attributes of the mundane world and so creates the wholeness of a geographical '*faithscape*' (Singh, 2013: 69).

Every region or place in India has its own sacred geography where humans meet with divinities and this reproduces the larger microcosmic web which is continuously regulated and expanded by the enactment of rituals, festivities and celebrations by the adherents of the faith system embedded within the *ritualscape*. By the reproduction of manifestations of sacrality in place and landscape, the associated territory converges to frame the greater unity of cosmic reality envisioned as *sacredscape* (*tīrtha-kshetra*). And this is maintained and transformed by the faith system that develops over time. There always exist distinctive forms of *genius loci* (spirit of place) which interconnect to generate the varying niches and intensities of the faith system. These deep associations between person and place-spirit results in a divine connection which crosses the boundary of space and time, and generates a world of meaning, feeling and revelation – a cosmic field of divine manifestation where humans and natural mystery meet, a *faithscape* (*āsthā-kshetra*). Faithscape thus encompasses sacred place, sacred time, sacred meanings and sacred rituals, and embodies both symbolic and tangible elements in an attempt to realise human identity in the cosmos (Singh and Haigh, 2015: 783). Mythological stories assemble the divinity's acts and life into a divine environment to make the faithscape more meaningful and grounded in place. Both the existential notion of the *sacredscape* and the experiential dimension of the *faithscape* have their roots in the cosmic order of sacred territory, which is constantly repeated, regulated and rejuvenated by rituals, rites and festivities that are embedded into resultant ritualscapes (cf. Singh, 2016: 118). *Faithscape* seeks to develop a two-way relationship between the pilgrim and the divine, howsoever conceived, and it offers two means of departure, one back to the mundane world and the other to the spiritual realms (Singh, 2011a).

In common Hindu cosmological belief, the physical and spiritual worlds are identified as two parallel dimensions of existence, and only by faith and revelation can one perceive their interconnectedness. Contemporary studies have shown that patterns of human movement in pilgrimage may take on cosmological significance, as journey to shrines are identified with macrocosmic cycles or movement among astral bodies (cf. Crumrine and Morinis, 1991: 5). In this context, the Hindu pilgrimage is a dialogue of 'imagining, memorizing and understanding' between an *outer* realm of vision and an *inner* realm of self. These two characteristics are the elements within a movement composed of two phases: Up from Body to Soul, and Up from Soul to Spirit. Through their active participation in the journey, pilgrims employ an active imagination which generates the phase, Up from Body to Soul. And by their ongoing active involvement pilgrims proceed to a deeper understanding – Up from Soul to Spirit. At the completion of their journey, pilgrims return, following the reverse steps, so that they proceed Down from Spirit to Soul and, subsequently, Down from Soul to Body. This process also generates the emergence of a *maṇḍala*, making a circular process where the point of return is synonymous with the point of departure. This conception of an archetypal mandala as the formal *a priori* structure is universal. No sacred place is without such an archetypal connotation (Singh, 1995: 98). As Edward Casey contends (1991: 290), 'They are a common place – a place of places, a place for places – and in this abstracted role threaten to become common places, taken for granted in their very universality'.

In addition to spiritual gains, Hindu pilgrimages have always concerned the gaining of social status and the relief of worldly cares (Haigh, 2011). In the ancient texts, very many of the blessings described concern the relief of sins or the fulfilment of wishes for health, wealth, success and so forth (cf. Jacobsen, 2013: 19–24). This is how these places are known for salvific qualities and soul-healing.

We form a sense of our-selves and the sense of our-place at a varying range of scales (*spatiality*), times (*sequentiality*), functions (*activity*), forms of mobility (*pilgrimages*), quests (*sacrality*) and our mental states (*belief systems*). We begin from the local scale and here we first experience the sacred message through the spirit of place or *genus loci*. According to theological notions, sacred places are consecrated or 'illuminated by faith', 'which because of their religious content, become the inevitable subject of auspicious visits' (Singh, 2011b: 6). Such places are considered sublime (*deva sthāna* in Hindu traditions) and their holiness is consolidated as a result of constant pilgrimage by devotees and their performance of rituals. In this way, such places are made constantly alive through the process of making, re-making and interfacing the sublime qualities and those understood to lead to soul-healing.

Place becomes sacred by the manifestation of sacrosanct processes and thus possesses 'power', here conceived as a spiritual energy of life-force that enables an individual or place to interact with the forces of the natural and supernatural world (cf. Radimilahy, 1994: 91). Sacredscapes are powerful places at which power is needed for transcendental energy or for protection from spiritual danger. Mythological description of these sacredscapes helps to develop a diverse range of images and imaginations (Singh, 1995: 97). Yet a sacredscape is more than an imagined product of the mind; it is also the landscape of faith, belief system, a state of transcendental consciousness and, as we have emphasized, is always becoming new, revived and transformed by the process of sacralization and ritualization.

The sacral bond between person and place is a reciprocal process, what Rudolf Otto (1923) describes as a *mysterium tremendum* associated with special power that a devotee perceives while crossing the mundane environment, a 'divine power' that Otto terms '*numinous*', which helps to achieve 'self-understanding' (*ātmasāta*) or 'self-awakening' (*ātmachetanā*). Eugene Walter (1988: 117) suggests that 'the quality of a place depends on a human context shaped by memories and expectations, by stories of real and imagined events – that is, by the historical experience located there'. In India, this notion is culturally rooted in Hindu traditions. The idea of *mysterium tremendum* also refers to a special rule of action, or customary form of making a place felt and experienced. There exist energies of place that shape the *genius loci*. What we suggest is that the quality of a sacred place depends upon the human memories, experiences, miracles and expectations that have been shaped by it. All these aspects are governed by rules of action that maintain continuity over a long span of time.

In elaborating this process Diana Eck (1996: 149) asserts that in the Hindu tradition, any place can become the sacred abode of the gods if the proper rites are performed. When a temple is consecrated and its image installed, the great rites of *pratishthā* serve to call the presence of the divine to that place. With any image fashioned of wood or stone or rudely crafted of clay, rites of invitation (*āvāhana*) are observed at the beginning of worship, inviting the deity to be present, and rites of dismissal (*visarjana*) are observed at worship's end, giving the deity leave to go.

The rituals performed by pilgrims are rooted in the theological-religious teachings of Hindu traditions and maintained through the passage of time by taboos, traditions and performances, passing from one generation to another while incorporating many contemporary beliefs and rituals, complexifying the belief systems. These rites and rituals are recorded in treatises and books of mythologies. Pilgrims' experiences lead to a sense of awe and wonder, and at the same time of relatedness. They believe that there exists an unseen order where the gods harmoniously look after us, helping pilgrims to gain relief from troubles, sorrows and sufferings, and provide peace and pleasure. In performing sacred journeys, visiting shrines and performing rituals, pilgrims 'proceed inwards for multiplicity to unity, just as in contemplation', and when they return to their home the process becomes reversed, thus developing a cyclic process between worldly activities and spiritual feeling. This profound sense of relatedness with cosmic nature can be understood only in the context of faith. While they approach certain shrines and divinities for specific needs and different purposes, above all the religious performance works as an 'instrumentality' of ritual by which one reaches the divine force (Singh, 2013: 138).

A taxonomy of Hindu sacred places

Hindu India is mythologized in its geographical features – the rivers, mountains, hills, forests, coastlines and unique landforms that all possess spiritual energies maintained and regulated by gods and associated divinities described in the ancient books of legends (*purāṇas*) (Singh and Rana, 2016: 69). The degrees and intensity of their power, together with aesthetic beauties, are descried with reference to hierarchy and distinctiveness. Classifying sacred places has been an important theme of geographic concern in terms of origin and location, motive, association and manifestation of power.

The most ancient parts of the *Vedas*, the *Rig Veda* (RgV, 1973) (*c.*13th century BCE), attach four chief connotations to the notion of *tīrtha*. It is: (1) a route or a place where one can receive power (RgV 1.169.6; 1.173. 11); (2) a place where people can dip in sacred waters as a rite of purification (RgV 8.47.11; 1.46.8); (3) a sacred site where God is immanent through possessing the power of manifestation (RgV 10.31.3); (4) a place associated with the religious territory (*kshetra*) that is sacralized on the basis of divine happenings and the work of the god(s) that took place there (*Śatapatha Brāhmaṇa* 18.9). According to the *Brahma Purāṇa* (70.16–19), a *c.*CE 7th-century text, pilgrimage sites may be classified into four categories: (1) *divine sites* related to gaining blessings from specific deities; (2) *demonic sites* associated with mythological demons who performed malevolent works and sacrifices there; (3) *sage-related sites* associated with the lives of important spiritual leaders; and (4) *human-perceived sites*, which are not believed to be 'chosen' but merely discovered and revered by humans. These categories overlap with, for instance, certain places being both important divine and sage-related sites.

Topographically, holy *tīrthas* may be classified into three groups: (1) *Jala Tīrtha* (water-sites), associated mostly with a sacred bath on an auspicious occasion, (2) *Sthān Tīrtha* (temple/shrines related to a place), related to a particular deity or sect and mostly visited by pilgrims attached to these, and (3) *Kshetra* (sacred grounds/ territory), areas usually shaped by the form of cosmic *maṇḍala*, passage along which brings special merit. The first exhaustive and annotated list of about 2,200 Hindu sacred sites, shrines and places was presented by Kane (1974) and other classificatory works on Hindu holy places include Dave's (1957/1961) four-volume work (in English) and the Gita Press's *Kalyāṇa Tīrthāṅka* (1957) with short and popular essays on 1,820 holy places of India (in Hindi). According to the *Kalyāṇa Tīrthāṅka* list, 35 per cent of all sacred places are associated with the god Shiva, 16 per cent with Vishnu, 12 per cent with various forms of the feminine divine and nearly three-fourths with water (Singh, 2009: 229).

In considering geographical scale, Hindu pilgrimage places may be conceived as falling into a four-tier hierarchy: as *pan-Indian*, those attracting people from all parts of India and glorified in the classical Hindu scriptures (see Map 6.1); *supra-regional*, referring to the chief places of the main sects and mostly linked to founders of various shrines (e.g., Pandharpur, cf. Mokashi, 1987);

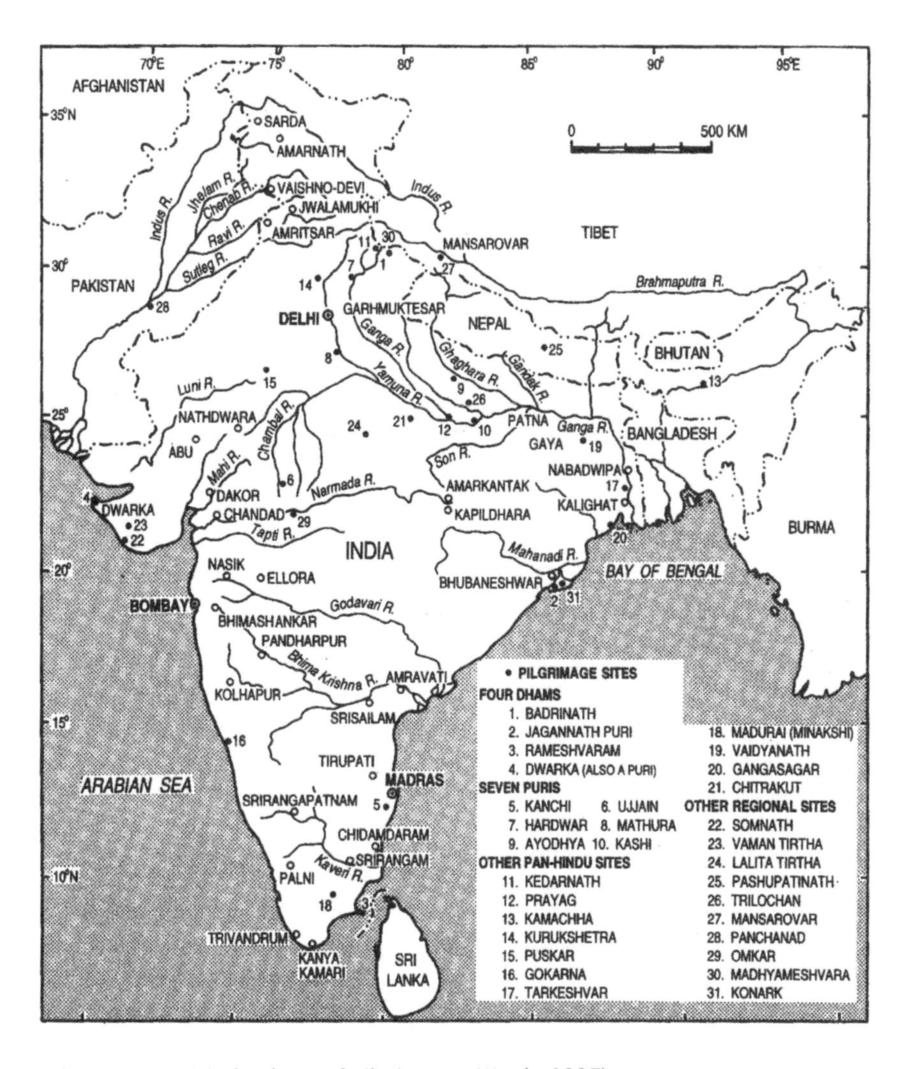

Map 6.1 Important Hindu places of pilgrimages (Singh, 1997).

regional, connoting the site's dominance in a particular culture or language group and perhaps narrated as *representative* of pan-Indian places; and *local spots* associated with attracting people from nearby villages or towns. Superimposition and transition among these differently classified places exist and their categorical status may change. Moreover, there also exist multilevel places whose identity changes according to the sacrality of time and the specificity of celebration; nevertheless, these sites maintain a loose hierarchy and traditionally generated appropriate relationship among them (Preston, 1992).

The seven sacred cities (*Sapta-purīs*) are Mathura, Dvaraka, Ayodhya, Haridvar, Varanasi, Ujjain and Kanchipuram (Singh, 2013: 7–48). Similarly, the twelve most important Shiva abodes are scattered all over India and are known as *Jyotir liṅgas tīrthas* (Shiva's site related to manifestation of light). The four abodes of Vishnu in the four corners of India are another group of popular pilgrimages (see Map 6.1). These are examples of pan-Indian pilgrimage places.

Goddess shrines are also pan-Indian examples. There are 51 special sites on the earth which symbolize the dismembered parts of the goddess's body (see Map 6.2); every region has its own tradition of varying forms of goddess (Feldhaus, 2003). The Tantric tradition symbolizes these sites as resting sites of pilgrimage by the goddess, resulting in a transformation of energy. These 51 goddess-associated sites later increased to 108 (Singh, 2013: 134–140). During the 10th–12th centuries CE, all these sites were replicated through their spatial transposition to Varanasi (Banaras), making this city a microcosm of India. These also remain active sites of pilgrimage and other rituals, attracting people from all corners of India.

With the belief that *devīs* at all Indian shrines are manifestations of the goddess in different forms, the tradition of 51 *Śakti Pīṭhas* is supposed to have come into existence sometime in the early medieval period, *c.*CE 7th century, and mythologically affirms a basic unity that embodies the goddess's power to procreate, produce and protect. The 51 *pīṭhas* have traditionally been classified according to their association with the (body) parts of the goddess they represent on the land, and sometimes based on major landforms (see Singh, 2013: 131), occupying hill/mountain tops or a distinct cave-like feature at the foothills (e.g., Shriparvat (3), Amarnath (2), Jvalamukhi (21), Manasarovar (22), Muktinath (23) and Uccaith (25), see Map 6.2).

These sacred sites are regulated, revived and sustained by *darshan* (auspicious glimpse) and *puja* (prayer and worshipping) that form the complex web of rituals. During the medieval period, many digests and treatises were written that elaborate the still practised rituals, describing the glories and merits, rules of performances and items used. Rituals play an important role in every sphere of humankind. In Hindu religion rituals start from birth (*janma*) and are completed upon death (*mrityu*) – though they may also extend after death (post-*shrādha*). The Indian ritualistic cities of Varanasi, Gaya and Ayodhya represent an aesthetic and unique type of cultural landscape that includes historical-religious traditional performances, mythology and faiths, custom, folklore, festivities, pilgrimage route and multi-cultural religious sites that are assembled in the formation of ritual landscapes.

Most Hindu pilgrimages and rituals are performed on auspicious occasions, sacred times defined in terms of astronomical-astrological correspondences that underpin their associated qualities of sacredness (*pavitrika*) and merit-giving capacity (*puṇya-phala*). These special occasions often coincide with the timing of sacred festivals and articulate the belief that, at such times, the spiritual benefits of visiting a particular *tīrtha* are most powerful. This can lead to the development of mass pilgrimages like those of the Kumbha Mela and Panchakroshi Yatra (Singh and Haigh, 2015: 785–786). Of course, the many and varied regional traditions of Hinduism, together with the rival claims of each *tīrtha*, contain many such occasions and festivals.

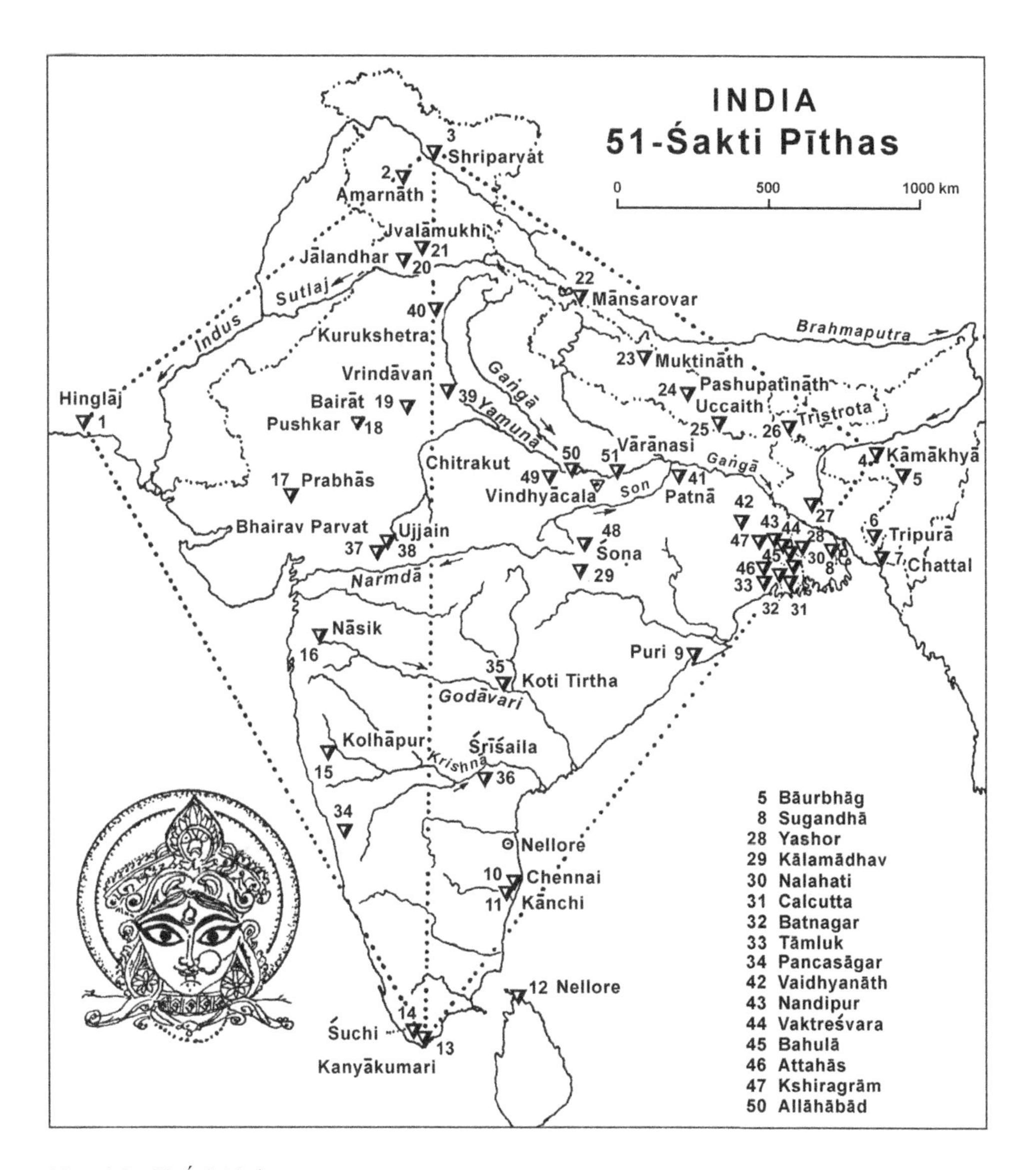

Map 6.2 51 *Śakti Pīṭhas.*

One of the hymns of the *Rig Veda* (7.9) states, 'The water in the sky, the waters of rivers, and the waters in well whose source is ocean, may all these sacred waters protect me'. Water is ubiquitously used as part of initiating rituals, is endowed with the power of sanctity and has many cosmological connotations in various mythologies, as we now discuss.

Sacred waters vis-à-vis sacredscapes

Among the many symbols of India endowed with spirituality, water is the most sacred, being both the purifier and the origin of mystery (Darian, 1978: 14). The Vedic text *Kaṭha*

Upanishad (IV.6) explains 'He who was born of old was born of water. Right from the waters, the soul drew forth and shaped a person'. Hindu mythology eulogizes the seven rivers of India, the Gaṅgā, Yamunā, Godāvarī, Sarasvatī, Narmadā, Sindhu (Indus) and the Kāverī (seeMap 6.3). This has powerfully promoted the development of a multitude of sacred places along these rivers, in addition to replicating other such sacred places from various parts of India through spatial transposition.

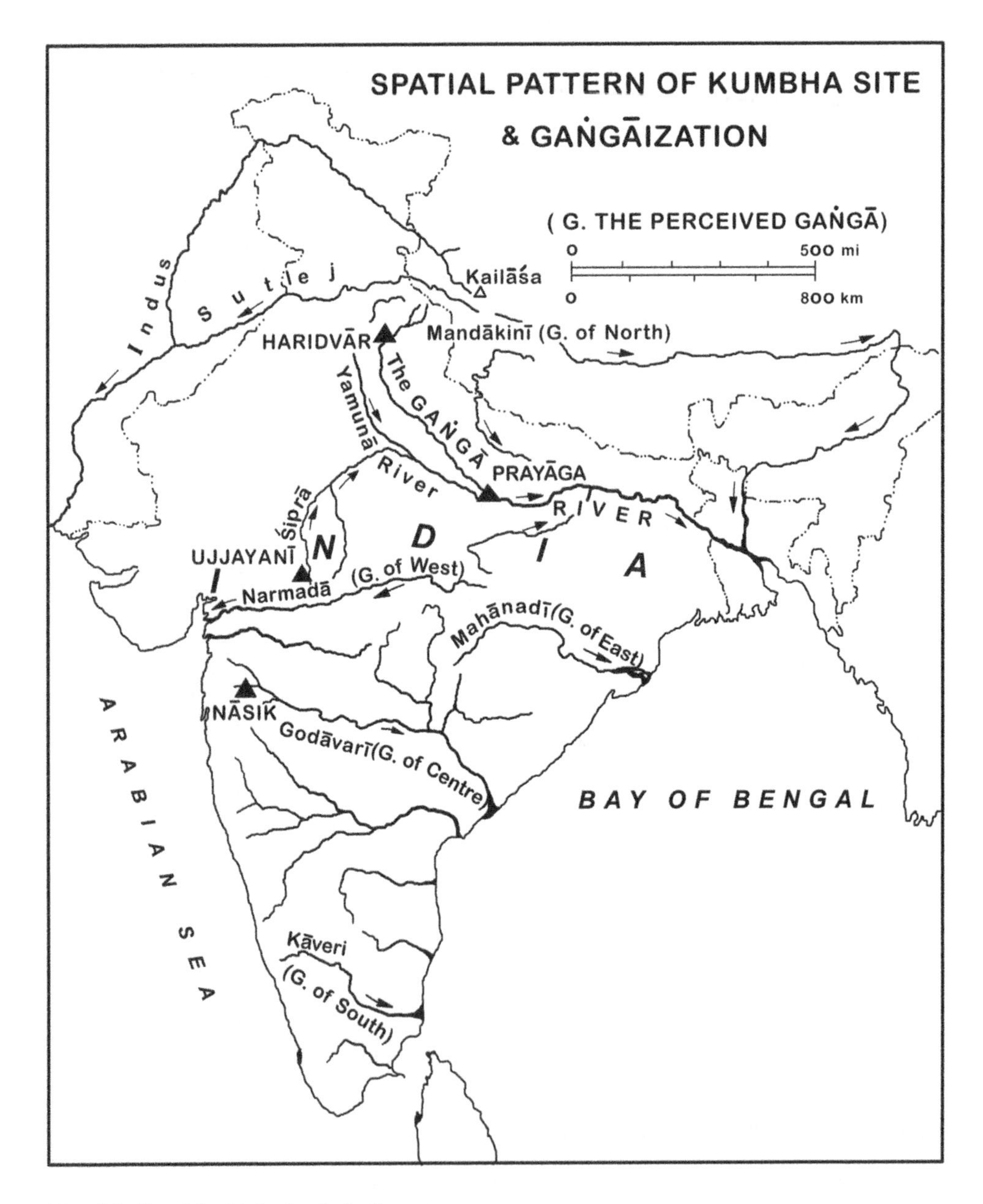

Map 6.3 Kumbha Melā sites in India.

Festivals (*melās*) at sacred sites are a vital part of Hindu pilgrimage traditions. Celebrating a mythological event in the life of a deity or an auspicious astrological period, *melās* attract around 450 million pilgrims from all over the country (Singh and Haigh, 2015: 783). The greatest of these, the Kumbha Melā, is a riverside festival held four times every twelve years, rotating between Prayagraj (Allahabad) located at the confluence of the rivers Ganga, Yamuna and the mythical Sarasvati (cf. Dubey, 2001), Nasik on the Godavari River, Ujjain on the Shipra River and Haridvar on the Gaṅgā (see Map 6.3). Bathing in these rivers during the Kumbha Melā is considered an endeavour of great merit, cleansing both body and spirit.

The Prayagraj (Allahabad) and Haridvar festivals are routinely attended by millions of pilgrims (13 million visited Allahabad in 1977, some 18 million in 1989, over 68 million in 2001, over 74 million in 2013, and over 105 million in 2019), making the Kumbha Melā the largest religious gathering in the world. It may also be the oldest. The 2019 event was spread over an area of 45 square kilometres and an 8-kilometre length of bathing ghats. Twenty-two million people resided in a temporary tent city divided into 22 sectors and infrastructural facilities included 122,500 toilets, 20,000 sanitation workers, 20,000 dustbins, 90 parking lots for 500,000 vehicles, 22 pontoon bridges on the two rivers, 500 shuttle buses and 22 hospitals (with 450 beds each).

Two traditions determine the origin/location and timing of the festival. Ancient *Purāṇas* tell of a battle between gods and demons wherein four drops of nectar (*amrita*) fell to earth on these *melā* sites (Singh and Rana, 2002; Map 6.3). And the timeframe is connected to astrological phenomena, determined by an alignment of planetary positions that generally happens in an eleven- or twelve-year cycle (Singh and Rana, 2002: 294).

The Gaṅgā in India functions as the archetype of sacred waters. By the process of the spatial transposition of her identity in different parts of India, the major rivers of a given part represent her. For instance, the Gaṅgā of the east is the Mahānadī river, of the south is the Kāverī, of the west is the Narmadā, of the north is the Maṅdākinī and in the centre is the Godāvarī river (see Map 6.3). In the source area itself there are 108 channels and tributaries named with the suffix Gaṅgā, with the two main tributaries the Bhāgirathī and Alakanandā, and where the two main tributaries meet is thereafter called the Gaṅgā.

Powerful places in Hindu history became cities and were responsible for generating entire civilizations. They seem to condense the culture and values of that civilization in one place. For example, Varanasi is known as the cultural capital of India, and microcosmic India, referred to by Diana Eck (1987: 2) as an 'orthogenetic' place that creates and sustains the ethos and moral order of the whole culture. Such places, or cities, reproduce the cosmological order and make it accessible on the human plane.

The process of spatial transposition of the holy places of India started in the 6th century and reached its climax by the 13th century Gahadavala period, by which time all the pan-Indian and regionally prominent sacred sites had been replicated in Varanasi. Mythological literature was created to manifest the power of the holiness of these sites, which culminated in making this city the 'holiest' for Hindus, preserving the '*wholeness*'. The sites of the four *dhāms* (abode of gods), the holy places in the four cardinal directions of the country (Badrinath in the north, Jagannath Puri in the east, Dvaraka in the west and Rameshvaram in the south), are re-established in Varanasi and are represented around the nuclei of the presiding deities at Matha Ghat (Badrinath), Rama Ghat (Puri), Shankudhara (Dvaraka) and Mir Ghat (Rameshvaram) (see Map 6.4).

Other religio-cultural places of India are also featured at different localities in Varanasi – Kedaranath at Kedar Ghat, Mathura at Bakaria Kund or Nakkhi Ghat, Prayag (Allahabad) at Dashashvamedha Ghat, Kamaksha (Assam) at Kamachha, Kurukshetra at Kurukshetra Kund near Asi and Manasarovar Lake at Manasarovar near Shyameshvara, amongst others (Map 6.4).

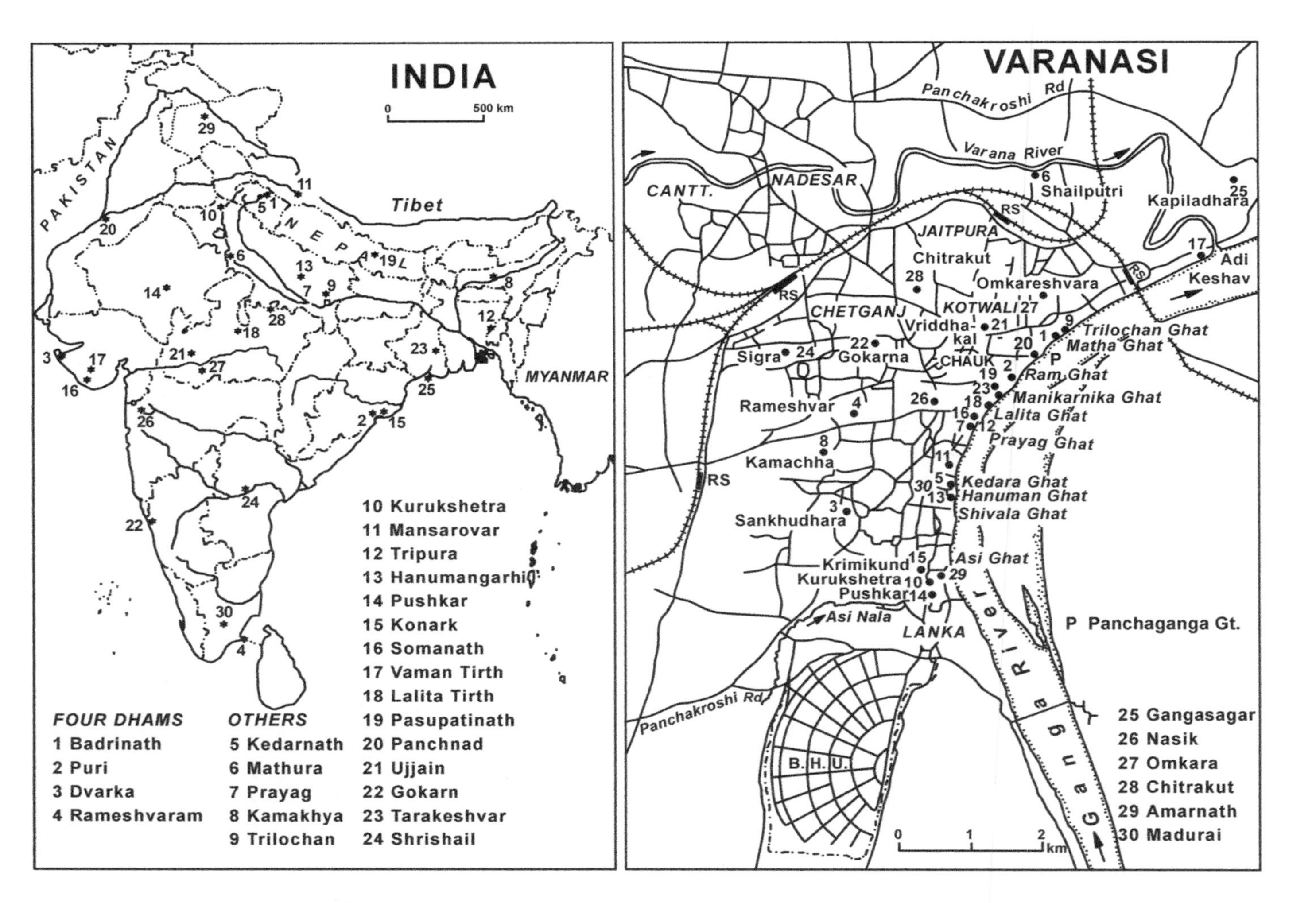

Map 6.4 Varanasi: an archetype of all-India.

Concluding remarks

Since the ancient past, the sacred place has been an intermediate zone created and explored in the belief that it has the ability to conjoin religious aspirants with their gods, thereby connecting the super-conscious mind of the human being with the celestial realm of divinities, serving an archetypal role as an *in-between* place within a cosmological frame (Barrie, 2010).

In India, while pilgrimage-tourism remains centred on informal devotion-based activities in pilgrimage centres, two aspects within religious travel remain. Religious-tourism (*dharma-kritya*) and spiritual-tourism (*moksha-dāyī*), while intertwined, have different infrastructural needs, require different services and have different driving forces, organizers, managers and modes. In practical terms, understanding these differences is a prerequisite for the effective development of Indian national strategies for sustainable development (Shinde, 2011).

Most Hindu sacred cities are examples of a complex whole, each regulated in function through an interlinking network of structural, morphological and cultural elements that together frame a 'cosmicized city'. The physical setting, the hilly territory and divine connotation mythologized in the historical past provide a basis from which to portray the emergent landscape into a complex form of cosmic geometry. Mythologies provide a cosmogonic outline that fosters an understanding of the archetypal nature of the earth spirit – and fuel the process of making and maintaining a sacred place. Born of the earth, of water, of fire, of space and of air (altogether called *pañchamahābhūtas*), the Hindu divinities are still here among us, though invisible. By performing rituals at these sites and temples, the sleeping gods are always awakened to bless the devotees. However, to perceive them we need a deeper vision of spirituality where humanity and divinity meet: this is exemplified in India's sacred geography.

Sacredscapes are distinct and unique cultural symbols of the creativity of human culture. If the sacredscape disappeared, there would be a loss of historical and cultural connections, of a heritage within Hindu psyche and cosmology. This would also result in the loss of human spirit and creativity in drawing near to the ultimate; humans would stop practising metaphysics and give up the search for cosmic interconnectedness between human being and earth-spirit (Singh, 1995: 108).

Carl Jung's provocation (1970, paraphrased in Swan, 1991: 304) expresses a moral and ethical concern for the sacred places and an associated faith system:

> People of our earth would never find true peace until they could come into a harmonious relationship with the sacred places where they live. Learning to encourage, harmonise with and perhaps even converse with the spirit of each place may be an essential survival skill to create a future world of peace where people live an ecologically sustainable lifestyle.

Injunctions to think *cosmically*, see *globally*, behave *regionally* and act *locally* but *insightfully* appeals for a cosmic vision, global humanism and self-realization. Jung's implicit promotion of an ecospiritual worldview, a spirit of wholeness and a sense of holiness grounded on an evolutionary cosmology is embedded in the Hindu conception of sacred place and sacred geography.

References

Barrie, T. (2010) *The Architecture of the In-between: The Mediating Roles of Architecture*. London & New York: Routledge.

Casey, E. (1991) *Spirit and Soul. Essays in Philosophical Psychology*. Dallas, TX: Spring Pub. Inc.

Crumrine, N. and Morinis, A. eds. (1991) *The Pilgrimage in Latin America.* Westport: Greenwood Press.

Darian, S. (1978) *The Ganges in Myth and History.* Honolulu: University of Hawaii Press.

Dave, J. (1957/1961) *Immortal India.* vol. 4. Bombay: Bhartiya Vidya Bhavan.

Dubey, D. (2001) *Prayāga, the Site of Kumbha Mela.* New Delhi: Aryan International Publisher.

Eck, D. (1987) 'The City as Sacred Center', in B. Smith and H. Reynolds (eds.), *The City as Sacred Center.* ISSSA vol. XLVI. Leiden: E.J. Brill, pp. 1–11.

Eck, D. (1996) 'Ganga: The Goddess Ganges in Hindu Sacred Geography', in J. Hawley and D. Wulff (eds.), *Devi: Goddesses in India.* Berkeley: University of California Press, pp. 137–151.

Eck, D.L. (2012) *India: A Sacred Geography.* New York: Harmony Books.

Feldhaus, A. (2003) *Connected Places. Region, Pilgrimages, and Geographical Imagination in India.* New York: Palgrave Macmillan.

Haigh, M. (2011) 'Interpreting the Sarasvati Tirthayatra of Shri Balarāma', *Itihas darpan, Research Journal of Akhil Bhartiya Itihas Sankalan Yojana* (New Delhi), 16 (2), pp. 179–193.

Jacobsen, K. (2013) *Pilgrimage in the Hindu Tradition: Salvific Space.* London: Routledge.

Kane, P. (1974) *History of Dharmashastra: Ancient and Mediaeval Religions and Civil Law in India.* vol. 3. Pune: Bhandarker Oriental Series. Originally published in 1953.

Mokashi, D. (1987) *Palkhi: An Indian Pilgrimage.* Albany: SUNY Press. reprint: Hyderabad: Orient Longman.

Otto, R. (1923) *The Idea of the Holy. An Enquiry into the Non-Rational Factor in the Idea of the Divine and Its Relation to the Rational* (Das Heilige: über das Irrationale in der Idee des Götlichen und sein Verhältnis zum Rationalen; Bresalau, 1923), translated from German by J. Harvey, London: Oxford University Press.

The Gita Press. (1957) *Tirthanka. Kalyana.* Annual no. 31. Gorakhpur: Gita Press. [Popular essays on 1820 holy places of India] <in Hindi>.

Preston, J. (1992) 'Spiritual magnetism: An organising principle for the study of pilgrimage', in A. Morinis (ed.), *Sacred Journeys: The Anthropology of Pilgrimage.* Westport: Greenwood Press, pp. 31–46.

Radimilahy, C. (1994) 'Sacred sites in Madagascar', in D. Carmichael, J. Hubert, B. Reeves and A. Schanche (eds.), *Sacred Sties, Sacred Places.* London: Routledge, pp. 82–88.

RgV, *Rig Veda, The Hymns of the Rig Veda.* (1973), ed. J. Shastri, Trans. R.T.H. Griffith, Rev. ed., 2 vols. Delhi: Motilal Banarasidass.

Shinde, K. (2011) 'Religious travel industry in India: Prospects and challenges', in World Tourism Organization (ed.), *Study on Religious Tourism in Asia and the Pacific.* Madrid: UNWTO, pp. 295–312.

Singh, Rana P.B. (2013) *Hindu Tradition of Pilgrimage: Sacred Space and System.* New Delhi: Dev Publishers & Distributors.

Singh, R.P.B. (1995) 'Towards deeper understanding, Sacredscape and Faithscape: An exploration in Pilgrimage studies', *National Geographical Journal of India,* 41 (1), pp. 89–111.

Singh, R.P.B. (1997) 'Sacred space and pilgrimage in Hindu society the case of Varanasi', in R. Stoddard and A. Morinis (eds.), *Sacred Places, Sacred Spaces.* Baton Rouge: Geoscience Publ., LSU, pp. 191–207.

Singh, R.P.B. (2009) 'Sacred space and faithscape', in R.P.B. Singh *Geographical Thoughts in India: Snapshots and Vision for the 21st Century.* Planet Earth & Cultural Understanding Series, no. 2. Newcastle upon Tyne, UK: Cambridge Scholars Publishing, pp. 227–265.

Singh, R.P.B. (2011a) 'Ritualscapes of Gaya, the City of Ancestors'', in R.P.B. Singh (ed.), *Holy Places and Pilgrimages: Essays on India.* [Planet Earth & Cultural Understanding Series no. 8]. New Delhi: Shubhi Publications, pp. 207–238.

Singh, R.P.B. (2011b) 'Sacredscapes & sacred places and sense of geography: Some reflections', in R.P. B. Singh (ed.), *Sacredscapes and Pilgrimage Systems.* [Planet Earth & Cultural Understanding Series, no. 7]. New Delhi: Shubhi Publications, pp. 5–46.

Singh, R.P.B. (2016) 'Understanding place & envisioning the cosmos', in B. Rodrigue, L. Grinin and A. Korotayev (eds.), *From Big Bang to Galactic Civilization: A Big History anthology, Vol. 2, Education & Understanding: Big History Around the World.* New Delhi: Primus Publications, pp. 117–126.

Singh, R.P.B. and Haigh, M. (2015) 'Hindu pilgrimages: The contemporary scene', in S. Brunn (ed.), *The Changing World Religion Map, Vol. 2: Sacred Places, Identities, Practices and Politics.* Dordrecht and New York: Springer Science + Business Media B.V., pp. 783–801.

Singh, R.P.B. and Rana, P.S. (2002) *Banaras Region: Spiritual and Cultural Guide.* Varanasi: Indica Books.

Singh, R.P.B. and Rana, P.S. (2016) 'Indian Sacred Natural Sites: Ancient traditions of reverence and conservation explained from a Hindu perspective', in B. Verschuuren and N. Furuta (eds.), *Asian Sacred Natural Sites: Ancient Philosophy and Practice in Conservation and Protected Areas*. London: Routledge, pp. 69–80.

Swan, J. (ed.) (1991) *The Power of Place. Sacred Ground in Natural & Human Environment*. Wheaton, IL: Quest Books.

Walter, E. (1988) *Placeways. A Theory of the Human Environment*. Chapel Hill: The University of North Carolina Press.

7

Colonial imaginaries, colonized places[1]

Uma Kothari

Introduction

This chapter critically examines the production of colonial imaginaries in the practice of Empire and how they were materially manifest – and contested – in colonized places. It considers the specificities of place in accounts of colonialism in order to appreciate the situatedness of colonial rule, and the entanglements of its material and discursive forces in actual places. I reveal the complex ways in which these imaginaries, and the practices and performances that sustained them, created segregated places that subsequently influenced the experiences of Empire for the colonized and colonizers. These reveal the complex and unstable meanings of places at different moments during British colonial rule (Roche, 2017).

The chapter goes on to show how the experiences of colonialists were not homogenous but mediated by place. I exemplify this by examining how individual colonial officers negotiated their administrative work in distinctive places. Indeed, the diversity of these colonial places reflects the contingent and heterogeneous character of the colonialist experience of Empire. Following this, the chapter demonstrates how these colonial imaginaries of, and practices in, colonized places were not transferred seamlessly but were resisted in various ways. Drawing on the connections and collaborations between anti-colonial leaders exiled by the British to the Seychelles, because the islands were imagined as remote and isolated, I demonstrate how these colonial depictions were exposed for their unfounded mis-representations of place.

I conclude by discussing how European imperialism, particularly as formulated in the British Empire, though it was diverse in practice, and fractured and ambiguous in meaning, and was simultaneously local and transnational, continues to inflect contemporary postcolonial legacies.

Colonial imaginaries of place

A colonial geographical imaginary refers to how places, spaces and landscapes are perceived, represented and interpreted through a colonial discourse. Indeed, Edward Said (1989: 218) claims that 'we would not have had an empire itself without the philosophical and imaginative processes' that enabled the acquisition, subordination and settlement of diverse places. The production of this imaginary has a long history.

The dawning of the 'age of exploration and discovery' in the 15th century was fuelled by technological advances in sailing, navigation and clocks. In 1492 Columbus reached North America and in 1497 Vasco da Gama sailed round the Cape of Good Hope. As these explorers travelled around the world they brought back fanciful stories about other places and Europe was subsequently flooded with accounts of 'new' lands. A key element of this knowledge was the process of classification through which animals, plants, particular 'races' and parts of the world were defined and categorized. These narratives produced a 'simulacrum of the Orient' that was reproduced materially in the West through 'regulatory codes, classifications, specimen cases, periodical reviews, dictionaries, grammars, commentaries, editions, translations, all of which together formed' (Said, 1978: 166). These classificatory ideas were also spread through the establishment of exhibitionary spaces. In London, the capital of Empire, these included the Great Exhibition at Crystal Palace in 1851, the British Museum, the Natural History Museum, Regents Park Zoo and Kew Gardens. Professional and academic institutions such as the Royal Geographical Society were also established to produce and disseminate these ideas.

Through these highly partial representations, distant places were identified as discrete spatial entities with their own particular fauna and flora, customs, cultures and landscapes, and through their transmission, people in Europe got to 'know' the culture and places of others, their styles of building, customs, religion, dress. These key characteristics and distinctions of place ascribed particular characteristics to different places through a colonial imaginary. This categorization of distinct, bounded places masqueraded as dispassionate scientific knowledge and became widespread, part of 'common' knowledge. Yet, it was inevitably informed by partial values and hierarchical assumptions, depictions unmediated by the people that were apparently being so expertly described by anthropologists, natural historians and geographers, amongst other specialists.

Through this imaginary, certain places were imagined and represented as landscapes without people, *terra nullius*. For instance, Africa became the 'Dark Continent', ready to be explored, made knowable and spatially classified. Where places were not seen as empty, their inhabitants were often described as people inherently predisposed towards wantonness, irresponsibility, immoral behaviour and childishness (Kabbani, 1986), with no ability for organized self-government (Mohanty, 1991). Such classificatory practices of knowing the realm of the other were deeply gendered. The explorer, scientist, hunter, soldier, administrator, missionary and geographer were invariably men, and the metaphorical notion was promoted that a colonized place was feminine, passive, awaiting its transformation into productive and regulated site. Places, landscapes and cultures were also seen as connected through essentialist discourses of environmental determinism. For instance, sub-Saharan Africa was identified as hot, tropical, disease ridden, chaotic, overgrown and wild and these characteristics of place were transposed onto its inhabitants and held to cause and reflect their particular characteristics.

Such places were conceived as suffering from a 'lack', implicitly an absence of those features found in Europe. For instance, writing of the Congo in 1907, explorer Richard Davis says,

> To tell what a place is like, you must tell what it lacks. One must write of the Congo always in the negative. It is as though you asked 'What sort of a house is this one Jones has built?' and were answered 'Well it hasn't any roof, and it hasn't any cellar and it has no windows, floors or chimneys. It's that kind of a house'.
>
> *(Quoted in Spurr, 1993: 96)*

Once a place was identified as 'empty' or 'lacking' and its people as 'wild and 'uncivilised', it seemed reasonable that it could and should be colonized and developed, for such places could be geographically re-envisioned. For example, writing for the *New York Tribune* in 1871 as he looks out at a valley near Lake Tanganyika, the explorer Henry Stanley re-imagines the landscape through western eyes:

> What a settlement one could have in this valley! See, it is broad enough to support a large population! Fancy a church spire rising where that tamarind rears its dark crown of foliage and think how well a score of pretty cottages would look instead of those thorn clumps and gum trees! Fancy that lovely valley teeming with herds of cattle and fields of corn, spreading to the right and left of this stream! How much better would such a state become this valley, rather than its wild and deserted aspect.
>
> *(Spurr, 1993: 30)*

These accounts of places, imagined as available for unhindered exploration and exploitation, provoked colonial ambitions and led to the implanting of settlements on distant territories. This involved the creation and maintenance of extensive spheres of interest, control and conquest through the appropriation, disciplining, aestheticizing and classifying of colonized places (Spurr, 1993).

Critically, they mobilized a series of discursive and practical strategies through which places were characterized as possessing a variety of attributes that represented them as devoid of cultural or productive value, but which could become productive and civilized through colonialism. This impulse was commonly expressed as 'the White Man's Burden', the idea that the colonizers had a responsibility to aid the development of 'less developed' areas, as is captured in the following quote from French colonial administrator Albert Sarraut in 1931:

> Without us, without our intervention... these indigenous populations would still be abandoned to misery and abjection; epidemics, massive endemic diseases and famine would continue to decimate them; infant mortality would wipe out half their offspring; petty kings and corrupt chiefs would still sacrifice them to vicious caprice; their minds would still be degraded by the practice of base superstition and barbarous custom; and they would perish from misery in the midst of unexploited wealth.
>
> *(Sarraut, 1931: 15)*

Though masquerading as universal, these imagined geographies were highly partial, emerging out of particular cultural, political and social perspectives. Indeed, they reveal more about colonial thinking and identities than the places they attempted to describe and classify. They conjured up particular images of these places as uninhabited, virgin territories and offered specific ways of talking about them as having no settled or indigenous population or inhabited by uncivilized people. These depictions legitimized colonial intervention and rule and justified how colonized people and places could be managed, manipulated and regulated.

Production and performance of colonized places

The imaginaries described above were materially manifest in how colonized places were planned, structured and organized, and through the on-going performative production of colonized places. Indeed, as Derek Gregory reminds us, in colonial settings as in other realms, people not only make history but geography in 'that their actions literally "take place"' (2004: xv).

Perceived as incomprehensible, unmarked and valueless prior to colonization, places were appropriated and cartographically mapped into seeming exactitude, carved into identifiable administrative districts, often irrespective of tribal, ethnic and other affiliations. Place names were often changed to reflect metropolitan cultural referents or amended through rough translations or transliterations of local names. Yet these practices, implemented without the mediation of the inhabitants, produced particular kinds of colonized places in which, as Albert Memmi (1965) writes, colonizer and colonized were situated in a special world with their own laws and situations.

Power was marked onto the fabric of colonized places through the functions of colonial rule. Colonized cities, for example, were sites for the establishment of key capitalist institutions such as banks and trading, shipping and insurance companies. They were also centres of administration with tax collecting points from which the wealth from the surrounding countryside was accrued and, as settings for political institutions, police and military bases, they were sites from which coercive power could be exercised.

Power was also inscribed on the colonial city through the creation of a binary spatial system that organized two separate sections that communicated with each other through a logic of violence, control and order. The reproduction of these spatial divisions led to the creation of a dual city in which separate European cantonments and native quarters were devised to inscribe the absolute cultural distinctions essential to supporting a rationale for colonialism. This demarcated the difference between the ruled and rulers through physical and geographical divisions and confirmed the absolute distinctions that colonialism sought to convey.

The cantonment had a function to ameliorate the huge death rate of the early colonizers, such as befell many of the British who initially travelled to India, by establishing systems of health, clean water, clean air and plenty of space. As such, cantonments were purified places, maintained and regulated to ensure adherence to strict social distancing, ordered and planned according to a grid pattern that functionally reinforced distinct areas. Shops, workplaces, factories and schools were policed to keep out undesirables, namely those 'natives' who were not workers and servants. This ordering of colonial places within the city highlights the anxious imperative for the colonial mind to demarcate, segregate and maintain boundaries between colonized and colonizer.

The cantonment was replete with parks, avenues, wide roads and compounds, and mobilized particular aesthetics based on specific urban values and architectural languages, with the erection of gothic and neo-classical churches, courts and administrative buildings. These were designed to contrast with the perceived otherness of mosques and Hindu temples as well as to dwarf them. In the domestic area of the cantonment, houses were saturated with signs of English/Britishness including ornaments and pictures in the interiors, and lawns and English flowers in the gardens, despite their unsuitability to the Indian climate.

The cantonment also frequently included an array of golf courses, horse paddocks, tennis courts and clubs. These clubs were places where rituals and practices of those who lived in the cantonment could be performed, such as the playing of croquet, tennis and cricket, and where 'sundowners' (a late afternoon alcoholic tipple) and dances could be enjoyed. These clubs or gymkhanas (as they were termed in British India) were the focal point of the community in which no 'natives' were allowed except to serve. In addition, the gendered separation of the spheres evoked a British domesticity away from the presumed uncertain morality and 'primitive' practices of the 'native quarter', and this was complemented by styles of dress, behaviour, manners, hobbies and pastimes, and other rituals and habitual performances that undergirded the reproduction of Britishness in place.

The 'native quarter' was conceived as the absolute opposite of the cantonment, as is depicted in Lord Russell's (Ching-Low Gail, 1860) pronouncement that here, 'the east grumbles, propagates, squabbles, sits in its decaying temples, haunts its rotting shrines, drinks its semi-putrid water'. Native spaces were alleged to be potentially dangerous, disease ridden and disorderly, and it was stridently expressed that their utter alterity must be kept at bay by measures to avoid social and cultural mixing. In the lurid imaginaries of the colonizers, the 'native quarter' was the

> out-of-bounds city where the living and the dead intermingle ... the carnivalesque world of the bazaar city where nothing is delineated but everything exists in a chaotic state of intermingling: a carnival of night and a landscape of darkness, noise, offensive smells and obscenities.
>
> *(Parry, 1993: 245)*

Indeed, this intermingling in which order was turned upside down, where there appeared to be few obvious divisions between public and private, or work and leisure, was disturbing to the colonial mind and confirmed ideas about complete difference. There was horror and disgust at poverty, disease, offensive smells, disfigurement and pollution, and the possibility of being touched by begging natives. Thus, the native quarter was seen as a source of danger that fed the feverish colonial imagination. As Dipesh Chakrabarty (1992: 544) writes, 'Indian chaos was contrasted with the immaculate "order" of the European quarters, where "pleasant squares", "white buildings with their pillared verandahs", and "graceful foliage" lent, to European eyes, "a fairy-like loveliness" to "the whole scene"'.

Paradoxically, however, the native quarter was also imagined as a site of desire, a place for sexual and aesthetic pleasure and adventure (Low, 1996), of brothels, gambling, dancing halls, hedonism, sensuality and decadence. As Stallybrass and White (1986: 191) remind us, colonial ideology, as with other articulations of power, 'continually defined and redefined itself through the exclusion of what it marked out as "low" – dirty, repulsive, noisy, contaminating'. And they continue, 'disgust always bears the imprint of desire. These low domains apparently expelled as "Other" return as the object of nostalgia, longing and desire'. The dual city was thus a place that reflected the intricate mix of desire, attraction, fear, loathing, adventure and repulsion that characterized the relationship between colonizers and colonized (Bhabha, 1994). It is these attributes that came to typify racialized and gendered colonial places.

Despite these clear demarcations of place and codes of behaviour, the practices of colonizers sent from the metropole to administer these peripheral places often blurred these dualistic boundaries. The dual city was variously negotiated and mediated by colonial administrators through their experiences of living in these places, producing further complexities and nuances to the kinds of imaginaries described above. The seemingly rigid ideologies of colonial discourse were thus compounded and transformed by its transmission into distinct local places, for, as Catherine Nash (2004: 112) observes, these ideologies 'did not travel out from the centre unchanged but were threatened, challenged, negotiated, made and remade in the encounters between those brought together through colonialism'. Thus, the meanings ascribed to certain ideas, behaviours and actions were not maintained wholly intact or homogenous across imperial places.

While the metropole–colony binary continues to dominate work on imperial history and historical geography, recent scholarship informed by postcolonial critiques challenges the notion of a singular, uniform colonial project implemented seamlessly across imperial places. Instead, a multiplicity of projects and discourses moved through various networks that were created and maintained between different parts of the empire. Indeed, Alan Lester (2006: 132) highlights the diversity of 'the agendas of colonial interests, their representations of colonised places and peoples and their practices in relation to them'. Thus, colonial discourses and practices were 'effective precisely because they were enormously flexible and adaptable' (Nash, 2004: 113). They were also open to individual interpretation and improvization as 'local textures of colonialism were immensely complex' (Potter, Binns, Elliot, and Smith, 2003: 38). Crucially, colonial administrators, while representatives of colonial power and institutions and acting as conduits for dominant colonial discourses, continually negotiated and mediated these conventions in accordance with their experiences on the ground, in particular places (Kothari, 2005).

As key agents in mobilizing the ideals expressed above, particular roles were strongly delineated by colonial officers in the rules that were embedded in colonial policies, hierarchies and codes, and enacted in, for example, collective performances, systems of recording and expatriate clubs. These colonized places were reproduced by the administrative tactics and protocols of colonial officers, the performance of authoritative roles towards colonized subjects, and the reinforcement of spatial divisions between colonizer and colonized. However, for many colonial officers, though they were informed by these powerful strictures and norms, being posted to 'remote' parts of the world, their very isolation forced them to improvize. These colonial officers were situated in specific cultural contexts where their individual subjectivities often supplemented the highly localized and place-based expressions of colonialism. Some experienced enduring and profound engagements with the places to which they were posted. While higher-ranked officers inhabited more enclavic (see Edensor, 1998) and highly regulated expatriate places that kept the surrounding otherness at bay, the colonial duties and interactions with local people of low-ranking officers were largely performed in rather heterogeneous places. With some remaining in one posting for at least 18 months, their practical situatedness shaped their relationship with local places and led to a conditional embeddedness. Though experiences varied according to the specificities of postings, as representatives of colonial authority 'in place', administrators performed a wide range of roles and activities that both partly reduced and reified the boundaries between themselves and the colonized, between their place of work and home, and between times for work and play. Continuously moving between the spaces of the colonizer and the colonized, some developed a more extensive involvement with places and a more substantive knowledge of their inhabitants (Kothari, 2006).

Colonial practices and performances in specific places, then, were simultaneously habitual and instrumental, contingent and individual. This illustrates how colonized places were not reified, bounded and singular, but were continuously reproduced through the activities and improvized performances of colonial administrators. These places were thus not solely defined by administrative infrastructures and protocols but were ceaselessly remade through the practices and performances of administrators, and were always in the process of becoming through the shifting social relations between colonizers and colonized. Such colonial places were therefore heterogeneous, permeable and continuously being transformed.

Challenging imaginaries of colonial place

While places were being remade by colonial practices and performances on the ground, colonial expansion continued to be justified and legitimized through the kinds of colonial imaginaries outlined above. These representations of distant places had to be continuously invoked to implement policies to sustain colonial rule. Yet some of these policies, founded as they were on mis-representations of places, inadvertently led to challenges to colonial authority and rule.

The British colonial government policies of sending anti-colonial 'agitators' into exile to the Seychelles provides such an example. At particular moments during colonial rule attempts were made to squash anti-colonial resistance by physically removing political leaders from their place of origin and relocating them to other colonized sites. This compulsory relocation of anti-colonial leaders, however, unintentionally fostered new networks and connections between colonized people and places that subsequently led to further acts of resistance to colonial rule. Indeed, from these places, exiles and their followers continued to mobilize against colonial control in innovative ways, producing new trans-imperial connections. The various forms of resistance that emerged and that connected different colonial places were enabled through the initial mis-characterization by the colonial administration as to the nature of the Seychelles. Imagined as a remote, isolated and bounded place, the colonial government mis-represented the Seychelles as far from being disconnected; however, the Seychelles has long been part of an extensive transnational network of people and ideas.

Geography makes a difference to the exercise of power, and the choice of where to banish dissidents was significant in shaping how colonial power was employed and experienced. Spatial isolation was a key characteristic in determining this geography of exile (Strange and Basford, 2003) and islands were seen as particularly good places for containment. Referring to the Andamans, Anderson (2007: 21) confirms that the landscape of the islands provided a 'natural prison' such that 'in the absence of secure places of confinement or sufficient personnel, the threats posed by the unknown jungle, sea and inhabitants comprised the convict guard'. The specific choice of the Seychelles as a destination for 'undesirables' was partly based on this perceived ease of containment, from and to which communication would be extremely difficult. Seeing the Seychelles as the least accessible place in the empire, from the late 1800s until the mid-1950s, the British authorities exiled nearly 500 prominent anti-colonial leaders and their followers from Egypt, Buganda, Somaliland, Ethiopia, Gold Coast, Palestine and other colonies to the Seychelles. The place was imagined as quintessentially remote, unconnected and largely devoid of 'civilized' and 'politicized' populations. Instead, it was simply understood as an isolated place to which exiles, from across the empire, could be safely contained with no fear of infecting others with their political sentiments. Considered a good 'dumping ground' (Lee, 1976) to 'resettle' people who challenged their authority, in the 'British colonial mind Seychelles was linked primarily with the concept of remoteness. As such they viewed it as an ideal place to exile some of the leading opponents of British imperial expansion' (Shillington, 2009: 33). Acutely aware of the potential for rebellion and revolt in other parts of the empire, having characterized the Seychelles as isolated, remote and inhabited by 'passive' peoples, colonial administrators were less concerned about the trouble the political exiles could stir up on the islands.

There was, however, much ignorance amongst the British administration about the Seychelles. This is exemplified in an exchange reported in Hansard in 1922 in which a Member of Parliament refers to the Indian Ocean islands as 'one of the most deadly places in the *Pacific*' (emphasis added, HC Deb 11 July 1922 vol. 156 cc1154-61). Importantly, far from being unconnected, the Seychelles has historically been constituted by global processes. Possessing no original population, it has been thoroughly shaped by various movements of people from other places, including French and British colonial administrators, traders, indentured labourers, slaves, plantation owners and exiles and thus has been very much constituted by a multiplicity of political, emotional, commercial, symbolic, cultural and familial connections (Edensor and Kothari, 2003). However, rather than producing placeless spaces within an abstract 'borderless' world, these connections are also historically and geographically specific.

Paradoxically then, while an imagined geography made the Seychelles an ideal place to banish political agitators, it exemplifies how places are produced through diverse connections and relations and, as such, became an unexpected nodal point in anti-colonial networks of resistance. These networks were produced through multiple connections between exiled anti-colonial leaders from different places in the empire who met in the Seychelles, and for encounters between exiles and the Seychellois, who, like them, were also colonized. These networks also enabled the maintenance of relationships with exiles' families and followers in their homeland. The islands then functioned as a place from which exiles created and accessed global networks.

In the Seychelles, exiles adopted a series of diverse strategies to confound the British authorities and to resist their exilic conditions. This resistance was multi-sited, simultaneously related to nationalist politics of independence in a distant homeland and forms of engagement in colonized Seychelles, their 'host' country. For example, the arrival and settlement of exiles connected the Seychellois to a geography of empire and forms of resistance taking place elsewhere beyond their islands. Furthermore, it raised consciousness of their own position within a larger colonial world, one replete with colonized people in other places similarly subject to British rule. The arrival of exiles was itself a significant event, one embodied in the idea and image of the ship that conjured up connections between distant places. In various ways and to differing degrees, the presence of exiles affected the mind set of Seychellois and was empowering in stimulating their own acts of resistance. Other small acts of resistance that connected colonized Seychelles to centres of imperial power were found in mundane, subtle forms of negotiation and communications between exiles and their colonizers. Places were connected through letters written by exiles to the colonial office in London demanding their release or expressing concerns over the adverse effects of the inclement weather, dietary and other requirements. As such, resistance was manifest across colonial places and in place.

These multiple sites of anti-colonial networks of resistance challenge place-based 'territorially embedded' (Tuathail, 1998: 82) understandings of politics. Indeed, the forms of translocal resistance that emerged demonstrate the importance of moving away from place-based politics (Amin, 2004; Featherstone, 2008) yet reveal how places, in this case the Seychelles, provided a new site in which identities could be reworked and the political activities of the colonized could take place. Exiles' various acts of defiance demonstrated that physical and geographical internment did not suppress their political motivations. Indeed, a wide network of ideas, cultures, resources and politics connected the 'remote' islands of the Seychelles to other places and cultural and intellectual traditions and produced anti-colonial solidarities between places (Routledge and Cumbers, 2009). Conceptualizations of the networks forged challenge perceptions of places as bounded, unchanging entities (Massey, 2004), as was also evident through the place-specific and contingent performances of colonial administrators described above.

Colonial legacies and postcolonial transformations

As shown above, Empire literally took place. Colonial imaginaries, practices and performances justified and sustained colonial rule but were also mediated, challenged and resisted in place. Yet, these imaginaries and practices endure. For example, they are inscribed on the landscape in particular places and continue to shape the movement of people between formerly colonized places and centres of imperial power. Furthermore, notions of progress and modernity, as exemplified by discourses and policies of international development, also persist. As such, forms of rule, cultures and practices travelled across colonial places and over time and have been subsequently reworked in the post-independence period.

Postcolonial analyses have shown how representations and articulations of colonial places did not come to an end at the time of decolonization but continue to be mobilized, reworked and mediated through ideologies, individuals and institutions in the post-independence period. By providing critical responses to the historical effects of colonialism and the persistence of colonial forms of power and knowledge into the present, postcolonial critics have challenged these orientalist discourses whereby the West continues to produce knowledge about other people and places (Said, 1989).

Many places in formerly colonized realms continue to reflect their colonial history through, for example, how places are organized. Bounded places comprising five-star hotels or guest houses, administrative buildings and leisure facilities such as sports clubs and gymkhanas continue to mark formerly colonized places. These realms are inhabited by highly paid Western international consultants, expatriate workers and national elites echoing distinctions across place that colonialism sought to impose through the purified spaces of, for example, cantonments in colonial places. By guarding the boundaries of these enclaves to ensure that the only local entrants are those admitted as low-paid workers serving Western and elite guests, the mingling and chaos that might upset the regulated environment is minimized. These enclavic places are further reproduced by an aesthetic control which puts a premium on manicured lawns and neat exteriors. Containing external and internal trimmings and decor which conform to European tastes, typically such places also combine icons, ornaments and pictures from Europe with contextualized signifiers of the local 'exotic'. This effect is consolidated by the cultural and leisure pursuits of the temporary inhabitants who perform drinking and sporting rituals often reminiscent of a colonial era. As familiar European havens, such places and the practices that take place within them are part of a neo-colonial geography inscribed upon the places of formerly colonized countries.

Contemporary tourist practices provide further examples of the resonances with the colonial imaginaries of formerly colonized places. Indeed, much contemporary tourism is entangled in colonial histories. For example, the remaining structure of the cantonment in certain postcolonial Indian cities, contemporaneously reconfigured as a venue for tourist accommodation, retain some of this aesthetic effect. For instance, as Barbara Weightman (1987: 235) writes, for contemporary European and North American tourists, 'perhaps the wide avenues and rational geometry … at one time symbolic of social order and control, soothe the nerves of those who confound confusion with danger in their perceived disarray of the maze-like indigenous city'. In the Maldives, maintaining high-end tourism in luxury resorts requires recreating a tourist imaginary of isolated and unpeopled island landscapes, an illusion of virgin territory, a pristine paradise on which tourists can re-map their identities and imagine themselves as castaways, albeit with luxuries and high-quality services. This particular vision of islandness is reproduced not only in travel writings and tourist brochures but on the landscape. Ensuring their continuing attractiveness to the high-paying tourists that seek out idyllic, isolated destinations also means keeping locals away and 'otherness' at bay. Many, therefore, have physical barriers and high fences around their perimeters to restrict entry and to insulate tourists from the very different surrounding environments. Indeed, the places outside the resort may be presented as dangerous by hotel staff, the realm of potential criminal acts or violence, hassle, chaos, dirt and disease and uncontainable difference. And, as such, hotel resorts create tourist enclaves that mimic the colonial (re) production of the kinds of places discussed above (Kothari, 2015).

These continuing legacies of colonial constructions and representations of colonized places have, however, also been reworked and challenged. After formal independence there was felt to be a need by some recently independent states to reconstitute the place as indigenous, as specifically Indian, Malaysian or Nigerian. Street signs were often changed to be named after nationalist figures rather than colonial characters. Statues were similarly replaced. Indigenous architectural styles were developed, national museums established and old nationalist myths, cultural traditions and heroes were celebrated. Simultaneously there was a felt need to use the place as a badge of modernity to show that colonial historical ideology was wrong, that, in fact, the newly independent country had been kept back by colonialism, stopped from developing and could be as modern as anywhere else, as industrial and sophisticated. For instance, the adoption and development of modernist architecture such as tower blocks and housing estates aimed to demonstrate that these places were intrinsically modern. As did the building of Chandigarh, designed by Le Corbusier, pre-eminent modernist architect, that provided a statement of India's desire to embrace modernity.

In this chapter I have shown how colonial imaginaries of other places enabled the acquisition and expansion of Empire and how colonial performances made and reproduced heterogeneous colonized places. Furthermore, I have highlighted how these representations informed policies that sustained colonial rule but were also mis-guided and so inadvertently enabled colonized people in and across different places to challenge this colonial authority. Yet, despite various forms of resistance, colonial imaginaries continue, albeit reworked, into the present shaping the structure, performances and practices in formerly colonized places.

Note

1 Parts of this chapter have been drawn from Kothari, U. (2011) Contesting colonial imaginaries: politics of exile in the Indian Ocean, *Geoforum* 43(4), 697–706.

References

Amin, A. (2004) Regions unbound: towards a new politics of place. *Geografiska Annaler: Series B Human Geography*, 86(1): 33–44.

Anderson, C. (2007) *The Indian Uprising of 1857–8: Prisons, Prisoners and Rebellion*. Cambridge: Anthem Press.

Bhabha, H. (1994) *The Location of Culture*. London and New York: Routledge.

Chakrabarty, D. (1992). Of garbage, modernity and the citizen's gaze. *Economic and Political Weekly*, 27 (10/11): 541–547.

Ching-Low Gail. (1996) *White Skins/Black Masks: Representation and Colonialism*. London and New York: Routledge, 160.

Davis, R. (1907) *The Congo and Coasts of Africa*. New York: C. Scribner's Sons.

Edensor, T. (1998) *Tourists at the Taj*. London: Routledge.

Edensor, T. and Kothari, U. (2003) 'Sweetening colonialism: a Mauritian themed resort', in D. Lasansky and B. McLaren (eds) *Architecture and Tourism*. Oxford, UK: Berg Publishers, 189–206.

Featherstone, D. (2008) *Resistance, Space and Political Identities: The Making of Counter-Global Networks*. Oxford: Wiley-Blackwell.

Gregory, D. (2004) *The Colonial Present*. Oxford: Blackwell.

Kabbani, R. (1986) *Europe's Myths of Orient*. London: Pandora.

Kothari, U. (2005) Authority and expertise: the professionalisation of international development and the ordering of dissent. *Antipode*, 37: 425–446.

Kothari, U. (2006) From colonialism to development: continuities and divergences. *Journal of Commonwealth and Comparative Politics*, 44(1): 118–136.
Kothari, U. (2015) Reworking colonial imaginaries in post-colonial tourist enclaves. *Tourist Studies*, 15 (3): 248–266.
Lee, C. (1976) *Seychelles: Political Castaways*. London: Hamish Hamilton.
Lester, A. (2006) Imperial circuits and networks: geographies of the British Empire 1. *History Compass*, 4 (1): 124–141.
Low, G. (1996) *White Skins, Black Masks: Representation and Colonialism*. London: Routledge.
Massey, D. (2004) Geographies of responsibility. *Geografiska Annaler: Series B, Human Geography*, 86(1): 5–18.
Memmi, A. (1965) *The Colonizer and the Colonized*, trans. Howard Greenfeld, vol. 21. London: Earthscan, 2003.
Mohanty, C. (1991) 'Introduction: cartographies of struggle: third world women and the politics of feminism', in Mohanty, C.T., Russo, A. and Torres, L. (eds) *Third World Women and the Politics of Feminism*. Bloomington, IN: Indiana University Press, 1–47.
Nash, C. (2004) 'Post-colonial geographies', in Cloke, P. Crang, P. and Goodwin, M. (eds) *Envisioning Human Geographies*. London: Hodder Arnold, 104–127.
O' Tuathail, G. (1998/1996) *Critical Geopolitics: The Politics of Writing Global Space*. London and New York: Routledge.
Parry, B. (1993) 'The content and discontents of Kipling's imperialism', in Carter, E. and Donald, J. (eds) *Space and Place: Theories of Identity and Location*. London: Lawrence & Wishart Ltd, 221–240.
Potter, R., Binns, T., Elliot, J., and Smith, D. (2003) *Geographies of Development*. 2nd edn. London: Prentice-Hall.
Roche, M.M. (2017) *(Dis)Placing Empire: Renegotiating British Colonial Geographies*. London: Routledge.
Routledge, P. and Cumbers, A. (2009) *Global Justice Networks: Geographies of Transnational Solidarity*. Manchester: Manchester University Press.
Said, E. (1978) *Orientalism*. London: Routledge.
Said, E. (1989) Representing the colonized: anthropology's interlocutors. *Critical Inquiry*, 15(Winter): 205–225.
Sarraut, A. (1931) *Grandeur et Servitude Coloniales*. Paris: Sagittaire, 15.
Shillington, K. (2009) *History of Modern Seychelles*. Basingstoke: Macmillan.
Spurr, D. (1993) *The Rhetoric of Empire: Colonial Discourse in Journalism, Travel Writing and Imperial Administration*. North Carolina: Duke University Press, 96.
Stallybrass, P. and White, A. (1986) *The Politics and Poetics of Transgression*. Ithaca, NY: Cornell University Press.
Strange, C. and Basford, A. (eds) (2003) *Isolation: Places and Practices of Exclusion*. London and New York: Routledge.
Tuathail, G.Ó. (1998) Political geography III: dealing with deterritorialization. *Progress in Human Geography*, 22(1): 81–93.
Weightman, B. (1987) Third world tour landscapes. *Annals of Tourism Research*, 14(2): 227–239.

8
Mobilities and place

David Bissell

Introduction

Over the past few decades, the social sciences have witnessed a profound 'turn' towards mobility. Mimi Sheller and John Urry (2006) have articulated this perhaps most forcefully through their declaration of a 'new mobilities paradigm' which acknowledges the 'immense' scale of contemporary travel. They write,

> all the world seems to be on the move. Asylum seekers, international students, terrorists, members of diasporas, holidaymakers, business people, sports stars, refugees, backpackers, commuters, the early retired, young mobile professionals, prostitutes, armed forces—these and many others fill the world's airports, buses, ships, and trains.
>
> *(2006: 207)*

Their suggestion is that previously, social science tended to be rather 'static', which had the effect of marginalising the significance of the movement of people that takes place for all sorts of different reasons.

What is undeniable is that acknowledging the significance of mobility changes how we understand place. As Tim Cresswell acknowledges, for some time, there was a tendency to think about places in rather static ways that marginalised mobility. For instance, he suggests that 'when people think and write about place they often fix on old small places that seem "authentic"' (2014: 79). Such places might on the face of it seem the very antithesis of mobility. Yet, given the immense rise of different sorts of mobility to which Sheller and Urry draw our attention, our understanding of even the most apparently 'static' of places needs to be much more sensitive to the different kinds of mobility that are happening. As Cresswell has argued, if place can be best understood as 'a way of seeing, knowing and understanding the world' (2014: 18), then 'mobility has always been part of place' (Ibid: 84).

This chapter synthesises some of the most significant academic literatures that have explored the relationship between place and mobility. To anchor these literatures, it draws on fieldwork from a research project that explores the lives of mobile workers in Australia whose lives routinely take them away from their places of home for days or weeks at a time. This phenomenon provides a useful lens through which to interrogate notions of place that are attuned to mobility. The overall argument accounts for how place is ultimately contingent on all kinds of different mobilities, and that 'place and mobility are not antithetical but co-constitutive' (Cresswell, 2014: 65).

Mobile work places

A key recent theme has been to explore employment-related mobility (Cresswell et al., 2016). Where previous studies of work focus on specific places, such as the work place itself, mobilities thinking has increased our sensitivity to the multiple kinds of mobility that are implicated in work lives, expanding our appreciation of the place of work beyond the confines of a building or room. Much attention has been devoted to exploring the mobilities that are required to get between home and work. Work on daily commuting, for instance, has highlighted how these mobilities constitute significant places in and of themselves, in which people undertake myriad meaningful activities (Bissell, 2018). The significance of understanding work places is also highlighted in research on the mobilities of work itself, where the work place might itself be mobile. Harry Ferguson (2009), for instance, shows the centrality of the car in performing social work, not only for undertaking work while in transit, but as a place where therapeutic transformations can happen between therapists and clients.

Our project set out to explore a related but distinctive phenomenon, where paid work takes people away from home for days and weeks at a time. Work-related mobility has become a key focus of policy concern in recent years. In Australia, one of the most significant forms of mobile work has been fly-in fly-out (FIFO) labour in the resource sector, where workers fly or drive from their homes in the coastal cities to live in 'host' mining communities for days or even weeks at a time. In zooming in on this population, our research targets a sizeable population that has been overlooked by research on mobility, which has tended to focus on either the to-and-fro of daily commuting (Edensor, 2011) or the much more episodic mobilities involved in infrequent circular migrations (Zhang, 2018). Although the definitions vary, long-distance commuting—or, in our terms, mobile work—generally refers to journeys of more than 100 km, often involving a temporary period of living away from the primary residence considered to be 'home'. Between 2006 and 2011, the number of mobile workers in Australia increased by 37% to around 213,773 or about 2% of the Australian workforce (de Silva et al., 2011).

Our project team was particularly interested in how this form of employment-related mobility changes the understanding of place for those involved. We interviewed 60 people, a combination of mobile workers and partners of mobile workers, in order to get a deeper sense of the impacts of this sort of work on households (Bissell and Gorman-Murray, 2019; Straughan et al., 2020). Here, I focus in on the experiences of one participant to provide us with a window onto exploring how we might consider place from the perspective of mobility. Any of our participations could have been selected, but Joe's interview was especially fascinating because his way of seeing place, in other words, his way of 'seeing, knowing and understanding the world' (Cresswell, 2014: 18), was characterised by mobility in intriguing ways. Joe is in his late 20s, single, and has been working as a technical operative on a mining site in the north west of Australia for almost a year. Like many people employed in the resource sector, his employment is characterised by a roster, working 12 hours on site for eight days, then taking six days off. The remainder of this chapter focusses on four aspects of Joe's experience and draw out implications for thinking about place.

Placeless in Perth

Early in the interview, Joe describes how once he had finished university in his home state of Queensland, he secured a position with a mining company in Western Australia. The condition was that he would have to spend three months working in Perth, the state capital, before starting at the remote mine site. However, he ended up having to work in Perth for a whole year because a decline in the economic fortunes of the resource sector coincided with his move. What we found particularly interesting is the way that he described his time in Perth. He says that 'I was quite a miserable person', which he first attributes to 'a mixture between lack of sleep or not having the time to have enough exercise or having to cook and everything every night or do all that type of stuff'. He later qualifies this time squeeze by pointing out his lengthy 1.5-hour commute to and from work, as well as the expectation that he would work longer than his scheduled eight-hour day. However, he subsequently suggests a further reason for his discontent. He says that

> it's not that it was hard, I was kind of constantly in a transitional period because I wasn't sure if I was staying in Perth or going to site, and it was very hard to make concrete plans. I just hated being there.

To evaluate this aspect of Joe's understanding of place, we turn to humanistic geographical writing concerned with exploring the relationship between place and mobility from the perspective of bodily experience. Yi-Fu Tuan's early writing is useful in this regard since he is interested in the ways that people make meaningful places. For Tuan, place is 'essentially a static concept. If we see the world as process, constantly changing, we would not be able to develop any sense of place' (1977: 179). So, for Tuan, place is an inherently sedentary notion, understood as 'a break or pause in movement—the pause that allows a location to become a centre of meaning' (Ibid: 14). Importantly, from this perspective, mobility undermines people's ability to form roots and attachments. Reflecting back to Joe's interview in this light, his time in Perth can be viewed as a 'break or pause in movement' of sorts, and therefore an opportunity to allow this place to become a centre of meaning for him. Yet his professed inability to 'make concrete plans' for instance, suggests a difficulty in forming meaningful attachments in the way that Tuan suggests stopping permits. Rather, his characterisation of his time in Perth emphasises transitoriness rather than stasis.

On one hand, this transitoriness can be understood in terms of Joe's daily mobility practices. The idea of living and working in the same city implies a sedentariness that obscures the significant mobility of Joe's long commute to and from work that restricts what he would like to do outside of work time. We might surmise that it is this daily mobility that might prevent him from forming more meaningful attachments. In this regard, we can supplement Tuan's writing with those of Edward Relph, his contemporary, whose writing highlights how the mobility facilitated by railways and highways 'imposed on the landscape' were a direct threat to places since they 'encouraged the spread' of what he terms 'placelessness' (1976: 90). However, on the other hand, Joe's transitoriness can also be understood in terms of his lack of desire to become rooted in Perth, instead wanting to get to work at the mine site itself, rather than being stuck in an office. Relph's writing is again instructive, since he writes that

> to have roots in a place is to have a secure point from which to look out on the world, a firm grasp of one's own position in the order of things, and a significant spiritual and psychological attachment to somewhere in particular.
>
> *(1976: 38)*

While Joe's physical mobility to and from work in the city is a practical encumbrance that prevents him from forming stronger attachments in Perth, it is also his dislike of the workplace itself that has contributed to this sense of transitoriness. Indeed, Joe implies that he did have a plan, 'a secure point from which to look out on the world', to use Relph's terminology, but this plan to begin working at a mine site after three months was thwarted by the economic uncertainty that hit the resource sector, giving rise to what we might call a more 'existential' mobility, a sense of a period of life currently being in transit, rather than stable. Both these forms of mobility are negatively coded in terms of their relationship to place. Cresswell (2006) labels this 'sedentarist' thinking, inviting us to think of the many domains of social and cultural life where mobility accrues negative connotations, as a perceived threat to place. From a sedentarist perspective, place and roots are seen to be morally superior to more mobile states of existence. The strong tones through which Joe refers to his 'miserable' time in Perth certainly seem to resonate with this disparagement of mobility, both practically and more existentially. However, the remainder of Joe's interview would suggest that mobility plays a much more affirmative role in his understanding of place.

Nomadic holidays

Joe recounts how, a year later, he finally got a position working at a mine site in the north west of the state, working an '8 and 6' roster. His love of his new job is unequivocal when he says that 'being up on site is bliss, absolute bliss'. His 12-hour work days are challenging, but he says, 'there's not really anything to complain about up here'. He likes the 'structure' that the work and rest times provide. He acknowledges that he might feel differently if he had a partner and children, but he is able to keep in regular touch with friends and family, calling his parents who live in Queensland a few times a week. He proceeds by telling me that his routine is different to that of many other people on site because 'every eight days I go off on holiday somewhere'. Since the cost of living in Perth is so high, and he would only be there for half the time, Joe says that it makes little sense for him to keep a base, preferring to be itinerant. He describes his situation as living 'a nomadic lifestyle'. Sometimes he'll plan where he's going a few months in advance, but at other times the choice of destination will be last minute. He recounts that this year,

> I went to Kuala Lumpur, I went to Singapore, I went to Bangkok, I went to Ho Chi Minh, I went to Brisbane, I went to Melbourne, I went to New Zealand. This coming break, I'm going to Hong Kong, the next break after that I'm going to be going to Fiji.

His choices are in part dictated by the 10-hour flight window that he has set himself.

In contrast to the theories of place developed by humanistic geographers Tuan and Relph and described in the previous section, where movement was negatively coded and being sedentary took on more positive connotations, Joe alludes to a much more mobile sense of place where his mobility away from the work site every eight days is described in a highly positive manner, framed through discourses of adventure and rootless pleasure. Echoing Joe's own language, his new life is better characterised by 'nomadic' rather than sedentary thinking. As Cresswell describes, in nomadic thought, displacement 'ceases to be a threat and becomes a virtue' (2006: 50). This more mobile sense of place that celebrates flow over fixity is particularly influenced by the social theories of poststructuralist thinkers such as Gilles Deleuze and Félix Guattari (1988). Their work on the figure of the nomad has been highly influential in highlighting how mobile figures wield considerable power through their capacity to evade the structures and organised powers of the state. However, for Deleuze and Guattari's nomad, points and nodes are subordinate to movement itself.

Joe's emphasis on the places that he visits during his time away from work suggest that a wholescale celebration of unmoored mobility is not an entirely faithful description of his sense of place.

Rather than seeing places as static and the people that move through them as mobile, recent work on mobilities has developed a more nuanced appreciation of how places are themselves better understood as constituted through mobility. As Sheller and Urry write, 'a clear distinction is often drawn between places and those travelling to such places. Places are seen as pushing or pulling people to visit. Places are presumed to be relatively fixed, given, and separate from those visiting' (2006: 214). So, from this perspective, Joe is not a 'nomadic' mobile figure that moves through the 'sedentary' cities that he lists. Sheller and Urry go on to describe how the new mobilities paradigm overturns 'this ontology of distinct "places" and "people"' (2006: 214). Instead, they suggest that there is a more 'complex relationality' where places and persons are connected through performances. Their argument encourages a focus on the actual practices and bodily performances that are enacted in different places, such that the distinctive qualities of even the most apparently 'fixed' places, such as a city square in any one of the places Joe visits, for instance, are actually much more faithfully captured through the different kinds of rhythmic, routine movement that go on in them at different times (Jensen, 2010).

In contrast to the figure of 'nomad', Joe's movements are remarkably patterned. His movements at the work site itself are regulated by the itinerary of his workday, and even his 'leisure' time is circumscribed by the institutional timetable which dictates key activities such as meal times. His work charter flights between Perth and the mine site are a similar routine requirement every eight days. David Seamon's (1980) description of 'place ballets' is an evocative way of capturing this sense of how places are composed of rhythmic, routine mobilities that can take place below the threshold of cognition. While influenced by humanistic notions of place which underscore the importance of stability, unlike Tuan and Relph, Seamon is interested in how such stabilities can be more mobile through people's routine practices. He argued that while people's daily activities include all kinds of small scale, largely unthought movements, which he calls 'body ballets', these take place in concert with those of others. As such, while Joe revels in his 'nomadic' lifestyle, preferring to holiday rather than going 'home', his movements are structured in highly specific ways.

Friendly flights

Much of the interview with Joe is preoccupied with discussion of the travel that he undertakes between work and the holiday cities to which he travels during his time off. When describing the best parts of his job, he exclaims unambiguously 'I love flying, so that comes in handy'. As part of his routine, before his two-hour charter flight to the mine site, Joe says that he always visits the airline lounge for breakfast and then he is able to sleep during the flight, especially if he has been 'sleeping at the airport' during the previous night to save on hotel costs. Referring to his work charter flight, he demonstrates that he is highly attuned to the type of plane he's on and what he will do often depends on whether it has wireless internet on board. The significance of airline lounges is a repeated feature of the interview. These are places in which he often waits for 'three or four hours' before getting on a plane to his next holiday destination. Indeed, he talks enthusiastically about the benefits of his frequent flyer programme and the 'rewards' that it provides him with, extending from lounge access to hotel upgrades and discounts, which 'makes that dollar go a little extra'. What is particularly striking is how Joe's flying experiences are thoroughly social. In spite of him being a 'solo' traveller, he talks warmly about how he uses his travel time to 'meet people on planes', even offering to take people into airline lounges. He recollects that 'I've met complete strangers and hang out with them for a day'.

So how does this further complicate our understanding of place? Referring back to the first section, Relph (1976) insists that technologies of transit spread the undesirable quality of 'placelessness', eroding the distinctiveness of individual places. Relph's concern is how the unique character of individual towns and cities is compromised by mobility. As the second section discussed, this perspective, which sees places as relatively bounded containers, has been extensively critiqued as overly 'sedentary', denying the significance of movement to the constitution of place. However, what is also missing in Relph's analysis is a concern for the character of spaces of transit themselves. Joe's interview suggests that a significant dimension of his sense of place is bound up with experiences that he has in the spaces and durations of passage themselves, rather than just the 'end points' of work and the cities that he visits.

Rather than imagining travel as an insignificant void between two significant places, some thinkers have been preoccupied with exploring the nature of spaces of transit. A prominent reference point here is Marc Augé's (1995) concept of the 'non-place', which he develops through reflections on the experience of transit spaces of which he suggests airports are particularly exemplary. For Augé, non-places are spaces of alienation and anonymity, since 'the space of non-place creates neither singular identity nor relations; only solitude, and similitude. There is no room for history unless it has been transformed into an element of spectacle' (1995: 103). Others have followed by describing a similarly de-peopled sense of transit spaces. Mark Gottdiener, for instance, suggests that airport departure lounges 'are not commonly a place for social communion' (2001: 34). However, while these thinkers encourage us to take more seriously these interstitial spaces themselves, they do not adequately capture Joe's thoroughly social experiences of flight which, far from characterised by an evaporation of social relations, is a place where he most keenly forms connections and has meaningful experiences that contribute to positive experiences of his mobile working life.

Recently, a rich seam of work has attempted to unpick such de-socialised assumptions by exploring the myriad ways in which interstitial spaces such as airports (Adey, 2008), cars (Edensor, 2003), train carriages (Bissell, 2016) and buses (Wilson, 2011) are sites of mobility in which distinctive socialities emerge and get reworked, in part according to the materiality of the spaces themselves. Other work on the figure of the passenger builds on this observation by showing how people on the move engage in myriad forms of work- and leisure-based activities, increasing our apprehension of how these are meaningful places (Adey et al., 2012). However, that Joe's professed love of flying is thoroughly bound up with the perks of lounge access reminds us that interstitial spaces such as airports can be experienced in very different ways by different people. As Mike Crang writes of Augé's non-place, 'this singular ego-ideal is sutured into the image of the forty-something, healthy male business traveller to the exclusion of other identities' (2002: 571). In some respects, Joe typifies the male, white, economically comfortable 'kinetic elite' that has been the target of critiques that have highlighted how the experience of others through airport spaces is not so friction free, as demonstrated by Peter Adey's (2004) work on the differential experiences of security and Kathy Burrell's (2011) work on the often uncomfortable experience of budget airlines. Such differential experiences are a reminder of the need to attend to the politics of mobility, where the easy movement of some comes at the expense of the constrained mobility of others.

Site transformations

It has now been a year since Joe began his FIFO working life. In reflecting on the changes over the past year, he affirmatively concludes that it has been 'everything I had hoped it would be'. What is interesting are Joe's reflections on the changes that have happened during the year, some of which refer to his own practices. For instance, he describes how he has 'culled' many of his possessions over the course of the year, including reducing his travelling luggage to a single carry-on case. While this reduction in possessions has enabled him to move more comfortably, it has also meant that he increasingly sees the work site as his home. 'I don't really have a spot', he says, recounting how when he comes back on site, he has to 'unpack everything and store things away', but often he has to 'move things from one place to another'. He concludes, 'I suppose home is basically wherever I dump my bag'. Other changes relate to his evaluation of the feel of the work site. He speaks fondly about the team that he works with, and how he has enjoyed getting to know new people as they have joined the work site, friendships cemented through activities such as the monthly cricket match. However, he has become increasingly sensitive to the quality of food, which has recently started declining owing to the company's finances. A final set of changes he reflects on concerns his dispositions. He says that his 'quality of life has definitely gone up, the quality of sleep and just overall, I suppose, my persona'.

In section two, Seamon's (1980) idea of a place ballet was discussed to describe how the coherence of places can be formed through the rhythmic movements that compose them. Joe certainly alludes to some of the more regimented daily rhythms that take place on site, where certain sorts of movements happen with clockwork regularity, such as the bus between the airport and the site, and smaller scale movements such as the movements to and from the mess hall. However, witness above how Joe also refers to a series of changes that have taken place since he began this job. This suggests that an emphasis on such patterned regularities risks obscuring the explicit and more subtle transformations taking place. One of the first scholars to appreciate the concept of place in a more open-ended manner was Alan Pred (1984) whose work highlighted that places are never 'finished', since the repetition of mobile practices means that they are always in a state of 'becoming'. Accordingly, while the practices that take place daily at the mine site have a rhythmic quality, all kinds of changes are taking place that mean that the mine site as a place is far from fixed, but is constantly changing.

Building on Pred's emphasis on process, one of the most influential theorists to consider place from the point of view of its mobility-induced transformations was Doreen Massey. Contrasting with previous ways of thinking about places as bounded, unchanging and fixed, Massey (2005) underscores how places can be better understood as 'events' involving myriad mobile practices that involve physical bodily movement, virtual communication, as well as the movement of capital and goods. Central to Massey's (1993) notion of a 'progressive' sense of place is the idea that places are constituted and transformed by the relationships, networks and connections which transcend a locale, rather than what might be apparently 'internal' to it. Guided by her approach, we can appreciate how the uniqueness of the mine site at which Joe works is not threatened by mobility, but is actually constituted through the mobility of people, possessions, goods, communication and capital that link the mine to other parts of Australia and beyond. As Joe suggests, his sense of place of the mine site is partly constituted by the comings and goings of all the different people that he has worked and socialised with, which over time has changed, rather like a kaleidoscope gradually turning.

Following from this notion of place being in ongoing transformation owing to mobility, Joe's reflections about how his dispositions have changed also indicate how mobile bodies can be understood in a more transitional manner. For instance, Joe's frequent travel to and from the mine over the course of the year has likely made him more skilled at living with fewer material possessions, changing his sense of self in the process. Such a point of view is championed by Brian Massumi who writes that

> Displacement is just the visible trail of qualitative changes in nature. Displacement is not just a shift of place. It's the index of a becoming: movement not just from one spatial location to another, but from one nature-changing entanglement to another. It's always a question of transformation.
>
> *(2017: 8)*

Work on mobilities has built on Massumi's notion of repetition as qualitative change to understand how routine mobilities such as the daily commute (Bissell, 2018) can give rise to different kinds of explicit and subtle transformations in the people involved, changing their capacities to act and to sense. As such, mobility is not just transforming place, but it is transforming the very bodies and objects that constitute these mobilities.

Conclusion: the fluidities and fixities of place

Scholarly debates have long grappled with the question of what mobility might be doing to place. Some thinkers have been concerned that mobility is antithetical to place, arguing that the specificity of place is eroded through movement. Certainly, there are many contemporary situations where this perspective can be discerned. However, this chapter has sought to demonstrate that, rather than antithetical, mobility and place are inextricably linked. As such, any understanding of place needs to be sensitive to the diverse ways in which different kinds of mobility make and remake place. In debating with some of the key literatures on place and mobility, this chapter has demonstrated the importance of meaning to movement. An understanding of place differs according to whether movements are evaluated as positive or negative, desirable or undesirable. From a 'sedentarist' perspective, mobility is viewed as a threat to place, whereas from a 'nomadic' perspective, mobility is celebrated and a more static notion of place is viewed with suspicion. In focussing in on Joe's reflections of working at a mine site for a year, this chapter has discerned some of the complexities relating to mobilities and place, showing that such conceptual distinctions are far from clear cut.

Reflecting on Joe's interview, on first blush we might surmise that the idea of place has little relevance to his self-professed 'nomadic' existence, especially given that he has no fixed home that he returns to after each 'swing' at work. Yet throughout the interview, we can discern a rather different appreciation of place emerging, one that is not premised on rootedness, immobility or authenticity. Instead, we witness Joe's appreciation of the multiple fluidities and fixities that together make up his evolving sense of place. The joys of his footloose city adventures during time off work are balanced with a similarly positive affection for the institutional 'moorings' that the mine site affords him, the personal relationships that such fixity has provided, while also retaining his connectedness with distanced family and friends. Indeed, zooming out to the wider context beyond Joe's own experiences, the mine site in Western Australia where Joe works is itself a complex place of fluidities and fixities. The work that happens at this site concerns the extraction and movement of iron ore to a port on the coast to be shipped overseas, facilitated by both physical technologies of transit and the mobility of capital; it is powered by a workforce that flies in from elsewhere; it is managed remotely from a capital city; it is changing shape each day. Yet, at the same time, this mobility is contingent on a host of things that are relatively immobile, such as fixed rosters, rules that govern conduct on the site, buildings that service the workers, and airport infrastructures that allow workers to come and go.

The field of mobility studies (Adey et al., 2014) encourages a nuanced consideration of place that is attentive to relative and contingent fluidities and fixities, rather than from the perspective of sedentarism or nomadism. While sedentarism and nomadism are powerful discourses, both see mobility and immobility in a dualistic manner. Instead, work undertaken in the 'new mobilities paradigm' attempts to survey a more intricately patterned terrain. Mobilities research asks crucial questions about the politics of relative fluidities and fixities, evaluating how they manifest and evolve in specific places for different people, shedding light on the politics of enablement and constraint in the process. In turning attention recently to the study of work and labour (see Cresswell et al., 2016), mobilities studies are asking pressing questions about how workplaces are contingent on mobilities of different kinds and invite renewed consideration of the role of mobility in the future of a variety of workplaces.

References

Adey, P. (2004) Surveillance at the airport: surveilling mobility/mobilising surveillance. *Environment and Planning A*, 36(8), 1365–1380.

Adey, P. (2008) Airports, mobility and the calculative architecture of affective control. *Geoforum*, 39(1), 438–451.

Adey, P., Bissell, D., Hannam, K., Merriman, P. and Sheller, M. (Eds) (2014) *The Routledge Handbook of Mobilities*. London: Routledge.

Adey, P., Bissell, D., McCormack, D. and Merriman, P. (2012) Profiling the passenger: mobilities, identities, embodiments. *Cultural Geographies*, 19(2), 169–193.

Augé, M. (1995) *Non-Places. Introduction to an Anthropology of Supermodernity*. London: Verso.

Bissell, D. (2016) Micropolitics of mobility: public transport commuting and everyday encounters with forces of enablement and constraint. *Annals of the American Association of Geographers*, 106(2), 394–403.

Bissell, D. (2018) *Transit Life: How Commuting Is Transforming Our Cities*. Cambridge, MA: MIT Press.

Bissell, D. and Gorman-Murray, A. (2019) Disoriented geographies: Undoing relations, encountering limits. *Transactions of the Institute of British Geographers*, 44(4), 707–720.

Burrell, K. (2011) Going steerage on Ryanair: cultures of migrant air travel between Poland and the UK. *Journal of Transport Geography*, 19(5), 1023–1030.

Crang, M. (2002) Between places: producing hubs, flows, and networks. *Environment and Planning A*, 34 (4), 569–574.

Cresswell, T. (2006) *On the Move: Mobility in the Modern Western World*. London: Routledge.

Cresswell, T. (2014) *Place: An Introduction*. Oxford: WileyBlackwell.

Cresswell, T., Dorow, S. and Roseman, S. (2016) Putting mobility theory to work: conceptualizing employment-related geographical mobility. *Environment and Planning A*, 48(9), 1787–1803.

de Silva, H., Johnson, L. and Wade, K. (2011) Long distance commuters in Australia: a socio-economic and demographic profile. Paper for 34th Australasian Transport Research Forum, 28–30 September 2011, Adelaide.

Deleuze, G. and Guattari, F. (1988) *A Thousand Plateaus: Capitalism and Schizophrenia*. London: Continuum.
Edensor, T. (2003) Defamiliarizing the mundane roadscape. *Space and Culture*, 6(2), 151–168.
Edensor, T. (2011) Commuter: mobility, rhythm and commuting. In T. Cresswell and P. Merriman (Eds.) *Geographies of Mobilities: Practices, Spaces, Subjects*. Aldershot: Ashgate, pp. 189–204.
Ferguson, H. (2009) Driven to care: the car, automobility and social work. *Mobilities*, 4(2), 275–293.
Gottdiener, M. (2001) *Life in the Air: Surviving the New Culture of Air Travel*. Lanham, MD: Rowman and Littlefield.
Jensen, O. B. (2010) Negotiation in motion: unpacking a geography of mobility. *Space and Culture*, 13(4), 389–402.
Massey, D. (1993) Power-geometry and a progressive sense of place. In J. Bird (Ed.) *Mapping the Futures: Local Cultures, Global Change*. London: Routledge, pp. 59–69.
Massey, D. (2005) *For Space*. London: Sage.
Massumi, B. (2017) *The Principle of Unrest: Activist Philosophy in the Expanded Field*. London: Open Humanities Press.
Pred, A. (1984) Place as historically contingent process: structuration and the time-geography of becoming places. *Annals of the Association of American Geographers*, 74, 279–297.
Relph, E. (1976) *Place and Placelessness*. London: Pion.
Seamon, D. (1980) Body–subject, time–space routines, and place-ballets. In A. Buttimer and D. Seamon (Eds.) *The Human Experience of Space and Place*. London: Croom Helm, pp. 148–165.
Sheller, M. and Urry, J. (2006) The new mobilities paradigm. *Environment and Planning A*, 38(2), 207–226.
Tuan, Y.-F. (1977) *Space and Place: The Perspective of Experience*. Minneapolis, MN: University of Minnesota Press.
Straughan, E. R., Bissell, D. and Gorman-Murray, A. (2020) Exhausting rhythms: the intimate geopolitics of resource extraction. *Cultural Geographies* (in press).
Tuan, Y.-F. (1978) Space, time, place: a humanistic perspective. In T. Carlstein, D. Parkes, and N. Thrift (Eds) *Timing Space and Spacing Time*, (vol. 1). London: Edward Arnold, pp. 7–16.
Wilson, H. F. (2011) Passing propinquities in the multicultural city: the everyday encounters of bus passengering. *Environment and Planning A*, 43(3), 634–649.
Zhang, V. (2018) Im/mobilising the migration decision. *Environment and Planning D: Society and Space*, 36(2), 199–216.

9

Place as human–environment network

Tree planting and place-making in Massachusetts, USA

Deborah Martin

The idea of place, and certain types of places, create certain imaginaries for people, as well as expressing values and understandings about geographies, and people. Place is often conceptualized as a site of human settlement, a named location where people live, work, and move around. In such a definition, places might be neighborhoods, or towns, or cities. But few people – except and originally Doreen Massey (1991) – often think of the whole globe as a place. Massey (1991) describes thinking about places as interconnected and global, calling for 'a sense of place which is extroverted, which includes a consciousness of its links with the wider world, which integrates in a positive way the global and the local.' Thinking of the whole globe as a place prompts a thinking about places as including human and non-human elements, expanding the word 'place' beyond what it might immediately conjure. This chapter uses a tree planting program in Massachusetts, USA, as an opportunity to explore how an instrumental, ecosystem-focused policy effort also produces places, through the deployment, and thus active networking, of trees.

Places are many configurations of things both fixed and moving; they are, therefore, networked, manifesting both conditions of a site, as well as relationships within and across that site. As John Agnew (1987) argued, places have three major dimensions. They are *sites* comprised, for example, of grassy hills or dusty plains, asphalt and concrete structures. Places are also *situations* (or processes) fostered by the activities of people meeting and traveling on roads or airplanes or dirt trails, exchanging goods and ideas. Finally, places are constituted by and repositories of the *feelings* that people develop in and about them, including satisfaction about the smell of a cherry tree in blossom, or emotions evoked by a bench where some people had a lovely chat (see also Massey, 2005). For many geographers, these three elements of place mean that places are *relational*, which means they develop out of, and physically manifest, relationships that are political, economic, and social (Amin, 2004; Massey, 2005; Pierce, Martin, and Murphy, 2011). Thinking about places as relationships as well as locations is helpful for understanding places as dynamic; they situate and connect people in relation to things and each other.

Cities provide a context for thinking about the full range of relationships that define and constitute places. Cities create and contain a host of human-built elements like roads, apartment buildings and houses, playgrounds and schools, and all sorts of workplaces, from office towers to factories and coffee shops or restaurants. Some building styles and materials express specific local cultures, climate, or physical characteristics; for example, in cities like Minneapolis or Houston, with extreme cold and heat (respectively), downtown buildings are connected by climate-controlled skyways and tunnels, so that people can travel through them without going outside. On the other hand, buildings and roadways worldwide exhibit globalized economic trends that foster similarities in architecture and signage, reducing the distinct sense of place which architecture and urban form sometimes expresses. All of these built things, however, essentially prioritize human lives; the concept of place tends to focus on human activities, movement, and feelings (which has a certain logic, since humans create the notion of place, and feelings we derive from them). But recognizing that the whole globe is a place (Massey, 1991), or that playgrounds are places, or that beaches are places, forces us to consider a host of non-human elements of places that help to make them meaningful. Oceans or water are essential elements of favorite vacation spots, for example. Oceans and water also connect and interact with fish, and soil, and plants. In cities, grass, trees, and sometimes water, are some of the most evident non-human elements in the landscape.

Trees offer a particular form of nature in the city that provides a multi-scalar sense of place; individual trees can delineate a street or yard-scape, and trees collectively make up an urban forest or canopy that shapes the ecosystem of the city. This multi-level or multi-scalar impact and role of trees means that they actively constitute all three of Agnew's (1987) place-elements of site, situation, and affect. Trees are part of places as sites by their physical presence, serving as markers – or memorials (Cloke and Pawson, 2008) – on a grassy park field, or as a border line between residential properties. Trees connect sites (places), shaping their situation, by chemically transforming carbon dioxide into oxygen, filtering rain water as it runs through the soil, linking parts of the earth to other parts by the extension of their root systems underground. Trees also serve to connect creatures such as birds and squirrels within their leaf canopies; and trees also link humans who walk among them or sit beside them. Finally, trees foster a sense of attachment and connection for individuals who encounter them (Jones and Cloke, 2002; Pearce, Davison, and Kirkpatrick, 2015).

Scholars have noted the significant ways that trees foster sense of place, generating both positive emotions of connection and well-being for people (Cloke and Pawson, 2008; Pearce, Davison, and Kirkpatrick, 2015). Trees also sometimes generate negative emotions, often related to the maintenance that they require in a landscape (Heynen, Perkins, and Roy, 2006; Kirkpatrick, Davison, and Daniels, 2012; Pearce, Davison, and Kirkpatrick, 2015). While research on trees recognizes their role in constituting place, Phillips and Atchison (2018) argue that much of the focus on trees as important sources of greening in cities has fostered a viewpoint of trees as primarily 'passive objects' (p. 2), rather than active participants in the production of urban space and places.

Certainly, urban trees do provide a host of environmental benefits, known as 'ecosystem services,' that improve air and water quality, human health, and serve economic, community, and aesthetic values (de Vries et al., 2003; Nowak and Dwyer, 2007). Many cities in the United States and elsewhere have created tree-planting programs to reap the benefits of urban trees, and to mitigate global climate change. Shade trees keep homes significantly cooler in the summer, and houses located in the lee of sheltering trees relative to prevailing winds enjoy reduced wind-chill in the winter (McPherson and Rowntree, 1993; Zölch, Maderspacher, Wamsler, and Pauleit, 2016).

In Massachusetts, an innovative tree-planting program provides trees to urban areas to improve energy efficiencies by creating summer shade and winter windbreaks. Called 'Greening the Gateway Cities' (GGC), the program explicitly addresses specific places: 'Gateway Cities' is a Massachusetts government term for mid-sized regional economic centers that historically provided good ('gateway') manufacturing jobs, but which today face a range of social and economic challenges in the post-industrial economy (MassINC, 2015). Since 2014, about 15,000 trees have been planted in fourteen of the twenty-six 'Gateway' cities in Massachusetts, in an effort to increase canopy cover in the Gateway Cities by 5–10% in the selected environmental justice neighborhoods (Mass.gov, 2019a). The GGC program has a paired focus on street and private trees; about 80% of the tree planting occurs on private property (yards of private homes, multi-family rental homes and commercial property), and 20% occurs on city streets or other public land.

The fact that the GGC state-level program is oriented to cities in the state that have certain characteristics highlights the policy value of a place identity and imaginary; the 'Gateway Cities' term is a policy category aimed at describing small and mid-sized urban areas in the state. Using the term Gateway Cities invites state agencies and politicians to imagine the need for policies – and create them – for regional economic centers that face social and economic challenges that are quite different from the vibrant and dominant Boston economy in the state. (Boston is the capital of Massachusetts and two thirds of the state's population lives in the Boston metropolitan area (World Population Review, 2019).) The idea of Gateway Cities creates a place category that draws attention to the particular and different characteristics of urban areas within a specific political and economic area; in this case, Massachusetts. For Massachusetts' GGC program, the focus is not only on social and economic characteristics, but on the physical environment of these cities.

The slogan 'Greening the Gateway Cities' suggests that these places are not already very green, perpetuating an historically based understanding of the gateway cities as former industrial centers, where mills and factories dominated. The 'greening' of the GGC program specifically focuses on the planting of trees. Since Gateway Cities are as small in population as 35,000 or as large as 250,000 people (Obar, 2012), the amount of greenspace – or tree canopy, to be specific – in the form of parks or forests or street trees varies considerably. Gateway Cities in the GGC program include very built-up Chelsea, adjacent to Boston, as well as coastal Fall River, which contains a higher proportion of suburban landscape. Although the amount of 'green' that already exist in the Gateway Cities varies, the focus of the program is on getting the ecosystem benefits of trees to more people in urban Massachusetts.

The GGC program is an environmental sustainability program because it has a central goal of reducing energy costs. Indeed, the Department of Conservation and Recreation (DCR), which runs the program, is a sub-unit of the Massachusetts Executive Office of Energy and Environmental Affairs (EEA). Funding for the program also comes from the Department of Energy Resources (DOER) and the Department of Housing and Community Development (DHCD). The program essentially seeks to change the costs of energy in nearby buildings through external changes to the environment, aiming to lower winter heating bills by reducing cold wind speeds, and lower summer cooling bills by providing shade. But, in addition to environmental and economic goals, the program is also social and political, because it targets only certain neighborhoods within the Gateway cities: those that contain a higher percentage of residents who rent than in other areas of the Gateway cities, those that contain housing that is decades and, in some cases, centuries old, and those that are not energy efficient (Mass.gov, 2019b). The targeted planting zones are also areas with less tree canopy than other parts of their cities, and relatively high wind speeds. These targeted areas are known as environmental justice neighborhoods.

Environmental justice refers to concerns about the uneven spatial distribution of environmental benefits, such as lakes, trees, and parks, as well as environmental harms, such as polluted soil, water, and air. Scholars of urban inequality have pointed to the important ways that degraded biophysical environments affect people unequally, in part because of the human processes (such as industrial production and its polluting by-products) that degrade the physical environment, and the resulting pollution that affects human health (Bullard, 1994; Lam, 2012). In particular, these scholars point to environmental racism as a fundamental environmental justice concern for urban policy (Bullard, 1994; Pulido, 2000). Environmental racism describes the disproportionate environmental dangers that affect low income and especially racial minorities in the United States. Although environmental racism tends to focus on pollution and other environmental harms, environmental justice also points to broader research on the positive effects for human health of access to biophysical amenities (Abraham, Sommerhalder, and Abel, 2010; Brownlow, 2006).

Thinking about environmental justice allows us to consider the full range of interactive elements that constitute places: people, economic processes, and environmental elements. The Greening the Gateway Cities program in Massachusetts highlights the importance of trees in cities as an environmental benefit which is not equally distributed to all. By trying to increase the number of trees in areas with a greater proportion of older, energy inefficient rental housing, the people who run this state program have to bring together a host of networks of people to make the program work. In doing so, they reshape neighborhoods into different places, both physically (because of the newly planted trees) and socio-politically, because of the networks of people who have to come together to make the program work.

In contrast to many residential tree distribution programs in which residents do the physical labor of planting (Nguyen et al., 2017), the Massachusetts DCR takes sole responsibility for tree planting, cooperating with different stakeholders to plan for tree watering and maintenance. Three categories of actors define the place and participants that work to achieve the goals of the GCC program. First is the state Department of Conservation and Recreation, which runs the program. When a Gateway city is designated as a site for planting, the DCR assigns one or two foresters to head up the program in that place. They are in charge of working with city and community partners, and establishing the locations for planting, and supervising the planting that occurs, using their own hired crews. The second set of actors is the city itself; usually a staff member in a department of public works or planning is a designated liaison and works with the DCR foresters to develop agreements and plans, and to ensure that the city is able to help designate planting sites and care for the trees planted on city property (such as street trees or plantings in parks or on school grounds). The third actors in the GGC are representatives of a community partner. Usually the partners are groups with some orientation to environmental issues; in some cases this could be an organization that already works to create tree stewardship and awareness (as in Fall River, Massachusetts, where the organizational partner had already been planting trees in the city), but it could be a more broad-based service organization (as in Chicopee, where the community partner is Valley Opportunity Council, which seeks to provide a wide range of services to low and moderate income families in its region). The community partners are important because they help to spread the word within the designated planting area about the free trees, and can encourage residents and local business owners to get involved.

Community partners such as the one in Chicopee, the Valley Opportunity Council (VOC), may find that their partnership with the DCR expands their understanding of their mission. The primary mission of the VOC is to provide opportunities to low income residents of Chicopee and Holyoke which focus on childhood education, food and nutrition, and energy assistance. It is through this latter mission that VOC and DCR identified a shared commitment through the GGC. The VOC assists the program by holding community events, such as tree planting, which bring awareness to area residents of the opportunity to have free trees planted on their properties (or those of a landlord). The VOC also created a yard sign that informs passersby that a nearby tree was planted by the DCR as part of the program, and that it is stewarded by the resident of the property. The yard sign provides a visual that demonstrates care of the neighborhood, and signals an investment in the community from the many new tree plantings in the area, both on the streets and in private yards. Through its involvement in the GGC program, the VOC can foster a community focus on trees and greening, which highlights the broader mission of the organization to improve the neighborhood. The physical presence of the trees themselves – and their continued need for stewardship – can help the VOC in bringing attention to the physical and social needs and relationships in the neighborhood.

These relationships and how they foster community awareness form part of a research focus by a team of students and faculty at Clark University that has been studying the GGC program. The research examines tree outcomes, and socio-political outcomes, of the GGC program. In the summer of 2018, the team conducted a survey of 2,271 trees, out of 2,393 planted in these two Gateway Cities (building on similar work in summer 2017 on the program in Holyoke, Chelsea, and Revere, MA). They found that almost 85% of trees planted in Chicopee and Fall River were located on private residential land, both single- and multi-family. (Multi-family residential sites comprised about 15% of the sites for planted trees.) The next most common planting site after residential land was maintained parks (8% of planting property types).

The preponderance of trees planted on private property raises three key issues for the realities of tree-planting programs, and for thinking about how tree planting relates to place and the networks that constitute place. First, post-planting stewardship is a key element of the success of the GGC program (and for any tree-planting program; Roman et al., 2015). Second, tree stewardship in the GGC relies upon individuals, or groups of individuals, rather than governments, since most trees are planted on private property. Finally, the introduction of new, juvenile trees into a landscape changes that place at the individual property level, as well as in the locale; in this case, in the neighborhoods. The combination of a changed landscape, and one needing a particular form of stewardship, brings together the practices that create and recreate places, through the interactions of people and their environments.

The places where trees are planted have an immediate change in landscape look (or, in the site). The trees respond to their environment by growing, either in ways that demonstrate stewardship and health, or by sending out distress signals such as basal sprouting, a signal of roots seeking water near the soil surface (and of inadequate watering). The healthy or distressed growth reactions of the juvenile trees point to their relations with area people and other aspects of nature, such as presence or absence of adequate rainwater as a supplement to watering by humans. Similarly, the interactions among tree stewards, DCR foresters, and VOC staff creates new relationships and senses of community, based around common interest and care of the trees.

Newly planted trees change the landscape, but the meanings they provide to that landscape depend on how they fare over time; dying trees can foster a sense of blight, whereas well-stewarded trees will thrive and provide green space, shade, and wind breaks for residents. Indeed, one resident, who lives in Chelsea, MA, commented after trees were planted, 'It's really pretty, it makes a big difference in the city, going down the street and seeing all the trees.' Another resident (in Fall River) said that, 'I just think it beautifies the neighborhood.'

The success of tree planting programs in actually beautifying neighborhoods and fostering such positive reactions depends in large part on what happens to the trees after they are planted. The addition of trees to a place creates a changed landscape, but those impacts will not necessarily be positive or beautifying: Dead trees are rarely a positive image, anywhere. In the GGC program, trees are stewarded by the owners or residents of properties where they are planted. In exchange for participating in the program and getting free trees planted on their properties, owners or their property managers (or tenants, in some cases) must commit to tree care. The DCR foresters who plant the trees try to make sure that property owners understand that trees need a lot of water, especially in the first few years, as well as good mulching.

Some participants in the program understand stewardship requirements and respond accordingly, such as this interviewee in 2018: 'I water in the morning, and water when the sun comes down. Two times a day. I give them probably about three to five gallons each.' Unfortunately for overall tree vigor and survivorship, however, it's not always clear to people how much they have to help urban trees to survive. For example, another tree recipient reported, 'I don't water my trees. Nature does that.' Despite these drastically different approaches, however, trees in the GGC program are thriving, by and large, with over 90% of the trees surveyed in Chicopee and Fall River in 2018 alive, and by far the majority of them were healthy.

The impacts of juvenile trees shape a place beyond their visible imprint on the landscape. Certainly, juvenile trees contribute to the physical scene of any place; they are part of the physical site and many material elements that make up the setting of place. Where they grow and thrive and send out new branches and leaves every spring, they invest in the health of their site. Where they struggle and decline and fail to leaf some branches, they may communicate weakness and struggle of the area as well. Where people are able to water trees and help them grow, both trees and waterers co-constitute a place.

The juvenile trees also shape less-seen elements of their place. For example, through the expansion of roots in the subsoil, trees take up water that would otherwise runoff into sewers or onto streets and sidewalks. Through the expansion of their branches their leaves or needles seasonally affect the dispersal of sunshine, creating oases of cool or areas where plants thrive where they need shade. Their branches also create habitat for small creatures and birds. All of these non-human elements form part of the physical setting of any place, and constitute elements in the landscape to which people respond, by gathering together on a summer day under shade, or working to clear leaves on a fall day.

The Greening the Gateway Cities program, too, constitutes a set of social relations and physical opportunities for networks of place elements to coalesce. One outcome of the program is that it creates new opportunities for a series of interactions; between city agencies and the state DCR, between these government entities and their community partners, and for all of these institutions to interact with property owners and residents in the environmental justice-defined tree planting zones. For residents of Chicopee and Fall River, the mere fact of newly planted trees generates opportunities for place identity. One resident described how tree planting constituted a sense of place around his home:

> These five houses here, these neighbors stick together. He's got trees in his yard, he got some in his yard, the guy across the street got some in his yard. So we're like affiliated like a little organization here with the trees, you know?

Another resident described greater interactions as residents shared information about the program with one another:

> I got involved in the program from my neighbor … So I asked the neighbor, I says, 'Those trees must have been expensive, no?' He says 'No, the city's providing them.' Then this [other] guy seen my trees, he asked me the same question, 'How much do these trees go for?' 'I got them for free.' He goes, 'Wow, heck you got his number?'

Place-making is not an explicit part of the GGC, and yet the program provides opportunity and incentive for people to come together, both formally (through the DCR, cities, and community organizations) and informally (on their streets and in their yards) in ways that constitute the place, materially and socially. Some residents see their neighborhoods also being transformed over time, as trees grow and children enjoy the enhancement of green spaces with more trees:

> I see a lot of the younger generation going into the park which wasn't happening before. And I see the ones with their little kids now going into the park, and that's a great thing. I did that with my kid, you know what I'm sayin'?

These social interactions and senses of place produced through the GGC indicate the ways that an energy-oriented, environmental justice program both imagines and produces places. The tree planting zones are imagined primarily through metrics relating to tree canopy, older majority rented housing stock, and relatively higher wind speeds. These factors emphasize environmental and physical elements with some social factors; the places are imagined and prioritized primarily in relation to potential energy savings. Yet the structure of the program also attends to existing and potential networks; the existing networks of the partner organizations, and the potential and emergent networks forged via the state and the city agencies which coordinate the program and create new lines of communication and interest among agencies, organizations, residents, and property owners.

Tree-planting programs highlight the important biophysical role of trees in urban environments, because of the impacts trees have on the indoor comfort and energy consumption of buildings, and health of residents (de Vries et al., 2003; Donovan and Butry, 2009). Socio-politically, tree planting requires communication, coordination, and funding among public and private organizations and individuals to ensure stewardship (Roman et al., 2015; Young, 2011). These actions, while oriented primarily towards trees, may, intentionally or otherwise, also create opportunities for community-building (Fisher et al., 2015). As a result, these environmentally oriented activities are also place-making activities. They change the biophysical material look and infrastructure of the impacted area, and foster new socio-political dynamics and relationships.

The case of the GGC program in Massachusetts illustrates the ways that multiple place imaginaries, formal and informal human actions, and non-human actions (in this case, of trees) co-constitute places. The state of Massachusetts politically constitutes a type of place, a Gateway City, as a site of both opportunity and struggle. The opportunities and the struggles of Gateway Cities derive from conditions within the cities, and crucially, their social, political, and economic relations with other places. Through constituting the Greening the Gateway Cities program, the state DCR conceptualizes these cities as sites of non-human opportunity and intervention, by emphasizing the ecosystem role of trees in fostering and constituting places. The program itself builds upon the attachments of people to trees, and the possibilities that networked relationships between trees, individual residents, community organizations, and state foresters, can create new places. Beyond its own program dynamics and particularities, the GGC demonstrates more broadly the ever-emerging character of places, which develop from ongoing social, political, and non-human actions and relationships over time and space.

Acknowledgments

I would like to acknowledge my collaborators John Rogan, Clark University, and Lara Roman, United States Forest Service. Our collaboration has helped to develop my thinking on all things urban-forestry-related. Thanks to Clark graduate student researchers Nick Geron and Marc Healy, and undergraduate HERO 2018 students who conducted fieldwork: Laura Cohen, Rachel Corcoran-Adams, Elizabeth Lohr, Rowan Moody, Andrew Pagan, and Yeannet Ruiz. Deep appreciation and thanks to Tim Edensor, who provided helpful comments on an earlier draft and demonstrated extraordinary patience with my writing schedule and pace. The work and efforts of Tim, and his co-editors Ares Kalandides and Uma Kothari, are much appreciated.

References

Abraham, A., Sommerhalder, K., and Abel, T. (2010) Landscape and well-being: a scoping study on the health-promoting impact of outdoor environments. *International Journal of Public Health*, 55: 59–69.

Agnew, J. (1987) *Place and Politics: The Geographical Mediation of State and Society*. Boston: Allen and Unwin.

Amin, A. (2004) Regions unbound: towards a new politics of place. *Geografiska Annaler B*, 86(1): 33–44.

Brownlow, A. (2006) An archaeology of fear and environmental change in Philadelphia. *Geoforum*, 37: 227–245.

Bullard, R. D. (1994) *Unequal Protection: Environmental Justice and Communities of Color*. San Francisco, CA: Sierra Club Books.

Cloke, P., and Pawson, E. (2008) Memorial trees and treescape memories. *Environment and Planning D: Society and Space*, 26(1): 107–122.

de Vries, S., Verheij, R. A., Groenewegen, P. P., and Spreeuwenberg, P. (2003) Natural environments—healthy environments? An exploratory analysis of the relationship between greenspace and health. *Environment and Planning A*, 35: 1717–1731.

Donovan, G. H., and Butry, D. T. (2009) The value of shade: estimating the effect of urban trees on summertime electricity use. *Energy and Buildings*, 41(6): 662–668.

Fisher, D., Svendsen, E. S., and Connolly, J. J. T. (2015). *Urban Environmental Stewardship and Civic Engagement: How Planting Trees Strengthens the Roots of Democracy*. Routledge Explorations in Environmental Studies. Abingdon, UK: Routledge.

Heynen, N., Perkins, H. A., and Roy, P. (2006) The political ecology of uneven urban green space: the impact of political economy on race and ethnicity in producing environmental inequality in Milwaukee. *Urban Affairs Review*, 42(1): 3–25.

Jones, O. and Cloke, P. (2002) *Tree Cultures: The Place of Trees and Trees in their Place*. Oxford, UK: Berg Publisher.

Kirkpatrick, J. B., Davison, A., and Daniels, G. D. (2012) Resident attitudes towards trees influence the planting and removal of different types of trees in eastern Australian cities. *Landscape and Urban Planning*, 107(2): 147–158.

Lam, N. (2012) Geospatial methods for reducing uncertainties in environmental health risk assessment: challenges and opportunities. *Annals of the American Association of Geographers*, 102(5): 942–950.
Mass.gov (2019a) Baker-Polito administration celebrates planting of 20,000 tree under greening the gateway cities program. https://www.mass.gov/news/baker-polito-administration-celebrates-planting-of-20000th-tree-under-greening-the-gateway. Last accessed 2 March 2020.
Mass.gov (2019b) Greening the Gateway Cities Program. www.mass.gov/service-details/greening-the-gateway-cities-program. Last accessed 21 January 2019.
Massey, D. (1991) A global sense of place. *Marxism Today* June: 24–29. http://banmarchive.org.uk/collections/mt/index_frame.htm. Last accessed 5 May 2019.
Massey, D. (2005) *For Space*. London and Thousand Oaks, CA: Sage Publications.
MassINC (Massachusetts Initiative for a New Commonwealth) (2015) About the Gateway Cities. http://massinc.org/our-work/policy-center/gateway-cities/about-the-gateway-cities/. Last accessed 4 August 2016.
McPherson, E. G., and Rowntree, R. A. (1993) Energy conservation potential of urban tree planting. *Journal of Arboriculture*, 19: 321.
Nguyen, V. D., Roman, L. A., Locke, D. H., Mincey, S. K., Sanders, J. R., Fichman, E. S., Duran-Mitchell, M., and Tobing, S. L. (2017) Branching out to residential lands: missions and strategies of five tree distribution programs in the U.S. *Urban Forestry and Urban Greening*, 22: 24–35.
Nowak, D. J., and Dwyer, J. F. (2007) Understanding the benefits and costs of urban forest ecosystems. In J. E. Kuser (Ed.), *Urban and Community Forestry in the Northeast*, 2nd ed., pp. 25–46. New York, NY: Springer.
Obar, S (2012) The latest trend: being a Gateway City. *CommonWealth Magazine*. https://commonwealthmagazine.org/economy/018-the-latest-trend-being-a-gateway-city/. 24 May. Last accessed 14 January 2019.
Pearce, L. M., Davison, A., and Kirkpatrick, J. B. (2015) Personal encounters with trees: the lived significance of the private urban forest. *Urban Forestry & Urban Greening*, 14(1): 1–7.
Pierce, J., Martin, D. G., and Murphy, J. T. (2011) Relational place-making: the networked politics of Place. *Transactions of the Institute of British Geographers*, 36: 54–70.
Phillips, C., and Atchison, J. (2018) Seeing the trees for the (urban) forest: more-than-human geographies and urban greening. *Australian Geographer*, 1–14. doi: 10.1080/00049182.2018.1505285
Pulido, L. (2000) Rethinking environmental racism: White privilege and urban development in Southern California. *Annals of the Association of American Geographers*, 90(1): 12–40.
Roman, L. A., Walker, L. A., Martineau, C. M., Muffly, D. J., MacQueen, S. A., and Harris, W. (2015) Stewardship matters: case studies in establishment success of urban trees. *Urban Forestry and Urban Greening*, 14(4): 1174–1182.
World Population Review (2019) Massachusetts Population 2018. http://worldpopulationreview.com/states/massachusetts-population/. Last accessed 14 January 2019.
Young, R. F. (2011) Planting the living city. *Journal of the American Planning Association*, 77(4): 368–381.
Zölch, T., Maderspacher, J., Wamsler, C., and Pauleit, S. (2016) Using green infrastructure for urban climate-proofing: an evaluation of heat mitigation measures at the micro-scale. *Urban Forestry and Urban Greening*, 20: 305–316.

Part II
The qualities of place

Tim Edensor

Introduction: the qualities of place

I am sitting in the garden of a small terraced house in Manchester. Damp pervades the muggy air. The tall oak trees in the adjacent park are starting to turn to autumn colours and in another month will be bereft of leaves as winter approaches. In the street, the evergreen dark hues of privet hedges complement the Victorian red brick facades of the houses. Bright blue rubbish bins sit outside each house awaiting collection on this Friday, as they will be in two weeks hence as part of the municipal collection schedule. On this mid-September morning, the sky is a predictably dull grey monotone. Periodic sounds of the smooth-running tram approaching and passing nearby puncture the ever-present drone of traffic coursing along the distant motorways and the hum of aeroplanes departing and arriving at the nearby airport. These mechanical sounds are accompanied by the rustle of the trees in the park next door, the frequent, harsh chatter of magpies and the chirrups of house sparrows. From an adjacent house, the clink of teacups and occasional bursts of pop music from the radio are heard and the smell of sizzling pork sausages fills the air. The mood will soon settle into a mid-morning lull after inhabitants have left for work and school. At the moment, this feels as if it couldn't be anywhere else other than the Manchester that I know so well, a backdrop to my daily routines that I usually neglect to consider. But it is always there, knowable, tangible.

This short passage identifies some of the qualities of the place in which I am currently located, and similar vignettes could be drawn for any other place. In this part, drawing on phenomenological, post-phenomenological and non-representational ideas, the focus is on the qualities of place. Consider a range of attributes that help us know when we are in place: the subtleties of weather, forms of everyday vegetation, the soundscape of birdsong or traffic and music, the ways in which people move and talk, the tastes and smells of local food, the affective intensities that accumulate in public spaces and the atmospheres of stations and forms of public transport, everyday rhythms of place and seasonal patterns, its colours and architectural forms, smells, vegetation, familiar institutions and places of gathering, and the mobilities that course through place.

Particularly influential in advancing an awareness of these qualities has been Tim Ingold's notion of the 'taskscape'. Ingold insists that place is continuously in process rather than static.

Workers and inhabitants serially encounter place as a perceptual realm, saturated with historical resonances from their own lives and those of others. In carrying out everyday tasks and ordinary practices, people habitually sense place and move through it unreflexively while possessing a competence borne of repeated practice in a familiar realm.

> A place owes its character to the experiences it affords to those who spend time there—to the sights, sounds and indeed smells that constitute its particular ambience. And these, in turn, depend on the kinds of activities in which its inhabitants engage. It is from this relational context of people's engagement with the world, in the business of dwelling, that each place draws its unique significance.
>
> *(Ingold, 1993: 155)*

Ingold's ideas mesh with those of other accounts that foreground sensation, habitual practice and everyday forms of inhabitation. Phenomenological and post-phenomenological accounts follow on from the seminal work of Yi Fu Tuan (1975: 152), who refers to 'distinctive odors, textural and visual qualities in the environment, seasonal changes of temperature and color, how (places) look as they are approached from the highway'. And focussing on the routine, mundane, unreflexive inhabitation of place, David Seamon (1979) terms the routine journeys that are repeatedly undertaken as 'place ballets', consolidating what David Crouch (1999) calls a 'lay geographical knowledge' that fosters a sense of belonging and informs how people know how to locate things and how tasks should be accomplished. Such habitual competencies include a practical knowledge of where to buy particular commodities, how to drive a car, place a bet, buy an alcoholic drink and catch a bus. These routines are not merely individual but intersect with those of others so that a sense of 'cultural community' may be co-produced by 'people together tackling the world around them with familiar manoeuvres' (Frykman and Löfgren, 1996: 10–11), strengthening affective and cognitive links and locating people in stable networks of relationships, objects and spaces. Shared habits also produce collective choreographies that delineate spatial and temporal constellations at which a host of individual paths and routines coincide. Local shops, bars, cafes, garages, parks and transport termini constitute such points of intersection, coined by Doreen Massey (1995) as 'activity spaces', spaces of circulation and meeting that collectively contribute to a shared 'common sense' of inhabitation. Deepened by time, such familiar realms are also embedded in sensory memory, as Lucy Lippard identifies in her claim that if 'one has been raised in a place, its textures and sensations, its smells and sounds, are recalled as they felt to child's, adolescent's adult's body' (1997: 34). To emphasise, a focus on such qualities undercuts the essentialist and classificatory conceptions of place to which we refer in the introduction to this book.

Returning to the scale of place discussed in the last part, these modest everyday arrangements merge a sense of the local with national and regional belonging. For most of the features encountered at an everyday level in the familiar environs of home and neighbourhood extend across national (Edensor and Sumartojo, 2018) or regional space (Stewart, 2013), including architectural, infrastructural, commercial, recreational, institutional, domestic and environmental commonalities (Edensor, 2002). Consequently, places are key sites in which 'individuals make sense of their relationship with the nation' (Jones and Fowler, 2007: 335), as well as of the other larger spatial contexts in which they are located.

Most of the time, the everyday qualities and features of place are unreflexively apprehended, yet there are times when they come into sharp focus as contested markers of place-identity. Most evidently, a heightened awareness about everyday place dawns when we enter unfamiliar spaces, perhaps travelling to an unfamiliar country. Mundane features and institutions seem peculiar, as do infrastructural details, foliage and architecture, along with strange smells, sounds, textures and climatic conditions, and the unfamiliar ways in which people move, speak, laugh and shout. Similar feelings may devolve when familiar place is rapidly transformed and new sensory features intrude into the everyday realm. For instance, Mikkel Bille (2019) explores how white Danish inhabitants of Copenhagen may recoil when they confront migrant houses in which the design of interior domestic illumination is brighter than the lighting practices that habitually compose a sense of *hygge*, loosely translated as cosiness, in their homes. Yet over time, new elements become integrated into place as part of its ever-changing fabric, and are likely to themselves become habitually and unreflexively sensed.

Sensation does not, however, provide unmediated access to the reality of place but is developed through habitual ways of inhabitation and according to conventions and values attributed to particular sensory experiences that circumscribe how they inform meaning. As Constance Classen emphasizes, 'we not only think *about* our senses, we think *through* them' (1993: 9). Accordingly, interventions to defamiliarize place can make us reassess how we perceive the worlds in which we habitually dwell, and thereby prove powerful in revealing the socially constructed nature of sensation. As Jacques Rancière submits, the distribution of the sensible 'revolves around what is seen and what can be said about it, around who has the ability to see and the talent to speak, around the properties of spaces and the possibilities of time' (2009: 13). Consequently, it is critical to pay attention to the ways in which power shapes place in distributing the sensible and yet, in particular performances, sound installations or light projections (Edensor and Sumartojo, 2018) for instance, 'new modes of sense perception' (Rancière, 2009: 9) can reveal familiar environments in a new and intriguing light. On such occasions, we might thus experience the 're-enchantment' (Bennett, 2001: 5) of place wherein heightened sensory awareness solicits how we 'notice new colours, discern details previously ignored, hear extraordinary sounds, as familiar landscapes of sense sharpen and intensify'.

A chapter by Michelle Duffy commences this part, tracing the origins and development of the notion of the soundscape and exploring how sound contributes to understanding the complex processes and relations that constitute place. In underlining the points made above about the values that are attributed to the senses and sensations, she shows how recent critical challenges draw attention to the 'natural' sounds to the neglect of other urban sounds, since these sonic elements are regarded as less valuable, despite being no less integral to the soundscape of place. She also insists that it is crucial to take account of what sound does for the various ways we listen, and the influence this may have on attributing meaning to sound.

This is followed by an account of the relations between weather and place by Phillip Vannini and Bradley Austin. The authors emphasize that everyday, ordinary and extraordinary weather events are profoundly revealing of the inseparability of all places from climate, and are part of the ways in which people sense and feel and respond to the emergent conditions of place. They demonstrate how inhabitants of places engage with and attune to weather in multiple ways – turning on a fan, adding ice to drinks or planning street snow removal – and through which they control, endure, adapt to, enjoy, or remove themselves from the weather-places they inhabit. They also explain how we represent weather places through a host of cultural practices, including sharing photographs on social media and engaging in small talk. They thus show weather to be an inescapable medium for our being-in-the-world.

Tim Edensor's chapter considers how the play of light and the distinct forms upon which it shines shape the everyday apprehension of place. Variegated patterns of sunlight and shade tone familiar space with colours, shadows and textures, revealing the vital qualities through which place continuously changes. These conditions are identified through considering human optics, via the quality of light that shines on place, according to the material affordances of place, and through the influence of cultural representations. The distinctive ways in which light and shadow fall on particular places is explored before an examination of how the light of place is apprehended and has been represented in painting and photography.

By contrast, in the following chapter, Nick Dunn investigates how place after dark has fallen is often conceived in binary opposition to how it is encountered and understood in the daytime. This, he underlines, does not account for the variations and qualities of darkness and light that may occur once the sun has set. This dichotomy, he argues, has been perpetrated through the ongoing development of artificial illumination, and infrastructures, policies and practices that seek to control and manage the night-time of place. The chapter particularly focusses on the potential of urban peripheries that may serve as sites for experimentation and imagination toward new conceptualizations. Drawing on experiences of the edgelands of Manchester, UK, Dunn illustrates the different coexistences between darkness and light, and reconsiders how we might design for place after dark.

Maarja Kaaristo's chapter focusses on the watery qualities of the linear place of the canal and the canal boat that cruises along these waters in the United Kingdom. First, she identifies the various boaters, walkers, joggers, anglers, cyclists, navigation authorities, local councils and volunteer organizations that use and maintain presence on and near inland waterways. She then focusses on the liquid, linear place of the canal, and on its geographical location, material setting and the particular relationships that circulate on and around waterways, to show how they are simultaneously relational, material and mobile, and produced by numerous entanglements, practices and interrelations. The chapter focusses especially on three key elements of life on the canal: slow mobilities, embodied watery materialities and convivial interactions.

In examining the apparently more solid materialities of stone, David Paton's account ranges across the worked landscapes of Cornwall, examining the relations between place and matter that have formed here over millennia. Focussing on one particular granite quarry, Paton looks at the storying of reciprocal, multiplicitous and conversant relationships that contest binary distinctions of living and non-living in the light of the insights generated in new materialist discourses. What emerges is a sense of how place is at once an already formed vessel within which bonds of many kinds are nurtured or abandoned, yet simultaneously place is the beating fleshy heart of an infinite array of more-than-human makers. In this case, people come and go, making their imprint on place while stone emerges, is fashioned and erodes over more extensive time scales.

Shanti Sumartojo focusses on how locating atmosphere at the centre of how we think about place can reinforce and advance an affective, experiential understanding of place. More specifically, Sumartojo contends that atmospheres help us to account for how places feel, what those feelings mean, and what might be possible as a result of their emergence and apprehension. By thinking in, about and through atmosphere, an approach to place is adopted that is contingent, excessive and slippery, that does not seek to delineate and fix place but to explore how place feels and what it means to people. This perspective also opens a speculative route to thinking about what implications place might carry into the future, and how it might stick with us long after we have left it.

In moving away from a focus on specific qualities Cristina Garduño Freeman, Beau B. Beza and Gloria Mejía investigate how language conditions how we apprehend and remember the textures of place. Moving away from the ways in which place is deployed as a concept in the Anglophone discourses of cultural geography, urban planning, heritage and environmental psychology, the authors explore how they are connected to distinctive places in Mexico City, and to how identity and culture are entangled in the physical environments we inhabit. In this Spanish-speaking, Central American context, they contend that the potency of the term 'place' is lost in its translation and application because of its ambiguity. Instead, more specific and less slippery terms are used: *lugar*, *sitio*, *ambiente*, *entorno*, each with their own semantic characteristics and, often, culturally specific meanings. We explore the translations, both linguistic and spatial, between the Anglophone and Hispanophone lexicons by examining literary discussions and memories to explore how language structures distinct understandings of places.

The final chapter in this part, by Sarah Wright, further decentres the ways in which Eurocentric and ethnocentric accounts can register the qualities of place by drawing on indigenous relationships with country on unceded Gumabynggirr land on the north coast of New South Wales, Australia. For its Aboriginal inhabitants, place here possesses agency and knowledge, is relational and agential rather than a passive backdrop on which human action occurs. In resisting notions that humans are separate from place, but are in a continuous state of connectedness and co-becoming with place, Wright shows how place is kin, a co-constituent of being, that is experienced in dreams, senses and feelings. Importantly, Wright asks us to consider the ways in which the rich understanding of place has been constrained by colonial and imperialistic thinking, and to grasp the importance of transcending this through ongoing processes of decolonization, an entreaty that could also be mobilized in many other contexts.

References

Bennett, J. (2001). *The Enchantment of Modern Life: Attachments, Crossings and Ethics*. Princeton, NJ: Princeton University Press.

Bille, M. (2019). *Homely Atmospheres and Lighting Technologies in Denmark: Living with Light*. London: Bloomsbury Publishing.

Classen, C. (1993). *Worlds of Sense: Exploring the Senses in History and across Cultures*. London: Routledge.

Crouch, D. (1999). Introduction: encounters in leisure/tourism. In D. Crouch (ed.) *Leisure/Tourism Geographies: Practices and Geographical Knowledge*. London: Routledge, pp. 1–16.

Edensor, T. (2002). *National Identity, Popular Culture and Everyday Life*. Oxford: Berg.

Edensor, T. and Sumartojo, S. (2018). Reconfiguring familiar worlds with light projection: the Gertrude Street projection festival, 2017. *GeoHumanities* 4(1): 112–131.

Frykman, J. and Löfgren, O. (eds) (1996). *Forces of Habit: Exploring Everyday Culture*. Lund: Lund University Press.

Ingold, T. (1993). The temporality of the landscape. *World Archaeology* 25(2): 152–174.

Jones, R. and Fowler, R. (2007). Placing and scaling the nation. *Environment and Planning D: Society and Space* 25: 332–354.

Lippard, L. (1997). *The Lure of the Local: Senses of Place in a Multicentered Society*. New York: The New Press.

Massey, D. (1995). The conceptualisation of place. In D. Massey and P. Jess (eds) *A Place in the World? Places, Cultures and Globalisation*. Oxford: Oxford University Press, pp. 45–86.

Rancière, J. (2009). *Aesthetics and Its Discontents*. Cambridge: Polity Press.

Seamon, D. (1979). *A Geography of the Lifeworld: Movement, Rest and Encounter*. New York: St Martin's Press.

Stewart, K. (2013). Regionality. *Geographical Review* 103(2): 275–284.

Tuan, Y.F. (1975). Place: an experiential perspective. *Geographical Review* 65(2): 151–165.

10
Soundscapes

Michelle Duffy

Introduction

The term *soundscape* is commonly used to refer to the sonic qualities that arise in a location, and how these then contribute to our understandings of a place. Some suggest that a soundscape can be understood as the acoustic equivalent of a visual landscape (Hall et al., 2013) and, therefore, is simply the collection of sounds in a place (Oliveros, 2005). Others argue that the soundscape is a 'perceptual construct' derived from the acoustic environment of a specific location, that then goes on to shape the range of responses an individual will have, that in turn contributes to place attachment, a sense of harmony and wellbeing (Brown et al., 2016). Some argue that the soundscape framework fails to capture what sound *does*, because 'what can be known through sound may not be accessible from the visible world' (Smith, 2000: 615). This chapter traces through the origins, development and critiques of *soundscape* as a concept and how this then influences our exploration of the sonic world. A critical examination of how the soundscape is conceptualised and subsequently examined can help us better understand the complex processes and relations that constitute place.

Schafer and the World Soundscapes Project

The term *soundscape* originated with the work of R. Murray Schafer in the late 1960s. Schafer, a composer, educator, artist and environmentalist, has long been concerned with the damaging effects of noise and the 'indiscriminate and imperialistic spread of more and larger sounds into every corner of man's life' (1977b: 3). His aim was to determine what sounds mattered to us and how we should listen to them, and he proposed a study of environmental acoustics in order to be able to nominate those sounds we want to preserve. Given his strong interest in environmentalism, it is the sounds that do not originate in 'nature' – sounds more likely to be human and more specifically industrial or technological in origin – that are to be minimised. While in some ways this could be interpreted as a nostalgia for a pre-industrialised world, Schafer's ideas arose out of scientific, social and artistic considerations of health and wellbeing. Better health, he suggested, would be achieved through an interdisciplinary acoustic design framework derived from acoustics, psychoacoustics, social behaviour and aesthetics.

Schafer's interest in composers like John Cage, who radically opened up what music could be, led to the suggestion that we conceptualise sound as part of 'a continuous field of possibilities lying *within the comprehensive dominion of music*' (Schafer, 1997: 5, italics in original). While perhaps largely descriptive and lacking a clear analytical model (Truax, 2001), connecting the soundscape to a musical context offered important ways to study soundscapes, and drew on particular listening practices that help determine significant aural features. Schafer proposed we start by determining the keynote of a soundscape, a term borrowed from music, and used to refer to the 'reference point' from which 'everything else takes on its special meaning … the keynote sounds of a given place are important because they help outline the character of men [sic] living among them' (Schafer, 1977b: 9). Yet, while fundamental to defining a specific soundscape, this keynote may not be heard consciously even as it may influence the behaviour and feelings of those present. Those sounds that we do consciously hear are called *signals*; for Schafer, these sounds are warning devices that can be used to communicate urgent messages, such as bells, sirens, whistles and horns. Schafer's final term is *soundmark*, a sound unique to a community and recognised as representative of that community.

From these listening practices, Schafer categorises soundscapes in one of two ways. The first, a hi-fi soundscape, is exemplified in rural settings with a 'quiet ambiance' where 'discrete sounds can be heard clearly because of the low ambient noise level'. In contrast, the modern city is a lo-fi soundscape, where 'individual acoustic signals are obscured in an overdense population of sounds' (1977b: 43). While an overly blunt instrument for defining soundscapes, embedded within these listening practices and characterisations is a careful consideration as to how we hear and listen, which Schafer linked to social bonding. Hearing and touch, he argued, are expressions of intimacy, and coincide when 'the lower frequencies of audible sound pass over to tactile vibrations (at about 20 hertz). Hearing is a way of touching at a distance' bringing about 'sociability whenever people gather to hear something special' (1977b: 11). Thus, Schafer indicates the importance of the body and sociality in apprehending the soundscape.

Schafer's (1997) methodological practice included first undertaking a soundwalk so that the researcher carefully listens to the sounds present in a locale. This initial step was not integral to Schafer's recording practice, but instead this was about the researcher preparing a detailed catalogue of the sounds present (such as classifying sounds according to frequency, duration, timbre) as well as their physiological impact on the listener (how the sound is perceived, in terms of the quality of the sound, its attack and decay). In addition, the researcher was to include careful consideration of the relationships between sounds and the listener, with acknowledgement that these perceptions would be shaped by the cultural context of the listener (Schafer, 1997). Analysis of these relationships was based on ideas of musical structure and relations. This is because we make sense of the soundscape through various physiological, physical and cognitive processes in which we unite various sounds as 'points of diffusion that in listening we attempt to gather' (Chow and Steintrager, 2011: 2). Thus, sounds operate as an organising force that may help people categorise, order and differentiate between places (Atkinson, 2011).

Influence and critique of Schafer's soundscape concept

Schafer's approach has had considerable influence. One area of research, particularly in the fields of urban planning and architectural design, continues to seek to minimise the deleterious effects of urban and industrial sounds and lessen sound's negative impacts on the body (Amphoux et al., 2004; Bijsterveld, 2008; Chelkoff, 2002; Grosjean and Thibaud, 2008). Other scholars have explored the impact of human sounds on non-urban environments (Caffyn and Prosser, 1998; Waitt et al., 2009). Studies concerned with the detrimental impacts of noise have led to the development of an ISO (International

Organisation for Standardisation) soundscape standard. For example, ISO 12,912–1 defines a soundscape as 'an acoustic environment as perceived or experienced and/or understood by a person or people, in context' (cited in Brooks and Schulte-Fortkamp, 2016: 2043). However, more recent work in this field seeks to incorporate perceptual responses to the physical evaluation of a soundscape, with the goal to develop a comprehensive guideline to collecting data for a soundscape study (Brooks and Schulte-Fortkamp, 2016). Schafer's work, along with that of the other members of the World Soundscapes Project, has also been influential in sound art and compositional practices, as well as providing important archival materials for historical studies. For example, Järviluoma et al. (2009) returned to the European villages first examined by members of the World Soundscape Project (published by Schaffer in *Five Village Soundscapes*, 1977a, 1977b) in order to capture and note any changes, particularly those arising out of the processes of urbanisation. However, critique of Schafer's approach includes questions about the ideological underpinning of how the soundscape is defined.

A key criticism of Schafer's work is that he privileges 'natural' sounds and rejects the sounds of the urban world in which most of us live. These 'sounds that matter' (Schafer, 1997: 12) readily align with views of 'a long dystopian history that descends from harmonious sounds of nature to the cacophonies of modern life' (Kelman, 2010: 213; see also Arkette, 2004). Schafer's nostalgia for the quiet of some pre-modern world (although Schafer fails to explain where or when his 'original soundscape' existed; Kelman, 2010) and his dismissal of man-made sounds simply as noise ignores the ways in which all sound serves as an acoustic marker of identity and difference.

In addition, even as Schafer does speak of the impacts of sound on a listening body, there is an implied assumption that a soundscape is considered a background to everyday life and activity, something that can be objectively contemplated (Helmreich, 2010) and thus this approach 'emphasises the surfaces of the world' rather than 'a phenomenon of experience' (Ingold, 2007: 12). Schafer's way of modelling the soundscape starts from a human-centred position, thus ignoring the complex and dynamic ways in which we are connected through sound to both the human *and* non-human elements of daily life. As Gallagher et al. (2017: 619) argue, the error of this human perspective is that it is often 'uncritically reproduced' and 'implicitly anthropocentric … other kinds of sonic encounters are frequently left out'. Thus, Schafer's approach to the soundscape maintains a separation of the human from the non-human.

Finally, the methodological practices suggested by Schafer fail to acknowledge the various ways in which we listen, and what influence this may then have on attributing meaning to sound. As Jean-Paul Thibaud argues, conceiving the soundscape in terms of a musical composition

> tends to enhance a contemplative perception of the world and exclude other kinds of more practical listening … Depending on the situation in which we are involved, we configure the surrounding one way or another: we can hear or listen, eavesdrop or heed, prick up our ears, notice or remark.
>
> *(1998: 2; see also Augoyard and Torgue, 2005)*

The suggestion that we listen to a soundscape as we would a musical composition therefore has implications as to how we engage with and interpret a soundscape. Critiques of these ideas lead to two broad topics for more detailed examination: what is meant by listening, and who is the subject that listens.

Sound and music

A more critical perspective of the soundscape needs to consider the qualities and processes of sound. Rather than defining sound as a static or neutral object, we need to acknowledge that distinctions made about the elements of this soundworld – how sound is classified as noise, as music and as silence, for example – are, as geographer Susan Smith notes, politically motivated, that 'the art of appropriating and controlling noise ... is in short, an expression of power' (2000: 616; see also Attali, 1992). The so-called cultural turn in human geography in the 1990s was important to a shift in thinking and broadened human geography's objects and agenda of study. Geographical approaches to culture had been aligned with anthropological ideas about culture as 'a way of life', but the work of geographers – in particular Peter Jackson (1989), Denis Cosgrove (1984) and Jim Duncan (1980) – was significant to challenging this notion of culture on ontological and epistemological grounds. What this shift enabled was a rethinking as to how place is constituted, something that had entered into human geographical thinking in the 1970s through phenomenological philosophy and its focus on how we experience place through the body and its senses (Saldanha, 2009). As with other areas of culture and everyday life, this shift influenced how we conceptualise sound and music. Rather than thinking of music as a cultural artefact or object and seeking its origins and subsequent diffusion across places, a geographical consideration of sound offers another dimension to thinking on our embodied encounters with the world. As argued elsewhere, 'paying attention to how sound prompts highly visceral experiences ... offers possible insights into how sounds are worked, reworked or silenced in everyday lives in meaningful ways' (Duffy and Waitt, 2013: 468).

The work of geographers exploring the relationship between music, place and identity resonates with studies in musicology, particularly major studies in popular music and ethnomusicology, which approached the construction of identity as something expressed through relationships between music and lifestyle. Such studies have explored the ways in which performers and audience perceive musical practices that help create a sense of group identity (Connell and Gibson, 2003; Dunbar-Hall and Gibson, 2004), as well as related studies such as the role of music in political action (Carroll and Connell, 2000) and as an economic commodity (Gibson and Connell, 2005). However, music is largely non-representational and therefore the manner in which musical sounds are put together does not result in a transparent and stable set of meanings. Nor are spaces or places the already present sites in which music occurs or through which music diffuses, but instead are formed and created through the temporal and spatial processes and practices of music and sound (Leyshon et al., 1998: 2).

In her earlier article, 'Soundscape', Smith questioned geography's lack of study into the imaginings of place through music and sound because they are 'inseparable from the social landscape' (1994: 238) and would help explore the emotional and intuitive qualities of social life, aspects she argued that are notably missing from the visual and rational modes of study. While Schafer had also looked to music as a framework for analysis of place through its sonic elements, Smith's use of performance brings to the fore 'a whole web of relationships ... anchored as much on those who listen as on those who make the sound' (2000: 633). Listening to sound, then, is more than an aesthetic experience; instead, we need to engage in what Smith calls 'the doing' of sound rather than undergoing a study of an assumed fixed, bounded and visualised object.

Conceptual and methodological considerations

Jean-Luc Nancy's examination of listening and hearing points to important differences between these two processes. Nancy contends that *hearing* is the cognitive process of understanding and comprehension, while 'to listen is to be straining towards the possible meaning, and consequently one that is not immediately accessible' (2007: 6). In this framework, listening (the French term *écouter*) operates in bodily, emotional, physiological and psychological ways through which we attempt to attribute meaning to sound (see also Simpson, 2009). This framing acknowledges that listening occurs in, around and through our entire bodies, and so draws our attention to ourselves as very much embedded in and through the human and non-human elements of a space rather than as separate entities situated within a place (Duffy et al., 2016). In addition, this framework acknowledges that our engagement with the sonic world happens not only through our ears but is an embodied set of responses. Immersion within the soundscape is therefore inherently a bodily experience.

One research trajectory examines this listening in terms of an interface between listener and the environment. For example, musicologist Holger Schulze proposes that listening is a material, physical and bodily process such that sonic thinking is a form of corporeal thinking,

> a form of listening immersed in the substance and the historical as well as sensational, fictional, and obsessive layers coating and entwining any sonic experience. Sensory critique in action.
>
> *(2018: 156)*

This 'thick' listening is informed by the idiosyncratic nature of the listener – his/her corporeality, historic experiences as well as the situatedness of actual listening, the material and technical aspects of these listening spaces. Schulze points out that all elements that constitute such thick listening shape how we, as individuals, respond.

Recent work inspired by performative forms of knowledge, including the more-than-representational (see Lorimer, 2005, 2007 and 2008), considers how sound exceeds representation, overcomes bodily boundaries, conceives of a multi-sensuous body and troubles the fixed, singular, coherent subject (Boyd and Duffy, 2012; Smith, 2000; Wood et al., 2007). As Nigel Thrift proposes, non-representational theory is 'about *the geography of what happens in the now*' (2008: 2, italics in original) and requires that attention is paid to the material and expressive forces through which sound is made and apprehended. This focus on sound as process and medium that shapes and reshapes people and places seeks to explore how sound moves through and is transmitted by bodies, things, technology, ideas, affective and emotional processes (Duffy et al., 2016). An example of such an approach is found in Boyd and Duffy's (2012) rhythmanalysis of the everyday soundscape of a busy university café. Their transcriptions of this soundscape demonstrate that what on first hearing might be received as noise can be used to elucidate the ways in which sound 'allows us to trace the affective relations and their impact on the constitution of place' (2012: online). In these sorts of studies researchers have explored the ways in which the elements of the soundscape – including rhythm, timbre and melody – constitute notions of subjectivity, space and place. Thus, this is an interest in the relationship between sound and the social production of meaning; 'sound must be considered as a public account of the social world. They can be observed and described as an expression of the way we live together and share our common daily environment' (Thibaud, 1998: 5; see also Kelman, 2010). Considering the soundscape in terms of the social world has facilitated nuanced studies of the relations of power in a range of settings, and the incorporation of soundscapes into multidisciplinary and multisensory projects offer compelling ways to uncover these relations (for example, Magowan et al.'s project, *Sounding conflict: from resistance to reconciliation* and Rebelo et al.'s project, *Understanding the role of music and sound in conflict transformation*). Yet this can also problematic, for what sometimes remains is an anthropocentric (and white) positioning linked to human consciousness, which has been critiqued not only 'for simplistic assumptions about sound and listening' (Gallagher et al., 2017: 619) but also because of assumptions around who or what listens.

This anthropocentric view is challenged in the work of artists and theorists such as Ros Bandt (*Hearing Jaara Jarra*, 2013) and Leah Barclay (*Biosphere Soundscapes*, 2017) who both work within acoustic ecology, a field of inquiry concerned with the ecological, social and cultural contexts of sonic environments. Others have sought to incorporate the non-human within compositional practice, underpinned by the theoretical approaches offered by more-than-representational and Actor Network Theory (ANT) (Westerlaken and Gualeni, 2016). In addition, some scholars have argued that we need to reconsider how we conceptualise sound itself, that

> any reconfiguration of 'sonic' ecology must inevitably acknowledge the invisible agency of sound as a force for revealing the possible assemblages that make up a place, and for reframing them in new and creative terrains for human and more-than-human negotiation.
>
> *(Pisano, 2017: 469–470)*

A more critical exploration of the soundscape shifts how we listen because we are asked to consider sound's materiality, as well as the ideologies and discourses in which it is located.

Doing soundscapes

Michael Gallagher and Jonathan Prior argue that valuable insights are lost if we do not consider the 'sonic aspects of organised spaces' (2014: 271) and they provide an overview of various methods that can help capture this, including listening, recording, playback, editing, distribution, broadcast, performance and installation. Indicative of the increasing interest in how sound mediates everyday experience is the growing number of studies on portable sound (Bull, 2000; Sterne, 2012), soundwalks (Butler, 2007), installations and experimental sound mappings of urban and wildlife spaces (Drever, 2009). Perhaps the most common way used in the examination of soundscapes is the soundwalk.

Although Schafer did not intend that an initial soundwalk be recorded, Hildegard Westerkamp, composer and member of the World Soundscape Project, developed this practice as a means for individuals to more actively listen to everyday environments and as a fundamental first stage to formalising the sound piece (McCartney, 2004). Her approach therefore requires us to consider the interdependence and indeed interplay between the individual and his/her environment. The use of portable audio devices such as MP3 players in soundwalks can, as Westerkamp (1974: 18) proposes, 'bring our existing position-inside-the-soundscape to full consciousness'. Devising soundwalks continues to be a means to carefully consider the sonic environment and is now commonly used to explore feeling, emotion and affect in the shaping and expression of our relationships to place. An example of this approach is found in Michael Gallagher's (2015) audio geography work, which uses a set of recordings created for a soundwalk at the ruined modernist St Peter's Seminary, Kilmahew (near Cardross, Scotland) as a means to 'seed stories about the past, present and possible future' and 'invite further engagement with the site, stimulate debate and perhaps attract additional visitors' (2015: 470). In this way, soundwalks are often interventionist: the emotions and affect generated have 'the potential to reconfigure listeners' relationships to place, to open up new modes of attention and movement, and in so doing to rework places' (Gallagher, 2015: 468). Thus soundwalks (as well as sound art installations) have become important tools and practices in thinking about spatial relations. For example, Michael Bull (2000) draws our attention to the practice of wearing personal stereos in order to enliven journeys by moving around the city to a personalised soundtrack. However, as in the audio geography practices of Gallagher (2015), such sound recordings can also disrupt connections to place, and instead produce an almost hallucinogenic experience as the sounds played out in the privacy of our head 'haunt' the place through which we move (and of course these personal devices can be a means to exclude ourselves from the everyday spaces we move through (Bull, 2000)). Thus, what Harris (2013: 13) describes as the 'deceptively simple process of recording sounds from a chosen environment' makes uncanny the listener's connection with his/her location, and challenges preconceived ideas about our spatial relations.

Since the 2000s social scientists have begun to seriously consider the material, sensory and sensual elements of sound. Lefebvre's (2004) notion of rhythm and method of rhythmanalysis have been particularly influential, originating in a form of ethnographic analysis that helps reveal the numerous spatial-temporal practices and struggles occurring in daily life (Edensor, 2010). Other researchers have called attention to the multisensory aspects of the soundscape, arguing that 'we make sense of sound in and through the situated body' and that 'a visceral approach facilitates a conceptualisation of sound/music as one of the interweaving forces that comprise the personal geographies of everyday life' (Duffy et al., 2016: 52). The impact of sound on bodies, feelings and emotions is also found in the work of scholars exploring ambiance (Thibaud, 2011) and affective atmospheres (Anderson, 2009). Inherent in these approaches is the recognition that sound is not something that can be perceived in an instant; rather it is a temporal and spatial event that unfolds around and through the materialities of place (Duffy, 2017).

Conclusion

The concept of soundscape is popularly used to describe the sounds present in a specific location, and thus is used as a form of representation of place. Yet this was not Schafer's original intent; rather, his concern was with the loss of 'nature' because of the 'cacophonies of modern life' (Kelman, 2010: 213). The practices of listening that he devised emphasised certain sounds that mattered to him, and thus were embedded within specific ideological and ecological discourses. Therefore, how we listen to the soundscapes around us and therefore use these sounds to define place is not a neutral practice. Nonetheless, Schafer's work, along with that of the WSP, has been influential in a range of disciplinary practices.

What is often missing from this approach is acknowledgement of sound as a process that unfolds over time and impacts upon the materialities and bodies in places. As Smith (1994, 2000) reminds us, the various elements of the soundscape offer a powerful way of knowing, being and inhabiting place. We listen in and through our situated bodies that then shapes orientations, identifications, choices, social interactions, as well as human and non-human relations that configure everyday life. What makes noise and sound meaningful is the sonic and cultural context out of which it emerges (Kelman, 2010). Thus, the soundscape is not simply a background to social life, and is more than the sonic qualities and relations that arise in a location.

How we make sense of sound, though, is not easily disentangled from our other senses, and scholars have cautioned using approaches that ignore the interconnectedness of the senses in human perception (Pink and Howes, 2010; Rodaway, 1994). Nonetheless sound has become increasingly important in the social sciences and humanities because of what it can tell us about the creation of place as it arises out of our social world and its intersection with that of the non-human. Yet, working with sound in ways that 'emphasises its being and doing – its nonrepresentational, creative, and evanescent qualities' (Wood et al., 2007: 868) can be challenging. Until relatively recently, studies of the relationship between sound and place have not fully engaged with the ways sound impacts on bodies and elicits feelings, emotions and affect; that as sound 'leaves a body and enters others; it binds and unhinges, harmonises and traumatizes; it sends the body moving, the mind dreaming, the air oscillating' (LaBelle, 2006: xi). Thus, sound is 'a perception of mutual engagement' (Stocker, 2013: xiii), something more than simply a background to social life because sound shapes and alters our perceptions of our world. More recent work in sound studies in a range of disciplines has sought to explore what such 'mutual engagement' might mean, with a number of scholars turning to a study of emotional and affective relationships. In these instances, the use of the term *soundscape* offers ways to capture these complex relations.

References

Amphoux, P., Thibaud, J-P., Chelkoff, G. (2004). *Ambiances en débats*. Bernin: À la croisée.

Anderson, B. (2009). Affective atmospheres. *Emotion, Space and Society* 2: 77–81.

Arkette, S. (2004). Sounds like city. *Theory, Culture and Society* 21(1): 159–168.

Atkinson, R. (2011). Ears have walls: the listening body in urban space. *Aether: Journal of Media Geography* 7: 12–26.

Attali, J. (1992). *Noise: The Political Economy of Music*. Minneapolis, MN: University of Minnesota Press.

Augoyard, J.-F., Torgue, H. (2005). *Sonic Experience: A Guide to Everyday Sounds*. Montreal: McGill-Queens University Press.

Bijsterveld, K. (2008). *Mechanical Sounds: Technology, Culture, and Public Problems of Noise in the Twentieth Century*. Cambridge: MIT Press.

Boyd, C., Duffy, M. (2012). Sonic geographies of shifting bodies. *Interference: A Journal of Audio Culture* issue 2, online: www.interferencejournal.com/articles/a-sonic-geography/sonic-geographies-of-shifting-bodies

Brooks, B., Schulte-Fortkamp, B. (2016). The soundscape standard. In W. Kropp, O. von Estorff, B. Schulte-Fortkamp (eds) Proceedings of the inter-noise 2016 – 45th international congress and exposition on noise control engineering: Towards a quieter future. Hamburg, Germany: Deutsche Geselischaft Fuer Akustik, pp. 2043–2047.

Brown, A.L., Gjestland, T., Dubois, D. (2016). Acoustic encounters and soundscapes. In J. Kay, B. Schulter-Fortkamp (eds) *Soundscape and the Built Environment*. London and New York: CRC Press/ Taylor and Francis Group, pp. 1–16.

Bull, M. (2000). *Sounding Out the City, Personal Stereos and the Management of Everyday Life*. Oxford: Berg.

Butler, T. (2007). Memoryscape: how Audio Walks Can Deepen Our Sense of Place by Integrating Art, Oral History and Cultural Geography. *Geography Compass* 1/3(2007): 360–372.

Caffyn, A., Prosser, B. (1998). A review of policies for 'quiet areas' in the National Parks of England and Wales. *Leisure Studies* 17: 269–291.

Carroll, J., Connell, J. (2000). 'You gotta love this city': the Whitlams and inner Sydney. *Australian Geographer* 31: 141–154.

Chelkoff, G. (2002). For an ecological approach to architecture: perception and design. First International Workshop: Architectural and urban Ambient Environment, Nantes, 6-7-8 Février 2002. Nantes: CERMA, Ecole D'architecture de Nantes, 2002.

Chow, R., Steintrager, J. (2011). In pursuit of the object of sound: an introduction. *Differences: A Journal of Feminist Cultural Studies* 22(2–3): 1–9.

Connell, J., Gibson, C. (2003). *Sound Tracks: Popular Music, Identity and Place*. London: Routledge.

Cosgrove, D. (1984). *Social Formation and Symbolic Landscape*. London: Croom Helm.

Drever, J.L. (2009). Soundwalking: aural excursions into the everyday. In J. Saunders (ed) *The Ashgate Research Companion to Experimental Music*. Aldershot: Ashgate, pp. 163–192.

Duffy, M. (2017). Re-sounding place and mapping the affects of sound. In T. Leppänen, P. Moisala, M. Tiainen and H. Väätäinen (eds) *Becoming with Music and Sound: Musicking Deleuze and Guattari.* London: Bloomsbury, pp. 189–203.

Duffy, M., Waitt, G. (2013). Home sounds: experiential practices and performativities of hearing and listening. *Social and Cultural Geography* 14(4): 466–481.

Duffy, M., Waitt, G., Harada, T. (2016). Making sense of sound: visceral sonic mapping as a research tool. *Emotion, Space and Society* 20: 49–57.

Duncan, J.S. (1980). The superorganic in American cultural geography. *Annals: Association of American Geographers* 70(2): 181–198.

Dunbar-Hall, P., Gibson, C. (2004). *Deadly Sounds Deadly Spaces: Contemporary Aboriginal Music in Australia.* Sydney: University of New South Wales Press.

Edensor, T. (ed). (2010). *Geographies of Rhythm: Nature, Place, Mobilities and Bodies.* Surrey and Burlington: Ashgate.

Gallagher, M. (2015). Sounding ruins: reflections on the production of an 'audio drift'. *Cultural Geographies* 22 (3): 467–485.

Gallagher, M., Kanngieser, K., Prior, J. (2017). Listening geographies: landscape, affect and geotechnologies. *Progress in Human Geography* 41(5): 618–637.

Gallagher, M., Prior, J. (2014). Sonic geographies: exploring phonographic methods. *Progress in Human Geography* 38(2): 267–284.

Gibson, C., Connell, J. (2005). *Music and tourism: On the road again.* Clevedon: Channel View Publications.

Grosjean, M., Thibaud, J-P. (2008). *L'espace urbain en méthodes.* Marseille: Éditions Parenthèses.

Hall, D.A., Irwin, A., Mark Edmondson-Jones, M., Phillips, S., Poxon, J.E.W. (2013). An exploratory evaluation of perceptual, psychoacoustic and acoustical properties of urban soundscapes. *Applied Acoustics* 74 (2): 218–254.

Harris, Y. (2013). Presentness in displaced sound. *Leonardo Music Journal* 23: 13–14.

Helmreich, S. (2010). Commentary: listening against soundscapes. *Anthropology News* 51(9): 10.

Ingold, T. (2007). Against soundscape. In A. Carlyle (Ed.), *Autumn Leaves: Sound and the Environment in Artistic Practice.* Paris: Double Entendre, pp. 10–13.

Jackson, P. (1989). *Maps of Meaning: An Introduction to Cultural Geography.* London: Routledge.

Järviluoma, H., Kytö, M., Truax, B., Uimonen, H., Vikman, N. (eds) (2009). *Acoustic Environments in Change.* Series A: Research Papers 13, Tampere: TAMK University of Applied Sciences.

Kelman, A. (2010). Rethinking the soundscape: A critical genealogy of a key term in sound studies. *Senses and Society* 5(2): 212–234.

LaBelle, B. (2006). *Background Noise: Perspectives on Sound Art.* New York: Continuum Books.

Lefebvre, H. (2004). *Rhythmanalysis: Space, Time and Everyday Life.* London and New York: Continuum Books.

Leyshon, A., Matless, D., Revill, G. (1998). *The Place of Music.* New York: Guilford Press.

Lorimer, H. (2008). Cultural geography: non-representational conditions and concerns. *Progress in Human Geography* 32(4): 551–559.

Lorimer, H. (2007). Cultural geography: worldly shapes, differently arranged. *Progress in Human Geography* 31(1): 89–100.

Lorimer, H. (2005). Cultural geography: the busyness of being 'more-than-representational'. *Progress in Human Geography* 29(1): 83–94.

McCartney, A. (2004). Soundscape works, listening and the touch of sound. In J. Drobnick (ed) *Aural Cultures.* Toronto: YYZ Books and Walter Phillips Gallery Editions, pp. 179–185.

Nancy, J-L. (2007). *Listening,* New York: Fordham University Press.

Oliveros, P. (2005). *Deep Listening: A Composer's Sound Practice.* Lincoln, NE: iUniverse.

Pink, S., Howes, D. (2010). Debate section: the future of sensory anthropology/the anthropology of the senses. *Social Anthropology/Anthropologie Sociale* 18(3): 331–340.

Pisano, L. (2017). Terrae Incognitae: crossing the borders of sonic ecology. Conference proceedings, Invisible places, 7–9 April 2017 São Miguel Island, Azores, Portugal, pp. 469–477.

Rodaway, P. (1994). *Sensuous Geographies: Body, Sense, and Place.* London: Routledge.

Saldanha, A. (2009). Soundscapes. In R. Kitchin, N. Thrift (eds) *International Encyclopaedia of Human Geography.* Amsterdam and Oxford: Elsevier, pp. 236–240.

Schafer, R. M. (ed.) (1977a). *Five Village Soundscapes.* Vancouver, BC: A.R.C. Publications.

Schafer, R. M. (1977b). *The Tuning of the World.* Toronto: McClellan & Stewart.

Schafer, R. M. (1997). *The Soundscape: Our Sonic Environment and the Tuning of the World*. Rochester, VT: Destiny Books.
Schulze, H. (2018). *The Sonic Persona: An Anthropology of Sound*. New York: Bloomsbury Press.
Simpson, P. (2009). 'Falling on deaf ears': A postphenomenology of sonorous presence. *Environment and Planning A* 41(11): 2556–2575.
Smith, S. J. (1994). Soundscape. *Area* 16: 232–240.
Smith, S. J. (2000). Performing the (sound)world. *Environment and Planning D: Society and Space* 18: 615–637.
Sterne, J. (2012). *MP3: The Meaning of a Format* Durham, NC: Duke University Press.
Stocker, M. (2013). *Hear Where We Are: Sound, Ecology, and Sense of Place*. New York, Heidelberg, Dordrecht, London: Springer.
Thibaud, J.-P. (1998). The acoustic embodiment of social practise. In *From Awareness to Action. Proceedings from 'Stockholm, Hey Listen!' Conference in Acoustic Ecology*. Stockholm: The Royal Swedish Academy of Music, pp. 17–22.
Thibaud, J-P. (2011). Sensory design: the sensory fabric of urban ambiances. *Senses and Society* 6(2): 203–215.
Thrift, N. (2008). *Non-Representational Theory: Space/Politics/Affect*. London: Routledge.
Truax, B. (2001). *Acoustic Communication*. 2nd edition. Westport, CT: Ablex Publishing.
Waitt, G., Gill, N., Head, L. (2009). Bushland walking: performing and managing nature in suburban Australia. *Social and Cultural Geography* 10(1): 41–60.
Westerkamp, H. (1974). Soundwalking. *Sound Heritage* 3(4): 18–27.
Westerlaken, M., Gualeni, S. (2016). Becoming with: towards the inclusion of animals as participants in design processes. *ACI '16 Proceedings of the Third International Conference on Animal-Computer Interaction*. Milton Keynes, United Kingdom. November 15-17, 2016. DOI: doi:10.1145/2995257.2995392.
Wood, N., Duffy, M., Smith, S. J. (2007). The art of doing (geographies of) music. *Environment and Planning D: Society and Space* 25(5): 867–889.

Websites

Sounding conflict: From resistance to reconciliation, online at: www.qub.ac.uk/research-centres/SoundingConflict.

Understanding the role of music and sound in conflict transformation, online at: www.qub.ac.uk/schools/ael/Research/Arts/ResearchImpact/Mozambique.

11

Weather and place

Phillip Vannini and Bradley Austin

The sun is shining brightly, but it is –63°C outside, with winds reaching in excess of 200 km/h. With a sense of marvel you gaze at the icy waters in the distance. You snap a picture to post on Instagram and because it's very warm by the window you reach for the small fan above you and twist it wide open. A voice interrupts your thoughts. 'Anything to drink, Sir?' You nod to the flight attendant and ask for a cup of juice. 'Ice with your juice?' 'Please,' you answer.

Few experiences of weather seem more ordinary and unremarkable to the contemporary global citizen than the one portrayed above. Yet, in actuality, moments like these are profoundly revealing of the inseparability of all places from their weather as well as of our continuous bodily and technological preoccupation with atmospheric conditions. How so? Unable to move on our feet across continents through the world's polar surfaces, humans have developed tools such as airplanes to carry out the task. Airplanes, ships, homes, tents, greenhouses, refrigerators, shopping malls are just some of the tools we have developed to keep us a step or two removed from atmospheric elements, so that we may travel, work, study, eat, consume, grow and keep food, and sleep in comfortable environs. All of these environments—indoor or outdoor—are weather-places characterized by distinct weather patterns. But these patterns do not just 'happen.' By turning on a fan, adding ice to our drinks, or perhaps by planning for street snow removal or shopping for beachwear we engage in *practices* through which we control, modify, endure, adapt to, enjoy, or remove ourselves from the weather-places we inhabit. Practices of course depend on *experiences*. We sense warmth so we ask for ice for our drinks. We feel a chill, so we slip under a blanket. We notice it's windy, so we go surfing. Finally, we *represent* weather-places by taking pictures for social media, by engaging in small talk with strangers, by sharing memories, by developing knowledge that can assist us in choosing when to go on holiday, and so forth.

Weather, place, and being are inseparable, and yet because they are so deeply knotted together it is easy to forget the multiple ways in which this entanglement matters. With growing interest in climate and long-term climate change around the world it is especially easy to forget about the importance of everyday weather events, no matter how seemingly insignificant, and thus all the more necessary to re-sensitize ourselves to the geographical, sociological, cultural, and historical meaningfulness of daily weather (see Behringer, 2000; Holland, 2013; Hulme, 2017). Unlike positivist and atomizing perspectives which treat the weather as an independent variable somehow external to the lifeworld, and yet capable of determining multiple behaviors, in this chapter we conceptualize the weather as an inescapable medium for our existence in the world. Following a phenomenological and relational perspective we argue that we are immersed within concrete events of weather as a condition of our being-in-the-world. The 'weather-world' (Ingold, 2007, 2008) is an all-encompassing universe of multiple weather processes unfolding in different places. These *weather-places* are the places we inhabit in our daily life, and as part of our inhabitation we continuously *weather* them through a variety of experiences, practices, and representations. In what follows we survey current research on weather, paying particular attention to geographical literature that deals with place. We exclude from our review research that deals with climate and climate change because it is with embodied and situated encounters of the weather as they are happening now that we are concerned. While resources from the field of geography are our primary focus, we also include relevant studies from cognate fields such as sociology and anthropology.

Experiencing weather-places

The notion of weather-place draws from the seminal work of Tim Ingold (2007, 2008) on weather worlds. Following a non-representational phenomenological approach and building on the ecological psychology of James Gibson, Ingold argues that the weather constitutes the medium for our existence in the world. Contrary to Gibson, who argued that the land was an interface separating earth and sky, Ingold writes that the land is a zone of 'admixture and intermingling' between the various substances of the earth and the air. It is in this zone that the weather constantly occurs, leaving its marks on the lives of creatures living in the land (Ingold, 2008). 'Instead of thinking of the inhabited world as composed of mutually exclusive hemispheres of sky and earth, separated by the ground, we need to attend,' Ingold (2007: 19) remarks, 'to the fluxes of wind and weather.' Contrary to dualist and atomized conceptions of weather, to feel the weather 'is not to make external, tactile contact with our surroundings but to mingle with them [...] in the continual forging of a way through the tangle of life-lines that comprise the land' (Ibid).

Following Ingold's perspective, experiencing the weather is not a matter of registering stimuli external to the body, but rather a precondition of our inhabitation of the lifeworld in its ongoing formation. Experiencing weather is thus a fundamental aspect of our immersion in place. Take for example a foggy environment. By changing the way in which sunlight travels through space, the opacity of fog is known to engender feelings of dislocation and disorientation. In doing so, Craig Martin (2011) writes, fog communicates a sense of inaction and stillness, conjuring feelings of fear, anxiety, and trepidation, pushing us to examine the human engagement with the tensions of proximity and distance. Fog prompts us to reconsider our relation not just with land but also with the sky. The sky is not something we see, but something we see in: luminosity itself (Ingold, 2007). So in this sense it is a mistake to say that in fog we cannot see the sky: fog is simply a reorganization of the sky, occurring through a change in the way luminosity unfolds during daytime. Fog leads us to attend 'to the dynamic processes of world-formation in which both perceivers and the phenomena they perceive are necessarily immersed' (Ingold, 2007: 28). This teaches us that we do not so much experience the weather, but rather experience *in* it.

Current empirical research on experiences *in* weather has focused on a wide range of events and conditions. Wind, in particular, has been the subject of much in-depth ethnographic analysis due to the ways in which winds overlap with notions of spirituality, health, and locality (see Low and Hsu, 2007). For example, writing in regard to the Foehn —a hot southerly wind occurring on the northern slopes of the Alps—Sarah Strauss (2007) remarks on the deep interconnectedness between wind and well-being in the town of

Leukerbad, Switzerland. In addition to the deep historical and cultural significance of the wind, Strauss finds that the Foehn seems capable of affecting the moods and atmospheres of the town, and even of having direct consequences on health and safety.

The relation between wind and health is also at the forefront of another study inspired by Ingoldian ideas, this time dealing with Zulu notions of pollution and spirituality. According to Rune Flikke (2013, 2018), winds are capable of blurring 'any clear distinctions between weather as an observable, objective fact and the internal, idiosyncratic, and personal realm of a spiritual encounter' (2013: 95). Flikke arrives at this conclusion after reflecting on the experiences of Thandi, one of her interlocutors. Following a quest for healing on a mountaintop—a ritual place removed from the pollution of the surrounding urban landscape—Thandi felt that the wind made the landscape come alive with the presence of spirits. This experience prompted Flikke to observe how winds 'physically blew through the body of the afflicted as through the caverns and caves of the landscape, filling them with enough "wind" (*umoya*) to heal and restore them, at least temporarily' (2018: 92). Thandi's body, in other words, became enwinded; rejuvenated in spirit, and restored in vital energy. While these experiences may seem 'exotic' and far-flung to a casual Western reader we do not need to think very hard to find parallels. We could cite, for instance, the case of health and wellness tourism, a business fueled by weary Northern hemisphere citizens' demand for rejuvenating and re-energizing health and wellness practices—ranging from open air meditation, to beach yoga and massage—enabled by sunny weather and balmy temperatures.

Many other weather events have been examined in relation to place. Gordon Waitt (2013), for example, has found that experiences of heat in the humid environment of Wollongong, Australia, are drenched in culturally coded notions of sweat. Sweat and sweatiness, as experienced by the young female participants to his study, were treated as embarrassing bodily dirt to be prevented, managed and hidden in the name of conventional gender ideologies of attractiveness and self-presentation. At the opposite end of the temperature spectrum, Jessica Finlay (2018) has found that snow and icy conditions were the source of several troubles for some residents of northern US cities. Snow, ice, and cold temperatures generated barriers to everyday activities, reducing general activity levels, and engendering feelings of isolation. Finlay's findings are in contrast to the work of Russell Hitchings (2010), who uncovered how London-based professionals were largely unaware of seasonal changes in weather due to their insulation in climate-controlled environments. What these cases go to show is that experiencing *in* weather—whether we are speaking of bodily sensations, community belonging, or one's perception of the environment—is a fundamental component of the way people inhabit their various lifeworlds. Experiences in weather, however, are variable and marked by sharp social and cultural differences.

Overall, research in weather-places has shown that experiences of weather are not passive responses to external weather stimuli. Air and meteorological events are constitutive of place-specific affective relations that enliven places through the attunement of bodies. For example, Gail Adams-Hutcheson (2017) has shown how the materialities and immaterialities of fluctuating weather conditions in New Zealand affect landscapes as vibrant practices of dwelling, both human and animal. Relatedly, Elizabeth DeVet (2017) has found that for her research participants weather was not a backdrop to everyday life, but instead a fundamental component of their lifeworld in Darwin, Australia. Her findings show that even in uncomfortable weather conditions people 'chose to remain weather-connected, working with and around the weather through the use of vernacular adjustment strategies (that is, through local culturally informed responses to weather)' (2017: 145). This is in line with what one of us (Vannini et al., 2012) found on the rainy west coast of Canada. Far from passive meteorological dupes, Vannini and colleagues found that individuals skillfully weather their lives by moving alongside atmospheric patterns, weaving original micro-climates within their dwellings, in search of a sense of balance and harmony with place. Weather is indeed a noun, but reflections on its experience across different places show quite clearly that it is also a verb. In the process of weathering people shape places and subjectivities dynamically, reflexively, and skillfully. This brings us to the notion of practice: the subject of the next section.

Practices

To weather place is to enroll a host of social, cultural, political, and economic practices in such a way as to make atmospheric conditions an integral component of what a place means and how it feels in both sensory and affective terms. As such, we can think of weathering in this way more as a complex set of practices. In this section, we illustrate weathering from different vantage points to show that, although meteorological phenomena may occur across broad scales, weather is itself a place-specific condition that is territorialized through sets of material practices.

One such set of practices that enable us to weather place are those associated with mobility. Taking rhythms of flooding in a suburb of Sydney, Australia, as a starting point, Neimanis and Hamilton (2018: 81) discuss human and nonhuman mobilities in terms of weathering as an embodied practice, writing that 'not all bodies weather the same,' and that 'weathering is a situated phenomenon embedded in social and political worlds.' For example, people with disabilities that require the use of wheelchairs may find themselves spatially and socially excluded as a result of interactions between the built environment and meteorological conditions, as Lindsay and Yantzi (2014) found among participants during Canadian winters. Such individuals weather place very differently than others without disabilities, often finding creative ways to negotiate their relationships with both the environment and other people (also see Chapter 27 by Wiesel and van Holstein in this book). As the experience of weather is contingent on mediating technologies (e.g., wheelchairs, or bicycles for Paul Simpson (2018)), so too are the spatial practices those bodies enact. In the Canadian context above, people bound to wheelchairs may find some places inaccessible and other places welcoming, based largely on the ways that material conditions (presence/absence of wheelchair ramps) enable or constrain movement under different atmospheric conditions. It is in this way that weathering place connects practices ranging from the implementation of construction codes to helping someone cross a street.

Such practices also productively make connections across scales from individual to national, creating further entanglements from which the atmosphere cannot be separated. Jacqueline Allen-Collinson (2018) describes 'weather work' in the context of a Welsh physical activity program known as *Mentro Allan* (Venture Out), which sought to increase physical activity among sedentary groups by promoting outdoor activities. Her ethnographic study of program participants found that 'weather work,' through an engagement with weather as a sense-making practice, also required 'weather learning' in order to cope and become comfortable with new kinds of interactions with the environment. *Mentro Allan* was a partnership between Public Health Wales and other organizations and was funded through the Big Lottery Fund (a non-departmental public body in the UK), which distributes funds raised by the National Lottery for various causes. While the participants in the program did indeed forge new relationships with weather, these practices were inherently connected to an assemblage of more distant quasi-governmental organizations, themselves subject to broader political forces and values. In other words, these participants did not enter into a pre-existing weather place but were, rather, singular components in a multifaceted meshwork of actors all working in concert to *produce* weather places.

The Polish winter of 1978–1979 provides an instructive example of the political dimensions of weathering in (re)making weather places. Leszek Koczanowicz (2007) illustrates how heavy snowfall and frigid temperatures effectively incapacitated the State's ability to administer emergency services and basic infrastructure (including heat), forcing people to rely on each other to meet basic needs. As people expressed anger in private homes and in lines for common goods, this event provided an opportunity to discuss the inefficiencies and inabilities of the administration. Furthermore, the event 'was considered a prelude to the "hot" summer of 1980 when [labor] strikes forced the Communist government to agree on the independent trade union Solidarity' (2007: 766). It is important to consider here that 'weather can reveal the true intentions of those in power, something that is always dangerous for politicians' (2007: 766). In this context, place was weathered through collective practices aimed at basic survival and ultimately united people, no matter how temporarily. While the meteorological event provided enabling conditions for economic, social, and political change, while also constraining mobility, it was the collective practice of weathering place that made use of these opportunities for broader social movements.

Relationships between social movements, political opportunity, and weather in New York City and Washington, D.C. were quantitatively analyzed by Huiquan Zhang (2016). Along with finding that social movements tend to be more violent in warmer weather, he highlights how activism takes place in particular environments, and how, in turn, weather is utilized differently among various actors with different goals. Importantly, Zhang (2016: 307) suggests that 'movements' sensitivity to weather conditions varies across different political opportunity contexts; movement activists are 'picky' about the weather when there are more political opportunities and resources; when society tolerates, even welcomes, movements; and when they have more options. Thus, the ways that places change are inherently linked not only to broader political and social structures, but also to the ways people enroll and act with weather as an enabling or constraining condition of place.

Economic practices of weathering are helpful in illustrating the ways that meteorological conditions can provide such possibilities for enabling different forms of human action. Focusing on 'tourist weather,' Rantala et al., (2011) undertook an analysis of wilderness guiding in Finland, finding practices such as 'anticipating' and 'coping with' weather to be central in the materiality of place. As an intrinsic component of place, weather presents opportunities (affordances) in such a way that people do not necessarily perceive the physical environment so much as they see meaning, 'for example, a snowdrift that affords the potential for skiing, tobogganing or a snowball fight' (2011: 291). In the broader context of tourism, these practices are rooted in economic and material conditions that affect how any number of possibilities for movement and action can be enrolled in place-making. The authors also address discursive practices that promote 'skiing weather.' In the most basic sense, it takes disposable income and time to take a skiing trip. Even if the weather is perfect for skiing, advertising it as such to lower-income groups without the economic means to go to the ski slopes renders 'skiing weather' almost meaningless for them. This cursory example underscores the relevance of weather as both a material and symbolic resource in economic practices that promote the production and consumption of place (see also Thornes and Randalls, 2007).

As we have shown here, weathering is a complex set of practices, at various times social (organizing space and people), cultural (constructing sense of place), political (including/excluding voices), and economic (promoting the production and consumption of place).

From individual bodies to nations, from persons with disabilities to wilderness guides, and from home and beyond, weather is not simply an external force to which people must adapt. Rather, it is clear that people weather place when they operate with, rather than against or in spite of, atmospheric conditions, because to weather place is to engage and appreciate more-than-human worlds. In this sense, weathering practices can become powerful and transformative. This, of course, depends on what weather means from the outset, which brings us to our next section: representation.

Representations

In this section, we consider the representation of weather through visual and textual discourse, identifying key processes through which weather is given meaning. Our aim here is twofold: first, to illustrate how representations of weather work to construct what places mean, and second, how dwellers of those places position their own identity within those representations.

Weather is commonly considered as a backdrop against which human activities are brought into focus, but as John Thornes (2008) finds in Monet's London Series (1899–1905), representing smoke and fog (smog) foregrounds urban weather as an emergent outcome of more complex interactions between social, economic, and environmental conditions—without a workforce, capital investment, consumption of natural resources, and fog itself, smog would not exist. Through representing turn-of-the-century London in this way, Monet provided a window through which we might now peer into the past to understand air quality in a time of rapid industrialization. Thornes (2008) then applies lessons learned from such analyses to ways that weather forecasters in the UK might conversely foreground societal impacts of winter weather on roadways through culturally distinct symbols, such as using 'traffic light' color schemes. 'Producing forecast images is not just about accuracy,' Thornes (2008: 579) writes, 'it is also about the material implications of constructing these images in a particular way.'

As textual and visual discourse, weather forecasts are key in helping shape everyday interactions with the atmosphere, which in turn help form place-based weather meanings. For example, while Stewart et al., (2012) found that survey respondents from drier climates in the US tend to find weather less salient than respondents in temperate or continental climates, they also found that how much weather matters in everyday life (namely, its salience) depends on confidence in the accuracy of forecasters. Taken together, it is important to consider that why and where weather matters (and what it means) is dependent on many factors ranging from physical perception, personal history, economic conditions, and social institutions.

The power of such institutions to influence the meaning of weather should not be discounted. As Michael Billig (1995: 117, emphasis original) succinctly writes in his work on nationalism:

> A homeland-making move transforms meteorology into *the* weather. And *the* weather – with its 'other places', its 'elsewheres' and its 'around the country's' … appears as an objective, physical category, yet it is contained within national boundaries … and all this, in its small way, helps to reproduce the homeland as the place in which 'we' are at home, 'here' at the habitual centre of 'our' daily universe.

As a site of cultural reproduction, the representation of weather in maps has been shown to both reinforce and repudiate existing power structures. For example, Yair Wallach (2011) shows how the exclusion of Palestinian toponyms in printed weather maps works to conceptually extend the Israeli state. Jaffer Sheyholislami (2010) similarly shows how the use of Kurdish place names in weather reports works to maintain Kurdish ethnic identity in his analysis of the satellite television channel Kurdistan TV, which lists Kurdish toponyms alongside official names (for example, Erbil, Iraq, is listed as *Hewlêr*).

Relatedly, Doherty and Barnhurst (2009) analyzed televised weather reports in Boston, Massachusetts, during the 2004 Democratic National Convention, a time when local residents were particularly attuned to weather conditions (that is, when weather was particularly salient and meaningful to everyday life in the city). Descriptions such as marching, migrating, keeping the upper hand, sneaking in, creeping up, struggling and failing were common ways of expressing meteorological conditions. They find that such framings tend to speak not only for weather, but also the natural world more broadly, despite using language that is typically used to describe more human movements. Through such framings, weather becomes discursively mobilized and is given a degree of agency in order to ascribe meaning.

Marita Sturken (2001) shows how multimedia representations of the 1997–1998 El Nino event helped produce apocalyptic narratives, which influenced how Californians saw themselves within the context of a global weather event. While some emergent narratives and representations 'may be read as a kind of collective humor' that attributed blame for anything from pest infestations to global economic trends, 'it was also part of a larger struggle for meaning' (2001: 186). Through newspaper cartoons and other discourse, weather representations became not only tools to communicate physical conditions in the atmosphere, but to more abstractly understand place itself. Similarly, Vannini and McCright (2007) find that weather tends to get its meaning through televised discourse in relation to middle-class leisure activities in places like golf courses and ski slopes. Such commodified representations do not simply reflect a need for information. Rather, they pull together various discourses in ways that produce what it means to be a Californian or a golfer, such that the product consumed is not only a representation of weather, but an identity in and of itself.

While commodified weather information does play a significant role in how people understand and position themselves in places, the emergence of social media also provides a digital space out of which physical weather-places can become actualized through a similarly relational process of meaning-making. For example, Bradley Austin (2014) highlights the material and discursive construction of bedrooms as important places where weather and society are co-productive. Through strategies of 'weather typing,' Twitter users in the US constructed and shared ideas about sex weather, cuddle weather, sleeping weather, and more, to the point that 'the reproduction of weather types implies the necessary reproduction of social conditions, which to some extent involves the reproduction of social space' (2014: 89).

What weather means is just as contingent upon meteorological events as it is upon the ways those events are represented, and those representations are often revealing of the cultural dynamics of a place. We have purposefully not set out to say 'weather means this' because, as Alan Stewart, 2009: 1840) remarks, 'the societal impacts of routine and extreme [weather] events may be as "fine-grained" as the individuals who experience them.' Place, as a concept, tends to be defined by uniqueness and particular characteristics. Weather representations inscribe different meanings for Palestinians and Kurds, Bostonians and Californians, Weather Channel viewers and Twitter users, and others. What we have illustrated here is that such difference, regardless of its source, is a fundamental element of place itself.

Conclusion

A great deal of positivist literature across the social sciences still attempts to tackle the analysis of weather events and weather-related practices and experiences in separation from the unique particularities of how weather takes place and makes place. Throughout this chapter we have argued that weather and place are inextricably entangled. To underscore our argument, in this short conclusion we want to stress the importance of *placing weather*. By placing weather we refer to the analytical necessity of contextualizing weather events in the precise spatialities in which they unfold (see Hulme, 2017). More precisely, we want to make two simple points. We believe that these two points about placing weather can lead researchers to become more sensitive toward the importance of weather across myriad contexts.

First, in placing weather we can become attuned to the environmental qualities of a particular environment and also become more alert to the personal, interpersonal, and more-than-human dimensions of weathering. In other words, when we place weather we uncover how its manifestations are never just 'good' or 'bad,' but always 'good' or 'bad' (or something in between) *for* a particular someone involved in doing something in particular. Take a snowfall in the city of Vancouver. While Vancouverites are not new to snow, accumulation of as little as 5 cm of snow is enough to cripple all forms of road transportation due to the many sloped streets of the city (and the fact that very few vehicles have studded snow tires). On days like these, weather is clearly a nuisance if you must get to work, but it is also a boon for the ski resorts surrounding the city, and a welcome excuse for some people to stay home instead of going to school or work. In placing weather within precise environs we situate weather events onto a complex map of social practices, experiences, and representations which ultimately shape the power of weather's occurrence in time and in place. In placing weather, in sum, we become attuned to the material and immaterial ways in which weather-related affects and atmospheres come to take place and make places.

Second, in placing weather we sensitize ourselves to the macrosocial and environmental consequences of weather events. As climate change continues to intensify around the planet, a better understanding of place-based weather experiences and practices becomes necessary in order to manage, predict, and manipulate energy demand. Research has shown that maladaptive thermal behaviors with consequences for increases in energy usage (for instance, using wasteful outdoor heaters or relying excessively on heating or air conditioning; see Hitchings, 2007) can be addressed through changes in social expectations and norms. Research has shown that people can adapt to novel or extreme weather conditions, especially when given the tools and the opportunities to do so (see DeVet, 2017; Fuller and Bulkeley, 2013; Hitchings, 2010; Opperman et al., 2018; Vannini and Taggart, 2014). Studies like these are sensitive to the notion of weather places, but more research of this kind is needed. By placing weather we can better understand how particular groups of people in distinct places deal in unique ways with specific weather conditions and events, both indoor and outdoor.

References

Adams-Hutcheson, G. (2017) 'Farming in the troposphere: drawing together affective atmospheres and elemental geographies', *Social and Cultural Geography*, published online before print: www.tandfonline.com/doi/abs/10.1080/14649365.2017.1406982?journalCode=rscg20

Allen-Collinson, J. (2018) 'Weather work: embodiment and weather learning in a national outdoor exercise programme', *Qualitative Research in Sport, Exercise and Health*, 10 (1): 63–74.

Austin, B. (2014) 'Perspectives of weather and sensitivities to heat: social media applications for cultural climatology,' Unpublished Dissertation. Kent State University. Available online at https://etd.ohiolink.edu/

Behringer, W. (2000) *A Cultural History of Climate*. New York: Polity Press.

Billig, M. (1995) *Banal Nationalism.* Thousand Oaks: Sage.

DeVet, E. (2017) 'Experiencing and responding to everyday weather in Darwin, Australia: the important role of tolerance', *Weather, Climate and Society*, 9 (2): 141–154.

Doherty, R. and Barnhurst, K. (2009) 'Controlling nature: weathercasts on local television news', *Journal of Broadcasting and Electronic Media*, 53 (2): 211–226.

Finlay, J. (2018) 'Walk like a penguin: older Minnesotans' experiences of (non)therapeutic white space', *Social Science and Medicine*, 198 (1): 77–84.

Flikke, R. (2013) 'Enwinding social theory: wind and weather in Zulu Zionist sensorial experiences', *Social Analysis*, 60 (3): 95–111.

Flikke, R. (2018) 'Healing in polluted places: mountains, air, and weather in Zulu Zionist ritual practice', *Journal for the Study of Religion, Nature, and Culture*, 12 (1): 76–95.

Fuller, S. and Bulkeley, H. (2013) 'Changing countries, changing climates: achieving thermal comfort through adaptation in everyday activities', *Area*, 45 (1): 63–69.

Hitchings, R. (2007) 'Geographies of embodied outdoor experience and the arrival of the patio heater', *Area*, 39 (3): 340–348.

Hitchings, R. (2010) 'Seasonal climate change and the indoor city worker', *Transactions of the Institute of British Geographers*, NS 35: 282–298.

Holland, P. (2013) *Home in the Howling Wilderness.* Auckland: Auckland University Press.

Hulme, M. (2017) *Weathered: Cultures of Climate.* London: Sage.

Ingold, T. (2007) 'Earth, sky, wind, and weather', *Journal of the Royal Anthropological Institute*, NS S19–S38.

Ingold, T. (2008) 'Bindings against boundaries: entanglements of life in an open world', *Environment and Planning A*, 40: 1796–1810.

Koczanowicz, L. (2007) 'Politicizing weather: two Polish cases of the intersection between politics and weather', *South Atlantic Quarterly*, 106 (4): 753–767.

Lindsay, S. and Yantzi, N. (2014) 'Weather, disability, vulnerability, and resilience: exploring how youth with physical disabilities experience winter', *Disability and Rehabilitation*, 36 (26): 2195–2204.

Low, C. and Hsu, E. (2007) 'Introduction: wind, life, and health', *Journal of the Royal Anthropological Institute*, NS S1–S17.

Martin, C. (2011) 'Fog-bound: aerial space and the elemental entanglements of body-with-world', *Environment and Planning D*, 29: 454–468.

Neimanis, A. and Hamilton, J. (2018) 'Weathering', *Feminist Review*, 118 (1): 80–84.

Opperman, E., Strengers, Y. Maller, C. Rickards, L. and Brearley, M. (2018) 'Beyond threshold approaches to extreme heat: repositioning adaptation as everyday practice', *Weather, Climate, and Society*, 10: 885–898.

Rantala, O., Valtonen, A. and Markuksela, V. (2011) 'Materializing tourist weather: ethnography on weather-wise Wilderness guiding practices', *Journal of Material Culture*, 16 (3): 285–300.

Sheyholislami, J. (2010) 'Identity, language, and new media: the Kurdish case', *Language Policy*, 9 (4): 289–312.

Simpson, P. (2018) 'Elemental mobilities: atmospheres, matter and cycling amid the weather-world', *Social and Cultural Geography*, published online before print: www.tandfonline.com/doi/abs/10.1080/14649365.2018.1428821?journalCode=rscg20

Stewart, A. E. (2009) 'Minding the weather: The measurement of weather salience', *American Meteorological Society*, 90 (12): 1833–1841.

Stewart, A. and Lazo, J. Morss, R. and Demuth, J. (2012) 'The relationship of weather salience with the perceptions and uses of weather information in a nationwide sample of the United States', *Weather, Climate, and Society*, 4 (3): 172–189.

Strauss, S. (2007) 'An ill wind: the Foehn in Leukerbad and beyond', *Journal of the Royal Anthropological Institute*, NS S165–S181.

Sturken, M. (2001) 'Desiring the weather: el Nino, the media, and California identity', *Public Culture*, 13 (2): 161–189.

Thornes, J. (2008) 'Cultural climatology and the representation of sky, atmosphere, weather and climate in selected art works of Constable, Monet and Eliasson', *Geoforum*, 39: 570–580.

Thornes, J. and Randalls, S. (2007) 'Commodifying the atmosphere: 'pennies from heaven'?', *Geografiska Annaler*, 89 A (4): 273–285.

Vannini, P. and McCright, A. (2007) 'Technologies of the sky: a socio-semiotic and critical analysis of televised weather discourse', *Critical Discourse Studies*, 4 (1): 49–74.

Vannini, P. and Taggart, J. (2014) 'Solar energy, bad weather days, and the temporalities of slower homes', *Cultural Geographies*, 22 (4): 637–657.
Vannini, P., Waskul, D., Gottschalk, S. and Ellis-Newstead, T. (2012) 'Making sense of the weather: dwelling and weathering on Canada's rain coast', *Space and Culture*, 15 (4): 361–380.
Waitt, G. (2013) 'Bodies that sweat: the affective responses of young women in Wollongong, New South Wales, Australia', *Gender, Place and Culture*, 21 (6): 666–682.
Wallach, Y. (2011) 'Trapped in mirror-images: the rhetoric of maps in Israel/Palestine', *Political Geography*, 30 (7): 358–369.
Zhang, H.T. (2016) 'Weather effects on social movements: evidence from Washington, D.C., and New York City, 1960–95', *Weather, Climate, and Society*, 8 (3): 299–311.

12

The luminosity of place

Light, shadow, colour

Tim Edensor

Introduction

In considering how place is sensed at an everyday level, inhabitants typically become accustomed to the range of affective and sensory qualities that play across space. The humidity in the air, the smells of flora or industry, the soundscape of birds or traffic, the textures of brick and stone all shape this everyday apprehension. And, as I discuss in this chapter, so does the play of light and the distinct forms upon which it shines. We live in places in which variegated patterns of sunlight and shade tone landscape with colours, shadows and textures that ceaselessly reconfigure our apprehension and attention. We are, as Ingold (2008) emphasizes, always immersed in the currents of a world in formation, and this dynamic play of light is an integral part of the everyday affective and sensory experience through which we inhabit and become attuned to place. Sedimented in mundane daily sensation, the luminous and shady qualities of place are unreflexively experienced most of the time, but they come to attention when they change (as occurred when the eruption of the Icelandic volcano *Eyjafjallajökull* produced highly colourful sunsets across Northern Europe in 2010) or when we visit other places that are suffused with unfamiliar qualities of luminosity. In this brief introduction, I identify the conditions that shape the apprehension of the light of place: first, through human optics, second, via the quality of light, third, according to the material affordances of place and, finally, through the influence of cultural representations. Subsequently, I discuss the distinctive encounter with light in a range of places. I first identify some distinctive ways in which light and shadow fall on places, create shadows and colours, and solicit certain everyday practices as well as creative place-specific designs. I then examine how the light of place has been represented in painting and photography before concluding with a brief discussion of how these distinctive lightscapes increasingly lure tourists to particular destinations.

The conditions that shape the experience of the light of place

Wherever we are, as humans we see with light in particular ways because we are bestowed with a particular visual apparatus. Though this varies, with some people being short-sighted or colour blind for instance, humans see the world enormously differently to non-humans, with their very diverse visual capacities. As a form of radiant energy, light is not perceived in itself but rather through how we visually discern the varying colours and intensities it produces as it shines upon space. The iris expands and contracts in controlling the amount of light admitted and the eye's convex lens focusses light to produce an inverted image of a scene on the retina. This image is sent via the optic nerve to the brain, which processes this information. Our distinctive human optical apparatus allows us to continuously adjust to qualities of brilliance, colour, intensity, radiance, shade and gloom; however, we diverge from those of the many non-humans that can discern objects that we cannot, such as birds of prey, who are able to see the ultra-violet part of the colour spectrum. To emphasize, we can only visually experience place with light through our distinctive optical capacities.

Seeing with the shifting quantities, patterns and intensities of light involves different modes of looking at the distinctive planes, angles and densities that constitute the places that we encounter (Wylie, 2006). Our response to this luminous fluidity is conceptualized by Alphonso Lingis (1998), who identifies how changing *levels* of light, characterized by depth of field and brightness, continuously play across the spaces within which we see things and with which we continuously adjust. He discusses how a red rose in a hospital room intensifies the whites of the sheets, thereby luring our gaze. This highlights how objects cast their colours on surrounding surfaces or become cloaked in shade, pale light or brilliant glare to reconfigure relationships across space. In continuously becoming attuned to these incessantly unfolding levels of light, it becomes apparent that we are never passive creatures that simply move across a pre-existing stage. As we walk through place, for instance, we might successively focus upon a striking feature, gaze into the distance, warily look at potential obstacles across the path, or glaze over in distraction. How we attend to and attune to these multiple features in a world of changing light shapes how we sense and make sense of place.

The distinctive ways in which light enables vision as it circulates between interior and exterior blurs the division between us as a viewing subject and the place we look upon, undergirding how we are an integral part of place, never distant onlookers. Despite this, places have often been reified by the conventions of scientific classification, touristic gazing, artistic representation and aesthetic appreciation that instruct us how to gaze upon, assess, understand and characterize them, recognize their supposed 'genii locus'. However, as our experience of the continuously changing light of place exemplifies how a seething vitality inheres in all settings and in our perceptions of them, such reifications are thus confounded by this dynamism. As the changing light exemplifies, places and landscapes are never preformed but are 'always in process … always in movement, always in making', vitally immanent and emergent (Bender, 2001: 3; Benedicktsen and Lund, 2010), and we are part of this, not distant onlookers.

This ever-shifting distribution of light across place is revealed by the different seasons and weather patterns that shape the distinctive light levels and register the passage of time through the day and the year. Levels of light depend upon the intensity and angle of the sun's rays, the presence or absence of cloud, and the kind and quantity of particles in the atmosphere that diffuse the light (see Chapter 11 by Vannini and Austin in this book). To emphasize, geographical location is important here in determining how light bathes place, and this changes throughout the year depending upon the position of the sun, with great global divergences in the diurnal and seasonal distribution of sunlight and the angle at which the sun hits place. For instance, Australia, Central Africa and eastern China receive much more solar radiation than California or the Mediterranean.

Our perceptual experience of light is also highly conditioned by the affordances of place, the ways in which light interacts with the distinctive surfaces, materials and objects upon which it reflects, deflects and is absorbed, and the capacities of landforms, fixtures, buildings and living things to attract and block luminosity. For example, light is reflected by water, absorbed by the pigment cells of plants and obstructed by large trees. And the distinctive ways in which things and material masses block the varying luminosity of the sun's rays create distinctive forms of shadow and shade. The way we experience light is also shaped by the ways in which it inflects the world with colour. As Diane Young submits, colours 'animate things in a variety of ways, evoking space, emitting brilliance, endowing things with an aura of energy or light' (2006: 173) across time and space, with the sunlight bestowing the material surfaces of place with distinctive hues that change according to season, time of day and the light's intensity.

Finally, although the light and shadow, textures, sounds, smells and tactilities reveal landscape's liveliness, we invariably apprehend and make sense of the light that falls on place, the colours and features that stand out – through cultural values and conceptions. As Constance Classen emphasizes, '(W)e not only think about our senses, we think through them' (1993: 9). Thus, distinctive cultures of looking ensure that seeing with light does not provide unmediated access to the reality of the world; rather, 'seeing involves movement, intention, memory, and imagination' (Macpherson, 2009: 1049). In considering the play of light in colouring space, purely scientific conceptions that construe the experience of colour as a mechanical or psychological response and colours as purely objective qualities cannot get at 'the emotion and desire, the sensuality and danger, and hence the expressive potential that colours possess' (Young, 2006: 174). The cultural values and understandings through which we make sense of how we see place with light veer from the cosmological to the moral, and from the aesthetic to the political, and vary enormously across time and space. As Veronica Strang insists, we formulate imaginative and symbolic concepts and categories that are aligned with our phenomenological experiences of space and place (2005). Moreover, these are often substantiated and supplemented by the ways in which landscapes are represented in photography, painting, film and literature and generate located cultural responses, as I discuss below. Most famously, with regard to shadow, Junichiro Tanizaki repudiates western obsessions with revealing everything with light and focusses upon the aesthetic potency of the multiple tones of grey that wash across urban interiors and exterior settings in Japan. Tanizaki finds beauty not in objects themselves but 'in the patterns of shadows, the light and the darkness, that one thing against another creates' (2001: 2).

Distinctive lightscapes: the light, shadow and colour of place

I now draw on selective examples to consider how light conditions the experience of place and, first of all, I turn to the lightscape of the UK. According to Greenlaw (2006), the temperate maritime climate of British landscapes generally take shape under cloudiness, mild shadows and weak sunlight that produce a distinctive tonal atmosphere typified by subtle and ever-changing patterns of light and shade. These undramatic continuous changes in the annual distribution of radiance and gloom are punctuated by infrequent hot sunny days and wintry gloom that stand out as exceptional, as textures, surfaces, folds and gradients interact with sunlight in unpredictable ways, as I have explored in a northern English upland setting (Edensor, 2017a). These are not merely aesthetic matters for, in adapting to this play of light, people shape routines that allow them to maximize the limited amount of sun that typically shines upon place, as exemplified by the increasing popularity of conservatories amongst British home-owners and the desire to live in accommodation that possesses a south or south-west facing aspect in order to bask in the morning sun in garden or lounge, and to stimulate houseplant growth. When the sun does break out for prolonged spells, patterns of social activity suddenly change as many inhabitants leave their homes to haunt beer gardens, go running, stroll in parks or drive into the countryside.

Yet Greenlaw's depiction is only partly salient for such patterns of light and cloud are not spread evenly throughout the UK. For instance, in North West Scotland, the play of light seems to be far more dynamic than in the muted lightscape of England, perhaps because of its closeness to the sea and plentiful lochs that reflect light, the somewhat sparse but sometimes effusive vegetation, the ancient, barren rock surfaces, and the effects of the North Atlantic Drift that generates dramatically changeable weather as it passes over the mountainous countryside. Passage through this landscape offers experiences of diverse hues across layered mountain vistas, silhouetted pine stands, silvery lochans, vibrant purple heather, the spreading russet of bracken, glaring mossy greens, brown-green deposits of kelp on seashores and the blues and greys of the sea. The unpredictable sky features a profusion of shifting clouds of manifold form and shade and, occasionally, extraordinary juxtapositions of murky clouds and brilliant patches of translucent sunlight that radiate parts of the landscape.

Also, in northerly European places, strong associations of light are embedded in Swedish, Norwegian, Finnish and Danish notions of belonging to place and landscape. While Henry Plummer recognizes the diverse topographies and ecologies of these landscapes, he submits that 'their skies share a subdued light that imbues the entire region with mystery' and evocatively describes this shared Nordic light as typified by the 'low slant of the sun ... long shadows and strikingly refracted colours' that pervade the winter months (2012: 6–7). During long twilights he describes how 'sky and snow are equally tinged with rainbow hues that linger for hours', and how on midsummer evenings 'the sun dissolves into an unreal haze that bathes the land in a fairy-like glow, its colours strangely muted and blurred'. Of course, there is considerable variation in Nordic light to the north and south of the region: towards the poles, darkness pervades diurnal experience in winter and light floods the day throughout summer, producing changing shadows over many hours. In these far northern realms, patterns of light and dark shape cultural rhythms, influencing when social practices take place across the year.

In contrast to these great seasonal differences, in the tropics the sun does not stray wildly throughout the year and so diurnal variations of light and shadow are less marked, instantiating more consistent social rhythms in place. These divergencies from northern lightscapes are also apparent in southern European settings, as Christian Norberg-Schulz (1980) points out in emphasizing the play of lucid light and deep shade in Mediterranean 'classical' landscapes. In the similarly Mediterranean climate of Los Angeles, perhaps the most striking, even infamous, elements of the lightscape are the lurid red sunsets that eventuate from toxic atmospheres saturated with a high content of aerosol and carbon dioxide. Norberg-Schulz (1980) also depicts the pared down Middle Eastern landscapes that are usually harshly lit by a cloudless sky, prompting denizens to seek shade in occasional tree-lined oases, with cities marked by deep colonnades, covered bazaars and arcades, high walls and shaded courtyards.

In considering how light inflects the surfaces of place to produce colour, the qualities of building materials mark out certain settings as renowned for their hue. For instance, Aberdeen is renowned as the 'Granite City' because, over several centuries, most of its coarsely crystalline silver-grey stone building material has come from the nearby Rubislaw Quarries, and the characteristically watery winter light and grey clouds of the Scottish northeast coast often compound the surrounding greyness. Bath is similarly typified by a monotonal colour since it is primarily composed out of local limestone, and this material constitution is underpinned by heritage legislation that requires the continuing use of this building material. These cities are comparable with Paris, which has primarily been supplied by limestone quarries in the Oise, 25 miles north of the city, which prompts Rebecca Solnit's description: 'Everything – houses, churches, bridges, walls – is the same sandy grey so that the city seems like a single construction of inconceivable complexity, a sort of coral reef of high culture' (Solnit, 2001: 196). Perhaps best known is the city of Siena, which has given its name to the uniform, distinctive earthy brown-yellow colour that clads the surfaces of its buildings.

The powerful radiance of Melbourne's summer light also generates particular qualities of light that shape place-specific activities. For instance, before leaving an interior environment to walk outside, most citizens equip themselves with sunglasses to minimize the glare and deploy hats or even parasols to reduce the harmful effects of the sun. Far more prosaically, exposure to this dazzling sunlight might potentially have other consequences: the prevalence of sun worshipping and sunbathing as popular cultural pursuits has culminated in high rates of skin cancer in Australia, joining the less severe impacts of impeded vision, sweating, headaches and thirst. Most evidently, the light and heat also inform the paths that pedestrians choose to take, where canopies appended to buildings or the thick foliage of trees on city-centre sidewalks provide shade that lure crowds not evident on sides of the road that lack such shady protection. In parks, large numbers of picnickers and gatherings of people eating lunch congregate under the thick shadow of plane and elm trees.

In Melbourne, the luminosity of the summer sunlight and the materialities upon which it shines produce especially distinctive forms of shadow. Everyday fixtures such as street lights, signposts, garbage bins and railings are bathed in bright light to produce highly delineated shadows that spread across space, soliciting an enhanced apprehension of their sculptural forms and transcending their typically utilitarian assignations. Elsewhere, especially when the angle of the sun becomes more acute, shadows become monstrously distorted or fantastically elongated. Furthermore, the shadows cast on vertical and horizontal surfaces bring surrounding elements of the environment into play, making infrastructural elements, foliage and architectural features more noticeable than usual. Most prominent are the shadows produced by the sunlight that beams upon the highly place-specific 19th-century domestic wrought ironwork of the inner suburbs, with their elaborate designs echoed in shadowy form, amplifying the intricacies of this unique feature (Edensor and Hughes, forthcoming).

Because shade is so necessary to maximize thermal comfort and minimize exposure to sun in Melbourne, this affords opportunities for designers and architects to devise inventive ways of manipulating the distinctive light, thus devising ways of making the built environment of place distinctive. The 19th-century canopies that line rows of city centre shops have recently been complemented by a range of pavilions and coverings that create dramatic patterns across floor surfaces. In addition, contemporary high-rise office and residential blocks are being designed so that shifting, geometric shadowy patterns play across their facades by the addition of small sills, decorative brickwork and window shades. These are recent examples of how architects through the ages have sought to inventively produce different aesthetic effects through manipulating light and shadow as discussed by Stephen Kite (2017) and to generate powerful atmospheres of place, as exemplified by Shanti Sumartojo in her discussion of the work of Tadao Ando in this volume (Chapter 16). Such manipulations not only apply to buildings for, as Böhme (1993) discusses, 18th- and 19th-century arboreal landscapes were orchestrated by landscape gardeners who 'tuned' space by managing the levels of light that filtered through woodland canopies on country estates. Similarly, the thick highly pollarded plane trees that afford thick shade in southern French town squares furnish public spaces in which people may gather in the heat of summer.

The production of effects of light and thick shadow may not necessarily be a deliberate practice, as epitomized by the inadvertent design wrought by soot in industrial Manchester. Prior to the Clean Air Act of 1956 which reduced airborne pollution, the coal fires and factory smoke covered the city's buildings with a thick black coat, producing sooty textures that absorbed daylight to produce a gloomy environment and a sense of almost permanent shadow (Dunn, forthcoming). Yet now that nearly all of this black coating has been removed, the city is revealed, unlike Paris, Bath and Siena, to be a polychromatic place, containing a great diversity of building materials that provide a plethora of textures and colours.

Having discussed how the distinctive relationship between light and place might be conditioned by the qualities of the light that are shaped by geographical position and weather conditions, material constitution and the skilful manipulation of light, I now turn to consider how particular representations have produced cultural understandings and heightened awareness of place-specific light.

Representations of light and place in painting and photography

I now examine how painters and photographers have represented light, shadow and colour in their work and consolidated the understanding and experience of place. A particularly pertinent example is provided by the paintings of English artist John Constable, who was famously engaged in a sustained attempt to identify the manifold play of clouds and light across southern England, with a particular focus on the skies of Salisbury, Suffolk, Hampshire and Hampstead Heath. The Heath was the location for more than 100 paintings created at different times of day and season that represented the enormous diversity of the dynamic sky. Constable's aesthetic concerns were allied to his acute scientific interest in the dynamic meteorological processes that produced such effects, termed by Kenneth Clark as 'the romantic conjunction of science and ecstasy' (cited in Thornes, 2008: 572), combining the painterly and the scientific in accounting for the specificity of place.

Returning to Nordic lightscapes, Henry Plummer asserts that their effects have 'permeated the arts' in these countries, stimulating representations of 'the frailest, most evanescent aspects of nature', as well as 'a mystical intensity to urban scenes'. He particularly draws attention to how 'quiet domestic interiors' are conditioned by an 'ethereal light washing into barren rooms to bring every surface under its spell, as it melts away contours and hangs in the air' (Ibid: 7). More specifically, Jan Garnert (2011) focusses on representations of midnight and twilight radiance in the paintings of the Swedish Anders Zorn, and the ways in which the people in his paintings, the rowers, strollers, farming folk, dancers, bathers and occupants of sitting rooms, followed singular social and seasonal rhythms that contribute to particular moods and atmospheres characteristic of very specific places and moments. By contrast, the northern Danish qualities of light are captured by the late 19th- and early 20th-century Skagen group of painters, including P.S. Krøyer, Anna and Michael Ancher, Holger Drachmann and Thorvald Bindesbøll, who established a colony at the northern coastal village of that name. They were attracted by the blend of seascape and the quality of the light at this place that provided fertile conditions for the emergent practice of *plein air* painting. The light represented in their paintings ranges from stormy skies and still summer radiance to midnight sun, vibrant sunsets and sunrises, moonlit nights and wintry, diffuse luminosity. Krøyer was especially inspired by the light of the evening 'blue hour', in which water and sky seem to merge. Skagen is now a popular tourist destination, marketed as the 'Land of Light' and an annual 'Blue September' festival is held to celebrate the twilight period celebrated by Krøyer.

The light, shade and shadow of these Northern landscapes starkly contrasts with the often harsh Australian sunlight, a quality that came to symbolically differentiate the nation from the British colonizing power (Miles, 2013). The distinctive luminosity of the landscapes of Victoria were ignored by 19th-century artists, many from Europe, who painted Australian rural scenes that resonated with the cloudy landscapes that Greenlaw characterizes. It is as if hemmed in by painterly conventions, their perceptions of the world, grounded in European modes of representation, limited their recognition of the lucid colours and light that they beheld. Towards the end of the 19th century such representations were challenged by the Heidelberg school of artists, a shifting collective who endeavoured to capture incandescent summer Victorian landscapes in their *plein air* painting. Key figures such as Tom Roberts, Fredrick McCubbin and Arthur Streeton rendered brilliant yellow fields, luminous trees, dazzling blue skies, long shadows and glistening beaches, scenes that often included rural workers, homesteaders and bathers but, tellingly, never indigenous Australians, those who formerly inhabited such expansive realms. Accordingly, these naturalistic, impressionistic landscapes were integral to the burgeoning assertions of settler colonial nationalism as well as regional identity. The sense of place that they fostered is afforded prominence in many art galleries and notably reinforced on the Bayside Coastal Art Trail, a 17-kilometre walk between the Melbourne coastal suburbs of Elwood and Mentone. Along the clifftops and in front of the sandy bays and former fishing settlements are information boards that feature many of the paintings they created in situ.

In further distinguishing forms of Australian light from the moody tones of European sunlight, Barbara Bolt questions the (Eurocentric) Heideggerian conception that light reveals truth, allows objective knowledge to be acquired and renders landscapes transparent to the onlooker. In Australia, she contends, there may be too much light, a light that bleaches out the details of the landscape that is beheld. In the shadowless glare of the midday sun, nothing is revealed: 'too much light on the matter sheds no light on the matter' (2000: 204). Here, the powerful effects of the sun upon human perception mark the limits of presenting place through realistic renderings. The fuzzy, blinding glow of the sun thwarts Bolt's attempts to paint Kalgoorie, a western Australian town roughly 700 kilometres from Perth, situated in a marginal desert landscape that in the glare of the sun is 'so fractured and messy that no form emerged', no clear distinction between foreground and background (Ibid: 210). Rather than illuminating the landscape with clarity, light makes it illegible. Forced to avert her gaze from the blinding radiance, the ground becomes the dominant manifestation of light in the landscape.

Melissa Miles shows how such representations of the glare of place have also been expressed through Australian photography. Although, like painting, photography represented the powerful sunlight to symbolize the promise, youth and energy of the young nation in distinguishing it from the grey British motherland, contemporary work has foregrounded the dazzling and volatile luminosity of the landscape. Miles (2013) particularly focusses on Tasmanian photographer Danielle Thompson, who photographs the light that filters through trees as layers of sharp luminous splinters, and the work of Paul Fusinato who directly photographs the light of the sun to disavow the possibilities of knowing it as a stable and legible entity.

The artistic representations discussed above only touch on the multitude of ways in which painters and photographers have sought to 'capture' the light of place, influencing the cultural understandings we mobilize when we encounter these and similar locations. These portrayals are also presented in art galleries, reproduced in advertising and loom large in tourist promotion. It is to this latter place-making practice that I now briefly turn.

Light, place-making and tourism

Finally, these qualities of light in place, fuelled by the kinds of representations I have discussed, are increasingly the focus of attraction for a range of tours to destinations that promote 'celestial tourism' (Weaver, 2011). Most obvious are tourist visits to Northern European destinations such as Iceland, northern Scandinavia and Canada, to experience the aurora borealis (Jóhannesson and Lund, 2017), and to the growing number of dark sky parks and reserves to witness the play of stars. Yet tourism has long focussed upon the effects of light in spectacular scenic landscapes and at iconic sites. For instance, sunset cruises at places as diverse as Santorini, Hong Kong and New York entreat tourists to gather on boats to experience and photograph the luminous skies around ports, mountains and cities and the prominent silhouettes against the glowing horizon. Perhaps most famously, the time at which the Taj Mahal is best appreciated is the subject of often contesting advice, with some advocating that its white marble tones are best brought out during the pale light of sunset, some contending that sunset is preferable, while others champion the view of the Taj by moonlight (Edensor, 1998). Such discourses underpin the global allure of the Taj and add to the plethora of images advertised in tourist promotion and photographically fashioned by tourists themselves. Another example of tourist desires to encounter illumined otherness is manifest in the growing popularity of tours to the Arctic Circle (Birkeland, 1999), to Tromsø, the Northern Cape, Hammerfest and other northerly places, to experience the 'Land of the Midnight Sun', and engage in fishing, walking, kayaking, cycling, whale watching and dining.

A more local attraction is *Manhattanhenge*, a recently created, highly place-specific event staged in New York that refers to ancient rituals associated with the seasonal arrival of light during the solstices. Occurring twice annually, in May and once in July, thousands gather to witness the setting sun radiate from west to east through the city's modernist, vertical, gridded streets and that, according to Andrew Wasserman, 'allows for a restructuring of one's experience of the city: it makes newly visible the relationship between the two-dimensional and three-dimensional structure of the city' (2012: 96). Similarly, in Melbourne each year, a multitude of spectators gather on the steps of Victoria's parliament, some armed with expensive camera equipment, to gaze down the linear stretch of Bourke Street, part of the Hoddle Grid, the rectilinear mesh of streets that form Melbourne's CBD designed by town planner Robert Hoddle in 1837. During the event, *Melbhenge*, the setting sun descends at an angle of 250 degrees west, briefly aligning with the street's furthest horizon, where confined by the canyon-like structure of the street's monolithic tower blocks, it brilliantly illuminates the asphalt surface and the gleaming steel tram tracks.

Conclusion

There have been few discussions about the themes discussed here: how daylight is a key element in the sensory apprehension and meaning of place. In recent years, a range of accounts explore the relationship between darkness and place (Dunn, Chapter 13, this book;

2016) twilight and place (Davidson, 2015) and illumination and place (Bille, 2019; Edensor, 2017b, 2018; Edensor and Bille, 2017; Ebbensgaard and Edensor, forthcoming; Isenstadt, 2018). This chapter thus attempts to address this lacuna.

I have emphasized that though the daylight may usually be unreflexively experienced, the qualities of tone, intensity, reflection, colour and shadow that it transmits to the look and feel of place in response to distinctive material characteristics condition everyday experience. I have also emphasized that place-specific, situated sensory attunements, affects and meanings resonate in artistic representations. Film-makers and television programme makers, as well as painters and photographers as discussed above, are often preoccupied with representing these luminous or gloomy specificities, and tourist marketing also lures tourists with the promise of the particular light effects of place. In addition, I have discussed how daylight has been harnessed and sculpted by architects and landscape designers as a crucial building material, deployed to bring out distinctive colours, textures and shadows, to produce diverse aesthetic, symbolic and sensory effects, and create relationships between external and interior spaces. Creative responses to the light of place continue to adopt new technologies and ideas, suggesting that ways of exploiting light in place-making will continue to evolve.

References

Bender, B. (2001) 'Introduction', in B. Bender and M. Winer (eds) *Contested Landscapes: Movement, Exile and Place*, Oxford: Berg, pp. 1–18.

Benedicktsen, K. and Lund, K. (2010) 'Introduction: starting a conversation with landscape', in K. Benedicktsen and K. Lund (eds) *Conversations with Landscape*, Farnham: Ashgate, pp. 1–12.

Bille, M. (2019) *Homely Atmospheres and Lighting Technologies in Denmark: Living with Light*, London: Bloomsbury.

Birkeland, I. (1999) 'The mytho-poetic in Northern travel', in Crouch, D. (eds) *Leisure/Tourism Geographies*, London: Routledge, pp. 17–33.

Böhme, G. (1993) 'Atmosphere as the fundamental concept of a new aesthetics'. *Thesis Eleven*, 36(1): 113–126.

Bolt, B. (2000) 'Shedding light for the matter', *Hypatia*, 15(2): 202–216.

Classen, C. (1993) *Worlds of Sense: Exploring the Senses in History and Across Cultures*, London: Routledge.

Davidson, P. (2015) *The Last of the Light: About Twilight*, London: Reaktion Books.

Dunn, N. (2016) *Dark Matters: A Manifesto for the Nocturnal City*, London: Zero Books.

Dunn, N. (forthcoming) 'Shadow', in P. Dobraszczyk and S. Butler (eds) *Manchester: Something Rich and Strange*, Manchester: Manchester University Press.

Ebbensgaard, C. and Edensor, T. (forthcoming) 'Walking with light in a changing area: Canning Town to Canary Wharf'.

Edensor, T. (1998) *Tourist at the Taj: Performance and Meaning at a Symbolic Site*, London: Routledge.

Edensor, T. (2017a) 'Seeing with light and landscape: a walk around Stanton Moor', *Landscape Research*, 42(6): 616–633.

Edensor, T. (2017b) *From Light to Dark: Daylight, Illumination and Gloom*, Minneapolis, MN: Minnesota University Press.

Edensor, T. (2018) 'Moonraking in Slaithwaite: making lanterns, making place', in L. Price and H. Hawkins (eds) *In Geographies of Making, Craft and Creativity*, London: Routledge, pp. 44–59.

Edensor, T. and Bille, M. (2017) 'Always like never before: learning from the lumitopia of Tivoli Gardens', *Social and Cultural Geography*, doi: 10.1080/14649365.2017.1404120

Edensor, T. and Hughes, R. (forthcoming) 'Moving through a dappled world: the aesthetics of shade and shadow in central Melbourne', in *Social and Cultural Geography*, [online] doi: 10.1080/14649365.2019.1705994.

Garnert, J. (2011) 'On the cultural history of Nordic light and lighting', In N. Sørensen and P. Haug (eds) *Nordic Light: Interpretations in Architecture*, Odense, Denmark: Clausen Grafisk, pp. 5–13.

Greenlaw, L. (2006) *Between the Ears*: 'the darkest place in England', BBC Radio 3, (25/3/06).

Ingold, T. (2008) 'Bindings against boundaries: entanglements of life in an open world', *Environment and Planning A*, 40(8): 1796–1810.
Isenstadt, S. (2018) *Electric Light: An Architectural History*, Cambridge, MA: MIT Press.
Jóhannesson, G. and Lund, K. (2017) 'Aurora Borealis: choreographies of darkness and light', *Annals of Tourism Research*, 63: 183–190.
Kite, S. (2017) *Shadow-Makers: A Cultural History of Shadows in Architecture*, London: Bloomsbury.
Lingis, A. (1998) *Foreign Bodies*, Bloomington, IN: Indiana University Press.
MacPherson, H. (2009) 'The intercorporeal emergence of landscape: negotiating sight, blindness, and ideas of landscape in the British countryside', *Environment and Planning A*, 41: 1042–1054.
Miles, M. (2013) 'Light, nation, and place in Australian photography', *Photography and Culture*, 6(3): 259–277.
Norberg-Schulz, C. (1980) *Genius Loci: Towards a Phenomenology of Architecture*, New York: Rizzoli.
Plummer, H. (2012) *Nordic Light: Modern Scandinavian Architecture*, London: Thames and Hudson.
Solnit, R. (2001) *Wanderlust: A History of Walking*, London: Penguin.
Strang, V. (2005) 'Common senses: water, sensory experience and the generation of meaning', *Journal of Material Culture*, 10(1): 92–120.
Tanizaki, J. (2001) *In Praise of Shadows*, London: Vintage Classics.
Thornes, J. (2008) 'Cultural climatology and the representation of sky, atmosphere, weather and climate in selected art works of Constable, Monet and Eliasson', *Geoforum*, 39(2): 570–580.
Wasserman, A. (2012) 'Street light: Manhattanhenge and the plan of the city', *Public*, 23(45): 94–105.
Weaver, D. (2011) 'Celestial ecotourism: new horizons in nature-based tourism', *Journal of Ecotourism*, 10 (1): 38–45.
Wylie, J. (2006) 'Depths and folds: on landscape and the gazing subject', *Environment and Planning D: Society and Space*, 24: 537–554.
Young, D. (2006) 'The colours of things', In P. Spyer, C. Tilley, S. Kuechler, W. Keane (eds) *The Handbook of Material Culture*, London: Sage, pp. 173–185.

13

Place after dark

Urban peripheries as alternative futures

Nick Dunn

Disentangling place and darkness

Place after dark is often conceived to be in binary opposition with how it is encountered and understood in the daytime. This limits the various ways in which place after dark may be understood as a dynamic that is simultaneously conceptual, experienced, material, and practised. One of the major obstacles in how we are able to consider darkness within the urban night is the significant weight of cultural meanings and values throughout history that still pervade our contemporary interpretations. Darkness remains misunderstood, bound up in negative associations, and frequently represented as both philosophically and physically inferior to light. This binary narrative has dominated much thought as Oliver Dunnett (2015: 622) explains,

> the idea of light, both in a practical and symbolic sense, has come to be associated with modernisation and the so-called "Enlightenment project" in various ways ... Here we can also see how the metaphor of light has taken on a moralising tone, seen as an all-encompassing force for good, banishing the ignorance of darkness in modern society.

This adversarial relationship has not gone unquestioned, with the considerable diversity and plurality of light and shadow examined through different critical lenses that suggest a counter-history of the significance of dark places (Dowd and Hensey, 2016; Gonlin and Nowell, 2018).

Perceptions of artificial lighting throughout history have also been subject to this entanglement of the symbolic and literal (Schivelbusch, 1988). Rather than a balanced reciprocity, the diminished status of darkness in relation to light in cities at night in these formulations has endured as Joachim Schlör (1998: 57) emphasises,

> Our image of night in the big cities is oddly enough determined by what the historians of lighting say about *light*. Only with artificial light, they tell us, do the contours of the nocturnal city emerge: the city is characterised by light.

This prevailing view negates the spatiality and physicality of urban darkness. It also occludes the considerable diversity of experiences and qualities to be found within darkness, drawing them into the widely held conception of the modern night as a consistent space and time. Concerning this variegation of dark places, their identities, and who they are constructed by and for, Robert Williams (2008: 514) reminds us, 'Night spaces are neither uniform nor homogenous. Rather they are constituted by social struggles about what should and should not happen in certain places during the dark of night'. Urban dark places are notable in this regard through ongoing strategies to 'deal' with them, typically through infrastructures, policies, and practices to control and manage them. In the context of many city centres in the West, darkness is unwanted, connected as it is to negative cultural and historical associations alongside contemporary perspectives of fear and crime. Although Western in their origin, values of light, clarity, cleanliness, and coherence have since been transferred across global cultural experience more widely, resulting in a worldwide decline of the 'nocturnal commons' to which urbanisation has significantly contributed (Gandy, 2017). This decline is not, however, a recent phenomenon. The changes in attitudes and beliefs toward the night between the sixteenth and eighteenth centuries in Europe were particularly important in shaping perceptions of darkness that have largely remained to the present day (Koslofsky, 2011). The opportunities for leisure and labour these societal transformations stimulated, in tandem with developments in both artificial illumination and street lighting, recast the night as an extension of the day through around-the-clock shiftwork and other forms of labour (Melbin, 1987). This framing has continued to be reinforced through the ongoing colonisation of the night, which has led to the present situation under the guise of 'progress'.

In the UK, the opportunities for plurality and diversity of urban lighting are currently limited by regulations imposed on street lighting to conform to *British Standard 5489 1:2013 Code of practice for the design of road lighting: Lighting of roads and public amenity areas* (British Standards Institute, 2013). Meanwhile, the complex demands for place management to ensure vibrant urban centres after dark is reflected in the Association of Town and City Management's (ATCM, n.d.) *Purple Flag Programme*, launched in 2012, as a means of designating those places which meet set criteria for clean and safe night-time economies. In addition, the appointment of Night Tsars in London and Manchester, in 2016 and 2018 respectively, illustrates the prevalent concerns for places after dark being able to enhance the night-time economy rather than support a broader opportunity for understanding the nuances and potentialities of urban darkness. Regulatory frameworks and codes of practice such as these are common in many countries, which along with the growing trend of appointing Night Mayors as ambassadors for cities at night, suggest that urban centres are highly constrained as official sites for experimentation in the dark given issues of safety and security. The question, therefore, is *where* sites for such exploration may lie if city centre locations are not viable? Given their accessibility, difference, and number, I propose that urban peripheries are suitable places for the diversity of gloom to be encountered and experienced. Such places are usually less regulated than inner urban areas and therefore more open to change, which makes them ideal places where different characteristics of, and relationships with, darkness can be experimented with. Further, I also suggest that at urban peripheries a form of 'composting' occurs concerning their identity. This is because they are allowed to settle and even decay on their own terms across a longer timeframe, tending to be more neglected than their city centre counterparts. Indeed, as Shaw (2018: 65) reflects, 'Paying close attention to the peripheries could produce a nocturnal city that is better for all'. Before encountering the edgelands of Manchester as sites for investigation, it is useful to consider how and why envisioning alternative futures for place after dark is increasingly important.

Envisioning alternatives and global challenges

How we understand place after dark is shaped by the different readings we are able to make of it. Envisioning alternatives to the way place after dark is portrayed is essential to being able to discern the plurality and diversity of its elements, materialities, and sensations. Visions for place carry and project the concepts and ideologies behind them and, in doing so, it is evident they are not neutral since what they omit can be as important, or even as controversial, as what they promote. The many visualisations produced for future cities are especially relevant to this latter point. Frequently constructed to depict the apparent virtues of coherence, cleanliness, efficiency, and light, visions for future cities rarely account for place after dark. Where darkness is present in such visions, it is typically used to shape the depiction of a foreboding future: dystopic, dirty, and dangerous. Why should this matter? Images such as these are critical in how we construct and share ideas for our collective future, providing portals for how the world might be (Dunn, Cureton, and Pollastri, 2014). They also make a vital contribution to our social imaginary, defined by Thompson (1984: 6) as, 'the creative and symbolic dimension of the social world, the dimension through which human beings create their ways of living together and their ways of representing their collective life'. In their expression of the 'not-yet' such imagery shapes our ideas of, and intentions toward, futures (Polak, 1973). A number of recent photography projects concerning place after dark signal an emerging body of alternative readings.

Thierry Cohen's *Villes éteintes* (*Darkened Cities*) (2012) is a series of photographs that form a vision of global cities without electricity. Through the composition of a photograph of each these cities with an accompanying one taken at a less populated location at the same latitude with greater atmospheric clarity, the images provide a provocation to reflect on our seemingly intractable reliance upon artificial illumination and its infrastructures. A key feature of these photographs is the star-filled sky that sits above the darkened cityscape in each location. This powerful motif across the series reminds us of our connection to the cosmos and the variations of darkness, relationships that are now lost to many urban populations due to light pollution. As a recurrent theme it also symbolises the artificial character of urban settlements, hinting at their precariousness, and highlighting the wider context of our planet and the challenges to its ecologies that urbanisation presents. The series is also suggestive of a future for cities where we reconnect with darkness via more responsible and less environmentally impactful ways of collective living.

William Eckersley's *Dark City* (2011) project meanwhile focusses on London's urban environment at night. All the images are unpopulated, bringing attention to a whole other set of time values and qualities of darkness rarely considered in such a busy city. The series portrays places that are outside of the touristic gaze and, through documenting them after dark, are highly affective of the atmosphere and evolving perception of darkness in cities that Edensor (2015: 436) suggests, 'might be conceived as an enriching and a re-enchantment of the temporal and spatial experience of the city at night'. The project is especially powerful in its description of the built environment with human life decanted from it, rendering urban places after dark as quietly humming with anticipation through these palpable absences (Figure 13.1).

Figure 13.1 Elevated view down a road with housing blocks, SE2. From the project and publication *Dark City*, 2011. © William Eckersley.

Work such as these two projects demonstrate the potentiality of images to support alternative readings on the present and enable speculation on the futures of place after dark. Conceived from a different perspective, the project *Through Darkness to Light* (2017) by Jeanine Michna-Bales provides an critical re-reading of how the present and the past of darkness may be reconsidered and understood. Following a route from the cotton plantations of central Louisiana, via the swamps of Mississippi and the plains of Indiana, to the Canadian border, the photographic series imagines a journey along the Underground Railroad as it might have appeared to a freedom seeker. Comprising a route of nearly fourteen hundred miles, the images evocatively depict a considerable array of different places after dark as they may have been encountered by those in search of freedom, an estimated one hundred thousand slaves, between 1800 and 1865. The photographs bring a different dimension to this important period of American history, drawing from written and oral historical accounts, they situate the viewer into the dark, mysterious and daunting possible journey to freedom. They collectively form a necessary document that reminds us of where we have been so that we might better understand both where we are and where we might go.

In projects such as these, as Bruno Lessard (2018: 63) observes, photography is able to 'question documentary media's perennial investment in the diurnal regime of visibility', which he suggests when considered from the perspective of darkness, 'reveals a dark photology fundamentally framed by pressing environmental, urban, and sociohistorical challenges'. This work is relevant in this context for its contribution toward how alternative conceptualisations for place after dark are made. It challenges the fallacy of binaristic framings of day and night (Gallan and Gibson, 2011) and reveals some of the differences to be found within darkness. Different readings of place after dark are essential in their role as stimulus toward being able to imagine alternatives. The use of images is vital to this process of reenvisaging darkness; its diversity and plurality. This is because of the agency of images to offer framings and interpretations of ideas effectively, as powerful communication devices that are both immediate and also rich in the level of information they can contain.

Such alternative formulations are imperative. The need to rethink how we live with darkness is a global challenge that requires urgent attention (Davies and Smyth, 2018; Pritchard, 2017). Key to this is the development of more effective strategies for urban lighting in order to reduce the deleterious and severely harmful effects of excessive artificial illumination upon our health, that of other species and the environment, and the attendant waste of valuable energy resources. Current patterns of use and trends in consumption suggest a very different picture globally as quantity appears to be prioritised over quality (Bille and Sørensen, 2007). Despite the various advancements made in the efficiency of light technologies, both energy usage for outdoor lighting and artificial night-time brightness have increased annually (Kyba, Hänel, and Hölker, 2014). The increasingly widespread installation of LED lighting for urban environments around the globe is a major factor here, being typically used to enable brighter lighting at no greater cost, rather than dimmer lighting at a lower cost. The impact of the blue-white colour of the light it commonly emits has also been the subject of scrutiny with regard to its harmful biological effects on humans and non-humans alike, though Pawson and Bader (2014) have demonstrated that LED lighting causes considerable ecological problems in all its variations. Clearly, there is much work still to be done in how we might better understand darkness and its complex relationship to place (Dunn, 2019), though the re-emergence of a consideration of the effects and aesthetics of urban lighting as a matter of public debate is encouraging (Meier et al., 2015). The next section considers a number of theoretical approaches through which place after dark may be apprehended and understood.

Making sense in and of the dark

Thinking about places after dark necessarily raises questions of belonging, identity, demarcation, and appropriation via the reciprocal relationship between the specific site conditions of the place and the behaviours these either promote or prohibit (Cresswell, 1996). The diversity and plurality of dark places, their differences and reinterpretations, recalls Doreen Massey's (1994) discussion regarding the identity of a place as being 'open and provisional'. There are two sets of dynamics at play here, sometimes in tandem and sometimes not. First are the processes and changes, ecological and otherwise, occurring within the place itself which results in its identity evolving. Second are the definitions and reformulations of the place's identity as a result of those that encounter it, move through it, perform or transgress within its boundaries. These two sets of dynamics are interesting in that they may be accretive, mutually interdependent, or unrelated. However, irrespective of their relationship with each other, they contribute to the continual emergence of place during the daytime and after dark.

Key here is the 'who' in such processes. Jacques Rancière (2009 [2000]: 13) explains how making sense of a sense is inherently political since it concerns, 'what is seen and what can be said about it, around who has the ability to see and the talent to speak, around the properties of spaces and the possibilities of time'. When transposed to a place after dark, this perspective enables the mobilisation of the variety of belongings that are situated in, relational to, and may also co-exist within a specific context. It is also important to remind ourselves of the non-human actions and routines that co-constitute the identity of a place. Nocturnal behaviours and rhythms of flora and fauna inscribe themselves, largely undetected, into the character of a place after dark, their patterns, frequency, and intensity dependent on the climate and ecology of a situation and its surroundings. Building upon this perspective and the implicit notion of non-human speakers in Rancière's work, Bennett (2010: 106–107) suggests,

> a political act not only disrupts, it disrupts in such a way as to change radically what people can "see": it repartitions the sensible, it overthrows the regime of the perceptible. Here again the political gate is opened enough for nonhumans (dead rats, bottle caps, gadgets, fire, electricity, berries, metal) to slip through, for they also have the power to startle and provoke a gestalt shift in perception.

It is because of elements such as these that the identity of nocturnal places often appears distinctly more contemplative, eerie and enchanting than in the daytime. They can feel strangely out of kilter with the rest of the world, where humans construct a sense of order through patterns and behaviours. This is because places are partly defined by the social relations that occur both in and across them, which at night are far less frequent as many spaces are devoid of people. In their absence, other creatures are freer to go about their routines undisturbed by humans or may take temporary ownership of such places if they are nocturnal species.

The originality and value of the nocturnal landscape for thought and creativity to flourish has been the subject of recent attention (Dunn, 2016; Foessel, 2017; Stone, 2018). Michaël Foessel (Ibid: 151) refers to the potential for a different 'regime of sensory experience' when out in the dark to support such processes. Walking in gloomy landscapes, urban or otherwise, offers a useful practise to experience and understand place (Bogard, 2013; Edensor, 2013; Yates, 2012). At the edge of the city, artificial illumination is less able to compete with the firmament of night. Due to the relative reduction in direct and diffuse light compared with inner-urban areas, there is typically a wider array of dimness and darkness, which in turn may render the identity of a place to feel smudged. This palpable looseness is co-constituted by the multi-sensory experiences of the nightwalker combined with the liminal zones of the city. This hinterland between what is known and visible and that which is not also supports walking in gloom as a means to speculate on the future. There is a distinct character to the outlying areas of cities, the not quite urban which blends effectively with the not quite dark. Within these peripheral urban places after dark the overlapping thresholds of personal identity and that of place can powerfully shape the notion and experience of the 'nocturnal sublime' (Stone, 2018). This entanglement is not simply physiological but also psychological. The rules and regulations of the city appear to dissolve here, a world beyond which always hovers at its edge: tangible but somehow always just out of reach. The process of becoming is never fulfilled but is one of continual emergence; a deep and tangled relationship between identity and place, open and provisional, in part due to non-human others. This 'staying with the trouble' (Haraway, 2016) amidst the silhouetted and crepuscular flora and fauna is very much concerned with being alive and alert to one's surroundings. By reconfiguring the sensible, edgeland places after dark provide multi-sensory experiences of dimness and darkness that offer glimpses of how we might better account for such diversity with design.

Walking after dark to experience and understand place

Over the last six years I have spent many hours walking through various urban centres and peripheries after dark, interested in how my physical and psychological relationships with place alter amongst variations of dimness and darkness. This set of experiences has led to a very specific and personal view. However, it is presented here as a contribution toward a growing body of work that has argued for a broader range of perspectives on gloom, daylight and illumination (Edensor, 2017; Gallan and Gibson, 2011; Shaw, 2018). In contrast to the more controlled and regulated night-time environments of city centres, urban peripheries are usually far less planned, managed or populated. As a result, I contend that these places after dark are able to offer what Tim Edensor (Ibid: 125) describes as, 'previously unanticipated ways of apprehension, soliciting perceptions that expand the capacities for imagining and sensing place otherwise, such approaches extend the compendium of ways of seeing'. In order to capture some of the different atmospheres and ambiances of place after dark, the following section presents a combination of my auto-ethnographic fieldwork and images of specific sites to assist the reader's understanding of the experience (Figures 13.2 – 13.5). These describe four places along a walk of Manchester's urban periphery after dark, taken on 4 April 2018. Leaving my house just after 11:00 p.m., the fourteen-mile walk lasted about six hours.

Figure 13.2 Sale Water Park, 4 April 2018. © Nick Dunn.

Figure 13.3 Banky Lane, Carrington, 4 April 2018. © Nick Dunn.

Figure 13.4 Bridgewater Way, Sale, 4 April 2018. © Nick Dunn.

Figure 13.5 M60 motorway underpass, Worsley, 4 April 2018. © Nick Dunn.

Walking out of the city's inner suburbs, I follow a rough arc between the meadows and the woods, less precise than the orbital motorway in the distance. Across Jackson's Boat Bridge, I become immersed again in woodland, the path moist and smeared underfoot. Emerging from the nearby field, the dark mirrored expanse of Sale Water Park stretches out ahead and either side. It appears to pull the surrounding landscape toward it, trees and hedgerows all hunkered down together in charcoal hues. The only exception to this gloomy forcefield are the electricity pylons; their angular filigree and overhead cables resisting the earth. The sky is awash with dark greys, murky blues and yellows. Waterfowl ruffle about near the water's edge, aware of my presence. What is evident is that, although very dim, there is enough light to be able to navigate around, to see and be seen. The vastness of the space makes me feel exposed, my straggling silhouette out of sync with the flat landscape. I walk along the concrete jetty and watch my own dark and oscillating reflection try to look back at me before threading under the motorway and into the roll and tumble of Priory Gardens.

Following the sodium-fused perimeter of Ashton upon Mersey replete with its rhythmic orangey glows, I head towards where the Box and Cox arrangement of outer suburbia yields to the landscape of agriculture, meadows, and sports clubs. Upon arrival, the nocturnal atmosphere along this seam feels uncanny, as if parts of two maps of different scales have been stuck together. At Banky Lane, the low buzz of the nearby substation hums steadily, a white noise lullaby for the two HGV drivers asleep in their cabins some yards away. The one illuminated signpost leaches the green from nearby flora into view, a sharp juxtaposition against the blanket of pinks and purples of the clouds above. The serene ambiance is disrupted by the oncoming noise of a motor car behind me. Shifting gears, it slows down to take the corner, crunching on the gravel underneath. The two occupants briefly staring dead-eyed at me before driving to the lock-up garages further down the lane. Retracing my steps, I head towards the canal.

As I walk along the Bridgewater Way, the canal sits calmly by my side, a parallel strip of sky framed in the ground. The sandy and stony path powders my boots in visual protest to their crunches. Geese slide silently against the canal's opposite edge. Launching out of the barge in front, two Jack Russells have a lot to say about my being here, guarding the invisible boundary of the territory they have decided is theirs to protect. Skirting around their domain, I am drawn into the undertow of where the canal passes beneath the motorway. The experience of illumination here is one of distinct contrasts; sharp yellows from the highway lighting slicing into the black underneath, the whole array suddenly reconfigured then reassembled by the headlights of a passing car overhead. The deep colour of the illumination, and lack of it, seems to be almost hewn from the materials it touches, a temporary sculpture carved by place and time as infrastructures overlap. Losing oneself in the slow time of such moments, the identity of place merges with the body, as if trapped in amber. It takes some determination to move on from such a beguiling display.

Continuing along the canal towpath, I weave through industrial estates, retail parks, and eventually back into the outer suburbs to the north of the city. Along the way, the firefly ends of cigarettes dance, belonging to a huddle of nightshift workers in the floodlit corner of a factory's car park. Beside this, all other movements and sounds aside from my own are either non-human in the foreground or the rumblings of unseen and distant vehicles. Striding up into Worsley Woods past the Old Warke Dam, the trees crowd into view on either side, leaves and mud slither against my boots, a trail of soft squelches left behind. Walking downwards between a field and woods, the lane pulls my feet towards the powerful glow of an underpass. Against the bruised peach-blues of the sky above, this otherworldly tunnel of light is solid and precise. Walking between the sentry of spindly trees and gloomy undergrowth either side, I enter the mass of colour, dissolving in its yellow-green tones as if it were a James Turrell installation. The unintended interplay here between darkness and light offering a different identity to the place and a new way of seeing it.

Designing with darkness

The methods presented here are part of a foray into examining and experimenting the reciprocity between our senses and place after dark. It is a nascent body of work that seeks to build different knowledges and understandings of the complex identities of place after dark, their distinct qualities and their coexistences. However, it is also important to remember that our senses are culturally conditioned, alongside with how we perceive darkness, since they are bound up in specific historical, geographical, and social contexts. Coincidental to the frequency of my walks after dark increasing in early 2014 was the announcement that Manchester City Council (2014) was to carry out a comprehensive replacement of fifty-six thousand street lamps with new LED lights. Several years later this process is well underway, and it is evident that the character of many areas of the city has already changed. Of relevance here is that this new layer of lighting technology will obstruct, or at least hinder, direct experience of a wider variety of sensible qualities of darkness that are currently accessible. This raises questions about the kind of experiences available in our urban landscapes and whether there are better ways to design with darkness.

In their book *Nightscapes*, Armengaud, Armengaud, and Cianchetta (2009) explore how night redefines the framework of thought and action in the realm of the imaginary, of territorial planning and of the practice of landscape. For their investigation, they ask (Ibid: 16), 'What is a specifically nocturnal public space? On the basis of which principle does one develop a project? If night means the ephemeral, the fragile, the spontaneous, how does one construct this element without distorting it?' As many cities like Manchester employ lighting techniques to extend their commercial offer, they promote particular atmospheres and ambiances that occlude alternatives. Our experiences of such places after dark powerfully shape our perceptions of gloom and in many ways reduce our ability to comprehend the diversity of elements, materialities, and sensations that darkness enriches. We know that such energy usage is not sustainable given the serious impacts urban lighting can have, as discussed earlier.

This situation has not gone unnoticed. *The Dark Art Manifesto* (2014) by Lowe and Rafael aims to create 'a lighting design philosophy that promotes a balanced view of light and darkness to enhance freedom whilst incorporating rational standardisation'. More recently, Roger Narboni (2017) has made a plea for cities to make use of 'dark infrastructures', to protect and preserve darkness and support green spaces and blue areas such as parks, canals, and rivers, by focussing their attention away from illumination toward a 'nocturnal urbanism'. Whether many cities are ready and willing to adopt the 'greenouts' as described by Nye (2010: 205–232), voluntary and temporary reductions in power consumption which provide a dim rather than dark urban environment, remains to be seen. What is clear is that faced with the challenges that are already changing the nature of place at a global scale, the need to better understand different histories, cultures, and practices for place after dark is more pressing than ever. Framed in this manner, I have aimed to show how, through the images of place after dark produced by an emerging body of contemporary photographers and my own ongoing experiences with the potential of urban peripheries as sites for experimentation and imagination, new conceptualisations and visions are possible. This is of critical importance if we are to develop wider and deeper knowledges of the situated, relational, and practised nature of place after dark, and evolve alternative futures for its preservation.

References

Armengaud, M., Armengaud, M., and Cianchetta, A. (2009). *Nightscapes: Paisajes Nocturnos/Nocturnal Landscapes*. Barcelona: Editorial Gustavo Gili.

ATCM. (n.d.). *Purple Flag Status: how It Fits Place Management Policy*. Available at: www.atcm.org/purple-flag

Bennett, J. (2010). *Vibrant Matter: A Political Ecology of Things*. Durham, NC: Duke University Press.

Bille, M., and Sørensen, T. F. (2007). An Anthropology of Luminosity: The Agency of Light, *Journal of Material Culture*, 12(3), 263–284.

Bogard, P. (2013). *The End of Night: Searching for Natural Darkness in an Age of Artificial Light*. London: Fourth Estate.

British Standards Institute (2013). *BS5489 1:2013 Code of Practice for the Design of Road Lighting: Lighting of Roads and Public Amenity Areas*. Available at: www.standardsuk.com/products/BS-5489-1-2013

Cohen, T. (2012). *Villes éteintes*. Paris: Marval.
Cresswell, T. (1996). *In Place/Out of Place: Geography, Ideology, and Transgression*. Minneapolis, MN: University of Minnesota Press.
Davies, T.W., and Smyth, T. (2018). Why Artificial Light at Night should be a Focus for Global Change Research in the 21st Century, *Global Change Biology*, 24(3), 872–882.
Dowd, M., and Hensey, R. (eds). (2016). *The Archaeology of Darkness*. Oxford: Oxbow.
Dunn, N. (2016). *Dark Matters: A Manifesto for the Nocturnal City*. Winchester: Zero Books.
Dunn, N. (2019). Dark Futures: The Loss of Night in the Contemporary City? *Journal of Energy History/Revue D'histoire De L'énergie*, Special Issue: Light(s) and darkness(es)/Lumière(s) et obscurité(s). 1(2), 1–27.
Dunn, N., Cureton, P., and Pollastri, S. (2014). *A Visual History of the Future*. London: Foresight Government Office for Science, Department of Business Innovation and Skills, HMSO.
Dunnett, O. (2015). Contested Landscapes: The Moral Geographies of Light Pollution in Britain, *Cultural Geographies*, 22(4), 619–636.
Eckersley, W. P. (2011). *Dark City*. London: Stucco.
Edensor, T. (2013). Reconnecting with Darkness: Experiencing Landscapes and Sites of Gloom, *Social and Cultural Geography*, 14(4), 446–465.
Edensor, T. (2015). The Gloomy City: Rethinking the Relationship between Light and Dark, *Urban Studies*, 52(3), 422–438.
Edensor, T. (2017). *From Light to Dark: Daylight, Illumination, and Gloom*. Minneapolis, MN: University of Minnesota Press.
Foessel, M. (2017). *La Nuit: Vivre Sans Témoin*. Paris: Editions Autrement.
Gallan, B., and Gibson, C. (2011). New Dawn or New Dusk? Beyond the Binary of Day and Night, *Environment and Planning A: Economy and Space*, 43(11), 2509–2515.
Gandy, M. (2017). Negative Luminescence, *Annals of the American Association of Geographers*, 107(5), 1090–1107.
Gonlin, N., and Nowell, A. (eds). (2018). *Archaeology of the Night: Life After Dark in the Ancient World*. Boulder, CO: University of Colorado.
Haraway, D. (2016). *Staying with the Trouble: Making Kin in the Chthulucene*. Durham, NC: Duke University Press.
Koslofsky, C. (2011). *Evening's Empire: A History of Night in Early Modern Europe*. Cambridge: Cambridge University Press.
Kyba, C., Hänel, A., and Hölker, F. (2014). Redefining Efficiency for Outdoor Lighting, *Energy and Environmental Science*, 7, 1806–1809.
Lessard, B. (2018). Shot in the Dark: Nocturnal Philosophy and Night Photography. In Cammaer, G., Fitzpatrick, B., and Lessard, B. (eds) *Critical Distance in Documentary Media*. Cham: Palgrave MacMillan, pp. 45–67.
Lowe, C., and Rafael, P. (2014). The Dark Art Manifesto, *Professional Lighting Design*, 91, 64–65.
Manchester City Council. (2014). *Street Lighting LED Retrofit Programme*, Executive Report. 12th February. Available at: www.manchester.gov.uk/meetings/meeting/2042/executive/attachment/16500
Massey, D. (1994). *Space, Place and Gender*. Oxford: Polity Press.
Meier, J., Hasenöhrl, U., Krause, K., and Pottharst, M. (eds). (2015). *Urban lighting, Light Pollution and Society*. New York: Routledge.
Melbin, M. (1987). *Night as Frontier: Colonizing the World after Dark*. New York: The Free Press/Macmillan.
Michna-Bales, J. (2017). *Through Darkness to Light: Photographs along the Underground Railroad*. New York: Princeton Architectural Press.
Narboni, R. (2017). Imagining the Future of the City at Night, *Architectural Lighting*. Available at: www.archlighting.com/projects/imagining-the-future-of-the-city-at-night_o
Nye, D. E. (2010). Greenout? In D. E. Nye (ed.), *When the Lights Went Out: A History of Blackouts in America*. Cambridge, MA: The MIT Press, pp. 205–232.
Pawson, S. M., and Bader, M. K-F. (2014). LED Lighting Increases the Ecological Impact of Light Pollution Irrespective of Color Temperature, *Ecological Applications*, 24(7), 1561–1568.
Polak, F. (1973). *The Image of the Future*. Trans. E. Boulding. Amsterdam: Elsevier Scientific Publishing Company.
Pritchard, S. B. (2017). The Trouble with Darkness: NASA's Suomi Satellite Images of Earth at Night, *Environmental History*, 22, 312–330.
Rancière. (2009 [2000]). *The Politics of Aesthetics*. Trans. G. Rockhill. New York: Continuum.

Schivelbusch, W. (1988). *Disenchanted Night: The Industrialization of Light in the Nineteenth Century*. Berkeley, CA: University of California Press.
Schlör, J. (1998). *Nights in the Big City*. Trans. P. G. Imhof and D. R. Roberts. London: Reaktion Books.
Shaw, R. (2018). *The Nocturnal City*. London: Routledge.
Stone, T. (2018). Re-envisioning the Nocturnal Sublime: on the Ethics and Aesthetics of Nighttime Lighting, *Topoi*. 10.1007/s11245-018-9562-4
Thompson, J. B. (1984). *Studies in the Theory of Ideology*. Cambridge: Polity.
Williams, R. (2008). Nightspaces: Darkness, Deterritorialisation, and Social Control, *Space and Culture*, 11 (4), 514–532.
Yates, C. (2012). *Nightwalk*. London: Collins.

14

Waterway

A liquid place

Maarja Kaaristo

Inland waters include tidal and non-tidal rivers, canals, lakes and some estuaries; they can be navigable for vessels of different sizes, but many are also disused, derelict, culverted, filled in or simply forgotten. The linear waterways can form networks, which run through landscapes, having sometimes been literally cut into them, as is the case with canals. They are physical and mythical, and can be sources of life and death, abundance and destruction (Mauch and Zeller, 2008). Before the modern period, rivers were essential as a means of transport for people and cargo and successful management and control of the water of the great rivers, such as the Nile, Amazon, Yangtze, Mississippi and Yenisei, have been fundamental to the development of human civilizations. As innkeeper and pamphleteer 'The Water Poet' John Taylor wrote in the 17th century, 'There is not any one Town or City, which hath a Navigable River at it, that is poore, nor scarce any that are rich, which want a River, with the benefits of Boats' (quoted in Willan, 1964: 7).

In addition to naturally occurring rivers, canals were already constructed for both irrigation and transport in the Mesopotamian, Egyptian and Roman empires, as well as in China. The latter's Lingqu Canal (the 'Magic Canal') in Guangxi (from 3rd century BC) is thought to be the first contour transport canal (that is, following the contour line of the land). China's Grand Canal (begun in the 5th century BC and thereafter mainly constructed between the 6th and 7th centuries AD) featured many important technical features, such as sluice gates, winches, pumps, round weirs and pound locks with vertical gates (Crompton, 2004). Nevertheless, drawing a line between a constructed and a 'natural' waterway is not simple, since improvements on rivers, connecting rivers and canalising them by dredging and making bank improvements has been a steady fixture of landscape alteration throughout human history, creating particular hybrid hydro-landscapes (Swyngedouw, 2015). Britain's first dead water canal with pound locks, the Exeter Canal (1566), was designed to bypass a complicated stretch of the river Exe. In France, Canal de Briare (1642) joined the Rivers Loire and Seine, and Canal du Midi or the Languedoc (1681) connected the Garonne with the Aude, which in effect meant connecting the Atlantic with the Mediterranean (Mukerji, 2009). The canals, constructed bodies of water, serve as a means of connecting natural bodies of water with each other and with various places, but they also connect human and non-human animal bodies, and it is in these relations that the waterways become places themselves.

In this chapter I discuss waterways as liquid places in the United Kingdom, with its current network of over 3,000 miles (5,000 km) of navigable and 1,800 miles (nearly 3,000 km) of non-navigable canals and rivers. The UK's 'Canal Age' started in 1761 with the construction of Manchester's Bridgewater Canal. The cargo-carrying capacity of the canals was one of the key enablers of the Industrial Revolution, as the waterways allowed reductions in the price of coal and subsequently carried various bulky cargoes easily across the country (Bagwell and Lyth, 2006). Their construction (itself a major driver of economic development), maintenance and use was a collective undertaking of a variety of human and non-human actors and actants, and collaborations of different groups, which resulted in increased social or physical mobility (or both) for many (Mukerji, 2009; Matthews, 2015). The key actors and actants included (but were not limited to) the workers who physically dug the canals, the engineers, the financiers, the 'boatpeople' living and working on them, the landscapes into which the canals were physically cut, the built infrastructure from lock chambers to bridges, the water in the canals, the narrowboats and barges and the horses towing them.

With the introduction of railways and over the 19th century, British canals largely fell into disuse and disrepair. However, they were transformed during the 20th century, repurposed from transport to tourism use, from an obsolete infrastructure into an experiencescape (Olsson, 2016); becoming a landscape of heritage, as well as a heritage landscape (Prideaux, 2018). After the 1944 publication of L.T.C. Rolt's widely read travelogue *Narrow Boat* (2014), waterways became part of a new narrative, that of leisure time spent in a simultaneously romanticised rural Arcadia as well as industrial heritage, thereby functioning as time machines, transporting boaters – slowly – back in time (Fallon, 2012; Boughey, 2013). Following extensive restoration projects, contemporary waterways are multi-use spaces for various individuals, groups and stakeholders, including boaters, walkers, joggers, anglers and cyclists, navigation authorities and volunteer organisations, but also for non-human animals, from wildlife at home on the waterways to domesticated animals and pets.

Canalised rivers and constructed canals are human fluvial modifications of the environment, resulting in the creation of socio-natural hydro-landscapes (Vallerani and Visentin, 2018). Throughout history, canals have passed through a number of stages, transportation, dereliction, dwelling and leisure, which have been foregrounded at different times. Inland waterways form networks connecting a variety of different places as they pass through various landscape types, forming 'a linked entity that includes the use of space, adjacent land, building forms and styles, agricultural and city life, natural ecosystems and cultural meanings that emerge from unique patterns of use, value and visualness' (Prideaux, 2018: 150–151). Yet this network itself is a place too, featuring a geographical location, the material setting of a locale and a particular emotional relationship with history and landscape (Agnew, 1987). Waterways are liquid, sometimes linear places that are simultaneously relational and material, but always in movement. Focussing on British canals, I discuss the inland waterways as liquid and hybrid places that come to exist through three key elements: the slow mobilities of boats, embodied watery materialities and convivial interactions on and by the water.

Slow boatmobility

'Waters take place. The movements of water in our daily lives link us to places and to each other' (Chen, 2013: 275). Yet these waters also constitute places. Mobility, socially produced motion (Sheller and Urry, 2006) on the waterways, is the result of a myriad of physical patterns of movement between different locations, various representations of movements, as well as the embodied ways of practicing and sensing the movement. Mobility, therefore, is not merely functional, allowing for the transport of bodies or cargo from one location to another, but it should instead be thought about in terms of the meaningful movement of people, goods, information and ideas (Cresswell, 2006). Water is all about mobility: human bodies always carry the water within them wherever they go (Neimanis, 2017) and, in turn, watery bodies carry human bodies and their boats.

One of the defining elements of the physical movement of people and boats on inland waters is the tempo, speed, pace and intensity of various activities (Adam, 2004). Both human and non-human mobilities in space–time are integral to the emergence of places, all part of 'the spatialisation of time and temporalisation of space' (Cresswell, 2006: 4). One of the most distinctive temporal characteristics of British inland waterways is that they are often perceived as slow. The material dimensions – the width and depth – of the mostly narrow canals determine the maximum speed of boats: 4 mph (6.4 km/h). Rivers allow for a slightly higher speed; for example, the limit on the non-tidal Thames is 5 mph (8 km/h). However, these formalistic rules only apply to an extent on the waterways where the code of conduct is shaped by lived experience. The official boaters' handbook states, 'Whatever the limit, if you make waves you're going too fast – slow down' (CRT, 2014: 52). This speed limit, equivalent to a walking pace, is punctuated with other practices such as stopping and temporarily mooring up in order to operate the locks and waiting for them to fill or empty, creating a rhythm rooted in habitual and routine boat-specific practices (Kaaristo, 2020).

The boaters who (semi)permanently dwell on the waterways tend to perceive time as simultaneously slow and unpredictable, close to and in harmony with the natural environment, where the presence of water elicits a sense of calm, quiet and tranquillity (Smith, 2007). The mobile and location-independent live-aboards, resisting dominant social temporalities, characterised by time–space compression and acceleration (Virilio, 2012), present a direct challenge to the terra-centrism of contemporary nation states – from the water, land is seen as the Other, not vice versa (Bowles et al., 2019). Their slow pace of life forms an important part of the boat dwellers' identity and results in a simultaneously political and utopian 'boat time' (Bowles, 2016). Boating as a practice, therefore, also relates to wider notions of slow living and, in the case of holiday boaters, slow tourism. As Fallon (2012: 149) contends, 'slowness is synonymous with canals, in fact, canal boaters are seen to be people who have chosen to slow down'. Slow living is a matter of choice and boat dwellers often make a conscious effort to change pace in order to 'slow' (Vannini, 2014), which has to be done in a correct and socially accepted manner. This includes following both written and unwritten boating rules, as well as a number of mundane practices, such as preparing food and drink or using the bathroom, for which boaters have to achieve and manage a number of bodily, technical and material alignments that attune their bodies to the slow tempo of waterscape (Kaaristo and Rhoden, 2017).

Mobility is also as much about immobility as about moving, as numerous 'stillnesses pulse through multiple ecologies with multiple effects' (Bissell and Fuller, 2011: 3). Immobility in terms of boating activities can be involuntary, as is the case if the boat's engine breaks down, the fuel runs out, or the boat becomes immobile because of the changing state of the water, for instance, if it freezes in winter. At the same time, stillness can become an opportunity, a potential and the boat a place to experience a subjective 'state of consciousness characterised by calmer mental rhythms and a shift in attention … towards the present moment' (Conradson, 2011: 72). The variety of these mundane mobile activities – steering the boat, sitting in the bow, resting in the kitchen, lying on the bunk, waiting around for the water tank to fill or a lock to empty or fill – can therefore be conceptualised as mobile stillness: a simultaneous mobility and immobility. Consequently, boating as an activity is a conglomeration of all the mobilities and immobilities discussed, and is best characterised by the notion of a stop-start tempo, where long and relatively peaceful stretches of time interchange with the more pressing and urgent activities (Bowles, 2016).

The watery slowscape is produced by a conscious deceleration, realised in a variety of embodied activities and practices. A locale can be associated with a particular pace of life, turning it into a timescape with specific temporal properties (Kaaristo and Järv, 2012). In phenomenological terms, both space and time arise from the experience of place (Casey, 1996) and the waterways become a 'liquid chronotope' (Peterle and Visentin, 2018), both symbolically-narratively in terms of a sense of place as well as materially-geographically as a location and locale (Agnew, 1987). The mobilities on and near water and the subsequent tempo of life lived on the water is of key importance here. However, in addition to the movement of people and boats in time, we also need to consider the movement of the key materiality of waterways – the water – since the capacity to successfully 'slow' (Vannini, 2014) can only emerge through cooperation with its properties and affordances.

Sensory and affective watery materialities

Water, the central materiality of what makes a waterway a place, is 'simultaneously an element, a flow, a means of transport, a life-sustaining substance, a life-threatening force, the subject, object, and often the very means of social and cultural activity' (Krause and Strang, 2016: 633). Water is a complicated and contextual substance: simultaneously embodying and embodied (Neimanis, 2017), ecological and socio-cultural; it is a process, equipped with a remarkable capacity to connect people, things and ideas (Linton, 2010). Waterways and their immediate surroundings, riverbanks, canal towpaths and lakesides, serve as bird and wildlife habitats. From a human perspective, freshwater ecosystems provide us with a water supply for drinking, essential for our survival. Our lives are inextricable from water; we need it for household and industrial uses, irrigation and aquaculture, and it supplies humans with sustenance in terms of fish, waterfowl, clams and mussels. Furthermore, water provides a variety of uses as diverse as transport, recreation, electricity generation, soil fertilisation, flood control, pollution dilution and enhancement of property values of waterside (re)developments.

Freshwater blue spaces also provide a number of opportunities for humans to act in various ways, including walking, running and cycling on towpaths and banks, angling, boating, kayaking, canoeing or paddle-boarding on the water as well as swimming in it. The linear waterways can be simultaneously rural, urban and wild, and in the context of both leisure and everyday life, 'water "reproduces" individuals and groups through creative engagement with a cultural land and waterscape' (Strang, 2009: 194) with sensory and aesthetic experiences playing a central role. Public spaces featuring water are seen as beneficial for social interaction as well as general wellbeing (de Bell et al., 2017), and the agency of water is central to how waterways emerge as places. The potency of watery materialities in place-making becomes apparent in their complex and cyclical relationships:

> We are literally implicated in other animal, vegetable, and planetary bodies that materially course through us, replenish us, and draw upon our own bodies as their wells: human bodies ingest reservoir bodies, while reservoir bodies are slaked by rain bodies, rain bodies absorb ocean bodies, ocean bodies aspirate fish bodies, fish bodies are consumed by whale bodies – which then sink to the seafloor to rot and be swallowed up again by the ocean's dark belly. This is a different kind of 'hydrological cycle'.
>
> *(Neimanis, 2017: 3)*

In discussions of experiences of water, the visual is often highlighted: 'like the hypnotist's flickering candle or swirling optical images, the visual qualities of water are indeed mesmerising' (Strang, 2004: 52). However, there is not just one generic (inland) form of water: the water of different waterways possesses different visual properties, and experienced observers may determine their location by the water's appearance. For example, a section of the Bridgewater Canal in Worsley (Greater Manchester) has a specific rusty orange colour due to the iron ore from the nearby abandoned mines. Some canals, for example the Rochdale and Ashton Canal in Manchester city centre, have an opaque, mud-coloured water, whereas in the Chesterfield Canal the water is particularly clear, like in a stream. The colour of the water can be a point of visual enjoyment, a geographical reference point and can signify a nostalgic, temporal modality (the past, the present or the imagined future); in short, the watery materialities are key contributors to place creation.

Equally importantly, water is experienced haptically; often our human bodies interact with and immerse themselves in water bodies, prompting sensations that broaden and amplify a sensory awareness of the environment. Outdoor (or wild) swimming is framed in emotional, affective and embodied ways, and has considerable health and wellbeing benefits (Foley, 2017) as the substance of water enters into direct and close relation with human bodies. Additionally, the visual consumption of water often depends on its perceived 'naturalness', the high concentrations of suspended particles in water that make its visual attributes bright and therefore attractive to humans (Smith et al., 1995). The canals, however, present a certain contradiction: whereas consuming the canal water visually and aurally is perceived as pleasurable, physical contact can evoke feelings of fear and disgust. While discussions on the sensory experiences of water often focus almost exclusively on its near universal enjoyability, still water in the canals can be perceived as a contaminant and a pollutant when it comes into direct contact with the human body. A recent study by Pitt (2018) highlights the need to think about the 'wateriness' of water and accept that luminescent blueness is not water's inherent property; sometimes, especially in urban areas, water can also be grey or brown, opaque or stagnant.

The human relationship with watery materialities is therefore twofold: it can be a provider of enjoyment and general wellbeing, but also represent an array of dangers or aversion and repulsion. These embodied experiences of water are both culturally and materially conditioned, emerging from relations between people as perceiving subjects and the environments they interact with through particular embodied actions (Edensor, 2006). The water is thus a source for enjoyment, affording the experience of tranquillity, as well as provoking emotions of anxiety and worry. These immediate and intense engagements with water are 'predisposed to enable affective relationships' (Strang, 2009: 32) and the combination of various sensory stimuli intertwine in shaping the experience of place (Crouch, 2000).

Conviviality on and by water

As a 'locale', the waterways are constituted through countless networks of everyday, routine social relations (Agnew, 1987), with different actors and stakeholders coming together. There is a sizeable community of 'live-aboard boaters' on British canals and rivers, the people who permanently live on their vessels: narrowboats, river cruisers or barges. Some own boats as second homes and some hire holiday boats for shorter periods. Current and former holiday boaters, live-aboards and those on board the occasional hotel and hostel boats, as well as the workboats, with their different levels of experience, knowledge and interest in the canals are all members of the boating community (Smith, 2007; Bowles, 2016; Roberts, 2019). They share a certain identity – that of a boater, which is realised through the numerous everyday convivialities of 'person-or-people-experiencing-place' (Seamon, 2013: 150) that occur on and near water and are centred on the practice of boating.

Boating is an affectual form of sociation and its participants form mobile groups, which are fluid, occasional and characterised by their sporadic gatherings and dispersals, as well as the shared experience, emotions and affect. Various intense interactions take place on board boats in the close physical proximity caused by the limited space where movement is restricted and privacy scarce (Kaaristo and Rhoden, 2017). This allows us to understand the numerous social relations as 'composite ways of spatialisation', attending to the internal relation between spatial territory and constant mobility (Brighenti, 2014: 3). The boundaries of the public and private dissolve, change and obtain new meanings on the water, and the slowly moving boat becomes an affective mobile place where numerous collaborative choreographies, patterns of movement and temporal schedules are constantly renegotiated.

Membership of this fluid community comes partly from self-identification, acceptance of other community members, and partly through practices of coming together on the waterways in numerous ways. Convivialities of a watery place are forged by the personal trajectory of each of its users, with the centres and peripheries continuously moving and shifting, depending on the subjective perspective. For instance, the live-aboard boaters' claim to the waterways is often spatio-temporal, stemming from dwelling on the boats and their imagined link with the 'boat people' who lived and worked on the canals from the Industrial Revolution onwards (Bowles, 2016). For others, group identity can derive from the physical act of navigating the boats on the waterways, the possession of the necessary skills and the knowledge and following of the unwritten code of conduct when boating. The membership can also emerge from volunteering on the waterways, contributing to the maintenance and upkeep of the canals (Trapp-Fallon, 2007), perceived as a form of commons. The waterways are produced through the mobile convivialities of all these actors and therefore cannot be defined merely through geographical, temporal or historical denominators.

Different places can co-exist simultaneously in the same space (Duarte, 2017), which can be territorialised consecutively by multiple actors. This results in a kaleidoscopic consumption of place where a riverbank, for instance, can be de-territorialised as an angling place and re-territorialised into a play area (Cheetham et al., 2018). This is not necessarily frictionless, since one group's territorialising processes can be contested by those appropriating the waterways in a different way. For example, from the boaters' perspective, anglers can sometimes be bothersome when they do not remove their fishing rods fast enough when boats cruise past. On the other hand, boats are supposed to slow down when passing anglers so as not to disturb the fish, and sometimes tensions arise when anglers feel that boaters have failed to do so. There are also long-standing conflicts between anglers and those on small non-powered vessels on the waterways, amplified by the UK's complex land and property rights (Church et al., 2007). The anglers hold a perception that canoeists disturb the fish, even though studies have concluded that this is not the case (Dudley, 2017). Likewise, conflicts can arise with the cyclists on the canal towpaths, who the boaters often perceive as going too fast and thus not subscribing to the much-valued ethos of slowness and the collective convivialities of shared space on the waterways, which can sometimes turn on itself and result in conflicts or tensions.

On the boats, on towpaths and riverbanks, lock-sides and canal-side pubs, people exchange information, acquire skills, complete their personal and individual trajectories and passages, and are socialised into fluvial norms and a value system. As a distinctive kind of place, waterways form through the constant territorialisation practices enacted by its multiple users. This results in an ephemeral and dynamic place where numerous relational micro-practices govern 'how spaces are actually used and what places mean' (Cheetham et al., 2018: 487). The waterway comes to exist as a convivial place through the diverse spatialities and mobilities of its various actors: the boats cruising on the canal network, the boaters observing (or failing to observe) the canal code of conduct, people walking, cycling or angling on the towpaths, wildlife as well as pets belonging to either passers-by or boaters. This fluid and perpetually mobile hydro-village is continuously co-constituted by numerous convivialities and sociabilities both on the boats as well as on the towpaths and riverbanks.

Conclusion: the hybridity of liquid places

In the 17th century the philosopher and mathematician Blaise Pascal wrote that 'Rivers are roads which move, and which carry us whither we desire to go' (in Middleton, 2012: 70). The linearity of the waterways is first and foremost relational – the lines and paths of waterways form numerous networks, connecting rural, urban and wilderness areas; they 'do not so much go back and forth as answer to one another, in a movement that goes not across but along' (Ingold, 2015: 151). The waterway, even if it often looks linear, is part of a network or a rhizome; the watery place is fluid and fluvial in its uniting of the physical, cognitive and socio-cultural. A whole host of many different actors come together on the waterways, all contributing to their 'throwntogetherness' (Massey, 2005), linking human to non-human and water to land. In the liquid place, land, water, human, animal, urban, rural, organic, non-organic, material and representational connect, intertwine and become interdependent, mutually constitutive and hybrid.

Therefore, as Chen (2013: 275) contends, 'it is neither that waters contain places nor that places contain waters: the specificity of situated waters articulated with places, with space and time, with dynamic bodies, materials and semantic contexts can enable a more thoughtful discussion of watery relations'. Thinking of the waterway as a place, we need to think about its spatial diffusion and linearity, and how it comes together as a 'fuzzy' place (Warnaby et al., 2010), where the boundaries (rural–urban, natural–constructed, water–land) are blurred and the place itself becomes practiced and performed, in other words, lived. Waterways should be conceptualised as liquid and relational occurrences generated by numerous dynamics, entanglements and interrelations between the convivial, material and mobile. Ingold maintains that '(T)he river will eventually open up into the sea, but it does not deliver its waters to a place' (Ingold, 2015: 151). Yet the waterway *is itself* a place – a liquid place, a socio-natural hybrid of convivial interactions, material relations and numerous mobilities.

References

Adam, B. (2004). *Time*. Cambridge: Polity Press.

Agnew, J. (1987). *Place and Politics: The Geographical Mediation of State and Society*. Boston: Allen and Unwin.

Bagwell, P. S. and Lyth, P. J. (2006). *Transport in Britain 1750–2000: From Canal Lock to Gridlock*. London and New York: Bloomsbury.

Bissell, D. and Fuller, G. (2011) Stillness unbound. In Bissell, D. and Fuller, G. (eds) *Stillness in a Mobile World*. London: Routledge, pp. 1–18.

Boughey, J. (2013). From transport's golden ages to an age of tourism: L.T.C. Rolt, waterway revival and railway preservation in Britain, 1944–54. *The Journal of Transport History*, 34(1): 22–38.

Bowles, B. (2016). 'Time is like a soup': boat time and the temporal experience of London's liveaboard boaters. *The Cambridge Journal of Anthropology*, 34(1): 100–112.
Bowles, B., Kaaristo, M. and Rogelja, N. (2019). Dwelling on and with water: materialities, mobilities and metaphors. *Anthropological Notebooks*, 25(2): 5–12.
Brighenti, A. M. (2014). Mobilizing territories, territorializing mobilities. *Sociologica*, 8(1): 1–25.
Casey, E. (1996). How to get from space to place in a fairly short stretch of time: phenomenological prolegomena. In Feld, S. and Basso, K. (eds.) *Senses of Place*. Santa Fe, NM: School of American Research Press, pp. 13–52.
Cheetham, F., McEachern, M. G. and Warnaby, G. (2018). A kaleidoscopic view of the territorialized consumption of place. *Marketing Theory*, 18(4): 473–492.
Chen, C. (2013). Mapping waters: thinking with watery places. In Chen, C., Janine, M. and Neimanis, A. (eds) *Thinking with Water*. Montreal and Kingston, London, Ithaca: McGill-Queen's University Press, pp. 274–300.
Church, A., Gilchrist, P., and Ravenscroft, N. (2007). Negotiating recreational access under asymmetrical power relations: the case of inland waterways in England. *Society and Natural Resources*, 20(3): 213–227.
Conradson, D. (2011) The orchestration of feeling: stillness, spirituality and places of retreat. In Bissell, D. and Fuller, G. (eds) *Stillness in a Mobile World*. London: Routledge, pp. 71–86.
Cresswell, T. (2006). *On the Move: Mobility in the Modern Western World*. New York: Routledge.
Crompton, G. (2004). 'The tortoise and the economy.' Inland waterway navigation in international economic history. *Journal of Transport History*, 25(2): 1–22.
Crouch, D. (2000). Places around us: embodied lay geographies in leisure and tourism. *Leisure Studies*, 19(2): 63–76.
CRT. (2014). *The Boater's Handbook. Basic Boathandling and Safety for Powered Boats*. Milton Keynes: Canal and River Trust in partnership with Environment Agency.
de Bell, S., Graham, H., Jarvis, S. and White, P. (2017). The importance of nature in mediating social and psychological benefits associated with visits to freshwater blue space. *Landscape and Urban Planning*, 167: 118–127.
Duarte, F. (2017). *Space, Place and Territory: A Critical Review on Spatialities*. New York and Abingdon: Routledge.
Dudley, M. (2017). Muddying the waters: recreational conflict and rights of use of British rivers. *Water History*, 9(3): 259–277.
Edensor, T. (2006) Sensing tourist spaces. In Minca, C. and Oakes, T. (eds.) *Travels in Paradox: remapping Tourism*. Lanham: Rowman and Littlefield, pp. 23–45.
Fallon, J. (2012). 'If you're making waves then you have to slow down': slow tourism and canals. In Fullagar, S., Markwell, K. and Wilson, E. (eds) *Slow Tourism. Experiences and Mobilities*. Bristol, Buffalo, Toronto: Channel View Publications, pp. 143–154.
Foley, R. (2017). Swimming as an accretive practice in healthy blue space. *Emotion, Space and Society*, 22: 43–51.
Ingold, T. (2015). *The Life of Lines*. London: Routledge.
Kaaristo, M. (2020). Rhythm and pace. The Diurnal aspects of leisure mobilities on the UK canals and rivers. In Amit, V. and Salazar, N. B. (eds) *Pacing Mobilities. Timing, Intensity, Tempo and Duration of Human Movements*. New York and Oxford: Berghahn, pp. 59–78.
Kaaristo, M., and Järv, R. (2012). 'Our clock moves at a different pace': the Timescapes of identity in Estonian rural tourism. *Folklore. Electronic Journal of Folklore*, 51: 109–132.
Kaaristo, M., and Rhoden, S. (2017). Everyday life and water tourism mobilities: mundane aspects of canal travel. *Tourism Geographies*, 19(1): 78–95.
Krause, F. and Strang, V. (2016). Thinking relationships through water. *Society and Natural Resources*, 29(6): 633–638.
Linton, J. (2010). *What is Water? the History of a Modern Abstraction*. Vancouver: UBC Press.
Massey, D. (2005). *For Space*. London: Sage.
Matthews, J. (2015). Mobilising the imperial uncanny: nineteenth-century textual attitudes to travelling Romani people, canal-boat people, showpeople and hop-pickers in Britain. *Nineteenth-Century Contexts*, 37(4): 359–375.
Mauch, C. and Zeller, T. (eds) (2008). *Rivers in History. Perspectives on Waterways in Europe and North America*. Pittsburgh: University of Pittsburgh Press.
Middleton, N. (2012). *Rivers: A Very Short Introduction*. Oxford: Oxford University Press.

Mukerji, C. (2009). *Impossible Engineering: Technology and Territoriality on the Canal Du Midi*. Princeton Studies in Cultural Sociology. Princeton: Princeton University Press.

Neimanis, A. (2017). *Bodies of Water: Posthuman Feminist Phenomenology*. London: Bloomsbury.

Olsson, A. (2016). Canals, rivers and lakes as experiencescapes – destination development based on strategic use of inland water. *International Journal of Entrepreneurship and Small Business*, 29(2): 217–243.

Peterle, G. and Visentin, F. (2018) Going along the liquid chronotope. The Po Delta waterscape through Gianni Celati's narration. In Vallerani, F. and Visentin, F. (eds) *Waterways and the Cultural Landscape*. New York: Routledge, pp. 122–139.

Pitt, H. (2018). Muddying the waters: what urban waterways reveal about bluespaces and wellbeing. *Geoforum*, 92: 161–170.

Prideaux, B. (2018). Canals. An old form of transport transformed into a new form of heritage tourism experience. In Vallerani, F. and Visentin, F. (eds) *Waterways and the Cultural Landscape*. New York: Routledge, pp. 145–159.

Roberts, L. (2019). Taking up space: community, belonging and gender among itinerant boat-dwellers on London's waterways. *Anthropological Notebooks*, 25(2): 57–70.

Rolt, L.T.C. (2014). *Narrow Boat*. Stroud: Canal and River Trust and History Press.

Seamon, D. (2013). Lived bodies, place, and phenomenology: implications for human rights and environmental justice. *Journal of Human Rights and Environment*, 4(2): 143–166.

Sheller, M. and Urry, J. (2006). The new mobilities paradigm. *Environment and Planning A*, 38(2): 207–226.

Smith, D. (2007). The 'buoyancy' of 'other' geographies of gentrification: going 'back-to-the water' and the commodification of marginality. *Tijdschrift Voor Economische En Sociale Geografie*, 98(1): 53–67.

Smith, D., Croker, G. and McFarlane, K. (1995). Human perception of water appearance. *New Zealand Journal of Marine and Freshwater Research*, 29(1): 45–50.

Strang, V. (2004). *The Meaning of Water*. Oxford: Berg.

Strang, V. (2009). *Gardening the World. Agency, Identity and the Ownership of Water*. Oxford: Berghahn.

Swyngedouw, E. (2015). *Liquid Power: Contested Hydro-modernities in Twentieth-century Spain*. Cambridge, MA: MIT Press.

Trapp-Fallon, J. (2007). Reflections on canal enthusiasts as leisure volunteers. In Jordan, F., Kilgour, L. and Morgan, N. (eds), *Academic Renewal: Innovation in Leisure and Tourism Theories and Methods*. Eastbourne: Leisure Studies Association, pp. 65–79.

Vallerani, F. and Visentin, F. (eds) (2018). *Waterways and the Cultural Landscape*. New York: Routledge.

Vannini, P. (2014). Slowness and deceleration. In Adey, P., Bissell, D., Hannam, K., Merriman, P. and Sheller, M. (eds), *The Routledge Handbook of Mobilities*. London and New York: Routledge, pp. 116–124.

Virilio, P. (2012). *The Great Accelerator*. London: Polity Press.

Warnaby, G., Medway, D. and Bennison, D. (2010). Notions of materiality and linearity: the challenges of marketing the Hadrian's Wall place 'product'. *Environment and Planning A*, 42(6): 1365–1382.

Willan, T. (1964). *River Navigation in England, 1600–1750*. London: Frank Cass.

15

Place-crafting at the edge of everywhere

David Paton

> We must resituate craft as a social instinct and an extension of environmental bonds.
>
> *(Shales, 2017: 22)*

> [Matter] is not mere stuff, an inanimate given-ness. Rather, matter is substance in its iterative intra-active becoming—not a thing, but a doing, a congealing of agency. It is morphologically active, responsive, generative, and articulate. Mattering is the ongoing intra-active differentiating of the world.
>
> *(Barad, 2012: 80)*

Figure 15.1 Masons working in the banker sheds at Trenoweth Dimension Granite Quarry near Penryn in Cornwall, UK.

Source: Photograph by David Paton, 2017.

In a quarry near you

Trenoweth Quarry is a small working granite quarry, slunk into a hillside, high above Falmouth's picturesque deep water harbour, in the far south west of Cornwall. Here, since the mid-1800s, the rolling topography has been dimensionally quarried, leaving deep angular cuts and 'benches' that follow the natural jointing of the granitic body. Today the saws whirr, forklifts move nimbly between stacks of granite and the corvids send out their calls against the roar of the wind. On hot sunny days a fine dust whips up around the tin sheds, wherein the rise and fall of the masons' hammers denotes the final shaping of lintels and quoins. When the weather blows in from the north and the rain drives horizontally into the sheds, all that matters is hot tea and the comforts of home. I have worked for and with this quarry for many years, and my life continues to revolve around and through it; and as someone at the quarry endearingly said, I 'would be fucked without it!'. How else could I talk about granite so confidently if it weren't for my time working in the quarry as a saw-man and mason. Indeed, what would I have to write about? How could I work on a twenty-five ton boulder, split it, lift it, shape it, if it weren't for the very placing of it all; if it weren't for the people and indeed the heavy machinery. To be honest, I can imagine no other place where matter has been so fundamentally unpacked through thoughtful and labouring bodies than this quarry.

This chapter attends to place and matter. I pay attention to place as being both analogous to craft, and that place can be crafted. As I guide you through some of the granitic landscapes of south west Cornwall, the quarry becomes a lens through which to highlight how acts of making align with the multiplicitous coproductions of place. The discrete specificities and mutabilities that transmit through maker and material are also present, I argue, within the always emergent identities of place. Place is always becoming; its specific natures assembling into part known and part unknowable narratives. In the western limits of Cornwall myth is woven through the landscape. Industrial practices resonate with ritual and Magick, sea–land binaries dissolve, and the literature of granite-working is evidenced in farms, mining tunnels, fogous, hedges and menhirs. These ancient features are tangled up with modern housing estates, factories, refuse tips and busy towns. Idealised promotions for artists' colonies and cosy fishing villages nestle alongside poverty and environmental degradation. All of this sits upon an immense granite body. A mineral-rich giant, whose once fluid mass emerged from the core of the earth and cooled just near enough to the surface to drive global economies and nurture a sense of belonging within its people.

Along with my quarrying experiences, I bring with me the rich textures of the 'thinking through making' perspectives that anthropologists such as Marchand (2008, 2010) and Ingold (2010, 2013) have pursued. Here, matter is vibrant and responsive and part of *the* conversation. In conclusion I also consider how new materialist work has profoundly extended crafted perspectives on the liveliness and socialities of matter towards the mythologies of place.

Attention has also recently been given to the geographies of craft, matter and making (Carr and Gibson, 2016; Edensor, 2013; Ferraby, 2015; Fox Miller, 2017; Gibson, 2016; Jakob and Thomas, 2017; Luckmann and Thomas, 2018; Price and Hawkins, 2018), demonstrating a renewed concern within cultural geography for how place-based narratives evolve through material relations and making practices—where the deeply personal continually transforms, and is being transformed, through in-place material interactions. I argue that, as a maker, I exist in a persistent threshold between matter and materiality, between raw exposure and itinerant meaning. Indeed, Lange-Berndt (2015), when bringing together critical perspectives on materiality and art practice, asserts that clarity has to be established between matter, material and materiality. Where Ingold made a profound case for these distinctions in 'Materials against materiality' (2007a), for the artists and critics that Lange-Berndt has collated, the need for an ontological specificity for matter to perform in certain ways is a persistent register through which ideas, and indeed places, are then negotiated. First, though, there is of course, the granite ....

Granite

For at least the past 5,500 years south west England's granites have provided mineralogical and spiritual sustenance. Scattered above and through the landmass are an entanglement of built structures that evidence the dismantling of the granite for tin and copper ores, shelter and ritual gatherings. These resources and potentialities are present within a crystal matrix which varies greatly as one travels west to east through the granite districts of Lands' End, Tregonning-and-Godolphin, Carnmenellis, St Austell, Bodmin and Dartmoor in Devon. The key components of granite are mica, feldspar and quartz; they come in many colours and patterns, and contain many variants and mutations. Some granites are soft and friable, and some are incredibly hard—like Trenoweth's very own fine-grained blue-grey 'Buckle and Twist'. Granite has been grown, it has a grain—gravitationally emplaced between 300 and 275 million years ago whilst the magmatic intrusions were heaving their way up through ancient sediments. The granite's horizontally bedded 'cleaving-way' or 'grain-way'—as the quarry workers term it—had also been exaggerated through the gradual erosion of rock above the granite body over millions of years, releasing pressure on the granitic mass and encouraging imperceptibly small gaps to grow between the crystals. In this way the granite performs in multiple axes and planes and, like following Ingold's tree (2010), the ripples of its 'growing' affect the bodies and lives of the people who work it.

To split a granite block is to ask nothing more than the universe to give up its secrets, where upon the 'Breath of the geologic' (Paton, 2016) is enacted as inhalations and exhalations from vastly differing 'Timespaces' (May and Thrift, 2001) merge. The old men, as Ernie the granite mason called them, worked deep in the quarry pit, reading the grain and jointing of the granite in order to extract block for the banker masons. The keener the banksman's knowledge, the less waste there would be to 'beat off' in order to square the block with hammer and point, and consequently the economy of the quarry would flow. Here, people and rocks tumble over each other and blur-out each other's material surfaces; and a social and cultural geology (Ferraby, 2015) evolves as a fleshy substrate in the world. Indeed, there is an incredibly rich array of quarrying terminology (Groundwork—*Tracing Granite*, Groundwork, 2018) which quite often refers to human and animal bodies. The granite is, as DeLanda proposes, a 'meshwork'; where the processual nature of rock is vitally present throughout all modes of life (1996). It is in this sense that I want to discuss place, where matter and the social are co-produced and, as such, the granite lies rumbling beneath this writing, teasing ideas along through its asymmetric junctures and fissures.

A worked landscape

I have come to know Trenoweth Quarry intimately over the past eleven years, as sculptor, quarry-worker and researcher. My relationship with this place has changed me forever, the trajectory of my life shunted, grounded and altered as I never thought I would be. My hammer-swinging arm has muscle-memories embedded deep in its fibres and my senses are alert to the sounds, patterns, densities and smells of the many different granites from the region. As I speak of these immense emplacements of granite, I also consider the freshly deposited topographies within the body. Through skilful labour the bone density increases, hard skin replaces the soft folds between thumb and forefinger, and the carcinogenic dust scarifies the alveoli of lungs, laying down meaty geologies the ramifications of which may well echo down through generations. That is to say, there are commonalities between place and matter that our labouring bodies resonate with and perform through—where reciprocating agencies disrupt and trouble any definitive binary conceptions of where life begins and ends, and the notion of place-crafting might also come into being.

Sometimes I ride my motorbike along the coast road between St Just and St Ives in the farthest westerly reaches of Cornwall, and as I speed along my eyes catch the pale granite boulders protruding from dark brown brackens and yellow flushed gorse. These boulders are scattered across the moorland and many show evidence of centuries of surface working and quarrying. Here, predating actual quarrying practices at the turn of the 19th century, the south west's early built environment was constructed from splitting and shaping surface boulders and weather-sculpted outcrops (Stannier, 1999). There is one substantial rock, though, that always catches my eye, with its sharp chiaroscuro definitions. It is plonked just off the roadside near Zennor on what is locally known as Quarry Corner—a site notable for numerous car crashes (Figure 15.2). I have filmed the boulder as strong winds bluster the clouds overhead, its plug-and-feather-split faces and angular forms morphing in the rapidly shifting light conditions, as if it were actually moving. This hard and sculpted clod maintains a melancholy presence yet is responsive and dynamic and shows its making through multiple affective orders. According to Mr Penhale and his family, the area just south of the boulder hosts a series of quarries and spoil heaps that only became apparent after a wildfire razed the gorse and bracken in the mid-1990s. The larger of these quarries was worked by a Jimmy Thomas in the 1930s, where he not only supplied the granite for the extension to the adjacent Tremedda Cottage, but also the remarkable, and now lichen-encrusted, war memorial in Zennor church yard sculpted by Ursula Edgecombe. The granite has been exposed by immense geological processes in order to be split by Jimmy and finely shaped with hammer and chisel by Ursula. There is always a story, always life, always death. Everywhere I feel the physiology of the marks left on the granite. I sense thoughtful bodies wielding tools, feeling their way through knowledge that is part acquired, part already present in the land. These imbricated tracings of labour etched across the land is the

Figure 15.2 Film still from *Breath of the Geologic ii* (David Paton, 2019) showing a stitch-spilt boulder near Zennor in Cornwall, UK.

language of place-making that physically resonates as some form of synesthetic response from my quarry work. The marks left by the tooled-up bodies are where the social and material manifest. It is indeed some kind of Magick that I now have this knowledge, where I can draw upon that moment of chisel smashing into granite, embracing some kind of phantasm of the past to inhabit my bones, muscles and organs.

One of the effects of quarry work has been to make me vigilant about the critical agency of matter. Whatmore describes this awareness as a reanimation of 'the missing "matter" of landscape, focussing attention on bodily involvements in the world in which landscapes are co-fabricated between more-than-human bodies and a lively earth' (2006: 603). Pitt (2015) aligns knowledge gathering with a *showing* of human and non-human natures. With reference to Ingold's work on knowing and learning, Pitt (2015) argues that plants and people can share their grown capacities in the context of exploring plant agency, given the right conditions that allow for a flow of mutual acknowledgement to emerge. Pitt suggests how this knowing and learning, or *showing* as she terms it, ties in with geographers' more-than-human exploration of livingness.

Also, the *practice* of being in the quarry has encouraged me to adopt labour as a crafting of place in very direct ways—the quarry is the place, the place is the quarry, and everything that happens there is quarryness and also just life. I do not want to claim the term labour as an overbearing political agenda, yet I find that it carries a heavy burden for a creative and critically aware practice fomented in the process of being a quarryman. It points directly at how acts of making correlate with a sense of belonging and contributes to the formation of place on a very slow, personal, and simultaneously universal, scale. The Brazilian architect and activist Sergio Ferro ('Sergio Ferro', spatialagency.net, 15/12/2014) provides a substantial critical context for the renegotiation of labour as an in-site creative ideology for dwelling. As with geography's cultural turn, it establishes the broader disciplinary territory of work/life human/nature practices, defined as the space where people's lives are constructed around the mutual adaptations of material conditions over generations. Dawkins and Loftus (2013) also examine praxis in the context of a Marxist reimagining of the senses, where an awareness of the contexts within which skills and labour practices foment can then be contextualised—as with my quarry work—enabling more complex perceptions to evolve through and beyond my own labouring body. This approach to reading place formed according to Whatmore, on the basis of a positional shift in social science research from '*discourse* to *practice* ... which relocates social agency in practice or performance rather than discourse—thinking and acting through the body' (2006: 604). These significantly repositioned relationships situate an exploration of place and matter very much within everyday life, and at the nexus of diverse life systems.

These merged properties of place and matter are commensurate with the multiple roles I continue to maintain within and beyond the quarry, and align with Massey's (2005) space-as-*trajectory*, a theory that proposes how multiple and mutable trajectories are exercised on a global scale of relatedness—localised specificities are at once rich with global or even cosmological significance. In line with what Pitt and others (Gibbs, 2009; Panelli, 2010) argue, political, ecological, gendered and economic thresholds have been gathered into a strategic restructuring of where, how and through what, life is lived. Being open to the *sharing* of grown properties is a matter of principle when working granite in the quarry, where a 'redistribution of energies puts the onus on "livingness" as a modality of connection between bodies (including human bodies) and (geo-physical) worlds' (Whatmore, 2006: 603). My fleshy knowledge of labouring with granite, as a base part of my weekly working commitments to family and home, render my senses always alert to the production of my life in this place, connecting to other coproduced livelihoods in this granitic and industrialised landscape. Frost states that when matter is

> worked upon and transformed by human labor, matter can be an agent by proxy, absorbing and translating the agency of individuals in ways that exceed each agent's deliberate intentions. The agency of matter, here, is an indirect extension and aggregate effect of the productive activities of the humans who work upon it
>
> *(Frost, 2011: 73)*

Equally, place has been, and is being, crafted whether one is aware or not of its machinations. Indeed, a political, cultural and social awareness of place-crafting is urgently needed if humans are to renegotiate a sense of belonging within this terran lifeworld.

Vitalities of land and sea

In *The Re-enchanted Landscape* (2017) White charts the origins and developments of occult, pagan and earth-mystery belief systems throughout the latter half of the 20th century in Cornwall. Filtered through the book are the methodologies of artists and ritual practitioners who folded themselves into the shadowed recesses of the land and celebrated its geological richness and proximity to the ocean. Often, it appears, these believers sought communion—with a nod and a dance—to the granitic monuments that have withstood millennia of social and cultural change. Here the stories of people are the stories of the land and sea. The land and sea are the host, the 'carrier bag' (Le Guin, 1989) where the liquidity of voice and rock register in the same space–time materialities, at once ancient and entirely present. At Le Guin's critical insistence, there are no individual heroes, but peoples that commune within their landed vessel for ritual union and birth without end. As White (2017) emphasises, there is something viscerally tantalising in the way the granite and ocean synthesise. On a hot and windless day I have sat on the busted slabs of granite below the imposing and exposed cliffs around the Gurnard's Head, again near Zennor. I peer through the seductive blue-green hue and follow the granite cliffs as they descend below the water's interface. At this moment it is easy to question where any hard threshold might exist, as two supposedly distinct habitats of sea and air merge. Water is becoming of rock, and rock of sea; and in this delicate moment there is a realisation of emergent and transitional material being ... and our fleshy bodies can indeed slip through these matters and commune with multiple belief systems, or none.

We are reminded that 'the ocean is classified as an object, a space of difference with a distinguishing ontological unity, the "other" in a land–ocean binary' (Steinburg in Anderson and Peters, 2016: XIV). In south west Cornwall this sense of the sea being 'other' plays a less potent social and cultural role due to the long history of industrialised transgressions of this sea–land binary through fishing and mining. As with Lahiri-Dutt's (2014) examination of the hybridised littoral of the Bengali floodplains, it is the entanglements of sea–land materialities, historically complex industrial strategies and labouring bodies that allows a fluid reading of the 'spongy' relations between water and rock. For centuries the far south west of Cornwall has been a hybridised lifeworld, where the differing materialities of water and rock are not resting in one state or another, but as a constantly evolving assemblage of resources and work/life hybridities (Martindale, 2012). Vickery (2016) also elaborates that there are more complex geologies and timescales in West Penwith that bring into being multi-axial registers of time and space, and that this again disturbs the boundaries between sea and land. What I want to emphasise here is that people's relationship to, and identifications with, a sense of place and home are malleable matters that are formed through immersive interaction. I, we, you, they instinctively craft belief systems from the matter surrounding us, responding to its material properties but also shaping it. In doing so, we shape our sense of what the conversation is or might be about. To submit to less bounded material constructs of sea and land is to see matter and place as a collaborative and lively constituent through which we move, and which moves through us. Celebrating the unpredictability of what time-grown entities might throw back at us allows the politics of matter to be engaged with and ultimately perhaps care will be afforded to animate and supposedly inanimate things.

Back at the intra-active sea–land threshold, communities that worked within, and grew to rely on, fishing and quarrying must have worked together to construct harbours and fishing havens. Sometimes their efforts were accepted by the sea–land entity, and sometimes not. However, as DeSilvey (2017) emphasises, the interactions of sea and land at Mullion Cove on the Lizard Peninsula are woven with socially and culturally complex narratives. The more-than-human labourings of the harbour wall have been tested over and over, and eventually superseded by the management of the site as a 'coastal village' and tourist destination. This framing of 'management' as a curation of material and cultural worth has now been called into question through DeSilvey's deeply thoughtful roam through many 'always-in-progress' structures. Although, I do have to say, perhaps, as someone who has masoned blocks of granite for the repair of the Penzance seafront after the 2014/2015 storms, there is always at the back of my mind a consideration of the extraordinary efforts of these labouring bodies that reside in a structure such as a harbour wall … and, as such, I might consider how to curate the materiality of that supposedly obdurate labour if the skills of granite masonry are also left to decay. It is indeed a case of my future history.

In West Penwith there are other monuments in a state of sea-sculpted precarity, these partially deconstructed buildings from the tin industry nestle on the wild and busted coastline near Botallack, their corresponding mining tunnels sinking deep down and spreading, nerve-like, out under the seabed—again traversing the supposed sea–land binary. Here labouring bodies have absorbed and excreted matter through discreet offerings and tooled-up rhythms. So we might also consider the relationship between industry and Magick, where an industrial animism (Leub, 2016) perforates any reductive reading of labouring bodies and craft persists in the darkest and harshest working environments. The world is not always made well as Barker and Pickerill (2019) and Yusoff (2018) highlight, where throughout history sea–land transgressions are littered with tragedies that migrate through time and space. Wrecked bodies have been deposited without any regard for life and the multiplicitous geographies and geologies of personhood.

The matter of craft

When I worked for the quarry as a banker mason, one of the most common architectural features we made were quoins. Quoins are part of the structural armature of buildings that feature on corners, around fenestration and doorways. Today we make quoins mainly from sawn block (Paton, 2015—*Making a Quoin*), although traditionally they were made from rough. Trenoweth has made thousands over the past few decades and they form part of the key economic foundation of the quarry. When a big order comes in, pallets of sawn block are placed in front of the masons' sheds and, over weeks, a rhythm evolves as quoin after quoin is dressed and stacked on pallets ready for delivery, and it did become tedious. Yet this is a hyper-aware tedium, where I grew to be more aware of the minute differences in the granite and the sensorial density of this process was felt in the thoughtful musculatures of my body. I also become acutely sensitive to multiple scales of affective slippages throughout the lifeworld of Trenoweth. All this takes place in the context of daily banter between quarry workers, enabling the harsh labour to be tempered by the unique capacities of each mason to riff off each other and the granite itself.

The skill of the craftsperson is to acknowledge the multiple hybridities of place and matter. To craft is social, it is awkward, sometimes dangerous and highly politicised … craft is not always beautiful. To craft is sometimes to destroy. Craft is about livelihood, craft is a verb, craft is a reciprocal and conversational formation that troubles the binary of living and non-living. To craft is not smooth or uniform, it is unpredictable. To craft is often the creation of a void, whose excavated matter is deposited elsewhere and used in unforeseen ways. The process of craft aligns with Barad's intra-action that

> queers the familiar sense of causality (where one or more causal agents precede and produce an effect), and more generally unsettles the metaphysics of individualism (the belief that there are individually constituted agents or entities, as well as times and places).
>
> *(Barad, 2012: 77)*

Indeed, in terms of place, the craft paradigm shows how the slow formation of the granite body, and its subsequent dimensional reduction for a quoin, are simultaneous co-productions across differing registers of material liveliness. Here we could again consider Le Guin's (1989) vessel—the first tool, brought into the world by women who moved through the vessel of the world—a vessel within a vessel ad infinitum. To craft, and to place, is the act of communication, within and without the body politic and as such aligns with the shifting parameters of new materialism.

The new materialist proposal, fleshed-out by Coole and Frost (2010) and given a chronological development by Dolphijn and van der Tuin (2012), is continually being extended and tested in multiple material assemblages (Bennett, 2010). New materialist ideas have also manifested very tangibly through arts and craft disciplines. Also, the maker, and/or craftsperson, has emerged as a very pragmatic critical model through which to examine the geographical entanglements of the social and material towards an understanding of place. Maker economies can become wrapped up in neoliberal agendas for the tidying of messy culture and 'dirty' societies (Price and Hawkins, 2018), and over-designing the handmade can be a route to localised and global commodification (Luckmann and Thomas, 2018). If care is not taken though, craft could become an over-convenient filter for a range of geographical and artistic dilemmas. Instead, as Gibson (2016) asserts, the societal and environmental imperatives of craft within evolutionary economic geographies demonstrates options for policy and livelihood change at community level.

Here, I am proposing that craft isn't something that happens in or to a place, rather craft practices are vital and co-productive features of material livelihoods moving through the placing of the body. From the perspective of my own labouring and research in the quarry, and in the face of the intense physical exhaustion and slow development of craft skills that have helped to pay my mortgage and support a family, the correlation between making and place is vitally clear. And in terms of the vibrant imperatives proposed by new materialism Gibson and Carr suggest how making becomes a process of iteration, and a maker works with this iteration prolifically. When the material pushes back, resisting the way it is being handled, a maker tries a different way. The material offers no reflection on ability in this moment; it is just an efficient way of working.

(2017: 303)

Here, the body is in constant negotiation with other matters, that in turn form part of wider socio-political entanglements that are simultaneously productive and disruptive (Clark and Yusoff, 2017). Any discourse on matter and the social will also be attending to the politics of the body (Davis, 2009), and consequently questions arise around the sites of agency and mutability, again forming within and without the body. As a species, as matter, we are in and of the world, 'we humans are all walking rocks' (Ellsworth and Kruse, 2013: 17) and as such we are responsible for, and answerable to, our own matters. Going back to Lange-Berndt's (2015) instance that matter, material and materiality require specific identities for artists, for people, for the world; actually, what requires acknowledgement here is that matter, material and materiality are not productive sites with their own bounded forms, rather they are the affective quantum productions of an ever shifting and decreasing set of variables.

We make in places with matter, places are matter made within matter. People are mattered throughout multiple coexistent places, and sometimes we get close enough to give these matterings a name … albeit temporarily on a geological scale. So, let us end back where we started, in Trenoweth Quarry, in Cornwall, right at the edge of everywhere. The quarry is real, it is harsh and beautiful. Full of magic and sweat, dust, broken machines and fibrous muscle. The quarry's articulated and busted surfaces are fleeting incidents on the way through the history of the earth. The quarry is a resource for making buildings, for walls, memorials and sculptures. Charles, Jamie and Liam make lintels and quoins and earn money in the quarry, and drink tea and coffee everyday in a small metal 'crib hut'. The quarry is local and was crafted at the beginning of the universe.

There is perhaps no finite answer to what place is, or what matter is, but we might say that they are interchangeable terms that perform as an always emergent and affective synthesis of form and formation. Through vigilant crafting, humanity can, or perhaps must, suppress notions of heroic progress and instead attend to this politics of careful making.

Bibliography

Barad, K. (2012) Intra-actions (Interview of Karen Barad by Adam Kleinmann). *Special dOCUMENTA (13) Issue of Mousse Magazine*, 34, pp. 76–81.

Barker, A. J. and Pickerill, J. (2019) Doings with the land and sea: decolonising geographies, indigeneity, and enacting place-agency. *Progress in Human Geography*, 10.1177/0309132519839863.

Bennett, J. (2010) *Vibrant Matter: A Political Ecology of Things*, Duke University Press, USA.

Carr, C. and Gibson, C. (2016) Geographies of making: rethinking materials and skills for volatile futures. *Progress in Human Geography*, 40 (3), pp. 297–315.

Clark, N. and Yusoff, K. (2017) Geosocial formations and the anthropocene. *Theory, Culture & Society*, 34 (2–3), pp. 3–23.

Coole, D. H. and Frost, S. (2010) *New Materialisms: Ontology, Agency, and Politics*, Duke University Press, USA.

Davis, N. (2009) New materialism and feminism's anti-biologism: a response to Sara Ahmed. *European Journal of Women's Studies*, 16 (1), pp. 67–80.

Dawkins, A. and Loftus, A. (2013) The senses as direct theoreticians in practice. *Transactions of the Institute of British Geographers*, 38, pp. 665–677.

DeLanda, M. (1996). *The 'Geology of Morals': A Neo-Materialist Interpretation* [online]. Available at www.t0.or.at/delanda/geology.htm [Accessed 05 June 2019].
DeSilvey, C. (2017) *Curated Decay: Heritage Beyond Saving*, University of Minnesota Press, Minneapolis, USA.
Dolphijn, R. and van der Tuin, I. (2012) *New Materialism: Interviews & Cartographies*, Open Humanities Press, USA.
Edensor, T. (2013) Vital urban materiality and its multiple absences: the building stone of Central Manchester. *Cultural Geographies*, 20 (4), pp. 447–465.
Ellsworth, E. and Kruse, J. eds. (2013) *Making the Geologic Now – Responses to Material Conditions of Life*, Punctum Books, Brooklyn, New York, USA.
Ferraby, R. (2015). *Stone Exposures: A Cultural Geology of the Jurassic Coast World Heritage Site*, PhD, University of Exeter, UK.
Fox Miller, C. (2017) The contemporary geographies of craft-based manufacturing. *Geography Compass*, 11, pp. e12311.
Frost, S. (2011) The implications of the new materialisms for feminist epistemology. In: Grasswick H. eds. *Feminist Epistemology and Philosophy of Science*. Springer, Dordrecht, pp. 69–83.
Gibbs, L. M. (2009) Water places: cultural, social and more-than-human geographies of nature. *Scottish Geographical Journal*, 125 (3–4), pp. 361–369.
Gibson, C. (2016) Material inheritances: how place, materiality, and labor process underpin the path-dependent evolution of contemporary craft production. *Economic Geography*, 92 (1), pp. 61–86.
Groundwork. (2018). *Tracing Granite* [online] Available at: https://groundwork.art/tracing-granite/ [Accessed 05 June 2019].
Ingold, T. (2007a) Materials against materiality. *Archaeological Dialogues*, 14 (1), pp. 1–16.
Ingold, T. (2010) The textility of making. *Cambridge Journal of Economics*, 34, pp. 91–102.
Ingold, T. (2013) *Making: Anthropology, Archaeology, Art and Architecture*, Routledge, London, UK.
Jakob, D. and Thomas, N. J. (2017) Firing up craft capital: the renaissance of craft and craft policy in the United Kingdom. *International Journal of Cultural Policy*, 23 (4), pp. 495–511.
Lahiri-Dutt, K. (2014) Beyond the water-land binary in geography: water/lands of Bengal re-visioning hybridity. *ACME: International Journal of Critical Geography*, 13 (3), pp. 505–529.
Lange-Berndt, P. (2015) *Materiality*, MIT Press, Cambridge MA, USA and Whitechapel Gallery Ventures Ltd, London, UK.
Le Guin, U. K. (1989) *Dancing at the Edge of the World*, Grove Press, NY, USA.
Leub, M. (2016) *A Renaissance of Animism: A Meditation on the Relationship between Things and their Makers* [online]. Available at https://designthinkingtank.at/a-renaissance-of-animism-a-meditation-on-the-relationship-between-things-and-their-makers/ [Accessed 05 June 2019].
Luckmann, S. and Thomas, N. (2018) *Craft Economies*, Bloomsbury Publishing Plc, London, UK.
Marchand, T. H. J. (2008) Muscles, morals and mind: craft apprenticeship and the formation of person. *British Journal of Educational Studies*, 56 (3), pp. 245–271.
Marchand, T. H. J. (2010) Making knowledge: explorations of the indissoluble relation between minds, bodies, and environment. *Journal of the Royal Anthropological Institute*, 16 (1), pp. 1–21.
Martindale, T. (2012). *Livelihoods, Craft and Heritage: Transmissions of Knowledge in Cornish Fishing Villages*, PhD Thesis, University of London.
Massey, D. (2005) *For Space*, SAGE Publications Ltd, London, UK.
May, J. and Thrift, N. (2001) Introduction. In: May J. and Thrift N. eds. *Timespace: Geographies of Temporality (Critical Geographies)*, Routledge, Abingdon, UK, pp. 1–46.
Panelli, R. (2010) More-than-human social geographies: posthuman and other possibilities. *Progress in Human Geography*, 34 (1), pp. 79–87.
Paton, D. A. (2015). *Making a Quoin/The Quarry as Sculpture: The Place of Making* [online]. Available at http://blogs.exeter.ac.uk/dapaton/2-making-a-quoin/ [Accessed 05 June 2019].
Paton, D. A. (2016) Stitch-split: the breath of the geologic. *Architecture and Culture*, 3 (3), pp. 267–270.
Pitt, H. (2015) On showing and being shown plants — a guide to methods for more-than-human geography. *Area*, 47 (1), pp. 48–55.
Price, L. and Hawkins, H. (2018) *Geographies of Making, Craft and Creativity*, Routledge, Abingdon, UK.
Sergio Ferro [online]. Available at www.spatialagency.net/database/sergio.ferro [Accessed 05 June 2019].
Shales, E. (2017) *The Shape of Craft*, Reakton Books Ltd, London, UK.
Stannier, P. (1999) *South Western Granite*, Cornish Hillside Publications, St Austell, UK.

Steinberg, P. E. (2016) in eds. Anderson, J. and Peters, K. *Water Worlds: Human Geographies of the Ocean*, Routledge, Abingdon, UK.
Vickery, V. (2016). *Fractured Earth: Unsettled Landscape Through Art Practice*, PhD, University of Exeter, UK.
Whatmore, S. (2006) Materialist returns: practising cultural geography in and for a more-than-human world. *Cultural Geographies*, 13 (4), pp. 600–609.
White, R. (2017) *The Re-enchanted Landscape: Earth Mysteries, Paganism and Art in Cornwall 1950-2000*, Antenna Publications, UK.
Yusoff, K. (2018) *A Billion Black Anthropocenes or None*, University of Minnesota Press, Minnesota, USA.

16

Thinking place atmospherically

Shanti Sumartojo

Introduction

On the Japanese island of Naoshima is the remarkable Chichu Museum. Opened in 2004 and designed by renowned architect Tadao Ando, it is built into a hillside above the Seto Sea, its exterior-less design maintaining the curve of the landscape and profusion of the surrounding woods. It houses works by only three artists – Claude Monet, James Turrell and Walter de Maria – all artists who, as the museum handbook explains, share an interest in 'producing work that confronts nature' through the elements of light, colour and shape (Chichu Museum, 2013: 5). The handbook lays out the museum's intention:

> Rather than merely gazing at each sculpture or painting, directly experiencing the space that arises out of this unique combination of elements is the best way to understand the aesthetic qualities of Monet, De Maria and Turrell.
>
> *(Ibid)*

The approach to the site builds slowly. On my visit, my companion and I walked a couple of kilometres from our hostel up a gentle slope, through the damp arboreal green, with glimpses of the grey sea on one side and the touch of mist on our faces and hair. After buying a ticket, an attendant gently directed us towards the entrance, a concrete slab covered in ivy a hundred metres further up the hill. The sharp angle of approach meant the shadowy opening was only discernible from very close, so our anticipation was tinged with uncertainty until just before we turned left into a sheltered walkway.

The building itself was a series of courtyards, passages, gardens and gallery rooms, in a material palette of grey concrete and muted white marble, composed with a confidence that nudged visitors from one space to the next. We could glimpse flashes of green reeds through an eye-level slit in one corridor, the organic profusion and colour of the plantings relieving the long, straight corridors and limited hues. Attendants padded around purposefully in off-white uniforms, guiding visitors through the rituals of entering an antechamber, removing shoes, putting on slippers, approaching the artworks carefully. It was quiet and the air was slightly humid and cool.

The artworks are arranged to be seen in a particular order, so the experience built and accreted as we moved from one room to the other, encountering each work in the context of what came before. The building, rituals, rhythms of movement, hushed sounds, gloomy light and moist air all contributed to the experience of the paintings and installations, thickening them beyond visual apprehension. As discrete artworks their impact was extended and enriched by the astonishing location and its immersive and seductive sensory and affective affordances.

This relates something of how the Chichu Museum felt when I visited it in the autumn of 2018, and its ineffable, at times almost overwhelming atmosphere. It acts as an introduction to those thoughts, feelings and sensations that are inextricable from the places where we experience them, that shimmer and pulse in and through our spatial encounters, that form a way of apprehending and understanding the world that cannot quite be described but that powerfully condition how we experience and remember things. Atmospheres imbue, seduce, immerse and envelop us, and we in turn are a part of them. They are absolutely emplaced, contingent on the particular qualities of the time, place and company, but cannot be reduced to them. When we call this 'atmosphere', and use this word to draw multiple elements together, something new comes into view about place.

Locating atmosphere at the centre of how we think about place does conceptual work that reinforces and advances our understandings of place by way of experience and feeling. Put simply, atmospheres help us to account for how places feel, what those feelings mean and what might be possible as a result of their emergence and apprehension. In the efflorescence of new scholarship on atmospheres, strongly linked to the affective, sensory and more-than-representational, place is foundational. In this chapter I explore the conceptual dialogue between place and atmosphere, tracing some of the entanglements that atmosphere can help us unravel, and use this to reflect on how atmosphere might frame place and how this is useful. By thinking in, about and through atmosphere (Sumartojo and Pink, 2018), I take an approach to place that is contingent, excessive and slippery, that seeks less to *define* place and more to understand how it feels and what this means to people. It also opens a speculative route to thinking about what implications place might carry into the future, and how it might stick with us long after we have left it.

Why atmospheres?

Even though they are commonplace, atmospheres resist definition, and this is reflected in the bulk of scholarship (my own included) that gets at them in terms of what they *do* rather than seeking to pin down what they *are*; indeed, Jean-Paul Thibaud (2015: 40) makes this point when he asks 'what does an ambiance make it possible to be, to experience, to do, to perceive and to share?' In part this is because atmosphere, inextricable from affective and sensory feeling, shapes how we understand and react to our worlds, and so is productive and generative of meaning.

Having said that, elsewhere I and others have made some attempt at definition. Griffero (2010: 6) calls atmosphere 'a something-more, a *je-ne-sais-quoi* perceived by the felt-body in a given space, but never fully attributable to the objectual set of that space ... [it is] a spatialized feeling, a something-more in a corporal sense'. Whilst not trying to determine such an 'objectual set', Sarah Pink and I have argued that we should understand atmosphere via what configures when we experience particular moods, resonances or impressions, elements that include 'sensation, temporality, movement, memory, our material and immaterial surroundings, and other people' (Sumartojo and Pink, 2018: 6). An emphasis on the subjective and experiential courses through scholarship on atmospheres, be it in attention to the senses, the body, or memory. Stephen Legg (2019: 3), for example, posits that 'atmospheres "braid" nature and culture together ... through all of the senses', whilst also tagging it to 'nostalgia, memory and forgetting'.

Thibaud's important contribution on *ambiance* brings into Anglophone scholarship a longstanding Francophone discussion of 'the situated, the built and the social dimensions of sensory experience' (2015: 40). This lends itself to designerly approaches that seek to understand how atmosphere can be created or staged, and how it might encourage particular encounters or forms of interaction. It follows that place is understood here in sensory terms, with the body playing a central role in constituting it: 'place emanates from a corporal engagement that is indissociable from its powers of guidance and expression' (Ibid: 41). That is, the body is always emplaced in specific empirical configurations, or 'tuned' spaces (Böhme, 2014) that guide its impressions, feelings and, presumably, capacities for action. Thibaud (2015) also engages with place via its sensory intensities such as brightness, heat or roughness, that might vary in importance and change intensity. Others have called these 'elements' or 'quasi-things' through which we perceive the world (see also Ingold, 2011). Writing on architecture and archaeology, and drawing on the work of Hermann Schmitz, Bille and Sørensen (2016: 13) explain that the experiential world is an 'atmospheric field of situations, potentialities or quasi-things ... such as feelings, voices or other corporeal forces'.

While this will be different with every situation or place in which atmospheres might be at work, and indeed changes as we dwell in and apprehend atmospheres in their emergence from moment to moment, thinking about the composition of atmospheres does remind us to check in with the affective, sensory and more-than-representational; the discursive and symbolic; the imagined, anticipated and remembered; and the complex relationalities amongst these things over time and in movement. This points to something of what an atmospheric framing can do for thinking about place – chiefly, to help us treat place as an ongoing configuration, akin to Doreen Massey's (2005) notion of 'throwntogetherness' that is empirically particular but always changing.

This relates to the notion that atmosphere is a constantly changing configuration that draws together people and their surroundings, a position that enables us to think about place as similarly emergent and ongoing:

> atmosphere emanates from connections between things and how they shift and align around and with us and with each other. While atmospheres are sensed, experienced and often understood in spatial terms, they are not limited by particular spaces – that is, they should not be thought of as bound or contained by space, or beginning and ending in clearly identifiable ways.
>
> *(Sumartojo and Pink, 2018: 56)*

Tim Edensor (2015) discussion of the atmosphere of a football match provides an engaging example of this. He details how attempts by club personnel to manufacture atmosphere with loud, recorded chanting and flashy graphics in a new stadium was met by fans with dismay and derision. These displays did not reflect longstanding experience and ritualised ways of travelling to the previous and more familiar ground, the regular interactions between fans on match day, the press of bodies and the raucous singing made almost deafening by the low roof. Instead, fans had their own strategies for heightening intensity at the new venue and making atmospheric meaning in their own terms. In this context of how people constituted, apprehended and valued atmosphere, Edensor describes the rapidly changing emotions of one crucial match that drew to a nail-biting conclusion and finally erupted into uncontrolled, carnivalesque euphoria at the end. Here, atmosphere was distinctive, powerful and memorable, and also contingent and unpredictable, ebbing and flowing over a brief period that 'melded affects, emotions, events, space and sensations that surged and waned before, throughout and after the game' (Edensor, 2012: 89).

If this account charts shifting atmospheric intensities, elsewhere Edensor joins others in attending to the more banal and unremarkable moods that can also be characterised as atmospheric (for instance, see Pink et al., 2014; Bissell, 2018). Even in everyday settings or routine activities, there is a dynamism to atmosphere, in its shifts, ripples, diminutions and gatherings that emerge ongoingly over time and in movement. This also helps us see how place is never frozen or still. Moreover, the configurations that constitute atmospheres beg the question of how these different elements relate to each other. Derek McCormack (2018: 20) insists that atmosphere 'holds in tension affective spacetimes that are both corporal and incorporeal' and 'emphasises the relational qualities of these affective spacetimes' between bodies and at different scales. This sense of dynamism is strongly reflected in its treatment as related to movement and mobility, as in David Bissell's (2010: 270) work on transport that accounts for 'how different affective atmospheres erupt and decay in the space of the train carriage', or Paul Simpson's (2017) emphasis on the atmospheres that emerge in and through the movement of urban cyclists.

This approach grounds atmospheres in the empirics of dwelling and moving in spacetime, asking how they are perceived by the people who are both enveloped in and help constitute them, and what has to configure in order for them to take hold (Sumartojo and Pink, 2018). This can be put differently as a question of noticing or attuning to atmosphere, and under what circumstances this is able to occur. This question of attunement is important because it brings the concept of atmospheres into direct dialogue with their empirical, emplaced settings, returning to the rough definition I sketched out above. It follows that because the configurations that make them apprehensible are so particular, atmospheres are always in some way place-specific, even if this is changeable. They not only make place, they make it meaningful in specific terms. This relationality and contingency make atmosphere 'good to think with' (Anderson, 2009) about place. This is the stepping-off point for the remainder of this chapter.

Thinking place atmospherically

How then might atmospheres contribute to our ways of thinking about place; or, what might be made possible by thinking place atmospherically? By thinking in, about and through atmospheres, we can put atmospheres to work conceptually and empirically to shed light on how we experience the world, and the sense and meaning we make of it (Sumartojo and Pink, 2018). Thinking *in* atmosphere means attending to it in our own and others' experiences as it is happening, and these are always place-based in some way. Methodologically, this takes the form of attuning to atmosphere sensorially and cognitively, and attending to how it emerges and changes, and in so doing, enriching accounts of place. Here, we recognise that atmosphere is already there all the time, even when we are not focussed on it.

Thinking *about* atmosphere asks us (and our research participants) to look back on it, or step outside it, to regard it at a remove and attempt to define it, or find the best terms with which to describe it. Here we must freeze it artificially for the purposes of understanding what it might be comprised of, for putting it in to words and working with it. This means stilling place temporarily, pausing its configurations long enough to examine them. Finally, thinking *through* atmosphere treats it as a means by which to understand something else conceptually. This is where atmosphere is revealed as linked to other, more foundational aspects of our worlds, such as how we understand our identities, our relationships with others or with ideas. In turn, this allows us to address the implications and potential of atmosphere, and what it makes possible to apprehend in new ways, including place.

Thinking in atmosphere + place

Thinking in atmosphere about place compels us to attend to its sensory qualities (Howes, 2005; Drobnick, 2014; Pink, 2015; Sumartojo, 2015). When we consider what makes a place 'stick' to us, what impressions it leaves most markedly, these are at least partially anchored in how we experience it: for example, a particular neighbourhood park understood in terms of the lemony smell of eucalyptus trees; the dry, scratchy grass on the back of the legs; the sound of dogs barking and scampering on the footpath; and the heightened attention to the proximity of those same dogs when they come closer and sniff at our hands and bodies. Our awareness of the park as a discrete place is therefore composed of these bodily sensations and variable intensities of somnolence, attention and concentration, for example, as dogs draw near and move away again. These feelings are contingent on the physical surroundings and the things in them, even if this is a very different view of the same park that we might have if we locate it on a map.

Thinking in place requires methodologies that enable researchers to attune to their surroundings as they dwell in and move through them (Sumartojo, 2018). Photography, for example, can work to help look at the detail of place and subsequently think through its effects and meanings. This is not so much a process of 'capture' – here the point is not to record experience or make aesthetically pleasing images – as it is a mode of visual attention that can open up new understandings. Using images, sound recordings or notes to consider place whilst in it is a useful way to attune atmospherically in the emergent flow of events.

Sara Ahmed (2010: 30) captures something of this contingency when she reminds us of the 'messiness of the experiential, the unfolding of bodies into worlds', writing of the role of affect in making the world meaningful. This focus on bodies in worlds can also be interpreted as emplacement, an experiential mode in which the *particular* sensory and affective affordances of our spatial surroundings make those surroundings distinctive and meaningful. Even the most banal or quotidian surroundings – in homes, supermarkets, cars – are conditioned by feelings that cannot be dissected out of embodiment and materiality. That is to say, when we think about place *in* atmosphere, it is lively and ongoing, very much in line with Massey's (2005) distinctive 'here-and-now', always moving and changing, and imbued with moods that also move and change. It follows that 'any spatial perception begins with a sense of presence or an attunement to a *place*', via the felt-body – in other words, place *is* atmosphere, from the point of view of 'human emplacement in the world' (Bille and Sørensen, 2016: 14).

Thinking about atmosphere + place

If thinking in atmosphere and place orients us to the processual and emergent, then thinking *about* their relationship means fixing this ever-changing here-and-now to describe it, understand what comprises it and what work it is doing, and to identify how it might be reproduced, manipulated or created. Perhaps the area in which we see this most clearly is in architecture and design and, as I have already touched on, considerable scholarship considers this dimension. The passage that opened this chapter, for example, narrated an encounter with the work of Tadao Ando, an architect renowned for his manipulation of material, void, light and shadow to powerfully moving effect, but who also considered his work on Naoshima to be about 'place-making' (Ando, 2014: 8).

Thinking about place as 'made' by atmospheric experience is also evident in the work of architect Peter Zumthor (2006), who orients his practice around the creation of sensory impressions that are perceived emotionally when people encounter the built form. Similarly, Juhani Pallasmaa (2014a: 230) argues that it is crucial for designers to move beyond visual perception and attend to the apprehension of built spaces by way of 'complex multi-sensory fusion of countless factors which are immediately and synthetically grasped as an overall atmosphere, ambience, feeling or mood'. To do so, he advocates that architects can cultivate a 'multisensory and fully empathic imagination' (2014b: 83) that is crucial to design a building from the point of view of how people may experience it.

Degen et al., (2017) take a different approach to the creation of place-specific atmospheres in their work on digital visualisations of architectural designs and how they are aimed to produce what the client will find visually appealing. This, they argue, is a significant departure from design work that is based on plans or other architectural approaches, instead selling a feeling that carefully crafted images are hoped to engender. The 'right' atmosphere was, for those who created architectural visualisations, impossible to define except by way of client approval. This shows that although designers can intervene in contributing certain elements, atmosphere *itself* cannot be designed, precisely because of its contingent and emergent nature. Indeed, many accounts of the staging or designing of atmospheres (Edensor and Sumartojo, 2015) detail the usually incomplete attempts to do so, and this is relevant as much for place and for any other designed outcome.

Using the concept of atmosphere to think about place also draws in imagined and remembered aspects, as places remind us of other places and times. Such nostalgic, excited, poignant or anticipatory moods link us to place in subjective and individual ways. But such place-specific feelings are also shared and circulate via the movement or stillness of bodies, the rise and fall of voices or music, temperature, the glitter or flicker of illumination and the light, shadow and colour that tincture places (for instance, see Edensor, 2012; Turner and Peters, 2015; Schroer and Schmitt, 2018; Sumartojo and Graves, 2018). This points to the complexity of atmosphere, and the importance of understanding what it evokes in the minds of people who experience and co-constitute it.

Thinking through atmosphere + place

The internal world of the 'perceiving subject' (Anderson, 2009) knots into how thinking *through* atmosphere can improve understanding of larger concepts. For example, there is an emerging body of work concerned with what atmospheres can tell us about politics. Tonino Griffero (2014) addresses this when he argues that 'an atmosphere still possesses and exercises authority or authoritativeness' not because 'I possess it ... but because it concerns me'. This is because its effects are comprised in part by the context in which an atmosphere occurs, the predispositions we have as we apprehend and contribute to it, including the degree to which we accept its normativity. However, at the same time that it connects to ideas, views or discourse, atmosphere's emplacement means that these concepts cannot be divorced from specific places and things. An account of another visit provides an example. During an

Australian national election, I was in London and went to Australia House to cast my vote. I queued briefly outside between metal barriers, took my turn to have my bag searched and entered the main building. I felt it a grand and imposing atmosphere. This feeling related to its ornate and expensive-looking marble interiors, high ceilings, columns and arches, glittering lamps and architectural built form typical of official 19th-century buildings. This was the specific spatial context, with an impact that also built through the security procedures and barriers at the entrance. My predisposition was coloured by my status as an Australian citizen that lent a slight (and slightly shameful) propensity to identify as a 'colonial' in London. My normative response rested on my scepticism of the governmental power that Australia House represented and manifested, especially because I voted on that occasion for the opposition party. Together, all these thoughts, attitudes and experiences were part of how the place felt, and its atmosphere was part of how the place worked on me, entered into my imagination and became a category of affective encounter with my surroundings.

On the one hand, this is because so often many of the elements that comprise atmosphere have been designed or planned; here, organisational power is evident. Christian Borch, 2015: 72) reminds us that the Nazi regime's 'political strategies were also based on the deliberate manipulation of previously non-politicised fields and objects', such as mass meetings and their 'politico-atmospheric staging' where people might be swept up in the feelings shared with the massed crowds. Nigel Thrift's (2004) call to attend to 'engineered affects' similarly gets at what this might mean, for example in public events' staging via light, sound amplification, crowd control or music.

Bille et al. (2015: 34), however, take a more nuanced stance. They invite us to consider 'atmosphere as a space of political formation that underlies the realm of discursive politics, but cannot be controlled in any simple and unambiguous way by political agents'. In other words, atmosphere might be intended to augment, enhance or magnify particular political agendas, but its very ambiguity means it does not pulse along predictable vectors. Indeed, when we attend to how atmospheres might be subject to design, manipulation, staging or production, we are also drawn to assess their effectiveness, and where they may fail, disperse or be interpreted differently. This is a politics of affect in which atmospheres might complicate or exceed the channels through which power is exercised. Indeed, if we treat atmosphere as a form of potential, because it is always in emergence, then we grasp that it is never completed and its work is never finished (Anderson, 2014; Massumi, 2015).

It follows that we can think about atmospheres and politics from another direction, here recalling Angharad Closs Stephens' (2015) insistence on 'taking feelings seriously' in the composition and circulation of shared identities. This means focussing on affective modes of perceiving and participating in state rituals and other expressions of the nation, which can help us see official attempts to compose publics by way of engendering 'moods'. Rather than thinking about how atmospheres might be designed to achieve particular ends, thus starting with political goals and ideologies, we might begin by thinking through atmospheres themselves and by trying to understand their political force and impact. In other words, if 'moods make publics', rather than the other way around, then we can refigure how we understand politics. This is akin to Ahmed's discussion of collective feelings and the impressions that others leave on us, feelings that align us with a collective, 'which paradoxically "takes shape" only as an effect of such alignments. It is through an analysis of the impressions left by bodily others that we can track the emergence of 'feelings-in-common' (Ahmed, 2004: 27). What this can reveal is that publics – groups of people composed around particular shared experiences or goals – can coalesce atmospherically; that is, they can become pulled into shared feelings and these come to foster the emergence of shared norms and understandings of the world. This matters because these shared dispositions, propensities, preoccupations and preferences nudge us towards action. Here, atmospheres are not neutral or inconsequential – instead they have real motivational capacities.

This enhances our understanding of the ways in which symbolic and representational aspects of place, which are so often related to political power or forms of identity, have potency. Such material symbols are everyday, emplaced reminders of political power. However, they are often not settled or latent, but can be hotly contested spatial signifiers of shared identities that work in part through the way they contribute to how place feels at multiple scales (see Rhys Jones, Chapter 3, this volume). This might include monuments or memorials, urban plans, buildings, street and place names, or borders and frontiers – and all of these have both symbolic weight and material form that we encounter in spatial contexts, as well as distinctive atmospheres to which they contribute. Moreover, because atmosphere is always in some way emplaced, the symbolic and discursive resonances of our surroundings are entangled with how we apprehend, feel and understand those surroundings. Atmosphere allows us to understand the representational meanings of our surroundings along with their sensorial affordances and how they connect to our memories and imaginations. It also helps us consider how place draws us together with others into shared experiences, meanings and identities. This capacity for foregrounding relationality is a key way in which atmosphere can enrich how we approach place, and can help us understand politics as constituted of many elements mixed together in dynamic combinations. Thinking through atmospheres is useful precisely because it homes in on what different aspects combine to imbue place with distinctive affective and sensorial feelings, how they ebb and flow in relation to each other, and what the larger political or social implications might be of these processes.

Place and atmosphere as potential

It follows that the relationality amongst different emplaced elements is also temporal – that is, place is made by specific times as well as spaces. This includes connections to other moments in time:

> Treating place as processual, open and unfinished configurations that change over time allows us to think of changes not just to different elements, but to their relationships with each other and the ways in which these formations reach backwards (memory, history) and forwards (anticipation, imagination) in time.
>
> *(Sumartojo and Pink, 2018: 56)*

In this sense, the unfinishedness of place is manifested atmospherically, precisely because when we think in place atmospherically, we must stay with it in emergence and ongoingness. Here, it is not possible to know what will happen next, even if we can anticipate that some of the conditions in which we dwell will shape this becoming. So while the atmospheres that help to comprise place are in some way unknown from one moment to the next, they are also contingent on what exists and is felt and experienced now. Attending to this contingency allows us to open up an orientation to the future that treats place as a form of potential, never completely actualised because it is always unfolding.

Accordingly, there are several implications of the drawing out and foregrounding of potential that atmosphere allows in thinking about place. The first is that the future of place is destabilised when we think it atmospherically, from the different elements that configure into it, to the way they relate to each other, to the ongoing change and unpredictability of these elements and relationalities. This may not herald a radical transformation, but possibilities for change are always there.

Second, this destabilisation does not have to run along the usual hierarchies of attention – in other words, a cooling shadow or a scent on the air might be felt as important as a historically significant place name or striking built form. It is not so much that atmosphere flattens the ways that we experience and make sense of the world, but rather it allows us to treat the minor, the sensorial or the fleeting as significant, and examine what these make possible in our understandings of place. As described above, the smell of the trees in a park or the quality of the dappled light when we sit under them might be as significant in how that place feels as the local politician it is named for. Thinking atmospherically brings these small details into the mix of place's meaning. This is important because it can unsettle hegemonic ways of understanding place and allow multiple understandings to not only co-exist but be granted importance.

Finally, thinking atmospherically about place as potential is a route for new things to come into being that weren't there before, because place is not closed or completed. This, I argue, is essentially hopeful, because while things might get worse, there is also always the possibility that they might get better. In this way, thinking place atmospherically gives us a future-orientation that can act to sustain and encourage us.

References

Ahmed, S (2004) *The Cultural Politics of Emotion*. London: Routledge.

Ahmed, S (2010) Collective feelings: or, the impressions left by others. *Theory, Culture & Society* 21(2): 25–42.

Anderson, B (2009) Affective atmospheres. *Emotion, Space and Society* 2: 77–81.

Anderson, B (2014) *Encountering Affect: Capacities, Apparatuses, Conditions*. Farnham: Ashgate.

Ando, T (2014) *Naoshima*. Paris: Le Bon Marché Rive Gauche.

Bille, M, Bjerregaard, P and Sørensen, T (2015) Staging atmospheres: materiality, culture, and the texture of the in-between. *Emotion, Space and Society* 15: 31–38.

Bille, M and Sørensen, T (2016) Into the fog of architecture. In M Bille and T Sørensen (Eds) *Elements of Architecture: Assembling Archaeology, Atmosphere and the Performance of Building Spaces*. London: Routledge, pp. 1–29.

Bissell, D (2010) Passenger mobilities: Affective atmospheres and the sociality of public transport. *Environment and Planning D: Society and Space* 28: 270–289.

Bissell, D (2018) *Transit Life*. Cambridge, Mass: MIT Press.

Böhme, G (2014) The art of the stage set as a paradigm for an aesthetics of atmospheres. *Ambiances* online https://journals.openedition.org/ambiances/315. Accessed 19 November 2018.

Borch, C (2015) *Architectural Atmospheres on the Experience and Politics of Architecture*. Basel: Birkhäuser.

Closs Stephens, A (2015) The affective atmospheres of nationalism. *Cultural Geographies* 10.1177/1474474015569994.

Chichu Museum (2013) *Chichu Handbook*, 2nd ed. Naoshima: Chichu Art Museum.

Degen, M, Melhuish, C and Rose, G (2017) Producing place atmospheres digitally: architecture, digital visualisation practices and the experience economy. *Journal of Consumer Culture* 17(1): 3–24.

Drobnick, J (2014) The museum as smellscape. In Levent, N and Pascual-Leone, A (Eds) *The Multisensory Museum: Cross-Disciplinary Perspectives on Touch, Sound, Smell, Memory and Space*. Lanham: Rowman & Littlefield, pp. 177–196.

Edensor, T (2012) Illuminated atmospheres: anticipating and reproducing the flow of affective experience in Blackpool. *Environment and Planning D: Society and Space* 30: 1103–1122.

Edensor, T (2015) Producing atmospheres at the match: fan cultures, commercialisation and mood management in English football. *Emotion, Space and Society* 15: 82–89.
Edensor, T and Sumartojo, S (2015) Designing atmospheres: introduction to special issue. *Visual Communication* 14(2): 251–265.
Griffero, T (2010) *Atmospheres: Aesthetics of Emotional Spaces.* London: Routledge.
Griffero, T (2014) Who's afraid of atmospheres (and of their authority)? *Lebenswelt* 4(1): 193–213.
Howes, D (Ed.) (2005) *Empire of the Senses: The Sensory Culture Reader.* Oxford: Berg.
Ingold, T (2011) *The Perception of the Environment: Essays on Livelihood, Dwelling and Skill.* London: Routledge.
Legg, S (2019) 'Political Atmospherics': The India Round Table Conference's *Atmospheric Environments, Bodies and Representations*, London 1930–1932, Annals of the American Association of Geographers, published online, DOI: 10.1080/24694452.2019.1630247.
Massey, D (2005) *For Space.* London: Sage.
Massumi, B (2015) *The Politics of Affect.* Cambridge: Polity.
McCormack, D (2018) *Atmospheric Things: On the Allure of Elemental Envelopment.* Durham, NC: Duke University Press.
Pallasmaa, J (2014a) Space, place and atmosphere: Emotional and peripheral perception in architectural experience. *Lebenswelt* 4(1): 230–245.
Pallasmaa, J (2014b) Empathic imagination: formal and experiential projection. *Architectural Design* 84(5): 80–85.
Pink, S, Leder Mackley, K and Roxana Moroşanu, R (2014) Researching in atmospheres: video and the 'feel' of the mundane. *Visual Communication* 14(3): 351–369.
Pink, S (2015) *Doing Sensory Ethnography*, 2nd. London: Sage.
Schroer, SA and Schmitt, SB (Eds) (2018) *Exploring Atmospheres Ethnographically.* London: Routledge.
Simpson, P (2017) A sense of the cycling environment: felt experiences of infrastructure and atmospheres. *Environment and Planning A* 49(2): 426–447.
Sumartojo, S (2015) On atmosphere and darkness at Australia's Anzac Day Dawn Service. *Visual Communication* 14(2): 267–288.
Sumartojo, S (2018) Sensory impact: memory, affect and photo-elicitation at official memory sites. In Drozdzewski, D and Birdsall, C (Eds) *Doing Memory Research: New Methods and Approaches.* London: Palgrave, pp. 21–37.
Sumartojo, S and Graves, M (2018) Rust and dust: materiality and the feel of memory at Camp des Milles. *Journal of Material Culture* 10.1177/1359183518769110.
Sumartojo, S and Pink, S (2018) *Atmospheres and the Experiential World: Theory and Methods.* London: Routledge.
Thibaud, JP (2015) The backstage of urban ambiances: when atmospheres pervade everyday experience. *Emotion, Space and Society* 15: 39–46.
Turner, J and Peters, K (2015) Unlocking the carceral atmospheric: designing extraordinary encounters at the prison museum. *Visual Communication* 14(3): 309–330.
Thrift, N (2004) Intensities of feeling: towards a spatial politics of affect. *Geografiska Annaler* 86B(1): 57–78.
Zumthor, P (2006) *Atmospheres.* Basel: Birkhäuser.

17

The urban Spanglish of Mexico City

Cristina Garduño Freeman, Beau B. Beza and Glenda Mejía

Introduction

Place is a potent and widely used concept in the Anglophone discourses of cultural geography, urban planning, heritage and environmental psychology. The concept refers specifically to the way we are connected to particular spaces; to our sense of attachment and the way identity and culture are entangled in the physical environments we inhabit. Yet when we move out of the Anglophone sphere, into the realms created by people who have different conceptual structures for their built environments articulated through their own languages and cultures, we find that the potency of the term 'place' is lost in its translation. For example, in Spanish, *lugar* (the literal translation of place) lacks ambiguity. Instead, in Spanish when referring to settings as 'places', the language is more specific. Less slippery terms are used: *sitio, ambiente, entorno*, where each of these terms conveys nuanced semantic characteristics and, often, culturally specific meanings. This phenomenon extends to other terms encountered within urban typologies of places. Neighbourhoods, streets, communities, suburbs and squares become *vecindarios, barrios, municipios, suburbios, colonias* and *plazas*.

In this chapter we explore the translations, both linguistic and spatial, between the Anglophone and Hispanophone discourses of place. We position 'place' as a relational term, used to describe the phenomenon whereby we distinguish some spaces as culturally significant that are imbued with attachment and implicated in our individual and collective sense of identity (Cresswell, 2005; Hubbard and Kitchin (2011 (2004)); Tuan, 1977). Our observations about the Anglophone roots of 'place' are prompted by our Latin-American backgrounds and consequent bilingualism to varying degrees coupled with our common academic interests in urban theory, identity and place. Cristina Garduño Freeman was born in Mexico City and grew up bilingual in Spanish and English. At the age of ten she migrated to Australia, losing connection with Mexico and her family for several years. She now has several projects and collaborations that reconnect her to Mexico. Beau B. Beza is Hispanic and was born and raised in California, USA. As an adult he migrated to Australia where he completed his PhD and entered academia. He pioneered study tours for built environment studios to Latin America, with a focus on Mexico (and Colombia). Glenda Mejía is from El Salvador. She migrated to Australia as an adult where she studied language and identity in Spanish-speaking communities from Latin America. Her relationship to Mexico is more distant; instead of a direct connection she knows Mexico as an academic and traveller, having taken students there in 2015 and doing research on Higher-Skilled Mexican Women. She is also connected through the common language of Spanish which is shared by much of Latin America. These connections have drawn us to explore how the language we use to *talk* and *write* about particular places has distinct implications for our perceptions, experiences and connections with them. In conversation we share experiences of Mexico by describing them in *both* Spanish and English, using phrases and terms that clearly articulate our shared connections to the country and the academic discourse of place.

From an Australian context, 'place', with its etymological relationship with the word 'plaza', suggests a connection that bridges linguistic and spatial divides. The public spaces described as 'plazas', with all the activity, symbolism and emotional attachment we have assigned to them, are often cited as exemplars of 'place' in academic literature (Low, 2000; Roa, 2008; Rojas and Timmling, 2008). However, the ambiguity of 'place', which in the English discourse is arguably a part of its theoretical value, cannot be replicated in Spanish because this language relies on more specific terms such as *lugar*, *sitio*, *ambiente* and *entorno* to convey meaning. As a result we argue in an earlier publication that discussions of places gain important meaning from being discussed in their *lingua franca*, as this then connects descriptions to the semantic structures embedded in language (Beza et al., 2019).

Here we want to extend our investigations to consider how the linguistic concept of 'Spanglish' can offer a metaphor/framework for understanding how the languages through which we write about places bridge cultural, physical and semantic divides. Spanglish is a cultural practice where bilinguals 'code-switch' between Spanish and English as a form of identity expression:

> The use of Spanglish creates another level of meaning where the hybridity of the Chicana/o experiences are negotiated. In this sense, Spanglish is a way to construct and reconstruct a third space of Chicana/o identity, a linguistic 'nepantla'.
>
> *(Sánchez-Muñoz, 2013: 440)*

Spanglish, on first examination, appears to be tied to the Hispanic experience in the USA, where *Chicanas* or *Chicanos*, as people with Mexican heritage are often colloquially referred to, share a hybrid American Mexican culture and language. While Spanglish is technically code-switching, rather than a language in its own right, as Sánchez-Muñoz argues, it operates as an alternative 'in-between' linguistic space that offers particular shared cultural meanings. It is in this sense that we want to explore how code-switching and the '*nepantla*' (third space) it creates can shed light on the intrinsic relationship between the language with which we talk about places and its effect on our perceptions and connections to them.

To do this we build on Yi-Fu Tuan's 1991 article 'Language and the Making of Place' to ground our approach to the personal, literary and semantic structures of place in Mexico City. First, we draw on *Coyoacân*, the suburb of Garduño Freeman's childhood memories and of Guillermo Sheridan's narrative critique (1998). Second, we discover *La Condesa*, a suburb which Beza came to know through academic analysis carried out in 2013 in collaboration with Munoz-Villers and López González Garza, colleagues from La Salle Universidad, Mexico City. A focus on *Coyoacân* enables us to discuss how language and affect are intrinsically connected, while looking at *La Condesa* prompts us to recognise how place is conceptually defined by the underlying linguistic structures of the language in which it is discussed. We use the Spanglish lexicon of Mexico City as a way of revealing how language structures distinct understandings of the same physical spaces.

Language and place

Yi-Fu Tuan writes on the relationship between language and place, describing their interdependence and, in particular, the way in which '[w]ords have the [...] specific power to call places into being' (1991: 686). Fundamentally, Tuan observes that without language (understood broadly to include text, speech, and visual representations) the reality of 'place' cannot be distinguished from 'space'. To elaborate on this premise, he brings forth examples of the way language renders and articulates places in many ways: from keeping indigenous people's rituals alive, to the appropriation through renaming of other's places as colonial explorers have done, or in conceptualising 'place' at various scales through geographical concepts such as 'the continent' or 'the local'. These sociological conditions of language are matched by Tuan's argument that cultural perceptions, experiences, memories and attachments are also important and that these are more readily shared and documented through narrative literature on place.

Forays into the direct relationship between language and place are uncommon. As Christine Higgins (2017) asserts, key scholars in the field such as Michel de Certeau, Edward Soja, Doreen Massey and Henri Lefebvre all focus on 'place' as a physically situated construct that overlooks its framing through language. Ron Scollon and Suzie Wong Scollon (2003) do study language in place, but do so through 'geosemiotics', a term coined to describe how signs come to have meaning once they are situated in the material world. Yet, as Barbara Johnstone (2010) demonstrates, language is intrinsically defined by place. It underpins many sociolinguistic analyses, which investigate the specificities of dialect or class, gender or race.

Although Tuan's paper is almost three decades old, his observations that '[...] geographers have focused almost exclusively on material processes and socioeconomic forces, without raising, explicitly, the role of language' (Tuan, 1991: 692) appears to continue to hold true. Perhaps this is because the characteristics of material and physical factors are much more straightforward to document and therefore discuss. Understanding the intangible and subjective interactions of language relies on investigating their reception both at an individual and at a cultural level. Language has power; it can influence our perception of place as Tuan (1991) asserts, a warm conversation in a kitchen (686) can engender affect for one's home, a literary description can frame a visit to a foreign place and the available words for analysis of a city can structure the potential meanings that can be communicated.

English, Spanish and Spanglish: code-switching as cultural identity

Spanglish is a contentious topic. Scholars from diverse fields that study its intersection with pedagogical, historical, cultural, linguistic, literary, demographic subjects have differing perspectives on Spanglish. To these areas we add the discourse of place and the field of urban planning to this conversation. Some scholars argue that Spanglish should either be discarded to avoid confusion or described as 'Spanish in the USA' (Dumitrescu, 2010, 2013; González Echevarría, 1997; Lipski, 2008; Osio, 2002; Otheguy, 2009; Otheguy and Stern,

2010) or as a form of Spanish-English language alteration (Rodríguez-González and Parafita-Couto, 2012). Other scholars are proud to use it (Morales, 2002; Stavans, 2003) and call it '*Chicana/o* Spanish' or '*Chicana/o* English' (Sánchez-Muñoz, 2013; Zentella, 2008). These scholars appreciate that while it is essentially a form of code-switching or code-mixing, it is used to express creativity; a hybrid/multicultural identity and a linguistic heritage of two worlds – the English and the Spanish (Casielles-Suárez, 2017; Fairclough, 2003; Morales, 2002; Rodríguez-González and Parafita-Couto, 2012; Zentella, 2008).

English has evolved over time. Historically, it was brought to Britain by the Anglo-Saxons, 'the Angles', one of the Germanic tribes, in 449 AD (Gelderen (2014 (2006)): 2). Like all languages, English has not developed in isolation; it has been influenced by, and borrowed from, both Latin and French. In contrast, Spanish has evolved directly from Latin, starting out as a dialect in Castile and Leon (Nadeau and Barlow, 2013: 1) in 200 AD. Both English and Spanish have spread well beyond their countries of origin in the process of colonisation. English is now the third most spoken language in the world and Spanish is the second (McCarthy, 2017). Spanglish is the result of the cross-fertilisation of Mexican culture with the United States and may have been practiced as far back as the early to mid-19th century when Mexicans and people from the (early) US began interacting. As this interaction and migration, by Mexicans, across the USA increased the mixing of the two languages proliferated (Villa, 2014: 391).

Speaking Spanglish is about identity, not about a lack of proficiency or negligence of linguistic rules. Indeed, as Dumitrescu contends, Spanglish is actually

> a well-known communicative strategy among bilinguals fluent in both languages, who alternate them for a variety of purposes, among which – contrary to popular belief – the lack of knowledge of one (or both) of the languages, or some form of mental laziness, is practically never the case.
>
> *(2012: 377)*

Spanglish is more readily understood as a form of cultural expression, one which gives speakers access to, as Ana Sánchez-Muñoz (2013) describes, a *nepantla* – a 'third space' – which exists in addition to Spanish and English.

When examining our built and natural environment we describe places through language. Language can enable us to 'see' places, but language is also an element of place. For example, part of the sense of place of Paris is the sensory experience of hearing people speak French. Whether or not one is bilingual, lived experience is not exclusively carried out in a single tongue. While thinking academically we (the authors) think in English, but at Latin-American events in Melbourne we switch to Spanish and Spanglish as a way of acknowledging our joint connection with the event and each other. While some scholars may claim that Spanglish is simply a code-switching phenomenon exclusive to the lower class or those with low levels of education in the USA, the authors of this chapter have observed, both through first-person and ethnographic observation, how Spanglish is also common with high-class Mexicans and regularly incorporated into everyday speech. Rodríguez-González and Parafita-Couto argue that '[r]egardless of which language is dominant, the use of two languages within the same utterances is a skill available not only to speakers when they produce code-switched utterances, but also to listeners and readers when presented with "mixed" language information' (2012: 472). Words, at times, come to us in 'the other' language simply because there is not always a direct equivalent in the context within which we are operating. It may be assumed that translating a term risks a loss of meaning. Using Spanglish can enable terms in English to hold their meaning when used as part of a conversation and their context to be recognised not only by the speaker but also by the reader or listener.

We argue that borrowing words from English (or Spanish) can often more appropriately express the meaning of the respective subject matter. It also has the effect of contextualising discussion in another 'world' and in a way that direct translation does not. This challenges the value of translation and acknowledges the linguistic appropriation that occurs when we write, describe and theorise places only within the Anglophone discourse. We ask, is there a value in using Spanglish when bilingual academics talk about places within a Mexican context? And how does that offer new readings for non-bilinguals of those places?

In the next section of this chapter we explore and draw on two suburbs in Mexico City connected with the authors' backgrounds and their recent involvement in these urban places. In order to explore the relationship of *this piece of written language* with these places we employ Spanglish to elucidate how language can structure distinct understandings of physical spaces. In recalling the sense of place of *Coyoacân*, Garduño Freeman utilises Spanglish to more authentically recall her childhood experiences of this place and to illustrate how Sheridan's use of Spanglish forces the reader to engage with Mexico's contentious relationship with the USA. In contrast, Beza's exploration of *La Condesa* reveals how it is impossible to truly distance urban analysis from the linguistic and spatial concepts of a city's *lingua franca*. While we employ Spanglish as a textual strategy, we do not aim to alienate the monolingual Anglophone reader and therefore we have bracketed our use of Spanish with written descriptions and literal translation of the words and phrases used. However, the intended or conceptual meaning we deploy, as argued in this chapter, comes from understanding the use of each word or phrase in the context of a third space or in-between-space of 'Spanglish'.

Remembering and reading place in *Coyoacân*

As a child I lived in *Coyoacân*, a historic suburb of Mexico City associated with bohemian lifestyles, cobbled streets and haciendas. My family and I lived on the edge of the *pueblo* (village) after which the *delegacion* (council, borough or municipality) was named, in a gated condominium made up of three fifteen-storey residential towers on Avenida Universidad. *Coyoacân* for me is associated with the *Mercado* (market) and *La Plaza* (an urban square, a plaza – in this instance it is a series of plazas). The Plaza always had street-vendors: during the day they sold *pepinos* (cucumber) and *jicama* (a type of turnip or yam bean that is crunchy, watery and slightly sweet) drenched in *limon y sal* (lemon and salt) or skewered *mangos* decoratively sliced into giant golden roses and sprinkled with *chile*. At night time the street vendors drew *hot cakes* (American-style pancakes) with batter into animals (mine was always a cat) spread with *cajeta* (a caramel spread made from goat's milk) over it and delivered piping hot on squares of butcher's paper. Other vendors sold *elote* (maize corn picked when young) boiled until tender, then dressed in *mayonesa, queso y chile* (mayonnaise – American style – with a salty white fresh cheese and powdered chile Piquín). In contrast, my memories of the *Mercado de Coyoacân* centre around the stalls, the fruit and vegetables, the cheese stall, the broom stall, the kitchenware stall, plasticware, meats and the *piñatas* (decorated papier-mâché vessels filled with fruit and sweets central to festivities). The *piñatas* hung from the concrete shells inspired by the *tianguis* (temporary markets typically constructed from steel scaffolding and bright coloured fabric awnings) that usurp the city, colonising a different street each day of the week. My favourite stalls when I was a child:

the stall which sold dolls clothes, of all kinds, including traditional Mexican dresses, handmade with miniature stitching; and in the lead up to *Navidad* (Christmas) the stalls that sold porcelain nativity sets, with evocative glass eyes, scaling a sea of Jesuses, Marys and Josephs as far as the eye could see. Of course, I have been back since migrating to Australia at the age of ten, and come to associate *Coyoacân* with Frida and Diego Rivera, and with *comida corrida* ('a sequence of food' – an inexpensive set meal eaten as the main meal of the day in the early afternoon around 2 to 3 p.m.) on the edge of the plaza. Also with *La Siberia*, a famous ice-cream chain producing *paletas de fruta* (fruit-based ice-block) of as many combinations as are imaginable. But I cannot think of *Coyoacân* without thinking in Spanish since the words used above draw out specific connotations, like flavours specific to this site, that allow my memories to be savoured fully rather than in English and in a partial way. This hybrid use of language, of a kind of Spanglish prompted by place-memories, triggers sensory memories that enable me to rekindle my connection to this place.

Guillermo Sheridan (1998) is a Mexican author and critic who wrote on *Coyoacân* for the Mexican literary magazine *Vuelta*, the article provocatively titled in English 'Yes, in *Coyoacân* you can' (Sheridan, 1998). While it is written in Spanish, the piece employs Spanglish as a literary device to convey sense of place and the Mexican ritual linguistic appropriation of English words and concepts. He peppers terms throughout the article such as *jipis* (hippies) and *pinfloi* (Pink Floyd) that capture the Spanish speakers' heavily accented pronunciation of English words. In doing so, Sheridan ties the text to the city, imploring the reader to enter into the urban and cultural space of *Coyoacân*.

Recently, Sheridan's 1998 original piece was republished in *The Mexico City Reader* (Sheridan, 2004). The anthology is a portrait of the city, often criticised for its crime, density, pollution and disorder, yet loved for its vibrant culture, food and people. In the anthology Sheridan's use of Spanglish has deliberately been retained despite being translated by Lorna Scott Fox into English. Sheridan's strategic use of Spanglish to connote the sensory qualities of *Coyoacân* remains, with *jipis* and *pinfloi* italicised to distinguish them from the English prose. Yet other Spanglish phrases borrowed from English, such as *casmir* (cashmere) *jarecrishna* (hare krishna) and *freaks* and *punks*, remain undistinguished from the text (we note that the text in Gallo's anthology actually reads 'Harry Krishna' but we have assumed this is a typographic error). These, while semantically clear in their Spanglish form, summon young people's cultural utterances; essential qualities of this place.

Translation is complex and decisions made about form and meaning of a translated text can feel like a betrayal to the original and its language. But Spanglish enables us to receive in more than one language, to understand place from multiple perspectives. The *Mercado de Coyoacân* does not connote the same sense of place that the Coyoacan Market does. Similarly, Sheridan's use of Spanglish in his original text helps to situate the narrative of *Coyoacân* in place. Scott Fox retains this linguistic strategy in much of the translation – although we would argue that all of the Spanglish would have enabled a more authentic experience for the reader. Spanglish in both cases structures a sense of place in Mexico City.

Urban lexicon of *La Condesa*

La Condesa is Spanish for 'countess'. A *colonia* (suburb) of Mexico City, *La Condesa* '[...] is considered a "ambiente urbano sobresaliente"/an outstanding urban environment [...]' (Beza, Munoz-Villers, and López González Garza, 2013: 33) by residents and visitors alike. *La Condesa*, like *Coyoacân*, is one of the city's most desirable places in which to live, visit and work. As a place, it exhibits many of the physical and societal elements associated with high quality places, such as walkability, vitality and diversity. Yet, it was not always so, and its transformation is grounded in a uniquely Mexican series of events, where Spanglish is written into its evolution.

As Tuan articulates, geographers have always constructed place through language: '[a] cademic geographers [...] have "named" entities on earth, from the climatic zones of the ancient Greeks, to the natural regions favored by nineteenth-century geographers, to the modern geographer's metropolitan fields' (1991: 693). Their task has been to articulate how places are made and unmade, to create analytical descriptions that enable a discourse about character, quality and transformation. But cities have their own lexicon, determined by their *lingua franca*, a form of hybridisation that is importantly embedded into a city's history and discourse: in Mexico City there are *colonias* and *vecindarios*; in New York there are *boroughs* and *neighbourhoods*; in Paris there are *banlieues* and *arrondissements*. To grasp *La Condesa*'s history, like grasping its regal name, requires terms and vocabulary that borrow from Spanish.

La Condesa and Mexican architecture/planning has been the focus of substantial scholarship (Carmona, 2010; Cobos, 2005; Cobos and López, 2005; Gortázar, 1996; Rangel, 2006). Beza, Munoz-Villers, and López González Garza (2013) and Beza (2016) suggest three phases to this *colonia*'s development, which are best referred to in Spanish:

1 *La Urbanización* – the urbanisation period from 1902 to 1985, where government policies, foreign investment, modern housing approaches (for instance, the installation of mains water) and newspaper coverage stimulated the emergence of this place;
2 *Despoblamiento y La Reconstrucción* – the outmigration and reconstruction period from 1985 to 1995 (resulting from the 1985 8.1-magnitude earthquake) where, because of the severe devastation to housing in this suburb, the residential population declined; and
3 *El Renacimiento* – the renaissance period from 1995–present where, because of land being cheap and government initiatives, small families, artisans and business entrepreneurs began moving back into the area.

Prior to *La Urbanización*, *La Condesa* was part of a private *hacienda* (estate) consisting of a *hipódromo* (racecourse) that to this day provides the key piece of open space infrastructure influencing the *colonia*'s spatial layout and urban development. In itself, and from this suburb's beginning, the *hipódromo* was '[...] enshrined in contractual urban development preservation arrangements [...]' (Beza, Munoz-Villers, and López González Garza, 2013: 30–1) and used by Jose Luis Cuevas (Mexico's first modern urban planner) to propose '[...] a series of elliptical and radial streets running outwards from the racetrack [to] ensure [...] that park and city came together through large green areas, parks, and tree-lined streets' (32). Interestingly, it is suggested that this thinking is borrowed from Ebenezer Howard's concept of the Garden City; Cuevas was possibly exposed to it while studying under Howard.

When the suburb was being redeveloped during *El Renacimiento*, it was a combination of 'new' ideas about urban development and pre-earthquake planning that has allowed it to be considered, as mentioned above, an *ambiente urbano sobresaliente*. For example, the area of *La Condesa* was part of Mexico City's decentralisation planning strategy, and was part of an interconnected network of small urban commercial centres where goods, people and vehicles could be dispersed by means of a tertiary road network (Beza, Munoz-Villers, and López González Garza, 2013; Cobos and López, 2005). The physical changes that resulted from the post-earthquake planning strategy during *El Renacimiento* helped to transform the perception of *La Condesa* residents, making it into the vibrant *colonia* it is today.

The Spanish language terms used above refer to and identify specific periods of *La Condesa*'s development, which importantly link it to Mexico City, rather than other parts of the world. For example, *El Renacimiento* when literally translated is the Renaissance and may conjure for those familiar with the historical context images of the European period in which the knowledge and art of classical Greece and Rome were rediscovered. The translation of *La Urbanización*, depending on context, may refer to urbanisation or development. The English language conceptualisations triggered by both these words depart from their contemporary Spanglish meaning where they identify a period when urban planning emerged in Mexico, here through Jose Luis Cuevas' design proposal for *La Condesa*. Additionally, when referring to *La Condesa*, the use of the term *colonia* typically solicits a mental picture of a quaint or walkable setting, whereas its translation into something akin to the English word 'suburb' may conjure images of a city's outer urbanisation, a realm very different to that of *La Condesa*. Hence, in the context of *La Condesa*, the *lingua franca* used above shapes the perception and conceptualisation of place in this distinctively Mexican setting.

Conclusion

Without language we cannot describe place. We cannot share it, discuss it and bring it forth into being. Language, through memories, critiques and analytical vocabularies, enables emotional connections, imagined experiences and academic knowledge to contribute to discourse. Place is not just constructed through physical matter; it is also appropriated, controlled and discussed across languages. Our exploration of Spanglish as a 'third-space', a '*nepantla*' that demands a more transparent and grounded connection, offers the writer, the reader and the academic new ways to ground their relationship with place and new ways to discuss its qualities. Critically, our argument also highlights that there are other ways place may be described, where local words are used to best convey the qualities of a setting. In this sense, Spanglish bridges a theoretical and spatial divide when the places we are discussing in English belong in a city where Spanish is the norm.

Using Spanglish ties place to experience, it draws in site-specific qualities. As Tuan (1991) observes, language frames both individual and collective experience. Our bilingual connection has made us realise that language can conjure specific relationships that have deep meaning and between the person and the setting. For example, *Cajeta* is more than 'goats milk caramel' – it is the taste of Mexican childhood and this single word conveys place to its enthusiasts. Similarly, *comida corrida* translates literally to 'food running', but in fact points to the cultural custom for a main meal in the middle of the day often taken at a restaurant with colleagues. A simple translation does not convey their larger contextual significance, but code-switch signposts that there is more to this term than its first sematic meaning. This is important in the way we analyse places as academics as well. In our analysis of *La Condesa*, for example, *programas ordenadores* (aka: *planes parciales de desarrollo*) literally translates into 'computer programs' or algorithms, but in fact one interpretation of the implied/conceptual meaning is 'Mexican government/community-inspired programs' which is a suggested part of the lexicon embodied by scholars and built environment professionals in Mexico.

It is perhaps more straightforward to discuss the physical characteristics of place as Tuan (1991) highlights, rather than, as mentioned above, the intangible and subjective interactions of language. However, this is no reason to ignore the potential and the implications of language for our perceptions of place. Urban Anglophone's Garden City, just like

Hispanophone's *La Plaza*, quickly conveys an urban type to others outside the concept's original linguistic field. Space and the language we use *is* culturally specific. This is a valuable frame; the lexicon of place should not be appropriated, but rather can prompt engagement with nuanced perceptions. Code-switching, through Spanglish or other hybrid ways of communicating, is to allow for multicultural and hybrid identities, for porous boundaries that recognise that knowledge and perception are grounded in language. Valuing Spanglish acknowledges that words are a form of appropriation and therefore Spanish words may more eloquently illustrate Mexican places, while code-switching assists mixed language readers and listeners, opening up a new pathway for understanding place.

References

Beza, BB 2016, 'Places for sustainability citizenship', in R Horne, J Fien, BB Beza and A Nelson (eds), *Sustainability Citizenship and Cities: Theory and Practice*, Routledge, Abingdon, pp. 139–149.

Beza, BB, Garduño Freeman, C, Fullaondo, D and Mejia, G, 2019, 'Place? Lugar? Sitio?: Framing Place and Placemaking through Latin American contexts', in J Hernández-Garcia, OB S. C., AG Jerez & BB Beza (eds), *Urban space: Experiences and considerations from the Global South*, Pontificia Universidad Press, Bogotá, pp. 19–42.

Beza, BB, Munoz-Villers, J and López González Garza, M, 2013, 'Finding common ground: Creating successful places in the redevelopment of Mexico City and Melbourne', *paper presented to 6th International Urban Design Conference*, Sydney Olympic Park, Sydney, Australia, 9–11 September. Paper available through Book of Proceedings: Peer Reviewed: https://urbandesignaustralia.com.au/archives/p/rp13.pdf.

Carmona, MDS, 2010, 'El trazo de las Lomas y de la Hipódromo Condesa', in *Diseño y Sociedad*, vols. 28–29, Otoño, pp. 16–23. Available at: https://archive.org/details/eltrazodelaslomasydelahipodromocondesa.

Casielles-Suárez, E, 2017, 'Spanglish: The Hybrid Voice of Latinos in the United States', *Journal of the Spanish Association of Anglo-American Studies*, vol. 39, no. 2, pp. 147–168.

Cobos, EP, 2005, 'Zona Metropolitana del Valle de México: megaciudad sin proyecto', *Ciudades*, vol. 9, pp. 83–104.

Cobos, EP and López, LM, 2005, 'Estancamiento económico, desindustrialización y terciarización informal en la Ciudad de México, 1980–2003, y potencial de cambio', in ACT Ribeiro, H Magallaes, J Natal, R Piquet (eds) & LM López (con), *Globalização e Território: Ajustes periféricos*, IPPUR: Arquimedes Edições, Río de Janeiro, Brasil, pp. 143–161.

Cresswell, T, 2005, *Place: A Short Introduction*, Short introductions to geography, Blackwell, Malden, MA.

Dumitrescu, D, 2010, 'Spanglish: An Ongoing Controversy', in S Rivera-Mills & JA Trujillo (eds), *Building Communities and Making Connections*, Cambridge Scholars, Newcastle upon Tyne, pp. 136–167.

Dumitrescu, D, 2012, 'Guest Editorial: "Spanglish": What's in a Name?' *Hispania*, vol. 95, no. 3, pp. 377–379.

Dumitrescu, D, 2013, '"Spanglish" and Identity within and outside the Classroom', *Hispania*, vol. 96, no. 3, pp. 436–437.

Fairclough, M, 2003, 'El (denominado) Spanglish en Estados Unidos: polémicas y realidades', *Revista Internacional de Lingüística Iberoamericana*, vol. 1, no. 2, pp. 185–204.

Gelderen, Ev, 2014 (2006), *History of the English Language*, John Benjamins Publishing Co, Amsterdam and Philadelphia.

González Echevarría, R, 1997, 'Kay Possa! Is 'Spanglish' a Language?' *The New York Times*, March 28, p. A29.

Gortázar, FG, 1996, *La arquitectura mexicana del siglo XX*, Consejo Nacional para la Cultura y las Artes, Federal District, Mexico.

Higgins, C, 2017, 'The Routledge Handbook of Migration and Language', in S Canagarajah (ed.), *The Routledge Handbook of migration and language*, Routledge., Abingdon, pp. 102–116.

Hubbard, P and Kitchin, R, 2011 (2004), 'Introduction: Why Key Thinkers?', in P Hubbard & R Kitchin (eds), *Key thinkers on space and place*, SAGE, London, pp. 1–17.

Lipski, J, 2008, *Varieties of Spanish in the United States*, Georgetown University Press, Washington, DC.

Low, SL, 2000, *On the Plaza: The Politics of Public Space and Culture*, University of Texas Press, Austin, Texas.

McCarthy, N, 2017, *The World's Most Spoken Languages*, viewed 19/01/19, www.statista.com/chart/12868/the-worlds-most-spoken-languages/.

Morales, E, 2002, *Living in Spanglish: The Search for Latino Identity in America*, St. Martin's, New York.

Nadeau, J-B and Barlow, J, 2013, *The Story of spanish*, Kindle edn, St Martin's Press, New York.

Osio, P, 2002, 'No se habla Spanglish: Useless hybrid traps Latinos in language barrio', *Houston Chronicle 8, diciembre 2002, IC.*, 8 December.

Otheguy, R, 2009, 'El llamado espanglish', in H Lopez-Morales (ed.), *Enciclopedia del español en los Estados Unidos: Anuario del Instituto Cervantes 2008*, Instituto Cervantes/Español Santillana., Madrid, pp. 222–243.

Otheguy, R and Stern, N, 2010, 'On So-called Spanglish', *International Journal of Bilingualism*, vol. 15, no. 1, pp. 85–100.

Rangel, RL, 2006, 'Ciudad de México: entre la primera y la segunda modernidades urbano-arquitectónicas', in P Krieger (ed.), *Megalópolis: La Modernización de la ciudad de México en el siglo XX*, UNAM, Instituto de Investigaciones Estéticas e Instituto Goethe-Inter Nationes, Mexico, pp. 179–186.

Roa, AS, 2008, 'The plaza de Bolívar of Bogotá: Uniqueness of place, multiplicity of events', in C Irazábal (ed.), *Ordinary places extraordinary events: Citizenship, democracy, and public space in Latin America*, Routledge., London and New York, pp. 126–143.

Rodríguez-González, E and Parafita-Couto, MC, 2012, 'Calling for Interdisciplinary Approaches to the Study of "Spanglish" and Its Linguistic Manifestations', *Hispania*, vol. 95, no. 3, pp. 461–480.

Rojas, RV and Timmling, HF, 2008, 'A Memorable Public Space: The Plaza of the Central Station in Santiago de Chile', in C Irazábal (ed.), *Ordinary places extraordinary events: Citizenship, democracy, and public space in Latin America*, Routledge., London and New York, pp. 84–102.

Sánchez-Muñoz, A, 2013, 'Who Soy Yo?: The Creative Use of "Spanglish" to Express a Hybrid Identity in Chicana/o Heritage Language Learners of Spanish', *Hispania*, vol. 96, no. 3, pp. 440–441.

Scollon, R and Scollon, SW, 2003, *Discourses in Place: Language in the Material World*, Routledge, USA & Canada.

Sheridan, G, 1998, 'Yes, in Coyoacán you can', *Vuelta*, vol. 22, no. 261, pp. 91–92.

Sheridan, G, 2004, 'Coyoacan I', in R Gallo (ed.), *The Mexico City Reader*, Kindle edn, The University of Wisconsin Press., Wisconsin, pp. Location 1196–222 of 3519.

Stavans, I, 2003, *Spanglish: The Making of a New American Language*, Rayo, New York.

Tuan, Y. F., 1977, *Space and Place: The Perspective of Experience*, University of Minesotta Press, Minneapolis.

Tuan, YF, 1991, 'Language and the Making of Place: A Narrative-Descriptive Approach', *Annals of the Association of American Geographers*, vol. 81, no. 4, pp. 684–696.

Villa, DJ, 2014, 'Spanglish', in C Tatum (ed.), *Encyclopedia of Latino culture: From calaveras to quinceañeras*, Greenwood, Westport, pp. 391–396.

Zentella, AC, 2008, 'Preface', in MN Murcia & J Rothman (eds), *Bilingualism and Identity: Spanish at the Crossroads with Other Languages.*, John Benjamins., Amsterdam, pp. 3–11.

18

Thinking, doing and being decolonisation in, with and as place

Sarah Wright

Introduction

Indigenous people, including many Indigenous scholars and community leaders (Wright et al., 2016; Coulthard, 2014; Graham, 2008; Jampijinpa, 2015; Larsen and Johnson, 2017), have repeatedly emphasised the need to attend deeply to place, and to work towards decolonising our relationships with it. This is an intensely important task, and one long overdue, in light of the devastating impacts of colonialism, modernism and patriarchal capitalism (Moreton-Robinson, 2015; Seawright, 2014). Engaging with place, talking and thinking about place, doing and being place is always political (Barker and Pickerill, 2019), although the ways it is political and what that politics means is different for all of us. For we are all differently positioned in relation to the call to deepen and rethink our engagements with place, we are differentially positioned in power relations, with different responsibilities (Noxolo, 2017). This is a journey and a call for all, to act and to know, from our place and with place.

In this piece I consider what it means to attend deeply to place, to live, be, know and act with and as place differently. I do this as a non-Indigenous person living on unceded Gumabynggirr land on the mid-north coast of NSW, Australia, and I both acknowledge this Country and its Elders, and recognise the privileges, complexities, limits, obligations and problems with my relationship to this stolen land, whose ongoing theft makes it possible for me, as a migrant-coloniser (Moreton-Robinson, 2015; Daley, 2020) to be here. While there is uneasiness around me writing this piece, I also acknowledge that it is the work of non-Indigenous people to address, where they can, their own responsibilities around decolonisation. My response, then, has two aspects. First is to attend to the call to nourish and reconfigure relations with and as place, taking the lead from Indigenous people who have generously shared diverse teachings about their worlds. Here, I seek to prioritise the cosmologies and struggles of Indigenous people, citing predominantly Indigenous scholars. Second is the need to attend to the power and politics that configure both Indigenous and non-Indigenous people's relations with, and knowledge about, place. I do this with particular attention to my place, my relationships to communities and the specific lands that I live on.

Attending to survivances and to Indigenous-led understandings means respecting place, understanding that, far from a backdrop to human action, place has agency, knowledge and law (Wright et al., 2016). Place is not an object of study (Daigle, 2016). Rather, it is a nourishing complex of relationships, of more-than-human kin (Coulthard, 2010). Place is relational and agential; it calls to action, it can teach (Larsen and Johnson, 2017). And in understanding that humans do not stand separate from place, that we are connected, differentially, through co-becoming, there is a need to understand the obligations and responsibilities of kinship. This means beginning or continuing to learn these responsibilities and to enact them not to a separate place, but as connected and co-emergent with place, acknowledging the thorny differential relations and responsibilities this entails (Suchet-Pearson et al., 2018).

Places are how and why we are who we are. Place brings us together, calls us and holds us within the shared quandaries, dilemmas and difficulties of co-existence. Place leads the ongoing survivances of Indigenous peoples; it suggests resistances, it teaches all diverse ways of being, differentially calling all into connection. And, through the specificities of place, come different worlds, the pluriverse, the world of many worlds (Larsen and Johnson, 2017). As we come into being in emergent ways, as we co-become, so we can become otherwise, more respectfully, with better relationships, so that there may be potential for different kinds of understandings, living our relationality as and with place.

Place as kin

Place, in Indigenous worldviews, is agential, a co-constituent of being. Indeed, according to Deloria and Wildcat (2001: 31) to be Indigenous is to "be of a place". Place nourishes and is nourished through affective more-than-human entanglements (Rose, 1996; Suchet-Pearson et al., 2018). It heals, guides and teaches with knowledge and Law. In doing so, it does not stand separate from humans, but comes about through diverse relationships with and as more-than-human kin (Graham, 2008).

Understanding place as sentient and part of a pattern of more-than-human kinship is foundational. Bakaman Yunupingu points out that this is one of the first learnings children are encouraged to understand. He says, "Land to us is a being. We feel bound to it. We feel related to it. That ties in with our singing, our dancing, our stories. It all relates. It is true. We believe in it. We know it's true" (The Yirrkala Film Project, 1986).

Place is neither a stage upon which actions play out, a passive container to hold "the environment", nor an object of study (Daigle, 2016). It, and the beings that co-constitute it, including humans, other animals, plants, soils, winds, spirits, waters, skies and dreams, hold story and song and knowledge. For some Indigenous people, place is an Elder, the "firstborn" that holds all other beings (Larsen and Johnson, 2017). Place actively personalises the energies of Creation, bringing beings into existence. Place creates life. It is the "active and ongoing unfolding of all creation into the situated and embodied forms of life that are all genealogical extensions of this place consciousness" (Ibid: 19). Place is never abstract.

Indigenous relationships with place, understandings of place, cosmologies and ontologies around place are necessarily deeply diverse. They must be, if such identities, relationships and understandings come into being with and as place because place, at the risk of sounding obvious, is different in different places (Yunupingu and Muller, 2009). It follows that more-than-human kinship and relationality looks, feels, is different in/as/with different places. Place-based understandings and emergent kinship, then, nourishes difference. Place in all its diverse relationality means the world is always plural, a pluriverse, a world of many worlds (Larsen and Johnson, 2017).

Indeed, conceptions and ideas of place are more diverse, more complex, more rich and beautiful and more powerful than can be expressed in words, particularly within English and in written form, and even more so as shared in these introductory ways by a migrant-coloniser such as myself. Indeed, importantly, Indigenous concepts of place go far beyond, far deeper than the introductory forms that may be shared outside the socio-legal obligations held by Indigenous cultures. There are always limits to what can be shared, what can be spoken, what can be understood (Kwaymullina, 2016). This is important to note. To say these sharings and insights are introductory, however, doesn't mean they are not profound. This is still more than enough to shake dominant conceptions of self and place and ownership quite rightfully to the core.

And it is not just the English language or processes of ongoing colonisation that make it difficult to communicate place. Coming into being with place is done with place, not with words, and in ways underpinned by diverse cosmologies and Indigenous Laws/lores, and, crucially, in ways embedded with differential relations of power, positionality and sovereignty (Hunt, 2014; Moreton-Robinson, 2015; Watts, 2013). Coming into being with place is affective, emotional and specific. Co-becoming place happens in dreams, feelings, thoughts, pre-cognitive moments and songs, through all senses (Jampijinpa, 2015). It is in the smell of drying mud from a waterhole and the song of the waterhole and the feel of mud on hands looking for waterlily root; in the rain on tarmac; the wind from the east in the afternoon on hot skin; and, in the sweat at night that tells someone who knows that fruits are ripe (Burarrwanga et al., 2013). It is in the love and the co-nourishment that comes from connection and knowledge, and in the energies that bind us to place through everyday thoughts, practices, feelings, beings and doings. What's more, there is much beyond human comprehension. There are things that only more-than-humans know about place, that no human can grasp; what or how the wind knows, makes, co-becomes place, or a fruit-bat or a rock. We co-become with and as place, but we do not know all that place knows.

Place is also always in emergence. It unfolds through relationships including through song, thought and practice (Murton, 2013). Brian Murton speaks of Māori conceptions of landscape and place, emphasising that creation is a process of continuous action that comes about through the unfolding of Te Kore, the primal power of the cosmos that already contains the form of every possible being. He draws on Stewart-Harawira's work (2005: 18) who shares:

> The cadences of ancient songs, of ritual calls, of sacred chants, through which the world is sung into existence, the flesh is sung onto the bones, and the relationships are sung which bind all together within the cosmos … (breathe) life into the network of subtle interconnections between human beings and the entire natural world.

For many Aboriginal people in Australia, place is Country. As Rose describes, Country is "a place that gives and receives life … a living entity with a yesterday, today and tomorrow, with a consciousness, and a will toward life" (Rose, 1996: 7). Within Country, everything comes about through emergent relationality through specific relationships (Wright et al., 2016). Country can and does communicate, nourish and is nourished in turn through relationships of care and responsibility. As Kwaymullina (2008: vi) puts it, "It is the whole of reality, a living story that forms and informs all existence. Country is alive, and more than love – it is life itself".

As Country comes about through relationships, the meaning and specificity of the relationships count. For Yolngu clans of north-east Arnhem Land, this is held by gurrutu, a kinship pattern that places all in relationship to each other and to more-than-human kin including ownership of particular lands, waters and seas (Wright et al., 2016). The specificity of our relationships to place also speaks to the need to be true to our own positionalities, in terms of relationships to place (as an Indigenous land owner or custodian of that Country or clan estate; as related to that Country; as migrant-coloniser). These specificities are ever important to the way we relate, understand, listen, learn, be and become. Yandaarra, a Gumbaynggirr and non-Gumbaynggirr research collective of which I am a part, speaks of the need to shift camp together. Together, we reflect (Smith et al., 2018: 11),

> Listening to Gumbaynggirr Country is to know one's place in Yandaarra. So we acknowledge the need to attend to our differences and positionalities, as also taught by Indigenous scholars and their allies, to be true to our own stories and to tell them, as we reach across them to connect where possible.

Place is to be attended to, respected for its teachings, its guidance and its Law/lore. Place calls, guides and teaches. Mary Graham (2001: 4), an Aboriginal scholar, Kombu-Merri and Wakka woman, explains that place defines and clarifies inquiry; it comes before and supersedes it. She says place, "informs us of where we are at any time, thereby at the same time informing us of who we are".

The agency of place is underscored by the work of the Bawaka Collective who cites Bawaka Country as the lead author on their work. Bawaka Country holds the collective and its research, it suggests topics, leads conversations, offers learnings, allows and disallows practices, thoughts and actions (Suchet-Pearson et al., 2018; Wright et al., 2016). Larsen and Johnson (2017), in drawing deeply from their entanglements and learnings with three places – the Cheslatta Carrier Nation traditional territory in British Columbia, Canada; the Wakarusa Wetlands in north-eastern Kansas, United States; and the Waitangi Treaty Grounds in Aotearoa/New Zealand – and from the First Nations people and allies that have heeded these places' call, speak to the ways place is profoundly pedagogical as it guides responsibilities and calls those who would listen, into dialogue, activism and ceremony. They also, movingly, point out that place teaches even when hurt.

Place and colonisation

And place is hurt. Relationships with place are hurt too; they have been hurt and continue to be hurt. This is important to understanding place. For, while it is important to acknowledge and support survivances and the rich everyday connections that continue to nurture Indigenous relationships with and as place, so, too, it is vital to acknowledge the damage, the violence and the ongoing attempts to disrupt them (Daley, Forthcoming).

For colonisation has long been underpinned by the theft of land, by the theft of place (Tuck and Yang, 2012). The way that colonialism has been able to proceed, capitalist expansion, life in settler colonial societies and in the countries of colonisers and ex-colonies – in short, the organisation of the world as we know it – has been built on the theft of land and the dislocation of people from place. In settler-colonial contexts such as Australia, Canada and the U.S.A, migrant-colonisers "settled"; they dispossessed

Indigenous people through a range of mechanisms including war, violence, attempted genocide, socio-legal machinations, and through those mechanisms they remain (Byrd, 2011; Todd, 2016; Veracini, 2015). As a result, the ongoing existence of these states, their very identity, the identity of those non-Indigenous people who live there, like myself, and their relationships with land, is premised on the theft and dispossession of others. And it is not just settler states, Indigenous people throughout the world face threats to their lands and their relationships to it, from settlers, from capitalist extraction, from state-run or sanctioned land-uses and from environmental change among myriad other causes. As Fredericks points out, "non-Indigenous territorialisation of sites and land holdings is only possible through dispossession and de-territorialising of Aboriginal people from that land" (Fredericks, 2013: 5). Land and place have been fundamental to the colonial relation, are fundamental to ongoing colonisation and are central to resistances to it.

The violences of colonialism not only mean land stolen, but violences that act to rupture people's relationships with places and disallow their obligations, in ways underpinned by very different modes of understanding place. This is "land-as-resource" (Graham, 2008), rather than land-as-relation; land is to be accessed, individually owned and used (Watts, 2013) in ways that prioritise rights and extraction over obligation. As Patrick Wolfe (2006: 388) points out, these are the ways that colonial societies "destroy to replace".

It's important to emphasise here that colonisation is not understood as an historical moment, an event, but as an ongoing structure (Veracini, 2015; Wolfe, 2006). Colonisation remains in the ongoing dispossession of lands, experienced through the cumulative impacts of generations dispossessed, the continuing-to-be-dispossessed, the ongoing issues of lack of access to land, the lack of control over decisions, the resource extraction, cultural disrespect and assimilative assaults, the racist conventions and laws, as well as the "force, fraud, and more recently so-called 'negotiations for recognition' on terms set by the State" (Coulthard, 2014: 7). These are deeply entangled capitalist, colonial and racist logics, the basis of how we are now; a matter of, at once, dispossession but of ongoing colonial possession (Moreton-Robinson, 2015).

This means that understandings of place must always be political. There can and should be no "cherry-picking" from Indigenous ontologies (Todd, 2016). Talking of place means talking of colonisation, Indigenous struggles and survivance, and acknowledging all of our positionality in that including, for non-Indigenous people, the ways we/they are implicated in and privileged by colonial relationships with place. Dispossession allows non-Indigenous people to wander through "our" national parks enjoying nature; violence and genocide has allowed enjoyment of "our" gardens and beautiful views.

While it is important to emphasise that colonisation is ongoing, it is also vitally important to acknowledge Indigenous survivances, strength and struggle (Vizenor, 1994). There is always, then, an incompleteness to colonisation. Indigenous people are still here. Indigenous land is still Indigenous land, everywhere. And culture and survivances continue to nurture people as place. For, while land and place are at the centre of colonialism, land and place are also at the centre of strength, resistance and survivance. This paradox has been emphasised by many Indigenous scholars and through myriad Indigenous movements (see, for example, Alfred and Corntassel, 2005; Coulthard, 2010, 2014; Fredericks, 2013; Snelgrove, Dhamoon, and Corntassel, 2014; Tuck and Yang, 2012). Even as place-as-resource provides the stimulus and foundation of the colonial-capitalist state, place calls and teaches and nourishes otherwise.

Place, Indigenous struggle and everyday self-determination

Place-based reciprocity provides an ethical foundation for many Indigenous peoples and their struggles. This is a point made by Glenn Coulthard of the Yellowknives Dene First Nation (2010: 79–80):

> Place is a way of knowing, experiencing, and relating with the world – and these ways of knowing often guide forms of resistance to power relations that threaten to erase or destroy our senses of place. This, I would argue, is precisely the understanding of land and/or place that not only anchors many Indigenous peoples' critique of colonial relations of force and command, but also our visions of what a truly post-colonial relationship of peaceful co-existence might look like.

For Coulthard and others, the reciprocal relationships held with place amount to an obligation to resist colonial modes of land ownership, and the organisational imperatives of colonial-capital accumulations with their concomitant abstracted ways of knowing and being, and relationships with place. Although heavily disrupted by colonialism, place and relationships with place nonetheless also provide the impetus and vision of Indigenous struggle and transformation, as well as what new, contemporary, just and reciprocal relationships with non-Indigenous people might look like. It is from place that the vision for change, the motivation for struggle and the hope and nourishment needed to continue might be found.

For Cree geographer Michelle Daigle, place also provides the basis for the everyday practices of self-determination and survivance that emerge from Indigenous place-based ontologies. Here, on a local scale, in everyday life, Indigenous practices of resurgence are guided and nourished by place. This is how many Indigenous people live self-determination often outside any formal state or intergovernmental structures (Alfred and Corntassel, 2005; Daigle, 2016; Vizenor, 1994). Daigle also points out that, while specific local more-than-human relationships are foundation to practices of self-determination, these practices also travel as they are mobilised by kinship relations that connect onwards to, through and with, other places and communities. Relational obligations are thus grounded but not fixed or static. Rather, accountability to place, land and kinship relations guides relationships with other clans, other human and non-human kin beyond specific lands.

Place also calls those who would heed it into activism, into struggle and into ceremony. Here, activism can take on an enlarged meaning through its ceremonial and spiritual aspects (Larsen and Johnson, 2017). In doing so, place may also lead to different kinds of decolonising subjectivities. While these alliances are led by Indigenous people, Indigenous communities and Indigenous ways of knowing and being, they also include those of settler/migrant-coloniser descent who may heed a call to fulfil their differential responsibilities. Here, place can challenge and call into dialogue, offering the gift of hospitality in ways that might encourage productive ontological relationships.

Place catalysing co-existence, doing decolonisation with and as place

Non-Indigenous people too have a part to play in the needed shifts. The burden should not be on Indigenous people alone. There has been more than enough of that. While place, as Coulthard points out, provides guidance for how Indigenous and non-Indigenous people might move towards more just decolonising relationships, such moves require non-Indigenous people to move, to listen deeply, to try and engage respectfully in ways that may not be easy to navigate. This means, in part, attending to place, staying with the difficulties of co-existence and taking up the challenges in place-based, ethical ways (TallBear, 2014).

Underpinning these calls, as Michelle Daigle points out, is a need for deeper engagements with the ontological underpinnings of place, including place's agency, and a greater attentiveness to geographies of responsibility. This is echoed by Larsen and Johnson who suggest "productive ontological relationships cannot occur without place or cannot occur within an impoverished understanding of place as backdrop" (2012). Place is not an object but a field of relationships, a teacher and guide. As Yuin Elder Uncle Max Dulumunmun Harrison (2009: 7) suggests, it is a matter of "really seeing what the land is telling you". He hopes his teaching may help remake connections between humans and non-humans, and so serve to honour his teachers and ancestors. These are never abstract but must continue to be brought into ourselves, whoever we may be, into our knowing and doing and acting.

This is in no way a call for non-Indigenous people to somehow "become Indigenous" or to appropriate Indigenous Law/lore, using it out of context. Neither is it a call for non-Indigenous people to race ahead (with a devastatingly familiar sense of entitlement) and just fix things. Uncle Max Harrison and others speak of the need for caution, for slowing down, learning and listening deeply (see also Rigney, 2006). As such, it is something both more modest and deeply fundamental, a call for non-Indigenous people to interrogate their own positionality, their relationships to the land including the way they have been and continue to be privileged by the colonial project even as they seek to begin to dismantle it, and to begin to build different kinds of relationships through learning to be differently with place.

The unevenness of the relationships is important to acknowledge; relationships between people and places are not symmetrical, not between different people or between people and places. As Noxolo et al. (2008: 164) point out, we are "all connected to structural processes that produce injustice", but "we are not all equally positioned". This means that decolonising relations to place is not easy, is not even possible in the current situation, contaminated as it is with colonialism and racism. While the vast majority of the world's geographers live on stolen land or in places made possible through stolen land and labour (here I'm thinking of Europe which wouldn't exist without its colonial histories/presents), to fully decolonise relations, as if this could be completed and moved on from, is not possible. What *is* possible is to move, take steps, lean inwards towards responsibility, contesting ignorance and to reflexive action and engagement, to *do*, but not in ways that seek to restore comfort, for the colonising remains and that is never comfortable, or innocence, for innocence cannot be found (Land, 2015; Tuck and Yang, 2012).

So that, to say the ethics are complex, the terrain is fraught; to say the end is not in sight; to say that non-Indigenous people must not, for once, think they can solve everything, lead everything, to say to non-Indigenous people "you are complicit", is *not* to say "do nothing". It is an important task, ever more important, and all can work, from their place, in recognition of their positionality. This is underscored by Daigle and Sundberg as they contend, "the discipline of geography will retain its Eurocentricity, coloniality and whiteness unless all geographers begin to do the anti-racist and decolonial work historically done by Indigenous, people of colour, women and queer faculty and students" (2017: 251).

That means that engaging with place in ways led by Indigenous people must be done carefully. The ethics are far from clear (Barker and Pickerill, 2019). Which brings me back to my positionality as a white, non-Indigenous woman living on unceded Gumbaynggirr land, back to me and my place in all this, and my blunders, white-women arrogances and misplaced fragilities. I write this piece from a deeply problematic place and I am unsure as I try to carefully navigate my path. While I am committed to the journey, I am in no way certain of where I am headed and have made many mis-steps along the way. I can't become innocent or unwhite, I can't ever have a simple or unproblematic relationship with place because I continue to live and benefit from, and have always lived and benefited from, the very colonial and exploitative relationships that I seek to change.

For me, one of the most important aspects is to try and make sure my learnings are not just academic, in my head or on the level of abstractions, words, and for professional advancement, but that they live in my life, my heart, my doings and beings over time. Within Yandaarra, both Aunty Shaa Smith, storyholder for Gumbaynggirr Country, and Neeyan Smith, her daughter, have spoken of the need to look to the question of purpose. Why are we doing this, they encourage us to ask as Indigenous and non-Indigenous people trying to respect Gumbaynggirr Law/lore from our different positionalities. What will we do with it? And how does it fit in our lives? Such knowings and doings link us inextricably to each other in and with place so that our purpose and responsibilities are not to something discrete and separate from us, but are part of our emerging, relational, more-than-human selves. For me, this means trying to unlearn and to learn, to challenge my colonised understandings and to bring this into practice in all parts of my life including my home and my parenting. It means acknowledging my positionality, my complicity in the theft and violences of colonisation, and being aware where things are simply not my place too, as I try to respect, but not appropriate, in ways that honour place and my ancestors. This includes supporting the struggles of others, including with time, resources or energy, walking together or stepping back (Land, 2015; Wright, 2018). This is sometimes destabilising and challenging, but it is important and beautiful too. In discomfort and awkwardness are ways power might shift, privilege be realigned, and assumptions might crack (Seawright, 2014).

Conclusion

As geographers, we are increasingly called to decolonise ourselves, our disciplines and institutions, our relationships with each other and place. For Indigenous people, including Indigenous geographers, that can mean decolonising themselves, pushing back against colonising institutions and processes, refusing engagements if and where necessary, and nurturing diverse survivances as and with place (Alfred and Corntassel, 2005; Daigle, 2016; Moreton-Robinson and Walter, 2009; Underhill-Sem, 2017). For non-Indigenous geographers, decolonising means reconfiguring relationships with place in ways that engage deeply and respectfully with Indigenous world views. Such engagements must seek, not to extract or appropriate, but rather challenge the foundations of dominant knowings and doings, even to the understanding of self. And underpinning all this, nourishing and challenging, is land, is place. For all, this is a difficult, even impossible, task. Yet it is one we must continue to navigate.

In contributing to this book, we have been invited to offer new conceptual frameworks and future directions in thinking about place. What I discuss here, however, is not new, as place and the understandings that have been shared here have come into existence with the world. Deep understandings of place and our relationships with it have long been offered by Indigenous people but they have also long been dismissed, ignored, or, if it suits, extracted. And, too, there have long been different engagements from social movements, groups of people coming together to journey with each other and with place in different, more respectful, ways. So these thoughts are not new. Neither are they just thoughts, ways of thinking about place. Rather, these are potential directions in thinking-doing-being place, in different kinds of ways. And in this, yes, there is a direction, a movement, a chance for a different kind of becoming and becoming-otherwise; an obligation, an imperative, to work at doing decolonisation with and as place, differentially, plurally, in our own ways, from our place and with place, together.

References

Alfred GT and Corntassel J (2005). Being indigenous: Resurgences against contemporary colonialism. *Government and Opposition*, 40(4), 597–614.

Barker A and Pickerill J (2019). Doings with the land and sea: Decolonising geographies, Indigeneity, and enacting place-agency. *Progress in Human Geography*, 10.1177/0309132519839863.

Bawaka Country including Suchet-Pearson S, Wright S, Lloyd K, Tofa M, Sweeney J, Burarrwanga L, Ganambarr R, Ganambarr-Stubbs M, Ganambarr B, Maymuru D (2018). Goŋ Gurtha: Enacting response-abilities as situated co-becoming. *Social and Cultural Geography*, 10.1177/0263775818799749.

Bawaka Country, Wright S, Suchet-pearson S, Lloyd K, Burarrwanga L, Ganambarr R, … Sweeney J (2016). Co-becoming Bawaka : Towards a relational understanding of place/space. *Progress in Human Geography*, *40*(4), 455–475.

Burarrwanga L, Ganambarr R, Ganambarr-Stubbs M, Ganambarr B, Maymuru D, Wright S, Suchet-Pearson S, Lloyd K (2013). *Welcome to My Country*. Crows Nest, NSW: Allen and Unwin.

Byrd J (2011). *The Transit of Empire: Indigenous Critiques of Colonialism*. Minneapolis: University of Minnesota Press.

Coulthard G (2010). Place against empire: Understanding indigenous anti- colonialism. *Affinities: A Journal of Radical Theory, Culture, and Action*, 4(2), 79–83.

Coulthard G (2014). *Red Skin, White Masks: Rejecting the Colonial Politics of Recognition*. Minneapolis: University of Minnesota Press.

Daigle M (2016). Awawanenitakik: The spatial politics of recognition and relational geographies of Indigenous self-determination. *The Canadian Geographer*, 60(2), 259–269.

Daigle M and Sundberg J (2017). From where we stand: Unsettling geographical knowledges in the classroom. *Transactions of the Institute of British Geographers*, 42(3), 338–341.

Daley L (2020). An urban cultural interface: (Re)thinking urban anti-capitalist politics and the city in relation to indigenous struggles. Unpublished PhD thesis. The University of Newcastle: Callaghan. Lara.daley@newcastle.edu.au.

Deloria V Jr and Wildcat DP (2001). *Power and Place: Indian Education in America*. Golden, CO: Fulcrum Resources.

Fredericks B (2013). "We don't leave our identities at the city limits": Aboriginal and Torres Strait Islander people living in urban localities. *Australian Aboriginal Studies*, *1*, 4–16.

Graham M (2001). Understanding human agency in terms of place: A proposed aboriginal research methodology. *PAN: Philosophy Activism Nature*, (6), 71.

Graham M. (2008). Some thoughts about the philosophical underpinnings of aboriginal worldviews. *Australian Humanities Review*, *45*, 181–194.

Harrison MD (2009). *My People's Dreaming*. Warriewood, NSW, Australia: Finch Publishing.

Hunt S (2014). Ontologies of Indigeneity: The politics of embodying a concept. *Cultural geographies*, 21(1), 27–32.

Jampijinpa Patrick W. S. (2015). Pulya-ranyi winds of change. *Cultural Studies Review*, *21*, 1.

Kwaymullina A (2008). Introduction: A Land of Many Countries. In *Heartsick for Country Stories of Love, Spirit and Creation*. S Morgan, T Mia and B Kwaymullina (eds). Freemantle: Freemantle Press. 5.

Kwaymullina A (2016). Research, ethics and Indigenous peoples: An Australian indigenous perspective on three threshold considerations for respectful engagement. *AlterNative: An International Journal of Indigenous Peoples*, *12*(4), 437–449.

Land C (2015). *Decolonising Solidarity*. London: Zed Books.

Larsen S and Johnson J (2012). *Heeding the Call of Place: Call for papers*. San Francisco: Association of American Geographers.

Larsen S and Johnson J (2017). *Being Together in Place: Indigenous coexistance in a more than Human World.* Minneapolis: University of Minnesota Press.
Moreton-Robinson A (2015). *The White Possessive: Property, Power and Indigenous Sovereignty.* Minneapolis: University of Minnesota Press.
Moreton-Robinson A and Walter MM (2009). Indigenous methodologies in social research (Online Chapter 22), Social Research Methods (Second Edition), Oxford, M. Walter (ed), South Melbourne, pp. 1–18.
Murton B (2013). From Landscape to Whenua: Thoughts on Interweaving Indigenous and Western Ideas about Landscape. In *A Deeper Sense of Place: Stories and Journeys of Collaboration in Indigenous Research.* J Johnson and S Larsen (eds). Corvallis: Oregon State University Press. 139–156.
Noxolo P (2017). Introduction: Decolonising geographical knowledge in a colonised and re-colonising postcolonial world. *Area*, 49, 317–319.
Noxolo P, Raghuram P, & Madge C (2008). 'Geography is pregnant' and 'Geography's milk is flowing': Metaphors for a postcolonial discipline? *Environment and Planning D: Society and Space, 26*(1), 146–168.
Rigney L (2006). *Indigenist Research and Aboriginal Australia.* Farnham: Ashgate.
Rose D B (1996). *Nourishing Terrains: Australian Aboriginal Views of Landscape and Wilderness.* Canberra: Australian Heritage Commission.
Seawright G (2014). Settler traditions of place : Making explicit the epistemological legacy of white supremacy and Settler Colonialism for place-based education. *Educational Studies, 50*(6), 554–572.
Smith A. S., Smith N, Wright S, Hodge P, & Daley L (2018). Yandaarra is living protocol. *Social & Cultural Geography*, 10.1080/14649365.2018.1508740.
Snelgrove C, Dhamoon RK, & Corntassel J (2014). Unsettling settler colonialism : The discourse and politics of settlers, and solidarity with Indigenous nations. *Decolonization: Indigeneity, Education & Society, 3*(2), 1–32.
Stewart-Harawira M (2005). *The New Imperial Order: Indigenous Responses to Globalization.* London: Zed Books.
TallBear K (2014). Standing with and speaking as faith: A Feminist-Indigenous approach to inquiry. *Journal of Research Practice*, 10, 17.
Todd Z (2016). An Indigenous feminist's take on the ontological turn: "Ontology" is just another word for colonialism. *Journal of Historical Sociology, 29*(1), 4–22.
Tuck E, & Yang KW (2012). Decolonization is not a metaphor. *Decolonization: Indigeneity, Education & Society, 1*(1), 1–40.
Underhill-Sem Y (2017). Academic work as radical practice: Getting in, creating a space, not giving up. *Geographical Research*, 55, 332–337.
Veracini L (2015). *The Settler Colonial Present.* London: Springer.
Vizenor G (1994) *Manifest Manners: Narratives on Postindian Survivance.* Lincoln, NE, and London: University of Nebraska Press.
Watts V (2013). Indigenous place-thought & agency amongst humans and non-humans (First Woman and Sky Woman go on a European world tour!). *Decolonization: Indigeneity & Society, 2*(1), 20–34.
Wolfe P (2006). Settler colonialism and the elimination of the native. *Journal of Genocide Research, 8*(4), 387–409.
Wright S (2018). When dialogue means refusal. *Dialogues in Human Geography*, 10.1177/2043820618780570.
The Yirrkala Film Project (1986). We believe in it … We know it's true. Filmmaker Ian Dunlop. National Film and Sound Archive of Australia.
Yunupingu D and Muller S (2009). Cross-cultural challenges for Indigenous sea country management in Australia. *Australasian Journal of Environmental Management*, 16(3), 158–167.

Part III
Identity and place

Samantha Wilkinson

Introduction: power, identity and place

Unprecedented levels of mobility and scales of globalisation have, on the one hand, constructed a celebratory discourse about globalisation and its promotion of the hybridity and the colourfulness of the "global village" (McLuhan, 1964). On the other hand, as is evident in current socio-political debates around Brexit and immigration in the UK, we are witnessing a rise in nationalism that romanticises an imagined past without migration and globalisation. At the heart of this cosmopolitan consciousness lies a deeper ontological contradiction between how place is conceptualised in these two discourses: a global village/contact zones, versus a geographical location owned by an imagined homogenous nation (see Badwan and Wilkinson, 2019). Nonetheless, as a consequence of increasing movements of people, there is no doubt that the notion of identity has become destabilised and reconfigured (Massey, 2004).

Moreover, place as a theoretical construct has attracted different interpretations in different disciplines. For instance, research in traditional sociolinguistics has tended to treat place as a space filled with norms, expectations and orders of indexicalities (Blommaert, 2010), whilst research in human geography has problematised and challenged static, homogeneous ontologies of place. Consequently, place becomes slippery (Markusen, 1996a), relative (Cele, 2013a), and spaces for meeting and sharing (Massey, 2004). By "meeting" Massey (2004: 3) means that there is a necessity of negotiating across and among difference "the implacable spatial fact of shared turf". Moreover, by "sharing", Massey (2004: 3) argues that places (localities, regions, nations) are necessarily the location of the intersection of disparate trajectories. The chapters in this part align themselves with liquid approaches to place that treat place as both a verb and a process. They ontologically and epistemologically draw on understandings of place as "meaning" rather than "location" (see also previous work by Entrikin, 1991), which emphasises the role of individuals' emotions, experiences and activities. By understanding place as meaning, and therefore as meaningful, we have to question whether places are always a prime source for the production of personal and cultural identity?

In people's everyday existence, there are many occasions when the term "place" is used. Some people may use the term to refer to a particular location or building. On other occasions, the term place is used in turns of phrase to represent the stance one should or should not maintain, such as "know your place", or "it is not my place to comment". In such expressions, place is much more than simply a spatial reference. It is clear that someone can belong, or not, in a particular place and that this is constructed relationally (Cresswell, 1996). Relational identities mean that we compare ourselves with others (Horton and Kraftl, 2014). As Crang (1998: 61, emphasis in original) writes: "identity can be defined as much by what we are *not* as by who we *are*". According to Cresswell (1996), there are expectations about behaviour that relate to a position in a social structure to actions in place. Put another way, people may be perceived to act appropriately or inappropriately in place, depending on their class, gender, race, sexuality and age, for instance. Of course, places themselves have their own power; Edensor (2002) contends that the congruence of affective, sensual and embodied effects, which inhere in relationships between people and place, strengthens the ontological power of place.

Place attachment, or a "sense of place", emerges from an increased depth of knowledge and association with a location, filling abstract space with meaning (Holton, 2015). From this perspective, those with limited connections to a place such as tourists are more likely to have a weaker sense of place in comparison with those with more historical connections, for instance, with those who were born or grew up in a particular place. This links with the identity of a place as constructed through marketing and place-promotion and the evolving individual and collective identities that emerge for those who live in a place. As Holton (2015) contends, recent conceptualisations have reconsidered place as the sedentary equivalent of mobility, to recognise the dynamic potential of place and its potential for evoking powerful emotional responses. A core theme running through the chapters in this part is identity, understood as both lived and imagined in ways that break down its contiguousness with a geographically bounded locality (Fortier, 1999). More of us now live in what Said (1979: 18) termed: "a generalized condition of homelessness"; that is, a world in which identities are at least differently territorialised, if not wholly deterritorialised (Gupta and Ferguson, 1992).

As is discussed in the Introduction to this book, Relph (1976: preface) argues that, in our modern era, an authentic sense of place is being gradually overshadowed by a less authentic attitude that he called placelessness; that is, "the casual eradication of distinctive places and the making of standardized landscapes that results from an insensitivity to the significance of place". Examples of such generic landscapes include hotels, motorways, service stations and airports. Such spaces may be considered to lack a "sense of place" and are typically kinds of what Auge (1995) terms "non-places". That is, places of transience where humans remain anonymous and that do not hold enough significance to be regarded as places. Places, by contrast, as the chapters in this part demonstrate, offer people a space that can potentially empower their identity, where they can meet other people with whom they share social references and histories.

People can belong or not in place. Belonging refers to the practices of group identity in manufacturing cultural and historical connections that mark out terrains of commonality that delineate the politics and social dynamics of "fitting in" (Fortier, 1999). Of course, it is worth thinking about those who "fit in" differently, or in contesting ways, and that there are individuals who may not want to "fit in" and others who cannot "fit in". According to Billig (1995), everyday representations of the nation, such as flags in everyday contexts, national songs, symbols embossed or printed on money and popular turns of phrase, are all aspects of what the author calls "banal nationalism", building a shared sense of national belonging amongst humans, a sense of tribalism through national identity. Similarly, at smaller scales, regional and local identifiers such as commonly shared slang words, place logos, the regalia of local football teams and folk songs similarly produce kinds of banal place identity.

Authors such as Read et al. (2003) have done well to bring to the fore how socio-economic status, age and ethnicity may influence feelings of place belonging. However, the authors do not pay attention to other key identity markers such as (dis)ability and gender. Fenster (2005) explores gendered feelings of belonging at the city scale, contending that there are different expressions of belonging in women's and men's narratives of their everyday life in the city; put another way, "the right to the city" is gendered. For instance, the author highlights how the "public" realm is perceived as the white middle- or upper-class heterosexual male domain. Sometimes, then, women in both Western and non-Western cities simply cannot wander around the streets, parks and urban spaces alone, and in some cultures cannot wander around at all (Fenster, 2005). Lefebvre (1991) based the concept of "the right to the city" not on formal citizenship status, but on inhabitance. As summarised by Purcell (2002), this evolves within it two fundamental rights: 1) the right to appropriate urban space, in the sense of the right of inhabitants to "full and complete use" of urban space in their everyday lives; 2) the right to participation, a right of inhabitants to take a central role in decision-making surrounding the production, or ideally co-production, of urban space. It is worth noting that authors such as Limonad and Monte-Mor (2015) have contended that the right to the city surpasses, by and large, the city itself; the authors contend that the city is not anymore an exclusive place for the urban and the countryside is not anymore an exclusive rural location. The chapters in this part explore many different facets of identity, including (dis)ability, gender, age, race, faith and class, and the role of both urban and rural places, in and beyond the UK, in shaping, and being shaped by, identity performances.

In the first chapter, Samantha Wilkinson and Catherine Wilkinson explore place, age and identity. They bring to the fore how young and older people use and occupy place to perform their identities, whilst highlighting how both young and older people can experience feelings of inclusion and exclusion from different places. The authors take inspiration from Kjorholt (2003b) and argue that there needs to be a greater understanding of "joint inter-generational geographies"; that is, we need to bring to the fore ways in which young and older people utilise spaces in mutual and supportive ways.

Gendering place is the focus of Dina Vaiou's chapter. Drawing from research conducted in Athens, she proposes three lines of argument to help think through the "gendering" of place. First, the everyday as a critical entry point into place-making through gendered labour and practices. Second, moving and settling through practices of care. Third, the multiplicity of often invisible borders and boundaries along the routes of mobility in the age of globalisation. Vaiou contends that these three lines of argument underline the relations of power involved, as they take shape along the lines of gender, ethnicity and other axes of inequality, and at different and interlocking scales.

Exploring another key identity marker, this time with a focus on race, Michele Lobo engages with the notions of encounter and co-belonging in the Northern Australian city of Darwin. The discussion poetically explores how racialised bodies that are easily overlooked mingle with the forces of a more-than-human world to create place as a lacework that touches and dances with the earth. Lobo argues that what she terms "shadowscapes" produce place as a site of new imaginations of co-belonging and earthly survival.

Paul Watt explores the identity marker of social class and place in the cities and suburbs of the global North. The crux of this chapter is place attachment and belonging to the residential neighbourhood with reference to the working class, middle classes and the "super rich". Watt contends that whilst community and place belonging were prominent aspects of working-class neighbourhoods up until the 1970s, since then there has been an epochal shift in the significance and nature of such neighbourhoods in European and North American towns and cities. With regards to the super-rich, Watt highlights that the relationship to space and place is decidedly ambivalent. On the one hand, the super-rich represent the epitome of globally rootless mobilities whilst, on the other hand, Watt highlights that the super-rich come down to terra firma, but only to places which are the dominant environments of people just like them.

Ben Rogaly's chapter foregrounds the contested politics of place, rethinking often taken-for-granted assumptions that there is an inevitable clash of interests between people categorised as "migrant" and people seen as "local". Reflecting in turn on the national, urban and workplace scales, he tells a story about how places change and the role of oral history in recording such change in an analysis of the often taken-for-granted binary of "local" versus "migrant" in the English midlands town of Loughborough. The chapter is also thought-provoking in terms of how oral history might itself contribute to changing places.

Rurality, place and the imagination are discussed in Rosemary Shirley's chapter, which centres on British ideas of the rural based in the UK. She explores how imaginary ideas of the rural feature in everyday life and are often connected to forms of consumption, arguing that this produces a set of rural mythologies – stories we are consistently told or tell ourselves about the rural. Shirley draws on examples from everyday life and popular culture to focus on two particularly prevalent rural mythologies: that the rural is healthy and that the rural is situated in the past.

Henrik Schultze explores community and place, providing a useful distinction between the two concepts; the author argues that community and place are not related inevitably but that place may be a necessary condition for the construction and maintenance of a symbolic community. Schultze draws on a case study on belonging conducted in Prenzlauer Berg, Berlin. We see how, for participants in Schultze's study, markers of difference may become salient and how sharp boundaries are drawn between "us" and "them".

Saskia Warren's chapter explores fashioning place, drawing on empirical examples from the British high street. The author looks at how women are using high street shopping to represent how minority faith followers can be contemporary, fashionable and cool. Warren explores how Muslim women, through their tactics and strategies around high street consumerism, use everyday forms of creativity – selecting, layering, mixing – to negotiate minority status in ways that add distinction and can become profitable. In everyday practices, Muslim fashionistas are challenging the meaning of public space as marked by masculine, white and secular cultural norms. As new Muslim female identities are formed, fashionable young Muslim women are fostering spaces of belonging and negotiating place in contexts where many once felt marginalised or "out of place".

Ilan Wiesel and Ellen van Holstein explore some of the key epistemological and ontological debates surrounding the relationship between disability and place. From an epistemological perspective, the authors consider tensions between structural and relational approaches to thinking about the production of place, its effects on people with disability and their agency within such processes. From an ontological perspective, the authors consider the systems of categorisation that underpin scholarly and other discourses on disability and place. Wiesel and van Holstein focus on the binary categories of "mainstream" and "specialist" places, and their central role in framing debates about the oppression of people with disability, deinstitutionalisation and social inclusion and exclusion.

In the final chapter of this part, Cuttaleeya Jiraprasertkun explores how the urban sprawl in Bangkok, formerly renowned as the Venice of the East with its distinctive houses situated along waterways surrounded by orchards, has created an intermingling of surviving and newly emerging elements. Jiraprasertkun draws on local vocabularies and knowledge to understand the complex processes through which Bangkok retains a distinctive identity rather than becoming a "city of nowhere".

The chapters in this part focus on a wide variety of places, from the micro-scale of the British high street to North American cities. The authors in this part do not present place as a passive backdrop to identity performances; rather, they present places as active constituents with the capacity to shape identity performances, and to be shaped by them. The chapters in this part highlight that identities are intersectional and forged in and through relations and, consequently, much like place, they are not rooted or static, but mutable, ongoing productions (Massey, 2004).

References

Auge, M. (1995) *Non-Places, Introduction to an Anthropology of Supermodernity*. Stanford: Stanford University Press.

Badwan, K. and Wilkinson, S. (2019) '"Most of the People My Age Tend to Move Out": Young Men Talking about Place, Community and Belonging in Manchester'. *Boyhood Studies*.

Billig, M. (1995) *Banal Nationalism*. London: Sage.

Blommaert, J. (2010) *The Sociolinguistics of Globalization*. Cambridge: Cambridge University Press.

Cele, S. (2013a) 'Performing the Political Through Public Space: Teenage Girls' Everyday Use Of A City Park', *Space and Polity*. 17(1): 74–87.

Crang, M. (1998) *Cultural Geography*. London: Routledge.

Cresswell, T. (1996) *In Place/Out of Place: Geography, Ideology and Transgression*. Minneapolis: University of Minnesota Press.

Edensor, T. (2002) *National Identity, Popular Culture and Everyday Life*. Oxford: Berg.

Entrikin, J. (1991) *The Betweenness of Place: Towards a Geography of Modernity*. London: Palgrave.

Fenster, T. (2005) 'The Right to the Gendered City: Different Formations of Belonging in Everyday Life'. *Journal of Gender Studies*. 14(3): 217–231.

Fortier, A.M. (1999) 'Re-membering Places and the Performances of Belonging(s)', *Theory, Culture and Society*. 16(2): 41–64.

Gupta, A. and Ferguson, J. (1992) 'Beyond 'culture': Space, Identity and the Politics of Differences'. *Cultural Anthropology*. 7(1): 6–23.

Holton, M. (2015) '"I already know the city, I don't have to explore it": Adjustments to "Sense of Place" for "Local" UK University Students'. *Population, Space and Place*. 21: 820–831.

Horton, J. and Kraftl, P. (2014). *Cultural Geographies: An Introduction*. London: Routledge.

Kjorholt, A. (2003b) '"Creating a place to belong": girls' and boys' hut-building as a site for understanding discourses on childhood and generational relations in a Norwegian community'. *Children's Geographies*. 1(1): 261–279.

Lefebvre, H. (1991) *Critique of Everyday Life*. London: Verso.

Limonad, E. and Monte-Mor, R. (2015) 'Beyond the Right to the City: Between the Rural and the Urban'. *Les Cahiers du Developpement Urbain Durable*. 13: 103–114.

Massey, D. (2004) 'Geographies of Responsibility'. *Geografiska Annaler: Series B, Human Geography*. 86(1): 5–18.

Markusen, A. (1996b) 'Sticky Places in Slippery Space: A Typology of Industrial Districts'. *Economic Geography*. 72(3): 293–313.

McLuhan, M. (1964) *The Gutenberg Galaxy: The Making of Typographic Man*. Canada: University of Toronto Press.

Purcell, M. (2002) 'Excavating Lefebvre: The right to the city and its urban politics of the inhabitant'. *Geojournal*. 22: 99–108.

Said, E. (1979) 'Zionism from the Standpoint of its Victims'. *Social Text*. 1: 7–59.

19

Place, age and identity

Samantha Wilkinson and Catherine Wilkinson

Introduction

This chapter considers how young people and older people use and occupy place to perform their identities, whilst highlighting how both young and older people can experience inclusion and exclusion in different places. In an increasingly globalising world, academic literature, both within and beyond the discipline of human geography, has challenged static, homogeneous conceptualisations of place (Dicken, 2010). As a consequence, place becomes a "process" (Massey, 1999), "slippery" (Markusen, 1996a) and "relative" (Cele, 2013a). In this chapter, we align ourselves with liquid approaches to place, drawing on understandings of place as "meaning", as opposed to "location" (Entrikin, 1991), emphasising the role of individuals' emotions, experiences and activities. Moreover, we seek to promote place not as a passive backdrop to the everyday lives of young and older people, but as integral to social relations (Wiles, 2005).

When young people and place are discussed in the academic literature, it is typically in relation to young people's appropriation of public spaces such as parks and streets (for instance, Cahill, 2000; Gough and Franch, 2005; Matthews et al., 1999). Conversely, older people tend to be narrowly represented in the literature as confined to their own homes, hospitals, residential or nursing homes (Andrews and Peter, 2006; Schneider et al., 2019; Tanner et al., 2008; Turner et al., 2018; Williams, 2002). This fails to recognise the importance of outdoor spaces (such as parks, allotments and garden centres) and a more diverse range of indoor spaces (including theatres and shopping centres) for older people's identity creation. This chapter argues that young and older people's places are part of an imagined geography that privileges certain places and thus inadvertently produces distorted notions of inclusion and exclusion. We therefore argue for a broadening of geographies of age that does justice to both young and older people's use of a diverse range of places.

Furthermore, this chapter highlights that there is a need for a relational geography of age and place (Hopkins and Pain, 2007). One way of achieving this is to explore intergenerational relationships in different spaces and places. To explain, whilst separate bodies of literature have discussed either young or older people's engagement with places, it is seldom that bodies of literature explore how younger and older people unite in places. Indeed, when intergenerational relationships are considered surrounding place, it is often in terms of conflict (e.g., adults attempting to move young people on from particular places) (Matthews et al., 2000; Pennay and Room, 2012). Consequently, we argue that there is a need for a much greater understanding of "joint inter-generational" geographies (Kjorholt, 2003a: 273); that is, how young and older people utilise spaces in harmony and togetherness. This chapter begins with an overview of literature on young people and place, before moving on to discuss older people and place. Before concluding, the chapter highlights a small body of literature on intergenerational relationships and place. In this section, we tease out what researchers can gain through adopting intergenerational approaches to exploring place. This chapter concludes by contending that there is a need for further research that uncovers the diverse places used by young people, adults and older people, both separately and together.

Young people and place

This section explores children and young people's experiences in a variety of places. The importance in doing so is clear when considering Kintrea et al.'s (2015: 666) argument that children and young people's aspirations can be "shaped by place". Historically, children have been conceptualised in two opposing ways, as devils and angels (Valentine, 1996). Dionysian and Apollonian views of childhood place children in the home. To explain, Dionysian views of childhood present children as troublesome and restrict them to the space of the home through curfews. Apollonian views place emphasis on exclusion from urban spaces, by emphasising rural idyllic childhood. Holloway and Hubbard (2001) support the notion that children's geographies are restricted, arguing that this is often related to assumptions that children are socially incompetent and unable to handle the rigours of navigating public space. As a consequence of the streets being depicted as inherently dangerous, children are largely restricted to occupying the space of the home or officially designated and designed playgrounds. Indeed, today's children have been nicknamed "cotton-wool kids" and the "bubble-wrap" generation, and it has been proclaimed that they grow up in walled gardens (Malone, 2007). Parents who seek to protect their children from life's dangers have been labelled "helicopter parents" (Coburn, 2006) and are accused of preventing their children from taking developmentally important steps towards independence. Young people are thus not gaining the experiential knowledge to become street literate (Cahill, 2000). It should be emphasised that by imposing such restrictions on children's engagements with a variety of places, adult geographies are simultaneously naturalised.

Whilst children are thus notable for their absence from public spaces, teenagers are often considered a ubiquitous presence in public spaces. Indeed, the popular press has the ability to create a moral panic surrounding young people's presence in these realms. The following sensationalist news headlines present young people, or "hooded youths", as "folk devils"; that is, as a bad influence on society (Oswell, 1998: 36): "Shocking Moment a Hoodie-Wearing Youth throws a Lit Firework into a 'Random' Car on a London Street" (*Daily Mail*, 2018); "Shocking Moment Hooded Thug Threatens Children with a Knife Outside School" (The *Sun*, 2018); and "Mother Dies after Attack by 'Hoodies' in City Centre' (*Daily Telegraph*, 2005). Holloway and Hubbard (2001) have proclaimed that the presence of teenagers congregating on street corners, in bus shelters and in parks may constitute both an unconscious and a deliberate contestation of adult space. The authors continue to assert that use of such places may represent teenagers' attempts to define their own ways of socialising and utilising space, free from the adult gaze.

However, for many young people, being together and walking together in streets is solely a gesture of care and responsibility for their friends, in terms of keeping them safe (Horton et al., 2014; Wilkinson, 2015). This display of care and concern thus contrasts markedly with popular representations of "anti-social" young people in public places (see Brown, 2013). Nonetheless, it is this "we-ness" of young people's mobilities – in which the focus of group members is directed inwards towards each other, rather than forming outward connection with others (Milne, 2009: 115) – that some police may find threatening in the "hanging out" behaviours of young people. Some police consider young people to be "unacceptable flaneur[s]" (Matthews et al., 2000: 279). Young people often feel stereotypically predefined as potentially deviant (Wilkinson, 2015); they believe they are perceived "as a potential threat to the moral fabric of society and up to no good" (Matthews et al., 1999: 1724).

As intimated above, police often seek to move on and separate young people who are in groups (Wilkinson, S., 2015). Consequently, many young people are not always mobile in public places through their own volition. Their walking practices are often characterised by an experience of always moving on (see Horton et al., 2014); young people are often "fixed in mobility", much like the young homeless people in Jackson's (2012: 725) study (see Chapter 8 by Bissell in this book). Whilst recognising that policing can sometimes constrain the scope for engaging playfully with place (Edensor and Bowdler, 2015), we contend that in some respects policing can enhance playful engagements with place. That is, many young people, rather than expressing frustration at constantly being ejected from parks (Townshend and Roberts, 2013), enjoy the "geographical game of cat and mouse" (Valentine, 1996: 594). Whilst young people appear to have the ability to actively resist policing by carving out new places (e.g., marginalised areas of parks) to assert their presence (see Hil and Bessant, 1999), it is important not to romanticise this. By reducing the visibility of young people in outdoor places, displacement may lead to young people occupying more covert and less safe spaces (Pennay and Room, 2012).

As can be seen thus far, there appears to be an imagined geography that privileges young people's use of certain places, such as streets and parks. In order to move beyond these distorted notions of inclusion and exclusion, this chapter now brings to the fore the importance of other everyday spaces to young people's identity construction. For instance, redressing the focus on public places, Hodkinson and Lincoln (2008) have explored the importance of micro-geographies of the home to young people's identities, with a focus on the personal space of the bedroom. The authors highlight that bedrooms hold extensive significance for children and young people, partly due to the increasing range of technologies within the bedroom. Interestingly, Hodkinson and Lincoln (2008) tease out similarities between the bedroom and virtual places of young people's online journals/blogs, arguing that online journals also function as a form of personal space for their users. Akin to the bedroom, Hodkinson and Lincoln (2008) highlight that the interactive, multi-dimensional place of online journals is safe and personally owned and controlled. Young people utilise such virtual places as part of negotiating their transitions to adulthood, for instance, through exploring identity and generating social networks.

Hollands (1995) likewise discusses the importance of the bedroom for young people, yet in the different context of going-out rituals such as finding outfits to wear, listening to music and crafting a positive affective atmosphere prior to entering bars, pubs and clubs. Indeed, Samantha Wilkinson (2017) has similarly highlighted how, due to their familiarity with the micro-geographies of the space of the home, some young people "socially sculpt" their experiences (Moore, 2013). For instance, young people can take control over their experiences of lightness and darkness in the home by staging atmospheres (Bille, 2015) in order to influence their experiences of drunkenness. In Wilkinson's (2017) study, one participant suggested that the practice of lighting candles is "a tool" (Bille and Sørensen,

2007: 263) to exercise a "gentle suggestion" (Sumartojo, 2014: 62) to friends that she desired the night ahead to be low-key. The work of both Hollands (1995) and Wilkinson (2017b) highlights that places are not fixed and static, but are continually (re)produced and performed.

As another example of moving beyond the privileging of parks and streets in the academic literature, Catherine Wilkinson (Wilkinson, C. 2015a), using a case study of a youth-led community radio station, KCC Live, based in Knowsley, neighbouring Liverpool, UK, finds that this place comes to feel like home for the young volunteers at the radio station. Writing later, Wilkinson (2019) explores the "hyperdiversity" of the young people at the radio station. She finds that, although young people who volunteer at the radio station live in a variety of neighbouring towns, some of which are positioned as rivals, at the place of the radio station they are united by their shared interests and put their differences aside. Thus, being in the place of the radio station stimulates the development of relationships across difference and diminishes binary distinctions between "us" and "them".

Another less commonly explored place in the literature, in terms of its importance to young people, is the shopping mall, a place blurring the boundaries between public and private. Matthews et al. (2000) highlight that the shopping mall provides a convenient place for hanging out, but they contend that this is not unproblematic, with many adults perceiving the public and visible presence of young people as uncomfortable and inappropriate. By locating themselves in shopping malls, Matthews et al. (2000) argue that the spatial hegemony of adulthood is brought into question. That is, for young people, the mall occupies as an important cultural space, going beyond its functional form. However, being present in, and occupying, a place beyond the realm of the home can create feelings of discomfort for some adults, who may seek to move such young people on. Nonetheless, Matthews et al. (2000) reveal that young people are not simply passive and can contest attempts to be moved on in attempting to protest against their marginality. Another author to explore young people's experiences in shopping malls is Tani (2015), who analyses hanging out as an interaction between the location and young people. She contends that the space of the shopping mall offers affordances to young people for hanging out, affecting their ways of being. Yet, equally, young people give new meanings to place, turning it into more than just a physical setting for hanging out. Here, the shopping mall has the capacity to shape young people's practices and experiences, and likewise the young people can contribute to the (re)production of place.

Building on the above body of literature, we contend that in order to move beyond the currently limited imagined geographies of young people, and thus restrictive ideas of inclusion and exclusion, future research must be conducted into young people's engagement with a diverse range of indoor and outdoor places. Having explored young people and place, this chapter now turns to engage with older people's use of place.

Older people and place

There is an assumption that as people get older and their mobility decreases, their geographical worlds shrink and they are less likely to be involved in outdoor activities (Wiles et al., 2009). It is thus unsurprising that, as with young people, there appears to be an imagined geography that privileges older people's use of certain places; in this case, home, residential homes, nursing homes or hospitals (Andrews and Peter, 2006; Schneider et al., 2019; Tanner et al., 2008; Turner et al., 2018; Williams, 2002). In this section we will explore the importance of home, the wider community, gardens and allotments for older people's identity construction. Moreover, this section will bring to the fore the impact of globalising processes on older people's relationship with place, along with discussing the significance of virtual communities.

Wiles et al. (2012) explore the meaning of the phrase "ageing in place" to older people in New Zealand. This term is used to refer to remaining living in one's own home with some degree of independence, as opposed to in residential care. The authors argue that ageing in place has the practical advantage of the security and safety of home. Home is seen as a refuge or base from which to go out and participate in activities. The authors highlight that ageing in place is connected to a sense of identity and this extends beyond the home to the wider community; for instance, the proximity and reliability of health or other services (such as police). Moreover, with a similar focus, Tanner et al. (2008) highlight that home is much more than a physical environment; older people and places are engaged in a reciprocal relationship through which home becomes a place of significant personal meaning. With a focus on recipients of a home modification service, Tanner et al. (2008) explore the impact of modifying the home on older people's experience of home as a place of meaning. The authors contend that home modification can both improve safety and comfort for the older person at home, whilst simultaneously strengthening the home as a place of personal and social meaning. By strengthening home as a place of security, safety and comfort, the continuation of habitual personal routines and rituals are enabled; it is through such mundane everyday practices that people are linked to their homes, and self-esteem and identity are consequently reinforced. However, the authors note that if older people are not involved in decision-making surrounding home modification, the impact on the meaning of home can be detrimental. Importantly, Wiles et al. (2009) highlight that the home is not a homogenous category; that is, even within the space of the home, older people tend to have particular places that they choose to spend most of their time, due to factors such as comfort and practicality. The importance of particular micro-geographies within the home is an area of research warranting further attention.

Moving beyond the boundaries of the indoor home, the work of Bhatti (2006) and Milligan et al. (2004) is important to discuss. Bhatti contends that whilst the indoor space of the home is now well explored in literature surrounding older people, the garden has been somewhat neglected. Bhatti seeks to redress this, contending that for individuals experiencing changes in later life the garden figures strongly as a place from which to engage the outside world and challenge ageist ideologies. Rather than being a separate sphere to the home, he argues that gardens are an important part of older people's sense of home, and gardening – as a form of bodily action and power – is significant in home making, providing psycho-social and health benefits in perpetuating control over their own physical space (Bhatti, 2006). Moreover, Milligan et al. (2004) draw on research conducted in northern England to examine how communal gardening on allotments and in domestic gardens may contribute to the maintenance of health and wellbeing amongst older people. They illustrate the sense of achievement, satisfaction and pleasure that older people gain from gardening. However, the authors recognise that whilst older people continue to enjoy gardening, the physical shortcomings attached to the ageing process mean they may increasingly require support to do so. They contend that communal gardening on allotment sites are inclusive places in which older people benefit from gardening activity in a mutually supportive environment that combats social isolation and contributes to the development of their social networks.

What is not brought to the fore in the above work is the impact of global processes on changing relationships to ageing in place. This is important to consider since older people tend to have high levels of attachment to place (Wiles et al., 2009). Indeed, Phillipson (2007) explores how globalisation is creating new social divisions between those who are able to exercise choice over their residential locations, which are consistent with their biographies and life histories, and those who experience marginalisation or who are rejected from their localities. Phillipson (2007) asserts that the impact of the global on the local may be of particular significance for older people given the duration of time they may have resided in the same community, along with the extent to which their mobility may at some point be somewhat restricted to defined territorial boundaries. Interestingly, with increased flows and connectedness facilitated between people by globalising processes, Kanayama (2003) explores how older people in Japan are becoming part of virtual communities. The author reveals that both the immediacy and asynchrony of computer-mediated communication helps older people to construct real human relationships in the virtual community, including social connectedness to others, as well as supportive and companionship relationships. Kanayama contends that older people are able to create a sense of greater propinquity by sharing stories and memories. Having explored older people's relationships with both physical and virtual places, this chapter turns to highlight the need for engaging with intergenerational relationships and place.

Intergenerational relationships and place

By segregating age categories rather than combining an interest in different generations, much of the existing literature can be accused of marking sharp distinctions between the "young" and the "old" (Hagestad and Uhlenberg, 2005), and thereby failing to consider how different generations occupy and perform in space together. We now turn to offer examples in which intergenerational approaches to age and place can be seen to allow for a more inclusive and holistic view of age and place, without fetishising the social-chronological margins (Hopkins and Pain, 2007).

Interest in harmonious intergenerational relations can be seen in the popular Channel 4 television programme *Old People's Home for 4 Year Olds*. The concept of this programme is that ten elderly volunteers, residents of the St Monica Trust retirement home, UK, are introduced to ten pre-school children for six weeks, as the retirement village opens a nursery where classmates range from three to 102 years old. Every day, the two groups spend time together, undertaking tasks aimed at fostering a relationship between them. The goal is to discover if the physical and mental health of older people may be improved by such an association. Viewers have been able to witness a widower, who formerly was very fixed and static in the home, sitting in his chair, become "very much alive". The show demonstrates that the intertwining of the lives, and by extension places, of the young and old has positive consequences. Furthermore, Boyd (2019) brought together young children, young people and the over 65s in an "intergenerational craft café" in Liverpool, UK. The intergenerational café concept is built on the idea that early childhood is a transformative period when young children start to develop attitudes and skills that can last a lifetime. The author discusses the café as a space for sharing knowledge as, each month, those over 65 display a different sustainable skill, such as sewing, knitting and baking. Boyd highlights that the place of the café offers an arena to challenge stereotypes and stigmas, such as that old people are "smelly or boring".

Further, Hopkins et al. (2011) draw on research conducted with young Scottish Christians and their guardians to explore the influence of intergenerationality on their religious identities and practices. The authors contend that intergenerational relations must be understood as part of place-based practices that are central to both the development and experience of young people's religious identities. The authors highlight how transmission in intergenerational religiosity is situated in particular settings, such as the home, but also in less obvious places where young people and their parents interact, including car journeys, on the route to church or walks in the countryside. This study is important in moving beyond stereotypical imagined geographies, highlighting the importance of banal, mundane, places for the development of intergenerational relationships. As can be seen, in comparison with studies highlighting generational discreteness, intergenerational studies can provide more nuanced understandings of everyday spatial processes (Hopkins et al., 2011).

Moreover, with a focus on an underexplored space of identity construction, children's huts in Norway, Kjorholt (2003a) proposes a "joint inter-generational" geographies approach. The author contends that children, in order to create a place to belong, need to craft their own special places during middle childhood. Children in Kjorholt's study made clear that their identities as autonomous individuals are constructed through a gendered generational relationship, comprised of males. That is, the young men's social practices in forests highlight continuity between generations of men in terms of how they use the space. As an integrated part of everyday life, knowledge of how to use the place is transmitted between generations. Continuity and integration with adults is thus a fundamental characteristic of the young men's stories. The work of Kjorholt is refreshing for making explicit intergenerational relationships in particular places that signal co-operation and reciprocity across generations, thereby redressing the emphasis in the literature and in the popular press, highlighted previously, on conflictual relationships between generations (Hopkins and Pain, 2007).

Conclusions

To conclude, as we have argued throughout this chapter, places are very important for young and older people to perform their identities. We have highlighted that there are many different experiences in places, and some of these may compete or conflict (Wiles, 2005). Moreover, we have sought to make clear that places are not static and passive, they are relational and active, with the ability to shape, and be shaped by, people's practices and experiences (Valentine, 2001). Nonetheless, this chapter has highlighted that certain places, such as streets and parks, have been prioritised in the academic literature when discussing young people's use of places (Cahill, 2000; Gough and Franch, 2005; Matthews et al., 1999). Meanwhile, particular places, such as the home, care homes and hospitals, have received most attention when older people's relationships with place are discussed (Andrews and Peter, 2006; Schneider et al., 2019; Tanner et al., 2008; Turner et al., 2018; Williams, 2002). This is not to downplay the importance of such places for young and older people, but to argue that young and older people's places are part of an imagined geography that privileges certain places, and thus inadvertently produces distorted notions of inclusion and exclusion. We thus urge future researchers to bring to the fore young and older people's complex relationships with a diverse range of often-overlooked places.

In this chapter we have highlighted the importance of understanding how intergenerational relationships are bound up with place. Largely, when intergenerational relationships surrounding place have been explored in the literature and popular press, it has been in relation to conflict between generations (Hopkins and Pain, 2007; Matthews et al., 2000). This chapter develops Kjorholt's (2003a) ideas to argue that there needs to be a much greater understanding of "joint inter-generational" geographies. That is, we must bring to the fore the ways in which young and older people utilise places in mutual and supportive ways. Put simply, there is a need for further research that uncovers the diverse places used by young people, adults and older people, both separately and together. Doing so may go some way towards challenging the stereotypes and stigmatisation associated with particular age groups, and will explore alternative ways of being a child, adult or older person (Hopkins and Pain, 2007).

References

Andrews, G.J. and Peter, E. (2006). 'Moral Geographies of Restraint in Nursing Homes'. *Sigma.* 3(1): 2–7.

Bhatti, M. (2006). 'When I'm in the Garden I can Create my own Paradise': Homes and Gardens in Later Life'. *The Sociological Review.* 54(2): 318–341.

Bille, M. (2015). 'Lighting Up Cosy Atmospheres in Denmark'. *Emotion, Space and Society.* 15: 56–63.

Bille, M. and Sørensen, T. F. (2007). 'An Anthropology of Luminosity: The Agency of Light'. *Journal of Material Culture.* 12: 263–284.

Boyd, D. (2019). The Legacy Café – A Trial of Intergenerational and Sustainable Learning in an Early Childhood Centre. In, Filho, W.L. (Eds). *Social Responsibility and Sustainability: How Businesses and Organizations Can Operate in a Sustainable and Socially Responsible Way.* Switzerland: Springer. pp. 73–388.

Brown, D.M. (2013). 'Young People, Anti-Social Behaviour and Public Space: The Role of Community Wardens in Policing the 'ASBO Generation''. *Urban Studies.* 50(3): 538–555.

Cahill, C. (2000). 'Street Literacy: Urban Teenagers' Strategies for Negotiating their Neighbourhood'. *Journal of Youth Studies.* 3(3): 251–277.

Cele, S. (2013a). 'Performing the Political through Public Space: Teenage Girls' Everyday Use of a City Park'. *Space and Polity.* 17(1): 74–87.

Coburn, K.L. (2006). 'Organizing a Ground Crew for Today's Helicopter Parents'. *About Campus: Enriching the Student Learning Experience.*

Daily Mail. (2018). 'Shocking Moment a Hoodie-Wearing Youth Throws a Lit Firework into a 'Random' Car on a London Street' [Online], available: www.dailymail.co.uk/news/article-6353619/Shocking-moment-hoodie-wearing-youth-throws-lit-firework-random-car-London-street.html [20 December 2018].

Dicken, P. (2010). *Global Shift: Mapping the Changing Contours of the World Economy.* London: Sage Publications.

Edensor, T. and Bowdler, C. (2015). 'Site-Specific Dance: Revealing and Contesting the Lucid Qualities, Everyday Rhythms, and Embodied Habits of Place'. *Environment and Planning A.* 47(3): 709–726.

Gough, K.V. and Franch, M. (2005). 'Spaces of the Street: Socio-Spatial Mobility and Exclusion of Youth in Recife'. *Children's Geographies.* 3(2): 149–166.

Hagestad, GO. and Uhlenberg, P. (2005). 'The Social Separation of Old and Young: A Root of Ageism'.

Hil, R. and Bessant, J. (1999). 'Spaced-Out? Young People's Agency, Resistance and Public Space'. *Urban Policy and Research.* 17(1): 41–49. *Journal of Social Issues.* 61. (2). pp. 343–360.

Hodkinson, P. and Lincoln, S. (2008). 'Online Journals as Virtual Bedrooms? Young People, Identity and Personal Space'. *Young: Nordic Journal of Youth Research.* 16(1): 27–46.

Hollands, R. (1995). *Friday Night, Saturday Night.* Newcastle: Newcastle-Upon-Tyne Press.

Holloway, L. and Hubbard, P. (2001). *People and Place: The Extraordinary Geographies of Everyday Life.* London: Pearson Education Limited.

Hopkins, P., Olson, E., Pain, R. and Vincett, G. (2011). 'Mapping Intergenerationalities: The Formation of Youthful Religiosities'. *Transactions of the Institute of British Geographers.* 36(2): 314–327.

Hopkins, P.E. and Pain, R. (2007). 'Geographies of Age: Thinking Relationally'. *Area.* 39(3): 287–294.

Horton, J., Christensen, P., Kraftl, P. and Hadfield-Hill, S. (2014). ''Walking ... Just Walking': How Children and Young People's Everyday Pedestrian Practices Matter'. *Social & Cultural Geography.* 15(1): 94–115.

Jackson, E. (2012). 'Fixed in Mobility: Young Homeless People and the City'. *International Journal of Urban and Regional Research.* 36(4): 725–741.

Kanayama, T. (2003). 'Ethnographic Research on the Experience of Japanese Elderly People Online'. *New Media & Society*. 5(2): 267–288.
Kintrea, K., St Clair, R. and Houston, M. (2015). 'Shaped by Place? Young People's Aspirations in Disadvantaged Neighbourhoods'. *Journal of Youth Studies*. 18(5): 666–684.
Kjorholt, A.T. (2003a). ''Creating a Place to Belong': Girls' and Boys' Hut-Building as a Site for Understanding Discourses on Childhood and Generational Relations in a Norwegian Community'. *Children's Geographies*. 1(1): 261–279.
Malone, K. (2007). 'The Bubble-Wrap Generation: Children Growing up in Walled Gardens', *Environmental Education Research*. 13(4): 513–527.
Markusen, A. (1996a). 'Sticky Places in Slippery Space: A Typology of Industrial Districts'. *Economic Geography*. 72(3): 293–313.
Massey, D. (1999). Spaces of Politics. In Massey, D., Allen, J., Sarre, P. (Eds) *Human Geography Today*. Cambridge: Polity Press. pp. 279–294.
Matthews, H., Limb, M. and Taylor, M. (1999). 'Reclaiming the Street: The Discourse of Curfew'. *Environment and Planning A*. 31(10): 1713–1730.
Matthews, H., Taylor, M., Percy-Smith, B. and Limb, M. (2000). 'The Unacceptable Flanuer: The Shopping Mall as a Teenage Hangout'. *Childhood*. 7(3): 279–294.
Milligan, C., Gatrell, A. and Bingley, A. (2004). ''Cultivating Health': Therapeutic Landscapes and Older People in Northern England'. *Social Science & Medicine*. 58: 1781–1793.
Milne, S. (2009). 'Moving Into and Through the Public World: Children's Perspectives on their Encounters with Adults'. *Mobilities*. 4(1): 103–118.
Moore, M. (2013). 'Studio 54 and the Production of Fabulous Nightlife'. *Dancecult: Journal of Electronic Dance Music Culture*. 5: 61–74.
Oswell, D. (1998). A Question of Belonging: Television, Youth and the Domestic. In Skelton, T. and Valentine, G. (Eds.) *Cool Places: Geographies of Youth Cultures*. London: Routledge. pp. 39–45.
Pennay, A. and Room, R. (2012). 'Prohibiting Public Drinking in Urban Public Spaces: A Review of the Evidence'. *Drugs: Education, Prevention and Policy*. 19(2): 91–101.
Phillipson, C. (2007). 'The 'Elected' and the 'Excluded': Sociological Perspectives on the Experience of Place and Community in Old Age'. *Aging & Society*. 27: 321–342.
Read, B., Archer, L., and Leathwood, C. (2003). 'Challenging Cultures? Student Conceptions of 'Belonging' and 'Isolation' at a Post-1992 University'. *Studies in Higher Education*, 28(3): 261–277.
Relph, E. (1976). *Place and Placelessness*. London: Sage.
Schneider, J., Pollock, K., Wilkinson, S., Perry-Young, L., Turner, N. and Travers, C. (2019). 'The Subjective World of Home Care Workers in Dementia: An 'Order of Worth' Analysis'. *Home Health Care Services Quarterly*. 38(2): 96–109.
Sumartojo, S. (2014). ''Dazzling relief': Floodlighting and National Affective Atmospheres on VE Day 1945'. *Journal of Historical Geography*. 45: 59–69.
The *Sun* (2018). *Shocking moment hooded thug threatens children with a knife outside school* [Online], available: www.thesun.co.uk/news/5292926/schoolchildren-threatened-knife-outside-school-hooded-thug [20 December, 2018].
Tani, S. (2015). 'Loosening/Tightening Spaces in the Geographies of Hanging Out'. *Social & Cultural Geography*. 16(2): 125–145.
Tanner, B., Tilse, C. and Jonge, D. (2008). 'Restoring and Sustaining Home: The Impact of Home Modifications on the Meaning of Home for Older People'. *Journal of Housing for the Elderly*. 22(3): 195–215.
The *Telegraph*. (2005). '*Mother Dies after Attack by 'Hoodies' in City Centre*' [Online], available: www.telegraph.co.uk/news/uknews/1496447/Mother-dies-after-attack-by-hoodies-in-city-centre.html [20 December, 2018].
Townshend, T.G. and Roberts, M. (2013). 'Affordances, Young People, Parks and Alcohol Consumption'. *Journal of Urban Design*. 18(4): 494–516.
Turner, N., Schneider, J., Pollock, K., Travers, C., Perry-Young, L. and Wilkinson, S. (2018). ''Going The Extra Mile' for Older People with Dementia: Exploring the Voluntary Labour of Homecare workers'. *Dementia: The International Journal of Social Research and Practice*.
Valentine, G. (1996). 'Angels and Devils: Moral Landscapes of Childhood'. *Environment and Planning D: Society and Space*. 14(5): 581–599.
Valentine, G. (2001). *Social Geographies: Space and Society*. Harlow: Pearson Education.

Wiles, J. (2005). 'Conceptualising Place in the Care of Older People: The Contributions of Geographical Gerontology'. *International Journal of Older People Nursing*. 14(8b): 100–108.

Wiles, J.L., Allen, R.E.S., Palmer, A.K., Hayman, K.J., Keeling, S. and Kerse, N. (2009). 'Older People and their Social Spaces: A Study of Well-Being and Attachment to Place in Aotearoa New Zealand'. *Social Science & Medicine*. 68: 664–671.

Wiles, J.L., Leibing, A., Guberman, N., Reeve, J. and Allen, R. (2012). 'The Meaning of 'Aging in Place' to Older People'. *The Gerontologist*. 52(3): 357–366.

Wilkinson, C. (2015a). *Connecting Communities through Youth-led Radio*. PhD Thesis. University of Liverpool.

Wilkinson, C. (2019). 'On the same wavelength? Hyperdiverse young people at a community radio station'. *Social & Cultural Geography*. 20(9): 1251–1265.

Wilkinson, S. (2015b). *Young People, Alcohol and Urban Life*. PhD Thesis. University of Manchester.

Wilkinson, S. (2017b). 'Drinking in the Dark: Shedding Light on Young People's Alcohol Consumption Experiences'. *Social and Cultural Geography*. 18(6): 739–757.

Williams, A. (2002). 'Changing Geographies of Care: Employing the Concept of Therapeutic Landscapes as a Framework in Examining Home Space'. *Social Science & Medicine*. 55(1): 141–154.

20

Gendering place

Mobilities, borders and belonging

Dina Vaiou

... Starting with a story

> Where am I to go back to? Back to Fieri? We don't have family there anymore. Besides, all those who somehow return, they come back here, you know. Things are not easy there either [...] and where do I take Mosa [the daughter]? She is not a package – she only knows this place, this neighbourhood, this language – and the same with me.
>
> *(From an interview with Elisa, in 2017)*

Elisa is a young woman from Fieri (Albania) who came to Athens in 1993, at the age of 9, together with her parents. The family settled in Kato Patissia, a densely populated neighbourhood in central Athens with strong migrant presence. She attended the local public school and went for higher education. She took a degree in regional planning in 2006 from the University of Thessaly, in the city of Volos, where she also worked for five years in bars and cafes in order to finance her studies. Upon graduation, she was very happy to find a job in a private planning consultants' office in Athens (2006–2013), she got married and had a daughter, Mosa, who was four at the time of interview. The young family found a flat not far from Elisa's parents in Kato Patissia where she has fond childhood memories, friends from school, knowledge of the local market and the everyday rhythms, good public transport and other services. In 2013, the neoliberal austerity policies hit both her and her husband, Sokol. The planning office closed down in 2013 and also, later, in 2016, the brokerage office where Sokol used to work. Small jobs and occasional day earnings were not enough to support them and Elisa decided to move to her parents' house:

> I had to decide what was more important, to have food on the table or to keep our "independence" – and my constant fear that we will not be able to pay the rent and be evicted [...] Sokol [her husband] felt humiliated, he could not somehow cope with the idea that he could not support his family [...] The situation became bad between us. We divorced about a year ago.
>
> *(From an interview with Elisa, in 2017)*

Elisa now works in a restaurant kitchen only weekends. She goes for food assistance to the municipality and to the neighbourhood solidarity initiative and collects vegetables for free when the weekly open market is about to close. She makes every effort not to depend entirely on her parents who, in any case, are in a difficult state as well since only Elisa's mother has a regular job as a live-out cleaner (she was a civil servant in Fieri).

Introduction

The story of Elisa highlights many of the issues that shape the debate about place and the ways in which it may be approached in the age of globalisation, elaborated in other sections of this book. It underlines in particular the ways in which significant numbers of people moving across the globe and crossing (different kinds of) borders have challenged understandings of place associated with settled communities and time-long cultures, localities with distinct character, more or less stable over time and set against the current fragmentation and disruption.[1] However, people moving across the globe, for reasons that differ dramatically along the lines of, for example, income and class inequalities, also engage in efforts to settle for longer or shorter periods of time, lay claims on territory and try to forge relations of belonging in places.

This chapter follows Doreen Massey's proposition to conceptualise place as a specific articulation, a particular unique point of intersection of social relations, social practices and materiality – a meeting place which integrates the global and the local, and is shaped by the experiences, narratives and symbolic meanings of different (and sometimes conflicting) groups and individuals (Massey, 1994). The accumulation of meetings and encounters, as they weave together "here", "now", "then" and "there", produce what is special or unique to a place, this "throwntogetherness" as Massey (2005: 140) calls it, leading to continuous negotiation with multiplicity. In this line of thinking, which combines material, social and emotional aspects, place is already necessarily gendered. Moreover, place is also bound to power geometries which combine gender to other axes of inequality and difference, like class, ethnicity, sexuality or age.

The story of Elisa[2] which runs through my text is inscribed in, and contributes to articulate, a dynamic conception of place as always becoming, constituted also through performances of gender at different scales, from the body to the global, and negotiation between previously unrelated processes, subjectivities and trajectories in space and time. In the following sections, I discuss three issues that may help think the "gendering" of place: the everyday as a critical entry point into place-making through gendered labour and practices; moving and settling through practices of care; and the multiplicity of often invisible borders and boundaries along the ways of mobility in the age of globalisation. Finally, I draw together some concluding points that underline the necessity to develop perspectives and methodologies that value people's diverse experiences as sources of knowledge production and permit one to "see" the power relations involved in making and gendering place.

Tesserae[3] in gendering place

The everyday and the neighbourhood as place

The everyday is a concept with a long history in philosophy and in the social sciences, in which the work of Henri Lefebvre holds a prominent position. His repeated elaborations (among many: Lefebvre, 1946, 1962, 1968, 1981, 1990, 1992) have raised the everyday to an object of critical reflection, linking it to the spatiality and multiplicity of urban living and to a radically different perspective on (urban) space and, I would argue, place. In this perspective, research interest and theoretical value include not only global processes, transgressions and "big" events, but also futile anxieties, the myriad things that need to be done every day for oneself and for others (also de Certeau,1984); all those "details" which are usually deemed humble, repetitive, taken-for-granted and, in any case, unworthy for the constitution of theoretical and explanatory frameworks.

The perspectives from those taken-for-granted activities in homes, neighbourhoods and communities which support the social, cultural and economic processes and make places habitable has drawn many feminists to explorations of the everyday, both theoretically and empirically, in concrete places (Peake and Rieker, 2013; Vaiou, 2013). It brings to the foreground of enquiry the ways in which, on the one hand, gendered power hierarchies are shaped through the mundane and ordinary practices, meanings and settings which constitute the everyday; while, on the other hand, it values such gendered experiences and practices as a basis for the production of different kinds of knowledge about, among other things, place and the multiple realities associated with it (among many: Harding, 1986; McDowell and Sharp, 1999). The theoretical and methodological emphasis on the everyday crosses over dichotomous conceptions and reveals continuities through different spheres of experience and interlocking spatial scales (Gardiner, 2000; Simonsen and Vaiou, 1996; Smith, 1987).

Among these interlocking scales, the neighbourhood, a highly contested concept in urban studies, particularly in its identifications with community and locality, emerges as an important reference in the everyday experiences and practices of different people (for a discussion, see Germain, 2002; Simonsen, 1997; Kalandides and Vaiou, 2012). Its meanings extend far beyond its spatial determinants and invest it with a renewed importance, distanced from conceptions of a bounded space. In this line of argument, the neighbourhood is an important socio-spatial scale at which movements of people, ideas, commodities, etc.,link the local with global processes; these processes support what Massey (1994) calls "a progressive sense of place": a conception of place as a particular moment in intersecting social relations, continuously in the making, through probable or unforeseen transformations and connections. All of these may materialise, or not, in the context of a system which includes heterogeneity, indeterminacy and openness to future restructurings, but also powerful geometries of power.

Migrant women, like Elisa, are caught in such geometries of power, as they seek to be included, to a greater or lesser degree, in the economic, social and political institutions and in the patterns of everydayness in the places where they settle (in her case, the neighbourhoods of Athens and Volos). At the same time, they transform and reposition these patterns and places, linking them to supra-local, even global, processes. In Elisa's case, the place/neighbourhood where she and her family have settled in Athens is very different from "then", when they were not "here" (see Massey, 2005); it is also very different from how mainstream narratives and some old residents' stories portray it. "Here", that neighbourhood/place in Athens with its layers of previous histories ("then"), is entangled with "now", with the different perceptions, memories and understandings generated by the multiple encounters that happen in material and imagined spaces. It is also entangled with allegiances to "there", where Elisa and her family have come from and seek to continue belonging as well. In this sense, neither the everyday nor the place can be identified exclusively with the local or with immobility in space and time.

Settling and practices of care

In Elisa's narrative, identifications with the neighbourhood/place are important for her, in order to sustain local and transnational networks and different opportunities for inclusion (see also Leitner and Ehrkamp, 2006; Salih, 2003). The latter have been strongly shaken in the long years of neoliberal austerity policies implemented in Greece and other countries since 2010. At the same time, the place has changed through the lifestyles, the ways of using material space and the everyday practices of "strangers" which inevitably intersect with those of "locals" – with all the controversies involved in such homogenising characterisations. In this changing landscape, care for neighbours, friends and family comes out as an important component in the process of familiarisation and the struggle for belonging. Elisa underlines, in this respect, how she has been assisted by neighbours to find a job (and previously a "good and affordable flat") and how they often look after Mosa "like their own grandchild". She also mentions how she reciprocates by caring for them in many informal but critical ways, for example with shopping, taking her elderly neighbour to the doctor or the bank, visiting and sharing time together. As part of neighbouring, she accompanies hers and the neighbour's children to kindergarten or to the playground, as she once was accompanied by a woman next door when she was a child and her mother had to leave for work very early.

These everyday acts are part of a bulk of caring labour at the scale of the body, the home and the neighbourhood and at the same time they are part of supra-local and even global processes and movements (Vaiou, 2013). Taking into account such small and repetitive acts of care, usually invisible in mainstream analyses, yet strongly associated with settling and belonging, becomes more than a theoretical conception. It is a major stake, a process of familiarisation with place, difference and otherness, all of which require investment of time and labour, both material and emotional. Here, gender differences come out prominently: migrant men are more likely to spend time with other men from their own community, or from work, and are only seldom involved in caring labour; migrant women are predominantly the ones who engage in contacts with neighbours, practically by caring and sharing common everyday routines.[4] Such differences reinforce notions of "naturalness" in the gendering of care, particularly under conditions of neoliberal austerity, and they form part of reluctant negotiations when common sense divisions of labour cannot be sustained (Dyck, 2005).

The dire conditions in which Elisa lives in Athens at times of neoliberal austerity are common to many people who have migrated after 1989 and have lived in the city for almost 30 years by now. For them, "here" is home, networks, everyday and longer-term life plans; it is also the place where their parents have struggled to "make life good". "There" is not part of their experience and returning is not an option; there is no "then" connected to it, and it would take a new migration project and the struggle to make yet another start. For Elisa, the repercussions of austerity, a supra local – even global – set of policies, have hit hard key aspects of everyday survival and longer term well-being as the safety net of the family can provide only marginal material resources. She lives in emotional strain and physical exhaustion, linked to little income, job insecurity and all the efforts to collect food, to identify services she could be eligible for free (e.g., health or childcare), as well as a divorce following the decision to move back to her parents' house.

Patterns of mobility and invisible borders

Raising mobility to a paramount value in the age of globalisation, and in the context of recent migrations, tends to undervalue the power relations involved in moving and the cost, both for those who move and for those who stay back (Massey, 1994). Part of this cost is only revealed when the different involvement of men and women in such practices is examined. Women and men moving across borders often live intense gendered experiences, as they face different behaviours and expectations in the different places they cross and/or settle for longer or shorter periods. Gender ideologies and practices are modified as subjects collaborate with or confront each other, their past and the changing economic, political and social structures linked with migration (Mahler and Pessar, 2001; Vaiou, 2012). In this context, gendered experiences of mobility as well as material and symbolic borders and boundaries are constituted at various geographical scales, from the national and the global to the local and familiar.

In Elisa's narrative, borders are used as a reference to institutional restrictions of passage between Albania and Greece; they are seen as lines, over mountains and seas, which set clear limits to the mobility of "others"(Allen, 2003) and remain embodied all the way along moving and settling. The incommensurate power of different states to determine borders and control, or channel movements, is played out in/on border zones and extended geographically along the trajectories of migrants and the places where they settle temporarily or permanently. It is further reinforced by a whole host of rules and legal barriers to mobility, which constitute migration policy beyond the border zone.

Boundaries, on the other hand, refer to socio-spatial and symbolic limitations which organise place and are shaped by such things as the level and quality of services one may receive; the diversity of local commerce; the quality of housing, and much more (Jess and Massey, 1995; Wills, 2013). Boundaries determine movement and access, and are constituted, at least in part, in relation to particular notions of gender and gender practices (Silvey, 2006). Mobility or immobility, access or barriers to particular places and activities, processes of inclusion/exclusion, are imbued with ideas and practices to do with gender. At the same time, they contribute to reformulate masculinities and femininities and often strengthen gender inequalities which permeate women's and men's attitudes, decisions and perceptions; they also have to do with conceptualisations of migration in which the model of migrant-traveller is already conceived as male (Kambouri and Lafazani, 2009; Pratt and Yeoh, 2003). Elisa underlines how she has overcome boundaries linked with language and "strangeness" at school, through the assistance of her teachers and of the neighbour who used to take her to the playground along with her own children; lying behind the words of her narrative is the insistence of her mother on education and the expected upward social mobility. In her effort to overcome boundaries, she feels that she has gradually (and perhaps unconsciously) been detached from "there" (Fieri), which remains strong in her parents' stories and in now occasional visits.

Her own everyday practices, feelings and perceptions have turned "here" (Athens) into a familiar place, the locale of her routines and interactions, which in turn contribute to (re) constitute and transform it (see also Cresswell, 1996). Employed below her skills for some years now, divorced and back to her parents' home, she feels stressed, devalued and somehow "trapped" in Athens and fears that if things get worse she will have "no (other) place to go". The expression of stress and depression, not uncommon in migrants' narratives, embodies accounts of self and family, and reveal important aspects of a place-making struggle.

Concluding comments

Thinking place along the lines explored in the previous section of this chapter emphasises openness, rather than a defensive putting up of barriers. However, openness (or its lack) is imbued with power relations, including ethnicity and gender. At various levels and scales in a geographically varied and uneven world, places are constituted, permeated/crossed, transformed, challenged by the practices of women and men. In this process, gender relations are continuously de- and re-composed (Nash, Tello, and Benach, 2005).

Women and men have different experiences of and inhabit places in different and unequal ways. These ways are connected with a whole range of displacements, inclusions and exclusions across borders and boundaries of various kinds. They are also connected with divisions of labour which take particular forms, often extreme, in the effort to make a home in unfamiliar places (a "here" and "now" in Doreen Massey's terms), while maintaining links and connections to a home "there" and "then" which is gradually becoming remote but remains always present (Vaiou et al., 2007). Behind such interlocking spatialities, one may find what Dyck and McLaren (2004: 521) call "women's landscapes of home"; that is to say the constitution of places within places, which are both material and symbolic sites stretching beyond localities and bridging over borders. Place in this context is not a refuge against insecure surroundings, but a dynamic terrain (Simonsen, 2008), in which quite often hybrid subjectivities are formed in, and through, the deployment of (even traditional) gender roles in paid work, leisure, consumption, and caring.[5]

Approaches that incorporate gendered experiences and perspectives pose difficult questions to conceptualisations of place and place-making, engagements with borders and boundaries, and perceptions of belonging. Gender, in its intersections with other axes of difference and inequality, determines to a large extent which bodies belong where; what kinds of spatial experiences different individuals and groups form; what techniques of exclusion correspond to what bodies, making them "out of place"; and who are "strangers" in a place – not because they are simply unknown to "us", but because they are already constructed as such even though they are painfully familiar (Ahmed, 2000; also Papataxiarchis, Topali, and Athanasopoulou, 2008). Such questions point to formal and informal socio-spatial arrangements through which place is constituted in and through unequal conditions of access and exclusion.

In order to unveil gendered aspects of place-making like the ones discussed in this chapter, it is necessary, on the one hand, to develop perspectives which would permit one to "see" such kinds of power relations and value people's diverse experiences as sources of knowledge production. On the other hand, it is also necessary to deploy methodologies that pay attention to usually "voiceless" subjects, acknowledging them as valid bearers of knowledge (among many, Dyck and McLaren, 2004). In the previous section I have attempted such a transfer of focus, through a discussion of the everyday and its places, of care and invisible borders, of moving and settling – in their connections with gender and feminist concerns. In this attempt, the story of Elisa (and the stories of all those women and men who contribute their experiences to research) is not an idiosyncratic example, but an entry point into the fields of knowledge and power within which migrant lives take shape and embody a web of relations extending at different scales. In this sense, it informs understandings of place-making based on global processes, and not only the reverse. Moreover, it contributes to de-naturalise common sense categories and ground "big pictures" and global analyses in the complexity of everyday practices, experiences and struggles which constitute place/s and make them habitable.

Notes

1 Also part of this disruption, and linked to the mobility of capital, ideas and information, as well as homogenisation of consumption patterns, are notions of "placelessness" or "non-places" also common in the literature (for a review, see Simonsen, 2008).

2 One of a significant number of stories of migrant and local women I have interviewed and talked with over the past 20+ years, in the context of several research projects and beyond.
3 In mosaic, a small piece of stone, glass, ceramic or other hard material cut in rectangular shape.
4 These shared practices modify earlier attitudes of non-migrants, now taking shape not by media representations but by reference to their own familiar neighbours – although racist language and aggression against migrants are not absent from the scene.
5 At the same time, images and representations of people who move, migrate, cross borders and boundaries are also gendered. Social attitudes differentiate migrant men and women, while migrant men and women themselves do not tell the same stories, nor do they recall the same memories of migration experiences. Gender also permeates the working of institutions (law, policies and practices of the administration) and of participation in civil society (Silvey and Lawson, 1999).

References

Ahmed, Sarah (2000) *Strange encounters. Embodied others in post-coloniality*, London and NewYork: Routledge.

Allen, John (2003) *Lost geographies of power*, Oxford: Blackwell.

Cresswell, Tim (1996) *In Place/Out of Place. Geography, ideology and transgression*, Minneapolis: The University of Minnesota Press.

de Certeau, Michel (1984) *The Practice of Everyday Life*, Oakland CA: University of Californa Press.

Dyck, Isabel (2005). ""Feminist geography, the 'everyday', and local-global relations: hidden spaces of place-making", Suzanne Mackenzie Memorial Lecture", *The Canadian Geogrpaher/Le Géographe Canadien* 49 (3): 233–243.

Dyck, Isabel and Arlene Tigar McLaren (2004). "Telling it like it is? Connecting accounts of settlement with immigrant and refugee women in Canada", *Gender, Place and Culture* 11 (4): 513–534.

Gardiner, Michael (2000) *Critiques of everyday life*, London: Routledge.

Germain, Annick (2002). "The social sustainability of multicultural cities: a neighbourhood affair?", *BELGEO* 4: 377–386.

Harding, Sarah (ed) (1986) *Feminism and methodology*, Bloomington: Indiana University Press.

Jess, Pat and Doreen Massey (1995) "The conceptualisation of place", in Doreen Massey and Pat Jess (eds) *A Place in the World? Places, cultures and globalisation*. Oxford: Oxford University Press in association with The Open University, 45–86.

Kalandides, Ares and Dina Vaiou (2012). "'Ethnic' neighbourhoods? Practices of belonging and claims to the city", *European Urban and Regional Studies* 19 (3): 254–266.

Kambouri, Nelli and Olga Lafazani (2009) "The house in Bulgaria, the collection of shoes in Greece: Transnational trajectories, migration and gender", in Dina Vaiou and Maria Stratigaki (eds) *The gender of migration*. Athens: Metaichmio (in Greek), 39–66.

Lefebvre, Henri (1946) *Critique de la vie quotidienne I: Introduction*, Paris: Grasset.

Lefebvre, Henri (1962) *Critique de la vie quotidienne II: Fondements d'une sociologie de la quotidienneté*, Paris: L'Arche.

Lefebvre, Henri (1968) *La vie quotidienne dans le monde moderne*, Paris: Gallimard.

Lefebvre, Henri (1981) *Critique de la vie quotidienne III: De la modernité au modernisme. Pour une métaphilosophie du quotidien*, Paris: L'Arche.

Lefebvre, Henri (1990) *Everyday Life in the Modern World*, New Brunswick, New Jersey: Transaction Publishers (first edition in French 1968).

Lefebvre, Henri (1992) (avec Catherine Régulier) *Eléments de Rythmanalyse. Introduction à la connaissance des rythmes*, Paris: éditions Syllepse. (English edition 1996, "Elements of Rhythmanalysis", in Eleanor Kofman and Elisabeth Lebas (eds) *Writings on Cities*, Oxford: Blackwell).

Leitner, Helga and Patricia Ehrkamp (2006). "Transnationalism and migrants' imaginings of citizenship", *Environment and Planning A* 38: 1615–1632.

Mahler, Sarah J. and Patricia R. Pessar (2001). "Gendered geographies of power: Analysing gender across transnational spaces", *Identities: Global Studies in Culture and Power* 7 (4): 441–459.

Massey, Doreen (1994) "A global sense of place", in Doreen Massey*Space, Place and Gender*. Cambridge: Polity Press, 146–156. (first published in *Marxism Today*, June 1991: 24–29).

Massey, Doreen (2005) *For Space*, London: Sage.

McDowell, Linda and Joanne Sharp (eds) (1999) *Space, Gender, Knowledge: Feminist readings*, London: Arnold.

Nash, Mary, Rosa Tello, and Nùria Benach (eds) (2005) *Inmigración, género y espacios urbanos. Los retos de la diversidad*, Barcelona: edicions Bellaterra.

Papataxiarchis, Efthymios, Pinelopi Topali and Aggeliki Athanasopoulou (2008) *Worlds of Domestic Labour. Gender, migration and cultural transformations in Athens of the beginning 21st century*, Athens: Alexandria and University of the Aegean (in Greek.

Peake, Linda and Martina Rieker (2013) "Rethinking feminist interventions into the urban", Introduction. in Linda Peake and Martina Rieke (eds) *Rethinking Feminist Interventions into the Urban*. New York: Routledge, 1–22.

Pratt, Geraldine and Brenda Yeoh (2003). "Transnational (counter) topographies", *Gender, Place and Culture* 10 (2): 59–166.

Salih, Rubah (2003) *Gender in Transnationalism: Home, longing and belonging among Moroccan migrant women*, London: Routledge.

Silvey, Rachel (2006). "Geographies of gender and migration: Spatializing social difference", *International Migration Review* 40 (1): 64–81.

Silvey, Rachel and Victoria Lawson (1999) "Placing the Migrant", *Annals of the Association of American Geographers*, 89 (1): 121–132.

Simonsen, Kirsten (1997) "Modernity, community or a diversity of ways of life: a discussion of urban everyday life", in OveKalltrop, IngemarErlander, OveEricsson and MatsFranzen (eds) *Cities in Transformation – Transformation in Cities. Social and Symbolic Change of Urban Space*. Aldershot: Avebury, 162–183.

Simonsen, Kirsten (2008) "Place as Encounters. Practice, conjunction and co-existence", in Jørn Øle Baerenholdt and Brynhild Granås *Mobility and Place. Enacting Northern European Peripheries*. Aldershot: Ashgate, 13–27.

Simonsen, Kirsten and Dina Vaiou (1996). "Women's lives and the making of the city. Experiences from "north" and "south" of Europe", *InternationalJournal of Urban and Regional Research*20: 446–465.

Smith, Dorothy (1987) *The Everyday World as Problematic: A Feminist Sociology*, Boston, MA: Northeastern University Press.

Vaiou, Dina (2012). "Gendered mobilities and border crossings: From Elbasan to Athens", *Gender, Place and Culture* 19 (2): 249–262.

Vaiou, Dina (2013) "Transnational city lives: Changing patterns of care and neighbouring", in Linda Peake and Martina Rieke (eds) *Rethinking Feminist Interventions into the Urban*. New York: Routledge, 52–67.

Vaiou, Dina, Anna Bacharopoulou, Theano Fotiou, Salomi Hatzivasileiou, Ares Kalandides, Rebecca Machi Karali, Olga Lafazani Kefalea, Rouli Lykogianni, Giorgos Marnelakis, Aleka Monemvasitou, Katerina Papasimaki, Fotini Tounta, and Eleftheria Varouchaki (2007) *Intersecting patterns of everyday life and socio-spatial transformations in the city. Migrant and local women in the neighbourhoods of Athens*, Athens: L-Press and NTUA (in Greek).

Wills, Jane (2013) "Place and Politics", in David Featherstone and Joe Painter (eds) *Spatial Politics. Essays for Doreen Massey*. Oxford: Wiley - Blackwell, 135–145.

21

Choreographing place

Race, encounter and co-belonging in the Anthropocene

Michele Lobo

Introduction

Experimental knowledges that move thinking beyond narrow human-centred conceptual vocabularies of place are urgent given debates on looming environmental catastrophes in the Anthropocene that threaten the survival of the planet (Gibson et al., 2015; Stengers, 2015; Latour, 2017). This chapter experiments with how we might live, breathe and dance with more-than-human worlds through a politics of place that proliferates imaginations of diverse planetary futures. Building on the work by feminist scholars Donna Haraway (2016), Gibson et al. (2015) and Kate Rigby (2015), who call for living earthly life intensely, joyously and ethically in the Anthropocene, I explore the possibilities for dancing with the earth body. Using participatory photographs and video clips of Darwin, Australia, I produce a choreography of human and more-than-human bodies that gently touches the skin of the planet in public spaces of the city informed by an experimental politics. Such a politics draws on embodied thought grounded in everyday urban worlds inhabited by racialised bodies whose identities cohere as Indigenous peoples, Aboriginal peoples, 'Blacks', 'multicultural people', 'Muslims', 'CALDs' (culturally and linguistically diverse), 'NESBs' (Non-English-speaking background), international students, ethnic/ethno-religious minority migrants, refugees and asylum seekers. Over the last 18 years I have interviewed and held focus groups with these hypervisible city dwellers in Paris, Detroit, Melbourne, Sydney and Darwin who inhabit the 'undercommons', a shared place of survival where the 'brokenness of being' (Harney and Moten, 2013: 5) gets under their skin (Ahmed, 2010; Simpson, 2011; Weheliye, 2014). In their everyday struggles with the slow, fleshy violence of racism, more-than-human encounters produce place as a site of survivance, nourishment, re-existence and resurgence (Rose, 1996; Vizenor, 2008; Lobo, 2014; Simpson, 2017; Mignolo and Walsh, 2018).

Ghassan Hage (2017: 125) argues that the emergence of such ethico-political spaces that enmesh 'multiple modes of existence' are crucial in unsettling dominant ways of inhabiting the world reproduced through western/white racism that exacerbate the ecological crisis. The advent of the Anthropocene suggests the dominance of such a racialised logics that is rooted in slavery, colonial capitalist domination as well as the domestication of the Other.

Rather than critique such racial dominance, this chapter focusses on an affirmative politics of place that centres performances of hope and responsibility. Such a politics is necessary so that our inheritance of 'trouble' in the Anthropocene moves away from despair, indifference, longings for Edenic pasts or 'salvific futures' in the city and beyond (Haraway, 2016).

Contemporary research on Anthropocene urbanism underlines the emergence of a political juncture that calls for urgency in thinking differently about cities (Derickson, 2018; Houston et al., 2018). Julietta Singh (2018), a (post)colonial feminist, participates in such knowledge production by drawing attention to the lingering of vital hope that is overlooked but can undo the mastery that governs webs of connections between humans, animals and the environment. For Indigenous scholar Leanne Simpson (2017), these webs embody relationships of gentleness, humility and carefulness as articulated by the thoughts about place of the Michi Saagiig Nishnaabeg, or salmon people, of Peterborough, Eastern Canada. Simpson (2017: 9) draws on these knowledges to describe place as an 'intellectual, political, spiritual, artistic, creative and physical' space that radiates responsibility. These insights from Indigenous and southern ontologies resonate with conceptual vocabularies of place that seek to expand the domain of the ethico-political through assemblage and multispecies thinking (Anderson et al., 2012; Haraway, 2016). Contemporary research within more-than-representational geographies, feminist science and technology studies, new materialisms, object-oriented ontologies and posthumanism distribute agency more widely by opening thought to propensities, affordances, and affectivities of nonhuman phenomena, planetary forcefields as well as technological objects (Bennett, 2010; Braun and Whatmore, 2010; Thrift, 2010b; Cresswell and Martin, 2012; McCormack, 2013; Amin, 2015; Stengers, 2015; Connolly, 2017). Ethnic Minority, Black, Indigenous and anti-racist scholars, however, interrogate western theoretical frameworks of liberal humanism, interspecies thinking and hierarchies of animacy for their failure to acknowledge 'unraced genealogies' (Chen, 2012; Nagar, 2014; Weheliye, 2014; Escobar, 2016; Chakrabarty, 2017; Puar, 2017; TallBear, 2017). Their refusal of racialised hierarchies of the human, inhuman, not-quite-human have the potential to decolonise the Anthropocene by proliferating diverse theoretical imaginations of place attentive to the identity of racialised bodies that cohere as well as dissolve through the passage of affects, sensations and intensities. The chapter has three main sections that focus on racialised logics and the skin of place, the laminations of place through experimental film making and choreographing more-than-human places.

The skin of place: shadowy bodies and everyday racialised logics

This section builds on seminal and emerging theoretical imaginations of place within Geography – a social space that is perceived, conceived and lived (Lefebvre, 1991; Soja, 1999); a contested political site produced through reiterative everyday practices (Cresswell, 2004); a 'critical interactional space' of responsibility (Massey, 2005: 11); a space of 'freedom and capacity' (Gibson-Graham, 2003: 50) as well as a place of belonging, togetherness and coexistence in the Anthropocene (Gibson-Graham, 2011; Larsen and Johnson, 2017; Howitt, 2018). I produce a choreography of shadow places from the 'undercommons' that is informed by my research agenda on race, affect, encounter and belonging in societies with white majority cultures over the last 18 years. When human encounters fatigue bodies who cannot or refuse to inhabit whiteness, a bodily orientation of mastery with material and symbolic privileges, this choreography that weaves together photographs as well as videos taken by residents in Darwin contributes to an energising affirmative politics (Hage, 1998; Ahmed, 2007; Lobo, 2013; McKittrick, 2015; Hage, 2017; Boittin, 2019).

In cities like Detroit, Paris, Melbourne and Paris, I felt and witnessed the palpability of mastery over place that unfolds through racism. Racism emerges as a material process, a machinic assemblage that entangles human/more-than-human entities, objects and technologies, as well as a brutal force that is subterranean or 'in your face' (Lobo, 2014). As it unfolds, racism increases pulse rates, churns guts, shackles bodies and sucks out the dignity of life for 'shadow citizenry' (Merrifield, 2015) that are either hypervisible or completely invisible in workplaces, parks, leisure centres, city streets, shopping complexes, public transport, airports, but also carceral spaces of prisons and asylum seeker detention centres. These marginalised and racialised urban dwellers who live on the periphery of attention are expulsed from the vortex of capital accumulation and widespread planetary urbanisation (Merrifield, 2015; Saldanha, 2015); see quotations from participants below:

JOSHUA: They want this white city, okay? White suburb, white city. Not Aboriginal city.
(male, >50 years, Aboriginal Elder, Darwin)

ALYA: The people also in the detention [centre] were saying they are having the experience that their dignity is crushed. I haven't gone through detention, but I feel the same thing.
(female, <50 years, humanitarian migrant, Darwin)

MARY: I don't know why I strain like I'm shackled in chains. They think I'm Islamically insane while I try to board an airplane … blood is coursing through my veins.
(female, <25 years, student/slam poet, Detroit)

HARRY: It's okay to have your life in your social ghetto, but now if you want to dress and own a flat in centre of Paris that becomes a problem. It's okay to dance and sing two days a year, because you have a national holiday. Exotic diversity, this entertaining diversity will be somehow tolerated to a certain extent and in specific areas, like a zoo, where that's your area.
(male, <50 years, activist, Paris)

GERARD: She was shouting 'fucking Indians don't you mind, don't you have any manners on the road', all that stuff. She was literally yelling at us using the country name, India and saying 'Go back to your country'.
(male, < 30 years, international student, Melbourne).

Affects of hatred, pain, anger and outrage circulate when the identity of bodies cohere in societies with different histories and geographies of white supremacy that have contributed to the advent of the Anthropocene. Puar (2017: 19) argues, however, that affect also dissolves identity so that bodies move from a diminished to an augmented capacity through a 'porous affirmation of what could or might be'. Such a focus on affective modalities contributes to thinking with place as a site of 'enfleshment otherwise' (Weheliye, 2014: 12) that unfolds through 'miniscule movements' and 'glimmers of hope' even in spaces of bare life where dreams of freedom are interrupted. These unbreathable places in which fleshy bodies of colour 'find themselves' (Stewart, 2011: 452) and are dehumanised, are also sites of 'thinking, feeling, sensing, being, doing and living otherwise' (Mignolo and Walsh, 2018: 102). When place is sentient it sheds its tough Euclidean skin through laminations that empty battle cries and 'breathe different atmospheres' (Thrift, 2014: 18) of human and more-than-human coexistence that are necessary if we are to think differently about cities and diverse planetary futures in the Anthropocene.

Laminations and shadowscapes

Laminations, a collaborative creative research event (2015, Darwin), and 'The Dark Matter of the Urban: forces, densities, velocities, affects, and more', a session at the Association of

American Geographers Conference, 2016, mobilised my thoughts on diverse planetary futures. Such futures are central to enacting a pluriverse or the co-constitution and entanglement of many different worlds (Escobar, 2016; Sundberg, 2014; Mignolo and Walsh, 2018). These worlds graft new skins and touch the earth through everyday events of responsible co-becomings. This section focusses on Greater Darwin (including Palmerston), a small tropical north Australian city at the centre of highly politicised national and international public debates on Indigenous wellbeing, racism and national security, but also, more recently, resilience to sea level rise, extreme weather events, bushfires and tropical cyclones (Loughnan et al., 2013; Spencer, Christie, and Wallace, 2016). Built on Larrakia or Saltwater Country in 1869, dystopian visions of Darwin are attributed to looming environmental crises, rather than continuing histories of white supremacist and interventionist policies that domesticate the worlds of Indigenous peoples as well as ethnic/ ethno-religious minority residents, migrant newcomers, refugees and asylum seekers. In this chapter these residents used participatory photography and videos of annual, weekly, bi-weekly, monthly social gatherings as well as everyday encounters that were part of work and leisure to express more-than-human vocabularies of place. As they moved, the shaky hand-held camera footage partially captured these multisensory resonances of place through events that included ethnic festivals (Indian Holi/Spring festival; India at Mindil), beauty pageants (Miss Africa Darwin), multicultural festivals (Seabreeze festival; Harmony Day celebrations), community dinners, film screenings/commemorative vigil services for asylum seekers (outside the detention centre, at community cafes), street marches/beach protests against racially discriminatory policies and irresponsible economic/environmental policies (March in March; Nightcliff beach protest), Indigenous celebration walks (National Aborigines and Islanders Day Observance Committee or NAIDOC March), Asian-style open-air markets (Parap Market, Mindil Beach Market; Nightcliff Market, Palmerston Market), cross-cultural football matches (Football Without Borders that engages asylum seekers) and Indigenous Caring for Country programs (Men and Women Larrakia Ranger programs). Shared activities of cooking, sewing and making art engaged ethnic-minority migrants, asylum seekers as well as Indigenous peoples, for instance, Sisters' Kitchen, Cooking for Senior Citizens, Art in the Underground, Painting in the 'Long Grass' for Indigenous elders who 'live rough'. I went on many walk-alongs with urban dwellers at beaches, open-air markets, community gardens, shopping malls and city streets. Contemporary research on walking informed by Indigenous movements such as the Zapatista Movement in Mexico and Idle No More in Canada draws attention to walking as a performative practice that enacts a pluriverse with multiple worlds (Sundberg, 2014). Videos taken by residents expressed these worlds through affects, sensations and intensities that were difficult to articulate in words but reanimated place deadened by everyday racism.

The outcome of participatory research in Darwin was a short collaborative film, a montage titled *Shadowscapes* co-produced through digital techniques of assembling, layering and juxtaposition. It involved carefully crafting and communicating affective relationships that centred responsible co-becomings with the earth through dark, moving shadows on the urban landscape. Such co-becomings laminate the earth through an affective, earthly pedagogy that expresses multiple modes of existence (Simpson, 2017). The participatory film shows how racialised bodies escape the containment of their skin and dance with more-than-human temporalities that constitute place. Images of outdoor spaces at morning, noon and evening lit by natural sunlight as well are supplemented by indoor community spaces with bright neon lights produce dancing shadows of co-belonging in the Anthropocene are lively rather than frozen in time and space. The film entangles the dark matter of bodily

Figure 21.1 A view of the Arafura Sea.

shadows with the shadowy lacework of diverse species of trees (banyan, eucalyptus, casuarina, mangroves), ambient environmental sounds, English/Aboriginal languages as well as accents, images of the coastline and archival maps of Darwin harbour. The film starts with my view of the Arafura Sea from a passenger ferry that plies between Cullen Bay (Darwin) and Wurrumiyanga, on the southern coast of Bathurst Island (Tiwi Islands) (Figure 21.1).

The ferry is used regularly by Tiwi Islanders and Aboriginal peoples of diverse backgrounds who make the two and a half-hour journey to work, visit family or access health facilities. This image of the Arafura Sea is juxtaposed with faint shadows on sandy Mindil beach, once a site for ritualised fights, corroborees and Aboriginal burials. The beach is marked by a small memorial that marks prior Aboriginal presence at Mindil beach, an open-air Asian-style market that celebrates contemporary Indigenous and ethnic minority encounters as well as paths that make up the Larrakia coastal Plant Use walk that focusses on Indigenous knowledges of Caring for Country (Figure 21.2).

This gentle touching of the thin sensitive sandy skin of the earth through Caring for Country programs that move beyond mastery and possession contrast with the slower violence but harder human touch that contributes to extreme weather events, ocean acidification, sea level rise and species loss (for instance, mangrove, seagrass, turtles, dugongs) in tropical coastal cities such as Darwin. Contemporary maps of the coastline and archival maps (prepared in 1937) of Darwin harbour that once had widespread mangrove forests are superimposed over images of the beach backed by coastal dunes (Figure 21.3, Figure 21.4).

These shadows of the past and present that entangle multiple temporalities create a choreography that laminates and enables possibilities for listening with the thin outer sensitive sandy skin of the earth. In her discussion of environmental justice projects, eco-feminist Val Plumwood (2008) argues that shadow places provide material and ecological support but elude knowledge and responsibility because they are often overlooked.

Figure 21.2 Larrakia coastal Plant Use walk/Mindil Beach, from participant video clips.

Figure 21.3 Mindil Beach.

Ambient sounds of crashing waves breaking on the shore, the whispering ebb and flow of tides, rippling creek waters, chirping crickets, buzzing mosquitoes and the strong sea breeze mingle with children's voices at the NAIDOC walk, animated voices at open-air Asian-style

Figure 21.4 Mindil Beach.

markets, loud contemporary western music, clicking Aboriginal clapsticks, footsteps that crackle forest undergrowth and the whirr of passing cars outside the asylum seeker detention centres. Larrakia Country emerges through the coexistence of diverse worlds performed through ambient sounds and voices that are foregrounded rather than silenced. In her exploration of philosophical animism informed by Indigenous ecologies, Deborah Bird Rose (2013: 103) argues that a place-based perspective that foregrounds active listening to 'multispecies, multi-cultural zones of inter-action' unsettles the mind/matter binary and is humbling as well as enriching. The montage materialises such listening through 'haptic visuality' (Marks, 2000: 3), a kind of haptic play that shows how the skin of racialised Indigenous as well as ethnic-minority bodies touch and are touched by the many skins of the city through the lens of the camera. Shapiro (2009: 1) argues that 'to touch something is both to act on the world and to be affected by it'. Touch is then an archetypal sense and a 'complex set of sensory practices and emotional intensities' (Dixon and Straughan, 2010: 453) that expresses an 'enfleshment otherwise' (Weheliye, 2014) in places that are always more than just human.

Choreographing more-than-human urban places: radiating responsibility

Shadowscapes produced by the intra-action (Barad, 2007) of bodies, light and the earth are often overlooked in films that centre upon the movements of human bodies. In this film, however, layered textured patterns of shadows produce multiple skins of the urban through affective resonances that often elude our grasp but show the eventfulness of the urban. This eventfulness that entangles life/non-life, animate/inanimate, mind/body and nature/culture emerges through a 'craftsmanship of the moment' (Thrift, 2010a: 142) made possible through technological mediation and digitisation of infra-thin shadows that capture gentle ways of touching the earth. Philosopher Erin Manning (2017: 97) draws on Marcel Duchamp's concept of the infra-thin to describe 'the most minute of intervals or the slightest of differences' expressed through contact that touches but is imperceptible, ungraspable and eludes definition. The film highlights these infra-thin moments that stay in the background of events but provide an insight into the thickness of the present that operationalises tendencies from more-than-human registers. It is barely perceived intensities or affective logics that are part of the textures of place that exceed capture and are only partially articulated as sound, colour and affective tonality that are necessary if we are to respond to Haraway's (2016: 1) call to stay with the trouble in the Anthropocene by learning to be 'truly present'. Like ephemeral shadows in public space that are infra-thin, never articulated in concrete form, these performances of being present highlight the urban as a place of earthly co-becomings. Manning (2017) theorisation of the politics of the infra-thin involves a pragmatics of the useless, where resonances from the eventfulness of the urban that are often excluded and might seem valueless come together to contribute to an ecology that has the potential to alter the habitual sensorium, bodily orientations, everyday practices and the flow of movements that assemble place.

The montage with infra-thin dancing shadows is an experimental as well as amateur art form that attempts to choreograph the city as a site of emergent ecological as well as socio-cultural diversity. But for this montage of moving bodies, maps and ambient sounds to continue to have value it must be 'activated anew' (Manning, 2017: 105) through infra-thin encounters or infra-individual experiential events that unfold every time it is shown at conferences, workshops, symposiums and creative exhibitions. At these events, dancing shadows are reactivated, echo, reverberate, become contagious and multiply to produce multisensory refrains. It is these multisensory refrains that have the capacity to affect routine decisions, habits and judgements that are part of everyday urban living. Amin (2015) regrets, however, that such incipient tendencies or potentialities are often barely recognisable and therefore rarely contribute to wisdom on the sociality of the urban. However, it is such sociality that has the potential to challenge dogmas of moral philosophy that centre the human agent as the bearer of responsibility and free will.

In the midst of struggle, place calls out, creates, stewards and teaches responsibility (Gibson-Graham, 2011; Haraway, 2016; Larsen and Johnson, 2017). Unlike 'responsibility for' (Manning, 2013: 68) that is heroic, sayable, reeks of liberal humanist forms of benevolence, generosity and agency in a 'passive and inert world' (van Dooren and Rose, 2016: 89), 'being responsible before' (Manning, 2016: 68) is a choreography or movement of thought that relays responsibility through openness to diverse planetary futures when the eventfulness of place calls out in the Anthropocene (Haraway, 2016). Simpson (2017) argues that place plays a crucial role in radiating responsibility through ecologies of intimacy with the earth that move beyond governance steeped in aggressive power, coercion, hierarchy and authoritarianism. This radiation of responsibility grounded in Nishnaabeg thought involves giving more than we take from the Earth and resonates with Larrakia protocols in Darwin performed by Indigenous people who 'live rough' as well as Land and Sea Rangers. Such knowledge steeped in the intelligence and brilliance of embodied thought is grounded in experience and is life propelling (Simpson, 2017). This is the force of embodied thought and ecologies of intimacy that the film tries to materialise through infra-thin shadows that gently touch, dance and hold the earth. This 'excess' that was expressed briefly by participants, but often escaped talk, captured affective attunements to the earth that nourished re-existence and resurgence among ethnic-minority migrants, refugees and asylum seekers whose lives were precarious and marked by frugality (Lobo, 2014, 2018).

Re-existence as well as resurgence are concepts central to embodied Indigenous thought on struggles that move beyond resistance or resilience, and can inform thinking about possibilities for diverse planetary futures in the Anthropocene. Walter Mignolo and Catherine Walsh (2018: 95) draw on Adolfo Alban Achinte's concept of re-existence to draw attention to 'agency, action, and praxis of the otherwise' or creative decolonial pedagogies that affirm dignity and hope in the midst of negation, violence and despair. For Leanne Simpson, resurgence is an everyday as well as collective process that draws on the kinetics of place-based creative practices grounded in Aki or the earth. These practices entangle landforms, living things, spirits, sounds, thoughts, feelings and energies into emergent ecologies and networks that are possible to partially capture through participatory films as shown in this chapter. Experimenting with digital visual technologies provides the potential to decolonise the Anthropocene by entangling Indigenous and ethnic-minority knowledges of place that refuse or show the inability to inhabit whiteness as a bodily orientation of mastery, possession and control.

Conclusion

The different colours of their skin and their haunting shadows cannot contain racialised bodies – skin is more than just a living membrane and a border that feels the pain, anger, outrage and fatigue of racism. Skin is also a multidimensional topological surface that folds subliminal bodily intensities from the eventfulness of the world (Connolly, 2010; Manning, 2016; Simpson, 2017). Through techno-social collaborations that focus on the immersive experience or 'thisness' of events, the skin of public space is reactivated through infra-thin moments that are elusive, difficult to articulate but capture generative forces that are propelling. *Shadowscapes*, an experimental creative intervention draws attention to these affective forces through an 'ontological choreography' (Kirksey, 2019: 17) of place that goes beyond the language of a racialised logics that has contributed to the advent of the Anthropocene. The film is a performative intervention grounded in an ontologically derived ethics of responsibility that entangles human and more-than-human worlds in webs of coexistence. Through a montage of videos that focusses on infra-thin laminations, place breathes and dances differently when we centre embodied thought grounded in the everyday lives of Indigenous and ethnic-minority peoples who inhabit the 'undercommons'. While such infra-thin laminations that choreograph public spaces may not necessarily result in revolutionary change in the Anthropocene, it seeds earthly co-becomings through everyday habits that attune us to how the more-than-human is 'materially constitutive' of everyday urban life (Ravaisson, 2008: 5; Yusoff, 2016).

Openness to generative more-than-human forces of the world that precede and exceed social relations is central to feminist geophilosophy or speculative theoretical frameworks that focusses on creative co-becomings emergent from the earth and cosmos attentive to difference (Grosz, 2011; Bosworth, 2017). Such co-becomings and coexistence with more-than-human worlds resonate with emerging Indigenous-led and ethnic-minority understandings of place in the Anthropocene within Geography (Saldanha, 2007, 2012; Nayak, 2010; Bawaka Country et al., 2016; Tolia-Kelly, 2016; Larsen and Johnson, 2017; Howitt, 2018; Lobo, 2018). Perhaps the whiteness of Anglophone Geography that Richie Howitt (2018) argues has so far produced a 'truly cacophonous silence' is beginning to be unsettled through new understandings of place.

Dancing shadows in public space embed bodies in the Earth. Bersani (2008: 183) argues that 'the only way we can love the other and the external world is to find ourselves somehow in it'. Public spaces then emerge as 'nodes of intensity' with human and non-human resonances that when fused together have the potential to create energy and incandescent light (Merrifield, 2014: 7). Indigenous peoples, ethnic-minority migrants, refugees and asylum seekers emerge as more than racialised shadow citizens when they feel the sunlight and learn how to stand the glare even though their eyes may sometime ache (Merrifield, 2015). In Darwin, aliveness to Earth or Country that is both land and sea nourishes racialised Indigenous and ethnic-minority migrants, refugees and asylum seekers with different histories of geographies of displacement and dispossession (Lobo, 2014). In their struggle for the freedom to be, to breathe and move in the urban commons, they are shadow citizens who draw attention to place in the Anthropocene as an ethico-political space of struggle but also re-existence and resurgence. These infra-thin laminations or ephemeral shadows of racialised bodies that are easily overlooked mingle with the forces of a more than human world to create place as a lacework that touches and dances with the earth. In comparison with practices that result in earthly exploitation, these shadowscapes produce place as a site of new imaginations of co-belonging and earthly survival.

Acknowledgements

This research was funded by an Australian Research Council Discovery Early Career Researcher Award (DE 130100250, 2013-2016). Special thanks to the participants, Larrakia Nation Aboriginal Corporation, Multicultural Council of the Northern Territory, Darwin Asylum Seeker Advocacy Network, Darwin Community Arts and the Northern Institute, Charles Darwin University, Australia. Thanks to Kaya Barry, Madeline Wilmot, Johanna Funk and David Kelly for creative video editing and ongoing conversations.

Bibliography

Ahmed, S. 2007. A phenomenology of whiteness. *Feminist Theory*, 8(2), 149–168.

Ahmed, S. 2010. *The Promise of Happiness*. Durham and London: Duke University Press.

Amin, A. 2015. Animated space. *Public Culture*, 27(2), 239–258.

Anderson, B., Kearnes, M., McFarlane, C., and Swanton, D. 2012. Materialism and the politics of assemblage. *Dialogues in Human Geography*, 2(2), 212–215.

Barad, K. 2007. *Meeting the Universe Halfway: Quantum Physics and the Entanglement of Matter and Meaning*. Durham: Duke University Press.

Bennett, J. 2010. *Vibrant Matter: A Political Ecology of Things*. Durham: Duke University Press.

Bersani, L. 2008. *Is the Rectum a Grave? Ann Other Essays*. Chicago: Chicago University Press.

Boittin, J. 2019. 'The Rot That Remains:' Aphasia and France's Multiple Temporalities of Race and Colonialism. *Postcolonial Studies* (forthcoming).

Bosworth, K. 2017. Thinking permeable matter through feminist geophilosophy: Environmental knowledge controversy and the materiality of hydrogeologic processes. *Environment and Planning D: Society and Space*, 35(1), 21–37.

Braun, B., and Whatmore, S. 2010. The stuff of politics: An introduction, In B. Braun and S. J. Whatmore Eds., *Political Matter: Technoscience, Democracy, and Public Life*. Minneapolis: University of Minnesota Press, ix–xl.

Chakrabarty, D. 2017. The future of the human sciences in the age of humans: A note. *European Journal of Social Theory*, 20(1), 39–43.

Chen, M. 2012. *Animacies: Biopolitics, Racial Mattering, and Queer Affect*. Durham and London: Duke University Press.

Connolly, W. 2010. *A World of Becoming*. Durham, NC: Duke University Press.

Connolly, W. 2017. *Facing the Planetary: Entangled Humanism and the Politics of Swarming*. Durham: Duke University Press.

Country, B., Wright, S., Suchet-Pearson, S., Lloyd, K., Burarrwanga, L., Ganambarr, R. et al. 2016. Co-becoming Bawaka: Towards a relational understanding of place/space. *Progress in Human Geography*, 40(4), 455–475.
Cresswell, T. 2004. *Place: A Short Introduction*. Malden, MA: Blackwell Publishers.
Cresswell, T., and Martin, C. 2012. On turbulence: Entanglements of disorder and order on a Devon beach. *Tijdschrift voor Economische en Sociale Geografie*, 103(5), 516–529.
Derickson, K. 2018. Urban geography III: Anthropocene urbanism. *Progress in Human Geography*, 42(3), 425–435.
Dixon, D., and Straughan, E. 2010. Geographies of touch/touched in geography. *Geography Compass*, 4/5, 449–459.
Dooren, T., and Rose, D. 2016. Lively ethography: Storying animist worlds. *Environmental Humanities*, 8(1), 77–94.
Escobar, A. 2016. Thinking-feeling with the Earth: Territorial struggles and the ontological dimension of the epistemologies of the South. *aibr. Revista de Antropología Iberoamericana*, 11(1), 11–32.
Gibson, K., Rose, D., and Fincher, R. 2015. Preamble, In K. Gibson, D. B. Rose, and R. Finche Eds, *Manifesto for Living in the Anthropocene*. Brooklyn: Punctum Books, pp. i–iii.
Gibson-Graham, J. K. 2003. An ethics of the local. *Rethinking Marxism*, 15(1), 49–74.
Gibson-Graham, J. K. 2011. A feminist project of belonging for the Anthropocene. *Gender, Place and Culture*, 18(1), 1–21.
Grosz, E. 2011. *Becoming Undone: Darwinian Reflections on Life, Politics, and Art*. Durham, NC: Duke University Press.
Hage, G. 1998. *White Nation: Fantasies of White Supremacy in a Multicultural Society*. Annandale, Victoria: Pluto Press.
Hage, G. 2017. *Is Racism an Environmental Threat*. Cambridge: Polity Press.
Haraway, D. 2016. *Staying with the Trouble: Making Kin in the Chthulucene*. Durham and London: Duke University Press.
Harney, S., and Moten, F. 2013. *The Undercommons: Fugitive Planning and Black Study*. Wivenhoe/ New York/Port Watson: Minor Compositions.
Houston, D., Hillier, J., MacCallum, D., Steele, W., and Byrne, J. 2018. Make kin, not cities! Multispecies entanglements and 'becoming-world' in planning theory. *Planning Theory*, 17(2), 190–212.
Howitt, R. 2018. Progress in Human Geography Lecture. 11 April 2018. New Orleans: Association of American Geography Conference.
Kirksey, E. 2019. Queer love, gender bending bacteria, and life after the Anthropocene. *Theory, Culture and Society*, 36(6), 197–216.
Larsen, S. C., and Johnson, J. T. 2017. *Being Together in Place: Indigenous Coexistence in a More Than Human World*. Minneapolis: Minnesota University Press.
Latour, B. 2017. *Facing Gaia. Eight Lectures on the New Climatic Regime*. Cambridge: Polity Press.
Lefebvre, H. 1991. *The Production of Space*. (D. Nicholson-Smith, Trans.). Oxford: Basil Blackwell Ltd.
Lobo, M. 2013. Racialised bodies encounter the city: 'Long Grassers' and Asylum seekers in Darwin. *Journal of Intercultural Studies*, 34(4), 454–465.
Lobo, M. 2014. Affective energies: Sensory bodies on the beach in Darwin, Australia. *Emotion, Space and Society*, 12, 101–109.
Lobo, M. 2018. Re-framing the creative city: Fragile friendships and affective art spaces in Darwin, Australia. Special issue article: Urban friendship networks: Affective negotiations in the city, In L Kathiravelu and T Bunnell eds., *Urban Studies*, 55(3), 623–638.
Loughnan, M., Tapper, N., Phan, T, Lynch, K, and McInnes, J. 2013. *A Spatial Vulnerability Analysis of Urban Populations During Extreme Heat Events in Australian Capital Cities*. Gold Coast: National Climate Change Adaptation Research Facility.
Manning, E. 2013. *Always More Than One: Individuation's Dance*. Durham: Duke University Press.
Manning, E. 2017. For a pragmatics of the useless, or the value of the Infrathin. *Political Theory*, 45(1), 97–115.
Manning, E. 2016. *The Minor Gesture*. Durham/London: Duke University Press.
Marks, L. U. 2000. *The Skin of the Film: Intercultural Cinema, Embodiment and the Senses*. Durham, NC, and London: Duke University Press.
Massey, D. 2005. *For Space*. London: Sage.

McCormack, D. 2013. *Refrains for Moving Bodies: Experience and Experiment in Affective Spaces*. Durham/ London: Duke University Press.
McKittrick, K. 2015. *Sylvia Wynter: On Being Human as Praxis*. Durham and London: Duke University Press.
Merrifield, A. 2014. *The New UrbanQquestion*. London: Pluto Press.
Merrifield, A. 2015. The Shadow Citizenry. *Eurozine*. www.eurozine.com/articles/2015-07-24-merri field-en.html.
Mignolo, W., and Walsh, C. 2018. *On Decoloniality: Concepts, Analytics, Prraxis*. Durham and London: Duke University Press.
Nagar, R. 2014. *Muddying the Waters: Coauthoring Feminisms across Scholarship and Activism*. Chicago: University of Illinois Press.
Nayak, A. 2010. Race, affect, and emotion: Young people, racism, and graffiti in the postcolonial English suburbs. *Environment and Planning A*, 42, 2370–2392.
Plumwood, V. 2008. Shadow places and the politics of dwelling. *Australian Humanities Review*, 44, 139–150.
Puar, J. 2017. *The Right to Maim: Debility, Capacity, Disability*. Durham and London: Duke University Press.
Ravaisson, F. 2008. *Of Habit*. London: Continuum.
Rigby, K. 2015. Contact improvisation: Dance with the earth body you have, In K. Gibson, D. B. Rose and R. Fincher Eds., *Manifesto for living in the Anthropocene*. Brooklyn: Punctum Books, 43–48.
Rose, D. 1996. *Nourishing Terrains: Australian Aboriginal views of Landscape and Wilderness*. Canberra: Australin Heritage Commission.
Rose, D. 2013. Val Plumwood's philosophical animism: Attentive interactions in the sentient world. *Enviironmental Humanities*, 3, 93–109.
Saldanha, A. 2007. *Psychedelic White: Goa Trance and the Viscosity of Race*. Minneapolis: University of Minnesota Press.
Saldanha, A. 2012. Assemblage, materiality, race, capital. *Dialogues in Human Geography*, 2(2), 194–197.
Saldanha, A. 2015. Scale, difference and universality in the study of race. *Postcolonial Studies*, 18(3), 326–335.
Shapiro, K. 2009. Reviving habit: Felix Ravaisson's practical metaphysics. *Theory and Event*, 12(4), 1–2.
Simpson, L. 2011. *Dancing on our Turtle's Back: Stories of Nishnaabeg Re-creation, Rresurgence and a New Emergence*. Winnipeg: Arbeiter Ring Pub.
Simpson, L. B. 2017. *As We Have Always Done: Indigenous Freedom through Radical Resistance*. Minneapolis: Minnesota University Press.
Singh, J. 2018. *Unthinking Mastery: Dehumanism and Decolonial Entanglements*. Durham and London: Duke University Press.
Soja, E. 1999. Thirdspace: Expanding the scope of the geographical imagination, In D. Massey, J. Allen and P. Sarre Eds., *Human Geography Today*. Malden, MA: Polity Press, 260–278.
Spencer, M, Christie, M, and Wallace, R. 2016. *Disaster Resilience Management and Preparedness in Aboriginal and Torres Strait Islander Communities in Darwin and Palmerston*. Final Report. Darwin: Charles Darwin University.
Stengers, I. 2015. *In Catastrophic Times: Resisting the Coming Barbarism*. Translated by Andrew Goffey. Paris: Open Humanities and Meson Press.
Stewart, K. 2011. Atmospheric atttunements. *Environment and Planning D: Society and Space*, 29(3), 445–453.
Sundberg, J. 2014. Decolonizing possthumanist geographies. *Cultural Geographies*, 21(1), 33–47.
TallBear, K. 2017. Beyond the life/not-life binary: A feminist-indigenous reading of cryopreservation, interspecies thinking, and the new materialisms, In J. Radin and E. Kowal Eds., *Cryopolitics: Frozen Life in a Melting World Mass*. Cambridge, MA: MIT Press, pp. 179–202.
Thrift, N. 2010a. Halos: Making more room in the world for new political orders. 1–36.
Thrift, N. 2010b. Slowing down race. *Environment and Planning A*, 2428–2430.
Thrift, N. 2014. The 'sentient' city and what it may portend. *Big Data and Society*, April–June, 1–21.
Tolia-Kelly, D. 2016. Anthropocenic culturecide: An epitaph. *Social and Cultural Geography*, 17(6), 786–792.
Vizenor, G. 2008. *Survivance: Narratives of Native Presence*. Lincoln: University of Nebraska Press.
Weheliye, A. 2014. *Habeas Viscus*. Durham and London: Duke University Press.
Yusoff, K. 2016. Anthropogenesis:Origins and endings in the Anthropocene. *Theory, Culture and Society*, 33(2), 3–28.

22
Class and place

Paul Watt

Introduction

This chapter examines the relationship between social class and place in the cities and suburbs of the global North. Place attachment – the affective, cognitive and social interactional sense of belonging to a place – can occur at a variety of spatial scales including the home, the neighbourhood, city, region and nation (Kusenbach and Paulsen, 2013; Lewicka, 2011; Manzo and Devine-Wright, 2014). The focus of this chapter is place attachment and belonging to the residential neighbourhood with reference to the working class, middle classes and the "super-rich".

It might seem perverse to focus on residential neighbourhoods given the argument from the mobilities' literature that globalisation and the accelerated flows of people have meant that residential location is less significant for social identities and interactions (Urry, 2007; Watt and Smets, 2014). There is certainly evidence regarding the accelerated spatial mobilities of the middle classes and the super-rich in relation to travelling for work and business (Andreotti, Le Galès, and Fuentes, 2015; Forrest, Koh, and Wissink, 2017). Nevertheless, these privileged classes are by no means unconnected to cities and their neighbourhoods, as discussed below. We will begin, however, by examining the shifting working-class relationship to the neighbourhood.

Place and the industrial working class: neighbourhood and community

Historically the industrial working class was spatially clustered into towns and cities, often close to their workplaces in factories, coal mines, steel mills and shipyards. These industrial working-class quarters were overcrowded and insanitary, as Engels (1844/1987) vividly describes with reference to 19th-century Manchester. Distinctive working-class neighbourhoods based around industrial and extractive employment developed throughout the 19th century and the first three quarters of the 20th century in European and North American towns and cities (Thrift and Williams, 1987). These neighbourhoods also formed the epicentres of the labour movement via the twin prongs of trades unions and labour/

social democratic/communist political parties (Parkin, 1967; Savage, 1987). Such neighbourhoods were still prominent in the post-1945 period as sociologists "discovered" tight-knit "urban villages" such as Bethnal Green in east London (Young and Willmott, 1957) and Boston's inner-city West End (Gans, 1962). These urban villages formed working-class communities based upon neighbouring, extended family ties, the church (in Boston's West End) and proximate sources of employment in local workplaces.

Harmonious pictures of the working-class neighbourhood as a singular, seamless "community" elide social differences and tensions, notably around gender, ethnicity and generation, as well as status distinctions between the "respectable" and "rough" working class (Blokland, 2003; Bourke, 1994; Watt, 2006). One example is the significance of ethnic differences within the male blue-collar workforce in Youngstown during its "Steeltown USA" heyday, differences that the steel companies exacerbated (Linkon and Russo, 2002). Nevertheless, the notion that post-war working-class neighbourhoods fostered a sense of community retains some veracity, even if this should not be uncritically mythologised or romanticised (Cohen, 2013; Pahl, 2005).

Many of these older working-class neighbourhoods, including Boston's West End (Gans, 1962), were eventually demolished under slum clearance/urban renewal programmes. Slum clearance began in the late 19th century but intensified under post-war urban renewal policies and the expansion of public/social rental housing. Urban renewal resulted in the forced relocation – displacement – of workers and their families to new modernist public/social housing estates, either in the inner city or suburban periphery (Clapson, 1998; Wacquant, 2008; Young and Willmott, 1957). Although urban renewal often improved housing conditions, it was also criticised since displacement could result in a diminished sense of place, as seen by how some relocated residents grieved for their lost homes and communities (Fried, 1966). Such mono-class public/social housing estates could, however, develop neighbourly social relations and positive place attachments over time (Clapson, 1998; Todd, 2014).

Place and the post-industrial working class: lost communities, ambivalent places

If community and place belonging were prominent aspects of working-class neighbourhoods up until the 1970s–1980s, the subsequent period has witnessed an epochal shift in the significance and nature of such neighbourhoods in Europe and North America. There are multiple reasons for this shift, including improved transportation, greater geographical mobility, the weakening of extended family relations and the increased prominence of privatised home-centred leisure, notably television. Probably most significant is deindustrialisation – the factories, coal mines and steel mills closed down or relocated thus removing the economic and organisational basis for industrial working-class neighbourhoods (Gest, 2016; Todd, 2014).

Factory closures in Western capitalist countries began as the post-war Long Boom stuttered to a halt during the late 1960s–early 1970s, but accelerated from the 1980s onwards due to widespread recessions and increased global competition. Previously industrial regions – for example the northern US states of Ohio, Illinois and Pennsylvania, northern France, northern and Midlands England – became "rust-belt" regions as their former industries declined and more or less disappeared. This decline was felt most acutely among those cities and neighbourhoods that were intimately associated with heavy industry and/or transport hubs. Gest (2016) has referred to such places as "post-traumatic cities" as in his research in deindustrialised Youngstown, the former "Steeltown USA" (Linkon and Russo, 2002), and east London with its now-closed docks and riverside industries (Cohen, 2013).

In-depth studies undertaken in formerly industrial/dockside working-class neighbourhoods have identified common narratives of place centred around community decline, especially among older residents (Blokland, 2003; Cohen, 2013; Gest, 2016; Jeffery, 2018; May, 1996; Watt, 2006, 2007). The present-day "community" is regarded as being a threadbare remnant of its former glories. Prominent themes in these "narratives of urban decline" include the loss of manual jobs alongside the depletion of the extensive and intensive social networks (based on trade unions, works organisations, pubs and bars) that used to exist in the industrial neighbourhood. Nostalgia has become a prominent lens through which places are interpreted (Cohen, 2013); when asked what kind of future they would like to see for their area, older residents reply "the past" (Gest, 2016).

Such narratives of urban decline are also informed by various social changes in addition to deindustrialisation and job losses. These changes include: the deterioration of public/social housing estates, often due to landlord neglect (Feldman and Stall, 2004; Karn, 2007); worsening leisure and consumption facilities (Jeffery, 2018; Watt, 2006); and increased crime and social disorder (May, 1996; Watt, 2007). Those public/social housing estates which relied on industrial jobs have suffered from increasing poverty as their residents were laid off and either became unemployed or worked in precarious post-industrial, service-sector employment (Smith, 2005; Todd, 2014). Elements within the white working class have also racialised community decline by symbolically associating it with the presence of minority-ethnic "incomers" (Blokland, 2003; Cohen, 2013; Gest, 2016; May, 1996; Watt, 2006, 2007). Thus, rather than place attachment, white working-class residents express *place detachment* to "their" neighbourhoods.

If narratives of urban decline are prominent, they are by no means all-encompassing or even dominant. In research on working-class public housing tenants in London, Watt (2006: 785) found that a sense of a lost community co-existed with a "discourse of belonging" based on the neighbourhood's location and accessibility to public transport, as well as how tenants had extensive social connections based on "knowing people" (family and neighbours) living at or near their estate. Such *ambivalent place attachments* have been noted in studies of deprived working-class areas in the US (Manzo, 2014) and UK (Jeffery, 2018). In so far as the working class are less geographically mobile than the middle classes and reside in a place over long periods of time, they are, if anything, more likely to develop neighbourly social relations (Young Foundation, 2010). Length of residence has generally been found to be "the most consistent positive predictor of attachment to residence places (usually neighbourhoods)" (Lewicka, 2011: 216). Morris (2019) found that longevity of residence was a key factor in generating a sense of community among public housing tenants in Sydney.

These working-class communal bonds can also cross ethnic differences over time (Watt, 2006). In her study of the St Ann's public housing estate in Nottingham, McKenzie (2015) identified social mixing between long-term white and Black Caribbean working-class tenants resulting in a shared "being St Ann's" place identity. Day-to-day neighbourliness and communal social bonds can also be enhanced by place-specific activities in working-class areas, such as carnivalesque Christmas decorations (Edensor and Millington, 2009).

An extreme version of urban decline and place detachment has been presented by Loic Wacquant (2008) in *Urban Outcasts*. In this book, Wacquant argues that the late-20th-century US black ghetto (South Side of Chicago) and French working-class banlieue (La Corneuve in the Parisian suburbs) are characterised by multiple reinforcing strands of disadvantage amounting to "advanced marginality". This marginality encompasses increased wage-labour insecurity indicating the formation of a post-industrial "precariat", as other research has indicated (Smith, 2005). In terms of place, Wacquant points out how these zones experience "territorial stigmatisation", as well as "spatial alienation and the dissolution of 'place'" (Wacquant, 2008: 241) involving the losses of home, security and cultural familiarity. The US black ghetto and French working-class banlieue have therefore shifted away from being "communal places" with shared meanings and mutual practices, to "indifferent 'spaces' of mere survival and relentless contest" (Ibid.). Wacquant also highlights how the hyper-ghetto of the US inner cities has no direct European equivalent in relation to the intensity of its racism, crime, poverty and welfare state withdrawal. The distinctive socio-spatial nature of the late-20th-century black inner-city ghetto is a prominent theme within US urban sociology, as seen in the work of Wilson (1987) which has a different emphasis from Wacquant. At the same time, research by Feldman and Stall (2004) on housing activism by the black female residents of Chicago's public housing "projects" has identified a greater sense of place belonging than Wacquant's uniformly bleak picture suggests (see Manzo, 2014). Qualitative studies of public/social housing estates in Paris (Garbin and Millington, 2012), London (Watt, 2013) and Sydney (Morris, 2013, 2019) have also provided nuanced accounts which query how far such neighbourhoods are merely epicentres of stigmatisation, place detachment and spatial alienation.

Place and the middle classes: mobilities and belonging

If the middle classes are more geographically mobile than the working class, does this therefore mean that the residential neighbourhood has limited importance for them in relation to place belonging? This question was addressed in *Globalisation and Belonging* by Mike Savage and colleagues (Savage, Bagnall, and Longhurst, 2005), a study of the northern English middle classes within a Bourdieusian theoretical framework. Savage and colleagues argued that the middle classes – unlike the working class who have a *traditional approach* to place based on residential longevity – have an *adoptive approach* to place predicated on moving to a neighbourhood and volitionally opting to belong to it – hence "elective belonging". Elective belonging involves middle-class households straddling the routes/roots antimony: flowing along routes via geographical mobility, but then *choosing* to put down roots in an area because of its aesthetic and ethical appeal – beauty, sense of authenticity, etc. (Savage, 2010). Hence the middle classes are not necessarily here-today-and-gone-tomorrow transients. In Bourdieusian terms, the middle classes choose specific neighbourhoods because they offer a socio-spatial fit between their "fields" (housing, education, consumption, etc.) and their "habitus" such that they feel "at home" and comfortable residing in such places.

The middle classes therefore choose certain neighbourhoods and this area choice can also reflect which particular *middle-class fraction* they belong to, as seen in research on gentrification in London (Butler and Robson, 2003). Hence the preference among students, artists and welfare professionals – typically low on economic capital but replete with cultural capital – to gravitate towards socially diverse, multi-ethnic areas such as Peckham in south London as opposed to East Dulwich, a nearby established middle-class area (Jackson and Benson, 2014). By contrast, the more economically well-off fractions of the middle classes are able to insulate themselves among people "just like them" in urban villages and suburban/rural villages, as research on the London and Parisian middle classes indicates (Bacques et al., 2015). Such "villages" become cocoons of class privilege whereby the presence of social difference – in terms of lower-class or minority-ethnic groups – is kept at bay via exclusionary socio-spatial practices. Such exclusionary practices are also highlighted by Duncan and Duncan (2004) in *Landscapes of Privilege*, their appropriately named study of an upmarket quasi-rural suburb of New York City.

The maintenance of class distinctions via spatial distancing is not, however, novel as seen by the "bourgeois utopian" nature of suburbia (Fishman, 1987). The Anglo-American model of suburbanisation in the UK and US meant that spatial movement out from the city to the suburbs was associated with a process of social mobility whereby those *moving out* also *moved up* the social hierarchy, notably via the intertwining of middle-class formation and suburban homeownership (Clapson, 2003; Jackson, 1973, 1985). One prominent example is the post-war US Levittowns such as Long Island (Gans, 1967). Culturally the Anglo-American suburbs involve a "suburban way of life" based on maintaining a tight social order, conforming to middle-class norms of respectability, and promoting a spatial distance from class and racialised "others" (Baumgartner, 1988). Racism was a key constituent part of post-war US suburbanisation via "white flight" from black inner-city areas (Jackson, 1985). More recently, minority-ethnic suburbanisation has expanded in the US and UK (Pattillo-McCoy, 1999; Watt, Millington, and Huq, 2014).

If suburbanisation has a long history, gentrification is a more recent middle-class socio-spatial trend. Gentrification was first identified by the sociologist Ruth Glass in relation to previously working-class areas in 1960s' north London (cited in Lees, Slater, and Wyly, 2008). Part of the appeal of such inner-city locales for first-wave "pioneer" gentrifiers – who tended to be left-liberal arts and welfare professionals – was that they were more socially mixed than the sterile suburbs, as Caulfield (1994) found in inner-city Toronto and Butler (1997) similarly noted in east London. Gentrification has subsequently morphed along multiple, complex lines and is no longer predicated upon the actions of artists and welfare professionals with modest incomes (Lees, Slater, and Wyly, 2008). Instead, gentrification is increasingly dominated by affluent banking and finance professionals who have displaced the pioneer gentrifiers, as in the case of "super-gentrification" in global cities such as London (Butler and Lees, 2006). Another recent trend is third wave "state-led gentrification" whereby urban regeneration policies demolish public/social housing estates resulting in the displacement of low-income tenants and their replacement by upmarket professional homeowners and private tenants (Lees and Ferreri, 2016; Watt, 2013; Watt and Smets, 2017).

These more recent affluent gentrifiers tend to live in inner-city areas via an oblique relationship to urban space, one which Atkinson (2006) refers to as "middle-class disaffiliation". This disaffiliation involves cultures and architectures of fear and mistrust; the affluent middle classes put physical and social barriers between themselves and spatially proximate working-class, minority-ethnic populations, for example by living in gated communities or separate apartment blocks (Butler, 2007; Davidson, 2010; Low, 2004). In a paper on suburban middle-class disaffiliation, Watt (2009) has reworked and critiqued the notion of elective belonging, discussed above, via introducing the concept of "selective belonging". This refers to a "spatially selective narrative of belonging that is limited to a given space within a wider area" (Watt, 2010: 154), whereby residents cocoon themselves within a middle-class "oasis" of order and security which represents a hierarchically privileged space in relation to nearby lower-class areas. Selective belonging narratives and practices have been identified among the middle classes in several cities including Paris and London (Bacqué et al., 2015) and The Hague and Amsterdam in The Netherlands (Pinkster, 2014).

Disaffiliation is, however, by no means ubiquitous. In a recent paper on Rotterdam, Bosch and Ouwehand (2018) found that middle-class Dutch and minority-ethnic newcomers to a deprived neighbourhood used local spaces such as shops and there was also evidence of place attachment to this ethnically diverse area. This raises the question as to what role ethnicity plays in relation to middle-class mobilities and place belonging? Barwick (2018) addresses this question in her study of the middle-class Turkish population living in Berlin. Among this upwardly mobile group, some remain living in predominantly lower-class Turkish areas, while others move to mixed-class areas. Nevertheless, as Barwick argues, both groups – stayers and movers – retain a strong sense of place attachment to their original Berlin Turkish neighbourhoods. Social mixing thus occurs regularly between middle-class and lower-class Turkish Berliners, but *within* their own minority-ethnic group. In this case, ethnicity forms a strong basis for social encounters and place belonging, irrespective of class differences.

It's debatable how far the notion of the restless, ever-mobile middle classes applies outside the Anglo-American context given that the continental European bourgeoisie has been less involved in suburbanisation or gentrification (Andreotti, Le Galès, and Fuentes, 2015; Bacques et al., 2015). As Andreotti et al. (2015: 11) note, "elite and middle classes did not systematically leave the city centres, and their urban presence in these areas has in fact become even more pronounced since the 1980s in most European cities" via "a continuous *embourgeoisement*" and reinvestment in traditional middle-class areas. In their study of upper middle-class managers in Paris, Lyon, Milan and Madrid, Andreotti, Le Galès, and Fuentes (2015) acknowledge the significance of routes – for example, transnational mobility for work and also moving to certain neighbourhoods for elective belonging reasons. At the same time, their research revealed considerable residential stability (even if broken by periods abroad), but especially among managers in Milan and Madrid. This stability reflects the continuing significance of extended family ties. As the authors say of the Milanese managers, "we found the highest density of interaction with friends and family, deep rootedness, low spatial mobility and very long-term relationships with best friends" (Andreotti, Le Galès, and Fuentes, 2015: 185). If Milanese managers remain embedded in traditional bourgeois neighbourhoods, what happens when such areas are subject to an arriviste group of the super-rich?

Place and the super-rich: crowded or lonely at the top?

There are relatively few academic studies on place and the rich, a reflection of the wider dearth of research on this elite group, protected as they are from the prying eyes of social scientists behind gated mansions and armies of staff. Nevertheless, there has been a recent academic focus on what are variously described as the elite, the upper class, the 1%, HNWI (High Net Worth Individuals),[1] UHNWI (Ultra-High Net Worth Individuals)[2] – or, most commonly, the "super-rich" (Dorling, 2015; Forrest, Koh, and Wissink, 2017; Hay and Muller, 2012). The source of this burgeoning interest lies in the massive increase in global wealth and income inequalities during the last 40 years as a result of neoliberalism and financialisation. Consequently, there is growing concern over social injustice and social cohesion which has prompted Danny Dorling (2015: 1) to ask "can we afford the superrich?"

The super-rich represent the apex of global rootlessness, sealed off from the places where the 99% live (Forrest, Koh, and Wissink, 2017; Watt and Smets, 2014). They travel in their exclusive private jets and yachts in order to visit their many homes around the world. When the super-rich do come down to terra firma, they only do so in certain exclusive places. Not only are the super-rich spatially concentrated in certain countries, notably the USA, Japan, Germany, China, France and the UK (Capgemini, 2018), but they tend to have homes in premier global cities such as New York, London, Tokyo, Paris, Singapore and Hong Kong (Forrest, Koh, and Wissink, 2017). The spatially uneven intra-national distribution of the super-rich can be gauged by how London is currently "home" to around 90% of the UK's

553,000 HNWIs (Atkinson, Parker, and Burrows, 2017). The super-rich also congregate together by buying homes within certain enclaves, including private islands such as Fisher Island in Florida (the wealthiest post-code in the US) and elite playgrounds such as Aspen, Colorado and Queenstown in New Zealand (Hay and Muller, 2012).

Within global cities, the super-rich are spatially concentrated within two main kinds of exclusive neighbourhoods. First are those already-established bourgeois neighbourhoods which have been the traditional epicentre of the upper and upper middle classes, such as the Upper East Side in New York City (Freeland, 2011; Hay and Muller, 2012) and Chelsea, Kensington, Hampstead and Belgravia in London (Burrows, Webber, and Atkinson, 2017). Second are new luxury private developments which are located in less salubrious areas, such as Vauxhall Nine Elms and The Shard in traditionally déclassé south London (Atkinson, 2018; Minton, 2017). These new private developments can also take advantage of waterfront locations, as in the case of Sentosa Cove, an exclusive gated community in Singapore which caters for wealthy Singaporeans and foreigners (Pow, 2011).

How do the super-rich relate to their residential neighbourhoods as places? Two emergent issues have arisen from the existing literature. The first is what impact the arrival of the super-rich has on traditional upper- and upper-middle-class areas. On this, Webber and Burrows (2016) have identified "old v. new money" conflicts over property extensions in Highgate Village, north London. This area has long formed the base for a traditional wealthy London elite (the old bourgeoisie) who have played a key role in local cultural organisations and events. This local paternalistic hegemony has been disrupted by the arrival of a "globally moneyed elite" (Webber and Burrows, 2016: 3148), who have proceeded to acquire old properties with the intention of extensively remodelling and enlarging them, or even knocking them down and rebuilding afresh. Thus, unlike Savage's elective belongers who choose a residential area for its established aesthetic appearance, the super-rich in Highgate want to dramatically alter this very same landscape because it does not suit their tastes – or those of their property advisers. This threat to the traditional character of the area has prompted planning disputes between the established and incoming elites as each group struggles for hegemony in relation to the aesthetics of home and neighbourhood. As Webber and Burrows (2016: 3152) note, while "the old elites drew status from their role as trustees of local charities and champions of community initiatives, the new elites identify far more with 'London' than they do with 'Highgate'". Compared with the established bourgeois elite, the super-rich have far less localist and far more metropolitan and global spatial orientations.

A second emergent issue in relation to place and the super-rich is that "the lights are on, but nobody's home", as seen in Atkinson's (2018) paper on new luxury developments in London. Many of these developments are either under-used or stand empty because they are investment opportunities rather than homes for living in. This creates a novel urban landscape – a "kind of dead residential space or necrotecture" (Atkinson, 2018: 1). Such half-empty spaces can generate social isolation – especially among new residents and wives when their husbands are away on business – as Pow (2011) has identified in Singapore's Sentosa Cove "golden ghetto".

Conclusion

This chapter has examined the relationship between social class and place in relation to residential neighbourhoods in cities and suburbs of the global North. The neighbourhood retains its hold as a source of place belonging and attachment for the working class, not least due to their greater residential longevity compared with both the middle classes and super-rich.

However, there is also evidence of greater working-class ambivalence to place, not least due to nostalgia for lost urban communities of the past associated with disappearing industrial employment. The middle classes have a complex relationship to their neighbourhoods which encompasses various mixtures of routes and roots, as encapsulated by the influential concept of elective belonging. There is also an ongoing tension between the middle classes and place, as captured by notions of disaffiliation and selective belonging, and which is most visibly expressed via architectures of fear and insecurity such as gated communities. The super-rich are the most globally mobile class and while their relationship to residential place is the most contingent and limited, it also throws up important research questions regarding how the super-rich and more established bourgeois elites vie for hegemony in their upmarket urban enclaves.

Acknowledgements

Thanks to Tim Edensor and Alan Morris for their insightful comments on earlier drafts of this chapter.

Notes

1 High Net Worth Individuals (HNWI) are those who own US$1 million or more in investable assets, but excluding primary residence, collectibles, consumables and consumer durables (Capgemini (2018: 3).

2 Ultra-High Net Worth Individuals (UHNWI) are those who own US$30 million or more in investable assets (Capgemini (2018: 7).

References

Andreotti, A., Le Galès, P. and Fuentes, F.J.M. (2015) *Globalised Minds, Roots in the City: Urban Upper-Middle Classes in Europe*, London: Blackwell.

Atkinson, R. (2006) 'Padding the bunker: strategies of middle-class disaffiliation and colonisation in the city', *Urban Studies*, 43(4): 819–832.

Atkinson, R. (2018) 'Necrotecture: lifeless dwellings and London's super-rich', *International Journal of Urban and Regional Research*. doi: 10.1111/1468-2427.12707.

Atkinson, R., Parker, S. and Burrows, R. (2017) 'Elite formation, power and space in contemporary London', *Theory, Culture & Society*, 34(5–6): 179–200.

Bacqué, M., Bridge, G., Benson, M., Butler, T., Charmes, E., Fijalkow, Y., Jackson, E., Launay, L. and Vermeersch, S. (2015) *The Middle Classes and the City: A Study of Paris and London*, Basingstoke: Palgrave Macmillan.

Barwick, C. (2018) 'Social mix revisited: within- and across-neighborhood ties between ethnic minorities of differing socioeconomic backgrounds', *Urban Geography*, 39(6): 916–934.

Baumgartner, M.P. (1988) *The Moral Order of a Suburb*, New York and Oxford: Oxford University Press.

Blokland, T. (2003) *Urban Bonds*, Cambridge: Polity Press.

Bosch, E.M. and Ouwehand, A.L. (2019) 'At home in the oasis: middle-class newcomers' affiliation to their deprived Rotterdam neighbourhood', *Urban Studies*, 56(9): 1818–1834.

Bourke, J. (1994) *Working Class Cultures in Britain 1890–1960: Gender, Class and Ethnicity*, London: Routledge.

Burrows, R., Webber, R. and Atkinson, R. (2017) 'Welcome to Pikettyville? Mapping London's alpha territories', *The Sociological Review*, 65(2): 184–201.

Butler, T. (1997) *Gentrification and the Middle Classes*, Aldershot: Ashgate.

Butler, T. (2007) 'Re-urbanizing London Docklands: gentrification, suburbanization or new urbanism?', *International Journal of Urban and Regional Research*, 31: 759–781.

Butler, T. and Lees, L. (2006) 'Super-gentrification in Barnsbury, London: globalization and gentrifying global elites at the neighbourhood level', *Transactions of the Institute of British Geographers NS*, 31: 467–487.

Butler, T. and Robson, G. (2003) *London Calling: The Middle Classes and the Re-making of Inner London*, Oxford: Berg.

Capgemini (2018) *World Wealth Report 2018*, available at: www.worldwealthreport.com/

Caulfield, J. (1994) *City Form and Everyday Life: Toronto's Gentrification and Critical Social Practice*, Toronto: University of Toronto Press.

Clapson, M. (1998) *Invincible Green Suburbs, Brave New Towns: Social Change and Urban Dispersal in Post-war England*, Manchester: Manchester University Press.

Clapson, M. (2003) *Suburban Century: Social Change and Urban Growth in England and the USA*, Oxford: Berg.

Cohen, P. (2013) *On the Wrong Side of the Track: East London and the Post-Olympics*, London: Lawrence and Wishart.

Davidson, M. (2010) 'Love thy neighbour? Social mixing in London's gentrification frontier', *Environment and Planning A*, 42: 524–544.

Dorling, D. (2015) *Inequality and the 1%*, London: Verso.

Duncan, J.S. and Duncan, N.G. (2004) *Landscapes of Privilege: The Politics of the Aesthetic in an American Suburb*, New York and London: Routledge.

Edensor, T. and Millington, S. (2009) 'Illuminations, class identities and the contested landscapes of Christmas', *Sociology*, 43(1): 103–121.

Engels, F. (1844/1987) *The Condition of the Working Class in England*, London: Penguin Books.

Feldman, R.M. and Stall, S. (2004) *The Dignity of Resistance: Women Residents' Activism in Chicago Public Housing*, Cambridge: Cambridge University Press.

Fishman, R. (1987) *Bourgeois Utopias: The Rise and Fall of Suburbia*, New York: Basic Books.

Forrest, R., Koh, S.Y. and Wissink, B. (2017) *Cities and the Super-Rich: Real Estate, Elite Practices and Urban Political Economies*, Basingstoke: Palgrave Macmillan.

Freeland, C. (2011) 'The rise of the new global elite, *The Atlantic*, Jan.-Feb., available at: www.theatlantic.com/magazine/archive/2011/01/the-rise-of-the-new-global-elite/8343/

Fried, M. (1966) 'Grieving for a lost home: psychological costs of relocation', in J. Q. Wilson (ed) *Urban Renewal: The Record and Controversy*, pp. 359–379, Cambridge, MA and London: The MIT Press.

Gans, H.J. (1962) *The Urban Villagers*, New York: The Free Press.

Gans, H.J. (1967) *The Levittowners*, London: Allen Lane, The Penguin Press.

Garbin, D. and Millington, G. (2012) 'Territorial stigma and the politics of resistance in a Parisian banlieue: La Courneuve and beyond', *Urban Studies*, 49(10): 2067–2083.

Gest, J. (2016) *The New Minority: White Working Class Politics in an Age of Immigration and Inequality*, New York: Oxford University Press.

Hay, I. and Muller, S. (2012) '"That tiny, stratospheric apex that owns most of the world" – Exploring geographies of the super-rich', *Geographical Research*, 50(1): 75–88.

Jackson, A. (1973) *Semi-Detached London: Suburban Development, Life and Transport, 1900–1939*, London: Allen & Unwin.

Jackson, E. and Benson, M. (2014) 'Neither "deepest, darkest Peckham" nor "run-of-the-mill" East Dulwich: the middle classes and their "others" in an inner-London neighbourhood', *International Journal of Urban and Regional Research*, 38(4): 1197–1212.

Jackson, K.T. (1985) *Crabgrass Frontier: The Suburbanization of the United States*, New York: Oxford University Press.

Jeffery, B. (2018) '"I probably would never move, but ideally like I'd love to move this week": class and residential experience, beyond elective belonging', *Sociology*, 52(2): 245–261.

Karn, J. (2007) *Narratives of Neglect: Community, Regeneration and the Governance of Security*, Cullompton: Willan Publishing.

Kusenbach, M. and Paulsen, K. (2013) *Home: International Perspective on Culture, Identity and Belonging*, Frankfurt am Main: Peter Lang.

Lees, L. and Ferreri, M. (2016) 'Resisting gentrification on its final frontiers: learning from the Heygate Estate in London (1974–2013)', *Cities*, 57: 14–24.

Lees, L., Slater, T. and Wyly, E. (2008) *Gentrification*, New York: Routledge.

Lewicka, M. (2011) 'Place attachment: how far have we come in the last 40 years?', *Journal of Environmental Psychology*, 31: 207–230.

Linkon, S.L. and Russo, J. (2002) *Steeltown U.S.A.: Work and Memory in Youngstown*, Lawrence: University of Kansas Press.

Low, S. (2004) *Behind the Gates*, New York and London: Routledge.

Manzo, L.C. (2014) 'On uncertain ground: being at home in the context of public housing redevelopment', *International Journal of Housing Policy*, 14(4): 389–410.
Manzo, L.C. and Devine-Wright, P. (2014) *Place Attachment: Advances in Theory, Methods and Applications*, London and New York: Routledge.
May, J. (1996) 'Globalization and the politics of place: place and identity in an inner London neighbourhood', *Transactions of the Institute of British Geographers, New Series*, 21: 194–215.
McKenzie, L. (2015) *Getting By: Estates, Class and Culture in Austerity Britain*, Bristol: Policy Press.
Minton, A. (2017) *Big Capital: Who is London For?* London: Penguin Books.
Morris, A. (2013) 'Public housing in Australia: a case of advanced urban marginality?', *The Economic and Labour Relations Review*, 24(1): 80–96.
Morris, A. (2019) *Gentrification and Displacement: The Forced Relocation of Public Housing Tenants in Inner-Sydney*, Singapore: Springer.
Pahl, R. (2005) 'Are all communities communities in the mind?', *The Sociological Review*, 53(4): 621–640.
Parkin, F. (1967) 'Working-class Conservatives: a theory of political deviance', *British Journal of Sociology*, 18: 278–290.
Pattillo-McCoy, M. (1999) *Black Picket Fences: Privilege and Peril among the Black Middle Class*, Chicago: University of Chicago Press.
Pinkster, F.M. (2014) '"I just live here': everyday practices of disaffiliation of middle-class households in disadvantaged neighbourhoods', *Urban Studies*, 51(4): 810–826.
Pow, C. P. (2011) 'Living it up: super-rich enclave and transnational elite urbanism in Singapore', *Geoforum*, 42: 382–393.
Savage, M. (1987) *The Dynamics of Working-Class Politics: The Labour Movement in Preston, 1880–1940*, Cambridge: Cambridge University Press.
Savage, M. (2010) 'The politics of elective belonging', *Housing, Theory and Society*, 27(2): 115–161.
Savage, M, Bagnall, G. and Longhurst, B. (2005) *Globalization and Belonging*, London: Sage Publications.
Smith, D.M. (2005) *On the Margins of Inclusion: Changing Labour Markets and Social Exclusion in London*, Bristol: Policy Press.
Thrift, N. and Williams, P. (1987) *Class and Space: The Making of Urban Society*, London: Routledge & Kegan Paul.
Todd, S. (2014) *The People: The Rise and Fall of the Working Class, 1910–2010*, London: John Murray.
Urry, J. (2007) *Mobilities*, Cambridge: Polity Press.
Wacquant, L. (2008) *Urban Outcasts: A Comparative Sociology of Advanced Marginality*, Cambridge: Polity Press.
Watt, P. (2006) 'Respectability, roughness and "race": neighbourhood place images and the making of working-class social distinctions in London', *International Journal of Urban and Regional Research*, 30(4): 776–797.
Watt, P. (2007) 'From the dirty city to the spoiled suburb', in B. Campkin and R. Cox (eds) *Dirt: New Geographies of Cleanliness and Contamination*, pp. 80–91, London: I.B. Tauris.
Watt, P. (2009) 'Living in an oasis: socio-spatial segregation and selective belonging in an English suburb', *Environment and Planning A*, 41(12): 2874–2892.
Watt, P. (2010) 'Unravelling the narratives and politics of belonging to place', *Housing, Theory and Society*, 27(2): 153–159.
Watt, P. (2013) '"It's not for us": regeneration, the 2012 Olympics and the gentrification of East London', *City*, 17(1): 99–118.
Watt, P., Millington, G. and Huq, R. (2014) 'East London mobilities: the Cockney Diaspora and the remaking of the Essex ethnoscape', in P. Watt and P. Smets (eds) *Mobilities and Neighbourhood Belonging in Cities and Suburbs*, pp. 121–144, Basingstoke: Palgrave Macmillan.
Watt, P. and Smets, P. (2014) *Mobilities and Neighbourhood Belonging in Cities and Suburbs*, Basingstoke: Palgrave Macmillan.
Watt, P. and Smets, P. (2017) *Social Housing and Urban Renewal: A Cross-National Perspective*, Bingley: Emerald.
Webber, R. and Burrows, R. (2016) 'Life in an Alpha Territory: discontinuity and conflict in an elite London village', *Urban Studies*, 53(15): 3139–3154.
Wilson, W.J. (1987) *The Truly Disadvantaged*, Chicago: University of Chicago Press.
Young Foundation. (2010) *Neighbouring in Contemporary Britain*, London: Young Foundation.
Young, M. and Willmott, P. (1957) *Family and Kinship in East London*, London: Routledge & Kegan Paul.

23

'Food-work city'

Oral history and the contested politics of place

Ben Rogaly

This chapter is prompted by two place questions posed by Doreen Massey: 'What does this place stand for?' (2007) and 'To whom does this place belong?' (Massey, 2011). The context is England in the Brexit era, the period from the 2007–2008 global financial crisis through to the uncertainty that followed the UK's 2016 referendum over EU membership. During this time, England witnessed heightened racisms, great ideological division, massive and growing inequality of wealth and power, and rapid changes in capitalist work and its spatial constitution. In the chapter I will use the lens of Peterborough, a provincial English city – where a majority voted to leave the EU in 2016 – to reflect on this conjunctural change and to raise the possibility of a more hopeful, anti-racist politics.

Massey consistently interweaved her research agenda with political interventions, collaborating with Stuart Hall in producing a sustained critique of neoliberalism (Hall et al., 2015). As Hall and his colleagues put it four decades earlier, the 'practical remedy' to the conditions that 'make the poor poor' … and 'the rich rich' 'involves taking sides'. This was explicitly presented as 'an *intervention* – albeit an intervention in the battleground of ideas' (Hall et al., 2013 [1978]: 4, original emphasis). Inspired by this perspective, the research on which this chapter draws has been an intervention from the start and not only through academic writing.

The politics in Massey's two questions lie in the way they highlight contests over place. However, their power is derived from Massey's more fundamental contributions to the theorisation of place. Places, as well as being contested, are, very importantly, multi-scalar and relational. So, if *World City* (Massey, 2007) is ostensibly a book about London, it is just as much about how what London stands for, and to whom it belongs, how it shapes and is shaped by the national territory of which it is the capital city, and by its unequal relations with places beyond the UK.

The relationality of place operates across time as well as space (Massey, 1991), since place is porous and dynamic rather than static and bounded. Places are built on the sediments of their own history and of the history of other places to which they are connected.

I use place in this chapter to rethink often taken-for-granted conceptions of an inevitable clash of interests between people categorised as 'migrant' and people seen as 'local'. Tens of thousands of people have travelled to Peterborough to find work since the Second World War. In recent decades, the food supply and warehouse distribution sectors have been important employers of newly arrived work-seekers as well as existing city residents. This is a story about how places change and the role of oral history in recording such change. It is also suggestive of how oral history might itself contribute to changing places.

In what follows, I approach Massey's two questions at three interconnected scales: the national, the urban and the workplace. In considering the national, I draw on my own critique (Rogaly, 2018) of a book published after the referendum by David Goodhart (2017) precisely because of its influential contribution to naturalising the categories I want to unpick, and in particular his idea of a national so-called 'mainstream' 'us' based on 'race' and migration history.

The book, *The Road to Somewhere,* was widely featured in the broadcast media. Although presented as, among other things, a robust study of social trends and an explanation of the majority vote for Brexit, the post-liberal, nationalist view it espouses can be seen as part of a Gramscian 'war of position', an ideological play aimed at creating a new 'common sense', or, in the terms of Massey's two questions, as an instrumental political intervention in contested national-level debates over 'who this place belongs to' and 'what this place should stand for'.

The book's central contention is that there are two prevalent ideological perspectives in England, which Goodhart refers to as 'Anywhere' and 'Somewhere', and which map in approximate and complex ways onto two groups of people in society, to which he gives the corresponding labels, 'Anywheres' and 'Somewheres'. While he attributes a range of views to each group, Goodhart's central point is that Somewheres are unhappy with the pace and extent of what he calls 'cultural change' since the 1960s, by which he means growing ethnic diversity, mass immigration and the legacy of multicultural policies. Anywheres, on the other hand, are comfortable with such change because they are mostly university graduates and, therefore, under the UK's residential university system, are much more likely to live away from the place they grew up in than Somewheres. The Brexit vote, Goodhart contends, was a 'revolt' by Somewhere people against the dominance of policies and social trends with which Anywheres are comfortable; he portrays it as Somewheres taking back a degree of relative power in the national polity.

It is the way Goodhart uses the 'us' word that most concerns me about his book because it promotes a racialised hierarchy in thinking about a national-scale response to our two 'place questions'. Goodhart believes that people whom he sees in cultural terms as outside 'the mainstream' should not receive equal treatment or status in the national polity. Although he does not define what 'mainstream' is in the English context, he implies that nationals of eastern and central European EU countries are further removed from it than western and southern European ones, as are international migrant people of colour from Africa and Asia. At some points in the book, Goodhart even places the descendants of these darker migrants (and citizen-migrants) (see Bhambra, 2016) further from his 'mainstream' too, regardless of their citizenship.

Goodhart's discussion of Muslims in Britain further illustrates the racialised hierarchy he constructs. He argues that 'mainstream public opinion is more wary of Muslims than other comparable minorities' (2017: 130). At the same time, *The Road to Somewhere* gives no sustained attention to hate crimes committed against people judged by their appearance and/or location to be Muslim nor to widespread structural discrimination against Muslims.

Goodhart exemplifies the power and reach of elite voices in a national-scale discursive battle. This is important because reactionary politics is often misattributed as having primarily, or even exclusively, working-class roots. Returning to the first of the two questions guiding this chapter, 'to whom does this place belong', Goodhart's power to shape debates over national-level belonging, by defining a 'mainstream' that is racialised white and Christianised, provides one kind of answer. Moreover, in terms of the second question, 'what does this place stand for?' Goodhart presents – as if it were an 'analysis' of two contrasting ideological perspectives – what is in fact a rhetorical argument, essentially his own, for what England and the United Kingdom more broadly *should* stand for.

This kind of positioning over place at the national scale has immediate implications for how people in England live together. As Stuart Hall wrote, what makes representations by the radical right dangerous is 'that they change the nature of the terrain itself on which struggles of different kinds are taking place' (2017 [1979]: 176). It is to place at the urban scale that I now turn.

The Peterborough project that I will discuss shortly emerged following research with Becky Taylor in three social housing estates on the edge of another English city, Norwich (see Rogaly and Taylor, 2009). Doreen Massey's theory of place as dynamic, porous and relational was critically important to the work that we undertook. The residential fieldwork, archival research and oral history interviews we conducted were all suggestive of much more spatial mobility in the lives of the mainly working class, mainly white, residents of the estates than were generally considered by national stereotypes of white social housing estate residents as fixed in place and threatened by spatially mobile outsiders and people of colour. Older narrators spoke with us, for example, about their own lives in the forces in remaining parts of the British empire in the 1950s and 1960s; about aunts who had married GIs and moved to the US after the war; and about sons and daughters who had moved to other parts of the UK and abroad.

As Massey put it, places are 'constructed out of articulations of social relations (trading connections, the unequal links of colonialism, thoughts of home) which are not only internal to that locale but which link them to elsewhere' (1995: 183). Through our interviews, we studied the meaning and experience of white British *emigration* as well as residential moves between social housing estates in Norwich, arguing that even short moves across space could be intensely felt.

In short, we were opening up the often racialised ideas, presented as common sense, about who was a migrant and who was local. It was the racialisation of the idea of the migrant that led Paul Gilroy to write that 'fascination with the figure of the migrant must be made part of Europe's history rather than its contemporary geography … We need to conjure up a future in which black and brown Europeans stop being seen as migrants' (2004: 165). My response to the problem identified by Gilroy was to look to the field of critical mobilities studies (Rogaly, 2015), which attend centrally to structure, including the reproduction of racisms, and class and gender inequalities. This helps move towards an understanding of the interplay of factors that led some people not to be able to afford to move, while others are forced to do so, and which enable still others to choose whether to stay put or go elsewhere. Yet, at the same time, critical mobilities scholarship often demonstrates a deep understanding of individual subjectivities, including how people characterise their own mobilities. Here, fixity is seen as equally important as mobility, as indeed is the relationship between fixity and mobility. For example, in his research on a Japanese inn, Chris McMorran (2015: 84) shows 'how some employees with "nowhere to go" besides [a specific type of inn] get stuck in a dead-end job, while simultaneously cherishing the freedom the [inn] provides, including freedom from domestic abuse'. He argues for a more nuanced analysis of how 'people's mobility at one scale can co-exist with their immobility at another', hinting at the importance of both classed and gendered processes in producing mobility and fixity.

Critical mobilities studies such as McMorran's (2015) therefore offer the potential to think about migration differently. They keep structural inequalities to the fore, including those associated with individual migration status in relation to national borders and with racisms. They do not claim that everyone is in the same position, nor that everyone is a migrant, a stance that is rightly critiqued by others (for instance, Back and Sinha, 2018). Importantly, however, critical mobilities studies offer a way of thinking about working-class identity that does not rely on a fantasy of primordial whiteness – and can therefore be, at the same time, inclusive of black and brown people, as well as people identified as 'migrants' (Valluvan, 2019). This helps provide a language for the potential for a 'common anger' (Massey (2011) shared, especially in conditions of neoliberal austerity, by people displaced by spatial mobility with others who have never changed their dwelling over a whole lifetime but may feel displaced by the passage of time during which the place around that dwelling may have become unrecognisable.

These ideas were central to subsequent research carried out in Peterborough in 2011 with Kaveri Qureshi and others, in a project called 'Places for All?' (see Rogaly and Qureshi, 2013). We built on the idea that being connected to places elsewhere, including through 'thoughts of home', was something that many people had in common, not only people who had recently crossed international borders to reach the city, but also those who had arrived from places within the UK and long-settled people of all ethnicities including the white British ethnic majority.

What Peterborough stands for, and to whom it should be considered as belonging, are actively contested. In late 2010, residents faced down a major far right rally composed mostly of outsiders to the city (Rogaly and Qureshi, 2013). Earlier, New Link, Peterborough's erstwhile support and advice organisation for newly-arrived migrants, won national recognition for its work (Rogaly, 2019: 229). A key aim at the start of the project was to use oral history to accompany local organisations already intensely engaged in increasing awareness of, and placing value on, first, Peterborough's multiple 'elsewheres' and how these contributed to making the city. These organisations' work was all the more urgent because of drastic cuts to local authority finances.

Peterborough, with a population of approximately 200,000 in 2019, is a multi-ethnic, multi-faith city. Moreover, the city has for over seventy years been the destination for international migrants responding to job openings there, including, in the 1950s and 1960s, large numbers of people arriving from Italy, south Asia and the Caribbean. Peterborough City Council welcomed east African Asian and Vietnamese refugees in the 1970s and early 1980s respectively; the city was also designated as a dispersal area for asylum-seekers at the turn of the millennium.

Yet the largest single flow of migrants to the city was made up of mainly white British Londoners and Scots at the time of the building of the city's satellite 'New Town' areas in the 1970s and 1980s. Biographical oral history interviews with long-term residents of Peterborough who remembered the arrival of the Londoners revealed both a range of mobilities experienced by those identifying as 'local', and a resentment produced through the sense that the city was making special efforts for new arrivals, all be they from within the UK. Some names have been changed in what follows.

Take the case of Sean Brennan, interviewed in 2012, a white man born in 1949, whose family background involved multiple mobilities. Sean's Scottish father grew up in working-class Glasgow, was de-mobilised at Connington Airfield in Cambridgeshire after World War Two and met Sean's mother on a trip to Peterborough's Embassy cinema where she was then working. Sean's maternal grandmother had travelled to England from Ireland to work in domestic service and it was *her* husband's job with Brotherhoods engineering that had brought the family to Peterborough. Sean grew up as part of a small Catholic minority in the city – and has been mobile in and out of the city his whole life through his work as a lorry driver across the UK. In spite of his Peterborough upbringing Sean was allocated one of the 'New Town' houses that were generally reserved for newcomers through a system he critiqued:

> The Union said I was a key worker and … I got a council house … originally they were built for overspill, for the cockneys as we used to call them and there was a lot of bad blood in town over that … if you lived in Peterborough you couldn't get a bugger ….

Sean felt that he and other existing Peterborough residents had been 'done to' by the system that the new Peterborough Development Corporation had set up – a discourse very similar to that of contemporary anti-immigrant sentiment, to which Sean referred directly in a comparison between the Londoners arriving in the 1970s and international migrants arriving in the 2010s. Yet this is complex. Sean had, after all, only *become* a Peterborough resident through his own family's move there. This complexity is suggestive of the potential to create greater solidarity across the city through an understanding of the commonality of having meaningful elsewheres – an *intervention* alongside the work of existing organisations and city residents seeking a pluralist notion of who the city belongs to, and thus what Peterborough, the place, should stand for.

Workplaces are the third and final scale of place I discuss. It was our hunch that bringing together individuals' experiences at work across their working lives contained the potential for people to find further common ground, rather than resenting fellow workers because of their ethnicity, faith identity, migration history or national heritage. We thought that creating such spaces would encourage people to look upward, at the employment practices of temporary work agencies and of the businesses that were end-users of labour that those agencies supplied workers to, as well as at neoliberal state policies that had produced a deregulated labour market that increasingly individualised risk.

Notwithstanding a longer history of migration to the city for manufacturing work (Tyler et al., 2018), many international migrant workers residing in Peterborough had come to seek work in the industrialised food supply chain. The city's location adjacent to the highly productive soils of the Fens region to the east continued to make it a hub for workers employed in temporary agency work in food production, packing and processing. In the UK, the period since the late 1980s saw major changes to capitalist work in the food supply chain, driven by the concentration of economic power in the hands of large-scale retail capital. Growers, packers and processors of food had ever-diminishing choices about where to sell their products: the vast majority of produce was sold to one of the large supermarket retailers. As a result those retailers were able to squeeze their suppliers – to apply pressure through exacting food quality standards and just-in-time supply requirements, as well as on price, with a step up in demands for suppliers to internalise additional costs associated, for example, with buy-one-get-one free offers.

These developments led, in turn, to a dramatic increase in demand for international migrant workers by the companies supplying the supermarkets, especially from the mid-1990s, further accelerating after EU enlargement in 2004. Some employers, who doubled as accommodation providers, sought international migrants as a temporarily captive labour force (Simpson, 2011). Others chose to employ migrants because they found them able and willing to work faster than British nationals, yet still taking the requisite care to meet the so-called quality standards.

The rapid growth in employment of migrant workers was accompanied by an intensification of workplace regimes (Rogaly, 2008). Developments included raising the bar for piece work – more boxes of strawberries picked for the same amount of money – tightening supervision and control of breaks, and insisting on working until an order was completed. Some companies combined ever-harder-to-achieve targets with fines and sanctions, including dismissal. This compounded the long history of job insecurity in the food production sector, manifest, for example, through agency working, uncertainty regarding daily working hours and zero hours contracts. As the food supply chain became ever more industrialised, mechanised and integrated, working conditions fed through into the rapidly expanding logistics sector (Beynon, 2016: 311), including companies with warehouses located on the edge of Peterborough, such as Amazon, Ikea and DHL.

Reid-Musson (2018: 895) has made similar connections between agriculture and the business of distribution and logistics in Ontario, Canada, referring to 'intimate and predatory forms of exploitation' that she calls 'management through algorithms' which are having 'profound effects on both consumers and producers – in logistics and transportation in particular'. Benvegnù et al. (2018: 95) recently went further, writing that 'changes in the modes of production and the development of global supply chains have put [storage and warehousing operations] at the heart of contemporary capitalism'.

The question of who workplaces belong to is an important part of the film *Workers* that Jay Gearing and I co-produced in 2018 following further interviews with current and former food factory and warehouse workers resident in Peterborough (Gearing and Rogaly, 2019). The film involves a diverse range of people, including both British nationals and international migrants. In revealing some of the harsh employment conditions in contemporary capitalist workplaces in these sectors, it confirms the power inequalities inherent in them. In that sense, these workplaces clearly *belong* to business rather than workers. But the film tells stories from multiple workers' perspectives and illustrates individual and collective agency and resistance, including demands for a say in the moral geographies of these workplaces (Rogaly and Qureshi, 2017). This shows that, in terms of the second of this chapter's two guiding questions, at times the workplaces stand for more than the exploitative processes that they are dominated by. Fruit-processing worker Agnieszka Sobieraj remembered a covert collective practice of eating the company's fruit while at work, and laughter in adversity:

> I remember this as a great time, the atmosphere, because we were in such a bad situation, it was summertime, we were wearing a lot of winter clothes because in the factory the temperature like fridge temperature, eight degrees, runny nose, I don't know, they're not nice some of the people who work there, they weren't nice, and you just try to do something in opposite just to keep yourself alive and happy … I think I never laughed so much because it was the only one way to survive. For example, we've learned how to eat fruits so no one can notice, so because you had runny nose all the time, so we were wiping our nose with the sleeve, so we thought if we put

> here ... inside your elbow if you put the fruit and you put then to wipe your nose, you quickly ... <Laughs> can eat pineapple or whatever we're doing.

In 2017, many non-British EU nationals working in the food supply chain or in warehouses faced uncertainty regarding their rights in the UK after Brexit. Moreover, the 2016 referendum had seen a spike in racist and xenophobic attacks (Burnett, 2016). Understanding workplaces as multiscalar and relational over space and time must, in the British context, cause researchers to look to connections with colonial divisions of populations whereby the 'human life of citizens protected by the state is bound to the denigration of populations cast in violation of human life' (Lowe, 2015: 6).

In *Workers* Joanna Szczepaniak tells of her van journeys to agricultural workplaces revealing sexist abuse of relatively powerless young women migrants unfamiliar with English. Capitalist work has long been connected with various forms of unfreedom. Studies of English rural workplaces in the 19th century (e.g., Griffin, 2012) attest to the historical entanglement of unfree employment relations in capitalist agricultural work. Oral history narrators in the 'Places for All?' research remembered how, following World War Two, racialised groups including temporary international migrants and Gypsies and Travellers, commonly formed part of the seasonal agricultural workforce alongside students, ex-miners and big-city dwellers. It can be more generally argued that work in food production, packing and processing has long been characterised by racial capitalism (Robinson, 1980; Bhattacharyya, 2018). For Lowe (2015: 149),

> [t]he term *racial capitalism* captures the sense that actually existing capitalism exploits through culturally and socially constructed differences such as race, gender, region, and nationality, and is lived through those.

Beyond the food supply chain, contemporary warehouse work in Peterborough can also be understood in terms of racial capitalism. People interviewed in Peterborough in 2017 spoke of inhumane treatment in warehouse workplaces. For example, Laura, who arrived from Poland at the age of 20 in 2005, has had years of experience as an agency worker and has at various stages taken on the role of supervisor. She reflected on her recent work at a warehouse:

> I don't know, it's not human really to treat people like that. They are not robots and they work, they try their best, they've got families, they want to have normal life. They work fast but they don't work fast enough, sometimes you might have stomach ache or worst day and because of this one day you have to work quicker, faster the next day because, for example on Thursday you do 80%, then next day you have to make sure you do 120, so for the whole week you got 100%.

Workers' accounts of this and other warehouses illustrated their experiences of Reid-Musson's (2018) 'management through algorithms'. Targets have become higher since the mid-2000s, and break-times are increasingly measured through monitoring the activity of workers' computerised packing guns. In one warehouse where Laura had recently worked, five late returns to work over several months of as little as one second each could result in dismissal.

In a different warehouse, another Polish woman, Sabina, spoke of how targets for pickers became more difficult to achieve as operations expanded. Walking times could increase by five or ten minutes with no reduction in productivity targets. Sabina narrated six years of work first as a worker, then supervisor and finally manager in this warehouse. She had been an anti-globalisation activist in Poland before she travelled to England in 2009 and her first thought on being sent by an agency to work for this large corporation was that she had sold out to the system. Her own critique is laced with solidarities with other supervisors and workers. For example, she spoke about her reaction when she realised a colleague was using his toilet break to work on a book he was writing:

> Yeah, generally I was very happy about people who actually were trying to fight with the system. I'm going to use a bad word but I always was saying to my workers, 'Fuck the system', and they did it, some of them, very well. So, one of them actually was going to write book in the toilet ...
>
> Obviously, I couldn't say, 'I'm not interested but, you know, let's do it', so I went to see that guy, Steve, and I ask him, 'What are you doing because I can see that you're disappearing', and they knew personally I completely didn't give a shit if they're going to achieve the productivity ... and he was so smiling at me and, 'Yeah, yeah, I'm writing a book'.

To return to this chapter's two questions, these portrayals of working conditions in Peterborough show warehouse workplaces to be changing in ways determined largely by those to whom they belonged: large business owners – demanding faster work, shorter breaks and ever higher productivity from workers. However, Sabina's background and her narration of her collusion as a supervisor with a rank-and-file worker who was subverting the rules of workplace show that, even under such intensified conditions, what the workplace stands for may be contested from below.

*

> It is important to ask ourselves what we can do to produce more prospects for hope, rather than trying to find hope in order to act.
>
> *(Brown and Littler, 2018)*

Many men and women interviewed in Peterborough about their workplace experiences and their lives in the city criticised supervisory regimes that prevented more than minimal interaction at work. Some of them explicitly valued the ethno-national diversity of the food factory and warehouse workforces. When strong connections were made across difference this meant greater potential to collude with others or even come together collectively to fight to be treated with dignity or to contest some of the worst conditions of exploitation in the workplace, including impossibly high targets, unjust fines or reductions in piece-rate payments leading to lower wages.

Of course this kind of informal agency rarely alters the structural injustices of the racial capitalism experienced in the workplaces of Peterborough's industrialised food supply chain or warehouse distribution sector, especially in a national context where average real wages in 2018 were worth less than they were in 2008 (TUC, 2018). However, although such moments of resistance may be short, they matter. As David Roediger argues, these moments contest the logic of capitalist management that entails the production of difference 'in its own interests'. Roediger (2017: 12) continues, attacking the idea in many liberal analyses following Trump's victory in the US that attention to racial injustice means sacrificing class justice: 'Struggles for racial justice are sites of learning for white workers, self-activity by workers of colour, and of placing limits on capitals' ability to divide workers'. They are thus struggles that can nourish, and be nourished by, class unity across racialised divides.

Hence the importance, in the case of Peterborough, of the actions of organisations alongside and in collaboration with which we did our oral history work – actions that seek to break down barriers between residents based on ethnicity, faith identity and whether someone is seen as a migrant or a local. When such efforts are successful they can turn our attention towards the more powerful forces that are invested in staying unseen, precisely through the promulgation of divisive discourses and practices, and move us away from a focus on the racial, ethnic or national difference we may see in the other. In this way the question of 'who does this place belong to?' cannot be separated from that of 'what does this place stand for?'

Note

This chapter is a revised and shortened version of a public lecture given at the University of Sussex, Brighton, England on 16 May 2018. The audio of the lecture is available at www.sussex.ac.uk/wcm/assets/media/303/content/54409.mp3 (accessed January 2019). Thanks to the Arts and Humanities Research Council (UK) for research funding (grant numbers AH/J501669/1 and AH/N004094/1).

References

Back, L. and Sinha, S. with Bryan, C., Baraku, V., and Yemba, M. (2018) *Migrant City*, London: Routledge.

Benvegnù, C., Haidinger, B., and Sacchetto, D. (2018) 'Restructuring labour relations and employment in the European logistics sector: unions' responses to a segmented workforce', in V. Doellgast, N. Lillie and V. Pulignano (eds) *Reconstructing Solidarity Labour Unions, Precarious Work and the Politics of Institutional Change in Europe*, pp. 83–103, Oxford: Oxford University Press.

Beynon, H. (2016) 'Beyond fordism', in S. Edgell, H. Gottfried and E. Granter (eds) *The SAGE Handbook of the Sociology of Work and Employment*, pp. 306–328, London: Sage.

Bhambra, G. (2016) 'Viewpoint: Brexit, class and British "national" identity', *Discover Society*, 34, July 5th.

Bhattacharyya, G. (2018) *Rethinking Racial Capitalism: Questions of Reproduction and Survival*, London: Rowman and Littlefield.

Brown, W. and Littler, J. (2018) 'Where the fires are: an interview with Wendy Brown', *Eurozine*, April 18th.

Burnett, J. (2016) *Racial Violence and the Brexit State*, London: Institute of Race Relations.

Gearing, J. and Rogaly, B. (2019) '"Workers": Life, creativity and resisting racial capitalism', *The Sociological Review Blog*, March 8th.

Gilroy, P. (2004) *After Empire: Melancholia or Convivial Culture?* Abingdon: Routledge.

Goodhart, D. (2017) *The Road to Somewhere: The Populist Revolt and the Future of Politics*, London: C. Hurst & Co.

Griffin, C. (2012) *The Rural War: Captain Swing and the Politics of Protest*, Manchester: Manchester University Press.

Hall, S. (2017 [1979]) 'The great moving right show', in S. Davison, D. Featherstone, M. Rustin, B. Schwarz (eds) *Stuart Hall Selected Political Writings: The Great Moving Right Show and Other Essays*, pp. 172–186, London: Lawrence and Wishart.

Hall, S., Massey, D., and Rustin, M. (2015) *After neoliberalism? The Kilburn Manifesto*, London: Lawrence & Wishart.

Hall, S., Roberts, B., Jefferson, T., Clarke, J., and Critcher, C. (2013 [1978]) *Policing the Crisis: Mugging, the State and Law and Order* 2nd ed., London: Palgrave.

Lowe, L. (2015) *The Intimacy of Four Continents*, Durham, NC: Duke University Press.
Massey, D. (1991) 'A global sense of place', *Marxism Today*, June.
Massey, D. (1995) 'Places and their pasts', *History Workshop Journal*, 39(1): 182–192.
Massey, D. (2007) *World City*, Cambridge: Polity Press.
Massey, D. (2011) Landscape/space/politics: an essay, available at https://thefutureoflandscape.wordpress.com/landscapespacepolitics-an-essay/ (accessed January 2019).
McMorran, C. (2015) 'Mobilities amid the production of fixities: labor in a Japanese Inn'', *Mobilities*, 10(1): 83–99.
Reid-Musson, E. (2018) 'Intersectional rhythmanalysls: power, rhythm, and everday life', *Progress in Human Geography*, 42(6): 271–284.
Robinson, C. (1980) *Black Marxism: The Making of the Black Radical Tradition*, Chapel Hill, NC: University of North Carolina Press.
Roediger, D. (2017) *Class, Race and Marxism*, London: Verso.
Rogaly, B. (2008) 'Intensification of workplace regimes in British horticulture: the role of migrant workers', *Population, Space and Place*, 14(6): 497–510.
Rogaly, B. (2015) 'Disrupting migration stories: reading life histories through the lens of mobility and fixity', *Environment and Planning D: Society and Space*, 33: 528–544.
Rogaly, B. (2018) 'Brexit writings and the war of position over migration, 'race' and class', *Environment and Planning C*, online early view. DOI: 10.1177/0263774X18811923.
Rogaly, B. (2019) 'Rescaling citizenship struggles in provincial urban England', in J. Darling and H. Bauder (eds) *Sanctuary Cities and Urban Struggles: Rescaling Migration, Citizenship and Rights*, pp. 217–241, Manchester: Manchester University Press.
Rogaly, B. and Qureshi, K. (2013) 'Diversity, urban space and the right to the provincial city', *Identities*, 20(4): 423–437.
Rogaly, B. and Qureshi, K. (2017) '"That's where my perception of it all was shattered": oral histories and moral geographies of food sector workers in an English city region', *Geoforum*, 78: 189–198.
Rogaly, B. and Taylor, B. (2009) *Moving Histories of Class and Community: Identity, Place and Belonging in Contemporary England*, Basingstoke: Palgrave.
Simpson, D. (2011) 'Salads, sweat and status: migrant workers in UK horticulture', unpublished PhD thesis, University of Sussex.
TUC. (2018) 'Real wages still down over £100 a week in some parts of the North West', available at www.tuc.org.uk/north-west/news/real-wages-still-down-over-%C2%A3100-week-some-parts-north-west-tuc-analysis-reveals (accessed January 2019).
Tyler, P., Evenhuis, E., and Martin, R. (2018) 'Structural transformation, adaptability and city economic evolutions, Case study report: Peterborough', Working Paper 10 (2018), p18, available at www.cityevolutions.org.uk/working-paper-peterborough-case-study/ (accessed January 2019).
Valluvan, S. (2019) 'The uses and abuses of class: left nationalism and the denial of working class multiculture', *The Sociological Review*, 67(1): 36–46.

24

Rurality, place and the imagination

Rosemary Shirley

Introduction

Picture a rural place. Does it have green rolling fields and a blue sky? Can you feel the breeze on your face and the cool fresh air in your lungs? Are there cows in those fields, or sheep, or llamas? Are those fields bordered by hand-built stonewalls, like a patchwork spread out across a verdant landscape? Or perhaps those fields are miles wide, surrounded by a wire fence with dusty brown earth. Are there mountains, hills, flatlands, dykes, extinct volcanos? Are there any people? Where do they live? In ancient thatched cottages, new housing developments, blocks of flats designed to house farmers or temporary pre-fabricated cabins. Is it peaceful? Can you hear that birdsong, that traffic hum, those machines in the quarry, the lorries dropping off at the Amazon distribution centre, the planes coming in to land at the airport, the open cast mining operation?

When we close our eyes and imagine, many of us will have a very strong picture of what a rural place looks like and, perhaps just as strongly, ideas about what should not be part of a rural landscape. At what point in the paragraph above did you find yourself running into a bump in the road, a disruption of your mental image of the rural? Was it the llamas? The dusty brown fields that were miles wide? Or was it as early as the green fields and blue sky? The answer to that question will be determined by your own geographical and cultural experiences of the rural. This chapter centres on the relationship between these two frames: the geographical specificities of rural places and the cultural constructions of rural places. It argues that these aspects are intertwined, each influencing the other in powerful ways.

What the rural looks like, who lives there and how the land is used varies enormously throughout the world and so when we are talking about 'the rural', as academics or cultural practitioners, it is important to be alert to the specificities of place and how they might counter our own culturally ingrained assumptions. However, these culturally ingrained assumptions, the ways in which we uncritically imagine the rural, are extremely powerful in informing our understandings of 'the real' rural. In fact, as this chapter will go on to discuss, they are an important part of what might be considered the elements that make up our understandings of the rural as a place and have meaningful effects on the reality of rural places.

Centring on ideas of the rural based in the UK, this chapter examines a number of theoretical co-ordinates as ways of conceptualising rural places. It explores how imaginary ideas of the rural feature in everyday life and are often connected to forms of consumption, arguing that this produces a set of rural mythologies – stories we are consistently told or tell ourselves about the rural. Drawing on examples from everyday life and popular culture the chapter ends with a focus on two particularly prevalent rural mythologies: that the rural is healthy and that the rural is situated in the past.

Conceptualising rural places

Even in a country as small as the UK, there are many different types of rural place. Living in a large new village close to retail parks and motorways in southern England is a different type of rural than living in a market town in the Lake District which is choked by tourist traffic for six months of the year, and is different again from living on a sheep farm in the Highlands of Scotland. In addition, there are many similarities between rural and urban experience in contemporary society. The currents of globalisation such as mass/global media and chain shopping connect rural and urban experience. For example, residents of town and country could find themselves sitting down to a meal bought in the same shops while watching the same US TV series on the same media streaming service. A closer look at this experience may reveal differences in the detail – the rural resident may have had to drive to some distance to the supermarket rather than walk or get a bus and broadband services maybe slower, but the similarities in everyday experience are clear.

The differences in rural places and the similarities shared between urban and rural places have led some geographers to questioned the value of the concept of the rural in thinking about place. In an article called 'Let's do away with the rural', Keith Hoggart (1990: 245) argues that 'the broad category "rural" is obfuscatory, whether the aim is description or theoretical evaluation, since intra-rural differences can be enormous and rural-urban similarities can be sharp'. I have suggested elsewhere that the term 'non-metropolitan' might be used instead of rural, in order to defamiliarise traditional notions of the countryside and allow room for thinking about places that are more usually excluded by the term rural; loaded as it is with the pressures and the picturesque and the peaceful, places that themselves complicate the polarities of the country and the city (Shirley, 2015: 4). Questioning received notions of the rural allows us to think more carefully about the typologies of places that fall into the bracket of rural today. We need to do this because, as demonstrated by the opening of this chapter, we all have strong ideas about what does and does not constitute a rural place and what does and does not belong there. This can lead us into a sort of blindness as to what rural places consist of. For example, instead of experiencing a rural place as consisting of fields, woodlands, motorways, electricity pylons, ancient earthworks and factories, we might think of the fields, woodland and ancient earthworks as belonging to the rural place and the motorways, electricity pylons and factories as being aberrations or pollutants, things that do not belong and are not rural. In addition to contributing to the idea that all rural places should be aesthetically pleasing, this way of thinking also situates the rural as being of the past or outside of modernity, as many of the elements deemed 'out of place' date from the twentieth or twenty-first century. Terms like non-metropolitan can help us re-evaluate the lived textures of contemporary rural places. I use the word textures here because it communicates something of the complex, simultaneous and multi-layered experiences of contemporary rural places. It helps us see how all these elements, both ancient and modern, are woven together as part of what constitutes the rural.

The usual visual characteristics we might understand as designating a place as rural, for example an open landscape of fields and trees, farms, twisty roads, isolated villages, country pubs, so important to cultural constructions of the rural are not relevant to the official, quantifiable understanding of what defines a place as rural. The definition of rural areas used by the British government is the Rural-Urban Classification system. This defines areas as rural if they fall outside of settlements with more than 10,000 resident population (DEFRA, 2017). It is interesting to note that the document detailing how this classification relates to the determining of local authority areas states that 'it should be noted that the classifications are based on populations and settlement patterns, not on how much countryside there is' (2).

Even with its single empirical measure of rurality – population density – the Rural-Urban Classification system alerts us to the presence of many different gradations of rural in the UK. As applied to the 2011 Census, the system identified six primary types of rural in England ranging from 'Rural hamlet and isolated dwelling in a sparse setting'– these were mainly situated in areas such as Cornwall and the far north of England – to 'rural town and fringe', which appear scattered in large numbers all over the country. Indeed, when we look at the map of England we can see that this system of classification produces the majority of the country, which is designated as either 'mainly rural' or 'largely rural'. It is perhaps surprising to learn that so much of the country is designated as rural. One of the reasons for this surprise maybe that the rural is represented in mainstream culture, not as the norm or the majority, but as something out of the ordinary and removed from everyday life. By this I mean that rural affairs are seen as a niche interest in national newspapers, radio programmes like BBC Radio 4's *Farming Today* are scheduled at 5.45 a.m. – when only farmers are thought to be awake – and that television programmes about rural places fall into specialised escapist genres like *Countryfile*, a rural magazine programme that is often criticised for its romanticisation of rural places, or landscape television where spectacular landscapes take centre stage (Bishop, 2017). This idea of the rural as being different from the norm of everyday experience, even though it takes up the majority of the UK's landmass, stems from the fact that most people live in urban locations, with recent governmental figures placing this as 83% of the population (DEFRA, 2018). As these statistics show, many more of us live in towns and cities than in rural places and this physical disconnection has increased the cultural importance of the countryside. Our relationship to the land finds expression in what we buy, how we choose to spend our leisure time and in conversations about global warming or plans for new power stations. Everyday decisions like choosing an organic chicken over a cheaper alternative or a fabric conditioner that claims to smell like a spring meadow reveal political, economic and imaginative connections to the idea of countryside. Our need for fresh air, to get out for a walk, to clear our heads, ties the countryside to ideas of physical and mental well-being. While large-scale protests against fracking in national parks or local petitions against housing developments in rural places point to our strong sense of how the countryside should look.

This means that while the rural is constituted as a series of geographically specific locations all with particular contexts, histories and needs, it is *also* significantly a place that is constructed in the imagination. Marc Mormont argues that 'rurality is understood as a social construct' – that is an imagined entity that is bought into being by particular discourses of rurality that are produced, reproduced and contested by academics, the media, policy-makers, rural lobby groups and ordinary individuals. The rural is therefore 'a category of thought' (Woods, 2011, referencing Mormont, 1990: 16).

An example of how the rural is constructed through forms of cultural imagination is the concept of the rural idyll. This category of thought imagines rural places as peaceful and timeless, a place of leisure rather than work, where humans and nature live in harmonious abundance. This way of conceptualising the rural has been woven throughout western culture from the third-century BCE poet Theocritus' *Idylls*, to Virgil's *Eclogues* and *Georgics*, into the pastoral landscape paintings of artists like Claude Lorrain and Nicolas Poussin in the seventeenth century. These images were used as inspiration for the large-scale landscaping of English country house estates by Lancelot Capability Brown and Humphrey Repton in the eighteenth and nineteenth centuries. As the wealthy tried to surround themselves with a supposedly lost classical idyll, these constructed landscapes often moved or disguised the homes of rural workers in order to remove any connection between the newly idyllic landscape and labour or poverty. The historian Jeremy Burchard (2002) examines the powerful influence of the idea of the rural idyll during the Industrial Revolution, with the shift in population from the rural to the urban, the development of model villages and garden cities and the nostalgia for past times in a rapidly developing society. We also find important critiques of the rural idyll in relation to class in the work of Howard Newby (1979) and John Barrell (1980) who demonstrate the complex interplay between the imagined or represented rural place and the realities of everyday rural life.

Today classical poetry and landscape painting find their contemporary antecedents in the popularity of cultural forms like new nature writing (see Smith, 2017), landscape-centred television programmes and consumer branding drawing on countryside themes. It is through such cultural expressions that we can begin to understand the function and power of the imaginary rural.

The necessity of attending to the rural as an imaginary place, as well as a myriad of rural spaces and lived experiences, can be seen in the work of Keith Halfacree, who argues that the rural should be understood through examining the tensions between three 'versions' of the rural: rural localities, everyday lives of the rural and formal representations of the rural. In Halfacree's model, the category of formal representations of the rural relates to documents and policies produced by the government which portrayed the rural in certain ways. This highlighting of the importance of representations of the rural in generating understandings of rurality can be extended to the representations of the rural which permeate everyday life through popular culture and consumer branding, becoming a pervasive imaginary landscape of the rural. Michael Woods (2015) extends this three-fold model by introducing the idea of the rural as an assemblage, a term that draws on the work of Bruno Latour (1993) (see Chapter 1 by Dovey, this book). Conceptualising the rural as an assemblage helps to highlight the connections and entanglements which characterise any account of place. For Woods (2015), the rural assemblage is composed of many different elements, including the material composition of a place – its geology as well as the plants, animals, people, histories and cultural representations.

What these different ways of conceptualising the rural make clear is that in order to understand rural places effectively we need to move beyond the physical geography of place and consider the rural as being constituted through discourse. Woods (2011: 31) states that

> a discourse … is not just a representation of reality, it creates reality by producing meaning and setting the boundaries of intelligibility. Neither does a discourse consist solely of written or spoken words, it includes images, sounds, bodily actions such as gestures, habitual thoughts and gestures and so on.

In addition to visual representations of the rural, the imaginary rural can also register as an embodied experience. Tim Edensor (2006) demonstrates how those engaging with leisure activities associated with rural places engage in a number of embodied performances, which relate to how they perceive that the countryside should be experienced. A small-scale embodiment of this kind that maybe particularly recognisable is the action of consciously taking a deep breath of 'country air' and perhaps remarking on its freshness or quality. Edensor (2006) conceptualises the British Countryside as a series of stages on which a number of contested performances take place that often have established scripts and costumes. For example, guidebooks and 'how-to' manuals give detailed behavioural instructions for groups as diverse as tourists, hunters, bird-watchers, climbers and ramblers. Each of these sub-sets don appropriate costumes dictated by differently oriented consumer marketing and tradition; tweed britches and shot gun for the grouse hunter, waterproof clothing and binoculars for the birdwatcher.

The imaginary rural is also embodied not only by visitors or those engaging in specialised country pursuits, but also the everyday practices of rural residents. The imaginary rural is not only something which influences the ideas about the countryside of those who live in urban areas, for the inhabitants of rural areas are also influenced by these representations. I will now go on to look more closely at how these imaginary rurals are constructed and circulated in contemporary society through a power set of what could be thought of as rural mythologies.

Rural mythologies

In *The Country and the City*, Raymond Williams ([1973]1993: 9) catalogues some of the discourses that circulate around these places as cultural concepts:

> On the country has gathered the idea of a natural way of life: of peace, innocence, and simple virtue. On the city has gathered the idea of an achieved centre of learning, communication, light. Powerful hostile associations have also developed: on the city as a place of noise, worldliness and ambition; on the country as a place of backwardness, ignorance, limitation. A contrast between country and city, as fundamental ways of life, reaches back into classical times.

In this description it is possible to recognise some of the most powerful stories we are told and continue to tell ourselves about rural places. These characteristics of rural places – peace, innocence, virtue, backwardness – are all linked to the idea that the rural is somehow divorced from modern society and therefore provides an alternative to it, in both positive and negative ways. Williams tracks the generation of these stories through the history of literature, however, as I have noted above, Woods insists that it is important to remember that discourses are not solely composed of written or spoken words, but also incorporate visual and aural elements, as well as embodied enactments. It is also important to note that discourses do not simply represent realities but that they also shape realities.

In contemporary society these stories circulate most effectively, in that they reach the widest audiences through encounters with aspects of everyday life such as television shows and consumer branding. Roland Barthes ([1957]1999) argues that the stories which circulate in everyday culture can be thought of as mythologies. Using a semiotic framework to analyse mid-twentieth-century French culture, Barthes draws attention to taken-for-granted or overlooked aspects of everyday life like soap advertisements, the popularity of steak and chips on restaurant menus, and the hairstyles of Romans in the film *Spartacus*. He wrote that:

> The starting point for these reflections was usually a feeling of impatience at the sight of the 'naturalness' with which newspapers, art and common sense constantly dress up a reality which, even though it is the one we live in, is undoubtedly determined by history. In short, in the account given of our contemporary circumstances, I resented seeing Nature and History confused at every turn, and I wanted to track down, in the decorative display of *what-goes-without saying*, the ideological abuse which, in my view, is hidden there.
>
> *(Barthes [1957]1999: 11)*

By highlighting the way in which representations of the world are constructed, but made to appear natural, Barthes argues that these mythologies are part of an ideological process in which what has happened (history) is made to seem as if it was always going to happen because it is the 'natural' state of things. In this way, mythologies as discourse have the ability not only to reflect realities, but also to produce realities. This suggests that certain discourses have ideological roots that justify certain conditions and prevent the imagining of alternatives. An example of this might be that the association of rural places with peace, as posited by Williams, and the communication of this mythology via the tourist industry, which frames the countryside as a place to get away from it all. This narrative supports a policy of under-investment in rural infrastructures such as public transport connections and broadband provision recently criticised in a House of Lords report (Lords Select Committee, 2018). Therefore, this imagined geography might further support the argument that such vital services for people who live and work in rural places could spoil highly valuable rural commodities such as a mythical (however misguided on a small island like the UK) remoteness or isolation.

In his recent re-working of Barthes' *Mythologies* to reflect contemporary western culture, cultural historian Peter Conrad (2016: 14) suggests that what Barthes was analysing is what we would now call branding. He argues that:

> In the affluent, contentedly secular society of the mid twentieth century, the meaning of myth underwent a change. The word once referred to stories that told hallowed truths, which as believers we took on trust. Now in common usage, it refers to a tissue of more or less amusing lies … The lost art was recovered by the writers of advertising copy, who had a sly awareness of its fictionality. Ancient myths were theological; although their contemporary equivalents are commercial, the products they tout still pretend to purvey spiritual benefits.

An example of the 'spiritual benefits' associated with using branding to closely link products to the powerful idea of the traditional farm and its associations of wholesomeness, nostalgia and being close to the land can be seen in the branding of a range of products by the supermarket chain Tesco. In 2016 the chain launched a new range of farm brands for fresh meat, fruit and vegetables, featuring the names of seven entirely fictional farms which had no relation to the produce being marketed. Complaints to the Trading Standards Institute were made by the National Farmers Union, claiming that the brands were misleading in that they appeared to indicate that the products had been farmed in the UK, when many had been imported from abroad (Rodionova, 2016a). The brand uses language which draws on a set of well-established rural mythologies in order to create associations between their products and a nostalgic, comfortable, picturesque image of farming. Rosedene and Redmere Farms, nominally responsible for Tesco's fruit and vegetables, both draw on old English place names for areas where rushes grow, giving both imaginary locations a pleasing heritage feel. Woodside Farm, representing pork products, conjures an image of happy pigs rootling around under beech trees. It cannot be a coincidence that Willow Farm and Nightingale Farm, used as brand names for poultry and salads respectively, are both farms that feature on the long-running agricultural radio soap opera *The Archers*. A move which enables the brand to draw on the deeply established mythological land of Ambridge, with its close knit community, green rolling hills and generations of farming experience. The appetite for such appealing, if entirely imaginary, images of the countryside is strong; Tesco recently attributed the popularity of their farm brands range as a significant factor in an impressive sales growth report (Rodionova, 2016b).

The following two sections explore the idea of rural mythologies further by exploring in more detail two prevalent rural mythologies: that the rural is a healthy place and the rural is aligned with the past.

The rural as a healthy place

The rural has long been imagined as a place of health and well-being. In Renaissance Italy, wealthy families sought refuge from plague-infested cities and created the practice of escaping to the comparatively fresh air of their country villas. In addition to physical good health, the rural has also been imagined as a place which is morally superior and better for one's mental health than the busy city full of morally dubious distractions like prostitution, drugs, alcohol and gambling. Early landscape painting often depicted St Jerome absorbed in the study of holy texts ensconced in a countryside setting, with the towers of the corrupting city sometimes visible on the horizon (Andrews, 1999). In the nineteenth century this idea was revived by writers such as Henry David Thoreau (1854) in *Walden: or Life in the Woods* and Ralph Waldo Emerson (1836) in his essay *Nature*, advocated for the moral properties of a life lived closer to nature.

Today, these positive associations between health and the countryside continue to circulate and have been most vividly mobilised by the consumer branding industry. Detergents and other cleaning products that make use of long-established associations between the countryside and cleanliness are familiar items on the supermarket shelves with products claiming to evoke the fragrance of a spring meadow or cherry blossom. However, one recent addition to their ranks shows how these imaginary versions of the rural are intertwined with the actuality of rural places. In 2014, a new collection of air fresheners was launched across the UK with a series of high-profile events, advertisements and supermarket promotions. A collection of eight fragrances each bore the name of one of the nation's National Parks. Some of the fragrances were loosely connected to characteristics of the parks themselves – *Snowy Mountain* for the Cairngorms and *White Roses and Pink Sweet Pea* for the Yorkshire Dales – white roses being the county's symbol. However, other fragrances appeared to bear little relation to their respective locations with, for instance, *Spring Breeze and Golden Lilies* use to signify the Peak District – which is not an area renowned for lilies. Together this range produced new imaginary landscapes from the *Mountain Sunrise* of Snowdonia to the *Midnight Berries and Shimmering Mist* of the Lake District.

Clearly, the product's branding was drawing on the centuries-old idea of the convergence of fresh air and rural places, in contradistinction to the stinking city. They also relate to the romantic landscape tradition seen in the work of artists such as Caspar David Fredrich whose lone wanderers stare into the unknowable distance from the top of mountains, or Samuel Palmer, who draws on the mystical druidical potential of landscapes in his dreamlike scenes of cornfields by moonlight. However, what makes these products particularly remarkable is in their association with the national parks themselves. The air fresheners were produced as part of a corporate partnership between the parks and the manufacturers. Such partnerships have been encouraged in response to a cut in public funding of around 40% and recommendations from the government that the shortfall should be generated through new partnerships with the private sector (Campaign for National Parks, 2015). The foundations for this partnership come from the USA where national parks cultivate relationships with brands like Coca Cola, Budweiser, Toyota and Disney. In 2016 new regulations were proposed to increase the visibility of corporate sponsors in US national parks, in a move designed to attract more lucrative sponsorship deals. This would include the opportunity for businesses to name certain features in the park and to display their logos on buildings, pathways and National Parks Service signage (Usbourne, 2016).

These air fresheners are not the only means through which there is large-scale circulation of imaginary landscapes that perpetuate simplified and romanticised ideals of rural places, but their success or otherwise also impacts on the funding and therefore the long-term management of the parks.

The rural is of the past

Popular representations of the countryside often situate rural communities and locations as alternatives to the tensions and pressures of modern everyday life. Drawing on the tradition of the rural idyll, discussed above, the British Sunday evening television schedule is well known for providing a quantity of relaxing rural-based escapism ahead of the working week to come (Sanghera, 2010; Woods, 2011). A long-running example from a mainstream channel is *Midsomer Murders*, in which crimes reminiscent of those featured in the novels of Agatha Christie are carried out in idyllic rural settings such as picturesque southern county villages with an abundance of thatched cottages, village ponds and ancient churches.

In almost an exact mirroring of the Sunday evening genre, there is the arena of rural horror, or the anti-idyll (Bell, 1997). Here, the countryside is represented not so much as of the past but as severely behind the times. Central to this is that the rural is isolated from modernity and this manifests itself in everything from a lack of mobile phone reception to slavish adherence to superstition, primitive sexual practices and violent hatred of outsiders. More recently, Adam Scovell (2017) has identified this fascination with the rural as a place of unsettling happenings as a genre he calls folk horror.

The popular portrayal of the countryside as outside of modernity is demonstrated by Tim Edensor (2002) in his analysis of the magazine *This England*, which claims to be Britain's best-selling quarterly magazine. He notes that it exclusively contains:

> photographs and sketches featuring little or no signs of modernity (no 'modern' buildings, hardly any cars and even television aerials are strangely absent). No youths are present in any picture, certainly no non-white locals or visitors are depicted and the urban is kept at bay. *This England* is located in the distant past, with little evidence of any post war development.
>
> *(2002: 42)*

Edensor's (2002) analysis together with the proliferation of the landscape/heritage television programming provides contemporary evidence of Williams' assertion that 'the common image of the country is now an image of the past' ([1973]1993: 297). These representations are not simply limited to consumption at home via screen or page, they also influence how rural places situate themselves as leisure attractions. The introduction of aspects of modernity such as new housing estates, wind turbines or solar farms trigger discourses around preservation; however, this also has a significant economic dimension, for rural places need to look like images of rural places in order to attract or maintain tourism. In his work on consumption and rural places, Paul Cloke (1993) observes that rural tourism sites, such as medieval castles, attract visitors by staging events such as vintage car rallies and craft fairs, which have no connection with the castle but are consumed along with it as part of a package of symbols of a kind of generic 'past'.

Recently scholars from different disciplines have been working to trouble the mythology that situates rural places as either outside of modernity or as somehow polluted by the evidence of modernity. These works move to re-situate the rural in narratives of modernity, not as a victim but as an active player central to many of the developments we associate with modern society. These range from the role of rural locations in large infra-structural projects like the national power grid and motorway system, to the influence of rural places on the development of modernist cultural practices in the visual arts and literature (Shirley, 2015; Bluemel and McCluskey, 2018; Stringer, 2018).

Conclusion

This chapter has argued that in trying to understand the rural in terms of place it is important not only to consider the specificities of real rural places, but to also examine how imaginary constructs of the rural are generated and circulate in society. It has asserted that these constructs or discourses can be found in the written word, visual culture, embodied or performative experience and habitual thoughts. It has introduced the concept of rural mythologies as a way of contextualising and analysing historical and contemporary manifestations of these discourses, and has focussed in particular on two popular rural mythologies: rural places are aligned with health and that rural places are of the past.

The circulation of these mythical images of the countryside situates the rural place firmly within the realm of consumerism. Through the purchasing of countryside-related products, the consumption of countryside-based programming or of countryside-themed experiences. However it is important to consider the real impact that these mythological relationships to the countryside have on rural places themselves. These discourses influence public and governmental opinion on how a diverse set of rural places with different histories, populations and needs should look, how they should feel, what sort of activities should and should not take place and what sort of infra-structure it requires. The imaginary rural is one aspect of what Woods refers to as the rural assemblage, however, in the UK, with the urban population far outnumbering the rural, these cultural mythologies become increasingly important in how the rural is understood as place.

References

Andrews, M., 1999. *Landscape and Western Art.* Oxford: Oxford University Press.

Barrell, J., 1980. *The Dark Side of the Landscape: The Rural Poor in English Painting 1730-1840.* Cambridge: Cambridge University Press.

Barthes, R., [1957] 1999. *Mythologies.* London: Vintage.

Bell, D., 1997. 'Anti-idyll: Rural horror'. In P. Cloke and J. Little (eds.) *Contested Countryside Cultures: Otherness, Marginalisation and Rurality.* London: Routledge, pp. 94–108.

Bishop, N., 2017. 'The nature of Nostalgia'. In V. Elson and R. Shirley (eds.) *Creating the Countryside: The Rural Idyll Past and Present.* London: Paul Holberton, pp. 100–104.
Bluemel, K. and McCluskey, M., 2018. *Rural Modernity in Britain: A Critical Intervention.* Edinburgh: Edinburgh University Press.
Burchard, J., 2002. *Paradise Lost: Rural Idyll and Social Change since 1800.* London: I.B.Tauris.
Campaign for National Parks, 2015. *Impact of Grant Cuts on English National Park Authorities: A Briefing based on Responses to Freedom of Information Requests.* [Online Accessed May 2019] www.cnp.org.uk/sites/default/files/uploadsfiles/Final%20national%20Stop%20the%20Cuts%20briefing%20July%202015.pdf
Cloke, P., 1993. 'The countryside as commodity: New spaces for rural leisure'. In Glyptis, S. (ed.) *Leisure and the Environment: Essays in Honour of Professor J. A. Patmore.* London and New York: Belhaven Press, pp. 53–67.
Conrad, P., 2016. *Mythomania: Tales of Our Times, from Apple to Isis.* London: Thames and Hudson.
DEFRA, 2017. *Defining Rural Areas.* [Online: Accessed May 2019] https://assets.publishing.service.gov.uk/government/uploads/system/uploads/attachment_data/file/597751/Defining_rural_areas__Mar_2017_.pdf
DEFRA, 2018. *Official Statistics: Rural population 2014/15, (Updated 28 February 2019).* [Online: Accessed May 2019] www.gov.uk/government/publications/rural-population-and-migration/rural-population-201415
Edensor, T., 2002. *National Identity, Popular Culture and Everyday Life.* Oxford: Berg.
Edensor, T., 2006. 'Performing rurality'. In P. Cloke, T. Marsden and P.Mooney (eds.) *Handbook of Rural Studies.* London: Sage, p. 7.
Hoggart, K., 1990. 'Let's do away with the rural', *Journal of Rural Studies*, 6: 245–257.
Latour, B., 1993. *We Have Never Been Modern.* Hemel Hempstead: Harvester Wheatsheaf.
Lords Select Committee on the Natural Environment and Rural Communities Act, 2006, 2018. *Rural Communities are being Failed by Government.* [Online: Accessed May 2019] www.parliament.uk/business/committees/committees-a-z/lords-select/nerc-act-committee/news-parliament-2017/nerc-act-report-published/
Mormont, M., 1990. 'Who is rural? Or how to be rural: Towards a sociology of the rural'. In T.Marsden P. Lowe and S.Whatmore (eds.) *Rural Restructuring: Global Processes and their Responses.* London: David Fulton, pp. 21–44.
Newby, H., 1979. *Green and Pleasant Land? Social Change in Rural England.* Hounslow: Wildwood House.
Rodionova, Z., 2016a. 'Tesco and other supermarkets using fake farm brands spark complaint from NFU'. *The Independent* 19 July. [Online: Accessed May 2019] www.independent.co.uk/news/business/news/tesco-and-other-supermarkets-using-fake-farm-brands-spark-complaint-from-nfu-a7144551.html
Rodionova, Z., 2016b. 'Tesco's range of 'fake farm' foods helps to boost sales'. *The Independent* 23 June. [Online: Accessed May 2019] www.independent.co.uk/news/business/news/tesco-results-profits-sales-fake-farms-willow-rosedene-boswell-dave-lewis-aldi-farming-a7097106.html
Sanghera, S., 2010. The Odd Appeal of Antique Britain on Sunday Night TV. *The Times* 20 October, p.3.
Scovell, A., 2017. *Folk Horror: Hours Dreadful and Things Strange.* New York: Columbia University Press.
Shirley, R., 2015. *Rural Modernity, Everyday Life and Visual Culture.* Abingdon: Routledge.
Smith, J., 2017. *The New Nature Writing: Rethinking the Literature of Place.* London: Bloomsbury.
Stringer, B., 2018. *Rurality Re-Imagined: Villages, Farmers, Wanders, Wild Things.* California: ORO Editions.
Usbourne, D., 2016. 'America's National Parks eye Corporate Sponsorship to plug Funding Shortfall'. *The Independent* 9 May. [Online: Accessed May 2019] www.independent.co.uk/news/world/americas/americas-national-parks-eye-corporate-sponsorship-to-plug-funding-shortfall-a7020786.html
Williams, R., [1973] 1993. *The Country and the City.* London: Hogarth Press.
Woods, M., 2011. *Rural.* Abingdon: Routledge.
Woods, M., 2015. 'Territorialisation and the assemblage of rural place: Examples from Canada and New Zealand'. In J. Dessein, E. Battaglini, & L. Horlings (eds). *Cultural Sustainability and Regional Development: Theories and Practices of Territorialisation.* Abingdon: Routledge, pp. 22–42.

25

The symbolic construction of community through place

Henrik Schultze

Introduction

In this chapter I will elaborate on the idea that community is not the same thing as place (Massey, 1991; Blokland-Potters, 2003; Blokland, 2017). I will take this claim as a starting point. Stating that place and community is not the same does not mean that they are not interrelated. What I am trying to do here is to highlight some aspects of this relationship between place and community.

There seems to be at least two main reasons for the assumption or rather the prerequisite for equating place and community which prevail in social science as well as in politics. First, a research agenda which looks only at the neighborhood level often assumes that physical proximity produces social proximity or that there is at least some cohesion to find. Whilst I use both terms interchangeably herein, it is worth highlighting that the equation of neighborhood and place can be problematic; this is not only a matter of scale, but also of putting meaning to spatial constructs. Put simply then, analytically, place and neighborhood is not the same thing; for more details see Agnew and Livingstone (2011). Second, an administrative or planning agenda needs to categorize neighborhoods as social units for the reasonable distribution of welfare as well as for planning purposes (Scott, 1998), which makes sense or, at least, cannot be avoided so far given the danger of over-simplification and reification (Brubaker, 2004; Jenkins, 2008; Scott, 1998). But to speak of deprived neighborhoods means to equate place and community. This may be a rational, sometimes convenient way for politicians to identify a concrete site for applying measures. For social scientists, it is not sufficient. We need to grasp the far more complex ways in which the sometimes uneasy relationship between place and community can be explained without a prior naturalization of both. What I want to discuss here is a third perspective on the link between community and place, not as an a priori connection, but as a construct where residents themselves build community through place.

So, my argument here is that community and place are not related inevitably but that place may be a necessary condition for the construction and maintenance of a symbolic community. For this purpose, I will briefly elucidate the debate of the relationship between place and community in social sciences, especially in sociology and urban studies. Then

I will turn to approaches on community which are able to sidestep some problems of this relationship and may offer an alternative to spatial determination. In this section I will also substantiate my argument that place becomes an important factor for the construction of a symbolic community with some empirical examples from my fieldwork.

Setting the scene

Empirically, I draw on my case study on belonging (see Schultze, 2017) conducted in Prenzlauer Berg, Berlin. Prenzlauer Berg is an area mostly built from the early 1870s – the so-called 'Gründerzeit' – and was part of the inner city of the former German Democratic Republic's (GDR) capital (East) Berlin. Due to physical neglect between 1945 and 1989 some of the critics of the political system of the GDR, like artists, writers or students, became aware of the opportunity to silently occupy the many empty flats, even whole buildings (Halbrock, 2004). From the 1970s onward, the housing authority – distracted by their large housing projects – lost track of the illegal occupations and more of those countercultural groups came, not only to stay off the grid but also to establish an alternative cultural space which was hardly possible somewhere else. What is important here is the distinction between a subculture and a counterculture. While we can define a counterculture as a 'set of norms and values of groups that sharply contradict the dominant norms and values of the society of which that group is a part' (Yinger, 1984, p. 3), a subculture does not entail this conflictual element (Yinger, 1960, p. 629). As part of this conflict, the counterculture developed their own infrastructures including hidden spots for illegal parties, concerts or off-theaters. It is important to mention this, because Prenzlauer Berg was not just a neighborhood; it was a place which symbolized an attitude of nonconformity, criticism and different alternative lifestyles vs. the mainstream and the state.

Prenzlauer Berg had already become a myth before the fall of the Berlin Wall in 1989 and was even celebrated in Western German commentaries (Felsmann and Gröschner, 2012; Haeder and Wüst, 1994). With the fall of the Wall the place became a hotspot for students, artists and all kinds of youth cultures from East, West and the whole world.

Nevertheless, Prenzlauer Berg has witnessed a fundamental change since 1989 both demographically and physically, as did other parts of East Berlin and East Germany. The reunification in 1990 initiated strong rent increases adjusting to the western housing market. Moreover, the restitution of nearly all buildings to their former owners (mostly Western Germans) led to a comprehensive urban renewal in which private property holders and the city administration were the main actors. With these economic investments and its strong cultural image as a bohemian site, Prenzlauer Berg became a desired place of residence for the better-off from 2002 onwards, while parts of the old counterculture kept living there.

Between 2011 and 2013 I conducted 35 in-depth interviews with both of these groups: the counterculture as well as better-off residents who moved in later. The interviews include arrival stories, narratives of belonging and reports on the use of the neighborhood. Both groups differ in length of residence and mostly in terms of economic capital whereas the cultural capital was quite similar. Here cultural capital is more related to a middle class which appreciate an urban environment and their ability to read the lifestyle codes of other urban middle class fractions, although the cultural capital of East Germans and West Germans may differ greatly due to their different biographies. Many writers in the GDR, for example, published their work in the underground while they were working as cemetery gardeners or boilermen.

Community as spatial concept

The urge to understand communal relationships has a long tradition in the social sciences and is – after all – the search for the answer to the question 'what binds society together?' This also touches upon the question of how groups are constituted and what the fabric of the boundary between different groups may be. One key term from the beginning to address these issues in sociology was community. Probably one of the first sociological attempts on community was Ferdinand Tönnies' (1887) differentiation between community and society, which was mainly a mapping on the rural/urban or pre-modern/modern divide. His notion of an analogy between the former rural community and the succeeding societal city was criticized implicitly and explicitly afterwards (König, 1958). Simmel (1995), as one of the more implicit critics, puts forward his idea of urbanity as lifestyle by emphasizing a different meaning of the urban as social space, rather than the rural and urban as mutually exclusive locations. Despite the contemporary denigration regarding life in the industrial city, he conceived the urban as a phenomenon of modernity in which communal relationships were not just destroyed but transformed in more fluid relationships. These relationships were no longer a matter of obligations and conformity but rather a matter of choice and diversity (see also Hampton and Wellman, 2018, p. 644). So, community ties seemed to be replaced by a more or less 'partial integration', as Bahrdt (1961, p. 39) puts it.

Maybe even at the beginning of the 20th century do we find this early divide between two perspectives in the sociological debate on community. On the one hand, we have Tönnies' normative lament of the lost (rural) community and, on the other hand, Simmel's observations of urbanites' 'blasé' attitude and reservation. Contrary to Tönnies, the loss of community was not the starting point for Simmel. What he stresses more is the progressing individualization as a result of an increasing division of labour and the liberation from narrowing social circles within the process of urbanization (Junge, 2012, p. 90). In short, the widening mobility and individualization seemed to be a trade-off against the 'good old' community and this was not necessarily a bad thing for scholars like Simmel or Bahrdt. On the other hand, the loss of community observed by Tönnies (1887) or Wirth (1938) was equated with social disorganization and anomy, especially in urban areas (Savage, Warde, and Ward, 2003, pp. 9–13). These two strands on theorizing community pervade also in today's thinking about community, reflected by the differentiation between the evidence of a 'lost' and a 'saved' state of community (Wellman and Leighton, 1979). Here, we can start to think about community and place in more detail.

In the early 20th century the Chicago School was probably one of the famous supporters for the concurrence of neighborhood and community reflected in the idea of 'natural areas'. Especially Robert Park was interested in how incoming people developed new forms of social relations in rapidly urbanized areas of Chicago. For Park, natural areas became the main source for social organization (Goist, 1971, p. 54). He, among others, conceived the modern city as an agglomeration of natural areas in which every incoming population would group themselves as a cultural unit in specific neighborhoods. The outcome of this sifting and sorting would be that 'natural areas and natural groups tend to coincide' (Zorbaugh, 1961, p. 47, quoted in Blokland, 2017, p. 21).

Maybe their search for an alternative community model in the form of 'neighborhood solidarities' which could replace eroding family and kinship ties in the modern city (Savage, Warde, and Ward, 2003, p. 13) was responsible for their unquestioned equation of place and community. But also many of the followers of the Chicago School oriented their research agenda that way. Wellman and Leighton (1979, p. 364) identified in the course of time even a substitution of the neighborhood concept in favor of the community concept. In their research agenda they discussed the relationship between community and society under the label of the 'community question', which is: 'the study of how large-scale divisions of labor in social systems affect the organization and content of social ties' (Wellman and Leighton, 1979, p. 365). In their review, community is often defined as: '*networks of interpersonal ties* (outside of the household) which *provide sociability and support* to members, residence in a *common locality*, and *solidarity sentiments and activities*' (Wellman and Leighton, 1979, p. 365, italics in original). The assumption that spatial proximity creates or at least fosters personal networks is in part responsible for a naturalized connection between place and community. Instead of such scientific naturalization, Wellman and Leighton (1979, p. 385, italics in original) proposed to study 'neighborhood *and* community rather than neighborhood *or* community'. The separation of the two concepts is not only helpful to avoid the equation of place and community but also in order to broaden the view for networks in a neighborhood as well as somewhere else.

The symbolic construction of community

Beside this network perspective, a third interesting strand of studying communities was developed after the spatial turn (Lossau, 2012). It was the notion of community as imagined and not structural, first advanced by the work of some British social anthropologists and sociologists (for example, Cohen, 1985, 1986), and later more prominently for the national level by Anderson (2006). These approaches also included concepts of identity and belonging to theorize community. This part of community research (rather at the margins of urban research) was probably part of the answer to the community question mentioned above because it investigated the impact of social change on communities. The idea of understanding community symbolically and not structurally started from the empirical observation that communities show a remarkable resilience, although their structural bases diminish due to macro social processes of change on the national and global level (Cohen, 1985, p. 76). By conceiving communities symbolically rather than structurally, scholars who investigated the symbolic expressions of communities were able to bridge the assumed divide between 'traditional' and 'modern' forms of social relationships in the early years of sociology (Hamilton, 1985, p. 9). Moreover, they could avoid the default connection of place and community. Cohen (1985, p. 12) identified a community when:

> the members of a group of people (a) have something in common with each other, which (b) distinguishes them in a significant way from the members of other putative groups. 'Community' seems to imply simultaneously both similarity and difference.

The membership in such a community does not necessarily depend on structural dimensions that express community, such as common location, class, ethnicity or networks. On the contrary, Cohen (1985) argued that such structural categories were challenged by another mode of establishing sameness. To be a member of a community depends on the symbolic construction of a 'veneer' of similarity (Cohen, 1985, p. 44): similarity within the community is emphasized while difference to the outside world is exaggerated (Cohen, 1985, p. 21). For this purpose symbols can be used, for example a football shirt, a wedding ritual, a flag or a lifestyle presented to the inside and outside of the community. The meaning which is assigned to a symbol may vary greatly within communities and to the outside. According to Cohen (1985, p. 19):

> The symbols of community are mental constructs: they provide people with the means to make meaning. In so doing, they also provide them with the means to express the particular meanings which the community has for them.

Those symbols not only mark similarity but also difference, even if we are not aware of it.

One crucial feature of symbolic communities is the boundary which demarcates the beginning and the end of belonging – the 'us' and 'them' – and is therefore an intrinsically relational construct (Cohen, 1985, p. 58). Symbolic communities identify themselves and are identified by others always in relation to each other:

> Since the boundaries are inherently oppositional, almost any matter of perceived difference between the community and the outside world can be rendered symbolically as a resource of its boundary. The community can make virtually anything grist to the symbolic mill of cultural distance.
>
> *(Cohen, 1985, p. 117)*

Cohen points here to the idea of sameness and difference again. To define who belongs does always imply to define who does not (Jenkins, 2008, p. 20).

If we think about the relationship between place and community we have to consider what definition of community is used. Scrutinizing the relationship between two or more theoretical concepts depends always on the definition of these terms. A slightly different angle on this definition may alter the nature of the relationship in question. If we define community as place-based, we may end up stating that place and community is one and the same thing. With the concept of a symbolic community I introduced a concept which reveals the possibility to track the process of the construction of community via place. In the next section I will put the concept of symbolic communities into the spatial dimension by providing some empirical examples from my research site.

The symbolic construction of community through place

As we saw in the description of Prenzlauer Berg, it isn't just a neighborhood like any other. The relationship between lifestyle and neighborhood seems to be very strong for some of my interviewees. Different lifestyles as a result of demographic changes may lead to very different ideas of how the neighborhood (and the residents) should be. Prenzlauer Berg (may be as an extreme case) is therefore a highly contested place whose meaning may differ profoundly between residents (and outsiders). This meaning making is what Tuan (1977, p. 6) conceived as the transformation from space to place:

> 'Space' is more abstract than 'place'. What begins as undifferentiated space becomes place as we get to know it better and endow it with value. (...) Furthermore, if we think of space as that which allows movement, then place is pause; each pause in movement makes it possible for location to be transformed into place.

One empirical example of how difference is marked is the evaluation of the newcomer's behavior in public space as improper. Dizzi is 37 years old. She is the mother to a 7 year old and a 20 year old, and was engaged with the counterculture in Prenzlauer Berg before she moved there in 2003. She travelled through her neighborhood often to pick up her younger son and encountered other mothers:

> That is something, I get this total aggression, with those MOTHERS, all those MOTHERS with their strollers. I don't know what motivates them, I don't know whether it just seems so extreme to me, but I think, I think this is really true, they have this entirely ignorant [attitude], something so … something so … something so really egocentric […] they walk with four of them next to each other and they don't give a shit whether someone needs to pass by, whether they should maybe step aside a little.

With the same irritation Uli, in his late 40s and father of a 10 year old, also recognizes signs of ignorance in the behaviors of newcomers:

> My impression is they have no self-doubts at all. At least they present themselves that way … and … yeah … I cannot relate to this. That is not my thing.

Interviewer: How do you notice that?

> Well, how they convey their stuff with such a firm conviction. So the thing with the baby strollers for example: We have children! The pavement is ours! We have rights, special rights if possible! That kind … and also I don't know if it is so healthy if you can afford a big condominium at 30–35 years of age.

It is interesting that Uli, working as an engineer, who labels himself a higher earner and draws a sharp boundary between 'us' and 'them', while he states at the same time: 'I am an executive employee and as such I have that thing [the income] in common with many of the folk around here'. We can see here that the construction of symbolic communities does not have to rely on, but rather crosscuts, key identity categories like class, gender or ethnicity, in this case class. What is more important here is the role of place for the observation and interpretation of the behavior of 'them' as marker of difference. Jointly used public space seems to be a crucial arena, not for establishing and maintaining communal relationships, but for strengthening the boundary between 'us' and 'them' and therefore for constructing community symbolically.

Another important way of reinforcing this boundary is the utilization and instrumentalization of the past. This points directly to the fact that the transformation from a structurally defined community to a symbolic one emphasizes the temporal dimension of this process. Members of communities can draw selectively on communal traditions or rituals for contemporary purposes:

> Symbols of the 'past', mythically infused with timelessness have precisely this competence, and attain particular effectiveness during periods of intensive social change when communities have to drop their heaviest cultural anchors in order to resist the currents of transformation.
>
> *(Cohen, 1985, p. 102)*

Yet how can we put place into play here? Saskia, a mother aged 35 with a 10 year old son, worries – as do many other interviewees who see themselves as part of the former counterculture in Prenzlauer Berg – about the closure of the old clubs in the neighborhood after the arrival of the better-off newcomers:

> I am very critical about how things turned out here. At Greifswalder Straße for example the Knaack[1] is closed now because of problems with new nearby flats complaining about

> the noise. I went there dancing when I was young. The Magnet-Club is also shut down because of such complaining. A friend has reopened it in Kreuzberg now. This is all changing, all are leaving and this place becomes rather a bourgeois family neighborhood.

Lofland (1998, p. 65) coined the term 'memorialized locales' for such locations, pointing out that those 'pieces of the public realm (...) can become lightning rods for feelings of "community" and for expressions of conflict' (Lofland, 1998, p. 66). Those clubs conceived as 'memorialized locales' symbolize a past meaning as locations which matched the lifestyle of the counterculture before. They did not lose their materiality but their meaning was changed, as Uli, whom we met before and who is also angry about the club-closures, puts it:

> The Knaack is gone, Magnet too and this is very symptomatic [for the change, HS]. It is full of symbolism that the 'Magnet' is an organic grocery store now.

If we take this account seriously, we see that it is not only about how one lifestyle follows the other, but how place is connected to community via symbols. The 'Bio-Company' is not just a successor as a store, it represents a whole new lifestyle in the eyes of my interviewees. It is remarkable that Uli and Saskia could barely remember when they used these clubs when they were still open, yet it becomes clear if we understand the location not just as a physical location, but as symbolic marker. Although the meaning of space has changed for there are creative ways to symbolize the community even within such a changing environment. One way is to create new material forms which symbolize a former community, as in the case of the 'Staatsgalerie'. That kind of exhibition venue – opened in 2010 – is one part of the memory repository used by the former counterculture and celebrated accordingly. Amalia is 48 and a kind of veteran of the early Prenzlauer Berg scene. She describes her sentiments regarding this and similar locations:

> Interestingly there are still openings of a few locations as the 'Staatsgalerie' which try to hit a stake in the ground again, like: we are still here and we still have a voice even if you don't care but we are still existing. So I would mention the 'Staatsgalerie' here absolutely but there were other spots, smoker's bars <laughs> ... put it this way ... open until dawn, always trouble with the neighbors, close from shut-down but still prevailing with a kind of hardiness. And this are locations where I feel most at home.

This is exactly the point when place and community – symbolically constructed – coincide. However, it is not natural as in the early conceptions of community; it is accomplished by experiencing community. It does not mean that community is local; rather, that it is constructed with location as a means. When places are changing and locations disappear there is the time for constructing community symbolically. Here the closed clubs serve not as locations per se but as 'memory figures' (Assmann & Assmann, 1988, p. 29), albeit their reversed meaning which goes far beyond the feelings of loss. The reverse meaning of these locations conjures the counterculture's community symbolically, while locations like 'Staatsgalerie' symbolize the boundary between symbolic communities. It is no coincidence that Amalia mentioned smokers' bars here as they contradict (if not spoil) the new meaning of Prenzlauer Berg as healthy, affluent and clean. What we observe here is how place is re-used through symbols and symbolic practices even if the habitus and the habitat don't match anymore (Bourdieu, 1991, p. 32). Moreover, the mutually exclusive appropriation is possible via symbolic neighborhood use at the very same place. The practices described above challenge the idea of a unified place-based community but rather suggest its symbolic deconstruction.

Conclusion

The early conceptualization of the relationship between place and community tended to insist on a strong dependency of both. Place seemed to be not possible without community. Urbanization was depicted as a dark era for community. The logical derivation was that changing places causes a lost community and paves the way for social disorganization and anomy, albeit with a few exceptions like Simmel who saw the diminishing of communities rather as a liberation than a curse. Relying on good and healthy communities and the quest for identifying and healing community problems persists also nowadays.

Beside the feel-good notion of community, the connection with place is also related to the idea of decoding social life. To glue community and place together seems to provide us with a straightforward view of the world. It provides us with the certainty of what we can expect in a given place. In that sense, place is used as a category enabling us to match people and specific attributes for making sense of a social world otherwise hardly readable. The same is true for various political or scientific endeavors: the use of the direct relationship between place and community seems to be a convenient tool for research, planning and governing. Knowing what to expect in a certain neighborhood in an unambiguous way makes it much easier to assess who residents are, and what they do or need.

The doings and thinking of my interviewees challenge the notion of one community at one place. Not because they do not relate to their place. It is because they use the place selectively and symbolically. We have seen how markers of difference may become salient, and how sharp boundaries were drawn between 'us' and 'them'. At this moment, the community is constructed symbolically; not because of structural differences but because of different meanings referring to a shared place. Places are not communities, but to think about places in terms of the symbolic construction of communities may help to highlight how their relationship is constituted. It may help us not only to challenge prior assumptions or images about specific places but also to explain processes of boundary work as analyzed here.

Note

1 The Knaack Club existed from 1952–2010 in Prenzlauer Berg and was shut down after a court order. It was a real club-institution especially for the counter culture in the early 1990s.

References

Agnew, J. A., & Livingstone, D. N. (2011). Space and Place. In D. N. Livingstone & J. A. Agnew (Eds), *The SAGE handbook of geographical knowledge* (pp. 1–34). Thousand Oaks, CA: Sage.

Anderson, B. R. (2006). *Imagined communities: Reflections on the origin and spread of nationalism* (Revised ed.//Rev. ed). London: Verso.

Assmann, A., & Assmann, J. (1988). Schrift, Tradition und Kultur. In W. Raible (Ed.), *Script-Oralia: Vol. 6. Zwischen Festtag und Alltag: Zehn Beiträge zum Thema "Mündlichkeit und Schriftlichkeit"* (vol. 6, pp. 25–50). Tübingen: Narr.

Bahrdt, H. P. (1961). *Die moderne Großstadt: Soziologische Überlegungen zum Städtebau. rowohlts deutsche enzyklopädie* Reinbek bei Hamburg: Rowohlt Taschenbuch Verlag.

Blokland-Potters, T. (2003). *Urban bonds.* Cambridge: Polity Press.

Blokland, T. (2017). *Community as urban practice. Urban futures* Malden, MA: Polity Press.

Bourdieu, P. (1991). Physischer, sozialer und angeeigneter physischer Raum. In M. Wentz (Ed.), *Stadt-Räume* (pp. 25–34). Frankfurt am Main: Campus.

Brubaker, R. (2004). *Ethnicity without groups* Cambridge, MA and London: Harvard University Press.

Cohen, A.P. (1985). *Key ideas. The symbolic construction of community* London: Routledge.

Cohen, A.P. (1986). Of Symbols and Boundaries, or, does Ertie's Greatcoat Hold the Key? In A.P. Cohen (Ed.), *Anthropological studies of Britain: Vol. 2. Symbolising boundaries: Identity and diversity in British cultures* (pp. 1–19). Manchester: Manchester University Press.

Felsmann, B., & Gröschner, A. (2012). *Durchgangszimmer Prenzlauer Berg: Eine Berliner Künstlersozialgeschichte der 1970er und 1980er Jahre in Selbstauskünften* (2nd ed.) Berlin: Lukas-Verl.

Goist, P.D. (1971). City and "Community": The Urban Theory of Robert Park. *American Quarterly, 23* (1), 46. doi: https://doi.org/10.2307/2711586.

Haeder, A., & Wüst, U. (1994). *Prenzlauer Berg: Besichtigung einer Legende*. Berlin: Ed. q. Retrieved from www.gbv.de/dms/faz-rez/F19950202RBUCH2-106.pdf.

Halbrock, C. (2004). Vom Widerstand zum Umbruch: die oppositionelle Szene in den 80er Jahren. In B. Roder & B. Tacke (Eds.), *Prenzlauer Berg im Wandel der Geschichte: Leben rund um den Helmholtzplatz* (pp. 98–124). Berlin: be.bra.

Hamilton, P. (1985). Editor's Foreword. In A. P. Cohen (Ed.), *The Symbolic Construction of Community* (pp. 7–9). London: Ellis Horwood and Tavistock Publications.

Hampton, K.N., & Wellman, B. (2018). Lost and Saved... Again: The Moral Panic about the Loss of Community Takes Hold of Social Media. *Contemporary Sociology: A Journal of Reviews, 47* (6), 643–651. doi: https://doi.org/10.1177/0094306118805415.

Jenkins, R. (2008). *Social identity* (3. ed.). *Key ideas*. London and New York: Routledge.

Junge, M. (2012). Georg Simmel. In F. Eckardt (Ed.), *Handbuch Stadtsoziologie* (pp. 83–93). Wiesbaden: VS Verlag für Sozialwissenschaften.

König, R. (1958). *Grundformen der Gesellschaft: Die Gemeinde. rowohlts deutsche enzyklopädie* Hamburg: Rowohlt Taschenbuch Verlag.

Lofland, L. H. (1998). *The public realm: Exploring the city's quintessential social territory. Communication and social order* Hawthorne and New York: Aldine de Gruyter.

Lossau, J. (2012). Spatial Turn. In F. Eckardt (Ed.), *Handbuch Stadtsoziologie* (pp. 185–198). Wiesbaden: VS Verlag für Sozialwissenschaften.

Massey, D. (1991). A Global Sense of Place. *Marxism Today*, June, 24–29.

Savage, M., Warde, A., & Ward, K.G. (2003). *Urban sociology, capitalism and modernity* (2nd ed.). Houndmills, Basingstoke, Hampshire, and New York: Palgrave Macmillan.

Schultze, H. (2017). Die Grenzen sozialer und räumlicher Zugehörigkeit (Dissertation). Humboldt-Universität zu Berlin, Kultur-, Sozial- und Bildungswissenschaftliche Fakultät, Berlin.

Scott, J.C. (1998). *Seeing like a State: How certain schemes to improve the human condition have failed. Yale ISPS series* New Haven, CT: Yale University Press.

Simmel, G. (1995). Die Großstädte und das Geistesleben (1903). In G. Simmel, R. Kramme, A. Rammstedt, & O. Rammstedt (Eds), *Aufsätze und Abhandlungen 1901 bis 1908* (vol. 1, pp. 116–131). Frankfurt am Main: Suhrkamp.

Tönnies, F. (1887). *Gemeinschaft und Gesellschaft: Abhandlung des Communismus und des Socialismus als empirischer Culturformen* (1. Auflage). Leipzig: Fues. Retrieved from www.deutschestextarchiv.de/book/show/toennies_gemeinschaft_1887.

Tuan, Y. f. (1977). *Space and place: The perspective of experience* Minneapolis, MN: University of Minnesota Press.

Wellman, B., & Leighton, B. (1979). Networks, Neighborhoods, and Communities: Approaches to the Study of the Community Question. *Urban Affairs Quarterly, 14* (3), 363–390.

Wirth, L. (1938). Urbanism as a Way of Life. *American Journal of Sociology, 44* (1), 1–24.

Yinger, J. M. (1960). Contraculture and Subculture. *American Sociological Review, 25* (5), 625–635.

Yinger, J. M. (1984). *Countercultures: The promise and peril of a world turned upside down* (1. Free Press paperback ed.). New York, NY: Free Press.

Zorbaugh, H. W. (1961). The Natural Areas of the City. In G. A. Theodorson (ed.), *Studies in Human Ecology* (pp. 45–49). Evanston, IL: Row, Peterson.

26

Fashioning place

Young Muslim styling and urban belonging

Saskia Warren

Introduction

The British high street is changing. Hardly a day passes without a news story about a retail chain facing collapse, threat of store closure and loss of jobs, or the heart of town and city centres failing. A vital aspect of this economic change is attributed to the growing might of on-line commerce in an increasingly globalised market-place. Much less attention has been paid to the changing dynamics of the high street amidst local contexts of social pluralism. Even more, the everyday practices of shopping in thinking through the relationship between minorities and place have been neglected in the wider narrative of *how* our high streets are changing. In this chapter I look at how women are using high street shopping to represent how minority faith followers can be contemporary, fashionable and cool. Muslim women, through their tactics and strategies around high street consumerism, use everyday forms of creativity – selecting, layering, mixing – to negotiate minority status in ways that add distinction and can become profitable. In everyday practices, Muslim fashionistas are challenging the meaning of public space as marked by masculine, white and secular cultural norms. As new Muslim female identities are formed, fashionable young Muslim women are fostering spaces of belonging and negotiating place in contexts where many once felt marginalised or 'out of place'.

A contested concept when first coined, modest fashion is now entering the mainstream. Modest fashion has mostly been attributed to fashions designed for minority faith groups from Abrahamic faiths – Islam, Christianity and Judaism. Studies have focussed on Muslim modest fashion, and to a lesser extent on Hasidic Jewish and Mormon forms of style (Lewis 2013). Different ideals and kinds of modest fashion have emerged that are subject to cultural, ethnic and religious differences over time, and changing cultural and political economies of fashion. What connects strands of modest fashion is consumer demand for clothing that provides bodily cover, originally rooted in the alignment of self-presentation with religious doctrine and its interpretation. Modest fashion choices bridge the sacred and everyday practices; meticulously studied through veiling practices in the geohumanities and religious studies around everyday 'lived' religion and material religion (see Dwyer 1999; Lewis 2015;

McGuire 2008; Secor 2002; Tarlo 2010; Warren 2019). Ethnic and Asian fashion styles popular in the 1990s can be seen as important precursors for the latter visibility and mainstream retailing of modest fashion along with the centring of questions of ethics and aesthetics in everyday wear. What started as a niche segment of the global fashion market thirty years ago has now grown exponentially with key high street stores such as H&M and Uniqlo launching their own modest fashion lines. Regular presence in fashion magazines, social media and luxury brands coupled with an expanding marketplace has situated modest fashion within mainstream fashion (Reuters 2017). Beyond the impacts of on-line commerce, the British high street has aesthetically altered through modest fashion and halal products in marketing campaigns and stock levels that spike in visibility across the year. For instance, the so-called 'Ramadan Economy' during the Islamic period of fasting and Eid-al-Fitr celebrations and gift-giving that follows are keenly targeted by high street retailers and brands in what is now the third biggest season for businesses after Christmas and Easter (Noor 2018).

In this chapter I reconsider geographical and social ideas of being in place/out of place through looking at practices of high street shopping by young Muslim women. By addressing the cultural economy of how power and agency are distributed in circuits of production, representation and consumption processes, it departs from a purely hierarchical framing of the power relation between production and consumption and between mainstream and ethnic (*or modest*) fashion (Jackson 2002). This responds to the findings of recent studies that social and economic exclusion of Muslim women remains acute in Britain (Karlsen and Nazroo 2016; Runnymede 2017). Everyday creativity of Muslim women in negotiating social and spatial discrimination is often overlooked, however, along with the kinds of subjectivity produced through forms of conformity and resistance. In this chapter I highlight the experiences of women in everyday tactics of shopping and self-styling as they attempt to build new Muslim female identities and spaces of belonging, and renegotiate a sense of place as part of a global Muslima *and* locally streetwise Mancunian urban youth. This is not an insular sense of place given embeddedness in global digital platforms dedicated to modest fashion in geographically dispersed locales, e.g., Indonesia, Malaysia, Dubai, UK, Canada, U.S. Nor is sense of place de-territorialised due to inflections by local creative scenes and quarters, including the aura of coolness and edginess in Manchester's Northern Quarter. Awareness of Muslim consumers around worker exploitation and consumer alienation from fashion production does not separate out a passion for fashion, or the significance of consumption processes for forging identities and belonging in Western states in finding new ways to fit or rebel. As has been argued elsewhere, symptoms of social and economic marginality have produced some extremely powerful arenas of cultural activity (Bhachu 1998; Jackson 2002). Everyday tactics on the high street and fashioning practices work as instances of 'the reflexivity employed in identity formation in the public negotiation of new subjectivities and axis of belonging that cross borders of nation states, piety and denomination, and racialised and ethnic difference' (Warren 2018, 2; Warren 2019). Fashioning place is thus advanced as key to understanding the performed and intimate relationships of being in place/out of place in changing dynamics of urban public space and the British high street.

In place/out of place: young Muslims and modest fashion

In the new cultural geography that has emerged in the last decade or so, a deeper emphasis has been given to the importance of religious and faith-based narratives, practices and beliefs in the production of place and space. Amidst changing geopolitical contexts of religious extremism and migration, calls have been made for the need to show greater awareness to aspects of religion, faith, spiritualism and morality in our everyday lives and the built and natural environment (Kong 2001; Oosterbaan 2014). The renewed importance of religion in public and civic life in secular societies has somewhat controversially been termed post-secularism (e.g., did religion ever really go away?) (Baker and Beaumont 2011; Dwyer 2016; Kong 2010). Whether this is considered a new turn or a re-turn, what is clear is that, especially in cities and neighbourhoods where there is a critical mass of people who identify with religious or spiritual groups, understanding systems of faith and belief is interwoven with the dynamics of place.

Convening multiple kinds of people as spatial entities, cities comprise difference with often subtle kinds of exclusion and inclusion marking out public spaces (Simonsen 2008). Ethnic exclusions and inclusions form part of the very construction of public space along with other intersectional markers of class, gender, race, religion, sexuality, age and disability (Ibid.; Warren 2017, 789). Public space in Western-liberal countries is typically no longer constituted through political participation, but rather by who has the rights and power to exercise social and economic agency in streets, squares and parks. Any attempt to build an accurate representation of a given place thus requires thinking through the material and perceptual ways in which people are in/excluded in space (Warren 2017). Beyond the built environment, we might also consider social-spatial structures of labour and leisure as well as the significant role of creative mediums in a critical investigation of the encoding of British cities, quarters and high streets with very different kinds of values; aesthetic, economic, social and moral.

At its simplest place is a space that is meaningful. Place is a relational concept constructed 'through the meanings that different individuals layer onto the world around them' (Warren and Jones 2015, 7). Writing on the experience of being in place/out of place, Tim Cresswell (2014a, 2014b) shows how it refers both to the geographical and to a position on a social hierarchy, e.g., what is deemed good or appropriate. Processes of place-meaning are subject to hierarchies between different social groups on who is included and permitted voice to shape the physical and narrative dimensions of a place. For power is spatialised as place is given meaning (Cresswell 2014b). Resistance to power, therefore, can take the form of organised disobedience (Ibid., 255). Whilst feeling out of place can be a sign of marginalisation and social isolation we might also recognise the potential of being 'out of place' when used as a visible means of protest or resistance to dominant forms of power – or challenging 'the status quo'.

Readings of place that fall into a bounded geographical approach giving emphasis to fixity and inwardness are flawed by what Connell and Gibson describe as a 'tendency to become too enmeshed in the detail of the local at the expense of recognising how the local is constituted within wider flows, networks and actions' (2003, 14). This can be about infrastructure but is also 'attributable to a process of mythologising place in which unique, locally experienced social, economic and political circumstances are somehow "captured"' (Ibid. 14). Writing on music, Connell and Gibson's insights also cast light on fashion scenes that are given rise in different places, and the significance of mobilities of people, ideas and tastes that travel across borders in the making of creative sub-cultures and places. In the co-creation of place with cultural scenes there are conflicts and tensions arising that are discursive – such as how varying identities are represented – and those that are concrete in struggles over public and private spaces.

Some of these struggles over recognition and representation are taking place on our high streets. Modest fashion on the high street first appeared from Muslim fashionistas shopping, or 'trawling', the high street in search of clothes that could be styled in modest ways (Warren 2019). Fashion blogs were created and trends dissected in commentary that increasingly took a political edge. Questions were posed on who mainstream fashion was for? Mainstream high street stores garnered critique for championing [Western] materialism and consumerism, and propagating an image that normalised white, secular and exposed bodies as fashionable (Lewis 2013; Pham 2015). The expansion into modest fashion lines of high street retailers and luxury brands signals awareness of the fashion industry of the wider bankability of diversity and, more specifically, the power of the Islamic marketplace. For geographers interested in geographies of embodiment and the performance and contestation of gendered identities in different spaces, the rise of modest fashion raises pertinent questions around the significance of dress in the demarcation of in-group and out-group boundaries and belonging in a shared society. As Clare Dwyer has written, studies on dress can reveal the active engagement by young women in constructing their identities through offering a challenge to meanings that are 'attached to different dress styles' and also 'in the reworking of meanings to produce alternative identities' (1999, 5). Contemporary playfulness with high street fashions and the insertion of modest lines in established retail stores point towards the instabilities of the binaries of 'Western' and 'traditional' (or indeed 'modest') forms of dress (Ibid.). As one of the few sectors and spaces where female bodies are valorised more than male bodies, I suggest that high street fashion and restyling has the potential to become a claim to and subversion of white, masculinist and secular dominance in public space.

Re-fashioning the British high street? Empirical examples

The British high street has been imagined as a mundane and ordinary place or, more recently, since the global recession, as a place of urban decline and vacancy. In these stories of the high street the ways in which it functions as a socially connective space have been attached to normative notions of white, commercial space and serving of local communities rather than tourist or visitor markets (Hall 2011; Hubbard 2017). The temporal and material comparisons between past and present 'summon forth memories of mundane and familiar acts of shopping and socialising, juxtaposing these with a present in which such rituals appear to have been destroyed and lost forever' (Hubbard 2017, 15). Attempts to revitalise the high street through pop-ups including micro-pubs have privileged class and gendered norms in economic restructuring and promoted questionable notions of shared cultural values (Ibid., 18). While these kinds of narrative are relevant to tracing perceived threat and opportunity for our high streets, they perhaps fail to adequately consider issues of social difference, especially faith, ethnicity and gender, as modalities of public experience. Alternatively articulated, *certain bodies* and *ways of seeing* the high street are often privileged over others. In addressing mainstream commercial space, it is assumed that the shoppers are also themselves mainstream, conventional, middle-of-road. Querying this, modest fashion and its consumers point instead towards how the high streets of city and town centres are negotiated and performed by discerning and often outspoken minority ethno-religious actors. Actively using tactics and strategies calls into play the work of De Certeau (1984) in his exploration of the use of derivés as a performed walk to disrupt the march of capital and strictures of urban planning. The ways in which the high street is selectively engaged, mixed up, resisted or championed is indicative of a much livelier set of dynamics and exchanges than we might ascertain from dominant representations of the urban commercial landscape in decline. Whilst moving away from any straightforward notions of everyday multiculture and conviviality, in a post-Brexit and 9/11 geopolitical context, on the high street the 'characteristics of different places driven by their histories, cultures and inhabitants act as a tactic of resistance against the production of a global generic' (Warren and Jones 2015, 10). For the young Muslim women in this study the material world offers a means through which to renegotiate aspects of social and cultural distinction and socio-spatial belonging.

In the sections that follow I trace the ways in which women are selectively engaging with the high street in order to create new fashionable Muslim subjectivities and relatable images of social difference. In walking this line, the young women in the study are seeking to foster a style that appeals to Muslims and non-Muslims alike. The empirics are taken from a major AHRC project on the roles of Muslim women in the British Cultural and Creative Economy. Over 45 semi-structured interviews were conducted and 6 focus groups (3 × Mixed; 3 × Muslim female; 42 participants in total) were held. I first explore how women make strategic plans in targeting particular shops for select items that disrupt normal fashion cycles in order to create distinctive looks. I then show how some women are utilising recombinations of high street and modest wear to build careers as hijabi bloggers and influencers. Next I discuss the symbolic importance of these fashion lines and fashionistas for perceptions of social integration and belonging, before showing how the mixing and assemblage of mainstream and Islamic influences is moulded around time and place. By changing appearance according to context, the practice of self-styling becomes entwined with everyday negotiations of being in place/out of place, along with gestures towards shared space.

Shopping

Over the past 10 years, instead of shopping at stores or online for clothing that specifically caters to Islamic needs, more young women are turning to the high street stores that their non-Muslim peers shop at (Lewis 2013). Across the study young Muslim women discussed their consumption at affordable stores such as Primark and H&M. Here they could selectively purchase clothes typically associated with 'disposable' fashion and reappropriate them at low cost for varied religious, professional and social purposes. Rather than shopping for trends – although this was discussed – the ways in which consumers attempted to create some space for distinction from their Muslim peers who looked for similar items in the same stores was striking. This was made particularly vivid in stories of buying headscarves. Headscarves are pinned for hijab-wearing, which means they develop holes and need to be replaced regularly. Oasis sells longer scarves that are ideal, but the store is more expensive than many other high street retailers. Primark was seen as a good option due to the low price of items, although its popularity amongst young Muslims means tactics are employed to avoid 'four people wearing the same one' (Haiza, November 2017). Haiza purchases her scarves from Primark but stores them away for a few months before introducing them to her look. Otherwise 'people are watching' and will copy her, as 'they know they cost £4 each' (Ibid.).

Instead of fast fashion that is trend-led, stores like Primark and H&M offer key wardrobe staples or basics that can be combined in ways that create a covered look. Maxi dresses and skinny jeans were bought in advance – 'they aren't going anywhere in terms of fashion' (Haiza, November 2017) – and then a temporal delay is placed on wearing them. Rather than a rush to intervene in fashioning place through high street shopping, the fashion rhythms are here slowed down. Saving shoppers from going abroad or to Jewish shops for modest clothes (deemed less stylish), these items could now be carefully pulled from the high street with more options for the buying of youthful, fashionable items than a decade previously. High street retailers and brands were not always effective in judging Muslim consumers, however. H&M's modest line launched in time for Ramadan and Eid 2018 was seen as 'almost cultural appropriation' (Clemantine, July 2018) for selling poor translations of traditional Asian styles such as the tunic. Instead, respondents wanted mainstream fashion styles tailored for greater coverage: 'Still give us the ruffled sleeves, we love that. We're into fashion, I want to see that style on me as well. But just a bit longer' (Clemantine, July 2018). No concern was articulated over whether pieces were purpose-designed hijabs or everyday scarves. The scarf took on sacred significance only through the practice of wearing; 'it is just a normal scarf tied differently' (Haiza, November 2017). Yet the accessibility and affordability of certain high street stores was seen as transformational in enabling the flexibility to easily combine religious, professional and youthful identities that was beneficial for Muslim women across the study.

Influencing

Successful careers have been forged through the visible roles of Muslim women in relation to high street brands and shopping such as creating new fashion influencers and bloggers. A crop of women with Dina Torkia as the figurehead have become well known with large social media followings through buying clothes and layering them in stylish, modest ways and, more recently, by buying modest fashion lines sold by high street and luxury brands (Warren 2018). The recent growth in specialist modest fashion lines serves as an exemplar of how fashion production is responsive to consumption, as influenced by digital feedback loops and media-savvy 'users', especially around how diversity sells in the marketplace (see Pham 2015). Safia observes, 'There's lots of Muslim bloggers on YouTube and Instagram that have masses of following and they are in the beauty industry. A lady that I follow on Instagram [Amena Khan] who wears the hijab was picked up by L'Oréal' (Focus Group, Muslim Heritage Centre, May 2018). As elsewhere in beauty and fashion, a large social media following can mean brands enrol actors or 'influencers' in order to reach their audiences. While models such as Khan 'get some criticism from the Muslim sector about how they're not representing Islam', Safia sees them as professional role models. Or, as another participant put it, 'Actually they're representing a really positive face of Islam and women progressing in the fashion industry and so-forth and within media' (Rohina, Focus Group 1, Oldham Library, June 2018). Similarly, Aaliyah noted how contemporary fashion appears more inclusive of Muslim women: 'I suppose the positive is that some of the big brands, they're using Muslim women as models as well. It's opening up, isn't it?' (Focus Group, Muslim Heritage Centre, May 2018).

Muslim women are engaging in the public life of Islam and religious discourse through everyday materialism in practicing shopping tactics and these strategies of fashion influencing. On the one hand, this has meant a widening of female voices in digital media discussing Islam and fashion and, on the other hand, it has resulted in more choice on the high street for modest fashion as retailers, magazines and vloggers together grow the Islamic marketplace. Beyond the rise of so-called hijabi YouTubers and vloggers attracting younger generations into consuming fashion and fashion design and media as careers, their prominence in the digital public sphere and influence on street fashion appears to be supporting young Muslims to express their changing identities. As Sabina noted, 'I think a lot of people are realising that they can express themselves through what they wear and it doesn't have to be boring and dull and just black, you can express yourself as a Muslim through different ways' (November 2017). This self-expression enabled by social media such as Instagram image-sharing works as testimony to the diversity of Islam where 'you can express the way you practice your religion in different ways' (Ibid.). In turn, it also serves to re-orientate narrow non-Muslim perceptions of Islamic dress in particular veiling fashions: 'There's this stigma attached to the scarf and the hijab and that stigma is slowly being broken apart' (Ibid.). Again, the entrepreneurial self-styling and branding of hijabi bloggers in particular was seen to be enabled by the fact that recombining and mixing high street looks was accessible and affordable to many. Accordingly the roles of Muslim fashion influencers and models are seen as educational, even where aims are more individualist in orientation: 'It has supported hijabi bloggers. I don't really care or think it is important. It is just more money for them [bloggers] though it still makes my life easier' (Haiza, November 2017).

Integrating and mixing

Everyday practices of shopping as a form of recreation enable closer relationships between Muslims and their non-Muslim peers. Fashionable clothing and styling practices work as identity markers that are shared providing common ground for cross-ethnic and religious friendships. This was discussed in detail by a mother, Parvina, in relation to her daughter during a focus group:

> Sometimes when I was growing up it was like 'well, I can't wear that because it's not covered ... ' and then it's like 'oh, well you don't buy in shops like us'. And *it separates you in a way. Whereas when you can buy things from the shop with your friends, the same shop as your friends, you feel more integrated and more accepted into society*. You know, there's a bit of a fine line between wanting to be accepted and having your individuality. But I think growing up, especially very young children and in your teens, that's your main thing. You want to be with your mates, you want to have similar things to them or go to the same shops. Or when you want to go out shopping. It's being able to say ... 'Oh they're going to that shop, oh yeah I can buy some things from there as well, that's great'.
>
> *(Parvina, Focus Group, Muslim Heritage Centre, May 2018;* italics my emphasis*)*

Parvina views her daughter as experiencing a stronger sense of belonging and integration than she remembers as a teenager, attributing this in part to shared high street shopping practices. She observes that it forges stronger bonds between her daughter and non-Muslim friends; 'there's no divide' (Ibid.). The everyday, often mundane high street is symbolically and materially important as a place in how it convenes young people across social markers of difference. Shopping as an activity might seem superficial but works as a performative act that bonds friends and friendships, and perhaps works as a connective tissue more widely in society. As Parvina's daughter Maha said: 'you can share'. And her mum responded: 'you both can wear it [Muslim and non-Muslim]'.

How consumers dress has revealed for some time that any binaries drawn between Asian and mainstream dress, and modest and fashionable clothing, do not reflect lived processes of mixing, hybridity and difference (see Dwyer 1999; Jackson 2002). This may be more necessary such as finding affordable clothing that is covered and appropriate, a playful expression of taste and attitude, or a recombination. For Sonia says: 'I just mix it up, yeah. Just, if I'm going out I might wear Asian clothes because I feel like wearing them' (January 2017). She notes that tolerance towards 'mixing it up' or wearing visibly Muslim clothing has become more acceptable over time in Britain: 'I think that's been a lot easier now, in Manchester especially, to be able to wear what you want to wear' (Ibid.).

Playing with image flexibly in order to 'fit' is significant here. This depends on particular places that are visited and the multi-sensorial interpretation and atmosphere of place apprehended by the subject. For instance, on visiting a cultural quarter on the edge of Manchester City Centre, the Northern Quarter, Clemantine says:

> Maybe if I'm wearing jeans and a long top, I might feel a bit more comfortable walking through the Northern Quarter because it's cultural and obviously there's lot of pubs, and stuff, and if I walk in to certain places I feel a bit out of place, and they're looking at me like 'what are you doing here?' Sort of thing. Even though the Northern Quarter is great, it's quite a friendly vibe, but I just feel a bit like an outsider. [Laughs] Do you know what I mean?
>
> *(July 2018)*

Clemantine perceives cultural norms in the Northern Quarter and adapts her image accordingly. On the days she is likely to enter the Northern Quarter she might tailor her image to wear a turban rather than conventional hijab, or wear jeans rather than an abaya, in order to mitigate against feeling 'a bit like an outsider' (Clemantine, July 2018). Fashion is here used as a barometer for feeling in place/out of place. It rises to the surface and writes itself across the bodies, streets and public spaces of the Northern Quarter. Being visibly Muslim also typically points towards other socio-religious encounters with difference in an area that is densely packed with bars, pubs and nightlife venues, such as drinking alcohol.

More complex still is how the playful and experimental aspects of mixing and hybridity belies pressure to assimilate into place. Discussing her friend, Laila said how on the Manchester spoken word scene certain sub-cultural styles had evolved. Over time, and with exposure to the scene, her friend's visual appearance changed in accordance – she started to wear a bandana underneath her chunky scarf, with large hoop earrings. Her whole physical appearance changed in order to 'fit in'; there was a shift, 'something had changed' (Laila, January 2018). She reflected that Muslim creatives often felt the need to have multiple lives and identities in relationship to place – but wanted herself to find a way of synthesising her creative and religious aspects as 'I don't want to lose my identity, and my sense of identity' (Ibid.). Pushing against an idea of identity as entirely fluid, where the post-subject endlessly recreates themselves through new styles, Laila resisted the pressure to have 'different performances for different people' (Ibid.). Alternative hybrid identities are fostered that subvert dichotomous meanings of Western and Islamic (Dwyer 1999; Hall 1993; Lewis 2013); identities that are contextual and relational positionings but which negotiate a sense of subjectivity with moral integrity. It reveals, too, how context might impact decision-making processes for self-styling with ethical and aesthetic implications for ethno-religious actors staking a claim to public space, and social and economic inclusion.

Conclusion

In this chapter I explore the use of the high street and shopping practices in a local context of social pluralism to interrogate notions of being in place/out of place. Cresswell critically assessed how being in place/out of place references that which is geographical and a position on a social hierarchy around what is good and appropriate. Fashioning place represents how subjects selectively and creatively build markers that situate themselves in relation to mainstream and high street fashion practices whilst maintaining a distinctive image as visibly Muslim. In negotiating a look that is modest *and* fashionable young Muslim shoppers enrol dress 'as a marker in the contestation of different constructions of belonging through which group boundaries are (re)produced' (Dwyer 1999, 6). Returning to Lewis (2013, 20), 'These consumers can be considered as creative co-producers, who "make do" with mainstream items and recombine them, engaging in forms of bricolage to put together an "Islamic" wardrobe'.

In matters of fashion and identity formation the sub-cultural is often presented as an 'authentic' space next to the commodified mainstream. Connell and Gibson (2003, 15) discuss this in relation to music where 'racial, gendered and socio-economic filters ... can act to polarise sections of society, marginalise groups of people from mainstream economic and political power, and silence oppositional or "alternative" cultural voices'. Sub-cultural scenes can become valuable and captured by the mainstream, however, as is evident in the ways that diversity and modest fashion are marketed by high street and luxury brands. New circuits of production and consumption mean Muslim bloggers and influencers are reworking high street looks that are relatively cheap to buy in order to launch careers, while brands have partnered with these fashionistas to reach their followers. Integrating aspects of ethnic and religious difference into mainstream fashion and street style, while market-driven, is impacting upon the dynamics of place and the very materialities of our high streets. Across the study, minority women discussed how it was making life easier for them. In various ways shopping affordable fashions, influencing, integrating and mixing though high street consumption enabled a stronger articulation of belonging across multiple, different groups at once.

This moves us away from place as marked by binaries of mainstream or marginalised, Western or traditional, modest or fashionable, to a relational space constituted by fluidity, heterogeneity and difference. Still, even as these flows are evident, recognition is given to how power structures space, influencing minority subjects and applying pressure to conform and assimilate in certain places. More broadly, endemic systems of exploitation in production processes that are crucial to understanding the global political economy of the fashion world do not undermine the desire to participate in consumption processes or the forging of new young Muslim identities in tactics around shopping and self-styling, and the remaking of places by fashionable and faithful subjects.

References

Baker, C., & Beaumont, J. (2011). Postcolonialism and religion: new spaces of 'belonging and becoming' in the postsecular city. In Beaumont, J. & C. Baker (eds), *Postsecular cities: Space, theory and practice*. London and New York: Continuum, pp. 33–39.

Bhachu, P. (2005). *Dangerous designs: Asian women fashion the diaspora economies*. Abingdon: Routledge.

Connell, J., & Gibson, C. (2003). *Sound tracks: Popular music identity and place*. Abingdon: Routledge.

Cresswell, T. (2014a). Place. In Cloke, P., Crang, P., & Goodwin, M. *Introducing human geographies*. Abingdon: Routledge. pp. 249–261.

Cresswell, T. (2014b). *Place: an introduction*. Chichester: John Wiley & Sons.

De Certeau, M. (1984). Walking in the city. In *The practice of everyday life* (Steven Rendell translator). Berkeley, CA: University of California Press, pp. 91–110.

Dwyer, C. (1999). Veiled meanings: Young British Muslim women and the negotiation of differences. *Gender, Place and Culture*, 6(1), pp. 5–26.

Dwyer, C. (2016). Why does religion matter for cultural geographers? *Social & Cultural Geography*, 17(6), pp.758–762.

Hall, S. (1993). Culture, community, nation. *Cultural Studies*, 7(3), pp. 349–363.

Hall, S. M. (2011). High street adaptations: Ethnicity, independent retail practices, and Localism in London's urban margins. *Environment and Planning A*, 43, pp. 2571–2588.

Hubbard, P. (2017). Enthusiasm, craft and authenticity on the High Street: micropubs as 'community fixers'. *Social & Cultural Geography*, 20(6), pp. 1–22.

Jackson, P. (2002). Commercial cultures: transcending the cultural and the economic. *Progress in Human Geography*, 26(1), pp. 3–18.
Karlsen, S., & Nazroo, J. Y. (2015). Ethnic and religious differences in the attitudes of people towards being 'British. *The Sociological Review*, 63(4), pp. 759–781.
Kong, L. (2001). Mapping 'new' geographies of religion: politics and poetics in modernity. *Progress in Human Geography*, 25(2), pp. 211–233.
Kong, L. (2010). Global shifts, theoretical shifts: changing geographies of religion. *Progress in Human Geography*, 34(6), 755–776.
Lewis, R. (Ed.). (2013). *Modest fashion: Styling bodies, mediating faith*. London: IB Tauris.
Lewis, R. (2015). *Muslim fashion: Contemporary style cultures*. Durham, NC: Duke University Press.
McGuire, M. (2008). *Lived religion*. Oxford: Oxford University Press.
Noor, O. (2018). *The Great British Ramadan* Available at: https://ogilvy.co.uk/sites/ogilvy-prelive/files/The%20Great%20British%20Ramadan%20%28Ogilvy%20Noor%29%20Summary.pdf
Oosterbaan, M. (2014) (Ed.) Special Issue: Public religion and urban space in Europe. *Social & Cultural Geography*, 15(6), pp. 591–682.
Pham, M-H. (2015). *Asians wear clothes on the internet: Race, gender, and the work of personal style blogging*. Durham, NC: Duke University Press.
Reuters, T. (2017). State of the Global Islamic Economy Report Available at: https://ceif.iba.edu.pk/pdf/ThomsonReuters-stateoftheGlobalIslamicEconomyReport201617.pdf
Runnymede (2017). *Islamophobia: Still a challenge for us all*. Runnymede Trust. [Online], available at: https://www.runnymedetrust.org/uploads/Islamophobia%20Report%202018%20FINAL.pdf.
Secor, A. (2002). The veil and urban space in Istanbul: women's dress, mobility and Islamic knowledge. *Gender, Place and Culture*, 9(1), pp. 5–22.
Simonsen, K. (2008). Practice, narrative and the multi-cultural city: A case of Copenhagen. *European Urban and Regional Studies*, 15, pp. 145–158.
Tarlo, E. (2010). *Visibly Muslim: Fashion, politics, faith*. London: Berg.
Warren, S. (2017). Pluralising the walking interview: researching (im)mobilities with Muslim women. *Social & Cultural Geography*, 18(6), pp. 786–807.
Warren, S. (2018). Placing faith in creative labour: Muslim women and digital media work in Britain. *Geoforum*, 97, pp. 1–9.
Warren, S. (2019). # Your Average Muslim: Ruptural geopolitics of British Muslim women's media and fashion. *Political Geography*, 69, pp. 118–127.
Warren, S., & Jones, P. (2015). *Creative economies, creative communities: Rethinking place, policy and practice*. London: Ashgate.

27
Disability and place

Ilan Wiesel and Ellen van Holstein

Introduction

The cumulative insights from geographical studies of disabilities have generated a small but rich and dynamic body of literature on disability and place (Butler and Parr, 1999; Chouinard, Hall, and Wilton, 2016; Hall and Wilton, 2017; de Vet, Waitt, and Gorman-Murray, 2012; Gleeson, 1999; Golledge, 1993; Hall, 2010, 2018; Holt, Bowlby, and Lea, 2017; Imrie, 2001; Kitchin, 1998; McClimens, Partridge, and Sexton, 2014; Metzel, 2005; Phillips and Evans, 2018; Wiesel and Bigby, 2015, 2016). Abilities such as being able to hear, to find one's way, to understand social cues or make oneself understood, and to follow social norms, are determined to some extent by the capacities and limits of human bodies and minds in all their diversity. But abilities and disabilities are also socially constructed and can take different form from one place to another.

A generalised discussion about people with disability inevitably risks obscuring the vast diversity within this population, with significant differences across the range of physical, cognitive, sensory and psychosocial abilities. Diversity among people with disability emerges also from differences in class, gender, age, ethnicity, as well as other identities. While some people might feel empowered by self-identifying as a person with disability, others do not wish to identify as such (Pionke, 2017). Nevertheless, the notion of disability, broadly conceived, is a necessary and useful lens for thinking about place.

Often driven by a social justice agenda, studies of disability and place have sought to understand where and how place has failed people with disability and deepened their oppression, and how place might be employed in their struggles for rights, equality and a better life. These studies revolve around questions of inclusion and exclusion, and enablement and disablement: what kinds of places can be accessed by people with different abilities? What abilities do people gain or lose in different places? And what is 'meaningful participation' and how might it be achieved in different places?

In this chapter we explore some of the key epistemological and ontological debates surrounding the relationship between disability and place. From an epistemological perspective, we consider tensions between structural and relational approaches to thinking about the production of place, its effects on people with disability, and their agency within such processes. From an ontological perspective, we consider the systems of categorisation that underpin scholarly and other discourses on disability and place. Specifically, we focus on the binary categories of 'mainstream' and 'specialist' places, and their central role in framing debates about the oppression of people with disability, deinstitutionalisation and social inclusion and exclusion.

Agency and structure

In recent years, debates about disability and place have been framed through two primary lenses that have conflicted at times. The first approach has focussed on place as the territorial manifestation of enduring societal structures of oppression (Gleeson, 1999; Imrie, 1996; Kitchin, 1998). The second approach has adopted a relational framework to focus on the more fluid possibilities that arise in encounters and relations between diverse people and non-human things, as they emerge in place (Chouinard, Hall, and Wilton, 2016).

A structural framework approaches disability as the outcome of a social structure, often referred to as ableism, which underpins the oppression of people with different kinds of impairments. Ableism serves and privileges the able-bodied person, who can be defined as someone who is 'cognitively, socially and emotionally able and competent; biologically and psychologically stable ... hearing, mobile, seeing, walking; sane, autonomous, self-sufficient ... economically viable' (Goodley, 2011: 79). Those who are judged as falling outside of these criteria are excluded, deprived, 'othered' and devalued by the mechanisms of ableism.

One of the key tenets of ableism is the widespread conception of disability as a personal tragedy (Oliver and Barnes, 2012: 11), a phenomenon that results from the impairments of an individual. Often described by critics as a 'medical model of disability', disability is understood as a personal deficiency that should ideally be 'cured' through medical intervention, so that the affected individual can function in society. In resistance to this individualised 'medical model', since the 1970s disability rights activists and scholars have proposed an alternative 'social model of disability' (Barnes, 1990; Finkelstein, 1980; Morris, 1991; Oliver, 1983, 1990; UPIAS, 1976). In its varied versions, the social model views disability as a form of collective experience of oppression. Rather than medical intervention to 'fix' individuals' impairments, the social model calls for collective social activism by people with disability and their allies, with the aim of transforming society itself and eliminating its ableism.

As a social structure, ableism is understood as pervasive and enduring; it is deeply embedded in all facets of a society's political and economic systems, social norms, culture, built environments and language. Ableism intersects with other social structures, such as capitalism (Gleeson, 1999), postcolonialism (Soldatic, 2015) and the patriarchy (Valentine, 2007). For instance, women with disabilities have lower labour participation rates than men with disabilities and are more likely to experience abuse, as an intersectional effect of ableism and the patriarchy (O'Hara, 2004).

Likewise, the shift from agrarian co-operative systems of production to urban capitalist systems organised around individualised waged labour excluded people with disability from emerging labour markets, and led to the disintegration of communal support systems. On the one hand, people with disability began to be regarded as unproductive and a social 'problem'. On the other hand, neoliberal politics and economics have shaped a new disability industry which approaches people with disability foremost as consumers, positioning disability itself as a commodity. This shift has been accompanied by the exploitation of disability support workers who are increasingly employed in precarious jobs (Gleeson, 1999; Oliver and Barnes, 2012, p. 55; Sothern, 2007). These processes reflect a distinct intersection of ableism and capitalism (Fritsch, 2015).

The structural understanding of ableism, and specifically the social model of disability, has inspired collective action by people with disabilities. Indeed, the social model is widely considered a key factor in the mobilisation of disability activism during the 1980s and 1990s. The focus on a disabling environment rather than impairments allowed people with diverse disabilities to organise their protests together with a shared purpose. These mobilisations led to notable achievements, including antidiscrimination legislation and a more recent shift to direct disability support payments in several countries. As claimed by scholar and activist Morris (2000: 1–3):

> The social model of disability gives us the words to describe our inequality. ... Because the social model separates out disabling barriers and impairments, it enables us to focus on exactly what it is which denies us our human and civil rights and what action needs to be taken.

At the same time, a structural view of ableism underscores a somewhat pessimistic stance that change can only be achieved through all-encompassing societal restructure. Such structural thinking underestimates the transformative potential of everyday practices through which people with disabilities challenge, resist and subvert oppressive and exclusionary attitudes and practices.

Disability geographers working within a relational framework understand place not as the manifestation of ableism as a social structure, but as constituted through 'seemingly mundane, habitual, non-reflexive practices' (Chouinard, Hall, and Wilton, 2016: 729). People's abilities, understood as capacity for action, emerge when they enter into relations with other people and non-human things. Therefore, rather than structurally predetermined ability, these geographers discuss a field of potential that emerges when a person interacts with other people and non-human things in place, and how they might enhance or constrain one another's ability to act.

Against structuralist pessimism, the open-endedness of relational thinking is 'an empowering perspective' (Jones, 2009: 492) because it recognises the potential for change and resistance that is embedded in everyday interactions and practices. However, it is problematic to assume that the fluidity of everyday practices is always empowering or that social and environmental structures are inherently oppressive. Furthermore, the evident persistence of certain forms of oppression of people with disability highlights the need to continue to recognise the systemic and enduring 'forces that restrict, constrain, contain, and connect the mobility of relational things' (Jones, 2009: 492).

Indeed, Hall and Wilton (2017: 729) acknowledge the 'enduring orders' that shape the possibilities of everyday interactions and relationships, but emphasise their embodied and performative nature. Ableism, from this perspective, is not an abstract, all-encompassing 'structure'; rather, it is the sum of many small encounters: a thoughtless gesture, a sensation of being out-of-place, a resurfacing memory of a past experience of humiliation.

'Shut in' specialist places

One of the primary 'enduring orders' that has shaped the experiences of people with disability has been their segregation from the so-called 'mainstream' places of society, and more recent movements towards reintegration. Consequently, the binary categories of 'mainstream' and 'specialist' places have come to play a central role in framing debates about the oppression of people with disability.

The mainstream place has often been conceived as a container of social goods from which people with disability are 'shut out' by a range of barriers; or, alternatively, as a dangerous wilderness in which they are most vulnerable. The specialist place has often been conceived as an impoverished, dehumanising asylum (literally or metaphorically) in which people with disability are 'shut in'; or, alternatively, as an oasis which shelters them from the dangers of an unaccommodating mainstream.

The legacy of the asylum is central to understanding the binary categories of mainstream and specialist. Since the 17th century, a large number of people with primarily intellectual and psychosocial disabilities have been incarcerated in institutions and asylums of various kinds. Inhumane living conditions, including overcrowding, disease, neglect, abuse, regimentation of daily activity and restriction of individual freedoms, were documented in numerous official inquiry reports, media campaigns and academic studies internationally (see Mencap, 2007).

The institutionalisation of people with psychosocial disability was central to the 17th-century 'Great Confinement', which, in Paris for instance, saw the incarceration of nearly one per cent of the city's population (Foucault, 1988). Nevertheless, the majority of people with disability have always lived 'in the community' – in ordinary residences, on their own or with relatives – rather than in institutions (Oliver and Barnes, 2012: 67). Thus, some analyses of the institution have focussed on its symbolic significance as spectacle, rather than its role as an actual place of residence. As conspicuous 'monuments' of modernity, institutions and asylums were seen as central to the project of building an enlightened civilisation (Foucault, 1988). These institutions were designed to discipline not only their inmates but also the general public at large, by showcasing the possibility of forced removal from the community to anyone who refused or was unable to conform with the principles of rationality and productivity (Foucault, 1988; Oliver and Barnes, 2012).

In the 1970s, nearly four centuries after the Great Confinement, a process of deinstitutionalisation began in various countries around the world, involving the closure or downsizing of institutions and the provision of support for people with disability to relocate to smaller and more dispersed housing in urban neighbourhoods. Deinstitutionalisation was driven by emerging welfare ideologies of the late 20th century, by medical advancements and by economic considerations. While fewer people now live in institutions, deinstitutionalisation remains incomplete and many institutions for people with primarily intellectual disability are still in operation around the globe, Sweden and Norway being notable exceptions (Wiesel and Bigby, 2015). In Australia, for example, over a period of four decades, the movement of deinstitutionalisation has progressed slowly and haphazardly. Most of the country's institutions have been closed and replaced with primarily smaller group homes. However, effective opposition by some relatives of residents, coupled with insufficient resourcing, have prevented the closure of several institutions that remain in operation. Furthermore, many people with intellectual or psychosocial disability experienced a process of 'trans-institutionalisation', moving out of (or avoiding admission into) state-run institutions only to enter other congregate settings such as shelters, prisons, nursing homes or psychiatric hospitals, where institutional practices and cultures thrive (Wiesel and Bigby, 2015).

Institutions in their varied forms are typically understood as sharing the attributes of the 'total institution' (Goffman, 1961), including the conduct of all spheres of life in a single place, a regimented daily routine carried out in groups of people treated alike and a strict hierarchy of 'inmates' and staff. In a setting that is separated physically and symbolically from the outside 'mainstream' world, these attributes produce a distinct sense of place which characterises many institutions. Although sheltered from capitalist competition and productivity, institutional time is structured as a strictly enforced routine. It is both a home for residents and a workplace for staff. It is defined by the hierarchical social relations between staff and residents but also by transgressions of the boundary with unexpected moments of conviviality and solidarity. It is a place of fear generated by both concealed and open abuse within the institution and, at the same time, for many, a safe haven or oasis (Philo, Parr, and Burns, 2005) of care and refuge from the outside world (Imrie, 2001).

The extent to which the institutional sense of place defines other specialist places is debated. In many countries, deinstitutionalisation saw the proliferation of small group homes, typically accommodating up to five residents with primarily intellectual disabilities, dispersed in residential neighbourhoods. Some critics (for instance, Sinson, 1993) have argued that, while smaller and more dispersed, group homes ultimately function as 'mini-institutions' with similar structures, routines and cultures. For example, residents' activities and routines becomes structured in relation to support workers' rosters. This makes it challenging for people to go out at night and on weekends, to make decisions spontaneously and thus to engage fully in community life (Milner and Kelly, 2009). It is argued that people continue to be compelled to participate in group activities, such as group outings in the community, which limit opportunities for encounters and conviviality with neighbours and passers-by. Others, such as Bigby et al. (2012), argued that while some group homes might still enforce a strict routine, resemblance between institutions and group homes is merely superficial because in group homes staff do not display the harsh treatment of residents that was common in institutions. Such debates illustrate how the legitimacy of specialist housing for people with intellectual disability is often framed within discourses contrasting an institutional sense of place with the notion of home.

Beyond accommodation, specialist services for people with disability are provided in various other domains, including special education, specialist (or 'sheltered') employment, specialist day centres, specialist sports clubs, leagues and events, specialist art and recreation groups and others. While no longer confined to a 'total institution' where all spheres of life are conducted and regulated, many people's lives still occur in primarily specialist places and services, even if these are more dispersed. Such 'institutionalised lives' (Hollomotz and Roulstone, 2014) can be critically understood as a simply more fragmented version of the asylum (Philo and Parr, 2000). Moreover, these specialist spaces are argued to reproduce various stigmas held in relation to people with disability (Kitchin, 1998). In writing about a specialist employment service for people with disability in Stockholm, Holmqvist et al. (2012: 207) argue that the practices and structures of the service actively encourage participants to internalise a 'passive identity – the identity of the occupationally disabled'. Beyond specialist places, the 'institutionalised lives' of people with disability expand to mainstream settings, when people with disabilities' access to, and experience of, such places are shaped by the practices and institutional constraints of support services (Wiesel and Bigby, 2016).

'Shut out' of mainstream places

Discussion about disability and place is often framed by a normative understanding that values the movement of people from specialist to mainstream spaces. Yet, as critics have pointed out, people with disability continue to be excluded from the social goods that are afforded to others in mainstream spaces, such as opportunities for employment and friendship, freedom from harassment and freedom to take risks. As expressed in the Australian Government report (2009: vi) that preceded its 2009 National Disability Plan:

> For many years people with disabilities found themselves shut in—hidden away in large institutions. Now many people with disabilities find themselves shut out—shut out of buildings, homes, schools, businesses, sports and community groups. They find themselves shut out of our way of life.

The concept of exclusion has gained political and scholarly currency since the late 1990s, shaped to a large extent by the policy rhetoric of the New Labour Government in the UK (Hall, 2010). The social inclusion agenda has focussed on removal of the material, symbolic, attitudinal and institutional barriers which prevent people with disability – alongside other marginalised groups – from accessing mainstream places and the social goods embedded in them. Yet, more critical interpretations have questioned the normative framing of access to the mainstream as a positive outcome in itself, shifting attention to the conditions under which access or inclusion are granted to people with disability and other marginalised people (Hall, 2010). From this perspective, even adjustments to increase mainstream participation of people with disability can be disabling if they fail to 'restore disabled people's self-esteem, dignity and independence' (Reeve, 2014: 99). Therefore, there is an increasing interest in the emotional nuances of people with disabilities' experiences in mainstream places (Imrie and Edwards, 2007).

A great deal of activism and literature has focussed on the exclusion of people with disability from built environments which are designed and managed with able-bodied people in mind, specifically 'a fit and able masculine body, the body as a machine, mechanical, fixed, taut, upright and pre-given to interaction' (Imrie, 2001: 234). Bodies failing to meet these standards – such as fat (Colls and Evans, 2014), ageing (Lager, Van Hoven, and Huigen, 2016) or disabled bodies (Imrie, 2001; Kitchin, 1998) – are ignored or forgotten. When the disabled body is accounted for in the design of built environments, it is typically imagined as a wheelchair user, discounting the vast diversity of impairments and disability experiences (Bredewold, Hermus, and Trappenburg, 2018; Reeve, 2014). Furthermore, some people feel they need to make their disabled identity known – for example by 'limping well' (Reeve, 2014: 111) – in order to use access features without harassment or suspicion.

Beyond physical barriers, a range of symbolic, attitudinal and institutional barriers underpin the exclusion of people with disability. When people with disability are unable to comply with the formal and informal norms of mainstream places, they are marked as being out of place, mocked as figures of fun (Imrie, 2001: 233) or frowned upon (Wiesel and Bigby, 2016). Such ableist norms are often communicated through the design and aesthetics of mainstream places, which mirror the relentless pursuit of 'a bombastic perfection of the body' (Siebers, 2005: 543) and the othering of people with disability and their diverse 'disability aesthetics' (Kitchin, 1998). People with disability often self-exclude from those places in which they are made to feel out of place (Burke, 2009).

When mainstream norms are adjusted through differential treatment of people with disability, even when the intent is to include, the effects are often oppressive and stigmatising (Holt, Lea, and Bowlby, 2012). For example, writing about a professional theatre company in the UK, Band et al. (2011) observed the frustration experienced by some actors with physical or sensory disabilities when they were demanded less rigorous discipline and performance standards than their peers. And in a mainstream college in Sweden, Zimmerman Nilsson (2015) found that teachers singled out students without disability as talented and addressed students with intellectual disability as grateful learners, to encourage able-bodied students to accept those with disabilities. Thus, mainstream places can prescribe different roles for people with and without disability resulting in experiences of exclusion within those places.While some forms of exclusion are subtle, others are crude. Not in My Backyard (NIMBY) opposition to community care facilities for people with disability represents a proactive erection of barriers to exclude people with disability from the 'mainstream' (Dear and Wolch, 2014; Gleeson, 1999). More recent literature has focussed on the impact of disability hate crimes – including both singular extreme events of violence and 'everyday low-level' harassment – on the social exclusion of people with disability, specifically intellectual disability (Hall, 2018). As noted by Chakraborti, Garland, and Hardy (2014) such experiences of being bullied and harassed are a common aspect of many people with disability's experience of mainstream places.

Some reflections

The notion of 'place' – and the specificity of investigating exclusion from, and ableism within, any particular place – has offered disability studies a powerful lens through which to approach some of its most pressing concerns. Since the 1990s, interest in disability and place aligned well with an emerging social model of disability that explained disability as a collective experience of oppression, as opposed to a personal medical misfortune. A more recent relational turn in thinking about place (e.g., Jones, 2009; Massey, 2005) has paralleled and reinforced critiques of the rigidity of the social model of disability, including its tendency to overlook diversity among people with disability and their experiences.

At the same time, geographies of disability have enriched the theorisation of place with some of the most vivid illustrations of how certain people are empowered, and others oppressed, through the design, aesthetics, regulation and social relations of place. Analyses of people with disabilities' experiences of being 'shut in' specialist places – their imaginary framed to a large extent by the legacy of the asylum – and 'shut out' of mainstream places and the social goods they treasure, are significant contributions to wider debates about socio-spatial exclusion and inclusion of marginalised groups.

These contributions lead to an understanding of specialist and mainstream places not as opposites, but as different manifestations of similar processes. Both the mainstream and the specialist place are spatial manifestations of ableism and its intersection with other social structures, such as capitalism, colonialism and patriarchy. Yet, as relational thinkers assert, notwithstanding more enduring constraints, there are also unexpected potentialities that can emerge in everyday encounters between diverse people as well as other non-human beings and things, which together constitute 'place'.

References

Australian Government. (2009) *Shut Out, The Experience of People with Disabilities and their Families in Australia*, Canberra: Australian Government.

Band, S., Lindsay, G., Neelands, J., and Freakley, V. (2011) 'Disabled students in the performing arts - Are we setting them up to succeed?', *International Journal of Inclusive Education 15*(9): 891–908.

Barnes, C. (1990) *Cabbage Syndrome: The Social Construction of Dependence*, London: Falmer Press.

Bigby, C., Knox, M., Beadle-Brown, J., Clement, T. and Mansell, J. (2012) 'Uncovering dimensions of culture in underperforming group homes for people with severe intellectual disability', *Intellectual and Developmental Disabilities* 50(6): 452–467.
Bredewold, F., Hermus, M., and Trappenburg, M. (2018) '"Living in the community' the pros and cons: A systematic literature review of the impact of deinstitutionalisation on people with intellectual and psychiatric disabilities", *Journal of Social Work* 20(1): 83–116.
Burke, S. (2009) 'Perceptions of public library accessibility for people with disabilities', *Reference Librarian 50*(1): 43–54.
Butler, R. and Parr, H. (eds) (1999) *Mind and Body Spaces: Geographies of Illness, Impairment and Disability*, London: Routledge.
Chakraborti, N., Garland, J., and Hardy, S. (2014) *The Leicester hate crime project: Findings and conclusions*, Leicester: University of Leicester.
Chouinard, V., Hall, E. and Wilton, R. (2016) 'Introduction: towards enabling geographies', in V. Chouinard, E. Hall, and R. Wilton (eds) *Towards Enabling Geographies: 'Disabled' Bodies and Minds in Society and Space*, Milton Park: Routledge, pp. 1–22.
Colls, R. and Evans, B. (2014) 'Making space for fat bodies? A critical account of 'the obesogenic environment"', *Progress in Human Geography 38*(6): 733–753.
de Vet, E., Waitt, G. and Gorman-Murray, A. (2012) '"How dare you say that about my friend": Negotiating disability and identity within Australian high schools', *Australian Geographer* 43(4): 377–391.
Dear, M. and Wolch, J. (2014) *Landscapes of Despair: From Deinstitutionalization to Homelessness*, Princeton: Princeton University Press.
Finkelstein, V. (1980) *Attitudes and Disabled People: Issues for Discussion*, New York: World Rehabilitation Fund.
Foucault, M. (1988) *Madness and Civilization*, New York: Vintage Book.
Fritsch, K. (2015) 'Gradations of debility and capacity: biocapitalism and the neoliberalization of disability relations.', *Canadian Journal of Disability Studies* 4(2): 12–48.
Gleeson, B. (1999) *Geographies of Disability*, London: Routledge.
Goffman, E. (1961) *Asylum*, New York: Anchor Books.
Golledge, R. (1993) 'Geography and the disabled: a survey with special reference to vision impaired and blind populations', *Transactions of the Institute of British Geographers* 18(1): 63–85.
Goodley D. (2011) *Disability Studies: An Interdisciplinary Introduction*, London: Sage.
Hall, E. (2010) 'Spaces of social inclusion and belonging for people with intellectual disabilities', *Journal of Intellectual Disability Research* 54(1): 48–57.
Hall, E. (2018) 'A critical geography of disability hate crime', *Area* (Early View): 1–8.
Hall, E., and Wilton, R. (2017) 'Towards a relational geography of disability', *Progress in Human Geography* 41(6): 727–744.
Hollomotz, A., and Roulstone, A. (2014) 'Institutionalised lives and exclusion from spaces of intimacy for people with learning difficulties', in K. Soldatic, H. Morgan, and A. Roulstone (eds) *Disability, spaces and places of policy exclusion*, London: Routledge, pp. 147–162.
Holmqvist, M., Maravelias, C., and Skålén, P. (2013) 'Identity regulation in neo-liberal societies: Constructing the 'occupationally disabled' individual', *Organization 20*(2): 193–211.
Holt, L., Bowlby, S. and Lea, J. (2017) '"Everyone knows me ... . i sort of like move about': the friendships and encounters of young people with special educational needs in different school settings.", *Environment and Planning A* 49(6): 1361–1378.
Holt, L., Lea, J. and Bowlby, S. (2012) 'special units for young people on the autistic spectrum in mainstream schools: Sites of normalisation, abnormalisation, inclusion, and exclusion.', *Environment and Planning A* 44(9): 2191–2206.
Imrie, R. (1996) "Ableist geographies, disablist spaces: Towards a reconstruction of Golledge's 'Geography and the Disabled"', *Transactions of the Institute of British Geographers* 21(1): 397–403.
Imrie, R. (2001) 'Barriered and bounded places and the spatialities of disability', *Urban Studies 38*(2): 231–237.
Imrie, R., & Edwards, C. (2007) 'The geographies of disability: Reflections on the development of a sub-discipline', *Geography Compass 1*(3): 623–640.
Jones M. (2009) 'Phase space: Geography, relational thinking, and beyond', *Progress in Human Geography* 33(4): 487–506.

Kitchin, R. (1998) '"Out of place','knowing one's place': Space, power and the exclusion of disabled people", *Disability & Society* 13(3): 343–356.

Lager, D., Van Hoven, B. and Huigen, P. (2016) 'Rhythms, ageing and neighbourhoods', *Environment and Planning A 48*(8): 1565–1580.

Massey, D. (2005) *For Space*, London: Sage.

McClimens, A., Partridge, N. and Sexton, E. (2014) 'How do people with learning disability experience the city centre? A Sheffield case study', *Health and Place* 28: 14–21.

Mencap. (2007) *Death by Indifference: Following Up the Treat Me Right!*, London: Mencap.

Metzel, D. S. (2005) 'Places of social poverty and service dependency of people with intellectual disabilities: 1a case study in Baltimore, Maryland', *Health and Place* 11(2): 93–105.

Milner, P. and Kelly, B. (2009) 'Community participation and inclusion: People with disabilities defining their place', *Disability and Society 24*(1): 47–62.

Morris, J. (1991) *Pride Against Prejudice. Transforming Attitudes to Disability*, London: The Women's Press.

Morris, J. (2000) 'Summary of Presentations', *Reclaiming the Social Model of Disability*. Accessed 6 April 2019 https://disability-studies.leeds.ac.uk/wp-content/uploads/sites/40/library/GLAD-Social-Model-of-Disability-Conference-Report.pdf

O'Hara, B. (2004) 'Twice penalized: Employment discrimination against women with disabilities', *Journal of Disability Policy Studies* 15(1): 27–34.

Oliver, M. (1983) 'Introduction: setting the scene', in M. Oliver, B. Sapey, and P. Thomas (eds) *Social Work with Disabled People*, London: Palgrave, pp. 1–6.

Oliver, M. (1990) *The Politics of Disablement*, Basingstoke: Macmillan.

Oliver, M. and Barnes, C. (2012) *The New Politics of Disablement*, Basingstoke: Macmillan.

Phillips, R, and Evans, B. (2018) 'Friendship, curiosity and the city: dementia friends and memory walks in Liverpool.', *Urban Studies* 55(3): 639–654.

Philo, C. and Parr, H. (2000) 'Institutional geographies: introductory remarks', *Geoforum* 31(4): 513–521.

Philo C., Parr H. and Burns N. (2005) '"An oasis for us': 'in-between' spaces of training for people with mental health problems in the Scottish Highlands", *Geoforum* 36(6): 778–791.

Pionke, J. (2017) 'Toward holistic accessibility: narratives from functionally diverse patrons', *Reference & User Services Quarterly 57*(1): 48–56.

Reeve, D. (2014) 'Part of the problem or part of the solution? How far do 'reasonable adjustments' guarantee 'inclusive access for disabled customers'?', in K. Soldatic, H. Morgan, and A. Roulstone (eds) *Disability, Spaces and Places of Policy Exclusion*, London: Routledge, pp. 99–114.

Siebers, T. (2005) 'Disability aesthetics', *Pmla 120*(2): 542–546.

Sinson, J. (1993) *Group Homes and Community Integration of Developmentally Disabled People: MicroInstitutionalisation?* London: Jessica Kingsley.

Soldatic, K. (2015) 'Postcolonial reproductions: Disability, indigeneity and the formation of the white masculine settler state of Australia', *Social Identities 21*(1): 53–68.

Sothern, M. (2007) 'You could truly be yourself if you just weren't you: Sexuality, disabled body space, and the (neo)liberal politics of self-help', *Environment and Planning D 25*(1): 144–159.

UPIAS. (1976) *Fundamental Principles of Disability*, London: Union of the Physically Impaired Against Segregation.

Valentine, G. (2007) 'Theorizing and researching intersectionality: A challenge for feminist geography', *The Professional Geographer 59*(1): 10–21.

Wiesel, I. and Bigby, C. (2015) "Movement on shifting sands: Deinstitutionalisation and people with intellectual disability in Australia, 1974–2014", *Urban Policy and Research 33*(2): 178–194.

Wiesel, I. and Bigby, C. (2016) 'Mainstream, inclusionary, and convivial places: Locating encounters between people with and without intellectual disabilities', *Geographical Review 106*(2): 201–214.

Zimmerman Nilsson, M. H. (2015) 'Inclusion functioning as exclusion: New students entering the Academy of Music in Sweden', *Teachers and Teaching: Theory and Practice 21*(3): 277–288.

28

Reading Bangkok

The transforming and intermingling city

Cuttaleeya Jiraprasertkun

Introduction

With the 1980s' policy to turn Thailand into a NIC (a Newly Industrialized Country), Bangkok has accelerated towards modernization and urbanization. Over the past three decades, the city has continually expanded to the outskirts, following the emergence of the *thanon* (road) and the expansion of new land-based settlements. Rapid development has caused the need to reduce traffic congestion in the inner city resulting in the building of the Outer Ring Road, expressways, and mass transit systems in 1978, 1981 and 1999 respectively. Such massive constructions have advanced the degree of urban sprawl along these transportation corridors. Consequently, the suburbs of Bangkok in popular understandings currently extend beyond the municipal boundary into the adjacent provinces of Nonthaburi, Pathum Thani, Samut Prakarn, Nakorn Pathom and Samut Sakhon.

While the city keeps expanding, middle-class people have increasingly moved to the suburbs in pursuit of a better life. The ensuing construction of amenities and service industries, such as shops, restaurants, malls and factories, in the peri-urban fringe area has drawn large numbers who work and live nearby. As a result, numerous rural villages are gradually gentrified by the settlements of newcomers and recent activities.

Thus, the phenomenon of urban invasion (urbanization replacing indigenous communities) has changed local characteristics from the traditional houses surrounded by greenery to a mixture of newly constructed buildings with the remnants of old elements. Water-based communities are threatened by the recently built land-based settlements and numerous villages (*muban*) are struggling to survive among the increasing gated housing estates (*muban chad san*). In each locality, the mixture is highly complex and intricate due to the proportion and degrees of various elements contributing to the uniqueness of each particular place.

This chapter aims to explicate the on-going gentrifying process of Bangkok. The complexity of places is analyzed through the reading of various typologies, scenes, and notions found in this city. While Bangkok is portrayed as a place one could perceive and define, it is the various sites that construct the diversity of Bangkok and contribute to the overall essence of the place.

Reading Thai place through the reading of Thai culture

In much recent literature, Bangkok has often been described by foreigners as a 'mixed' and 'diverse' city (Bartlett, 1959: 67; Fournereau, 1998: 26; Mulder, 1996: 134). Yet, many visitors and scholars who have attempted to express these intermingling characteristics have mostly failed to portray the deeper qualities of the place. Here then, there is a need to consider how we can read this place in a Thai context and be able to illustrate the delicacy of each place.

The uniqueness of Thai culture has strongly framed the ways in which Thai people perceive, conceptualize, and mentally construct their world. It is proposed here that the reading of culture is the key to comprehending the reading of space and place. Hence, local knowledge of places is needed in order to yield deeper ways of understanding a particular place and its distinctive qualities. In order to conduct an experimental reading of places in present-day Bangkok, one therefore must seek specific knowledge of place in the Thai context. This chapter explores ideas and images of place conceived in Thai culture in both the past and the present.

Through such reading, one must acknowledge the quality of unboundedness which has been expressed in all sorts of Thai representations, such as physical characteristics of blurred and cantilevered elements along the edge of canals and streets, interdependency and flexibility in social behavior, and also notions about the inseparability of humans, nature, and supernature in the ways that Thai people frame their worldviews. These particular qualities are integrated in the ways in which this chapter seeks to understand the intermingling characteristics of the places of Bangkok.

Reading changes of place

The idea of 'reading and interpreting place' – how people perceive, shape, and value their environment – is the starting point for this chapter. Much literature indicates the importance of dynamic and changing qualities in conceptualizations of place and landscape (O'Hare, 1997: 21). Place connotes the idea of dynamic processes where physical and social constituents are continually produced and reproduced by the interactions of humans and their surroundings over a period of time (Gustafson, 2001). Through such constitutive processes, social identity and meanings are constructed and attached to existential space (Brindley, 2003). This conceptual understanding is consonant with the idea that place is changing with cultural and social transformations (Walmsley and Lewis, 1984). At the same time, meanings and connotations of place from people's perspectives are modified and reconceptualized alongside transformations of physical existence.

The above understandings stimulate the idea that we need to read place through its constitutive processes, which are continuous from the past through the present into the future (Eckbo, 1969). How then can we read the constituting process of community? The dimension of time or temporal effects is an important factor, highlighting how community formations, meanings, and affiliations are constructed through longitudinal and on-going processes (Rivlin, 1987), supplementing the reading of change through space with change through time (Tuan, 1977). This chapter displays the transformations of Bangkok over time (from the 1950s to the present day) via the observation of remnants of the past that coexist with present-day characteristics. Here, the gentrifying process is portrayed through the phenomenon of urban invasion by the middle and working classes over local settings and traditional ways of life.

Bangkok in nostalgia: before the 1950s

> The approach to Bangkok is equally novel and beautiful. The Meinam is skirted on the two sides with forest-trees, many of which are of a green so bright as to defy the powers of art to copy. Some are hung with magnificent and fragrant flowers; upon others are suspended a variety of tropical fruits ... A few huts of bamboo, with leaved roofs, are seen; and in the neighboring creeks, the small boats of the inhabitants are moored. Here and there is a floating house, with Chinese inscriptions on scarlet or other gay-coloured paper.
>
> *(Bowring, 1969: 392)*

The scenes and stories of Bangkok in 'the old days', with houses along waterways surrounded by orchards (*Ban suan rim khlong*), often portrayed in the literature as Bangkokian characteristics, appear to have been conceived around the idea of an archetypal 'Thai space' in the collective imagination. The repeated image of the 'old' Bangkok displayed in various media stimulates the perception that such traditional qualities have survived and still exist, at least in people's minds, in the present day.

It is nevertheless important to note that the ways in which the 'old' and the 'new' Bangkok are classified here differs from other literature. Here, environmental changes are taken as a major consideration; thus the 'new' image of Bangkok for present purposes only began in the 1950s (King Rama IX's period) when the phenomena of urban invasion and modernization most observably began. It is noted that this also coincides with socio-economic changes of Bangkok: the city's population has grown from approximately 1 million in 1950 to 9.5 million in 2018, and the economic structure has been increasingly developing towards industrialization (Ouyyanont, 2000).

The idea of a water-borne city with its prosperity in agriculture was commonly perceived by tourists and readers. The notion 'Venice of the East', originally ascribed by a German visitor (Mandelslohe) to describe Ayutthaya in 1537, was still an apt description of the city of Bangkok up to the 1950s, because of its extensive networks of interlacing rivers and canals (Eliot et al., 2001; Fournereau, 1998; Hinshiranan, 2000; Nawigamune, 2003; Plynoy, 2001a). Additionally, there are some interesting aspects of the city's names and their meanings, with many originally referring to river, canal, or sea, that support the notion of Bangkok as an aquatic city. For example, the term 'Bang' (from Bang-kok) originally meant a dead-ended canal which was dug to let the water flow into the field and illustrates a common geographical characteristic of this area. Moreover, the fertility of the land is commonly portrayed through a familiar phrase for Thai people – 'there is fish in the water and rice in the fields' (*nai nam mi pla nai na mi kao*) – which has been applied to Thai places ever since the Sukhothai period (1238–1350) (Basche, 1971; Plynoy, 2001a).

Nowadays, even when Bangkok's physical characteristics have been almost completely replaced by modernization, these archetypal notions of Bangkok in the 'old' days still thrives in the minds of travellers who frequently seek classic scenes of life along the river in a green and peaceful atmosphere. One wonders if such environments can survive in the bustling Bangkok of the 21st century.

Transforming Bangkok in the 21st century

> Klongs abounded. Bangkok was then known as the Venice of the East and the City of Angels. Someone recently noted it is now known as the Vice of the East and the City of (fallen) Angels
>
> *(Klausner, 2000, p. 42)*

Since the transitional period in the reign of King Rama V (1868–1910), Bangkok has continually evolved to come closer to the common understanding of what constitutes a modernized city. Inspired by the scenery of European countries, many projects were created in Bangkok during this period in order to achieve the goals of 'civilization' (*khwamchroen*) and 'betterment' (*khwamkaona*), referring to the principle of 'civilization', which is described as somewhat opposite to the concept of 'nature' (O'Connor, 1989). The subsequent major trigger for change was transportation development, with the construction of canals, roadways, trams, and railways, causing an immense expansion of urbanized areas, particularly to the north and the east of the city (Figure 28.1a–f).

The huge growth of urbanization, as well as major influences from modernization affected by global factors such as mass transportation and construction technology, have had major impacts on the environment and on the emergence of newly intermixed characteristics of places. The 'merging phenomena' of visual and social worlds, with their superimpositions and intermixtures of past and present scenes, old and new elements, village and city senses, and spiritual and practical engagements, have created 'an intermingled essence', which has been perceived as expressing distinctively Thai characteristics in modern Bangkok. The perception of Bangkok as a city of muddle, chaotic and messy, has become part of a stereotypical understanding for tourists or even the city's inhabitants.

This intermingling characteristic has arguably underlain the Thai way of negotiating and adapting to new things, elements, concepts, or attitudes in culture and society (Bunnag, 2003). Thais have been able to 'mix' and 'combine' local and new ideas, bringing betterment into their lives yet still retaining their cultural roots. This unique mixture is expressed in several cultural

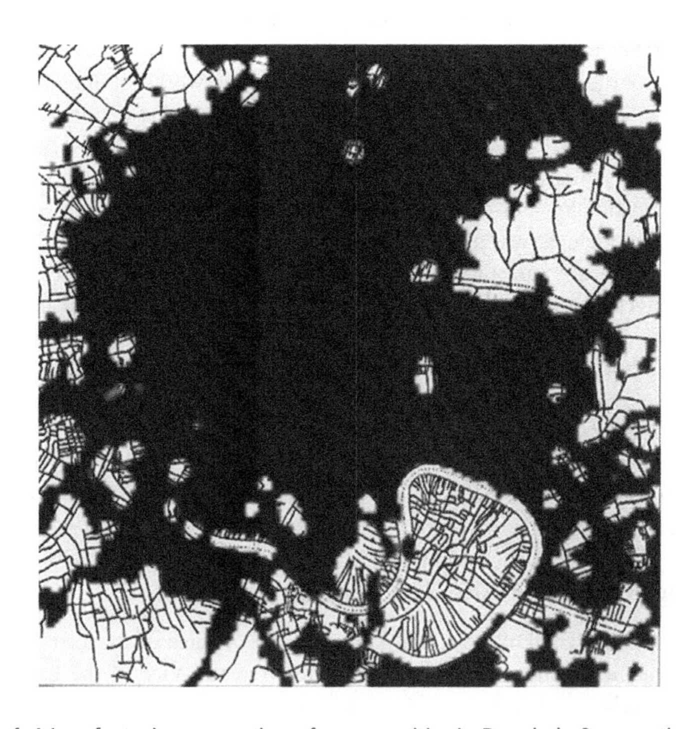

Figure 28.1a–f Maps featuring expansion of communities in Bangkok. Source: the author.

Figure 28.1a–f (Cont.)

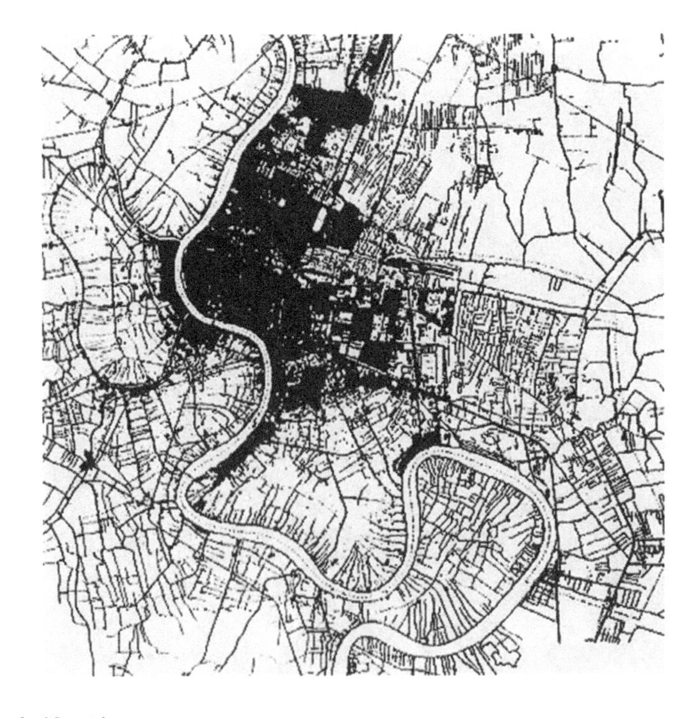

Figure 28.1a–f (Cont.)

Figure 28.1a–f (Cont.)

representations, including the palace architecture which combines the Thai and European architecture together, and the way Thai people utilize and integrate modern spaces such as the railway station, the shopping mall or the footpath with local lifestyle, by sitting on the floor and selling food. Bunnag described the flexibility of 'Thainess' with the idea of 'co-evolution' that integrates 'external forces' of globalization and modernity with 'internal forces' of local culture, tradition, environment, and context (Ibid: 11).

There is however a question of how these integrations have changed and been reflected in the patterns of settlements, human daily activities, and people's attitudes towards place and modern concepts of the nation state. The following sections discuss the merging processes and hybridizing characters of Bangkok's environment via the co-existence or reciprocity of water-based and land-based settlements, *chonnabot* (rural) and *muang* (city) lifestyles, and images of *muangkao* (old city) and *muangmai* (modern city).

Water-based and land-based cities

You mustn't judge Bangkok by New Road and Bangkapi or Rajadamneon Avenue. On the other side, the floating market as the tourist books call it, you'll see thousands of people living almost as the people of Bangkok and Ayuthia have lived for hundreds of years. There are still busy klongs this side of the river too, if you know where to look for them (Bartlett, 1959: 39).

Contemporary Bangkok is spatially constituted by the two parallel networks of *khlongs* (waterways) and *thanons* (roads). The impression of the water-based city, which was once prominent for its labyrinthine *khlongs* and its extensive waterscape, coexists with the more recently emerging image of the automobile and the road-oriented city (Jumsai, 1997; Yantrasast, 1995). Over the past seventy years, a huge system of roads of all sizes (*thanon-trok-(sog)soi* – a familiar phrase among Thai people) has rapidly been constructed and bridges have been superimposed on the existing layer of *khlongs*. Urban development which once concentrated on *khlong* excavations has increasingly shifted to road and recently rail constructions, while several canals have been filled in to provide space for wider roads to carry the larger numbers of vehicles (Askew, 1993; Basche, 1971). The economic boom in the late 1980s had impacts on the growth of businesses and, subsequently, on building construction and transportation developments. Roads have been used as a tool to open up the countryside and nature, thereby to spoil them according to some perceptions (Basche, op. cit.; Mulder, 1996). Consequently, numbers of roads and bridges have been constructed to link the inner area to the outer parts of Bangkok in all directions, followed by the expansion of urbanized areas linked with infrastructure, support facilities, and land-based settlements.

Since the 1950s, the emergence of roads has increasingly overshadowed the uses of *khlongs* and railways as major routes of transportation (Askew, 1993; Keyes, 1987). In earlier times when *thanons* were first constructed, roads and avenues were still a recently acquired luxury in Bangkok and not many vehicles could be seen (Fournereau, 1998). *Khlongs* on the other hand were commonly used and involved in many aspects of people's everyday lives. But this seems to have gradually reversed along with the modernizing processes of Bangkok over the past century. Roads have, in turn, become a vital element in people's lives. Like any other large city, traffic congestion, often described by both locals and visitors as a 'nightmare', became a serious problem in the 20th century (Figure 28.2b) (Askew, 2002; Basche, 1971; Bello et al., 1998).

The dilemma of *khlong* and *thanon*, implying natural and built environments (although, of course, both are constructed), portrays contrasting qualities that have become ingredients of a new 'merging phenomenon' and intermingling characteristics that Bangkok today seems to express. The riverine notion is argued to have survived not in the form of water per se but in the form of aqueous imagery deeply involved in traditional Thai ways of life (Jumsai, 1997). Jumsai suggests that the essence of water, represented through the symbolic forms of the 'naga' and 'Mount Meru', a serpent worship and cosmological model representing the water symbol in Hindu-Buddhist mythology, still exists in many forms of cultural elements in present-day Thai society. Besides, the remnants of traditional Thai houses built on stilts (Figure 28.2a), the boating system, rites and festivities relating to water, also remind us of the 'amphibious' characteristics: how Thai people were able to live 'with' water and its fluctuations as if this was the native element of their everyday lives (Bowring, 1969; Fournereau, 1998; Jumsai, op. cit.). As people were formerly flexible and adaptable in their ways of living, the coming of water through flooding was not perceived as the coming of disaster, but rather as auspicious.

Physical and behavioural changes have had significant impacts on people's perceptions of the environment. The idea of constructing roads was initially accepted as a 'new' technology, bringing only advantages to people's lives, including prosperity (*khwam rungruang*), civilization (*khwam charoen*), a sense of being abreast of the times (*khwam tan sa-mai*), of stepping forward (*khwam kaona*), and convenience (*khwam sa-duak*) (Plynoy, 2001b). Influenced by these positive attitudes towards the emergence of roads, Bangkok has slowly developed to be a more land-based city by absorbing the Western idea of dry land development while the *khlong* system and the flooded nature of this low-lying land still persist. Several flooding situations, including the most severe deluge in 2011, indicated that there have been serious attempts to control floods and provide drought or flood relief to Bangkok. Additionally, the increased popularity of roads underlies how modernization has influenced the development and direction of the city. The grand, wide, sweeping, ornamental boulevards like Ratchadamnoen Avenue, the Bangkok

Figure 28.2a A contemporary Thai house built on stilts located along KhlongBangluang in Thonburi.

Figure 28.2b A common scene of traffic congestion in Bangkok.

version of the Champs Elysees, are a good example of an attempt to construct the city following the Western concepts of standardization, cleanliness, and the visual aesthetics of formality (Bartlett, 1959).

Indeed, modern development has treated floods as a problem and a cause of damage or disaster to people, in contrast to the auspicious sense of the coming of water perceived in the old days. This suggests that the Bangkokian's life is moving away from nature: instead of negotiation or subjugation to it, they attempt to manipulate and control the environment. Besides the evidence of concrete dams and embankments along most of the urban *khlongs*, air-conditioned buildings, cars, indoor plants, interior lighting and piped water also provide evidence of this shift (Broman, 1984; Vichit-Vadakarn, 1989).

The lives of *chonnabot* (countryside) and *muang* (city)

> I was charmed by Bangkok and I propose to be aggressively syrupy about it in the most buckeye travelogue manner ... Its character is complex and inconsistent; it seems at once to combine the Hannibal, Missouri of Mark Twain's boyhood with Beverly Hills, the Low Countries and Chinatown.
>
> *(S. J. Perelman, 1948, cited in Warren, 2002, p. 9)*

Over the past seventy years of the transformation of Bangkok, the phenomenon of replacing *chonnabot* (countryside) with *muang* (city) has been continual, stretching from the inner city to the urban fringe and reaching into the surrounding provinces. The urban sprawl of Bangkok has caused modifications of physical characteristics as well as changing people's ways of living and worldviews. Nowadays, the city of Bangkok has 'mixed up' the senses of rural and urban, traditional and modern, and village and city contexts to various degrees, thereby creating the notions of 'complexity' and 'the unfamiliar' in the social environment (Mulder, 1996: 134).

Since 1782, Bangkok has functioned as a 'primate city' and national capital in Siam, currently Thailand (Askew, 2002; Keyes, 1987). Besides being an administrative hub, it is also a center of business, trade, social activities, cultural intermixture, and the diverse movements of contemporary life. Bangkok's newly intermixed characteristics of rural and urban worlds have influenced the development of other large up-country towns, such as Chiang Mai, Khon Kaen, Khorat, and Hat Yai, so that they too are becoming alien worlds to villagers (Keyes, op. cit.). This replicating phenomenon highlights the important changes to the environment, not only in Bangkok but also in the whole country.

Accordingly, the notion of '*muang*' has been conceptualized as an 'ideal city' in the Thai geographical imaginary, yet its expansion, through the construction of roads, has threatened and gradually overtaken the survival of '*ban*' or village and traditional ways of living (Figure 28.3b). From modernizing processes, distinctions between sophisticated city people (*khon muang*) and simple country 'hicks' (*khon ban-nok*), including their lifestyles, values, and tastes, have become more apparent (Askew, 2002). Consequent on the economic boom in the late 1980s, the popular image of the Bangkokian (*chao krung*) is largely framed around the lifestyles of two main groups, a middle class or elite group who live the urban dream of big houses in *muban chad san* (commercial housing projects) and a lower class of laborers who live in *chumchon ae-at* or slums (Askew, 1994, 2002; Koanantakool and Askew, 1993). A research project conducted in an urban locale in Bangkok, Dalad Plu (author's spelling), which described people's values as affection, enlightenment, power, skill, wealth, rectitude, respect, and well-being, led to the comprehension that urban society is hybridized and at the same time getting more superficial and materialistic.

While the debate on urban society (*sangkhom muang*) reinforces the urban–rural distinctions, the evidence shows that countryside (*ban-nok*) personality traits still persist even in the city (Figure 28.3a) (Askew, 2002: 102–103). The lifestyle of villagers (*chao ban*) who live outside the capital is seen as paralleling that of Thai peasants (*chao suan* or *chao tung* (*chaona*) or orchard growers or farmers) (Nartsupha, 1994). In these contexts, social relationships and interdependency within the extended families and neighborhood are significant. Evidence of people willing to help and be considerate to each other, known as *nam chai*, is more rare though it is still valued in Thai society (Nartsupha, 1991). An economic system based on self-sufficiency (*tam-ma-ha-kin*), which has long existed in Thai society, has slowly shifted to a commerce-based economic system (*tam-ma-ka-kai*), in which people function primarily in a money economy (Wongtes, 2001). This would seemingly indicate a lessening of the rural–urban distinction due to the transformation of rural ways to accord with the 'rational' and 'money-oriented value-system' of urban society (Cruangao, 1962, cited in Askew, 1994: 41, 42; Klausner, 2000).

Due to the deterioration and transformation of traditional qualities in the modern era, the importance of *watthanatham chumchon* (culture of community) has been stimulated in Thai society over the past twenty years (Nartsupha, 1997). *Muban* (village), or *chumchon* in Nartsupha's term, has actually developed and reproduced a harmonious 'collective identity' over a long period of time (Nartsupha, 1991). Hence the study of *chumchon* (or *muban*), and

Figure 28.3a A surviving chonnabot lifestyle in Thonburi.

Figure 28.3b Public transporting systems in the city of Bangkok.

the acknowledgment of such qualities, are argued to yield solutions for reviving cultural identity and traditional lifestyle in Thai society (Nartsupha, 1991).

The notions of *muang kao* (old Bangkok) and *muang mai* (modern Bangkok)

> [It is] difficult to take Bangkok seriously. It is the most hokum place I have ever seen, never having been to California. It is the triumph of the 'imitation' school; nothing is what it looks like; if it's not parodying European buildings, it is parodying Khmer ones; failing anything else, it will parody itself.
>
> *(Gorer, 1986, cited in Warren, 2002: 8)*

Within the concept of *muang*, there is also another layer of perception that has accumulated and been reconstructed through time. While Bangkok and its environment have developed and transformed, the term *muang* and its connotations appear also to have changed. Due to the continual changes and boundless expansions of Bangkok's environs, the issue of 'no hard definition' could well be extended into the synthesizing concepts of *muang kao* (the old city) and *muang mai* (the modern city) as well as to people's perceptions towards these notions. The shifting concepts and connotations of *muang* reflect the surviving or emerging qualities of Bangkok as a place.

The Chao Phraya River divides Bangkok, yielding the Thonburi bank (the western side) and the Phranakorn bank (the eastern side). Thonburi, once briefly the capital city of Siam, has been recognized as the old city (*muang kao*) of Krungthep (Bangkok) whereas Phranakorn is known as a newer city (*muang mai*), established at the beginning of the Chakri dynasty in 1782. Prior to the King Rama V period (1868–1910), Thonburi was the center for the urban communities; most people at that time lived on the west bank of the river whereas the Phranakorn bank consisted only of governmental buildings, some business districts, residences for foreigners and vast fields far behind (Saksri et al., 1991; Wallipodom, 2000). The notion of '*muang mai*' in that time had a sense of lifelessness and fewer activities than '*muang kao*'.

Since modernizing processes began in the early 1900s, along with the expansion of roads and thereby community settlements, Krungthep (Bangkok) has been promoted mostly on the eastern side. People have, however, started turning back to Thonburi following the construction of bridges and roads connecting to the west after 1932 (Silapacharanan, 2000). A new layer of *muang mai* has been superimposed on Thonburi and, in the same way as Phranakorn, is also continuing its development to be more *muang mai* and to enlarge its boundary to the area previously perceived as *chonnabot* (Chanchareonsook, 1970; Yantrasast, 1995). The meanings and characters of '*muang kao*' that people held at a previous time have transformed into those of the emerging '*muang mai*', which is viewed as obliterating older senses of '*muang kao*'. Therefore, Thonburi is both an ancient (forgotten) capital and a very new city – a seeming confusion, even contradiction, that runs through the ambiguities of the contemporary identity of the place as both very ancient and utterly new.

Due to urban sprawl, which has underlain the intermingling characteristics of old and modern Bangkok, it is now hard to read any boundary of *muang*, or some sense of a city center (Basche, 1971). For historians and conservationists, the core of Bangkok is the inner Rattanakosin Island where the grand palace has been set as the center of the royal Capital City. For the revolutionaries and the founders of democracy, the city's main axis is Thanon Ratchadamneon, the first grand avenue in Siam and the historical street of democracy. Moreover, while the city center for business and commerce might have moved to Thanon Silom, teenagers may instead consider that their city center is the area called Siam Square and the shopping complexes around. In brief, we could argue that the Bangkok city center is and has always been fragmented and shifting.

The transforming identity of Bangkok

The research indicates that the urban invasion phenomenon has threatened local settlements and thereby transformed several localities on the fringe to be more *muang* (city) as shop-houses, housing estates which are normally gated communities, factories, and offices have been built along extended transportation routes. The evidence shows that plenty of orchards and rice fields taken over by developers have typically been flattened and allotted into several pieces of land for sale. These new developments often ignore the old networks of canals and walkways as well as historical and social contexts. Nowadays, both traditional linear settlements along waterways and modern housing estates co-exist without knowing or interacting with each other. Consequently, as gentrification has increased, local people are increasingly moving out of their locale because they feel uncomfortable living among strangers.

These phenomena of urban invasion, land development and city transformation have repeatedly occurred on the fringes of Bangkok, along both the Eastern and Western Ring Road. As a result, the landscapes along these highways turn out to be somewhat homogeneous with scenes of shop-houses forming the edge, alternating with modern housing estates and scattered signage. While the old ways of life along the canal with greenery still exist, they are hidden behind these strips of shop-houses, forming another world.

Nowadays, the boundary of the city keeps extending further with seemingly no limit – the third ring road is going to be constructed in the near future. With a concern that Bangkok seems to be moving towards the 'city of nowhere' (with similar patterns of development and gentrification occurring all around the city), it should be acknowledged here that remnants of the past are what actually constitute the special characteristics of each locale. The complex levels and variety of integration between water-based and land-based features, *chonnabot* (rural) and *muang* (city) lifestyles, and the notions of *muang kao* (old city) and *muang mai* (modern city) in each place has constructed its identity.

Rethinking place and identity

The above narratives illustrate well the above discussed theory that place is not static; it keeps changing, transforming, integrating past and present, memory and existence, remnants and emergence. The reading of place therefore needs to be revisited from time to time because of this never-ending movement.

Pictorial recollections of old Bangkok as the city of canals will always exist in the minds of people, though the reality is one of an increasing number of highways and railways which dominate the city's visual image. This indicates that the understanding of place identity is more complex than what we see. What people perceive to be 'a place' is the accumulation of experiences they have with it, together with what they observe today, and how they imagine or foresee that place in the future. Place identity reinforces the idea that place is truly a representation: it represents the assimilation of past, present, and future. Hence place identity is also changing, following people's perceptions of a place.

The coexistence of rural and urban senses as well as the old and the new of Bangkok today shows that the city of Bangkok (or elaborated further to 'the Thais') lives with the 'unboundedness' – there is no separation or boundary for the differences; thus it is understandably described by foreigners as mixing, intermingling, and complex. The fact that there is never a contradiction for a variety of things to be together like Buddhists, Muslim, and Christian; or human, spirit, ghost, and god shows that these intermingling (or more precisely 'unbounded') characteristics indeed represent the identity of Thai places.

References

Askew, M. (1993) 'The making of modern Bangkok: state, market and people in the shaping of the Thai metropolis', in the proceedings of: Who gets what and how?: challenges for the future, Vol. Background report, at, The 2010 Project.

Askew, M. (1994) *Interpreting Bangkok: The Urban Question in Thai Studies*, Bangkok, Thailand: Chulalongkorn University Press.

Askew, M. (2002) *Bangkok, Place, Practice and Representation*, London; New York: Routledge.

Bartlett, N. (1959) *Land of the Lotus Eaters: A Book Mostly about Siam*, London; Melbourne: Jarrolds Publishers.

Basche, J. R. (1971) *Thailand: Land of the Free*, New York: Taplinger Pub. Co.

Bello, W., Cunningham, S. and Poh, L. K. (1998) *A Siamese Tragedy: Development & Disintegration in Modern Thailand*, Thailand: White Lotus Co., Ltd.

Bowring, J. (1969) *The Kingdom and People of Siam*, Kuala Lumpur, New York: Oxford University Press.

Brindley, T. (2003) 'The social dimension of the urban village: a comparison of models for sustainable urban development', *Urban design international*, Vol. 8, 53–65.

Broman, B. M. (1984) *Old Homes of Bangkok, Fragile Link*, Bangkok, Thailand: Siam Society: DD Books.

Bunnag, D. (2003) 'Co-evolving heterogeneity', In R. Powell (ed.), *The New Thai House*, Singapore: Select Publishing Pte., Ltd, pp. 10–22.

Chanchareonsook, A. (1970) 'Housing in Bangok-Thonburi, A Talk at USAF Meeting', May 12, Bangkok.

Eckbo, G. (1969) *The Landscape we See*, New York: McGraw-Hill.

Eliot, J., Bickersteth, J., Buranakarnkul, N., Buranakarnkul, S., Hosaini, H. E. and Saville, J. (2001) *Thailand Handbook*, ((Eds, Dawson, P. and Fielding, R.), Footprint, 3rd ed.), Bath, England: Footprint Handbooks.

Fournereau, L. (1998) 'Bangkok in 1892', Translated by Tips, W. E. J., Originally published under title Bangkok in Le Tour du Monde, 1894, Vol. 68, pp.1–64, Bangkok: White Lotus Press.

Gustafson, P. (2001) 'Meanings of place: everyday experience and theoretical conceptualizations', *Journal of environmental psychology*, Vol. 21, 5–16.

Hinshiranan, N. (2000) 'Chao Phraya River and canal network: nostalgia of aquatic traditions', in the proceedings of: Chao Phraya Delta: historical development, dynamics and challenges of Thailand's rice bowl, at.

Jumsai, S. (1997) *Naga: cultural origins in Siam and the West Pacific*, 3rd ed., 3rd Printing in paperback reprint ed., Bangkok: Chalermnit Press and DD Books.

Keyes, C. F. (1987) *Thailand, Buddhist Kingdom as Modern Nation-State*, Boulder: Westview Press.

Klausner, W. J. (2000) *Thai Culture in Transition*, Collected writings of William J. Klausner, 3rd ed., Bangkok, Thailand: The Siam Society Press.

Koanantakool, P. C. and Askew, M. (1993) 'Urban life and urban people in transition', in the proceedings of: Who gets what and how?: challenges for the future, Vol. II - Synthesis papers, at, The 2010 Project.

Mulder, N. (1996) *Inside Thai Society: An Interpretations of Everyday Life*, 5th updated and enlarged ed., Amsterdam, Kuala Lumpur: Pepin Press.

Nartsupha, C. (1991) *Watthanathamthaikabkabuankarnplianplangthangsangkhom* [Thai culture and the transforming process of Thai society], 5th (2004) reprint ed., Bangkok, Thailand: Chulalongkorn Press.

Nartsupha, C. (1994) *Watthanathammubanthai* [The culture of Thai village], 2nd (1998) reprint ed., Bangkok, Thailand: Sangsan Publishing Co., Ltd.

Nartsupha, C. (1997) *Ban kabmuang* [Ban and muang], 2nd reprint ed., Bangkok, Thailand: Chulalongkorn Press.

Nawigamune, A. (2003) *Thintharn ban chong* [Tales of the river and places], Bangkok, Thailand: Pimkam Publishing.

O'Connor, R. A. (1989) 'From 'fertility' to 'order', in the proceedings of: Culture and environment in Thailand: a symposium of the Siam Society, at, DuangKamol, 393–414.

O'Hare, D. J. (1997) 'Tourism and small coastal settlements: a cultural landscape approach for urban design', Ph.D. Thesis Dissertation, Oxford Brookes University.

Ouyyanont, P. (2000) 'Transformation of Bangkok and concomitant changes in urban-rural interaction in Thailand in the 19th and 20th centuries', in the proceedings of: Chao Phraya Delta: historical development, dynamics and challenges of Thailand's rice bowl, at.

Plynoy, S. (2001a) *Chiwit tam khlong* [Life along the khlong], 4th reprint ed., Bangkok, Thailand: Saitharn Publication House.

Plynoy, S. (2001b) *Lao ruang bang-kok* [Telling stories about Bang-kok], vol. 2, 6th reprint ed., Bangkok, Thailand: Saitharn Publication House.

Rivlin, L. G. (1987) 'The neighborhood, personal identity, and group affiliations', *Neighborhood and community environments*, (Eds, Altman, I. and Wandersman, A.), New York: Plenum Press, 1–34.

Saksri, M. R. N., Tiptus, P., Worawan, M. R. C., Sajkul, V., Sathapitanonda, L. and Julasai, B. (1991) *Ongprakorpthangkaiyaphap Krung-Rattanakosin* [Physical components of Rattanakosin], Bangkok, Thailand: Chulalongkorn University Press.

Silapacharanan, S. (2000) 'Thonburi: conservation of agricultural area and urban development', in the proceedings of: Chao Phraya Delta: historical development, dynamics and challenges of Thialand's rice bowl, at.

Tuan, Y.-F. (1977) *Space and Place: The Perspective of Experience*, Minneapolis: University of Minnesota Press.

Vichit-Vadakarn, J. (1989) 'Thai social structure and behavior patterns: nature versus culture', in the proceedings of: Culture and environment in Thailand: a symposium of the Siam Society, at, DuangKamol, 425–447.

Wallipodom, S. (2000) 'The Chao Phraya society: development and change', in the proceedings of: Chao Phraya Delta: historical development, dynamics and challenges of Thailand's rice bowl, at, 7–33.

Walmsley, D. J. and Lewis, G. J. (1984) *Human Geography: Behavioural Approaches*, London; New York: Longman.

Warren, W. (2002) *Bangkok*, London: Reaktion.

Wongtes, S. (2001) *Maenamlamkhlongsaiprawatsart* [River and khlong: the historical stream], 3rd ed., Bangkok, Thailand: Matichon Press.

Yantrasast, K. (1995) 'Bangkok's water logic: a study on the phenomenon of water and urban transformation', Doctor of Engineering in Architecture Dissertation, Department of Architecture, Faculty of Engineering, The University of Tokyo.

Part IV

Power, regulating and resisting in place

Paul O'Hare

Introduction: power, regulating and resisting in place

Places are regulated through the interaction of an array of social, economic, cultural, legal, administrative and bureaucratic forces. These forces are themselves influenced – to varying degrees – by a great multiplicity of statutory, commercial and business interests, and voluntary actors and agencies. Regulation takes many different modes which can be codified and enforced through legal arrangements, contracts and orders, social regulation, or through norms, practices and cultures. Regulation can be hierarchical – a form of command and control. Yet, in other guises, regulation can be much more diffuse. What emerges is a complex interplay of formal and informal regulatory practices, from different sources that are performed across spatial scales and temporalities combining to cumulatively shape and reshape the places in which we live out our daily lives.

Regulations – and regulatory forces more broadly – stem from their own rational imperative. But what fundamentally unites them is their attempt to lend a greater sense of order and a greater sense of control over place. On the surface, the regulation of place can seem rather mundane; a byword for restriction. Indeed, regulation is often criticised as being prescriptive, creating uniform, standardised and profoundly uninteresting places. Worse still, it can be suggested that the over-regulation of place can subject places to design and management that is not motivated by optimism but, instead, and rather pessimistically, to plan for a whole host of worst-case scenarios. However, despite anti-regulatory sentiment, regulation can make places safer, more civilised, more ordered; more inherently liveable. When functioning as they should, regulatory orders and practices keep our buildings standing, keep our workplaces safe and – we hope – keep places functional. Moreover, it is difficult to contest the laudable public health origins of contemporary spatial regulation that enhance contemporary human life.

The power to define and manage place in accordance with particular values, whether through policing, surveillance, legal procedures, commemoration, architecture and managing risk, contrasts with efforts that resist such attempts or that experiment with establishing alternative ways of living in place. In these ways, power is continuously inscribed, deleted and reinscribed on places. Regulations and regulatory interventions can endure across time, producing something of a palimpsest of past efforts to regulate a place that remains traceable. On other occasions, places are cast anew, with limited if any reference to previous efforts to order, control and standardise. Such ordering practices fundamentally seek to secure place against processes that would blur its established meanings, functions and physical and environmental qualities.

Some levers of regulatory control are much more overt than others. In the United Kingdom in 1947 the right to develop land was effectively nationalised, introducing a plethora of rules and regulations on land speculation and building development, and establishing a precedent for state intervention in hitherto private activities. Consequently, what has subsequently become known as spatial planning has had an enormous physical impact on places, particularly with regard to the flow of people, money, goods and traffic across scales. In this way, spatial planning represents an effort to create more orderly and efficient places. It does this in a multitude of ways. From a strategic perspective, land use planners establish the broad frameworks and principles that determine spatial decision-making, shaping what is permissible and lending a degree of certainty for those involved in, or subject to, these bureaucratic procedures. Here, the rational design and segregation of spaces that are allotted for housing, roads, state institutions and businesses demarcate functional, single-purpose spaces and the modes of regulation that reinforce such delineations. At alternative spatial scales, development control systems seek to regulate the correct use of land, including the appearance of proposed developments, their function and functionality, and their potential impact on neighbouring properties. Beyond this, a further tier of regulatory intervention in the form of building control is responsible for ensuring that construction takes place in a manner that meets safety and quality standards. These frameworks, and checks and balances, place the onus on the developer, the construction industry, suppliers and eventually the owners, occupants and users of buildings to adhere to regulatory standards. To another degree, regulation is also enacted by the ongoing repair and maintenance that seeks to secure places against wear and tear and the loss of their functional and physical coherence. Such procedures also take place across scales, from large-scale repairs to transport infrastructures that enter, leave and traverse through place, to the everyday household chores that keep domestic spheres 'regularised' – the vacuuming, washing, mowing and sweeping that keep the home ordered (Graham and Thrift, 2007).

As noted earlier, regulation is fundamentally about taking control of space – of claiming and controlling territoriality. It also includes those measures of policing, surveillance and directing people in order to regulate who enters, departs and moves through place and what they do whilst there. In this respect, place regulation is marked by exclusionary and inclusionary practices that foreground who is out of and who belongs in place, demonstrating how regulatory practices have significant impacts on society and people. Perhaps the most notorious and maligned form of territorial regulation is that constituted by the encroachment of agencies and technologies of surveillance that have proliferated across place. CCTV technologies have been considered as especially emblematic of systems that monitor activity, conveying the impression that certain places are designed to accommodate only particular practices and those that violate such designations will be subject to removal or admonishment. The arguments that these and similar surveillance cultures and practices render place akin to a modern-day panopticon are well rehearsed and will be revisited in this part. Of course, regulatory practices through surveillance and security are by no means limited to technologies such as CCTV systems or electronic sensors. Security guards, bouncers, door-keepers and the police serve to enforce regulatory environments, maintaining order – or at least conveying an impression of order over places.

It is worth noting that surveillance systems are the technological manifestations of a much broader – and quite troubling – increase in the neo-liberalisation and privatisation of public space. The proliferation of gated realms further intensifies the ascribed functions of places, compounding the saturation of some spaces with security technologies and personnel, asserting a blatant place territoriality in a deliberate, conspicuous manner. It has similarly been proposed that planners and built environment professionals have been co-opted into the creeping, insidious securitisation of public space through their engagement in counter-terrorism strategies (Coaffee et al., 2009).

Steven Flusty (1994) developed a useful typology of fortress urbanism and describes how this has infiltrated the contemporary city. He calls this 'interdictory space'; space that is designed to exclude and determines place not just by dictating the function but by influencing the sensibilities of the users of public space. Similar ideas have been developed by the urban theorist Oscar Newman (1972) who referred to how space should become 'defensible'. Taken to an extreme, it has been suggested, such is the tightness of regulation, securitisation or even militarisation of these urban environments that they have been likened to 'fortress cities' (Davis, 2006). Yet despite a proliferation of neo-Foucauldian and neo-liberal critiques of surveillance and its widespread manifestation in contemporary space, the owners, managers and the users of public space accept – and often even welcome – its omnipresence. Many people choose to live and consume in the confines of these highly regulated environments and are usually paying a premium for the privilege.

Despite the proliferation of these technological strategies of control, the most powerful process that secures the meaning and practice of place is that of self-regulation and the surveillance of fellow citizens. These mobilising gazes, facial expressions and overt interventions reveal how we are complicit in securing and regulating place. For instance, Oscar Newman's work on territoriality is often used to justify the ever more intensive securitisation of public spaces, yet this was only one element of his conceptual development. Critically, Newman advocated improved opportunities for surveillance – that, combined, can bring environments under the closer control of residents, so that they could be controlled not by the police but by local communities (1972). Herein lies a paradox. According to Tony Bennett (1995: 48), we have changed from flaneurs who simply gaze tolerantly on the diverse people who enter place and carry out unorthodox practices to 'citizen-policemen' who monitor ourselves and others to ensure the codes of place are followed. Such places contrast with the kinds of inclusive, heterogeneous streets championed by Jane Jacobs (1961), where there is mixing of different people, and it is this very inclusivity that is maintained by the mobilisation of 'eyes on the street', a mode of regulation that seeks not to restrict activities to a narrow range but to ensure their continuance.

This leads into a deliberation of the considerable geographical diversity through which different kinds of places are regulated. For instance, the bazaars of India are especially multifunctional places, mixing together small businesses, shops, street vendors, public and private institutions and domestic housing. Diverse activities of loitering, sitting and observing, meeting friends, and carrying out domestic activities such as collecting water, washing clothes, cooking and childminding are part of daily experience (Edensor, 2000). Moreover, mobile hawkers, beggars, musicians and magicians, political and religious speakers move across space, joining the flow of different bodies which crisscross the street in multi-directional patterns, with vehicles paying little heed to formal traffic rules as they jostle for position. Here is a variegated sensual experience of space in which particular smells, sounds, sights and tactilities have not been regulated away, yet spaces are still rational and ordered in their own ways.

Though people tend to inhabit places that have been shaped largely by forces over which they may have little control, other places are exceptions to this. One example is the freetown of

Christiania, an autonomous enclave in the centre of Copenhagen, Denmark, established in 1971. Here most regulatory and governmental procedures are locally governed. Ntounis and Kanellopoulou (2017) term this a 'jurisdictional heterotopia' in which the 'Laws' are collectively policed or regulated by specialised groups of Christianites and are largely observed. If they are flouted, this may result in a communal ostracising of the offender. These may be marginal places, outside the purview of hierarchical power, or may involve forms of tactical urbanism through which a range of actors attempt to reclaim place. These vary in scale from high profile actions such as those carried out by Extinction Rebellion, the occupation of the financial districts of cities that occur during anti-capitalism protests, democratisation marches or the Reclaim the Night marches, to the more contingent, fleeting colonisations of place by skateboarders, graffiti artists and parkour exponents.

Clearly, therefore, regulatory practices in their different guises have a significant influence on the spaces and places in which we live, work and socialise. Moreover, regulatory influences, nested against political and public policy initiatives, permeate our places at manifold spatial scales. Regulation, then, as an exertion of power, is always inextricably entangled with place in dynamic and diverse ways, affording scope for some to act and claim space while limiting the power of others to stake their claim to belonging. However, perhaps more critically, they have a defining impact on the engagement and interaction of citizens. Accordingly, this part examines the various contrasting and even competing ways in which places are regulated, managed, policed and ordered. Moreover, they engage with the regulatory conventions, forces and practices that – in their own ways, and across varying spatialities and temporalities – shape place.

The first chapter, written by Nikos Ntounis, Dominic Medway and Catherine Parker, places the part into context by providing a broad introduction to managing places. The authors refer to the essential plurality and interdisciplinarity of places to detail the contradictions and paradoxes that come to the fore in contemporary place management.

In their chapter, Paul O'Hare and Iain White turn attention toward how risk and resilience influence place. They detail how risk is perceived and encountered differently across society and space, though they note how many risk management and resilience strategies tend to be spatially blind and often focus on coping with risks rather than adaptation away from threats. Moreover, policy often ignores the wider political economy that embeds risk and disadvantage in a place. They conclude by arguing that, to be efficacious, contemporary risk management and resilience policies must be linked more strongly to place-based social contexts.

In 'Mapping place', Chris Perkins looks at how maps, in their various formats and presentations, tell the story of a place. Despite this, he contends, mapping simultaneously takes its relationship to place for granted, reducing the subjective diversity of places to the spatial anonymity of the map. Ultimately, cartographic practices render the complexity of places to usable knowledge space, but one where diversity and difference is 'off the map'.

Jenny Kanellopoulou's chapter introduces the complex relationship between law and place. The chapter emerges from the fundamental premise that 'there cannot be law without a place of application', but that this is often not understood. The chapter addresses this gap by identifying particular aspects of the law of pertinence to geography.

In her chapter, Rachel Woodward details how understanding the influence of military power across place requires a comprehension of complex, connected practices and ideas around control. Drawing upon the case study of Newcastle upon Tyne, the chapter describes how the process of militarisation, rather than its presence, is often overlooked but is vital to understand the places we live in today.

Jon Coaffee develops themes that have become inherent to regulatory place management across contemporary cities and beyond: that of policing and securing place. Coaffee proposes that place safety is essential to creating successful, functioning places in two fundamental ways: first through the creation of social and regulatory forces to maintain order and control, and, second, more tangibly through the integration of 'fortification' of space. Given these dual pressures, the management of public space for safety and security must be cognisant of broader needs to preserve and enhance the vitality of public realm.

Quintin Bradley examines the bond between place and participation, adopting the proposition that the devolution of power to local neighbourhoods through participatory practice is a fundamentally more democratic form of governance. Bradley first outlines the distinction between what he refers to as the 'domestic realm' of neighbourly care and the formal arena of neighbourhood governance, before introducing the concept of localism as the vehicle for political participation. The concept of community identity is then proposed as a mobilising force through which participative democracy can be advanced.

In exploring how alternative places that escape normative regulation can be devised, Alexander Vasudevan's chapter investigates the experimental geographies produced by activist communities living in cities across the global North. Urban squatting movements in Europe and North America provide the focus for this interpretation of place, particularly in terms of their everyday practices and spatial imaginaries, and with regard to how the site of occupation emerges as an alternative habitus where occupiers create new and often experimental forms of shared city living.

In 'A monument in the city', Hélène Ducros examines how a single building becomes fundamental to how a place – in this case configured as a memoryscape – is perceived and interpreted. The chapter reflects on how Ducros had her own preconceptions of a familiar space challenged, critiquing how place is a conduit for 'unintended' monumentality.

In the final chapter of this part, James Lesh examines how efforts to respect heritage in place have been dominated by pragmatism at the expense of the more diverse conceptualisations of place that are recognised across other disciplines. In particular, Lesh highlights the distinction between what is tangible and what is intangible in places: the physical fabric of a place, in contrast to social and human constructions of place. In so doing, the chapter draws attention toward more critical and more sensitive interpretations of place conservation, in the hope of broadening the practice and scope of those involved in protecting and enhancing heritage regulation and planning.

References

Bennett, T. (1995) *The Birth of the Museum*, London: Routledge.

Coaffee, J., O'Hare, P. and Hawkesworth, M. (2009) The visibility of (in)security: The aesthetics of planning urban defences against terrorism. *Security Dialogue* 40(4-5): 489–511.

Davis, M. (2006) *City of Quartz: Excavating the Future in Los Angeles*, London: Verso.

Edensor, T. (2000) Moving Through The City, In D. Bell and A. Haddour (eds) *City Visions*, London: Palgrave, pp. 121–140.

Flusty, S. (1994) *Building Paranoia: The Proliferation of Interdictory Space and the Erosion of Spatial Justice*. Los Angeles: LA Forum for Architecture and Urban Design.

Graham, S. and Thrift, N. (2007) Out of order: Understanding repair and maintenance. *Theory, Culture & Society* 24(3): 1–25.

Jacobs, J. (1961) *The Death and Life of Great American Cities*, New York: Random House.

Newman, O. (1972) *Defensible Space - Crime Prevention Through Urban Design*, Macmillan, New York.

Ntounis, N. and Kanellopoulou, E. (2017) Normalising jurisdictional heterotopias through place branding: The cases of Christiania and Metelkova. *Environment and Planning A*, 49(10): 2223–2240.

29
Managing places

Nikos Ntounis, Dominic Medway and Catherine Parker

> Current place management policy is struggling to resolve the paradoxes and contradictions that revolve around notions of localism/globalism, hierarchies/networks, heterogeneity/homogeneity, competition/cooperation, equity/efficiency and the like … the desire of places to be unique and different confronts a practice which more often leads to similitude and uniformity. It is also obvious that place management policies operate within societies of increasing plurality of cultures, life-styles, expectations and interventions. Place management has thus become more difficult, complex and unpredictable but equally more necessary, demanding, and indeed fascinating.
>
> *(Ashworth, 2008: 18)*

Back in 2008, in the inaugural issue of the *Journal of Place Management and Development*, the Editorial Advisory Board reflected on what place management is, how it can be defined, and what it means for both academics and practitioners. The breadth of definitions and interpretations offered in that first issue provides an excellent starting point for understanding the various and varied potential applications of 'place management'. Thus, the term can refer to special place-based institutions that organise, promote and represent places and extend beyond the public sector; it can also include the adoption of marketing and management models that are extrapolated to the policy field and fitted to places for specific purposes (e.g., regeneration and sustainability projects, public space management, conservation of culture and heritage); and it may also extend to the implementation of retail and land use planning policies for the successful development of the built environment in cities and towns. Such associations highlight the inherent complexities that surround this interdisciplinary field. Indeed, over a decade after the first edition of the *Journal of Place Management and Development*, the central subject of this publication is still trying to find its place in the academic field. Consequently, the paradoxes and contradictions that the late Greg Ashworth then highlighted – and as he recounts in the opening excerpt to this chapter – are still taking centre stage in the development of place management theory, policy and practice.

Of course, both academics and practitioners can be held liable for the ambiguity and vagueness of place management as both a theoretical field that is constituted by knowledge sources from a multitude of disciplines (Coca-Stefaniak, 2008), and as a 'loose' process that can be applied in places in a variety of ways. This situation is also a likely reflection of the fact that there is no 'one right way' to manage a place or space in order to produce the laudable but rather vague outcome of making places better (Parker, 2009). Ultimately, there is even difficulty and complexity in defining what place management is, such that the concept can include any process, tool, design, intervention or practice that aims to contribute in place. Furthermore, the means by which it is practised being open to anyone's interpretation of what is right and wrong for the place in question.

Whilst this complexity and plurality might initially be perceived as a major drawback, we argue that a more singular approach towards the study and practice of place management would prevent the incorporation of core knowledge from a plethora of theoretical, conceptual and methodological choices that are 'whirling' in its heart. Indeed, as it will be shown in this chapter, place management theory, policy and practice is influenced by inputs from many fields of knowledge, and from a multitude of (often conflicting) stakeholders who have 'place' as their common denominator. In this vein, we will highlight place management's emergence into academia from a geographical perspective, with emphasis on its operational and strategic practices that suggest a turn towards an engaged, smart pluralist approach towards managing places.

The historical emergence of managing places across multiple scales

Since the beginning of the first settlements and the creation of the first ancient villages, towns and cities, places have been developed as a result of people's organising and managing efforts (Parker, 2011). In ancient Greece, city-states laid the blueprint for the idea of *polis* – a bounded, territorial, administrative and politically autonomous city that nurtured intimate relationships with its citizens (Agnew, 1994). In the English context, place-based forms of organisation were the standard process that the Church, the parish, the market, or the electorate would run in the proximity of their territories (Stewart-Weeks, 1998). Such forms of management and organisation were explicitly rooted in the idea of place as a 'terrestrial surface that is not equivalent to any other' and 'that cannot be exchanged with any other without everything changing' (Farinelli, 2003: 11). This also highlights the significance of a geographical perspective on managing places since the very beginning.

Of course, society has come a long way since then, and so has the practice of place management, as the rapid urbanisation of cities necessitated more formal and cohesive structures of management that mirrored the organisational structure of the private sector with the adoption of rules, legislations, and functional forms of organising as a more efficient way of managing places (Adams, 2008; Parker, 2011). New narratives of constant competition appeared, which 'brought places face to face with capital without the intermediation of the state' (Dirlik, 1999: 45), and necessitated new forms of governance and organisation with the aim of attracting capital from elsewhere. As such, an understanding of place management from the neoliberal perspective started to emerge, which suggests that management of cities can be improved by entrepreneurial modes of urban governance. The shift towards entrepreneurship seeks to promote a range of 'capacity-building' initiatives and to establishing public–private partnerships that value private enterprise and free-market economics as drivers of change in places (Hall and Hubbard, 1996; Harvey, 1989a; Sassen, 1991). This focus on competitive positioning has led to a new urban politics (Cox, 1993) of place that aimed to enhance the possibilities for better economic conditions by promoting and managing the city as an urban growth machine that caters for the resourceful private sector (Irazábal, 2009).

However, the negative outcomes from passively adopting an entrepreneurial stance towards economic development that hit many cities in recent years have resulted in a series of homogenising trends. These include developing a similar mix of mega-projects, regeneration projects and buildings, thereby applying similar urban policy and planning solutions which can eventually result in urban monotony (Harvey, 1989a) and to the creation of corporatised non-places (Auge, 1995). Additionally, the ongoing processes of urban transformation deprive places of their distinctiveness and extract valuable resources via exploitative practices. By linking the above outcomes with the process of place management, it can be argued that the destructive mantra of place competition 'not only contributes to an undermining of local distinctiveness but also weakens the power of civic institutions to affect local change as they become subject to remote decision-making' (Millington and Ntounis, 2017: 368).

As places are constantly undergoing 'dynamic market-led and planning-led change, even in times of crisis' (Salet and Savini, 2015: 448), and their spatial organisation becomes increasingly polycentric and discontinuous (Hall, 1997; Massey, 2005), there is an increasing demand to address the problems of globalisation not only from top-down initiatives, but also from local forms of governance and organisation. This suggests an interdependency between local institutions and networks of 'partners' and communities in the formulation and implementation of local policies and strategies for urban transformation, which vary considerably in different localities, despite being influenced by the same global circumstances (Parés et al., 2014). As such, the management of places relies 'increasingly on the instruments of soft regulation and network management, as local government becomes ensnared in its reliance on other actors' (Blanco et al., 2014: 3133).

It follows from the above that the task of managing places in the era of 'network local governance' (Peyroux et al., 2012: 112) becomes a 'rather haphazard affair' (Parker, 2011: 5). This is because a variety of place stakeholders can exert different leadership styles at the same time during the place management process, thereby facilitating a dynamic interaction that can shape place development on an everyday basis. Among these stakeholders, local people and communities emerge as an important group with new rights and powers, exercising flexible, performative forms of citizenship that aim to safeguard mutual and equal benefits, and equivalency in participation and decision-making, for all place stakeholders (Lepofsky and Fraser, 2003). In short, citizenship becomes 'a hybrid between a given status and a performative act' (Fraser et al., 2016: 863) that becomes material through the actions and behaviours (Pine, 2010) of the people involved in place management.

From this brief historical review, it is evident that existing place management configurations have in part originated from a need to deviate from managerial and competitive market solutions in favour of a just decision-making process regarding place commons, the emergence of public participation, and the shift from government to governance. Early research from a public and urban policy perspective advocated the abolishment of departmental silos in favour of multidisciplinary management teams that try to achieve equity by customising services and allocating resources based on locational needs. From this perspective, place management is seen as an outcome-based approach that can be viewed as a central responsibility in the new governance era, since it:

- facilitates the fundamental restructuring of public and administrative sectors and state and local government;

- offers promise as a policy framework for re-conceptualising community relationships with the state and the markets;
- delivers improved community outcomes, particularly in the context of place-based or spatial policies, for particular groups of people in particular communities (Mant, 2002, 2008; Reddel, 2002; Walsh, 2001).

This acknowledges the fact that place management is likely to be an intensive, continuous process involving a coordinated effort across a range of agencies (Walsh, 2001). In fact, because place management is increasingly concerned with tackling local problems that are usually an outcome of broader market forces, collaboration between civil society, the private sector, markets, the state, and other governmental bodies is an essential requisite for everyone who is involved in the process. In this sense, success in place management is directly related to the effectiveness of people and partnerships that are engaged in the process, and on how much influence they have in the construction of new urban policy initiatives and practices, the creation of new economic and spatial imaginaries, the achievement of improved outcomes for their local areas, and so on (Jessop, 2013; Ward, 2003). It is therefore unsurprising that place management is progressively considered 'as a symbiotic element of strategic significance in the long-term impact and sustainability of towns/cities', which needs to be 'at the heart of the planning, design and overall placemaking processes' (Coca-Stefaniak and Bagaeen, 2013: 532).

From this brief review, it is evident that a number of interdependent factors across multiple scales are driving the current place management debate. These include the turn to more localised forms of stakeholder participation and engagement; the increasing influence of global economic and market trends that influence a place's competitiveness in multiple arenas; the restructuring of governance mechanisms into networked forms that lead towards collaborative, supra-local arrangements; and the shift towards multidisciplinary management teams that aim to tackle siloed forms of place governance. Based on the interdependencies above, Adams (2008) argues that place management, as a form of urban public policy, is increasingly influencing the social, economic, human and natural capitals in place(s). In this sense, the view of place management as an emergent form of urban public policy is acknowledged.

The practice of place management

The multiple activities associated with place management accentuate the need for their itemisation, in order to understand what can be broadly considered as 'place management practice'. In this respect, Yanchula (2008) provides a comprehensive hierarchical framework that lays out place management activities in a way that allows local partnerships and organisations to make a conscious decision regarding which activities they want to be in charge of (see Figure 29.1).

Such activities must usually start from the most trivial tasks (but mostly forgotten by macro-level planning strategies), such as picking up litter and making sure that the place is clean and safe. As cleanliness and safety directly affect people's lived experiences and lead to negative place attitudes and perceptions in relation to phenomena like crime (Medway et al., 2016), place maintenance is seen as a 'place management 101' practice that shifts place managers' efforts towards 'a more micro-marketing perspective. Here, the attitudes of individuals to a given place, and those factors which may directly affect such attitudes, drive place marketing [and management] activity' (Parker et al., 2015: 1106). Apart from the basic janitorial activities, place management is also concerned with systematic attempts to re-

Figure 29.1 Hierarchy of place management activities and their expected outcomes.
Sources: Yanchula, 2008: 93.

familiarise place and give a renewed sense of pride to local citizens via a series of festivals, events and promotions (Richards and Wilson, 2004; Rota and Salone, 2014). These 'soft' temporal place interventions rely mostly on place marketing and branding practices to drive regeneration in certain places, contrary to more traditional economic development initiatives that require hard infrastructure, but have long been important and reliable attractors of visitors that add vitality and market a place as fun and pleasurable. Marketing and branding here should be understood in their strictest sense, describing practices of place selling that comprise of place management organisations' attempts to find the right place consumers, create favourable images, organise simple promotions of certain events, attractions and urban functions and sell the place as a more or less pre-defined and pre-packaged product (e.g., Ashworth and Voogd, 1990; Warnaby and Medway, 2013).

In the higher levels of place management practice, we encounter its increasing function as a strategic vehicle for long-term interventions in places. For example, partnerships and organisations involved in place management can affect a place's design and physical place development by making streetscape improvements, creating new public spaces and public art, providing public amenities, adding street furniture, greenery, and so on (Biddulph, 2011; Whyatt, 2004). Beautification practices respond to problems such as social and urban decline (Mayer and Knox, 2010), enhancing pleasure and quality of life for the people living in them (De Nisco and Warnaby, 2014), and can help businesses to attract and retain customers and employees in an area, thus giving them the incentive to protect the quality of their environment by being a part of local partnerships, and by engaging with public realm management.

Another traditional activity in place management is capital attraction, business retention, and recruitment. Such activities attempt to attract industries, promising businesses low production costs and a friendly environment to operate in (smokestack chasing) (e.g., Kavaratzis and Ashworth, 2008; Wood, 1993). However, place management does more than attracting this one type of capital. Resident, tourist, and investment attraction are equally important in the place management agenda. Niedomysl and Jonasson (2012) purport that places which adopt place marketing strategies are not necessarily trying to attract all forms of capital simultaneously, but instead focus on those forms of capital to which they are more accustomed. Such activities are also in line with large-scale projects, such as the development of new areas; property development; the restructuring of current districts; the revamping of town centres and high streets; the delivery of flagship projects such as sport stadia, art galleries and convention centres; the gentrification of older areas and housing; the creation of new transport hubs and road infrastructure; and so on (e.g., Biddulph, 2011; Smyth, 1994; Warnaby, 2009).

Although local partnerships are not directly involved in the actual construction and delivery of these 'hard' infrastructure and long-term development projects, they often mobilise and create short-term regimes that can steer public opinion and decision making in order to secure national or European funding for specific developments (e.g., Geddes, 2006; Irazábal, 2009; Stone, 1993). Consequently, urban planning practices are often focussed on satisfying narrow urban policy agendas for local growth coalitions (formed from land owners, business leaders and local government representatives, among others) that aim to maintain or improve the competitive standing of places.

However, these narrow approaches to strategic planning may lead to the development of new urban spaces and place identities that are more accustomed to 'the outsider, the investor, developer, businesswoman or – man, or the money-packed tourist' (Swyngedouw et al., 2002: 550–551) rather than local people. This aligns with the entrepreneurial city thesis (Cox, 1998; Hall and Hubbard, 1996; Harvey, 1989b), favouring urban entrepreneurial policies and growth-oriented agendas (McCann, 2004) rather than developmental initiatives that are concerned with the redistribution and provision of public goods (Mayer and Knox, 2010). According to Oliveira (2015), such developmental practices must be framed through communicative, participatory, and relational approaches to strategic spatial planning, which incorporate political debates and democratic participation in innovative and creative ways that can lead to more effective place development interventions. Therefore, local partnerships involved in development projects can adopt these planning approaches in order to nurture dialogue and participation between place stakeholders, which makes strategic spatial planning an important element of the place management process in terms of collective strategy-making (Pasquinelli, 2014).

Once place management organisations and partnerships have run through the spectrum of activities and practices mentioned above, they should have enough influence to become effective advocates for sustained transformational change and place repositioning (Yanchula, 2008: 95). In this sense, place management organisations do not differ a lot from 'policy networks', 'policy communities', or 'advocacy coalitions'. Advocacy implies a certain accumulation of power prior to lobbying, but this does not preclude that this power will only be used for exploitative purposes (systemic power), which favours those groups concerned with economic growth at the expense of those concerned with redistribution (Hidle and Normann, 2013). This is highlighted through the attempts of advocacy planners to tackle growing citizen opposition and inequalities to decision-making processes, typically by facilitating public participation and collaboration in planning (Sorensen and Sagaris, 2010). Such an approach can lead to the generation and reproduction of network power and emancipatory forms of strategic spatial planning (Irazábal, 2009).

In these instances, place management partnerships are 'power alliances that favour cooperation rather than competition' (Naughton, 2014: 15) and promote a discourse that showcases advocacy as inclusive and accommodating to varying viewpoints, without losing sight of the goals of social equity (Irazábal, 2009). In most cases, however, there is a tendency to sustain the power relationships between elite forms of local governance and participatory practices (Geddes, 2006; Guarneros-Meza and Geddes, 2010). Therefore, place management needs to be thought of as a kind of leadership that can leverage a place's assets and mobilise the community, co-ordinate and utilise the potential of new and existing networks (van den Berg and Braun, 1999) and bypass political pressures, while also 'operating through the contingent intersections in which most places are necessarily embedded' (Collinge and Gibney, 2010: 476).

A relational understanding of place management

It is evident – from the breadth of practices associated with it – that place management can have a wide societal and financial impact and act as a key process for organising area-based regeneration. It should be noted, however, that prescriptive place management still adheres to a nomothetic and mechanistic approach that fails to address the interdependencies and traverses between different initiatives and practices, as well as between place management's theoretical underpinnings. Indeed, place management practices intersect through all spectra of the so-called place management pyramid. In reality, place management processes do not follow a hierarchical pattern as Figure 29.2 suggests. Instead they reflect the 'messy' realities of places and the need to move past specific hierarchical structures and strategies and take into account the external and internal complexities of 'the relatively autonomous systems, sectors and organisations that need to be coordinated if places are to maintain and develop their competitiveness' (Omholt, 2013: 29).

Plurality in place management

Based on the above, it follows that decision making and management must become 'less hierarchical and myopic and more place-based and "porous" to allow more intelligence and input from the location' (Millington et al., 2015: 5). The difficulties of addressing the fuzzy problems of

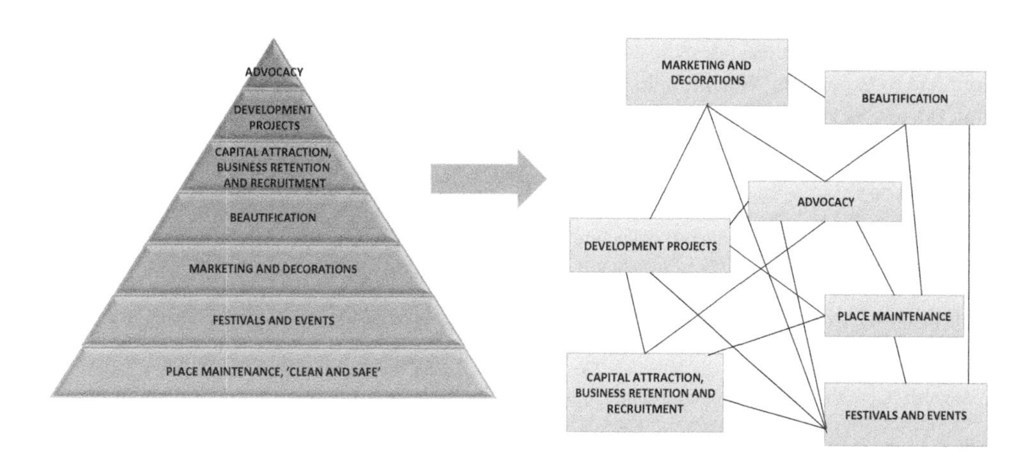

Figure 29.2 From hierarchical to networked-relational place management models.

places require spaces in which a plurality of interests, opinions, conflicts, different values, and power relationships are addressed, and consequently challenge existing knowledge, conventional wisdom, and practices (Brand and Gaffikin, 2007; Forester, 2010; Hillier, 2007). Therefore, other approaches towards collaboration are needed, such as a focus on knowledge exchange between stakeholders, by considering the shifting behaviours and interactions of these in local partnerships (Le Feuvre et al., 2016), and by embracing conflict and 'social untidiness', if it is for the greater good (Brand and Gaffikin, 2007).

It can be argued that a participatory, pluralist and relational approach to place management can infuse it with a spatial and social emphasis rather than a business focus. This suggests that in order for place management to gain theoretical legitimacy, a wider conceptualisation of its adjacent fields is needed, which will allow a fuller appreciation of the complexity and pluralism that is inherent in places. As such, it is argued that place management, as a synthesised, place-based process of strategic significance that aims to solve complex problems and produce specific outcomes for places and people, has the potential to act as an organising buzzword that is open enough to allow different fields and theoretical traditions to contribute in its development (Miettinen et al., 2009).

Conclusion: place management as a boundary concept

Therefore, a future way forward for the field and practice of place management is its conceptualisation as an interdisciplinary boundary concept. This understanding draws similarities with the notion of 'boundary objects' (Star and Griesemer, 1989: 393), which highlight the simultaneous plasticity and robustness of those scientific objects that can have 'different meanings in different social worlds but their structure is common enough to more than one world to make them recognizable, a means of translation'. Place management, as shown in this chapter, has similar properties that allow its conceptual development across intersecting social worlds, as it 'can operate as a concept in different disciplines or perspectives' (Mollinga, 2008: 24) and is 'imprecise and open enough to allow people from different traditions to join without renouncing their respective worldviews' (Miettinen et al., 2009: 1313).

Viewing place management as an interdisciplinary boundary concept parallels with Parker's (2008: 5–6) interpretation of place as something that is shared and understood by many disciplines, but also as a wilderness that cannot be claimed by any particular subject area. This is similar to the idea of 'marches' – vague, imprecisely defined spaces that exist around borders and highlight the possibility of multiple plotlines (Anzaldúa, 1987; Clandinin and Rosiek, 2007). It suggests opportunities for further research where place management is concerned. Such research can continue to explore the vibrant multiplicity of theoretical underpinnings of the place management concept, as well as its continuing application in practice.

References

Adams D (2008) Place management: collecting definitions and perspectives: reflections from the Editorial Advisory Board. *Journal of Place Management and Development* 1(1): 17–28.

Agnew J (1994) Timeless Space and State-Centrism: The Geographical Assumptions of International Relations Theory. In Rosow SJ, Inayatullah N and Rupert M (eds), *The Global Economy as Political Space*, Boulder, CO: Lynne Rienner, pp. 87–108.

Anzaldúa G (1987) *Borderlands/La Frontera: The New Mestiza*. San Francisco: Aunt Lute Books.

Ashworth GJ (2008) Place management: collecting definitions and perspectives: reflections from the Editorial Advisory Board. *Journal of Place Management and Development*, 1(1): 18.

Ashworth GJ and Voogd H (1990) *Selling the City: Marketing Approaches in Public Sector Urban Planning*. London: Belhaven Press.

Auge M (1995) *Non-places: Introduction to an Anthropology of Supermodernity*. London: Verso.
Biddulph M (2011) Urban design, regeneration and the entrepreneurial city. *Progress in Planning* 76(2): 63–103.
Blanco I, Griggs S and Sullivan H (2014) Situating the local in the neoliberalisation and transformation of urban governance. *Urban Studies* 51(15): 3129–3146.
Brand R and Gaffikin F (2007) Collaborative Planning in an Uncollaborative World. *Planning Theory* 6(3): 282–313.
Clandinin DJ and Rosiek J (2007) Mapping a Landscape of Narrative Inquiry: Borderland Spaces and Tensions. In: Clandinin DJ (ed.), *Handbook of Narrative Inquiry: Mapping a Methodology*, Thousand Oaks, CA: SAGE Publications Ltd, pp. 35–76.
Coca-Stefaniak JA (2008) Place management: collecting definitions and perspectives: reflections from the Editorial Advisory Board. *Journal of Place Management and Development*, 1(1): 17–28.
Coca-Stefaniak JA and Bagaeen S (2013) Strategic management for sustainable high street recovery. *Town and Country Planning* 82(12): 532–537.
Collinge C and Gibney J (2010) Place-making and the limitations of spatial leadership: reflections on the Øresund. *Policy Studies* 31(4): 475–489.
Cox KR (1993) The local and the global in the new urban politics: a critical view. *Environment and Planning D*, 11(4): 433–448.
Cox KR (1998) Spaces of dependence, spaces of engagement and the politics of scale, or: looking for local politics. *Political Geography* 17(1): 1–23.
De Nisco A and Warnaby G (2014) Urban design and tenant variety influences on consumers' emotions and approach behavior. *Journal of Business Research* 67(2): 211–217.
Dirlik A (1999) Globalism and The Politics of Place. In Olds K, Dicken P, Kelly PF, et al. (eds), *Globalisation and The Asia-Pacific" Contested Territories*, London: Routledge, pp. 37–54.
Farinelli F (2003) *Geografia. Un'introduzione ai modelli del mondo*. Turin: Einaudi.
Forester J (2010) Foreword. In Cerreta M, Concilio G, and Monno V (eds), *Making Strategies in Spatial Planning*, Dordrecht: Springer, pp. v–vii.
Fraser J, Bazuin JT and Hornberger G (2016) The privatization of neighborhood governance and the production of urban space. *Environment and Planning A* 48(5): 844–870.
Geddes M (2006) Partnership and the limits to local governance in England: institutionalist analysis and neoliberalism. *International Journal of Urban and Regional Research* 30(1): 76–97.
Guarneros-Meza V and Geddes M (2010) Local governance and participation under neoliberalism: comparative perspectives. *International Journal of Urban and Regional Research* 34(1): 115–129.
Hall CM (1997) Geography, marketing and the selling of places. *Journal of Travel & Tourism Marketing*, 6(3–4): 61–84.
Hall T and Hubbard P (1996) The entrepreneurial city: new urban politics, new urban geographies? *Progress in Human Geography* 20(2): 153–174.
Harvey D (1989a) *The Condition of Postmodernity*. Oxford: Blackwell.
Harvey D (1989b) From managerialism to entrepreneurialism: The transformation in urban governance in late capitalism. *Geografiska Annaler. Series B, Human Geography* 71(1): 3.
Hidle K and Normann RH (2013) Who Can Govern? Comparing Network Governance Leadership in Two Norwegian City Regions. *European Planning Studies* 21(2): 115–130.
Hillier J (2007) *Stretching Beyond the Horizon: A Multiplanar Theory of Spatial Planning and Governance*. Aldershot: Ashgate.
Irazábal C (2009) Realizing planning's emancipatory promise: Learning from regime theory to strengthen communicative action. *Planning Theory* 8(2): 115–139.
Jessop B (2013) Recovered Imaginaries, Imagined Recoveries: A Cultural Political Economy of Crisis Construals and Crisis-Management in The North Atlantic Financial Crisis. In Benner M (ed.), *Before and Beyond the Global Economic Crisis*, Cheltenham: Edward Elgar, pp. 234–254.
Kavaratzis M and Ashworth G (2008) Place marketing: how did we get here and where are we going? *Journal of Place Management and Development* 1(2): 150–165.
Le Feuvre M, Medway D, Warnaby G, et al. (2016) Understanding stakeholder interactions in urban partnerships. *Cities* 52: 55–65.
Lepofsky J and Fraser JC (2003) Building community citizens: Claiming the right to place-making in the city. *Urban Studies* 40(1): 127–142.
Mant J (2002) Place management as an inherent part of real change: a rejoinder to walsh. *Australian Journal of Public Administration* 61(3): 111–116.

Mant J (2008) Place management as a core role in government. *Journal of Place Management and Development*, 1(1): 100–108.
Massey D (2005) *For Space*. London: Sage.
Mayer H and Knox P (2010) Small-town sustainability: Prospects in the second modernity. *European Planning Studies* 18(10): 1545–1565.
McCann EJ (2004) 'Best places': Interurban competition, quality of life and popular media discourse. *Urban Studies* 41(10): 1909–1929.
Medway D, Parker C and Roper S (2016). Litter, gender and brand: The anticipation of incivilities and perceptions of crime prevalence. *Journal of Environmental Psychology* 45: 135–144.
Miettinen R, Samra-Fredericks D and Yanow D (2009) Re-turn to practice: An introductory essay. *Organization Studies*, 30(12): 1309–1327.
Millington S and Ntounis N (2017) Repositioning the high street: evidence and reflection from the UK. *Journal of Place Management and Development*, 10(4): 364–379.
Millington S., Ntounis N., Parker C. and Quin S. (2015) *Multifunctional Centres: A Sustainable Role for Town and City Centres*. Manchester: Institute of Place Management.
Mollinga P (2008) *The Rational Organisation of Dissent: Boundary Concepts, Boundary Objects and Boundary Settings in the Interdisciplinary Study of Natural Resources Management*. Bonn: ZEF.
Naughton L (2014) Geographical narratives of social capital: Telling different stories about the socio-economy with context, space, place, power and agency. *Progress in Human Geography* 38(1): 3–21.
Niedomysl T and Jonasson M (2012) Towards a theory of place marketing. *Journal of Place Management and Development* 5(3): 223–230.
Omholt T (2013) Developing a collective capacity for place management. *Journal of Place Management and Development* 6(1): 29–42.
Oliveira E (2015) Place branding as a strategic spatial planning instrument. *Place Branding and Public Diplomacy* 11(1): 18–33.
Parés M, Martí-Costa M and Blanco I (2014) Geographies of governance: How place matters in urban regeneration policies. *Urban Studies* 51(15): 3250–3267.
Parker C (2008) Extended editorial: place – the trinal frontier. *Journal of Place Management and Development*, 1(1): 5–14.
Parker C (2009) Making Places Better: An International Perspective. In *International Cities Town Centres & Communities Society Conference*, 27–30 October, Geelong, Victoria, Australia: Deakin University.
Parker C (2011) Place management: An international review. Manchester.
Parker C, Roper S and Medway D (2015) Back to basics in the marketing of place: the impact of litter upon place attitudes. *Journal of Marketing Management* 31(9–10): 1090–1112.
Pasquinelli C (2014) Branding as urban collective strategy-making: The formation of Newcastle Gateshead's organisational identity. *Urban Studies* 51(4): 727–743.
Peyroux E, Putz R and Glasze G (2012) Business Improvement Districts (BIDs): the internationalization and contextualization of a 'travelling concept'. *European Urban and Regional Studies* 19(2): 111–120.
Pine AM (2010) The performativity of urban citizenship. *Environment and Planning A* 42(5): 1103–1120.
Reddel T (2002) Beyond participation, hierarchies, management and markets: 'New' governance and place policies. *Australian Journal of Public Administration*, 61(1): 50–63.
Richards G and Wilson J (2004) The impact of cultural events on city image: Rotterdam, cultural capital of Europe 2001. *Urban Studies* 41(10): 1931–1951.
Rota FS and Salone C (2014) Place-making processes in unconventional cultural practices. The case of Turin's contemporary art festival Paratissima. *Cities* 40: 90–98.
Salet W and Savini F (2015) The political governance of urban peripheries. *Environment and Planning C* 33 (3): 448–456.
Sassen S (1991) *The Global City*. Oxford: Blackwell.
Smyth H (1994) *Marketing the City: The Role of Flagship Developments in Urban Regeneration*. London: E.&. F.N.Spon.
Sorensen A and Sagaris L (2010) From participation to the right to the city: Democratic place management at the neighbourhood scale in comparative perspective. *Planning Practice and Research* 25(3): 297–316.
Star SL and Griesemer JR (1989) Institutional ecology, `Translations' and boundary objects: Amateurs and professionals in Berkeley's Museum of Vertebrate Zoology, 1907–39. *Social Studies of Science* 19(3): 387–420.

Stewart-Weeks M (1998) *Place management: fad or future?*, A presentation to an Open Forum, Institute of Public Administration Australia (NSW Division), 20th August 1998, Dixson Room, State Library, Sydney, The Albany Consulting Group, Sydney.

Stone CN (1993) Urban regimes and the capacity to govern: A political economy approach. *Journal of Urban Affairs*, 15(1): 1–28.

Swyngedouw E, Moulaert F and Rodriguez A (2002) Neoliberal Urbanization in Europe: Large-Scale Urban Development Projects and the New Urban Policy. In Brenner N and Theodore N (eds), *Spaces of Neoliberalism*, Chichester: John Wiley & Sons, Ltd, pp. 194–229.

van den Berg L and Braun E (1999) Urban competitiveness, marketing and the need for organising capacity. *Urban Studies* 36(5–6): 987–999.

Walsh P (2001) Improving governments' response to local communities - is place management an answer? *Australian Journal of Public Administration* 60(2): 3–12.

Ward K (2003) Entrepreneurial urbanism, state restructuring and civilizing 'New' East Manchester. *Area* 35(2): 116–127.

Warnaby G (2009) Non-place marketing: transport hubs as gateways, flagships and symbols? *Journal of Place Management and Development* 2(3): 211–219.

Warnaby G and Medway D (2013) What about the 'place' in place marketing? *Marketing Theory* 13(3): 345–363.

Whyatt G (2004) Town centre management: how theory informs a strategic approach. *International Journal of Retail & Distribution Management* 32(7): 346–353.

Wood A (1993) Organizing for local economic development: local economic development networks and prospecting for industry. *Environment and Planning A* 25(11): 1649–1661.

Yanchula J (2008) Finding one's place in the place management spectrum. *Journal of Place Management and Development*, 1(1): 92–99.

30

Risk, resilience and place

Paul O'Hare and Iain White

Introduction

Risk, we are told, has emerged to become a defining feature of our modern world (Beck, 1992; Giddens, 1991). Whether it encompasses fears regarding climate change, meteorological and geophysical hazards, financial crises, warfare, terrorism, threats to public health, or precarious employment, threats are part of the background soundtrack of everyday life. This range of risks intersects with people and places across temporal and spatial axes, from the global to the local, and spanning immediate threats and longer-term intergenerational risks. Their consequences for places are equally diverse, from the incremental stresses associated with a prolonged drought or economic recession, to the more sudden, uncertain and unpredictable impacts of events associated with earthquakes or terrorist attacks. Even where hazards appear relatively localised or geographically contained, the interconnected nature of the contemporary world means that the shockwaves of a particular risk ripple outwards, potentially affecting places and people who initially appear unconnected from the primary hazard. The Global Financial Crisis, for example, resulted in a restriction of credit that affected businesses far removed from the housing sector and a lowering of purchasing power that hit those in service sectors who did not even own homes (Squires and White, 2019). Moreover, in this instance, the impacts of the financial shock extend from the immediate disruption for individuals and families dealing with the impacts of unemployment, retracting incomes or the inability to access credit, to a more insidious influence on the flows of capital and labor that shape current development, future trajectories, and ultimately influence the 'success' of places.

The nature, distribution, impacts and even the very existence of risks is socially constructed, strongly correlated to the ability of societies to manage them, and the strategies that are adopted in response. Dependent upon the nature of the threat, risk management is a task that may be shared, either formally or informally, between a number of stakeholders across scales, including the international community, the nation state, citizens, and the private, voluntary and community sectors. Where the cause and effect of risks seem to be readily identifiable, and there is an agent with a relevant mandate and ability to take executive action, then implementing measures to try to mitigate or manage risk can be relatively straightforward. This is particularly the case when 'solutions' to manage risk have been developed. For example, the compelling evidence concerning the benefits of wearing seatbelts in cars made it relatively easy for governments to act. Yet, even in this instance, the mandatory wearing of seatbelts took some time to be adopted across the world, and 'belting-up' initiatives were sometimes met with a degree of resistance on the part of lobby groups and car drivers and passengers themselves. As with so many risks and risk management strategies, it is proposed that individuals should have the ability to decide whether to take risks without interference from the state, whether it concerns their health and safety or whether they want to live in a 'high-risk' flood zone.

This is partly related to wider perceptions of the role and scope of the state, which differs between places. For example, many in the US prefer the 'freedom' of deciding whether or not to purchase expensive, individualised healthcare, rather than the universal 'socialised medicine' of the UK's National Health Service, which is funded by general taxation. Often, too, risk assessments can be debated, with some contesting the appraisal of the severity or extent of hazards, or contentions that assessments of the benefits of interventions might be overstated. This is particularly noticeable in issues relating to uncertain risks that may have an economic impact, such as how the future impacts of climate change may impact on property prices (Haughton and White, 2017). These examples also demonstrate how risks that may be calculated in isolation, and perceived very differently, intersect. While it may be convenient scientifically to imagine risks in such a way, in reality health risks or climate risks may be seen through the lens of more intangible political or economic risks. There are, too, suggestions that risk management strategies, rather ironically, might drive other risk-taking behavior. Insurers refer to this as 'moral hazard' or the 'Peltzman Effect' whereby the impact of 'safety' features is undermined, paradoxically rendering systems as a whole more vulnerable (see Peltzman, 1975).

Recognising the complex and constructed nature of risks has been a key research challenge over recent years, and has broadened the disciplines involved from the initial focus on statisticians, modellers, and engineers, to social scientists, geographers or architects. It is now acknowledged that underlying any discussion regarding responses to risk are broader governance issues such as the balance of responsibility between the state and the individual, or how risk may be perceived and encountered differently across society and space and their relative sensitivity and vulnerability.

In part response to this complex context of risk, the concept of resilience has rapidly emerged as part of the lexicon of risk management across policy and practice (c.f. Coaffee and Lee, 2016; White and O'Hare 2014). It offers a normative, anticipatory perspective to risk management and risk governance that has potential to manage multiple hazards by promoting the ability of people and places to cope with turbulence or enable systemic change to be less vulnerable. Politically, resilience is ultimately presented as an effort to govern and to seek to control uncertainty. Indeed, resilience has become a staple feature of public and social policy initiatives and offers the potential for place-based responses to a plethora of anticipated and unanticipated risks alike that may be difficult to predict, remove, or reduce.

Both risk and resilience have strong connections to place. However, the ways that this becomes operationalised in practice has significant implications for places and the people that inhabit and use them. In this chapter we draw from both the international literature and from our own research to offer insights into how place is understood and incorporated in risk and resilience policy, practice and governance. In doing so we disassemble the broad notion of risk across place, in particular examining how risk is uncertain and is imbued with intricate complexities across scales, temporalities and intersectionalities. We deal with each of these themes in turn, across three recurring conceptual themes that are necessary to understand the place-based characteristics and distinctions of risk:

- First, that while risk and resilience strategies are specific to place and people, their reliance upon technocratic assessments rooted in historical calculative practices means that these tend to be spatially blind, homogeneous, and focus on coping with risks rather than adaptation to the threat.
- Second, risk management strategies tend to be siloed, ignoring the wider political economy that embeds risk and disadvantage in a place.
- And third, following on from these assertions, we argue that to better manage risk and better apply resilience we need to link these more strongly to place-based social, cultural and economic contexts.

Risky places

Two classic cases can be deployed to introduce key ideas relating to risk and place. The first is the early research on health and environmental risk conducted by noted English physician John Snow (1813–1858). 19th-century London – like many cities – hosted frequent outbreaks of cholera, which at the time was thought to be caused by 'bad air'. Skeptical of this theory, Snow talked to residents and mapped cholera cases, eventually isolating the source of an outbreak to the Broad Street neighborhood water pump. The pump was disabled to demonstrate that the risk was not connected to a general miasma but was *place-specific*. His geographical analysis, though quite simple in mapping terms, fundamentally changed how this risk was perceived and managed and had significant implications for the medical profession (see Figure 30.1).

The second example is the work of pioneer social researcher Charles Booth (1840–1916). Determined to establish the true extent of London's poverty in the late 19th century, Booth surveyed and mapped the income and social class of the inhabitants of the city (Figure 30.2). The maps were a forerunner of the advanced spatial analysis commonly used today. They were color-coded on a street-by-street basis with a series of classifications ranging from 'Lowest class. Vicious, semi-criminal' to 'Upper-middle and upper classes. Wealthy'. The analysis concluded that 31% of citizens were living in extreme poverty, influencing both the burgeoning socialist movement at the time and government policy (Hall, 2002).

These examples introduce key concepts relating to risk and place which we will now elaborate upon:

1. Some risks are ***certain***, stemming from a clear, identifiable hazard, while others are more uncertain and diffuse, such as those relating to poverty or climate change;
2. Some can be addressed quite easily with a low-cost solution at a neighborhood scale, whilst others are ***complex*** and relate to the broader political economy;
3. From a ***scalar*** perspective, some map well onto existing institutions and policies, while others span multiple administrative issues and boundaries;
4. Risk has important ***temporal*** dimensions: some are retrospective, occurring after the detriment has been suffered, while others are more anticipatory;
5. Lastly, some stem from a single, distinct risk, while others are subject to ***intersecting*** issues and disadvantages.

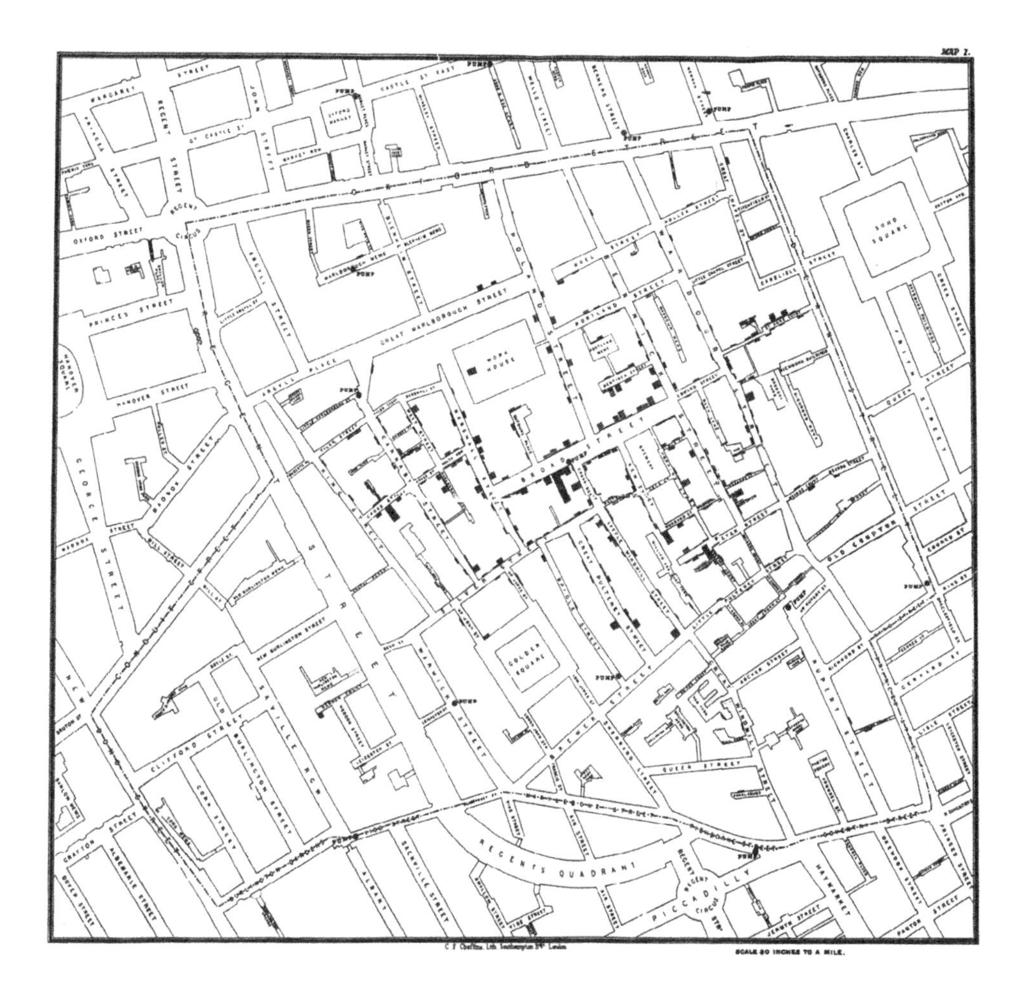

Figure 30.1 Original map by John Snow (1855) detailing the place clustering of cholera cases in the London epidemic of 1854.

We will now discuss each of these in turn in an effort to elaborate upon the challenges and nuances of managing risky places.

Certainty and place

Significant effort is invested in efforts to fix a risk to a 'place' or, in other words, to spatialise risk. For example, hazard maps delineating varying degrees of risk, such as those associated with flood risk, tsunami lines, or coastal erosion, are regularly published into the public domain with increasing degrees of technical precision or confidence. Demonstrating this, like most countries throughout the world, the United Kingdom publishes maps identifying the extent and severity of flood risk (see https://flood-warning-information.service.gov.uk/long-term-flood-risk/map for England-specific flood risk maps). These are useful in multiple

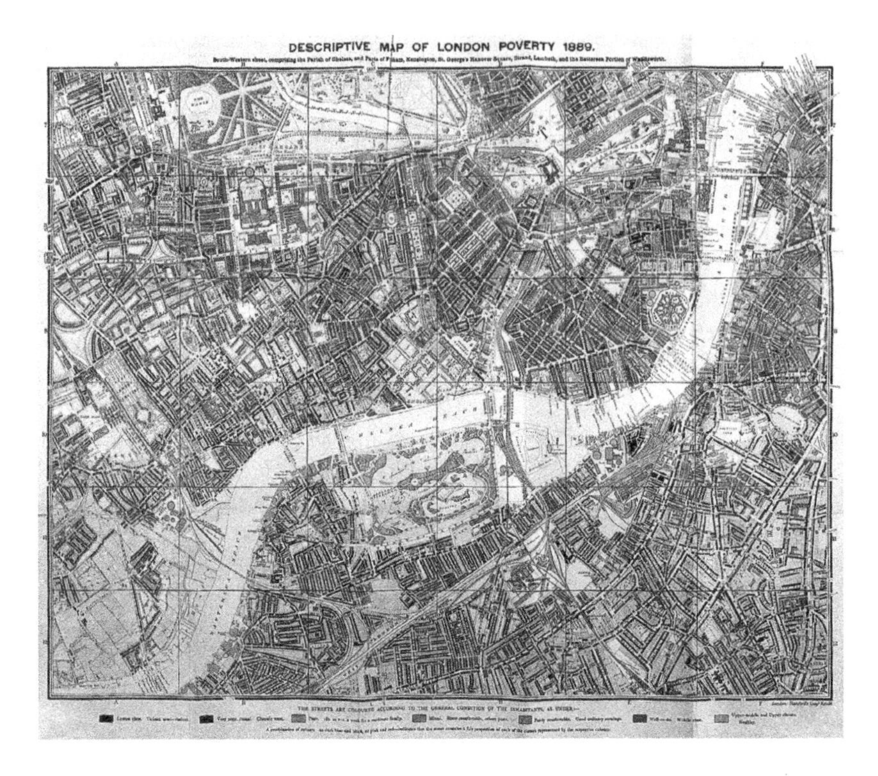

Figure 30.2 Charles Booth's map, Descriptive of London's Poverty.

ways, from helping decision makers allocate resources or policy attention to areas of 'higher' risk, or for the individual home purchaser. Risk assessments perform a similar dual function, providing information on the threats to places and their potential impacts, as well as supplying part of the evidence base that enables decision makers to decide how to respond. While appearing scientifically authoritative, they have however been accused of providing a 'false precision', given the extent to which they are subject to frequent revision (White, 2013).

More broadly, these approaches aid resilience in a number of ways, such as by shaping development and markets, or the logics and decisions of developers, businesses or individuals who may wish to be exposed to less risk. As such, technical devices by which societies make risk visible hold real power. They are political instruments that help form perceptions of places as 'risky' or 'safe'. Depending upon where you put the line on a map, or whether scientific judgment says a place is low, medium, or high risk, investment or capital may flow differently, and people or businesses may move. There are even periods of negotiation where, before publication, technical lines may be redrawn to trace property boundaries in an effort to provide clarity over zoning and development rights, or be subject to legal challenge as communities produce alternative models that see the spatiality of risk differently (Haughton and White, 2017).

While mapping, zoning or risk assessing are common risk management techniques, several threats struggle to fit into this place-oriented technical approach. The term 'deep uncertainty' can help explain risks that entail interactions that may be so intricate and dynamic that system changes and the timing and impact of any consequences lack any clarity. For instance, traditional quantitative assessments of probability of occurrence and impact are difficult to ascertain with regard to long-term climate change scenarios as they rely heavily on how society responds to the threat. Here uncertainty is not something that can be quickly reduced. Knowledge accumulates, but slowly, as social aspects of risk resist traditional methods of risk management that are designed to provide precise lines, figures, or probabilities for decision makers.

Discourses connected to certainty and place provide an important context in two ways: they demonstrate an engagement with complex systems thinking and simultaneously – and perhaps paradoxically – they highlight the scientific limits and managerial constraints of managing risk. For instance, public policy decisions require evidence on risks to proceed, but while this reduces political liability and increases the defence of selected options, it doesn't reflect the ways that the dynamism of risks resists such a techno-rational approach. If there is one thing we can say with certainty, it is that certainty regarding risk is misplaced. These will not only be revised as more scientific endeavor reveals anew their inaccuracy, but society is considerably more complex and messy than scientific models or assessments can acknowledge. In cases like these, the logical response may not be to quantify, 'solve' or 'manage' the problem but rather to be more adaptive or 'resilient' to the impacts. The pursuit of resilience here entails efforts to reduce what is known as path dependencies; in this instance this might be developments that lock risk into places over the long term. For example, through the construction of new housing or new infrastructure, development patterns are set for decades into the future, and so too are risks. In contrast, adaptive decision making would entail effort to work with communities to help them understand, select options and development trajectories, and ultimately to make the transition to a more resilient future (Haasnoot et al., 2013; O'Hare et al., 2015).

Complexity and place

Any effort to understand the impact of risk on a place requires an appreciation of how risks fit within existing technical or managerial remits and jurisdictions. In cases where a clear causal link between the driver of a risk and a place is demonstrable, executive and authoritative action can be easily executed. In the case of John Snow's effort to demonstrate patterns of cholera outbreak and their proximity to users of a water pump, an agency with responsibility for managing the water supply infrastructure and for public health was able to decommission the facility. A further example is arguably the most successful instance of environmental risk management of modern times. In the 1970s scientists discovered the risk of increased UV radiation stemming from a growing hole in the ozone layer. The cause was clear: the release of chemicals, mainly chlorofluorocarbons (CFCs) used in refrigeration and aerosols, into the atmosphere. Politicians acted swiftly. A global ban of CFCs and other ozone-depleting gases ensued, and now the ozone layer is recovering. Readers may reflect upon how this successful situation appears to differ to other global environmental risks, such as climate change or biodiversity loss. A key factor was the nature of the problem, which was more simple to solve in comparison. For example, an alternative technology was almost available and only a few countries produced CFCs, so the environmental risks did not represent an intractable political and economic risk ... in simple terms, it highlights the links between the characteristics of certain problems and the potential for them to be managed. In contrast, Booth's observations regarding poverty arguably has much more complex drivers, from the broad and often seemingly insurmountable political, economic and social conditions, through education, social and public health policy, to the employment circumstances of particular families and individuals. Consequently, there is little agreement on, or understanding of, the causes of poverty and disempowerment, never mind consensus on how these societal ailments can be resolved.

In 1973 Rittel and Webber used the phrase 'wicked problems' to refer to public policy dilemmas that are hard to understand, complex, and dynamic. Although the concept of 'wickedness' has been subject to significant critique in more recent times, particularly regarding the loose way it becomes assigned to, or frames, policy problems that are simply difficult (for instance, see Peters, 2017), it emphasises how risks must fit within existing decision-making frameworks in order to be effectively managed. Here we can consider the differing managerial problems presented by risks from a depleting ozone layer and climate change. While both are global, for climate change the managerial pathway for resolution is disputed, in part due to the economic reliance on fossil fuels and the varied industries that generate greenhouse gas emissions, not to mention the technological and economic difficulties in transitioning to alternative energy systems.

Just as the previous section highlighted the scientific limits of managing risk, this one emphasises the political, institutional, and managerial constraints of managing risk. The emergence of resilience as a managerial response is a direct function of the types and complexity of risks existing in a place. Even if there is agreement that 'something must be done', there may be little agreement on the precise action that can or should be taken, and uncertainty regarding the ability of people and places to respond.

Scale and place

The previous section indicated the challenges of risks that might transcend geographical boundaries or that may extend beyond administrative and political boundaries. However, the perennial quest to manage risk and to foster resilience has long had a considerable influence on the material form and the construction of built environment across multiple spatial scales. The management of risk in this respect is complex, ranging from spatial planning through to the minutiae of building elements. Risk can determine what gets built where, the design, layout and materiality of our buildings, and how and when a building is used by its inhabitants. Risk management in the construction industry is also big business, driving innovation in the development of materials and research in developing better and safer building systems and techniques.

Such interventions have a long history. The safety and security that gathering in places can bring helped develop our early cities in the first place, while the very first formal building regulations in London were devised in the wake of the Great Fire of 1666 (Imrie and Street, 2011). The management of risks in places is associated with the rise of institutions, regulations, and aspects such as best practice. Today this is perhaps most notable through the codification of construction standards and the drafting and enforcement of planning legislation that governs what can be built where, for what function, and with what form. Buildings are regularly claimed to be earthquake proof or flood resistant, while on a larger scale planners and urban designers are encouraged to make the built form more resilient to take greater account of the threat of terrorism in the design of public places (Coaffee et al., 2009) or to 'make space for water' (White, 2010).

Research is also starting to reveal how the development of societies, particularly the forms of bureaucracy associated with the effective functioning of the typical nation state, means that risks tend to be more easily managed within particular scales and sectors. In today's risk society, the many professionals involved in the construction industry and those with responsibility for installing, maintaining, and controlling key services that we all rely on have a responsibility – usually legally mandated – for managing public safety and public risk. Yet, these risk and resilience policies have increasingly been subject to critique. For example, in a study of resilience practice in the UK, it was found that resilience within spatial planning was characterised by a simple return to normality that is more analogous with neo-liberal norms and techno-managerial modes of risk governance rather than potentially more transformative alternate interpretations of the concept (White and O'Hare, 2014). This underpins existing behavior and normalises risk, ignoring wider sociocultural concerns through concentrating on technical and reactive measures at building scales. Therefore, existing ways of 'knowing and doing' influence the uneven scalar distribution of both risk and resilience across places, which also helps to foreground our next question: to what extent does risk and resilience engage with temporal considerations?

Time and place

Risk management is conducted in many stages, ranging from the management of a specific threat or preparedness to a disruptive event, through to anticipatory risk management: the assessment and even embracing of risky futures in decision making. While the immediacy of an emergency crisis response provides a visible reminder of how well we can react to a present shock or stress, to a certain extent the future is *always* a part of risk management for places. For instance, environmental audits or impact assessments that are often a required component of spatial planning, development control and routine place management strategies, influence decision making and enable risks and rewards across time and into the future to be deliberated. These can have significant implications for the design and function of places. One way to picture the influence of temporal risk on places is via the concept of 'riskscapes', an idea that emphasises the various ways in which risk becomes manifest over space and time (Müller-Mahn et al., 2013). Just as decisions we have made in the past shape the real and imagined geographies of risk, so too do decisions about possible futures.

Questions of central interest here is when and why does the voice of the future become articulated or silenced? And what does this mean for the risk to and resilience of places? One way to explore issues connected to time is through the application of the precautionary principle, which argues that where there are threats of serious damage, a lack of scientific understanding should not be used as a reason to justify inaction across a period of time. It is a feature of public policy that many risks only become apparent after detriment has occurred, but should we always be so reactive? The risks presented by genetically modified crops provide a good illustration; while they may bring benefits to crop yield, we may not yet fully understand the impact of genetic modification of plants across longer periods of time. So should society go ahead and adopt innovation, wait until science provides more certainty, but potentially create new risks for places, or should the proponents have to prove they are *not* a risk, and so places forego their benefits before they can be used? Does the burden of proof exist to prove it is safe, or to prove it is risky?

Against the backdrop of apocalyptic visions of an uninhabitable future world, efforts to understand our ability to more effectively govern future risk is a burgeoning academic field. Anticipatory risk management is framed and understood through both indeterminacy and uncertainty, meaning that decision making focused on futures is an immensely challenging endeavor. In part this helps explain some of the logic and rationale for resilience: 'whereby a future becomes cause and justification for some form of action in the here and now' (Anderson, 2010: 778). Time does, however, challenge the mechanisms of risk management, which tend to codify and discipline phenomena (Lane, 2014; White and Haughton, 2017). Just as we create boundaries spatially, we also bound temporally. For example, cost benefit analyses are defined temporally, say for 5 or 10 years, and future detriment may be heavily discounted in comparison with current concerns. This is not necessarily a bad thing, however. Scientific and economic growth may mean that future generations and places are more able to manage risks than we are in the present. So, as with all risk, it is a trade-off that will differ across cultural, political and societal contexts that will – as it has been demonstrated here – vary across time.

In many respects risk and resilience policy fundamentally involves choices: how much is acceptable, to whom, and why? Considering temporal risk perspectives across spatial considerations and the longer-term evolution of place is vital because it reveals boundaries to decision making with regard to what places, generations, or times are given voice in the present, and the ways that our methods and approaches may exert hidden influences on outcomes. Our final layer of complexity develops this strand of thinking by exploring the extent to which risks such as these are considered in isolation. How does risk management consider people or places subject to intersecting environmental and societal risks?

Intersectionality and place

The idea that across society people can be disadvantaged by where they live is an important one for this chapter (Fainstein, 2011). But it is increasingly acknowledged that risk cannot be wholly explained through a simple analysis of spatial proximity – often referred to as exposure – to hazards. Instead, other dimensions of risk such as vulnerability and sensitivity are essential to holistic understandings of risk, particularly in terms of understanding why some individuals, communities and places are more affected by hazards than others. Some people and places are not only subject to more hazards, they may have less ability to cope with shocks or stresses. Beyond assisting us in assessing how risk affects place and society, understanding these broad profiles is vital to understanding how places react and respond to hazards when they strike; often referred to as the capacity to respond to a threat. In a similar sense, the composition and profiles of places at risk become critical to understanding risk in a more complete sense. The term intersectionality has gained considerable momentum in recent years, demonstrating the interplay between and across the various characteristics and social circumstances of people. It can be similarly applied to places.

Under these more nuanced interpretations of risk, the type of hazard – whilst remaining a critically important variable of a risk profile – lies adjacent to a broad range of complex, intricate and related social, economic, and cultural characteristics. Demonstrating this, literature focusing on environmental justice and environmental racism illustrates how the distribution of risk is related to wider social factors such as deprivation or race (Bullard, 1999; Walker, 2012). In a study of flooding, we argue that while there are compounding interactions between flooding and other social disadvantages across multiple public policy areas and scales, this goes largely unacknowledged (O'Hare and White, 2018). For example, people can be made less resilient to the effects of a natural hazard by losing their job, or by simply living in an area with low economic investment which may struggle to attract scarce resources. This highlights the existence of public policy silos and how risk management practices are based in the technical professions that tend to see people and communities as homogeneous entities in a given spatiality. Such accounts, for instance, prompt risk managers to be cognisant of and to accept that the impacts of a hazard are rooted with complex social and economic disadvantages that stem from beyond the specific area of the hazard itself.

Helping to demonstrate this point, the term climate disadvantage has emerged as a concept that associates risk management approaches to broader socio-economic factors and conditions that frame vulnerabilities and capacity to respond to hazards (Lindley et al., 2011). Such work – residing at the intersection between hazard research and broader understandings of societal disadvantage and vulnerability – is useful in illuminating more sophisticated and holistic understandings of the risk profile of a place. How risk is encountered, understood, endured and responded to is immensely uneven across society, but often this is not recognised by risk analysis or policy and practical efforts to address risk. Consequently, the resilience of people and places is not just related to mapping, zoning, or emergency response, it also is connected to the capabilities and capacities of communities and aspects such as social networks or empowerment (Mehmood, 2016).

Conclusion

Risks, clearly, are an inevitable facet of modern society. They are multifaceted, dynamic, and affect people and places differently. They may also have diverse scientific and managerial facets relating to politics and uncertainty. Fundamentally, place has a dialogical relationship with risk. Risk maps onto places and – by extension – the people that inhabit these places in very different, often unanticipated, ways. Spatiality is equally important to understanding risk and the associated framing of risk management initiatives. But it is vital to be cognisant that spatiality – just like other dimensions of place-making – extends beyond exposure or proximity to a hazard. Ultimately, such observations stem from concerns that predominantly technical and static risk management decision-making frameworks are frequently not reconciled with the broader dynamic and complex political-economy contexts within which they reside, not to mention across environmental, cultural and socio-economic place characteristics. More specifically, risk management approaches are bound within their own calculative rationalities, for instance, by often bundling people and places together into 'neat' administrative units. But these are often conducted with limited attention to the sensitivities and capabilities of inhabitants whilst specific risks are nested against other much broader factors and forces. Risk mapping exercises all too often focus on exposure rather than taking account of these broader characteristics. Moreover, risk is dynamic and mobile, cascading across space and time. Any effort to capture the complexity of risk is infused with uncertainty and unknowns.

In an effort to illustrate these tensions, this chapter has disassembled the phenomena of risk and its relationship to place across social, spatial and temporal dimensions of analysis. The chapter has also noted how, at the same time, resilience has emerged as an influential rhetorical device; a mobilising concept for policy and political efforts to offer hope in the face of risk. The term permeates public policy, particularly policy relating to place. But the articulation of resilience policy can similarly take place through rather opaque processes that are simultaneously technically complex and imbued with complex power dynamics. As a wider range of people – including citizens and civil society – are expected to assume greater responsibility for resilience, the contrasting exposure, vulnerability and capacity to respond of different places must be recognised and addressed. But a greater recognition of the compounding socio-spatial-temporal dimensions of risk and resilience will more accurately reveal why certain places are more vulnerable to risk and better placed to pursue resilience than others.

References

Anderson, B. (2010) Preemption, precaution, preparedness: anticipatory action and future geographies. *Progress in Human Geography*, 34, (6), 777–798.

Beck U. (1992) *Risk Society, Towards a New Modernity*. London: Sage Publications.

Bullard R. (1999) Dismantling environmental racism in the USA. *Local Environment*, 4, (1), 5–19.

Coaffee J., Lee P. (2016) *Urban Resilience: Planning for Risk, Crisis and Uncertainty*. Basingstoke: Palgrave Macmillan.

Coaffee, J., O'Hare, P. and Hawkesworth, M. (2009) The visibility of (in)security: the aesthetics of planning urban defences against terrorism. *Security Dialogue*, 40, (4-5), 489–511.

Fainstein, S. (2011) *The Just City*. Ithaca, United States: Cornell University Press.

Giddens A. (1991) *Modernity and Self-Identity: Self And Society in The Late Modern Age*. Cambridge: Polity. Giddens.

Haasnoot, M., Kwakkel, J.H., Walker, W.E., Ter Maat, J. (2013) Dynamic adaptive policy pathways: a method for crafting robust decisions for a deeply uncertain world. *Global Environmental Change*, 23, 485–498.

Hall, P. (2002) *Cities of Tomorrow*. Oxford: Blackwell Publishing.

Haughton, G. and White, I. (2017) Risky Spaces: creating, contesting, and communicating lines on environmental hazard maps. *Transactions of the Institute of British Geographers*, 43, (3), 435–448.

Imrie, R. and Street, E. (2011) *Architectural Design and Regulation*. Oxford: Wiley-Blackwell.

Lane, S.N. (2014) Acting, predicting and intervening in a socio-hydrological world. *Hydrol. Earth Syst. Sci.*, 18, 927–952.

Lindley S., O'Neill J., Kandeh J., Lawson N., Christian R., O'Neill M.(2011) *Climate Change, Justice and Vulnerability*. York: Joseph Rowntree Foundation.

Mehmood, A. (2016) Of resilient places: planning for urban resilience. *European Planning Studies*, 24, (2), 407–419.

Müller-Mahn, D., Everts, J., Doevenspeck, M. (2013) "Making Sense of the Spatial Dimensions of Risk." Pp. 202–07 in *The Spatial Dimension of Risk. How Geography Shapes the Emergence of Riskscapes*, edited by D. Müller-Mahn. London: Routledge.

O'Hare, P. and White, I. (2018) Beyond 'just' flood risk management: the potential for-and limits to-alleviating flood disadvantage. *Regional Environmental Change*, 18, (2), 385–396.

O'Hare, P., White, I. and Connelly, A. (2015) Insurance as maladaptation: resilience and the 'business as usual' paradox. *Environment and Planning C: Government and Policy*, 34, (6), 1175–1193.

Peltzman S. (1975) The effects of automobile safety regulation. *Journal of Political Economy*, 83, (4), 677–726.

Peters, G. (2017) What is so wicked about wicked problems? A conceptual analysis and a research program. *Policy and Society*, 36, (3), 385–396.

Rittel, H.W.J., and Webber, M.M. (1973) Dilemmas in the general theory of planning. *Policy Sciences*, 4, 155–169.

Squires, G. and White, I. (2019) Resilience and Housing Markets: who is it Really For?. *Land Use Policy*, 81, 167–174.

Walker G. (2012) *Environmental Justice: Concepts, Evidence and Politics*. London: Routledge.

White, I. (2010) *Water and The City: Risk, Resilience and Planning for A Sustainable Future*. London: Routledge.

White, I. (2013) 'The more we know, the more we don't know: reflections on a decade of planning, flood risk management and false precision. *Planning Theory and Practice*, 14, (1), 106–114.

White, I. and Haughton, G. (2017) Risky times: hazard management and the tyranny of the present. *International Journal of Disaster Risk Reduction*, 22, 412–419.

White, I. and O'Hare, P. (2014) From rhetoric to reality: which resilience; why resilience; and whose resilience in spatial planning?. *Environment and Planning C*, 32, (5), 934–950.

31
Mapping place

Chris Perkins

Approaches to mapping

This chapter focusses on maps and mapping as objects and processes. As such it explores thinking, making, doing and enacting mapping. Every map, whether analogue or digital, tangible and permanent or intangible and ephemeral, static or mobile, and whether it is deployed to navigate, control, inform, inspire or imagine, tells a story of place. However, whilst maps are obviously about places, the detail of their relationship to place is frequently taken for granted and largely under-analyzed.

Map making and using maps have often been seen as practical activities, conforming to Cartesian logics, in which the map is separate from and represents certain qualities relating to the territory that it charts. Map making has been seen as a technical activity in which knowledge relating to place accumulates over time. Cartography in this scientific view works to improve the ways in which maps might communicate information about a place. As such, cognitive processes demand research, as against cultural approaches. Quantitative methods dominate in the accumulation of mapping knowledge, which is seen as incrementally improving shared best practice. Research agendas articulated by the International Cartographic Association (see Griffin, Robinson, and Roth, 2017) reflect this scientific orthodoxy.

Mapping reduces the subjective diversity of places to the spatial anonymity of the map, in which the object and practice is cast as neutral and which is deployed to work as a tool linking geography to theme. It does so by relying upon frequently unquestioned logics of coding, simplifying, classifying and symbolizing the world, depicting and inscribing certain aspects of a place, disseminating a shared cultural image, locating us and things in a common framing, but also carrying this knowledge across places.

This functional view of mapping dominated cartographic thinking for much of the twentieth century, during which cartography emerged as a discipline concerned with the production of useful knowledge (Fernández and Buchroithner, 2013) and it continues to imbue the day-to-day practices of GI science (see Wilson, 2017) and the practices of map design (see Field, 2018), as well as strongly influencing approaches to map use such as usability studies (see Nivala, Brewster, and Sarjakoski, 2008). In popular culture mapping still connotes veracity – maps place media stories and as such make them more real. Mapping in language continues to be used as a synonym for order and truth. We all take the Google route planning algorithm for granted and trust the navigation that it facilitates. However, these instrumental tool-like qualities impact upon how we interpret places and how we move around the city. Maps alter places. Every place is changed by the navigational logic available in the various Google Map interfaces, with their single instantly recognizable and globally standardized style, and their egocentric placing of you as a blue dot in the centre of a world reduced to a screen (Abend and Harvey, 2017). And the map's capacity to standardize arguably distances the form from the uniqueness so frequently associated with place, and explored throughout this handbook, reducing complexity, diversity and experience to a knowable and usable knowledge space in which diversity and difference is 'off the map'.

It is precisely this unspoken but two-way relationship that explains the inclusion of this chapter in a section of the Handbook focussing on Power, for since the 1990s mapping has been seen as reflecting and reifying social norms and is increasingly critiqued. Strongly influenced by the work of Brian Harley, Denis Wood, Denis Cosgrove and John Pickles, critical approaches to cartography and GI science increasingly saw the history of cartography as a history in which maps were deployed as tools serving the interests of powerful groups (Perkins, 2018a). In this view maps constitute places and change the world, they don't just work as functional neutral tools. This constitutive power can be clearly seen in the ways in which land use and lines of communication follow surveying practice, with lines on the ground echoing lines on the map. Rich places continue to be better mapped than poor places. Cadastral mapping plays an important role in the commodification of land (Fogelman and Bassett, 2017). The nation state continues to use maps to fix borders and legitimize its control over territory. Mapping helps make territorial claims over place more real (Boria, 2008). So maps in this view are authoritative documents designed to achieve political ends.

Theorists have drawn on diverse approaches to interpret the powerful nexus through which places come to be fixed (see Dodge, Kitchin, and Perkins (2009a) for an overview of these changing intellectual currents). More relational approaches have increasingly gained influence, interrogating social difference mapped out across long-recognized divides such as colonial power, gender relations, wealth, race, social class, educational difference, disability and sexuality. Foucauldian, Derridean and Saidian approaches have been used to help understand widely different mapping contexts. So the simple relationship between map and territory is increasingly complicated: sometimes a map is made by distilling information from a place; in other contexts the map changes the place and sometimes maps are made from other maps, with no link to the place at all (Janelle and Goodchild, 2018).

Towards the end of the twentieth century new challenges have gained ground as theorists increasingly came to question the certainty that maps embody particular political ideas. A post-representational orthodoxy has emerged in which thinkers have come to question the so-called 'ontic security' of the map, focussing instead on performative aspects of mapping as a process. Mappings in this view become emergent, ontologically insecure and mutable, and ideas about place that they facilitate are always developed in relation to the context in which they are situated (Kitchin, Gleeson, and Dodge, 2013). In part this change in emphasis stems from technological change described below, in which maps are no longer fixed and in which people can create their own changing images of places. But it also reflects the zeitgeist of performative ways of understanding place charted throughout this volume.

For some researchers these insights stem from detailed ethno-methodological exploration in which everyday relations between maps and people are charted (Brown and Laurier,

2005). Others ground performative approaches in technology studies, focussing on the transduction and technicity (Kitchin and Dodge, 2007). Others see the potential of feminist embodied geographies as informing approaches to mapping (D'Ignazio and Klein, 2015). The challenge of non-representational theory has also been taken up by researchers such as Gerlach (2018) and Deleuzian thought has been particularly influential in the remapping of cartography as performative. Together these performative approaches redirect attention towards assemblage thinking, relational thought, emergence and events instead of spatial structures, hybridity, affects enacted through mapping and towards more than human ways of approaching mapping and place.

Epitomising these trends has been the recent popularity of 'deep mapping' of different places (Wood, 2015). Deep mapping constitutes an interdisciplinary approach in the spatial and digital humanities that focusses upon narrative qualities of mapping place. As such it enrols a place and the activities in that place, together with the affective qualities associated with particular encounters, to deliver a multi-layered, multi-dimensional spatial account. So deep maps become performative and multi-mediated spatial anthropologies of place, attending to possibilities instead of fixing one particular track and delivering a richly qualitative and subjective remapping (Roberts, 2016). There is also a growing appreciation of the importance of the materiality of mapping, with research focussing on the difference that paper maps make (Hurst and Clough, 2013), as well as a growing recognition that mapping works in a more than visual fashion, dependent on haptic encounters as well as vision (Rossetto, 2019).

Situating the idea: mapping modes

In 2009 Martin Dodge, Rob Kitchin and I reflected on the medium, through the alliterative lens of modes, methods and moments (Dodge, Perkins, and Kitchin, 2009b). As such, the intent was to draw attention to the temporality of all mapping, and its active constitution in digital mappings, which can stitch time together, enrol ephemerality and mobility, or in different contexts chart (in)formalities (see Lammes et al., (2018) for a more detailed exploration of these processes). We also used the notion to signal possible directions for geographic research.

Mapping modes offers a very appropriate framing through which to highlight recent interventions relating mapping to place. The notion of mapping modes was first deployed by Edney (1993) as an assemblage of technologies, things, people, ideas and meanings that coalesce at different times. Modes bring together thought and practice that all too frequently have been seen as separate. The framing enables mapping to be deployed in different places, but at different times modes might coexist – for example, even in the era of digital mapping on mobile devices the printed map continues to be published. Every mapping mode inevitably privileges different kinds of map – place relations. These are delivered through changing affordances that flow from the assemblage characteristic of that mode. For example, medieval mapping in Europe tended to emphasize very generalized narratives of place, in which temporality was more important than detailed depictions of particular places – mappae mundi operated as Christian chronological stories, bringing the known world together to reinforce a particular morality, instead of serving as a depiction of reality structured by a coordinate system (Woodward, 1985).

By way of contrast, the hegemony of national mapping agencies and of printed topographic surveys which began with the Cassini mapping of France in the eighteenth century and continued through to the end of the twentieth century was usually associated with the mapping of places at unprecedented scales, but with a standardized view of place conforming to the interests of the nation state (Rankin, 2016). Military interests usually underpinned these projects and state secrecy often characterized the mapping of place, with citizen access to mapping frequently limited. Feelings and subjectivities were outwith the map. The expert knowledge of the surveyor emphasized notions of accuracy and precision, but ignored the context of social forces.

Towards the second half of the twentieth century increasing adoption of neoliberal policies weakened the power of these state monopolies and emerging suites of technologies gradually rendered the fixed paper format increasingly redundant. These technological changes can usefully be summarized in terms of infrastructures, algorithms and interfaces. Some infrastructures make mapping possible. Cloud-based computing, for example, certainly strongly facilitates mapping in an age of big data (Wilson, 2017). And the pervasive but hidden links to infrastructures of military power that underpin earth imaging and GPS lead to us all using maps that are a side-effect of military expenditure. A second kind of infrastructure is offered by the mapping itself. Geo-browsing via mapping interfaces is central to the mission of non-state actors such as Google and Apple whose 'free' mapping platforms garner users and offer new affordances, in which place can be made anew each time a user accesses an online map. Posting and geotagging make new places during our online interactions (Zook and Graham, 2007). Accessing the web feeds information to these corporations who use this to target needs of users.

The algorithms that underpin these systems were seen as a black box in the early years of the new millennium, but in the last decade significant research has started to chart how the code underpinning new mappings of place actually works. Emerging interest in digital geographies focusses attention on the power of these algorithms to script events and place (Ash, Kitchin, and Leszczynski, 2018). The availability of API code allows a mass sharing of information in mash ups, delivering geo-referenced data against a common backdrop. Algorithms power mapping but also change what the maps afford.

These algorithms present mapped information on multiple platforms through interfaces. Interfaces work as technological and visual artefacts through which we can access the world and, as such, Galloway (2012) argues that they frame but also exclude. Interfaces focus attention as well as affording different tasks. In the past mass production of the paper map made possible limited interaction, but fixed images could be shared across cultures and offered a large area across which eyes could roam. Screen spaces offer different affordances to their users – digital mapping delivered to a desktop or laptop supporting mouse-based interaction offers a 'double click' level of interaction and a reasonably large area of display. Different modes of existence become possible when a digital map is displayed on a screen which supports a 'double tap'-based manipulation, where direct haptic encounter with the screen interacts with location-aware capability embedded in the device, and where the interface itself moves with the person across space (Hind and Lammes, 2015). A major consequence of this shift is that places can now seem more embodied and personal, as they appear to be both accessible and knowable through the fingertips. We appear to have more power over places, even though the mapping technologies that afford us this power paradoxically distance us from direct sensory encounters with places, and also disempower us by feeding information about us to unseen strangers.

It is tempting to see technological change as the fundamental difference between the current mapping modes and those which characterized the twentieth century, but in practice changing mapping infrastructures, algorithms and interfaces are also mediated by the assemblage of visual culture and changing notions of authorship in the context of an ongoing raft of neoliberal policies. Visual studies until recently has focussed largely on media beyond mapping. Artistic practice and cartography have until recently been seen as separate fields. But in the last thirty years a profusion of map art has begun to blur this formerly separate boundary (Harmon, 2010) and to draw attention to the narrative qualities of mapping, including the capacity to embody memories. Authorship has also been profoundly changed in the contemporary mapping mode. Crowdsourcing of mapped information grew apace from the early years of the new millennium, with so-called Volunteered Geographic Information (VGI) increasingly being merged with other data sets. This new model clearly enrols new voices into a co-produced mapping of place, which potentially supplants the top-down mapping of the older mode. A proliferation of participatory systems attests to the potential of Open and Free mapping and a potential democratization of the mapping of place (Rød, Ormeling, and Corné, 2001). However, in-depth investigation of the practices of VGI suggests that we should be more cautious in welcoming the emancipatory potential of shared mapping – Haklay (2013) suggests that *OpenStreetMap* and other neo-geographic systems tend to reproduce socio-economic divides.

Contemporary examples of mapping moments across place

The shift towards a more performative understanding of mapping and the combinations of technologies in the contemporary mapping mode described above make it a rich time to be mapping place. Dodge, Perkins, and Kitchin (2009b) called for increasing attention to be given to 'mapping moments' – the particular and detailed practice through which mapping is called into being, whether it is being imagined, surveyed, coded, digitized, drafted, displayed, discussed, deployed, stored or destroyed. In this section I summarise recent mapping in literature, art, indigenous cartographies, emotional and sensory cartographies, counter-mapping and open-source mapping, which exemplify some of these moments, highlighting how place might be reimagined in existing mapping, but also reflecting on newer possibilities emerging as creative mapping encounters. Together the examples highlight the subjective, creative and collaborative potential of mapping place.

Re-imagining the dominant mode

A performative approach draws attention to the possibilities of re-reading mapping crafted with different intents in mind. In 1987 Brian Harley offered a subversive reading of the Ordnance Survey six-inch quarter sheet that charted the official authoritative view of the town of Newton Abbot in Devon where he and his family had made their home (Harley, 1987). Instead of the neutral depiction of the built environment he recounted the associations and memories evoked by experiences and events that took place across the mapped space and explored the potential of a biographical re-reading. This brief and at times touching piece highlights how we might all reclaim the map. Another strategy is to bring together maps that are usually separate, to reveal ideas that would be unnoticed were these to be considered in isolation. Offenhuber (2018) shows the potential of this comparative re-imaging in his analysis of different media mapping of the Islamic State, and he reveals the symbolic value of mapping as against the taken-for-granted indexicality of the separate forms.

Map art also offers a refashioning of established mapping norms, and frequently pokes fun at the capacity of official mapping to capture the world, by taking established forms and enhancing or changing their appearance. Artists frequently juxtapose one mapped place against another, play with map scale, change aesthetic conventions, reframe mapping, or break up and reassemble the official map. The turn towards digital and socially networked mapping has strongly encouraged this ongoing encounter. New artistic forms, such as locative art, have emerged to repurpose accepted mapping modes (MacDonald, 2014). Ethnographic approaches to the affordances that maps provide are another form of reimaging dominant mapping modes. The taken-for-granted ways in which people deploy maps in their day-to-day navigation are thus revealed as deeply cultural and associated with everyday taken-for-granted practice, instead of being read as cognitive exercises. Map skills in this context are situated and enacted (Duggan, 2018).

Crowdsourced alternatives also afford changes to established mapping. We can all add layers or pins onto a Google Map backdrop, remaking a new version of the world. A profusion of apps facilitate user interaction with mapping, and the gig economy depends upon the mapping bringing these layers together, whether it is using an *Uber* taxi or ordering food through *Just Eat*. Cyclists can now record their own tracks using the *Strava* app www.strava.com and share these individual mapped journeys with others in their social network, incorporating a personal experience of the world into the map. We can play location-based games such as *Pokémon Go*, in which digital and virtual imagery becomes augmented and overlain on top of a conventionally mapped backdrop. We can swipe left or right in dating apps such as *Tinder* or *Happn*, with the capacity to meet people according to the places that are shared and, again, grounded in location-based technology deploying the Google map. Formerly separate media can be brought together when geocoded video, photos and sound are shared via a map interface, as in the *Snapmap* offered with *Snapchat*.

So across different contexts dominant modes of mapping place are being resisted and these moments of difference are frequently enacted in subjective and individual creative acts.

Subjective imaginings

The map's capacity to make links to place has long been recognized. Certain genres of written work have exploited this capability to render their imagined worlds more real – fantastic worlds are mapped as a terrain across which the plot takes place (Lewis-Jones, 2018). Sometimes this is to help an author craft words; sometimes it is part of a publicity machine and sometimes it involves post-hoc justification of literary invention (Cooper, Donaldson, and Murrieta-Flores, 2016). This may be in the hands of the author, or from a fan base, to share and enhance reading. This creative imagining of a fantastic world parallels academic work linking the written work and its associations with place. Thus literary atlases map out associations between author's referencing of place and the real world, whilst literary GIS explores the corpus to mine potential textual geo-referencing (see, for example, Cooper and Gregory, 2011).

Artists and community groups have also reimagined local contexts in unique community mapping, varying in context from local planning protest to 'green mapping' of alternative futures, and from counter-mapping resources in the face of multinational interests to remapping of place names as a strategy for reasserting cultural heritage (Perkins, 2018b). In some cases this has been facilitated as part of initiatives to reclaim the uniqueness of places – affirming the importance of local voices against outsider and powerfully competing claims to these places. Thus Common Ground facilitated the development of over 2,500 Parish maps across the UK in the period from 1984, in an initiative that has subsequently spread across Europe and beyond (Wood, 2018). Published endeavours to remap terrains are sometimes read as attempts to construct deep maps of communities, such as the interpretation of Tim Robinson's mapping of the West of Ireland by Cronin (2016), or Denis Wood's (2013) mapping of his neighbourhood, Boylan Heights in Raleigh, North Carolina. These local views cast a new light on places. Wood, for example, maps what has been seen as unmapped or unmappable, such as radio waves, or a paperboy's route, or the light cast by street lamps, or the location of Halloween pumpkins on porches. Contested and liminal places are a rich terrain for this approach, such as the mapping compiled by Garrett Carr to accompany his walk through the Irish borderland (Carr, 2017). Many of the mapping moments evoked in these personal re-mappings relate to memories of places. Others evoke the sensory geographies bound up in different senses of place. Graphic artist Kate McLean's many smell maps of urban environments explore the unique smellscapes of these different places, translating one sensory modality into another, to communicate a shared embodied response to smell captured during group smell walks, translated into pastel-toned visual contour maps (McLean, 2018). Her projects chart the ephemerality of perceptions of smell, evoking the animated qualities of encounters across the smellscape. Other affective mapping also attends to feelings about places that are beyond vision. From 2004 Christian Nold published emotion maps by capturing bodily reactions with bio-sensors and translating these into mapped equivalents (MacDonald, 2014). In the second decade of the twenty-first century emotion mapping has been widely applied as a data collection device, bringing people's feelings to the attention of people with the power to manage built environments. Less policy-oriented approaches to places have adopted an explicitly playful approach to mapping (see Perkins, 2009). Many different genres of computer games deploy mapping as a gameboard across which players can compete. Examples of many different affordances in these games are described in Playful Mapping Collective (2016).

Other subjective re-mappings of place deliver significant political challenges. Maps can be subversive vehicles for protest (Firth, 2014). The 2018 parallel publication in hard copy, as an electronic book and as a website of *Not an Atlas* from Kollektiv OrangoTango+ (Orangotango Kollectiv, 2018) gives a very good flavour of the rich diversity of alternative mapping practices, drawing on global sources. These counter-maps deliver political possibilities, at times deploying sophisticated technology, and in other contexts celebrating a handcrafted creativity. They are usually collaborative and frequently deploy the mapping in an educative mode. Several contributions in this collection deliver practical advice about how to map. *Not an Atlas* shows how critical cartography has grown in the two decades since the term was first used. Maps created here assert women's rights, they form a central part of struggles for justice in land rights, they combat multinational power, they reveal secret military sites and they celebrate the authorship so often hidden in the current mapping mode. The global coverage in the collection also highlights a potential escape from the global northern domination of mapping, and gives indigenous groups a voice.

Conclusions

The subjectivity of personal mapping of place outlined above demands novel geographic methodologies. Some of these methods deploy the map as a tool to investigate aspects of place in a participatory fashion. Sustainable development is facilitated if there is community buy-in and mapping techniques can make this more likely (Craig, Harris, and Weiner, 2002). Hazards can be more carefully managed when community mapping takes place, because it is local communities that are impacted by risk and it is local people who have the mundane experiences of living with a hazard (Haworth, Whittaker, and Bruce, 2016). Community cohesion can be charted using spatial technologies to map real-time movement and interaction (Huck et al., 2019). A sense of place can be explored by participants mapping what they value in the form of sketch maps (Boschmann and Cubbon, 2014). But the diverse, paradoxical and mutable qualities in the mapping of place also invite us to change how we regard mapping. Instead of focussing upon the potential of the form for offering a mimetic success – a perfect capturing of place – we should rather focus on its partiality, and the mapping failures that play out across different cartographic modes. Failure offers a rich potential for processual interpretation, instead of serving as a dystopian downside of progress. Hind and Lammes (2015) reflect on the potential of failure in the *OpenStreetMap*-based mapping of flooding in the Somerset Levels in 2014 and on mobile-based protest mapping deployed in public demonstrations. In both of these cases the mapping practices play out and are ongoing – when floods took place and a route across the Levels became impassable new crowdsourced mapping was created to offer a more accurate reflection of routes across the flooded areas, even though advice existed against mapping temporary features. This mapping worked whilst the floods were in place, but increasingly failed once they receded – but amidst this 'failure' the temporary remapping drew attention to the mapping policies of the website and also to tensions around the rules that govern how a place might be mapped.

This kind of research emphasizes the emergent and emancipatory potential of digital mapping. It also shows how making maps can be part of a process for changing places.

References

Abend, Pablo, and Francis Harvey. 2017. "Maps as Geomedial Action Spaces: Considering the Shift from Logocentric to Egocentric Engagements." *GeoJournal* 82 (1): 171–183.

Ash, James, Rob Kitchin, and Agnieszka Leszczynski. 2018. *Digital Geographies*. London: SAGE.

Boria, Edoardo. 2008. "Geopolitical Maps: A Sketch History of a Neglected Trend in Cartography." *Geopolitics* 13 (2): 278–308.

Boschmann, E. Eric, and Emily Cubbon. 2014. "Sketch Maps and Qualitative GIS: Using Cartographies of Individual Spatial Narratives in Geographic Research." *The Professional Geographer* 66 (2): 236–248.

Brown, Barry, and Eric Laurier. 2005. "Maps and Journeys: An Ethno-Methodological Investigation." *Cartographica: The International Journal for Geographic Information and Geovisualization* 40 (3): 17–33.

Carr, Garrett. 2017. "Land and Power: Making a New Map of Ireland's Border." *Cartographica: The International Journal for Geographic Information and Geovisualization* 52 (3): 251–262.

Cooper, David, Christopher Donaldson, and Patricia Murrieta-Flores. 2016. *Literary Mapping in the Digital Age*. London: Routledge.

Cooper, David, and Ian N Gregory. 2011. "Mapping the English Lake District: A Literary GIS." *Transactions of the Institute of British Geographers* 36 (1): 89–108.

Craig, William J., Trevor M. Harris, and Daniel Weiner. 2002. "Community Participation and Geographic Information Systems." In *Community Participation and Geographical Information Systems*, edited by William J. Craig, Trevor M. Harris and Daniel Weiner, 29–42. Boca Raton, FA: CRC Press.

Cronin, Nessa. 2016. "'The Fineness of Things': The Deep Mapping Projects of Tim Robinson's Art and Writings, 1969–72." In *Unfolding Irish Landscapes*, edited by Derek Gladwin and Christine Cusick, 53–72. Manchester: Manchester University Press.

D'Ignazio, Catherine, and Lauren F. Klein. 2016. "Feminist Data Visualization." In *Workshop on Visualization for the Digital Humanities (VIS4DH), Baltimore. IEEE.*

Dodge, Martin, Rob Kitchin, and Chris Perkins, eds. 2009a. *Rethinking Maps: New Frontiers of Cartographic Theory*. London: Routledge.

Dodge, Martin, Chris Perkins, and Rob Kitchin. 2009b. "Mapping Modes, Methods and Moments". In *Rethinking Maps*, edited by Martin Dodge, Rob Kitchin, and Chris Perkins, 220–243. London: Routledge.

Duggan, Michael. 2018. "Navigational Mapping Practices: Contexts, Politics, Data." *Westminster Papers in Communication and Culture* 13 (2): 31–45.

Edney, Matthew H. 1993. "Cartography without Progress': Reinterpreting the Nature and Historical Development of Mapmaking." *Cartographica: The International Journal for Geographic Information and Geovisualization* 30 (2-3): 54–68.

Fernández, Pablo Iván Azócar, and Manfred Ferdinand Buchroithner. 2013. *Paradigms in Cartography: An Epistemological Review of the 20th and 21st Centuries*. Berlin: Springer Science & Business Media.

Field, Ken. 2018. *Cartography*. Redwood, CA: Esri Press.

Firth, Rhiannon. 2014. "Critical Cartography as Anarchist Pedagogy? Ideas for Praxis Inspired by the 56a Infoshop Map Archive." *Interface: A Journal for and about Social Movements* 16 (1): 156–184.

Fogelman, Charles, and Thomas J. Bassett. 2017. "Mapping for Investability: Remaking Land and Maps in Lesotho." *Geoforum* 82 (June): 252–258.

Galloway, Alexander R. 2012. *The Interface Effect*. Cambridge: Polity.

Gerlach, Joe. 2017. "Mapping as Performance." In *The Routledge Handbook of Mapping and Cartography*, edited by Alexander J. Kent and Peter Vujakovic, 114–124. Abingdon: Routledge.

Griffin, Amy L., Anthony C. Robinson, and Robert E. Roth. 2017. "Envisioning the Future of Cartographic Research." *International Journal of Cartography* 3 (Supp 1): 1–8.

Haklay, Mordechai. 2013. "Neogeography and the Delusion of Democratisation." *Environment and Planning A* 45 (1): 55–69.

Harley, J. Brian. 1987. "The Map as Biography: Thoughts on Ordnance Survey Map, Six-Inch Sheet Devonshire CIX, SE, Newton Abbot." *The Map Collector* 41: 18–20.

Harmon, Katharine. 2010. *The Map as Art: Contemporary Artists Explore Cartography*. New York: Princeton Architectural Press.

Haworth, Billy, Joshua Whittaker, and Eleanor Bruce. 2016. "Assessing the Application and Value of Participatory Mapping for Community Bushfire Preparation." *Applied Geography* 76: 115–127.

Hind, Sam, and Sybille Lammes. 2015. "Digital Mapping as Double-Tap: Cartographic Modes, Calculations and Failures." *Global Discourse* 6 (1–2): 79–97.

Huck, Jonny J., J. Duncan Whyatt, John Dixon, Brendan Sturgeon, Bree Hocking, Gemma Davies, Neil Jarman, and Dominic Bryan. 2019. "Exploring Segregation and Sharing in Belfast: A PGIS Approach." *Annals of the American Association of Geographers* 109 (1): 223–241.

Hurst, Paul, and Paul Clough. 2013. "Will We Be Lost without Paper Maps in the Digital Age?" *Journal of Information Science* 39 (1): 48–60.

Janelle, Donald G., and Michael F. Goodchild. 2018. "Territory, Geographic Information, and the Map." In *The Map and the Territory: Exploring the Foundations of Science, Thought and Reality*, edited by Shyam Wuppuluri and Francisco Antonio Doria, 609–627. The Frontiers Collection. Cham: Springer International Publishing.

Kitchin, Rob, and Martin Dodge. 2007. "Rethinking Maps." *Progress in Human Geography* 31 (3): 331–344.

Kitchin, Rob, Justin Gleeson, and Martin Dodge. 2013. "Unfolding Mapping Practices: A New Epistemology for Cartography." *Transactions of the Institute of British Geographers* 38 (3): 480–496.

Lammes, Sybille, Chris Perkins, Alex Gekker, Sam Hind, Clancy Wilmott, and Daniel Evans. 2018. *Time for Mapping: Cartographic Temporalities*. Manchester: Manchester University Press.

Lewis-Jones, Huw. 2018. *The Writer's Map: An Atlas of Imaginary Lands*. Chicago: University of Chicago Press.

MacDonald, Gavin. 2014. "Bodies Moving and Being Moved: Mapping Affect in Christian Nold's Bio Mapping." *Somatechnics* 4 (1): 108–132.

McLean, Kate. 2018. "Mapping the Invisible and Ephemeral". In *The Routledge Handbook of Mapping and Cartography*, edited by Alexander J. Kent and Peter Vujakovic, 509–515. London: Routledge.

Nivala, Annu-Maaria, Stephen Brewster, and Tiina L. Sarjakoski. 2008. "Usability Evaluation of Web Mapping Sites." *The Cartographic Journal* 45 (2): 129–138.

Offenhuber, Dietmar. 2018. "Maps of Daesh: The Cartographic Warfare Surrounding Insurgent Statehood." *GeoHumanities* 4 (1): 196–219.

Orangotango Kollectiv. 2018. *This Is Not an Atlas*. Bielefeld: Transcript Verlag. www.transcript-verlag.de/978-3-8376-4519-4/this-is-not-an-atlas/

Perkins, Chris. 2009. "Playing with Maps". In *Rethinking Maps: New Frontiers in Cartographic Theory*, edited by Perkins C. Dodge, M., Kitchin, R., 167–188. London: Routledge.

Perkins, Chris. 2018a. "Critical Cartography". In *The Routledge Handbook of Mapping and Cartography*, edited by Alexander Kent and Peter Vujakovic, 80–89. London: Routledge.

Perkins, Chris. 2018b. "Community Mapping." Oxford Bibliographies. Oxford: OUP. www.oxfordbibliographies.com/view/document/obo-9780199874002/obo-9780199874002-0184.xml.

Playful Mapping Collective. 2016. *Playful Mapping: Playing with Maps in Contemporary Media Cultures*. Amsterdam: Institute of Network Cultures.

Rankin, William. 2016. *After the Map: Cartography, Navigation, and the Transformation of Territory in the Twentieth Century*. Chicago: University of Chicago Press.

Roberts, Les. 2016. "Deep Mapping and Spatial Anthropology." *Humanities* 5 (1): 5.

Rød, Jan Ketil, Ferjan Ormeling, and Corné van Elzakker. 2001. "An Agenda for Democratising Cartographic Visualisation." *Norsk Geografisk Tidsskrift - Norwegian Journal of Geography* 55 (1): 38–41.

Rossetto, Tania. 2019. "The Skin of the Map: Viewing Cartography through Tactile Empathy." *Environment and Planning D: Society and Space* 37 (1): 83–103.

Wilson, Matthew W. 2017. *New Lines: Critical GIS and the Trouble of the Map*. Minneapolis: U of Minnesota Press.

Wood, Denis. 2013. *Everything Sings*. Catskill, NY: Siglio Press.

Wood, Denis. 2015. "Mapping Deeply." *Humanities* 4 (3): 304–318.

Wood, Denis. 2018. "Mapping Place". In *The Routledge Handbook of Mapping and Cartography*, edited by Alexander J. Kent and Peter Vujakovic, 401–412. London: Routledge.

Woodward, David. 1985. "Reality, Symbolism, Time, and Space in Medieval World Maps." *Annals of the Association of American Geographers* 75 (4): 510–521.

Zook, Matthew A., and Mark Graham. 2007. "Mapping DigiPlace: Geocoded Internet Data and the Representation of Place." *Environment and Planning B: Planning and Design* 34 (3): 466–482.

32
Of place and law

Jenny Kanellopoulou

About this chapter

This chapter aims to introduce the reader to the intertwining relationship between law and place from a variety of perspectives, including but not limited to doctrinal legal analysis. Legal geography, an interdisciplinary academic project that has gained popularity in the past decades, will also feature in the discussion, aiming to bridge a gap between two seemingly unrelated fields of study, which are, however, built on common conceptual ground: there cannot be law without a place of application. In this vein, law and geography go hand in hand in the dedicated study of the rules that shape, create, and govern place, and it is the present chapter's intention to posit that this relationship should be acknowledged in theory as well as in practice. Unambiguous as it may be, this relationship remains largely overlooked in the study of the law, which is seen as 'insensitive' to the place-specificity of human activity.The chapter will commence with a short introduction to the existing academic thought on law and place by briefly touching on the interdisciplinary endeavours of legal geography, whilst highlighting the lacuna in 'traditional' legal theory's appreciation of place and space. From there, the chapter will delve into the role of place and space in doctrinal legal thinking more explicitly, identifying areas of law that merit geographical attention, and advocating for an alternative interdisciplinary viewing of law and society through the geographical lens. Consequently, this chapter attempts to offer fertile ground for the cross-pollination of ideas with respect to the geographical appreciation of the law, and vice versa.

An overview of place in law

Traditionally, the law has constituted its own academic discipline, distinct from the social and political elements that lead to the creation of legal rules. Under this prism, reading and applying the law appears to remain immune to non-doctrinal or black letter analysis, for fear that the impartial application of rules be jeopardised. It follows that any prejudiced or biased application of legal rules can affect legal certainty, which is incompatible with any democratic system of governance. This traditional viewing leaves little room for expansion into adjacent and related disciplines when it comes to the way the law is read and applied; the law remains a rigid and impermeable construct.

This is a sought-after quality of the law in a democratic legal system, as the law needs to be seen and recognised as such, and to be applied equally; however, this also constitutes the law's harshest critique and constraint, since phenomena such as social inequalities and geographical particularities do not seem to prima facie affect legal decision making. To address this lacuna, academic scholarship has expanded into socio-legal and geographical dimensions. Nevertheless, such projects remain mainly on the theoretical plain as, indeed, the study of the letter of law, alongside its application and enforcement, remain doctrinal in nature. For the purposes of the present, attention is paid to the expansion of law into geography (and vice versa), in an attempt to 'map out' how law and place configure each other in 'traditional' and 'non-traditional' legal thinking.

Hardly a newcomer in legal thinking and legal literature, the relationship between law and the notion of place is continuously entering new paradigms and is constantly being enriched with new works of academic merit. A lot has changed since Blomley's *Law, Space, and the Geographies of Power* (1994), with legal geography entering the debate and establishing itself as a distinct field of scholarship, mainly as a human geography sub-project. Indeed, legal geography as an academic pursuit has caught the attention of critical geographers wishing to focus on the spatial dimensions of the law, its institutions, as well as other non-state actors (Bennett, 2011; Hubbard, 2012; Konsen, 2013, indicatively). At the same time, socio-legal and critical legal scholars are constantly re-discovering geography's appeal, addressing issues ranging from the spatialities of injustice in the broadest sense (Delaney, 2016) to more traditional disputes related to a designated place (e.g., housing, landlord/ tenant disputes, property transactions inter alia), and perhaps most famously, to issues related to the regulation of the public versus the private place.

Despite the above, the present contribution does not aspire to become a précis of the legal geography project per se; this would fall outside the scope of a compendious discussion on law and place. At this point, it suffices to say that legal geography has been the most dedicated attempt to provide the law with a spatial interdisciplinary reading, emphasising the role of place in the making, the application and the interpretation of legal rules. For those interested, the development of the academic discourse on legal geography is periodically being documented by David Delaney (2015, 2016, 2017), who reports on the progress made in the study of the field both thematically and methodologically, by offering a critical assemblage of academic literature from both the geographical and the legal perspective. Rather, the aim of this short chapter is to consider the relationship between law, place, and space from a variety of perspectives, including but not limited to 'traditional' doctrinal legal thinking, which might not be as familiar to non-lawyers. It is interesting to observe how traditional legal theory has more recently attempted to answer questions predominantly left to the hands of socio-legal scholarship, including issues regarding the relationship of place and space with the law.

Whilst it is interesting and indeed inspiring to witness the expansion of lawless geography into anti-geographic law (Blomley, 1994; Clark, 1989; Delaney, 2000; Pue, 1990) and vice versa, it is important to explore why this mutual understanding had not always been evident. The claim 'place matters', which has long been advocated in geography, has reached both adjacent and more remote disciplines, and rightly so. On the other hand, however, it is not surprising to see that the law does not seem to 'matter' to the same extent, given its traditional static and reactive nature. Plainly put, the law is traditionally seen as something to be dealt with 'later' and by lawyers, despite its omnipresent nature and its place-making qualities.

Indeed, the relationship between law and place is undisputed, even though it has not been always documented as such. Not only does the law make place (as stated previously), it also constitutes place, indicating one of the most interwoven relationships between disciplines: geography aside, law is perhaps the only discipline across humanities and social sciences that is so closely related to the notion of place; place is a sine qua non condition for the existence and the application of the law.

The law shapes and defines its place of application to the extent that the two become inextricably interwoven to form a sphere of authority. Once this sphere of authority has been established, the study and the application of the law can begin, as people, institutions, actors and disputes are brought together to form the legally relevant place. This place 'ends' where the power of the law to apply ceases: outside the borders of the sovereign state the laws of other jurisdictions may apply (speaking, for instance, of the domestic jurisdiction) or perhaps none, should we speak of supralegal or infralegal issues (extralegality or issues of non-legal relevance), as per Johns (2013) in Blomley (2014). As such, the law has both inherent spatial relevance and place-making capabilities; it affords meaning to a well-defined place that the law itself creates. It follows that under this prism this interwoven relationship is a facet of legal certainty, as place and legal order constantly configure each other, and can always be asserted and interpreted judicially.

Place and space in legal theory

Such is the unambivalence of the law's spatial relevance that a broader debate on the significance and the role of the place in law per se is almost a non-issue in doctrinal legal thinking, which sets out to answer questions in law rather than about the law. From this perspective, the place of a given legal order is the place or the territory where a people establish their self-governing will. Consequently, from a constitutional perspective, law represents a spatial declaration of sovereignty and manifests itself within the borders of a given territory, but also outwards, i.e., internationally, for as long as it remains rooted in the same sovereign mandate (international representation of states and participation in international institutions). Hence, in international law terms, this so-called Westphalian sovereignty is respected and acknowledged as a principle of non-interference in matters of domestic jurisdiction, as recognised in the Charter of the United Nations (United Nations, Charter of the United Nations, 24 October 1945, 1 UNTS XVI, available at: www.unweb site.com/charter).

So far, the law's traditional spatial manifestation visualises as the constant flow or dialectic between the national and the international. In normative terms, outside the geographical and physical borders of the domestic jurisdiction lies the universal foundation of international law. The legal continuum remains unbroken, and so does the uninhibited spatial presence of the law itself. In this vein, Teubner has commented on 'the historical unity of law and state' (1997: xiii), reaffirming the territorial relevance of a legal system.

More recently, however, traditional legal theory has examined this inherent spatial element of the law under post-Westphalian, modernist perspectives that question the law's origins, and considers to what extent and in what way norms that bypass this traditional perspective will (or can) be recognised as law (Walker, 2010). And this is something that merits the attention of the non-legal scholar: law has the ability to create metaphorical borders (co-ordinates as per Walker) in order to establish itself, thus dismissing issues that have not been legislated or adjudicated, as mentioned previously albeit briefly. What remains outside these borders are issues of justice/injustice, or rather issues that have the ability to foster relevant legal debates, meaning questions on what should and could be legal/illegal and is not yet. This border-crossing between law and non-law is not just a theoretical exercise in what separates legal from social sciences, it is also an indication of the difference between a place where the law exists and applies, and a space comprising of various human and non-human actors where questions relevant to the law are raised.

To illustrate, in normative terms the law is situated in space and time, a quality that Walker refers to as situatedness of 'qua law in space and time' (Walker, 2010: 16), signalling an act of framing the space–time continuum in order to assert legal meaning and thus secure legal certainty. Not only does this quality point to the law's static nature, thus affording the law place-making qualities, it also allows for further practices of complementarity and compartmentalisation within the law itself. Once established and situated in space–time, the law provides its own frame of reference:

> The law of the sovereign state [...] may conceive itself in finite terms, as limited in space and marked by a beginning (...) in time; yet it recognises no space within the place of its jurisdiction and no time [...] where the activities of those located within that expanse of space and time can escape its jurisdiction. The law of the sovereign state, then, pertains, potentially at least, to the whole of law and the whole of life under the law who fall under its perpetual jurisdiction.
>
> *(Walker, 2010: 56)*

We can see, therefore, that the law requires both physical and metaphorical borders. Physical, geographical borders are required to separate the legal orders of different sovereign states and the transition from the national to the international, and metaphorical borders to separate the legal from the non-justiciable. The two are not mutually exclusive, as the legal continuum, once delineated or situated geographically, affords meaning to its surrounding places and concepts in perpetuity, as seen in Walker's writings.

More recently, legal scholarship has faced the need to engage with the post-national era of legal globalisation, where the traditional legal viewpoint or dichotomy of national and international has required revisiting. Even in the globalised world where the lines between private actors and international institutions (and thus obligations) become blurry (i.e., from the WTO to the ICANN), legal certainty still needs to be achieved. In this vein, two concepts re-enter the debate, the need for boundaries-setting in the legal hierarchy, and the role that geography (as a discipline) is called to play in achieving legal certainty.

Aiming to address the first issue, Lindahl (2010) reaffirms that the regulation of human behaviour by the law occurs through the setting of boundaries that determine 'who ought to do what, when and where' (2010: 35). Lindahl posits that even though legal orders appear de-territorialised in a global environment, they still manage to become accumulated by the national jurisdictions, leading to some form of emplacement rather than displacement of global legal orders (internalisation and application of international law or EU law serving as an example). Therefore, he contests that 'no legal order is conceivable unless it is bound spatially' (2010: 44).

What Lindahl actually points to (perhaps unintentionally) is the inevitability of entering into a debate about place and geography if we wish to address issues that challenge the traditional normative legal perspective. We need to delve into the roots of the law in order to address more perplexing issues about the law in a globalised environment. In other words, it is time to revisit qua law, something that Walker sets out to consider. Ultimately, it appears that this is something that cannot be done un-geographically, especially since the law's traditional borders and boundaries are constantly being challenged: what makes the law is tied to where the law happens.

Thus, considering what makes the law reiterates Lindahl's and Walker's assertion that, among other things, the law is made by and in a certain place and a certain time. In the globalised post-Westaphalian world, what remains to be answered is if the place where law happens should be the sovereign territory that subsequently assigns meaning to the national and the international, the law and the non-law, or an alternative notion of place and space, where borders and boundaries remain fluid, and so does the very idea of what constitutes law.

The fluid borders of legal certainty

As Antonia Layard (2010: 412) has pointedly observed, speaking about the 'law of place' means to separate and delineate 'places from spaces, applying different legal rules either side of an often-invisible boundary line'. It is true therefore that for the legal scholar this distinction between place and space denotes more than a theoretical exercise: as law happens and exists somewhere there can be no legal certainty until that place is prescribed. Consequently, it can be argued that in the relevant legal space, relationships, actions, actors, and institutions carry legal weight, and as such require a closer examination by legal scholars, practitioners, and lawmakers. Should a certain action occurring in the relevant legal space be legal or illegal? Should people and interactions in the relevant legal space attract the attention of the legislator and the judiciary? In other words, is there or should there be law in its normative sense? If so, what should that law be?

Ultimately, this exercise of delineation is an exercise in affording legal meaning or an exercise in legal certainty. This is the legal equivalent to the definition provided by Carter et al., (1993) that 'place is space to which meaning has been ascribed'. Even though disputed in critical geography (e.g., Massey, 2005), this distinction between place and space holds merit when talking about the law and its situated qualities in normativity. The most evident manifestation of this invisible legal boundary line between the legal place and the relevant legal space can be found somewhere where the traditional jurisdiction is either absent or in limbo, meaning somewhere where power relations, resources and spatiality are all much more able to play their part in determining (legally informed) decisions, absent the traditional jurisdiction (Collis, 2004).

Examples of this can be numerous, ranging from outer space to Antarctica, and from unregulated urban squats to illegal activities in public spaces (e.g., cannabis festivals, graffiti, parkour). In such situations that seem to fall outside a predetermined place of authority, legislators and legal scholars have struggled with ascribing legal specificity. In the cases of outer space and Antarctica national appropriation and claims of sovereignty are expressly excluded, by virtue of Art. II of the Treaty on Principles Governing the Activities of States in the Exploration and Use of Outer Space, including the Moon and Other Celestial Bodies (18 UST 2410, 610 UNTS 205, 6 ILM 386, 1967) and Art. IV of the Antarctic Treaty (402 U.N.T.S. 71, 1961).[1] This means that the law of no single sovereign state applies to these particular spaces, whereas international law is evoked for the remainder of human activity with respect to scientific exploration and international cooperation.

To illustrate, a dispute with respect to human activity will be treated as a dispute arising between two or more nationals of specific sovereign jurisdictions, willingly subject to the rules, norms, and conventions that make up the international legal order. It follows that, notwithstanding its geographical significance, the place of a dispute can become completely bypassed by a sovereign legal system. This does not necessarily mean that such spaces remain irrelevant to the law, as this unique interaction between law, human activity and place specificity should be factored in to the way the law is applied and enforced, nevertheless. In other words, these places manifest the departure from the standard Westphalian viewpoint, challenging the notion of space and place in traditional legal thinking; their study advances the need to read the law in a geographical and place-specific manner.

Similarly, in spaces of ambiguous legal status within a specific jurisdiction, we encounter paradoxical relationships with the law. Such spaces within a recognised jurisdiction can challenge the way the law is made and applied 'from within', pushing the limits of law-making once again. In these terms, an urban squat remains an illegal occupation that often borders legitimacy and acceptance by the authorities (Ntounis and Kanellopoulou, 2017), despite any claims for 'sovereignty' on the part of the squatters (as in the case of Freetown Christiania in Copenhagen, Denmark). Equally, but on a smaller scale, graffiti or parkour (Mould, 2009) can be appreciated as forms of art and sport, adding value to the urban environment (Iljadica, 2017; Layard and Milling, 2015; Mubi Brighenti, 2010, inter alia), even though they may contradict municipal regulations. From a criminal law perspective, Cowley et al. (2015) examined the alternative penal system that governs prisons in Kyrgyzstan, identifying a quasi-sovereign legal ordering running in parallel with the Kyrgyz penal code (the law 'outside').

On the antipode, we encounter pseudo-public spaces, meaning spaces where the public element is present in geographical terms (i.e., a place for public use) but the legal element has been withheld. These consist of spaces such as shopping centres and privatised town centres where legal rights such as freedom of assembly and freedom of speech remain restricted (MacSithigh, 2012) by virtue of the rights deriving from private property. Even though such cases are not aligned with the notion of sovereignty, they still manifest a spatio-legal oxymoron in the modern era of astute privatisation.

It follows that spaces such as these either remain outside the confines of a traditional jurisdiction (literally or metaphorically) or represent an alternative reality thereof, something that shifts the onus from norms to alternative guarantees of societal integration (Jalava, 2003). In consequence, we can see how the law's boundary lines can remain fluid in both substance and enforcement. The relevant legal space can bear qualities and characteristics that challenge the law's normative certainty. At the same time, the law remains present as a frame of reference, a juxtaposition between here and there. This shows that the law's normative confines still need to be acknowledged, and unsurprisingly so: 'that which is placed outside of law is not simply a space of not-law but is itself a product of law' as Blomley explains (2014: 136).

Hence, should we agree that in the relevant legal space power dynamics, human interaction, human and non-human actors, all carry equal weight in achieving legal results (e.g., regulation of human activity) whilst bypassing the rules of the jurisdictions outside, what remains to be seen is how this is possible and what this could mean for the sought-after legal certainty. Taking, for instance, the example of a group of squatters wishing to engage in urban planning activities or operate a business out of the illegally occupied premises: how can any resulting endeavour gain legitimacy, having itself become delineated from the laws of the outside jurisdiction? And should it ultimately gain such legitimacy through societal acceptance, what does this mean for the production of law and the establishment of legal certainty?

Legal bracketing as an exercise in legal certainty

Trying to find common ground and deduce acceptance by the legal source, as well as by the law's institutions and relevant actors, can be a challenging exercise, predominantly since it requires revisiting what makes law or qua law (going back to Walker's writings once again). I believe that Blomley's concept of legal bracketing can help bypass this theoretical impasse. Blomley refers to the law's ability to 'stabilize and fix a boundary within which interactions take place more or less independently of their surrounding context' (2014: 135). There exists a difference, however, between what Blomley defines as bracketing in legal terms and what normative scholarship describes as the law's situated borders. Blomley's concept of legal bracketing aims to bring a specific situation to the foreground and afford legal meaning to a certain interaction's diverse and heterogeneous constituent parts. Bracketing in this sense means to afford legal meaning by putting an interaction under the microscope and by simultaneously disentangling it from those elements that lack legal relevance, in a set place and time.

Hence, bracketing is not about crossing some fluid boundary lines from lawlessness to certainty; it requires observing an interaction closely. As such 'it entails complex and subtle calculations that govern what is, and what is not to be included within a particular setting' (2014: 136). He explains that bracketing does not (and should not) equal simplification, as the law is not a simple thing (the law matters!), but is a performative frame, carrying effect and real-life consequences. Delineated in space and time as it may be, Blomley's legal bracket is as temporal as the interaction in hand requires (e.g., a land dispute, a contract, a planning application). This temporality does not require legal isolation and re-interpretation of actors and interactions. Rather, it affords an interaction with legal clarity and certainty by accepting its heterogeneity and by focussing on how the law happens. Elsewhere, Blomley has referred to such a space as a splice (1994, 2003). Thus, the concept of legal bracketing accepts the law as relational, a coming of being of 'entities, a temporary touching down of heterogeneous and spatially stretched relationships', borrowing from Massey's writings on space (2005).

Arguably, the best quality of the legal bracket is that it remains meaning- and space-specific without being prima facie tied to a pre-defined legal order. Thus, bracketing an interaction and subsequently aiming to appreciate how the law comes into being, allows for an alternative viewing of the spatio-legal conundrum that could potentially address the issues encountered in spaces of legal ambiguity. Within the splice, law is both performed and relational: this enables the social and spatial elements that 'make law' to become accentuated and examined as such. Consequently, interactions within a splice can be further categorised into areas of legal practice (e.g., contract, tort, criminal law, etc.), something that complies with legal certainty.

This results in the relevant legal space of ambiguity becoming a space of legal relevance, a space where the law has meaning beyond the scalable hierarchies that the traditional jurisdictional order prescribes (Valverde, 2009). Indeed, in normative terms the law recognises vertically defined hierarchies; beginning from the international, down to the national and the local (Ferguson and Gupta, 2005). This vertical viewing has been found to exclude rather than include socio-spatial relations and interactions in the name of legal ordering: an example of this can be seen in the Canadian cases of *Federated Anti-Poverty Groups of BC* v. *Vancouver (City)* 2002 BCSC 405, *Victoria (City)* v. *Adams* 2008 BCSC 1363, and *Victoria (City)* v. *Adams* 2009 BCCA 563, where the panhandling and camping-out practices of homeless people claiming their rights to livelihood were deemed objectionable by municipal regulations. Commenting on these cases, Blomley (2013) argues that these regulations are designed to exclude the poorest from the public space and to micro-manage space within city limits and raises the question of the municipality having acted ultra vires. Bracketing these activities, however, allows for an equal representation of the claims of both parties, bypassing the restrictions of the legal hierarchy. Within the bracket, human activity is seen in conjunction with the place where it occurs, and its legality or the illegality is consequential to both social and spatial factors. Ultimately, and going full circle to where this short piece started, within Blomley's splice the law does not have to shed its social and geographical dimensions.

It follows that Blomley's and other scholars' writings on the legal space (e.g., Delaney's nomosphere 2004, 2010; Graham, 2010; Philippopoulos-Mihalopoulos, 2007 lawscape) remain heavily related to critical geography as they seek to 'investigate the co-constitutive relationship of people, place and law' (Bennett and Layard, 2015: 406). Besides, legal geography's distinctive feature is the elegant and detailed attention to 'the complex processes of legal constitutivity' (Delaney, 2015: 98) and the analysis of the reciprocal or mutual constitutivity of law, society, and space. In addition, Braverman (2011) lays down the questions a legal geographer should contemplate on, namely where is this place, event, or dispute located, what do we see, and how do legal and spatial meanings combine. By following these proposed methodologies, the result would be an evaluation of the legal continuum, since the legal, the spatial, and the social come under investigation.

As Holder and Harrison suggest (2003: 4), '"doing law in geography" helps our understanding of how law shapes physical conditions and legitimates spatiality, and makes clear that law has a physical presence, or even many presences'. The work of Braverman et al. (2014) notes the relationship between the where and the how of law and space, pinpointing that the legally significant place is anything but static; a constantly flowing dialogue can be observed, an ability of the legal space to influence how law happens. This realisation is crucial to the present analysis as, given the ability of the where and the how to converse, their interdependence can also be deduced.

Conclusion

To conclude, even though this present chapter promised not to become a report on legal geography, it did result in arguing that geography can provide theoretical and more practical answers to questions that traditional Legal Theory has started to raise. Legal geography stresses 'the normative appeal and institutional significance of place' when examining local legal cultures, and argues for the 'imbrication of the legal, the social and the spatial' (Blomley, 1994: 28, 63). This shall not be done at the expense of the legal element, as the law too matters. One last thing that needs to be pointed out is that the law's static and spatial qualities do not have to be mutually exclusive. We still need to be able to delineate the law from its surroundings and afford legal meaning to interactions. This is a prerequisite in any democratic sovereign state, where the rule of law prevails. Nevertheless, legal certainty needs to be achieved in a manner that does not exclude spatial and social concerns from its narrative. This can be achieved should attention be paid to how and where law happens, following practices such as Blomley's bracketing.

By emphasising the triadic element brought forward by such a geographical viewing of the legal space (legal, social, spatial), geography can be employed not only to enrich legal thinking, but also to bypass practical impasses encountered in traditional legal hierarchies (the rights of the homeless to livelihood, of the activists to protest and so on).

Ultimately, geography can aid groups and individuals 'voice' spatial concerns and rights associated with places by introducing an alternative to the traditionally a-spatial legal framework.

Note

1 Note that by virtue of para.2 Article IV '*No acts or activities taking place while the present Treaty is in force shall constitute a basis for asserting, supporting or denying a claim to territorial sovereignty in Antarctica or create any rights of sovereignty in Antarctica. No new claim, or enlargement of an existing claim, to territorial sovereignty in Antarctica shall be asserted while the present Treaty is in force*', whereas several states have either claimed or reserved the right to claim territories in the future.

References

Bennett L (2011) A pub, a field and some signs – a case study on the pragmatics of proprietorship and legal cognition. Paper presented at COBRA 2011 – Royal Institute of Chartered Surveyors International Research, Manchester 12-13 September.

Bennett L and Layard A (2015) Legal geography: Becoming spatial detectives. Geography Compass 9(7): 406–422.

Blomley N (1994) *Law, Space, and the Geographies of Power.* New York: Guilford Press.

Blomley N (2003) Law, property, and the geography of violence: The frontier, the survey, and the grid. *Annals of the Association of American Geographers* 93(1): 121–141.

Blomley N (2013) What sort of legal space is a city? In Brighenti (ed.), *Urban Interstices: The Aesthetics and Politics of Spatial In-Betweens*, pp. 1–20, Farnham, UK: Ashgate.

Blomley N (2014) Disentangling law: The practice of bracketing. *Annual Review of Law and Society* 10: 133–148.

Blomley N (2016) The territory of property. *Progress in Human Geography* 40(5): 593–609, doi: 10.1177/0309132515596380.

Braverman I (2011) Hidden in plain view: Legal geography from a visual perspective. *Law, Culture and the Humanities* 7(2), pp. 173–186. doi: 10.1177/1743872109355579.

Braverman I, Blomley N, Delaney D, et al. (2014) *The Expanding Spaces of Law: A Timely Legal Geography*. Redwood City, CA: Stanford University Press.

Carter E, Donald J and Squires J (1993) *Space and Place: Theories of Identities and Location.* London: Lawrence & Wishart Ltd, xii.

Clark CL (1989) Law and the interpretative turn in the social sciences. *Urban Geography* Routledge 10(3): 209–228.

Collis C (2004) The proclamation Island moment: Making Antarctica Australian. *Law Text Culture* 8: 39–56.

Cowley A, Ryan C and Dunn E (2015) The law, the mafia and the production of sovereignties in the Kyrgytz penal system. *Ab Imperio* 2: 183–208.

Delaney D (2000) Semantic ecology and lexical violence: Nature at the limits of law. *Law Text Culture* 5: 77–112.

Delaney D (2004) Tracing displacements: Or evictions in the Nomosphere. *Environment and Planning D: Society and Space* 22(6): 847–860.

Delaney D (2010) *The Spatial, the Legal and the Pragmatics of Place-Making: Nomospheric Investigations.* Abingdon: Routledge.

Delaney D (2015) Legal geography I: Constitutivities, complexities, and contingencies. *Progress in Human Geography* 39(1): 96–102.

Delaney D (2016) Legal geography II: Discerning injustice. *Progress in Human Geography* 40(2): 267–274.

Delaney D (2017) Legal geography III: New worlds, new convergences. *Progress in Human Geography* 41 (5) 667–675.

Ferguson J and Gupta A (2005) Spatializing States: Toward an Ethnography of Neoliberal Governmentality. In Inda JX (ed), *Anthropologies of Modernity: Foucault, Governmentality, and Life Politics*, Oxford, UK: Blackwell Publishing Ltd., pp. 105–131.

Graham N (2010) *Lawscape: Property, Environment, Law.* Abingdon: Routledge.

Holder J and Harrison C (2003) Connecting Law and Geography. In Holder J and Harrison C (eds), *Law and Geography*, Oxford: Oxford University Press, pp. 3–16.

Hubbard P (2012) Kissing is not a universal right: Sexuality, law and the scales of citizenship. *Geoforum* 49: 224–232.

Iljadica M (2017) Copyright and the right to the city. *Northern Ireland Legal Quarterly* 68(1), 59–80.

Jalava J (2003) From norms to trust: The Luhmannian connections between trust and system. *European Journal of Social Theory* 6(2): 173–190.

Johns F (2013) *Non-Legality in International Law: Unruly Law*. Cambridge: UK: Cambridge University Press.

Konsen L (2013) *Norms and Space: Understanding Public Space Regulation in the Tourist City*. Lund Studies in Sociology of Law 41. Lund: Media-Tryck, Lund University.

Layard A (2010) Shopping in the public realm: A law of place. *Journal of Law and Society* Blackwell Publishing Ltd 37(3): 412–441.

Layard A and Milling J (2015) Creative Place-Making: Where Legal Geography Meets Legal Consciousness. In Warren S and Jones P (eds), *Creative Economies, Creative Communities: Rethinking Place, Policy and Practice*, Farnham: Ashgate, pp. 79–102.

Lindahl H (2010) A-Legality: Postnationalism and the question of legal boundaries. *The Modern Law Review* 73(1): 30–56.

MacSithigh D (2012) Virtual walls? The law of pseudo-public space. *International Journal of Law in Context* 8(3): 394–412.

Massey D (2005) *For Space*. London: Sage.

Mould O (2009) Parkour, the city, the event. *Environment and Planning D: Society and Space* 27(4): 738–750.

Mubi Brighenti A (2010) At the wall: Graffiti writers, urban territoriality, and the public domain. *Space and Culture* 13: 315–332.

Ntounis N and Kanellopoulou E (2017) Normalising jurisdictional heterotopias through place branding: The cases of Christiania and Metelkova. *Environment and Planning A* 49(10): 2223–2240.

Philippopoulos-Mihalopoulos A (2007) Introduction: In the Lawscape. In Philippopoulos-Mihalopoulos A (ed), *Law and the City*, Abingdon: Routledge-Cavendish, pp. 1–20.

Pue WW (1990) Wrestling with Law: (Geographical) Specificity vs. (Legal) Abstraction. *Urban Geography* 11(6): 566–585.

Teubner G (1997) Foreword: Legal Regimes of Global Non-state Actors. In Teubner G (ed), *Global Law Without a State*, Aldershot: Ashgate, pp. xiii–xvii.

Valverde M (2009) Jurisdiction and Scale: Legal `technicalities' as resources for theory. *Social & Legal Studies* 18(2): 139–157.

Walker N (2010) Out of place and out of time: The law's fading coordinates. *Edinburgh Law Review* 14: 13–46.

33

Militarisation and the creation of place

Rachel Woodward

How do military power and military phenomena create place? What does it mean if we say that a place is militarised? Are those effects permanent or temporary, cause for celebration or resistance? It seems obvious perhaps that military power works through places by the exercise of military force. If we follow Weber's suggestion that the military is a complex of functions and capabilities to which the nation-state grants authority to use violence, then military violence is central to place creation as a capacity of the nation-state. The temporal and spatial politics of the military creation of place are, we usually assume, reducible to an assessment of the legitimacy or otherwise of military deployment or presence.

Yet the military creation of place happens through multiple practices and through an enormous variety of military phenomena, expressed through place with different outcomes, and constituted too through the places where they come into being. Understanding the constitution and expression of military phenomena and military power across place requires an understanding of some complex, interlocking practices and ideas around control (Woodward, 2004). These include the controls exerted by physical military presence, the controls exerted by the availability or otherwise of information about military phenomena, the modes of governance through which control is exerted over geographical space and processes, and the discourses of national and international security through which military control over place is legitimated.

In this chapter I want to suggest an approach to understanding the military creation of place that rests on a conceptualisation of place as physical location, place at its most basic, most common-sense. This is not to deny the availability (as this Handbook shows us) of a range of ways of conceptualising place, but rather to narrow the conceptual focus down in order to enable a fuller understanding of militarisation and place to emerge. Conceptualising place as physical location prioritises two ideas. The first is of scale, for physical locations can be identified at a range of scales, and considering militarisation and the creation of place necessarily requires some thought about scale. The second is that prioritising physical location prioritises tangible presence and actual materialities. This capacity of place, as something that can be visible, observable, touched, walked and sensed, as I hope to show in this chapter, is significant when trying to understand militarisation and the creation of place. It is not that the non-visible and intangible are unimportant when considering militarisation; militarised places may induce fear in the absence of any clear indicators of military power, and military power may be exerted in non-material, non-physical ways – cyber warfare would be an example. But prioritising the tangible and material capacity of militarised places draws our focus towards the generative capacity of militarisation and its capacities for producing change, and seeing and touching the effects (or perhaps relics, in some instances) directs our attention to military agency (for an overview, see Pearson et al., 2010). In other words, the militarisation of place is not just something that happens, but rather is the consequence of innumerable conscious, deliberate acts.

Militarisation in this chapter is conceptualised as 'a multifaceted social process by which military approaches to social problems gain elite and popular acceptance' (Kuus, 2008, p. 625). This definition emphasises process and dynamism, and opens up opportunities for the identification of multiple points of engagement and purchase across social formations. Crucially, Kuus's definition is underpinned by an understanding of militarisation as an integral part of social life, with purchase across social formations from high politics to the mundane routines of the everyday, beyond the military institutions responsible for military phenomena (see Woodward et al., 2017). As this chapter will show, ready recourse to phrases like 'the militarisation of place' is resisted when we think more carefully about what militarisation actually is, conceptually. Conversely, prioritising an idea of militarisation as processual means that our attention is then focussed on how processes actually work as an object of empirical investigation, rather than just assuming that there is some kind of malign or benign – depending on your viewpoint – agency at work that is the responsibility of structural forces over which we have no control. In addition, conceptualising militarisation in these terms draws our attention to both the ease with which we can resort to the simplicities of the binaries inherent in the terms 'military' and 'civilian', and the necessity of resisting simplicity and interrogating exactly what those binaries might mean. Drawing on Birgitte Refslund Sørensen and Eyal Ben Ari's (2018) analytic approach to civil–military relationships from an anthropological perspective, this chapter uses their terminology of civil–military entanglements to speak to the idea of simultaneously recognising distinctions between civilian and military phenomena, whilst also recognising that they are entangled, co-constituted and co-dependent.

In focussing on multiple practices and phenomena through which military power is expressed in place and constituted through it, and mindful of the importance of scale and materiality to understanding militarisation and the creation of place, this chapter examines one particular place, the northern English city of Newcastle upon Tyne. Newcastle is as specific or as general as any other smallish city and wider conurbation in the UK. Teaching about the militarisation of place in the context of an undergraduate Geography degree here in the city, we start to think about this by leaving the lecture theatre and walking around the city. This experience underscores for me the utility of researcher engagement with the places where we situate our enquiries. The traditional-sounding idea of field-walking contains a single but essential point, that understanding the constitution of place follows from sensory engagements with place. The visual, aural, haptic, affective and olfactory are as important as factual information and conceptual framing when trying to understand place, and these senses are engaged when we go and look at the thing we're investigating. In this chapter, I explore some facets of militarisation and the creation of place through observations from field-walking in Newcastle upon Tyne. The military phenomena engaged with in this chapter are selective. These are conquest, defence, the production of weapons, the basing of military forces and modes of memorialisation. These are chosen because they are prevalent in Newcastle; other phenomena will demand prioritisation in other places, and I return to this point in the conclusion.

Conquest

The very existence of the city of Newcastle upon Tyne is a consequence of military conquest and occupation. Situated at the lowest feasible crossing point of the River Tyne, which runs eastwards to the North Sea across this part of northern England, although there is some evidence of pre-Roman settlement, the city's origins are conventionally dated back to the Roman establishment of fortifications and a crossing point, Pons Aelius, in about 120 CE under the Emperor Hadrian. Initiator of a built territorial boundary at the northern edge of the Roman empire from about 122 CE, Hadrian's Wall defined the city and leaves its mark in the contemporary city's urban morphology in the present. Hadrian's Wall is most commonly understood as a physical means of controlling population movement rather than an absolute barrier, in ways that have resonated at some points with the US–Mexico and other contemporary borders (Madsen, 2011). Hadrian's Wall when built marked a regular west–east line across the narrowest part of the island of Britain. Its eastern terminus is some 8 km east of the centre of the city of Newcastle at the fort known to the Romans as Segedunum and since then to the region's inhabitants as Wallsend.

Hadrian's Wall within Newcastle is no longer visible as a masonry structure or set of earthworks, but continues to resonate as a feature in the landscape because of its shaping of the street pattern of the city. The line of the wall is traceable – and obviously visible – in the straight line of a major road into the city from the west (today's Westgate Road). Standing at a point in the centre of the city and looking due west, this road is an obvious and enduring marker of this past episode of military infrastructural development for the purposes of territorial control. Marked out in concrete of two contrasting colours, the line of the wall is identified in a 4-metre section of enclosed pavement in front of a Victorian building in the city centre, with this concrete and an accompanying explanatory plaque only noted by those who care to stop and look (see Figure 33.1). The point about the military landscape here is the endurance in the urban fabric of military infrastructure, which in turn leads to questions about what can be seen and what cannot. The point about civil–military entanglements, though, is more opaque. Acts of entanglement – the relationships between a military power and the civilian populace in this place – can only be guessed at. What remains are traces. But these traces still remind us that civil–military entanglements have a spatiality, a physical form – they take place, literally. They leave their mark, even if those marks, like a barely visible difference in a concrete pavement, are obscure.

Although the origins of the city are Roman, the naming of the city derives from the construction of a new castle in the 12th century (which is still very much visible). This new castle replaced an older post-Norman Conquest 11th-century motte and bailey structure, which itself comprised a wooden fortification (the motte) surrounded by an enclosed area (the bailey) on the same site as an older Roman fort. The site on the edge of the gorge looking down on the river and its crossing point, bordered on another two sides by ground that falls away steeply to a river and a tributary burn, was clearly chosen for its defensive geography. The subsequent stone structure remains, visible as a dominant 12th-century castle keep and a more modest tower known as the Black Gate. The two now stand as individual structures, now separated by the railway on a viaduct. These traces of military control may be clearer in their physical form than those of the Roman era but the point here is about time and change in the militarisation of place. To visit this part of the city in the present is to visit an interesting example of medieval engineering, a tourist attraction, a place to sit and take a rest. The life of the city does not revolve around the castle as it once would have done. A visit to the castle is not a military encounter. But the dominance of this structure in

Figure 33.1 The line of Hadrian's Wall marked out in coloured concrete, Newcastle upon Tyne. © Rachel Woodward.

this place, and its absolutely specific relationship to the site's morphology, makes the point about enduring marks that military power etches onto place, and the ways that physical location enable military power to take place. The point is also made, encountering the castle, about the temporalities of militarisation; militarisation, so often understood as a contemporary force negatively affecting a place or a populace otherwise untrammelled by its effects, something to be resisted, is so often always already present (Jauregui, 2015).

Defence

Sometimes the impacts of military power in shaping place are obvious and visible, woven into the fabric of settlements, with a profound effect on their function and appearance (for an overview of these 'ecologies of power' see Bélanger and Arroyo, 2016). Although usually on a far more modest scale, compared with the ecologies of power of the contemporary US armed forces, for example, the traces of past imprints of military power may still be evident centuries after their instigation. Newcastle upon Tyne, like many long-established urban centres, had defensive military features for much of its history and these are still evident in the present in the traces of the city walls which date to the early 14th century. Much of the city's medieval and early modern history is as a border town and although the modern border between England and Scotland is now stabilised far to the north, traces of earlier defensive functions are still present. These traces speak to civil–military entanglements because they remind us of the use of military power in determining what lies within and beyond a zone of defence and protection. Newcastle was for much of its history a walled city, and although much of the masonry of this wall and its gateways was removed in the later 18th and 19th centuries as the city grew (a function of its emergence as an economic centre built on trade, particularly coal, and heavy industry), traces remain. These are evident in standing medieval walls within the city centre, in remaining gateposts and towers, in the street pattern in places and in the naming of streets (Westgate Road, Newgate Street, Sandgate). The remaining walls are now subject to statutory protection as historic structures, protections not afforded in previous centuries when demolition allowed the city to grow. The morphology of the city has grown around them, with successive rounds of construction and demolition of the urban fabric such that in the present they sit as isolated blocks of masonry with, in some sections, no visible relationship to their surroundings in terms of form or function. One section backs on to the long straight line of Stowell Street (see Figure 33.2), now portrayed by inhabitants and city tourism promoters as Chinatown, focus of New Year celebrations early in February each year, of nights out in the restaurants and daytime business in the Chinese medical practitioners' clinics, solicitors' offices and food shops. One of the turrets in the wall is now used as a small arts performance venue. Another section of the old medieval wall, running perpendicular to the gorge through which the River Tyne flows, is surrounded by new office space, an older pub, the remains of a central depot for the Royal Mail, a casino and a carpark. Down by the river at Sandgate, a modern

Figure 33.2 Portion of city walls backing onto Stowell Street, Newcastle upon Tyne. © Rachel Woodward.

road embankment has been faced with stone on which is carved a representation of the river as it flows across this section of northern England, and a commemorative plaque signals the former existence of the town walls in this part of the city.

The remaining segments of the city's walls, as with the castle, remind us to consider the temporality of militarisation; the city may have built walls as a military response to a perceived social threat (invasion), and this would have, we assume, gained both elite and popular acceptance as an appropriate response, but this is of course historical. The walls show in material form the way that militarisation creates place – the walls from their inception shaped the morphology of the city – and also that these priorities fade. An imposing physical barrier, a piece of infrastructure engineered to assert military control, is now reduced to small sections of minor curiosity. Militarisation may create place but in ways that can later be amended or adapted. Other traces of military defence in the city are entirely absent. For example, look at the Tyne Bridge, an imposing structure spanning the gorge through which the River Tyne flows 60 metres below. Opened in 1928 as a major traffic artery linking Newcastle on the north bank of the Tyne to Gateshead and beyond to the south, we see it in the present as communications infrastructure. No traces remain of its planned role during the early years of the Second World War as a point at which anticipated invading German armed forces coming from the north would be stopped in their tracks by the demolition of the bridge. The bridge was part of a network of points which comprised the Tyne 'stop line', one of many such lines of defence across the landscape of Britain comprising points where the momentum of an invading force could be slowed (Collier, 1957). This feared invasion (widely anticipated in 1939/1940, mostly forgotten by 1943) never happened and the explosive charges wired to the Tyne Bridge were removed, leaving (as far as I am aware) no trace. To think, in the present, of the Tyne Bridge being blown up is almost impossible. Militarisation can sometimes leave no trace in place.

Armaments production

The defence industries responsible for armaments production are distinctive shapers of place (for an overview, globally, see Tan, 2009; Kurç and Neuman, 2017). The 19th-century wealth of the Tyneside region lies with heavy industry. A confluence of engineering innovation, the availability of capital investment, the proximity and accessibility of sources of power (coal) and of land, the existence of the River Tyne to facilitate import and export, and the availability of an industrial workforce (not least because of changes in agricultural practices and rural–urban migrations through the 18th and the 19th centuries) all contributed to the industrial development of the region, much of which involved the production of the means of violence. Examples include the Armstrong-Whitworth company's manufacture of warships, aircraft and armaments across the 19th and 20th centuries at Elswick, about 5 km upriver from the centre of the city. Industrial arms production has now all but disappeared but leaves its traces in the landscape. The Swing Bridge across the river in the centre of the city, the only crossing point over the Tyne gorge at low level, was built in 1876 and designed to pivot horizontally on its central axis to enable warships (amongst others) to pass downstream from the Elswick works to the sea. Away from the river and up by the university, a memorial statue to the company's founder, the engineer and industrialist Lord Armstrong, is flanked by bronze panels depicting the Swing Bridge and the fitting of cannon to a warship. That this memorial is situated by Newcastle University is appropriate. The university traces its origins in part back to a 19th-century College of Physical Science founded by Lord Armstrong in 1871. The civil–military entanglements through which military phenomena create place include, therefore, not just the production of armaments but also the development of the educational and training infrastructure to enable technical innovation. The university where I sit and write this today is as implicated in the militarisation of place as the last remains of industrial manufacture of warships.

The remains of the Armstrong-Whitworth works at Elswick are unmarked by any identifying plaque, and a prototype Challenger tank, which for a couple of decades was parked outside the factory on the Scotswood Road, a major traffic artery running along the north bank of the River Tyne, has now been moved. It sits outside a local science and natural history museum. The tank has been made safe, both literally and metaphorically, through its partial encasement in wooden packing; a symbol of military violence reconstructed as public art with the added benefit that children cannot climb on it, as they would otherwise be likely to do. But does it militarise the place in which it is now parked? If we consider the idea of militarisation introduced at the start of this chapter, which points to its processual capacities for enabling military solutions to social problems, it would appear not. This in turn prompts the idea that militarisation may not be identifiable even where visible indicators of military phenomena exist. Because what exactly is the social issue being addressed by the parking of a redundant tank, reconstituted as art, outside a civic museum?

Basing

After the signs of war itself, military installations are the clearest indicator that military phenomena are expressed across place – indeed, they literally 'take place'. Place appropriation for military purposes happens for a host of functions and these invariably leave their mark (Gillem, 2007; Lutz, 2009). Across the River Tyne, in Gateshead, sits one such site, the Royal Navy's HMS *Calliope* (see Figure 33.3). This small establishment is flanked by two symbols of the regeneration of the Tyne's quayside, the Sage concert hall and the BALTIC centre for contemporary art. HMS *Calliope* is currently home to the Royal Navy Reserve, the Northumbrian Universities Royal Naval Unit (NURNU) and the Royal Marines Reserve Tyne detachment. A ship from the Royal Navy fleet, the small P2000 fast inshore patrol vessel HMS *Example* used by the NURNU, is moored alongside when not at sea on exercises.

Civil–military entanglements are clearly visible at HMS *Calliope*. Obvious military presence is indicated via signs for the Royal Navy and Royal Marines, the white ensign flag flying outside the main building, and the battleship-grey HMS *Example* moored alongside (see Figure 33.3). HMS *Calliope* is a small establishment and the site often looks deserted. The visibility of the military presence is muted but still discernible, entangled with signs of civilian engagement. What looks like a carpark within the base has a large letter H painted on it, a helicopter pad. HMS *Example* has fixed upon it a plaque looking very similar to the Newcastle University crest. This naval base is normalised within the landscape; however, the two iconic architectural structures which flank it, Sage and BALTIC, draw the eye from what is otherwise a fairly nondescript glass-clad, two-storey rectangular structure which looks like what it is: office accommodation, training and leisure facilities for an organisation. This is a naval base in the heart of a city which, if not hidden in plain sight, is certainly less visible than what sits around it.

This (in)visibility is aided by the importance of the site as a Reserves base. Visibly military personnel are hard to spot at this site, because this is the home base for personnel who are more usually elsewhere at their civilian places of work. The story of the role of the Reserves in cementing (or otherwise) civil–military relations in the UK is beyond the scope

Figure 33.3 HMS *Calliope* with HMS *Example* berthed alongside, with the Sage Gateshead, BALTIC centre for contemporary art, Tyne Bridge and Millennium Bridge. © Rachel Woodward.

of this chapter (see Edmunds et al. 2016 for more detail). What is evident though in the buildings here is the tenacity of defence planners in recognising the importance of location for current UK defence personnel policies reliant on developing Reservist numbers and roles. HMS *Calliope* has been on this site since the late 1960s, moving into what had been a Royal Mail depot at a time when the surrounding quayside was exclusively industrial. The surroundings have changed due to deindustrialisation and arts-based regeneration strategies, but still HMS *Calliope* remains. Successive conversations between urban planners and defence land managers about alternative accommodation, which would surely suit ambitious plans for complete regeneration of this part of the urban area, have come to nothing. It is hard not to speculate that the continued location here of the Royal Navy is not just a consequence of convenience (the site is very accessible to road and public transport for those reporting for training) but also for its symbolic value, asserting the continued existence of the Royal Navy even when the total regular force itself now numbers under 30,000 people (Ministry of Defence, 2017) and has a surface fleet (including HMS *Calliope*) of under 90 vessels (Axe, 2016). The role of militarisation in the creation of place is symbolic as well as material.

Memorialisation

Civil–military entanglements are also evident in war memorials. It is a truism worth repeating that war memorials are often simultaneously prominent in places, and ignored.

Post-1918, when grief was sharp and funds forthcoming, many war memorials and associated sites of memorialisation were built across Britain. Tyneside is no exception. For example, in Gateshead, south of the river, a central municipal war memorial dominates a busy suburban traffic intersection. A couple of kilometres away a public park contains at least four memorials to Britain's 20th-century wars and many of Gateshead's suburbs have their own modest war memorials. Within the Newcastle University precinct, a central quadrangle comprising manicured gardens and flanked by the university's older red-brick buildings is home to a memorial to staff and students who lost their lives during the Second World War (see Figure 33.4). Inside the university's oldest building (named after university benefactor, engineer and armaments manufacturer Lord Armstrong whose industrial activities did much to enable the arming of the First World War's belligerents) is a marble plaque listing the names of the 223 students and staff who were killed in action as members of the British armed forces during the First World War. The carved inscription on the plaque was updated recently to cement this as a memorial to all those students killed whilst on active service in armed conflict.

The literature on war memorials and their functions in landscapes reminds us of the cultural, political, regimental and individual invocations that these memorial landscapes invoke (examples include Hoffenberg, 2001; Johnson, 2004; Muzaini and Yeoh, 2005). War memorials bind together ideas about personal loss with ideas about nationhood, 'just war', sacrifice. They are about local, regional and national identities and about the provision of a narrative explaining and often justifying loss in war.

Figure 33.4 Newcastle University Second World War memorial. © Rachel Woodward.

In the urban spaces of Tyneside, a number of these sites become animated once a year on Remembrance Sunday, usually the first Sunday in November and a date chosen after the end of the First World War for national commemoration services because of its proximity to the date of the Armistice on 11 November 1918. This animation, in the centre of the city at Eldon Square, at a suburban road intersection in Gateshead, and within the university's precincts, comes in the form of short commemorative services and a march-past of military personnel (including Reservists and young cadets). Sound and the experience of sound is important to these events, with the stilled traffic (roads usually open to traffic are closed for the event), the silence of the crowd at particular moments in the ceremony, the music of military bands, the tramp of marching boots on tarmac, the solo bugle playing *The Last Post* to mark the closure of proceedings. There is an expectation of solemnity at these events, an expectation of the wearing of the Royal British Legion's poppy pinned to a lapel. Poppy wreaths are laid and small wooden crosses amenable to personal inscription are planted or placed where the ground allows. The animation of these memorial landscapes is periodic and specific to this point in time in November, but the poppy wreaths and crosses stay on through the year, gradually fading as the effects of wind and rain weather synthetic fabric and thin wood. These poppy wreaths mark out memorials until they disappear, and the memorials return to their habitual state of anonymity.

As sites of civil–military entanglement and places where militarisation becomes evident, the war memorials of the city encapsulate the experiential, contradictory and processual nature of those relationships. These places become locations for visible, public displays of support for the armed forces. Presented as services of commemoration, the march-past and the men and women in uniform are infrequent occurrences in British cities; there is a strong element of spectacle here and, as Debord (1967) and others remind us, the spectacle is implicated in processes of social control. The solemnity and order behind these performances of commemoration, officiated by leaders of the church and enacted by those granted state legitimacy for the bearing of arms, and their element of spectacle, remind us of the endurance and power of state narratives of the legitimacy of its military operations in processes of memorialisation for the dead (see, for a similar example from Israel, Lomsky-Feder, 2005). But these are also personal events for many people, occasions when private grief and remembrance can be publically shown and recognised. These are places where civil–military entanglements become evident at the level of the personal and experiential, just as much as they are places for displays of state narratives (Basham, 2016). Military phenomena – personnel, martial music, modes of personal deportment – create these as militarised places through the controls they exert over what these places mean, albeit for a limited period of time. The visible displays of public acceptance for military modes of remembrance show the dynamism of militarisation in the creation of place.

Conclusion

This chapter has prioritised an idea of militarisation that foregrounds process over presence. Thinking in these terms focusses our attention on the practices through which military approaches to issues in social life come to be accepted – in short, how militarisation operates. Encounters with various places in the city of Newcastle upon Tyne provide opportunities for engagements with physical locations such that the generative, dynamic capacities of military phenomena to create place become visible or evident, not least through sensory means. How other military or militarised practices and their place-shaping capabilities in other places would bring different conclusions is of course an open question.

Certainly, the perceptions of military personnel themselves about their roles in the creation of place and ideas about place would merit more sustained examination than currently exists, as Chitukutuku and Maringira (2019) suggest. It is also the case that military constructions of virtual space in the contemporary period complicate our thinking about what military places might in fact be (Cristiano, 2018). The phenomena discussed here – conquest, defence, arms production, basing, memorialisation – are not specific to this city, and may be relatively unimportant in other contexts. Equally, in other cities and other places, different phenomena, processes and militarised practices will be evident or more prominent. As a growing literature on urban securitisation makes clear, the causes, costs and consequences for urban life, planning, design and living can be endlessly variable (Coaffee, 2009; Coaffee et al., 2009; Graham, 2010; Pullan, 2011). We should also be mindful of the invisibility – or rather, muted presence – of military power via information and communication technologies. But the point about thinking through military phenomena in terms of the physical locations in which they occur – where they take place, in other words – is that these acts of observation and engagement help explain the generative capacities of militarisation as a process that is profoundly geographical in the ways that it is constituted by and expressed through place.

References

Axe, D. (2016) What the US should learn from Britain's dying navy. *Reuters news* website www.reuters.com/article/us-uk-military-navy-commentary/commentary-what-the-u-s-should-learn-from-britains-dying-navy-idUSKCN10L1AD Accessed 19 March 2020.

Basham, V. (2016) Gender, race, militarism and remembrance: the everyday geopolitics of the poppy. *Gender, Place & Culture* 23 (6): 883–896.

Bélanger, P. and Arroyo, A. (2016) *Ecologies of Power: Countermapping the Logistical Landscapes and Military Geographies of the US Department of Defense*. Cambridge, MA: MIT Press.

Chitukutuku, E. and Maringira, G. (2019) Spirituality in African military geography: Soldiers on deployment and spiritual engagements with landscapes. In Woodward, R. (Ed.) *A Research Agenda for Military Geography*. London: Edward Elgar, pp. 133–146.

Coaffee, J. (2009) *Terrorism, Risk and the Global City: Towards Urban Resilience*. Farnham: Ashgate.

Coaffee, J., O'Hare, P. and Hawkesworth, M. (2009) The visibility of (In)security: The aesthetics of planning urban defences against terrorism. *Security Dialogue* 40 (4-5): 489–511.

Collier, B. (1957) *The Defence of the United Kingdom*. London: HMSO.

Cristiano, F. (2018) From simulations to simulacra of war: game scenarios in cyberwar exercises. *Journal of War and Culture Studies* 11 (1): 22–37.

Debord, G. (1967) *The Society of the Spectacle*. New York: Red and Black.

Edmunds, T., Dawes, A., Higate, P., Jenkings, K.N. and Woodward, R. (2016) Reserve Forces and the transformation of British military organization: soldiers, citizens and society. *Defence Studies* 16 (2): 118–136.

Gillem, M. (2007) *America Town: Building the Outposts of Empire*. Minneapolis: University of Minnesota Press.

Graham, S. (2010) *Cities Under Siege: The New Military Urbanism*. London: Verso.

Hoffenberg, P.H. (2001) Landscape, memory and the Australian war experience, 1915-1918. *Journal of Contemporary History* 36 (1): 111–131.

Jauregui, B. (2015) World fitness: US Army family humanism and the positive science of persistent war. *Public Culture* 27 (3): 449–485.

Johnson, N. (2004) *Irish Nationalism and the Geography of Remembrance*. Cambridge: Cambridge University Press.

Kurç, C. and Neuman, S.G. (2017) Defence industries in the 21st century: A comparative analysis. *Defence Studies* 17 (3): 219–227.

Kuus, M. (2008) Civic militarisation. In Flusty S, Dittmer J, Gilbert E, Kuus M (2008) Interventions in banal imperialism. *Political Geography* 27: 617–629.

Lomsky-Feder, E. (2005) The bounded female voice in memorial ceremonies. *Qualitative Sociology* 28: 293–314.
Lutz, C. (Ed.) (2009) *The Bases of Empire: The Global Struggle Against US Military Posts*. New York: NYU Press.
Madsen, K.D. (2011) Barriers of the US-Mexico border as landscapes of domestic political compromise. *Cultural Geographies* 18 (4): 547–556.
Ministry of Defence. (2017) *UK Armed Forces Monthly Service Personnel Statistics*, 1st September 2017. London: Ministry of Defence.
Muzaini, H. and Yeoh, B.S.A. (2005) War landscapes as "battlefields" of collective memories: reading the Reflections at Bukit Chandu, Singapore. *Cultural Geographies* 12 (3): 345–365.
Pearson, C, Coates, P. and Cole, T. (2010) Introduction: Beneath the camouflage; revealing militarized landscapes. In C. Pearson, P. Coates and T. Cole (Eds) *Militarized Landscapes: From Gettysburg to Salisbury Plain*. London: Continuum, pp. 1–18.
Pullan, W. (2011) Frontier urbanism: the periphery at the centre of contested cities. *The Journal of Architecture* 16: 15–35.
Sørensen, B.R. and Ben-Ari, E. (Eds.) (2018) *Rethinking Civil-Military Relations*. New York: Berghahn.
Tan, A. (Ed.) (2009) *The Global Arms Trade: A Handbook*. London: Routledge.
Woodward, R. (2004) *Military Geographies*. Oxford: Blackwell.
Woodward, R., Jenkings, K.N. and Williams, A.J. (2017) Militarisation, universities and the University Armed Service Units. *Political Geography* 60: 203–212.

34
Policing place

Jon Coaffee

Today, inter-related strategies to deal with insecurity are widespread and have become an essential part of place-making efforts, with successful places being those that are seen as safe, well maintained and well managed. Such approaches to 'policing place' commonly incorporate two key elements. First, place *management* whereby a series of spatial and temporal rules and regulations are socially enforced or dictated by the forces of law and order. Second, the *fortification* of space, where the introduction of defensive design measures such as surveillance cameras, walls, barriers, and gates cause physical segregation of the landscape and can have significant implications upon place qualities.

In today's fast paced, increasingly globalised and insecure world where risk, crisis and uncertainty abound, such strategies of policing place are often highly visible and politicised, and have an important and influential genealogy. In this chapter, the policing of place is viewed through two chronological lenses that highlight how the police have increasingly become intertwined with urban planners and other stakeholders in the creation of the built environment and in the governance of places. The first lens highlights how over time, initial urban planning ideas of defensive architecture and its influence on human behaviour have been increasingly utilised by the police in attempts to 'secure by design' as the perception of fear in public places has increased. In the contemporary period this has led to an increasingly sophisticated array of safety and security measures, and techniques of crime prevention being deployed, with the police emerging as an increasingly important actor in the construction, management and use of places. The second lens focusses upon how these defensive design ideas and the role of the police in place-making has further been adapted in attempts to counter the threat from urban terrorism. Here, the policing of place has become a delicate balancing act between ensuring safety and security and maintaining the vitality of the public realm.

Enhancing safety through policing and design

Policing has, since the early nineteenth century, been used as a means to secure cities. In 1814, Sir Robert Peel, the Chief Secretary for Ireland, introduced the Peace Preservation Act, which established a Peace Preservation Force to prevent crime, protect property and deal with unrest and rioting. When back in the UK, Peel was appointed Home Secretary and in London he founded a civil police force, the Metropolitan Police, in 1829. Before this, some London squares were secured by private 'guardians' armed with canes, who kept order and made sure the squares were only used by residents and their guests. In time, similar police forces became established across the country for the preservation and maintenance of law and order.

As industrialisation and associated urbanisation continued apace through the nineteenth and twentieth centuries, policing continued to be a key element of managing insecurity. In time though, the form and function of the built environment was also increasingly seen as central to effective safety and security and to place-making through the purported ability of design to determine behaviour (often referred to as design determinism). Over fifty years ago, writer and journalist Jane Jacobs (1961) highlighted how poor urban design could contribute to diminishing community safety and developed the concept of 'eyes on the street'. The logic underpinning her idea was simple: the more people in the streets, the safer they become through the enhancement of informal surveillance. Such observations also elucidated the importance of the design of the city in producing or mitigating potential criminogenic environments, influencing human behaviour and affecting quality of life. Specifically, Jacobs (1961: 35, emphasis added) argued that

> A city street equipped to handle strangers and to make *a safety asset* … must have three main qualities:
>
> First, there must be *a clear demarcation between what is public space and what is private space.* Second, there must be *eyes upon the street, eyes belonging to those we might call the natural proprietors of the street.* And third, the sidewalk *must have users on it fairly continuously*, both to add to the number of effective eyes on the street and to induce the people in buildings along the street to watch the sidewalks in sufficient numbers.

For Jacobs, one of the main characteristics of a thriving place is that people feel safe and secure, despite being among complete strangers. Jacobs' community-centred viewpoint on planning and the importance of safety sought to ensure that everyday experiences were represented in public spaces that she saw as central to the dynamics of city life. Essentially, if such locations are accessible, attractive, and safe they encourage a range of uses and activities. Conversely, when public spaces are neglected, they can create a feeling of insecurity where people fear to tread. The logic was simple: the more people in the streets, the safer they become. Their 'eyes on the street' provide constant monitoring of the urban environment for signs of danger. Urban security was therefore not simply a matter of policing routines but was related to the physical design of the built environment, the quality of public spaces and their ability to attract people onto the streets. This was further seen in relation to ideas of clear borders or demarcated territory that gave a sense of ownership of communal spaces. As Jacobs exemplified,

> the presence of buildings around a park is important in design. They enclose it. They make a definite shape out of the space, so that it appears as an important event in the city scene, a positive feature, rather than a no-account leftover.
>
> *(Ibid: 54)*

Jacobs' pioneering work was inspirational in stimulating research and thinking on the relationship between certain types of environmental design and reduced levels of violence.

This led, from the early 1970s, to urban design and planning tools commonly being used to address the causes of crime, disorder and incivility, and its impacts. In America, defensive architecture and urban design were increasingly used as a direct response to the urban riots that swept many US cities in the late 1960s, as well as the perceived problems associated with the physical design of modernist high-rise housing blocks that were viewed by criminologists of the day as breeding grounds for criminal activity. The strategic manipulation of the built environment, it was argued, could create places that would discourage unwanted behaviours (particularly opportunistic crime) and encourage 'good citizenship'. Crime Prevention Through Environmental Design (CPTED) approaches, it was claimed, could create environments and arrange buildings to deter crime. As its pioneer, C. Ray Jeffrey, argued 'in order to change criminal behaviour we must change the environment (not rehabilitate the criminal)' (1971: 178). Such ideas were also progressed when, after a study of large public housing estates in St Louis and New York, architect and planner Oscar Newman published *Defensible Space – Crime Prevention Through Urban Design* (1972) provoking an intense debate on the relationship between crime and the built environment.

Defensible space encompassed a 'range of mechanisms – real and symbolic barriers ... [and] improved opportunities for surveillance – that combine to bring the environment under the control of its residents' (1972: 3). Defensible space was, in Newman's words, offered as an alternative to the target-hardening measures that were rapidly being introduced to American housing developments at this time as a way of stimulating greater community involvement in crime management and overall place quality. In essence, defensible space became a means by which the residential environment could be more subtly redesigned to 'become liveable and controlled not by the police, but by a community of people who share a common terrain' (1972: 2). Newman did not rule out the use of security fences or electronic surveillance technologies but relying on these measures was seen as a last resort if subtler design solutions were unsuccessful (Coaffee, 2003). Echoing Jacobs' central tenet of successful place-making that 'policing' does not necessarily take place just through organised police forces but also through informal community surveillance, the notion of defensible space fundamentally incorporated ideas of civic responsibility:

> The designs catalyse the natural impulses of residents, rather than forcing them to surrender their shared social responsibilities to any formal authority, whether police, management security guards, or doormen. [They call] for the return of participation and control to the local level.
>
> *(Newman, 1972: 11)*

In practice, Newman proposed that places could become more defensible, or safer, if they are clearly demarcated and shrunken, and proposed four interrelated design features that contribute to secure residential environments:

1 *Territoriality* – creating a sense of ownership and of community of shared space amongst local residents by the zoning or demarcating of public space;
2 *Natural surveillance* – the use of environmental design to improve the ability of residents to observe their locality;*Image* – alteration of the structure of buildings to counteract the 'stigma' of public housing;
3 *Milieu* – alteration of environmental surroundings of residential areas to merge with areas of the city considered safer and to enhance their security.

Newman's ideas, like those of CPTED, were of the late 1960s and early 1970s and, as such, reflected the growing interest of the architectural profession in the relation between environment and behaviour, with some influence from anthropological ideas of territoriality. Defensible space was considered attractive at this time, as it emphasised the use of the environment to promote residential control and the potential to return to a more human and less threatening environment.

Since their incorporation into mainstream urban planning, defensive design measures and their potential for policing place have been further advanced due to rising crime rates, the escalation of social conflicts related to material inequality, intensifying racial and ethnic tensions and the heightened fear of crime. This has led to an increasingly sophisticated array of fortification, surveillance and security management techniques being deployed by urban authorities and the police to protect perceived urban vulnerabilities based on the principles of Defensible Space/CPTED (Cozens et al., 2001).

Since the work of Jeffrey and Newman, 'second and third generation' schemes have evolved to increasingly include measures such as access control, tactics such as 'target hardening' and advanced technologies such as CCTV, to enhance the surveillance of space. As a result, the police have increasingly become incorporated as an influential stakeholder in planning and place-making processes in the same way as planners became enmeshed in the fight against crime. For instance, in the late 1980s, the UK Association of Chief Police Officers (ACPO) and the Home Office pioneered a new approach to reduce crime – Secured by Design (SBD) – which sought to embed CPTED and defensible space ideas into new housing developments (Cozens, 2002). In the UK, this meant that police-led SBD approaches became integral to overarching safer city approaches and led to the police increasingly taking on a risk management role as strategic advisers to planners, project managers and other built environment professionals, on the basis that good design and layouts reduce criminal activity.

Normalising crime prevention through design

By the 1990s, new security approaches based on defensive design principles were significantly influencing the design and management of the urban landscape, with it being implied by some that 'form follows fear' (Ellin, 1997). For many commentators, this was part of a broader social shift towards a 'culture of fear' that pervaded urban life (Furedi, 2006; Garland, 2001).

In the 1990s, in some cases the response of urban authorities to insecurity was dramatic, especially in North America and, in particular, Los Angeles, where it was argued that the implementation of crime displacement measures and the surveillance of particular places was taken to an extreme. Los Angeles assumed a theoretical primacy within urban studies, with an overemphasis on its militarisation that portrayed the city as an urban laboratory for anti-crime measures. Fortress urbanism was highlighted as the order of the day, as an obsession with security became manifested in the urban landscape with 'the physical form of the city ... divided into fortified cells of affluence and places of terror where police battle the criminalized poor' (Dear and Flusty, 1998: 57). The emergence of the fortress city also appeared to be about transforming the city in the mirror of middle-class paranoia combined with economic vibrancy. As Haywood (2004: 115) noted, 'this was the corporate Los Angeles manning the ramparts in a bid to protect its economic interests by excluding those individuals and groups no longer necessary for (or dangerous to) the perpetuation of profit in the city's new globalised economy'. Here, as in many other locations, safety and security became a key leitmotif in place marketing campaigns that sought to attract inward investment and new wealthy residents.

Mike Davis is perhaps the most cited author on 'Fortress LA' in depicting how city authorities and private citizen groups responded to the increased fear of crime by 'militarising' the urban landscape and the multitude of public spaces therein. His dystopian portrayal of LA (1990, 1998) provided an alarming indictment of how increasing crime trends are effecting the development and functioning of the contemporary city through the radicalising of territorial defensive measures and with the increased role played by the Los Angeles Police Department (LAPD) in the development process. As Davis starkly highlights, 'in cities like Los Angeles on the hard edge of postmodernity, one observes an unprecedented tendency to merger urban design, architecture and the police apparatus into a single comprehensive security effort' (1990: 203). As the boundaries between the two traditional methods of crime prevention – law enforcement and fortification – become blurred, defensible space and technological surveillance, once used at a micro-scale level, were being used across the city to protect an ever-increasing number of city properties and residences.

Territoriality, as we saw earlier, was a central tenet of Jacob's and Newman's defensible space approaches and for many decades has been used to help explain the impact of defence as key in shaping the contemporary urban landscape (Gold and Revill, 1999). In the 1990s' policing perspective, how the police strategically use the notion of territoriality to control the city was seen as paramount. Territoriality was the central idea in Steve Herbert's ethnographic study (1997a, 1997b) which indicated the significance of territoriality for the LAPD, who actively seek to control the spaces they patrol, which intensifies as power is resisted – 'contested spaces preoccupy the police most' as they create boundaries and restrict access as they seek to regulate space (1997b: 399). Herbert's work provided vital insights for understanding how the police were increasingly active in the planning process, forming partnerships with local authorities or community groups to reduce crime and the fear of crime in the public realm. It further highlighted how the actions of the police can be interpreted through the control of space and, more broadly, how the 'power of the police is inserted within the fabric of the city' (1997a: 6/7).

Securing an urban renaissance

In the twenty-first century, the rhetoric of not just creating better public space management 'but safer too' has become part of the broadening 'designing out crime' agenda with managing and regulating the design aspects of crime prevention viewed as integral to high-quality places. In the early years of the new millennium, the policing of place emerged as an increasingly 'holistic' approach to security, which became progressively embedded within the latest round of government initiatives concerning community safety and the development of 'safe' and sustainable communities. Echoing Jacobs' 'eyes on the street', Rogers and Coaffee (2005: 323) highlighted explicit links between attempts to promote new sustainable urban realms founded upon the principles of social mixing, connectivity, higher densities, walkability, and high-quality streetscapes, and broader issues of urban safety:

> In the UK, recent policies of design-led urban renaissance have been concerned with making the environment of cities as a whole more attractive whilst at the same time improving the safety, management and governance of public spaces.

Table 34.1 Attributes of sustainability relevant to crime prevention and community safety

Attribute	*Descriptor*
Access and movement	Places with well-defined routes, spaces and entrances that provide for convenient movement without compromising security
Structure	Places that are structured so that different uses do not cause conflict
Surveillance	Places where all publicly accessible spaces are overlooked
Ownership	Places that promote a sense of ownership, respect, territorial responsibility and community
Physical protection	Places that include necessary, well-designed security features
Activity	Places where the level of human activity is appropriate to the location and creates a reduced risk of crime and a sense of safety at all times
Management and maintenance	Places that are designed with management and maintenance in mind, to discourage crime in the present and the future

Other reports and policy statements on the state of the built environment have also made it clear that general issues of community safety had become a political priority, with acceptance of the idea that successful places are safe, well maintained and well managed. For example, the 2004 publication in the UK of *Safer Places: The Planning System and Crime Prevention* argued that 'safety and security are essential to successful, sustainable communities', often with a focus upon designing out crime. Not only are such places well-designed, attractive environments to live and work in, but are also places where freedom from crime, and from the fear of crime, improves the quality of life. This guide also identified seven 'attributes of sustainability' to be considered 'as prompts' to promoting community safety (ODPM, 2004: 13). These 'attributes' are highlighted in Table 34. 1 and draw significantly on ideas of CPTED and defensible space that had been utilised by built environment professionals and law enforcement agencies since the 1970s, as well as practices of SBD.

For many, this approach puts a specific responsibility on planners and developers to use design for controlling human behaviour and thwarting crime in public spaces, and that:

> The principle that urban design can be used to control and order mobility is embedded in the design guidelines for sustainable communities planning ... Designers are called on to use environmental tactics and practices to shape the form and character of public spaces and development areas so that they become less attractive to potential criminals and undesirables.
>
> *(Raco, 2007: 50)*

Ongoing urban regeneration and safety concerns have foregrounded concerns over the changing nature of 'publicness', which might be accused of being diluted or purified by design and management processes that focus upon creating a clean, safe and accessible public environment. These physical and managerial realities of ongoing urban rejuvenation, that have inscribed neoliberalism into place management, have been subject to critiques that raise concerns about the negative impacts of security design and practice on everyday public urban experience (Atkinson and Helms, 2007; Coaffee, 2005) and the continued coalescing of planning and policing in place-making activities.

From policing place and designing out crime to designing-in counter terrorism

Increasingly, such security and place-making approaches have entered the current discourse about countering terrorism and the protection of public places. Since 11 September 2001, concerns about the likelihood and impact of terrorist attack against crowded public locations have heightened the sense of fear in many urban locations as future attacks against these so-called 'soft targets' appear more likely (Coaffee et al., 2008). Crowded places such as shopping areas, transport systems, sports and conference arenas were seen as at high risk and could not be subject to military-urbanism with its guns, gates and guards posture, or traditional security approaches such as searches and checkpoints, without radically changing public experience. The creation of an environment which is inherently more resilient and less likely to suffer attack through 'designing-in' counter-terrorism to physical and managerial systems was seen to offer hope of improving security in an acceptable and effective way.

In the new millennium, the actuality and fear of terrorist attacks, often conducted by suicide attackers or hostile vehicles and tactically aimed at crowded places, led to considerable ongoing research on how the police can seek to make such places more secure whilst retaining their vibrant character. These crowded areas have features in common such as their lack of access control but may be bounded (a stadium) or unbounded (a shopping area). While iconic buildings and specific hubs such as airports have long been identified as targets for terror attacks, a more general reading of public places is also required if resilience is to be enhanced. For example, recent counter-terrorism design guidance for crowded places issued in Australia refers to policies that embed features into the urban environment that counter or mitigate the effect of a successful attack, utilising the language of resilience. This symbolises a protective, proactive and necessary approach that creates a human environment that is difficult to attack, modifying the nature of terrorist targets by lessening their physical vulnerability:

> A resilient crowded place has trusted relationships with government, other crowded places, and the public. It has access to accurate, contemporary threat information and has a means of translating this threat information into effective, proportionate protective security measures commensurate with the level of risk they face.
>
> *(Australian National Government, 2017)*

Challenges of policing places in the age of terrorism

Together, this blend of changing terror methods and targets, especially directed against crowded places in urban centres, has provided a challenge for policing and security professionals in keeping such places safe. Protecting crowded public places is one of the key policing priorities in the ongoing fight against crime and terrorism in many Western countries, with periodic calls for counter-terrorism measures to be embedded within the design, planning and construction of public places given their importance economically and in the vitality of urban life. The fear is that vibrant public places will become sterile locations if targeted by terrorists. Equally, if security measures are too militaristic they can discourage the public from frequenting such locations. There have therefore been calls for planners and the police to consider security solutions that blend more seamlessly into the built environment, making them more aesthetically pleasing and more in keeping with the principle of attracting people into the public realm (Coaffee, 2018; Coaffee, O'Hare, and Hawkesworth, 2009).

In essence, spatial design, material choices, aesthetics and many other 'design' factors can influence the vulnerability of a location to attack but also change the fundamental character of places. It is not suggested that design intervention is a stand-alone defence against a terror attack, but it does form one part of a multi-layered and multi-pronged system of security. Equally important are the governance arrangements that ensure that all relevant stakeholders – not just the police and other security professionals – are informed and consulted about the development of counter-terrorism measures and strategies.

Current approaches to place-making highlight the importance of engaging with multiple stakeholders and local communities but, until recently, key issues related to the social impacts of counter-terrorism strategies and of community engagement have been largely absent from official discussion on counter-terrorism. Yet the policing of place cannot occur in a professional silo. For counter-terrorist design solutions to be successful, they must not only be *effective* (incorporating the design and managerial solutions previously mentioned) but also *acceptable to* the owners, inhabitants and users of particular places (Coaffee et al., 2008). Here, acceptability encompasses complex financial, social and aesthetic considerations. Recent scholarship in the humanities, urban studies, planning and architecture has highlighted the risks that counter-terrorism measures pose for the functional integrity of places. These can be their potential to contribute to an atmosphere of fear, a culture of surveillance, consequences for social control and freedom of movement, and a reduction in democratic involvement in urban planning and construction, possibly leading to the increasing militarisation of urban design (Graham, 2010; Coaffee, 2010). There is also a particular risk that counter-terrorism measures may alienate specific members of the community that feel singled out as threats, which in extreme cases could lead to radicalisation. There is clearly a need to address the problem of terrorism while remaining attuned to these social concerns. Such concerns, however, have proved difficult to quantify within the overly technocratic counter-terrorism design solutions based on secured by design principles – notably security bollards – remaining commonplace.

Balancing security and place quality

Whilst ongoing urban revitalisation and cultural renaissance have increasingly emphasised inclusivity, liveability and accessibility, these 'quality of life' values often sit uneasily beside concerns to 'design-out' terrorism, as security becomes an integral part of the design process and an important ingredient in public realm improvement strategies (Coaffee, O'Hare, and Hawkesworth, 2009). Crowded public places have great potential to benefit from design changes offering deterrence or protection from terrorism, but their essential role means that security cannot be allowed to detract from their primary function. In other words, whatever design, engineering or management changes are implemented, predominantly by policing measures, this must be proportional to the ongoing threat of terrorism and seek to minimise disruption to civic life.

In recent years, the frequency, complexity and death toll of terror attacks – particularly those using hostile vehicles to target public spaces – have increased the need for the use of specialised street furniture and security bollards. These design elements, if integrated appropriately, can help reduce the risk of these atrocities in ways that balance the effectiveness of intervention measures with their acceptability by users of those spaces. The predominant view that has emerged is that security features should, where appropriate, be as unobtrusive as possible. In some locations this has led to security features that are increasingly camouflaged and subtly embedded within the cityscape. For example, balustrades or artwork erected as part of public realm improvements or 'hardened' benches, lampposts or other streetscape elements that still provide a 'hostile vehicle mitigation' functionality, with designs capable of stopping a seven-ton truck traveling at 50 miles per hour, have been utilised in selective locations as a stealthy security measure (Coaffee, O'Hare, and Hawkesworth, 2009).

Further, in response to such challenges a number of countries – notably the UK, the US, Europe, Australia and Abu Dhabi – have advanced strategic guidance on how the owners and operators of public spaces, in conjunction with the police, can respond to the latest wave of vehicle-borne terror attacks. This has been done through embedding security and policing practices into design plans in ways that reflect upon and turn threat information into effective, protective security measures, which are considered at the earliest opportunity within a design process and which are proportionate with the level of risk faced. For example, New York's Times Square – one of the densest and most visited public areas in the United States – has been transformed from a congested vehicular space into a largely pedestrianised location in the name of enhancing security. The Times Alliance has also sought to embed security in its redesign of the public realm and to replace the large blocks of concrete that are painted white with the initials 'NYPD' in blue.

Through the use of strategically placed and crash-rated steel bollards and granite benches, the area has, as far as possible, sought to limit the opportunity for vehicle rammings whilst not detracting from the area's vibrancy. As noted by designers Snøhetta, the security functionality of the plaza was to be integrated as far as possible into the overall design:

> Our method has been to protect the plaza areas while also using design elements that don't overwhelm the public experience. We wanted to be sure safety measures did not define the public space while also creating highly effective protective features in the most populated areas. Bollards, in connection with other integrated security features, form the basis of the security design for the plaza. These elements allow for fluid and intuitive circulation between the plazas. This was a fundamental concept of the redesign as a whole, which focused on reducing visual and physical clutter and confusion in the Square, creating a simplified surface that allows people to move comfortably and naturally through the space.
>
> *(Snøhetta Press Release, 2017)*

What kind of public space do we want to live in?

As we have seen, over many decades, complex relationships have existed between human behaviour and the material design of the environment, and the interaction between ideas from planning, criminology and policing practices. The contemporary policing of space therefore emerges as a complicated and negotiated settlement based on perceptions of crime (including terrorism), the criminological potential of particular locations, the changing techniques of policing and fluid criminal practices and motivations. In many ways the historic ideas of defensive design and eyes on the street have evolved, with design concepts and security techniques having become mainstreamed within both planning and security practices and, more broadly, embodied in the praxis of policing place.

Ultimately, how our public places are designed and managed in response to risk and threat tells us a lot about the type of society we are and the type of society we would like to be (Coaffee, 2017). In this sense, providing prescriptive guidelines to the police on protecting against crime and terrorism in public places is a difficult task, especially in societies that value freedom of movement and expression but are fearful of crime or under threat of attack. More broadly, crime and counter-terrorism measures deployed in public places must seek to balance security effectiveness with social and political acceptability. We live in dangerous times, but how we react to the threat of crime and terrorism will impact on our public realm and civic sense for many years. In many ways, the threat to cities also comes from our policy responses to such risks; both have the potential to harm the freedom of movement and expression that define a vibrant place.

If we want a humane and accessible public realm and a genuinely open society we should not let exceptional security measures become the norm as we seek more adaptable and effective ways of coping, in a calm and measured way, with crime and terrorism. This involves more than a police-centric response and requires those that use, secure, design and manage places to work collectively. We need to think innovatively about how we can police places effectively whilst retaining the qualities that make them accessible, friendly, walkable and welcoming, places that are attractive and sustainable as well as safe.

References

Atkinson, R. and Helms, G. (eds.) (2007) *Securing and Urban Renaissance: Crime, community and British Urban Policy*, Bristol: The Policy Press.

Australian National Government (2017) Australia's Strategy for Protecting Crowded Places from Terrorism, at www.nationalsecurity.gov.au/Securityandyourcommunity/Pages/australias-strategy-for-protecting-crowded-places-from-terrorism.aspx

Coaffee, J. (2003) *Terrorism, Risk and the city: the making of a contemporary urban landscape*, Ashgate, Aldershot.

Coaffee, J. (2005) Urban renaissance in the age of terrorism – revanchism, social control or the end of reflection? *International Journal of Urban and Regional Research, 29* (2) 447–454.

Coaffee, J. (2010) Protecting vulnerable cities: the UK resilience response to defending everyday urban infrastructure, *International Affairs*, 86 (4) 939–954.

Coaffee, J. (2017) Urban terrorism isn't going to stop. Can city planners help reduce its lethal impact? *The Washington Post*, 22 June.

Coaffee, J. (2018) "Designing Proportionate Security Responses To 'Soft Target' Terrorism", *Europe's World* at www.friendsofeurope.org/publication/designing-proportionate-security-responses-soft-target-terrorism

Coaffee, J., Moore, C., Fletcher, D. and Bosher, L. (2008) Resilient design for community safety & terror-resistant cities, *Proceedings of the Institute of Civil Engineers: Municipal Engineer*, Vol 161, Issue ME2, 103–110.

Coaffee, J., O'Hare, P., and Hawkesworth, M. (2009) The visibility of (In)security: The aesthetics of planning urban defences against terrorism, *Security Dialogue*, 2009. 40 489–511.

Cozens, P. (2002) Sustainable urban development and crime prevention through environmental design for the British city. Towards an effective Environmentalism for the 21st century, *Cities*, 19 (2) 129–137.

Cozens, P.M., Hillier, D., Prescott, G. (2001) Crime and the design of residential property. Exploring the theoretical background - Part 1, *Property Management*, (paper 1 of 2), 19 (2) 136–164.

Davis, M. (1990) *City of Quartz: Excavating the Future in Los Angeles*, Verso, London.

Davis, M. (1998) *Ecology of Fear: Los Angeles and the Imagination of Disaster*, Metropolitan Books, New York.

Dear, M and Flusty, S. (1998) Postmodern urbanism, *Annals of the Association of American Geographers*, 88 (1) 50–72.

Ellin, N. (1997) (ed.) *Architecture of Fear*, Princeton Architectural Press, New York.

Furedi, F. (2006) *Culture of Fear Revisited*, 4th ed., Continuum, Trowbridge.

Garland, D. (2001) *The Culture of Control: Crime and social order in Contemporary Society*, OUP, Oxford.

Gold, J.R. and Revill G. (1999) Landscapes of defence, *Landscape Research*, 24 (3) 229–239.

Graham S. (2010) *Cities under Siege: The New Military Urbanism*, Verso Books, New York.

Haywood, K.J. (2004) *City Limits: Crime, Consumer Culture and the Urban Experience*, Glasshouse press, London.

Herbert, S. (1997a) *Policing Space: Territoriality and the Los Angeles Police Department*, University of Minnesota Press, Minneapolis.

Herbert, S. (1997b) On prolonging the conversation: some correctives and continuities, *Urban Geography*, 18 (5) 398–402.

Jacobs, J. (1961) *The Death and Life of Great American Cities*, Peregrine, London.

Jeffery, C.R. (1971) *Crime Prevention through Environmental Design*, Sage, Beverly Hills.

Newman, O. (1972) *Defensible space - crime prevention through urban design*, Macmillan, New York.

Office for the Deputy Prime Minister. (2004) *Safer Places: The Planning System and Crime Prevention*, ODPM/Home Office, London.

Raco, Mike. (2007) Securing sustainable communities: citizenship, safety, and sustainability in the new urban planning, *European Urban and Regional Studies*, *14* (4) 305–320.

Snøhetta Press Release, April 19, 2017: https://snohetta.com/news/362-snohettacelebrates-opening-of-times-square-redesign

35

A passion for place and participation

Quintin Bradley

This chapter is about the bond between place and participation, or the idea that devolving decisions to local neighbourhoods makes for a more participatory democracy. It concerns the practices of 'community localism' or the devolution of statutory governance to local neighbourhoods. The key assumption underpinning the state rationality of localism is that the smallest geographical unit of governance – the local neighbourhood or place – provides the greatest opportunities for citizens to participate in decisions (Lowndes & Sullivan, 2008). The local becomes in this rationality a metaphor for empowerment and democratisation. It is implied that devolution to neighbourhoods automatically makes decision-making more participatory. This has been dubbed 'the local trap' by scholars who point out that there is nothing intrinsic to local-scale decision-making to guarantee greater popular participation (Purcell, 2006). This chapter argues against that common-sense view. It maintains that places can be more democratic simply because they are more local.

The argument is constructed over three sections. In the first I explore the role of place in regulating social relations and establishing norms of behaviour, and I discuss the spatial divide between the behaviours associated with the domestic realm of neighbourly care – the place of community – and those associated with the formal economy of public and political life – the place of governance. In the second section, I introduce the state rationality of localism as a breach of spatial boundaries across this division of labour. Localism provides the legislative and institutional permissions that enable people in domestic spaces to practice democratic governance and to 'scale-up' an economy of reciprocity and neighbourly care into political participation (Smith, 1993). In the third section, I discuss the spatial technologies of community localism that enable places and their associated behaviours to be changed and rendered more participative. I advance the concept of community identity frames to explain the practices through which place can be co-produced as neighbourly and neighbourly relations can be transformed into more formal processes of participative democracy (Bradley, 2014; 2017). The chapter concludes that the identity work done by communities is a political project through which participatory governance is established as a social relationship of place and places become both local and more participative.

Place and subjectivity

Places have meaning and purpose. Physical and social settings are inscribed with meanings that govern expected behaviour and social interaction (Goffman, 1969; Jenkins, 2008). It is through place that the roles and categories that define social identity are learned and internalised (Manzo & Perkins, 2006). Dominant place meanings are established through authorised discourses and practices. The role of statutory town and country planning systems, for example, is to allocate particular land uses to specific places. Planning decrees the activities of place in broad terms: it identifies land for residential uses, or retail or for light industrial use. This allocation of use to land attributes specific behaviours to places (Lefebvre, 1991). Shopping malls expect visitors to shop. Often there is an element of coercion in place to ensure that people abide by these spatial codes. Security guards may exclude young people from shopping malls if they are conspicuously failing to consume. Some of these forms of coercion are subtle and are exerted through peer pressure and through everyday social relations. Friends may tell you to 'lighten up' at a nightclub or 'quiet down' in a library. In many ways our subjectivity, what we think as well as what we say and do, and how we present ourselves, is actively performed by the meanings of the places we occupy. In company with the feminist theorist Judith Butler (1997), we can understand this as a form of citational practice. Places are encoded with normative prescriptions about how they should be used and what values, social behaviour and relationships are to be associated with them. In our social relations we cite, or reference, as if it was a body of literature, this spatial code. In the course of everyday life, we 'read' places and reference their invisible rules in our behaviour and comportment (Moisio & Luukkonen, 2015). Place is then an active agent in cementing social distinctions and hierarchies and in maintaining inequality and injustice. The most evident impact on equality is in the distinction between private and public places where spatial codes map onto a gendered division of labour between the domestic and formal economies (Staeheli, 2002). The encoding of place as public or private, residential or economic, assigns different status to the labour of women and men. It marks out notional divisions in society between the 'economy' and the 'community' and depoliticises as it moralises the domestic sphere (Roy, 2001; Spivak, 1988).

In her theory of performativity, Judith Butler argued that socio-spatial positioning (her focus was on the gendered body) is made concrete through the repeated citation of regulatory norms. As we repeatedly reference the invisible rules of place and adapt our behaviour accordingly, our behaviour becomes increasingly constrained and predictable. This is an active process and we may do it slightly differently each time, although Butler is at pains to point out that our behaviour is 'regularised' (Butler, 1993: 95). Place becomes part of our subjectivity in this way. We are the embodiment of regulated space. Butler (1997: 10) maintained that this regulation acted on, and through, our relationship to place; she wrote: 'Individuals come to occupy the site of the subject (the subject simultaneously emerges as a "site")'. The subject of regulation is our relationship with place; we can be put in our place, or we can be out of place. We are always a subject-in-place. It might follow from this that any change to a place may also present opportunities to change the behaviours associated with that place. Just as what is expected of us, and what we are expected and authorised to do and say changes as we move from one place to another, we might argue, along with Butler, that we can subvert the meaning of spaces and occupy them in ways that expand and potentially transform their normative use and restrictions. The accent here is on the active and emergent nature of place (Jupp, 2008: 334). Places and our relationship with them are continuously subject to reproduction and reinterpretation that projects 'the instability and incompleteness of subject-formation' (Butler, 1993: 226).

In *The Production of Space* Henri Lefebvre (1991) famously classified our relationship with place into three dialectically entangled spatial elements, as conceived, perceived and lived. This spatial triad can provide a conceptual model for theorising how the regulatory power of place may be challenged. In conceptualising his triad Lefebvre located the motor for reiterative change in lived space or representational space that 'the imagination seeks to change and appropriate' (Lefebvre, 1991: 39). He recognised that the 'living' of space is coloured by the imagination, and by memories and emotional associations that have the potential to produce variances in the normative processes through which the subject-in-place is reproduced. The meanings that are attached to place through residence, and through familiarity and routine 'living', may be in conflict with the spatial codes of formal practices, such as those of the town and country planning system. Everyday reiterative practices make space familiar and malleable, so that its spatial codes and its meaning can be transformed. Repeated experience, familiar daily routines and established paths transform space so that it 'gets under the skin' and becomes a 'field of care' (Tuan, 1975: 418). We can understand this as a practice of domestication, of making space familiar, so as to enlarge the range of possibilities encoded within it. This is a practice associated with residence, with place attachment and with the concept of community. As Butler stressed it involves repetition and familiarity but, as Lefebvre points out, it is also about the imagination and about a passion for place. We can transform a place in our minds and we can see how it could be changed through our actions. Becoming attached to a place might give us licence to act differently; it might offer us a greater range and freedom of action, and more room to breathe.

We associate place attachment with community and the work of community and neighbourhood groups or associations. While community groups are largely confined to the private and the domestic sphere, they have become increasingly central to political strategies. In the next section I discuss the state rationality of localism and the licence it offers communities to challenge the subjectivity of place.

Localism, place and democracy

A promise to devolve decision-making to local communities has been a constant theme in the political strategies of localism that have been central to the restructuring of state power since the 1970s. State relations of government have been widely transformed into new assemblages of distributed governance that have promised 'a reordering of public space' (Mohan & Stokke, 2000: 250), attributing political content to a particular spatial form in their conflation of the local with better and more democratic governance (Painter et al., 2011). In constituting the local as a metaphor for democracy, community localism foregrounds the pivotal role played by place in cementing social differentiation and in naturalising power relations (Marston, 2000). By conflating community governance with empowerment, localism makes socio-spatial positionalities visible and, at the same time, makes them the object of political attention and therefore vulnerable to change (Leitner, Sheppard & Sziarto, 2008).

Community localism is presented as a transfer of responsibility from the state to the community; it builds on, and seeks to co-opt into governance, a long tradition of grass roots activism and neighbourhood campaigns focussed on the local welfare state (Hall & Massey, 2010; Williams, 1993). The mobilisation of urban social movements around place as community has, if anything, been renewed by the ubiquity of the concept in government discourse. The ability of community groups to move fluidly from campaigning for improvements in local services, to running them, and back again, has been a subject of particular commentary among feminist scholars (Newman, 2012). Community action manifests itself as an ethic of care extended into the public sphere of governance. It mobilises household reproductive labour as a model of co-operation on which to reconstruct the local welfare state (Abel & Nelson, 1990). It borrows from an economy of reciprocity exemplified by the informal provision, most often by women, of material and immaterial help through extended family and neighbourhood networks (Williams & Windebank, 2000). The high levels of trust necessary to support this economy of care were founded on the geographical immobility of women (McCulloch, 1997) and developed in the absence of alternative means of surviving 'as an extended subterranean chain' of services and good deeds (Bulmer, 1986: 112). This is what Raymond Williams called 'the positive practice of neighbourhood' that aims to foster the social relations of community as a model for the collective organisation of society; 'the basic collective idea' that 'the provision of the means of life will, alike in production and distribution, be collective and mutual' (Williams, 1958: 326). It is also, and paradoxically, a relationship that can be commodified in the form of social capital, an adaptable coinage that awards an exchange value to economies based on use value, and that resonates with governmental discourses of responsibility, enterprise and active citizenship (Portes, 1998).

The rationality of localism extends an invitation to community and neighbourhood organisations to take part in the governance of place and to take responsibility for the delivery of statutory services. Localism relocates the domestic norms of a gendered private space to the public sphere, promising that politics can be brought within reach and made subject to the rhythms of daily interaction. This is a breach of spatial boundaries that superimposes the public space of formal governance onto the private sphere of place attachment and neighbourly care. The exercise of formal governance in a residential and domestic setting suggests that the public space of democracy can be enacted as domestic and familiar and that power and decision-making can be made neighbourly and brought within reach. In doing so, it locates political space within familiar patterns of place attachment and neighbourliness and suggests that questions of power and governance can be decided on a domestic scale (Smith, 1993). To change the meaning of space is to create the possibility that people and their social relations can change. The engagement of neighbourhood groups and community organisations in the devolved governance of localism appears as a political struggle over the value and meaning of place and the social relations it prescribes (Mihaylov & Perkins, 2015). It is a conflict over knowledge and power and the role that place plays in the creation, maintenance and transformation of social identity (Haraway, 1991).

The political technology of localism gives license to community action at the same time as it seeks to embrace communities as the embodiment of a responsible and governable public. It provides institutional forms and regulatory permissions through which the boundaries of political space may be moved into new alignments and confers the privileges of statutory authority on a moral economy of reciprocity and neighbourly care. In its rhetorical conflation of place and participation it provides the spatial codes through which resident-led organisations can attempt to construct place as both local and participative.

Constructing place as participation

The adoption of methodologies of participation by resident organisations and neighbourhood groups has been widely perceived as a response to a crisis of legitimacy in representative democracies. Formal processes of delegation and electoral accountability are seen as insufficient for the construction of a democratic society (Della Porta et al., 2017). Participatory theory envisions the maximum participation of citizens in their own governance, especially in sectors of society beyond those that are traditionally understood to be political (Pateman, 1970). By taking part in decisions directly, people are expected to acquire political competencies and experience a heightened sense of political efficacy and empowerment. This is a theory of democratisation; of the extension of democracy into civil society, into the economy and into the neighbourhood (Hilmer, 2010).

The political ubiquity of public participation has transposed the legitimacy concerns of representative democracies onto the new publics empowered by the political rhetoric of localism. Participatory theory is associated with the pursuit of political equality as a 'process where each individual member of a decision-making body has equal power to determine the outcome of decisions' (Pateman, 1970: 71). In practice, the resource inequality intrinsic to the operation of market societies presents an almost insurmountable obstacle to attempts to achieve political justice through the direct popular representation of interests (Dahl, 1998; Freeman, 1970; Mansbridge, 1973). The absence of traditional procedures of authorisation and accountability has provoked criticism that participatory initiatives are open to capture by private interests, and that participants are drawn inequitably from those with existing attachments and the means to influence decision making.

Groups that lack a convincing electoral mandate may construct other forms of democratic legitimacy in their claim to represent a defined constituency or locality. Pitkin (1967) identified 'descriptive representation' as the system of legitimacy most associated with public participation and direct democracy. In descriptive representation the absence of formal methods to enable accountability are mitigated by a correspondence between a specific public, their experiences, interests or demographic make-up, and those who represent them as their political surrogates (Mansbridge, 2003). There is an unavoidable ambiguity in the precise nature of this correspondence. In her influential work 'Can the subaltern speak?' the post-colonial theorist Gayatri Chakravorty Spivak (1988: 276) pointed to the distinction between representation as proxy and representation as an act of signification. A definition of representation as 'acting for' or 'standing for' ignores 'the constitutive dimension of representation' (Saward, 2010: 9). The claim to correspondence with a specific public, a place and a set of interests entails an act of signification in which a portrait of a community of place and interests is created. The idea of a constituency that can be spoken for, and a set of interests that can be represented, is a work of identification that manifests a specific public and that awards it voice, needs and preferences. To understand the act of descriptive representation as signification is to attend to identity work as a democratic practice: the framing of collective identities around place and constituency, and the democratic contentions that arise as identities are challenged, debated and transformed.

A collective identity is framed around shared residence in a defined territory and the proximity of representatives to their constituency, signifying their physical closeness to the designated public (Houtzager & Lavalle, 2010; Piper & von Lieres, 2015). Participation is here understood as a condition of 'nearness' in which routine interaction and face-to-face encounters provide the mechanism for accountability between represented and representative (Kearns & Parkinson, 2001). Peter Somerville (2005: 122) identified the democratic processes associated with living in nearness:

> The high probability of repeated interaction within a community means that members have a strong incentive to act in socially beneficial ways to avoid retaliation next time.

> Frequent interaction lowers the costs and raises the benefits associated with discovering more about the characteristics, recent behaviour and likely future actions of other members.

The routine social interactions that constitute nearness as participative democracy have to be actively constructed through 'neighbouring' work (Bulmer, 1986). Transforming place into nearness means bringing decision making within reach by embedding it in the rhythms of reciprocity and neighbourly care. In addition to the implied accountability of physical proximity, the claim to democratic legitimacy through nearness conveys possession of a shared local knowledge. The representatives claim to know the interests and concerns of their constituents, because they are in regular contact with them and because they share the same familiar space of community, or nearness in the neighbourhood (Bradley, 2014). The claim to local knowledge is a necessary adjunct to the substantive action of representation which entails speaking for, and acting in, the known interests of the public (Pitkin, 1967). The notion that shared turf entails shared knowledge presumes a set of interests that come attached to nearness and that have to be actively constructed as a collective identity through the practice of neighbouring.

The connection between place and the construction of collective identities has been captured in the concept of community identity which describes emotional connections to the locality and to its cultural context (Long & Perkins, 2007; Puddifoot, 1995). Community identity encapsulates the attribution of distinctive meaning to place and the association of place with enhanced behaviours and changed social relations (Puddifoot, 2003). It transcends the individual responses to place attachment and place identity familiar to the literature of environmental and community psychology and signals instead a process of collective identification, in which the passions of place translate into statements of social and political purpose (Dixon & Durrheim, 2000; Kyle, Mowen & Tarrant, 2004). This collective work of identification has been dubbed 'place framing' by Martin (2003) and is typically associated with community groups and neighbourhood organisations who integrate the multiple values that residents ascribe to place into a convincing narrative to mobilise collective action.

Martin drew on the social movement concept of collective action frames (Benford & Snow, 2000) to explain how organisational discourses are assembled by neighbourhood groups to inspire and legitimise place-protective action. In social movement studies, the technique of frame analysis (Snow et al., 1986) has become a key diagnostic tool for interpreting the discursive assemblage of shared identities that is necessary for groups to mobilise collective action and generate plans for change. Martin's thesis was limited to a discussion of the place definition work carried out by community activists to build local organisations. Further development of the concept of place frames is required to understand the impact of identity work on place and its social relations. One approach to this analysis is through the social movement concept of collective identity frames. Collective identity frames (Melucci, 1995) are emotional and often passionate constructs that are negotiated, elaborated and developed in group relationships and acquire their resonance through widely-shared and familiar symbols, interpretations and self-definitions (Polletta & Jasper, 2001). They represent an assemblage of three collective identity processes: the demarcation of group boundaries, the production of a repertoire of shared values, and the promotion of collective efficacy or belief in the ability of the group or organisation to bring about change (Taylor & Whittier, 1992). Applied to the place-based work of community organisations, the theory of collective identity broadens Martin's outline of place framing to connect place to participation and to show how place framing can encode space with the social relations of nearness.

My concept of *community identity frames* enables connections to be drawn between place attachment and the practices of identity work diagnosed by social movement theorists. A community identity frame does not only assert a common identification of place but forges a connection between place characteristics and particular social relations. It seeks to encode place with specific social meanings that legitimise and normalise a defined set of spatial practices. These, in turn, provide the rationale for the claims of community organisations to represent their neighbourhood and for the conditions of nearness that enable the practices of participative democracy. A community group and their representatives must assemble a resonant frame of community identity from the diverse and potentially conflicting place meanings expressed and felt by residents so that it amplifies a shared sense of place and enhances feelings of belonging and capability. Community identity frames are a negotiation rather than a defined agreement, and the test of participatory democracy is the extent to which publics are able to organise and articulate their competing interests in nearness, and either reach a negotiated settlement or continue to contest the issue (Marres, 2005). In highly diverse neighbourhoods, it can be argued that community identity frames are a project of manufactured unity in which a coherent vision of place is distilled from discordant views, potentially to mobilise a population despite tensions and divisions. In small rural communities, the community identity frame might serve to amplify a shared sense of place and enhance feelings of belonging and capability (Bradley, 2017).

A community identity frame provides a sense of place and the social relations associated with it. It invokes shared interests and local knowledge and promotes a sense of living in 'nearness'. This identity work roots participatory governance in the everyday labour of reciprocity and care. It lays the foundations for the descriptive representation essential to participatory democracy and produces place as both neighbourly and democratic, as the final section explains.

Participation and place: a conclusion

The transposition from neighbourhood as place to neighbouring as an activity and on to nearness as a condition of being is the root of the concept of community and its bond with participatory democracy. Neighbourhood organisations aim to construct a resonant frame of community identity that assembles a public around a characterisation of place as neighbourly and participative. This collective identity frame is distilled into a spatial code to invoke the subjectivities of distinctive environments and establish norms of social relations. The spatial norms assembled in a community identity invoke a particular public and implicitly align place with an emplaced culture or, to borrow a phrase from organisational theory, 'a way of doing things around here' (Bower, 1966: 4). Places are rendered as capable of managing their own affairs, of holding duties of care and stewardship, and are represented as civic polities where collective decisions can be taken. Although this collective identity appears as a discursive construction it is inscribed on place through the repeated citation of regulatory norms. Community representatives are rendered representative through active proximity and everyday interaction. The routine labour of neighbourly care provides the communicative networks that sustain participation in community decisions. To this community identity of participative nearness, the state rationality of localism adds the permissions and opportunities of legislative framework and institutional form. Community groups may assume devolved responsibility for devising a statutory plan for neighbourhood development, build and manage community housing or take over the delivery of local services. Localism gives licence to the democracy of nearness and the local becomes a place of empowerment and democratisation through the practice of participatory governance as neighbourly care. In this way, despite all evidence to the contrary, the smallest geographical unit of governance provides the greatest opportunities for citizens to participate in decisions.

References

Abel, E.K. & M. Nelson (eds.) (1990) *Circles of care: work and identity in women's lives*. Albany, NY. SUNY Press.

Benford, R. & D. Snow. (2000) Framing processes & social movements: an overview and assessment. *Annual Review of Sociology*. Vol. 26: 611–631.

Bower, M. (1966) The way we do things around here. *A New Look at the Company Philosophy. Management Review*. Vol. 55, No. 5: 4–11.

Bradley, Q. (2014) Bringing democracy back home: community localism and the domestication of political space. *Environment & Planning D: Society & Space*. Vol. 32, No. 4: 642–657.

Bradley, Q. (2017) Neighbourhood planning and the impact of place identity on housing development in England. *Planning Theory & Practice*. Vol. 18, No. 2: 233–248.

Bulmer, M. (1986) *Neighbours: the work of Philip Abrams*. Cambridge. Cambridge University Press.

Butler, J. (1993) *Bodies that matter: on the discursive limits of sex*. London. Routledge.

Butler, J. (1997) *Excitable Speech: a politics of the performative*. London. Routledge.

Dahl, R. (1998) *On democracy*. London. Yale University Press.

Della Porta, D., F. O'Connor, M. Portos & A.S. Ribas. (2017) *Social movements and referendums from below: direct democracy in the neoliberal crisis*. Bristol. Policy Press.

Dixon, J. & K. Durrheim. (2000) Displacing place-identity: a discursive approach to locating self and other. *British Journal of Social Psychology*. Vol. 39: 27–44.

Freeman, J. (1970/2013) The tyranny of structurelessness. *WSQ: Women's Studies Quarterly*. Vol. 41, No. 3 & 4: 231–246.

Goffman, E. (1969) *The Presentation of the Self in Everyday Life*. Harmondsworth. Penguin Books.

Hall, S. & D. Massey. (2010) Interpreting the crisis. *Soundings*. Vol. 44, Spring: 57–71.

Haraway, D. J. (1991) *Simians, cyborgs and women: the reinvention of nature*. London. Free Association Books.

Hilmer, J. (2010) The state of participatory democratic theory. *New Political Science*. Vol. 32, No. 1: 43–63.

Houtzager, P. & A. G. Lavalle. (2010) Civil society's claims to political representation in Brazil. *Studies in Comparative International Development*. Vol. 45: 1–29.

Jenkins, R. (2008) *Social Identity*. London, New York. Routledge.

Jupp, E. (2008) The feeling of participation: everyday spaces and urban change. *Geoforum*. Vol. 39: 331–343.

Kearns, A. & M. Parkinson. (2001) The significance of neighbourhood. *Urban Studies*. Vol. 38, No. 12: 2103–2110.

Kyle, G., A. Mowen & M. Tarrant. (2004) Linking place preferences with place meaning: an examination of the relationship between place motivation and place attachment. *Journal of Environmental Psychology*. Vol. 24: 439–454.

Lefebvre, H. (1991 [1974]) *The production of space*. Trans. D. Nicholson-Smith. Oxford. Blackwell.

Leitner, H., E. Sheppard & K. Sziarto. (2008) The spatialities of contentious politics. *Transactions of the Institute of British Geographers*. Vol. 33, No. 2: 157–172.

Long, D.A. & D. Perkins. (2007) Community social and place predictors of sense of community: a longitudinal analysis. *Journal of Community Psychology*. Vol. 35, No. 5: 563–581.

Lowndes, V. & H. Sullivan. (2008) How low can you go? Rationales and challenges for neighbourhood governance. *Public Administration*. Vol. 86, No. 1: 53–74.

Mansbridge, J. (1973) Time, emotion and inequality: three problems of participatory groups. *The Journal of Applied Behavioural Science*. Vol. 9, No. 2/3: 351–368.

Mansbridge, J. (2003) Rethinking representation. *The American Political Science Review*. Vol. 97, No. 4: 515–528.

Manzo, L. & D. Perkins. (2006) Finding common ground: the importance of place attachment to community participation and planning. *Journal of Planning Literature*. Vol. 20, No. 4: 335–350.

Marston, S. (2000) The social construction of scale. *Progress in Human Geography*. Vol. 24, No.2: 219–242.

Marres, N. S. (2005) *No issue, no public: democratic deficits after the displacement of politics*. Amsterdam. Ipskamp Printpartners.

Martin, D. (2003) Place-framing as place making: constituting a neighbourhood for organising and activism. *Annals of the Association of American Geographers*. Vol. 93, No. 3: 730–750.

McCulloch, A. (1997) You've Fucked Up the Estate & Now You're Carrying a Briefcase! In Hoggett, Paul (ed.) *Contested Communities: experiences, struggles, policies*. Bristol. Policy Press.

Melucci, A. (1995) The Process of Collective Identity. In: Johnston, H. & B. Klandermans (eds.) *Social movements & culture*, pp. 41–63. London. UCL Press.

Mihaylov, N. & D. Perkins. (2015) Local environmental grassroots activism: contributions from environmental psychology, sociology and politics. *Behavioural Sciences*. Vol. 5: 121–153.

Mohan, G. & K. Stokke. (2000) Participatory development & empowerment: the dangers of localism. *Third World Quarterly*. Vol. 21, No. 2: 247–268.

Moisio, S. & J. Luukkonen. (2015) European spatial planning as governmentality: an inquiry into rationalities, techniques and manifestations. *Environment & Planning C: Government and Policy*. Vol. 22: 828–845.

Newman, J. (2012) Making, contesting & governing the local: women's labour and the local state. *Local Economy*. Vol. 27, No. 8: 846–858.

Painter, J., A. Orton, G. Macleod, L. Dominelli & R. Pande. (2011) *Connecting localism & community empowerment: research review and critical synthesis for the ARHC connected community programme*. Project Report. Durham. Durham University, Dept. of Geography & School of Applied Social Sciences.

Pateman, C. (1970) *Participation and democratic theory*. London. Cambridge University Press.

Piper, L. & B. von Lieres. (2015) Mediating between state and citizens: the significance of the informal politics of third-party representation in the global South. *Citizenship Studies*. Vol. 19, No. 6-7: 696–713.

Pitkin, H. (1967) *The concept of representation*. London. University of California Press.

Polletta, F. & J. Jasper. (2001) Collective identity and social movements. *Annual Review of Sociology*. Vol. 27: 283–305.

Portes, A. (1998) Social Capital: its origins and applications in modern sociology. *Annual Review of Sociology*. Vol. 24: 1–24.

Puddifoot, J. (1995) Dimensions of community identity. *Journal of Community and Applied Social Psychology*. Vol. 5: 357–370.

Puddifoot, J. (2003) Exploring 'personal' and 'shared sense of community identity in Durham city, England. *Journal of Community Psychology*. Vol. 31, No. 1: 87–106.

Purcell, M. (2006) Urban democracy and the local trap. *Urban Studies*. Vol. 43, No. 11: 1921–1941.

Roy, A. (2001) A 'public' muse. *Journal of Planning Education and Research*. Vol. 21: 109–126.

Saward, M. (2010) *The representative claim*. Oxford. Oxford University Press.

Smith, N. (1993) Homeless/Global: scaling places. In Bird, J., Curtis, B., Putnam, T., Robertson, G. & L. Tucker (eds.) *Mapping the futures*. London. Routlege. 87–119.

Snow, D., E. B. Rochford Jnr., S. Worden, & R. Renford. (1986) Frame alignment processes, micromobilisation and movement participation. *American Sociological Review*. Vol. 51, No. 4: 464–481.

Somerville, P. (2005) Community governance and democracy. *Policy & Politics*. Vol. 33, No. 1: 117–144.

Spivak, G. C. (1988) Can the Subaltern Speak? In Nelson, Cary & Lawrence Grossberg (eds) *Marxism & the interpretation of culture*, pp. 271–316. Chicago. University of Illinois Press.

Staeheli, L. (2002) Women and the *work of community*. *Environment & Planning A*. Vol. 35, Part 5: 815–831.

Taylor, V. & N. Whittier. (1992) 'Collective identity in social movement communities: lesbian feminist mobilization', pp. 104–130 in A Morris and C. Twigger-Ross, C. & D. Uzzell (1996) place and identity processes. *Journal of Environmental Psychology*. Vol. 16: 205–220.

Tuan, Y. (1975) Place: an experiential perspective. *The Geographical Review*. Vol. 65, No. 2: 151–165.

Williams, C. & J. Windebank. (2000) Helping each other out? Community exchange in deprived neighbourhoods. *Community Development Journal*. Vol. 35, No. 2: 146–156.

Williams, F. (1993) Women & Community. In Bornat, J., C. Pereira, D. Pilgrim & F. Williams (eds) *Community Care: a reader*. Basingstoke. Macmillan.

Williams, Raymond. (1958) *Culture and society*. London. Chatto & Windus.

36

Experimental places and spatial politics

Alexander Vasudevan

Introduction

At the heart of this chapter is a longstanding preoccupation with the experimental geographies produced by activist communities living in cities across the global North. Over the past decade I have been conducting detailed research on the history of urban squatting in Europe and North America (see Vasudevan, 2015, 2017). This is a project that has highlighted the everyday practices and spatial imaginaries of squatters from the late 1960s to the present, exploring what squatters actually did, the terms and tactics they deployed, the ideas and spaces they created. At stake here, for me, has been a commitment to developing an understanding of squatting that focusses on its 'world-making potentialities' (Muñoz, 2009: 56). As I have tried to argue, the very site of occupation – from squatted house to autonomous social centre – should be seen as a place of collective world-making – a place to quite literally build an alternative habitus where the act of occupation became the basis for creating new and often experimental forms of shared city living whether it be architectural, social and/or political.

In this chapter I examine the wider significance of these practices for how we – as geographers – come to conceptualise and research the relationship between experimentation and place-making. We often describe experimentation as a process that involves the 'testing of a scientific hypothesis in a secluded space distinguished by an artificial set-up, the inducement of change, and the measurement of effects' (Evans, 2016: 430). While experimentation is therefore associated with a 'community of specialists' working in the carefully calibrated boundaries of a laboratory, a number of scholars have demonstrated that the 'place' of experimentation cannot be reduced to controlled 'scientific' settings. Geographers have, in this context, drawn particular attention to the spatiality of scientific experimentation and the diverse set of practices and sites that it encompasses (Davies, 2010; see Jellis, 2015; Kullman, 2013; Last, 2012; Livingstone, 2003; Powell and Vasudevan, 2007). Laboratory environments, they argue, should not be seen as isolated 'placeless' sources of knowledge formation. Rather, as David Livingstone reminds us, they bear the imprint of their location. 'Place matters', he writes, 'in the way scientific claims come be regarded as true, in how theories are established and justified, in the means by which science exercises the power that it does in the world' (Livingstone, 2003: 14; see also Gieryn, 2006; Kohler, 2002).

Set against this backdrop, it is perhaps unsurprising that the relationship between science and the city, experimentation and urbanism, has become an increasingly important source of scholarship in its own right. Research has zoomed in on the co-production of science and the city and the role that the urban has played as both a setting for experimental venues and as an object of experimentation (Dierig, Lachmund and Mendelsohn, 2003; Lachmund, 2013; Reid-Henry, 2010; Vasudevan, 2006). At the same time, cities have also become a test-bed for new modes of administration, planning and policy-making. According to this view, cities are now key laboratories for the testing of new and often risky innovations in policy and governance (Bulkeley and Castan-Broto, 2013; Edwards and Bulkeley, 2018; Evans and Karvonen, 2014; Halpern et al., 2013; Karvonen and Van Heur, 2014; Peck and Theodore, 2015). Urban 'experiments' – whether it be a living laboratory, a science park or a local community initiative – are widely seen as an adaptive politico-technical response to the changing nature of contemporary urbanisation (Evans, 2016: 437).

This chapter offers, however, a rather different reading of urban experimentation and place. While recent scholarly attention has focussed on the ways in which cities are governed and managed through the practice of experimentation, my own approach draws attention to the *micropolitical* experiments *with* place adopted by housing activists – and squatters in particular – in cities across Europe and North America over the past 50 years. The widespread occupation of empty flats and other vacant buildings that coincided with the rise of new social movements in the late 1960s was, in no small part, a product of a housing crisis that condemned significant numbers of people to misery and prompted many to seek alternative and often informal forms of shelter.

Squatters thus acted out of an urgent desire to meet basic housing needs but they also experimented with new forms of shared living. More often than not, squatters were confronted with abandoned or run-down spaces that required significant renovation and, for many, the very act of occupation was understood as a constructive mode of repair and transformation. It is the modest 'experimental politics' (Lazzarato, 2017[2009]) cultivated by squatters that I seek to re-trace across three interconnecting registers: 1) as a form of *worlding* through which an alternative urban infrastructure was continuously made and re-made; 2) as a form of *commoning* that connected people and places through new geographies of care and solidarity; 3) and, in concluding, as a form of *dwelling* and endurance that pointed to the possibilities of re-defining what it means to live in a city. Taken together, these 'experiments' depended on an expansive understanding of the places they produced. Place, in the eyes of squatters, was less a strictly delineated location than a radically open and contested source of potentiality and promise.

Building experiments

In September 1971 a group of activists broke into the recently abandoned Bådsmandsstræde Barracks, a former military base that was located alongside parts of Copenhagen's seventeenth-century defensive ramparts. The activists explored the sprawling network of barracks, workshops and halls and documented their 'adventures' in one of Denmark's leading alternative newspapers, *Hovedbladet*. 'Here was', in their own words, 'the framework for an alternative city that could be produced through a range of shared experiments' (*Hovedbladet*, October 2–3, 1971). The squatters appealed to other activists to join them and the base was quickly overrun by hundreds of young people (Karpantschof, 2011: 39–40).

As I have argued elsewhere (Vasudevan, 2017), local authorities as well as the Danish Ministry for Defence were unprepared for the actions of the squatters. The occupation quickly grew and, within a few months, there were over 500 squatters living in Christiania. A police eviction was ruled out by the authorities who chose to negotiate with the base's new inhabitants. On 31 May 1972 an preliminary agreement between the Danish state and the squatters in Christiania was signed. Within a year, the government had designated the site as a temporary 'social experiment' until a decision could be made on the future use of the barracks. For the residents of Christiania, this became an experiment in autonomous forms of decision-making and self-organisation that were, in turn, supported by a thriving underground economy that included the widespread sale of drugs (Christiania, 1971; Karpantschof, 2011: 41; Zinovich, 2015: 87).

Squatted spaces across Europe and indeed North America were often seen as sites of experimentation by both their occupants and local municipal authorities. While many people became squatters to take control of their own lives and respond to basic housing needs, they also found, in their actions, new possibilities for re-habilitating and re-imaging the spaces they occupied. Squatting was, in other words, far more than a simple act of survival. It was indelibly shaped by the new social movements that flourished in the 1960s and 1970s and the experimental geographies they produced in response to the violent predations of late capitalism. As Michel Foucault (2009) later recalled, 'the revolution in modern Europe was not only a political project but also a form of life' (169).

This was, moreover, a 'revolution' that promoted a vision of the built form as a key laboratory for new forms of living, organising and working. The repertoire of contention adopted by the protest cultures that first emerged in the late 1960s depended on a 'fundamental revaluation of forms of political action and interaction' (Slobodian, 2012: 171). From teach-ins and happenings to experiments in communal living and working, the protest techniques adopted by a new generation of activists were anticipatory. They actively prefigured the shape and structure of the alternative that they imagined and, in many cases, constructed.

The relationship between housing activism and architecture played a central if understudied role in the emergence of squatter movements in Europe and North America. In the 1960s, architecture departments were early sites of radical agitation. In Milan, faculty at the Polytechnic (Politechnico di Milano) were involved in the formation of study groups that highlighted the alternative political possibilities of architecture, the failures of urban planning across Italy and the need to build high quality working-class housing. In 1967, in opposition to government attempts to 'modernise' the study of architecture, students occupied the faculty for fifty-five days. The occupation led to the introduction of a new experimental curriculum which blurred the boundaries between student and professor, campus and city, but also anticipated the major role that the department would play in 1971 in supporting and collaborating with a large group of squatters based on the Via Tebaldi (see Wright, 2002: 89).

In the case of West Berlin, it was the work of Aktion 507, a group of young architects and planners based in the Faculty of Architecture at the Technische Universität (TU-Berlin) that challenged the planning policies that had been adopted by the city. Aktion 507 were responsible for *Diagnose zum Bauen in West-Berlin*, a ground-breaking 1968 exhibition which drew attention to the experiences of inner-city residents who had been re-located to new satellite communities on the outskirts of the city. The same architects also turned to more engaged forms of participatory research. They began work in the Märkisches Viertel, a large satellite estate in the district of Reinickendorf whose ongoing construction was part of West Berlin's First Urban Renewal Programme initiated in 1963. They were joined by over 100 students and 8 lecturers from the Pädagogisches Hochschule in Berlin who had received funding from the Volkswagen Foundation to set up a five-year research scheme on grassroots organising and community self-empowerment (see Gribat, Misselwitz and Görlich, 2018).

The Märkisches Viertel quickly became a laboratory for alternative forms of community organisation which, in turn, treated the city, the neighbourhood and the built form as both the setting and stage for the articulation of new political practices. Most of the students that became involved in the Märkisches Viertel had already been active in the extra-parliamentary opposition of the late 1960s where they had been exposed to an action repertoire whose disruptive character belied a commitment to building a 'new way of living' (Dutschke-Klotz, 1996: 97). While local residents were often suspicious of the students' intentions, they worked closely with and used a range of tactics developed by the New Left in West Germany to highlight the plight of the neighbourhood. It is in this context that in May 1970 the very first squat in West Berlin was set up in an abandoned factory hall in the Märkisches Viertel (Vasudevan, 2015).

The factory occupation in the Märkische Viertel was short-lived and remains a largely neglected chapter in the history of housing activism in Berlin. The major wave of squatting that would later follow at the end of 1970s provided, however, a much larger platform for squatters who often saw the houses they occupied in light of the creative possibilities that they offered. In West Berlin, squatters famously proclaimed that 'it is better to squat and mend than to own and destroy (*lieber instand(be)setzen als kaputt besitzen*)' (Laurisch, 1981: 34). The same squatters adopted the term *Instands(be)setzung* as a slogan to describe their movement; the term is a combination of the German for renovation *(Instandsetzung)* and squatting (*Besetzung*). Occupying empty apartments offered, in the words of one squatter, 'a new commitment to repair and maintenance'. 'The constructiveness of our approach', they added, 'lies in *what we do* with flats' (Klein and Porn, 1981: 112, emphasis added).

Practically, whether it was in West Berlin in the 1970s and 1980s or in countless other cities in Europe and North America during the same period, squatting's so-called 'constructiveness' was marked by a slow and incremental process of renovation. From Hamburg to New York, London to Amsterdam, squatted houses were makeshift sites that reflected a modest ontology of mending and repair. Occupation went hand-in-hand with renovation as joists, pipes and floorboards were replaced, electrical wiring fixed, plumbing restored, windows installed and roofs repaired (see Vasudevan, 2015).

Squatters were, in this way, able to cultivate a radical form of *place-making* that transformed the ways in which people, materials and ideas came together. While squatters often relied on the most rudimentary of tools, they were quick to share the skills and techniques that they learned. Throughout the 1970s and 1980s in particular, squatters across Europe and North America produced their own handbooks and DIY manuals. In Amsterdam, for example, a *Guide For Squatters* (*Handleiding voor krakers*) was published in 1969. The rough and ready guide included instructions on how to repair a house with information on planning and housing policy in Amsterdam. For New Yorkers, it was *Survival Without Rent* that served as a step-by-step guide for would-be squatters in the Lower East Side in the mid-1980s. Housing activists in Berlin relied, in turn, on a series of detailed articles that were published in one of their own widely circulated magazines, the *Instandbesetzerpost*.

As squatters slowly and painstakingly brought vacant buildings back to life, they also experimented with the form and substance of the spaces they occupied. Just as the skills they learned spoke to pressing housing needs, they also reflected a desire to create new environments that prefigured how one might come to know and live the city differently. Many squatters were therefore drawn to the *malleability* of a building as a physical expression of their political commitments. As the history of feminist architectural practices in London, for example, shows, squatting offered a point of intersection between theory and practice.

Groups such as the Feminist Design Collective and Matrix were closely connected to the actions of squatters in the 1970s and their desire to rupture traditional social values and gender relations. As one historian recently concluded, 'urban squatting in London of the 1970s enabled a generation of feminist women to engage directly with the built environment: to shape it and adapt it at the level of the household and the community' (Wall, 2017: 139).

For urban squatters, experimentation was thus characterised by a 'physical interaction with the materiality of housing' (Wall, 2017: 139). This was a 'hands-on' process that led to the creation of new social configurations that often challenged traditional understandings of domesticity and kinship (Cook, 2013; Di Feliciantonio, 2017; Stanley, 2018). Walls were demolished, communal spaces (kitchens, meeting rooms, performance spaces) were added and alternative building materials were used to meet the changing needs and wishes of squatters. The squat represented, in this context, a place of collective world-making – a place to imagine alternatives, to express anger and solidarity, to explore new identities and different intimacies, and to experience and share new feelings (Gould, 2009: 178).

It is, of course, the case that many, if not most, squatted houses were carved out of settings of extreme iniquity that were themselves shaped by wider cycles of capitalist accumulation and the creative destruction of urban landscapes. The activists that I worked with, and the archives they meticulously assembled, pointed if anything to an acute awareness of the 'housing question' and the use value of squatted houses as places of care, refuge and solidarity. This was, moreover, a critical awareness that extended far beyond the walls of squats. It also was responsible for the creation of a radical urban infrastructure and the dense web of social spaces, practices and interactions that it supported. Whether it be London or New York, Amsterdam or Berlin, this was an infrastructure that encompassed alternative bookshops and cafés, cinemas, community gardens, concert venues, cycle repair shops, day-care centres, galleries, neighbourhood social centres and workshops. At stake here, as I hope to show in the section that follows, was the making of an urban commons and the challenges of building cultures of resistance in otherwise hostile urban worlds.

Placing the urban commons

Much of my thinking on the history of urban squatting has focussed on an understanding of squatted houses as key sites through which a radical urban commons was precariously developed and experimented with (see SqEK, 2014). In recent years the 'commons' has come to offer 'radical collectivity as a mode of living against the bounded present'. According to this view, the commons is best understood, in the words of one commentator, 'as a place, a structure of feeling, and an idea [that] provides refuge in the ruins of capital's totality' (Stanley, 2018: 489). As an experiment in commoning, squatting, therefore, draws particular attention to the 'reciprocities of infrastructure and sociality' and the relationship between local articulations of autonomy and self-determination and a wider ecology of protest and resistance (Amin, 2014).

In their efforts to create liveable communities, squatters are, to paraphrase the geographer Amanda Huron, 'practicing the urban commons' (2018). Huron is one of many scholars who has, in recent years, attempted to theorise what we mean by the urban commons. For Huron, the commons can be conceived of in three interlocking ways: as a resource, a community of people using that resource and, finally, as a set of institutions (and infrastructures) that have been developed to manage that resource.

In the specific case of squatters, it is housing – and perhaps the built form generally – that serves as the main source of a political imaginary that seeks to conjoin the built form with a more expansive transformation of urban politics. Urban commoning, in this context, treated the various vacant sites occupied by squatters as a medium through which 'institutions of commoning take place' (Stavrides in Huron, 2018: 1065). I have developed this point in greater detail elsewhere and showed how the emergence of squatter movements across Europe and North America in the 1970s was connected to the development of new trans-local networks that brought activists together and played a crucial role in the assembling and circulation of an urban commons (see Vasudevan, 2017).

At a local scale, many squatted houses – most notably in Italy, Spain and the UK – doubled as 'social centres' that also focussed on other forms of political organising (see Chatterton and Hodkinson, 2006; Mudu, 2004). These centres, as Lõpez (2013) has shown, occupied a prominent place within local neighbourhoods. In some cases, they offered meeting and exhibition space for activists and artists respectively. In other cases, they hosted initiatives associated with other political campaigns (anti-fascist and anti-racist organising, migrant solidarity, precarious workers' rights, community gardening schemes, etc.). In other cases still, they became key centres for the development of feminist, queer and trans collectives.

The association of squatting with 'commoning' has, admittedly, a much longer history as the anarchist and historian Colin Ward has shown (2002). The belief that squatters are simply exercising a common customary right remains, in the eyes of many, a seductive view though it also represents an increasingly romanticised framing of anti-capitalist praxis. While squatters often saw their actions as a form of expropriation and separation that removed the places they occupied from the circuits of capital in which they were nothing more than exchangeable commodities, their status as urban commons was not without its own contradictions and inconsistencies. As some commentators have shown, urban commons – including many housing projects set up by squatters – were exclusionary in their own right and often sites of racialised and sexualised violence (Amantine, 2011).

The recent history of radical housing struggles in North American and other settler colonial societies was also marked by a rather different understanding of 'occupation', where squatters were not only responsible for the occupation of indigenous lands but also provided a pretext for the obliteration of longstanding native geographies and the complex life-worlds they supported. Seen in this way, occupation was as much a 'methodology of the settler state' as it was an experiment in urban commoning. Or, to put it somewhat differently, the urban commons should not be seen as an innocent place (or process) that is somehow 'immune to the force of capital's colonial violence' (Stanley, 2018: 490).

In Vancouver, for example, the 2001 occupation of the Woodwards Building (what became known as 'Woodsquat') was a product of an intensifying housing crisis in the city's Downtown Eastside. It was also characterised by the tensions surrounding the demands raised by native residents of the Downtown Eastside, who made up over 40% of the neighbourhood's inhabitants and reminded the other occupiers that the squat was located on unceded territory (Krebs, 2003/4). A few years later and during the Occupy movement in the United States, efforts were made to recognise the settler epistemologies that were unwittingly reproduced by the encampments that appeared in numerous US cities. This culminated in the unsuccessful attempt by Occupy Oakland to rename itself as 'Decolonise Oakland' (see Stanley, 2018).

To conceive of the urban commons as an object of critique therefore highlights some of the limitations that have accompanied recent political injunctions to 'form communes' and create new experimental sites of political action (Invisible Committee, 2009). To do so is to also recognise that the experimental nature of the commons cannot be reduced to a strict place or location but is better understood as an ongoing activity and process. While the systematic attack on alternative forms of living and working in cities across Europe and North America over the past few decades has often prompted housing activists and squatters to retreat into and defend the places they created, it has also encouraged them to recognise the enduring possibilities of experimenting *with* place and the wider alliances and solidarity this necessitates. Here, Marx's own framing of the commune in the *Grundrisse*, as Eric Stanley reminds us (2018), is especially instructive. 'The commune', Marx writes, 'appears as a coming-together [*Vereinigung*], not as a being-together [*Verein*]' (Marx, 1993: 483). The relationship between experimentation and place remains, if nothing else, a process, a coming-together, an opening to the possibilities of an another world.

From dwelling to survival

Much of my own research has focussed on re-tracing the recent history of urban squatting in settings squarely located in the global North. The alignment of experimentation and place that this chapter has examined is predicated, in many respects, on a particular reading of spatial politics that has until recently tended to overlook the innovative and often improvised ways in which residents in the global South – many of whom are themselves squatters – have developed to survive and extend new forms of life in settings that are widely viewed as 'inhabitable' (Simone, 2019). These are settings, according to AbdouMaliq Simone and Edgar Pieterse (2017), where there is 'no obvious programme or stable ideological edifice that can be invoked to establish an orderly modern politics of city-making or rights' (151).

In response, Simone and Pieterse have called for a rather different commitment to 'experimentation' and to recognise its significance for re-thinking how we conceptualise and inhabit cities (2017: 151). For Simone, in particular, this involves a different kind of attentiveness to the 'intense volatility and uncertainty' that has come to define urban districts across the global South (2019: 16). Simone is at pains, in this context, to caution against a reading of the intense forms of precarity and toxicity at work here that relies, in turn, on a crude celebration of resilience and resourcefulness. Rather, he draws attention to the rhythms of endurance and improvisation that create sites of operation that 'cannot be stabilised as points of anchorage, as settlements to inhabit' (2019: 20). These are experiments in urban living that may not go anywhere. They do not remain or endure in one place but find a way of inhabiting and connecting places however short-lived they may be. And yet, Simone is at pains to remind us that this often depends on its own kind of 'deep relationality' which may point to other ways of occupying space beyond existing impositions (2019: 59).

I would, nevertheless, like to highlight the significance of this approach to the relationship between experimentation and place and the need to re-calibrate *how* we approach the history of housing activism in the global North. I am not advocating a retrospective decentring of the kind of concept work undertaken by squatters living in cities across Europe and North America, but rather a further recognition of the unruly nature of this work and its enduring significance for how we might know the city differently. To this end, I have begun to carve out a problem-space that recognises the cardinal significance of migrant struggles – both empirically and conceptually to the history and geography of squatting I have been working on (see final chapter in Vasudevan, 2017).

From Amsterdam to Frankfurt, Barcelona to New York, migrant communities have played an important role in the organisation and transformation of housing struggles (see Mudu and Chattopadhyay, 2016). These are stories that remain largely untold but were ultimately decisive in the emergence of squatting as an *experimental politics*. This is, moreover, an urgent undertaking given the recent revival of squatting in Europe as a necessary alternative to dominant anti-immigrant policies that seek to deny asylum seekers and refugees the agency to inhabit and shape the city on their own fragile terms. Across Europe, a number of squats have been set up as temporary spaces of refuge and hospitality for forced migrants who often find themselves in political limbo or unable to access local services. These are spaces that reflect the efforts of squatters to secure their own right to housing and the basic fundamentals of survival. These are, moreover, spaces that may acquire a certain stability though they are, more often than not, temporary staging-posts that bring bodies, materials and things together in shifting configurations of dwelling and endurance. At stake here, in the end, is a rather different experimentation with place but also, more importantly perhaps, a set of experiences that represent a renewed experimentation with *the place of urban theory* not to mention a more expansive articulation of what it means to inhabit the city.

Conclusion

I would like to conclude by briefly reflecting on the wider significance of this chapter to our understanding of the relationship between experimentation and place-making. The nature of experimental practice, as Steven Shapin and Simon Schaffer remind us, is ultimately the product of a specific arrangement of bodies, materials and knowledges in which matters of fact were 'made into the foundations of what counted as proper scientific knowledge' (1985: 3). According to Shapin and Schaffer, this was a process that increasingly depended on a physical, social and epistemic site – the laboratory – in which experiments were 'performed and witnessed' (1985: 334).

While the modern laboratory is often understood as a strictly controlled environment, geographers of science, in particular, have demonstrated that experimentation is not confined to the boundaries of a given place but is, in fact, a *generative* process through which particular places were continuously assembled in the search for scientific truth. It is perhaps unsurprising, in this context, that the 'city itself has long been an important locus for the advancement of science' (Evans, 2016: 432). Urban experimentation has not, however, been confined to specific venues such as clinics, hospitals, laboratories and universities. It has also played a decisive role in the development of new forms of urban planning and governance.

At the heart of this chapter is a rather different and modest reading of experimentation that seeks to translate and transpose these concerns to the actions of urban squatters for whom the city represented a site of dwelling and necessity as much as it acted as a source of political action and claim-making. For the same squatters, to 'see like a city' (Amin and Thrift, 2017) provided an opportunity to experiment *with* and *through* place and recognise the possibilities that emerging forms of collective life produced. If nothing else, these are practices that further enlarge our understanding of experimentation as a mode of place-making.

Acknowledgement

Research for this chapter was supported by the British Academy Mid-Career Fellowship [RA15AZ]. The chapter builds on arguments developed in my recent book on the history of squatting (*The Autonomous City: A History of Urban Squatting*).

References

Amantine. (2011) *Gender und Häuserkampf*, Münster: Unrast Verlag.

Amin, A. (2014) 'Lively infrastructure', *Theory, Culture & Society*, 31: 137–161.

Amin, A. and Thrift, N. (2017) *Seeing Like a City*, Cambridge: Polity.

Bulkeley, H. and Castan-Broto, V.C. (2013) 'Government by experiment? Global cities and the governing of climate change', *Transactions of the Institute of British Geographers*, 38: 361–375.

Chatterton, P. and Hodkinson, S. (2006) 'Autonomy in the city? Reflections on the social centres movement in the UK', *City*, 10: 305–315.

Cook, M. (2013) '"Gay times": Identity, locality, memory, and the Brixton squats in 1970s London', *Twentieth Century British History*, 24: 84–109.

Davies, G. (2010) 'Where do experiments end?', *Geoforum*, 42: 667–670.

Dierig, S., Lachmund, J. and Mendelsohn, A. (2003) *Science and the City*, Chicago: University of Chicago Press.

Di Feliciantonio, C. (2017) 'Spaces of the expelled as spaces of the urban commons: Re-emergence of squatting initiatives in Rome', *International Journal of Urban and Regional Research*, 41(5): 708–725.

Dutschke-Klotz, G. (1996) *Rudi Dutschke – Wir hatten ein barabrisches, schönes Leben*, Cologne: Kiepenheuer & Witsch.

Edwards, G.A.S. and Bulkeley, H.A. (2018) 'Heterotopia and the urban politics of climate change experimentation', *Environment and Planning D: Society & Space*, 36(2): 350–369.

Evans, J.P. (2016) 'Trials and tribulations: Problematizing the city through/as urban experimentation', *Geography Compass*, 10(10): 429–443.

Evans, J.P. and Karvonen, A. (2014) '"Give me a laboratory and I will lower your carbon footprint!" – Urban laboratories and the governance of low-carbon futures', *International Journal of Urban and Regional Research*, 38(2): 414–430.

Foucault, M. (2009) *Le gouvernement de soi et des autres. Le courage de la vérité*, Paris: Gallimard.

Gieryn, T. (2006) 'City as truth-spot: Laboratories and field-sites in urban studies', *Social Studies of Science*, 36(1): 5–38.

Gould, D. (2009) *Moving Politics: Emotion and ACT UP's Fight against AIDS*, Chicago: University of Chicago Press.

Gribat, N., Misselwitz, P. and M. Görlich (eds) (2018) *Vergessene Schulen: Architekturlehre zwischen Reform und Revolte um 1968*, Leipzig: Spector Books.

Halpern, O. et al. (2013) 'Test-bed urbanism', *Public Culture*, 25(2): 272–306.

Huron, A. (2018) *Carving out the Commons: Tenant Organising and Housing Cooperatives in Washington, D. C.*, Minneapolis: University of Minnesota Press.

Invisible Committee. (2009) *The Coming Insurrection*, Los Angeles: Semiotext(e).

Jellis, T. (2015) 'Spatial experiments: Art, geography, pedagogy', *Cultural Geographies*, 22(2): 369–374.

Karpantschof, R. (2011) "Bargaining and barricades: The political struggles over the Freetown Christiania, 1971-2011", in H. Thörn, C. Wasshede and T. Nilson (eds) *Space for Urban Alternatives? Christiania, 1971-2011*, Gothenburg: Gidlunds Förlag, pp. 38–67.

Karvonen, A. and Van Heur, B. (2014) 'Urban laboratories: Experiments in reworking cities', *International Journal of Urban and Regional Research*, 38(2): 379–392.

Klein, J. and Porn, S. (1981) 'Instandbestzen', in I. Müller-Münch et al. (eds) *Besetzung – weil das Wünschen nicht geholfen*, Hamburg: Rowohlt, pp. 108–125.

Kohler, R.E. (2002) 'Labscapes: Naturalising the lab', *History of Science*, 40(4): 473–501.

Krebs, M. (2003/4) 'Demands', *West Coast Line*, 41(2/3): n.p.

Kullman, K. (2013) 'Geographies of experiment/experimental geographies: A rough guide", *Geography Compass*, 7(12): 879–894.

Lachmund, J. (2013) *Greening Berlin: The Co-production of Science, Politics and Urban Nature*, Cambridge, MA: The MIT Press.

Last, A. (2012) 'Experimental geographies', *Geography Compass*, 6(12): 706–724.

Lazzarato, M. (2017[2009]) *Experimental Politics: Work, Welfare and Creativity in the Neoliberal Age*, trans. A. Bove, et al., Cambridge, MA: The MIT Press.

Livingstone, D. (2003) *Putting Science in its Place: Geographies of Scientific Knowledge*, Chicago: University of Chicago Press.

Laurisch, B. (1981) *Kein Abriß unter dieser Nummer*, Berlin: Anabas Verlag.

Lõpez, M.M. (2013) 'The squatters' movement in Europe: A durable struggle for social autonomy in urban politics', *Antipode*, 45: 866–887.

Marx, K. (1993) *Grundrisse: Foundations of the Critique of Political Economy*, London: Penguin Classics.

Mudu, P. (2004) 'Resisting and challenging neoliberalism: The development of Italian social centres', *Antipode*, 36: 917–941.

Mudu, P. and Chattopadhyay, S. (eds) (2016) *Migration, Squatting and Radical Autonomy*, London: Routledge.

Muñoz, J.E. (2009) *Cruising Utopia: The Then and There of Queer Futurity*, New York: NYU Press.

Peck, J. and Theodore, N. (2015) *Fast Policy: Experimental Statecraft at the Threshold of Neoliberalism*, Minneapolis: University of Minnesota Press.

Powell, R. and Vasudevan, A. (2007) 'Geographies of experiment', *Environment and Planning A*, 39: 1790–1793.

Reid-Henry, S. (2010) *The Cuban Cure: Reason and Resistance in Global Science*, Chicago: University of Chicago Press.

Shapin, S. and Schaffer, S. (1985) *Leviathan and the Air-Pump: Hobbes, Boyle and the Experimental Life*, Princeton: Princeton University Press.

Simone, AM. (2019) *Improvised Lives: Rhythms of Endurace in an Urban South*, Cambridge: Polity Press.

Simone, AM. and Pieterse, E. (2017) *New Urban Worlds: Inhabiting Dissonant Times*, Cambridge: Polity Press.

Slobodian, Q. (2012) *Foreign Front: Third World Politics in Sixties West Germany*, Durham, NC: Duke University Press.

SqEK (Squatting Europe Kollective). (eds) (2014) *The Squatters' Movement in Europe: Commons and Autonomy as Alternatives to Capitalism*, London: Pluto Press.

Stanley, E. (2018) 'The affective commons: Gay shame, queer hate, and other collective feelings', *GLQ*, 24(4): 489–508.

Vasudevan, A. (2006) 'Experimental urbanism: *Psychotechnik* in Weimar Berlin', *Environment and Planning D: Society and Space*, 24: 799–826.

Vasudevan, A. (2015) *Metropolitan Preoccupations: The Spatial Politics of Squatting in Berlin*, Oxford: Wiley.

Vasudevan, A. (2017) *The Autonomous City: A History of Urban Squatting*, London: Verso.

Wall, C. (2017) '"We don't have leaders! We're doing it ourselves!": Squatting, feminism and built environment activism in 1970s London', *Field Journal*, 7(1): 129–141.

Ward, C. (2002) *Cotters and Squatters: Housing's Hidden History*, Nottingham: Five Leaves Books.

Wright, S. (2002) *Storming Heaven: Class Composition and Struggle in Italian Autonomist Marxism*, London: Pluto Press.

Zinovich, J. (2015) 'Christiania: How they do it and for how long', in A. Moore and A. Smart (eds) *Making Room: Cultural Production in Occupied Spaces*, Barcelona: Los Malditos, pp. 84–94.

37

Monumentalizing public art through memory of place

Place-based interpretation and commemorability

Hélène B. Ducros

From the Cosmos, like a soft bouquet, fell these meteorite flowers …

Mireille Dreisine

As you stroll along Paris' Boulevard Raspail, an unexpected modernist building grabs your attention, as might also on the sidewalk in front of it a series of large granitic slabs showing deeply carved parallel grooves. Depending on time of day, these capture light to compose different atmospheres, from dismal on grey rainy days to radiant at sunset on long summer evenings, but always primarily mineral. If the presence of those scattered blocks seems fortuitous to distracted passersby, a turn around the corner into the Rue du Cherche-Midi quickly suggests that they belong to a larger artistic ensemble comprising a central cluster of much taller blocks located on the side of the building. As I grew up in the neighborhood, this sculpture has been part of my quotidian landscape. I have always heard that it was meant to honor what once stood on these grounds: a military prison that was demolished in the 1970s. An inquisitive eye might also discern on the ground, sharing space with the imposing structure, two inconspicuous plaques evoking individuals' suffering at various times in the history of the prison. Perhaps that day wilted flower bunches left behind from an anniversary date celebration might even cover the plaques.

Concerned with heritage and memory of place, I envisaged the artistic creation as part of what I considered a clear memorial landscape, setting out to examine the story of the familiar sculpture as a commemorative artifact. Originally answering Young's (2016) prompt that "contemporary inquiries into material culture and history demand an exploration of how questions about the aesthetic, social, political, and performative dimensions underlying these sites resonate with the public" (p. 38), I was interested in unveiling the ways in which mnemonic objectives intervene in the creative process, the manner in which different parties negotiate over the nature and shape of memoryscapes in public spaces, and the ensuing reactions by neighborhood residents and other implicated actors. I embarked into preliminary archival research, only to be faced with unanticipated findings, which led me to deepen my investigation

Figure 37.1 Shamaï Haber, Maison des Sciences de l'Homme, Paris. Main cluster.

into the layers of the monumentalization process. What I did not find took over my inquiry: there was no trace of explicit commemorative intent. In his benchmark and still influential 1903 essay, Aloïs Riegl (1982) defined a monument as "a human creation erected for the specific purpose of keeping human deeds alive in the minds of future generations". Further refining this classic understanding, he categorized the concept by questioning purpose and intent as necessary prerequisites. In this lineage, through the installation in the 1970s of an unnamed sculpture in the Parisian public space (Figure 37.1) on the ground where once stood a prison, I advance that, in monument-making, explicit purpose in meaning-making is not as important as place. By unveiling the process and context of the sculpture's creation, I examine the role of place as a conduit for making unintended monuments.

Methods

I highlight the relationship between place, memory and monument by considering what it takes for public art to function as commemorative device. Rather than apprehending memorialization as an agent of place-making as it has often been done (for example, in Post, 2018), I conversely question the role of place in the construction of mnemonic value. Under study is a work by Shamaï Haber (Łódź 1922–Paris 1995), a Polish-born sculptor established in Paris in 1949 and acclaimed for his inventive interaction with urban spaces. The massive unnamed sculpture located at the intersection of the Rue du Cherche-Midi and Boulevard Raspail wraps around the first all-metal and glass building ever built in France. Entirely renovated in the 2010s for asbestos removal and modernization, the building is home to the Maison des Sciences de l'Homme (MSH) Foundation and the Ecole des Hautes Etudes en Sciences Sociales (EHESS). One originality of Haber's 300-ton piece dwells in its integration within the urban architectural landscape and its function as a fluid hyphen between the street and the building's interior. Encompassing 25 vertical and horizontal striated blue granite blocks dispersed around and inside the building with a highest point in the main cluster at 8+ meters, the sculpture testifies to Haber's integrative, monumental and mineral approach to public art that is palpable throughout his oeuvre, for example in the gigantic granitic disc from which water flows, designated as *Le Creuset du Temps* (Place de Catalogne, Paris).

To better understand the historical, spatial and social influences on the production and reception of the monument, I consulted a mix of primary and secondary sources. I delved into the national archives of the Ministry of Culture to collect correspondence, architects' plans (by Lods, Malizard, Depondt and Beauclair) showing the integration of the sculpture into the design of the building, and press articles concerning the design, financing, construction and public reception. I scrutinized multi-lateral correspondences between the architects, the funder (the state) and the artist to appreciate how different actors argued their choices. I also obtained documents from the MSH's institutional archives—press articles and commentaries—and questioned MSH employees and researchers (including retired scholars) about the sculpture and the military prison. To analyze the making of Haber's creation into a commemorative object, I examined how selected scholars approached the concept of monument in relation to place memory. My findings led me to posit that it is the illegibility of the non-figurative sculpture that fostered meaning-making through viewers' place-based interpretation, conjointly with explicit—but separate—markers of place memory.

Monuments

Monuments reveal societies' connection with the past, present and future. Since the advent of nation-states in particular, monuments have been central to public debates about contemporary representations of history. Constituting effective tools of transmission for nationalist ideals, they advance cohesiveness and continuity in one version of the nation (Lowenthal, 1985; Nora, 1997). At the intersection of history, memory and identity-making, monuments propose an answer to the question of what people and ideas from the past are worthy of representation, while simultaneously underlining the moral justification behind choices made (Howells, 1866). Understood as a subset of the broader category of "memorials", monuments constitute "the material objects, sculptures, and installations used to memorialize a person or set of events" (Young, 2016, p. 38), sometimes also substituting for tombs and places of mourning (Sherman, 2006). Early in the twentieth century Riegl (1982) identified three types of monuments. First, the intentional monument—the most evident and deliberate type—explicitly recalls moments from the past. While the second type—the historical monument—also fulfills this function, it does so within a particular temporality chosen subjectively according to viewers' preferences, so that historical monuments may be unintentional. Finally, monuments "of age-value" comprise all artifacts regardless of original significance and purpose providing they confirm "the passage of a considerable period of time"(p. 24). Many have commented that a quasi-obsessive concern with memory—whether individual or collective—characterizes the modern era, leading to an affirmed tendency towards systematic and excessive patrimonialization (or "heritagization") (Drouain, 2006; Jeudy, 2008; Heinich, 2009; Lowenthal, 2015) and a broadened patrimonial field that now includes landscapes (Poulot, 2006). As post-modern societies are driven by melancholia and a longing for an often reimagined past, they become more likely to imbue with value artifacts from the past, representations of the past, or sites of past events. In urban environments in particular, monuments have been linked to commemoration, becoming "parts of the city landscape, spatial points of reference or elements founding identity of a place" (Caves, 2005, p. 318). They function as symbolic landmarks vital in shaping how local residents identify with place and historical discourses about place (Krzyżanowska, 2016).

Memory and place

The tug between individual and collective memory is pivotal in memory studies, a field described as engaging in how individuals' memories acquire social and tangible form and gain political significance through cultural texts, images, architecture, or citizens' associations (Sherman, 2006). For memory theorist of reference Maurice Halbwachs (1950), all memory is collective and social, whereas other scholars focus on individual aspects of memory-making, with a specific focus on place as mnemonic agent. For David Lowenthal (1985), collective memory yields to collective and present understandings of the past. In these various understandings, history, heritage, individual and social memories all come together in place as a node. However, for Michel de Certeau (1984), memory is not localizable. He envisages it as the "anti-museum"—unattached to place. Place being external to time (and vice-versa), it is incompatible with memory. Making memory place-dependent or constraining it spatially risks hollowing it of its meaning. If fixed somewhere, memory loses the plasticity needed to preserve remembrance. Nonetheless, even for de Certeau, memory intervenes in place as a set of possibilities that can not only change place, but also be altered in place. Pierre Nora's (1997) seminal *Lieux de mémoire* is one of the most comprehensive attempts at theorizing the relationship between place and memory—*loci memoriae*. In his multi-volume opus, Nora analyzes both memory *of* place and memory *in* place through the concept of *lieu de mémoire*. Going beyond a simple *site*, his expanded understanding of *lieu* (translated as "realm") produces complex and multi-layered meanings. Place becomes the embodiment of polyvocal material producing symbolic and functional value through a limited number of signs, which are made. In the context of memory, *lieu* becomes "any significant entity, whether material or non-material in nature, which by dint of human will or the work of time has become a symbolic element of the memorial heritage of any community" (Nora, 1996, p. vii). Turned into an abstract, place becomes an opportunity for self-discovery, of which the construction requires examination.

Metaphoric communication and counter-monument

Commemorative spaces create meaning and experiences calling for interpretation. Commemoration consoles, recognizes loss, consecrates places, honors courageous deeds, and creates pedagogical moments (Sherman, 2006). Its language must resonate with the public. As a tool of communication and a subset of public art, commemorative monuments trigger a dialogue with viewers, who in turn produce meaning through their personal pasts, their engagements with memorial structures and the spaces they occupy, and their personal stories in relation to them, even when those narratives are at odds with official or explicit interpretations (Young, 1994). Public monuments were long envisaged as comprehensive, integrative, lasting, and fixed artifacts stabilizing physical and cognitive landscapes. As such, their vocation was to promote consensus and conflict resolution, providing historical closure while merging polysemic understandings of historical characters and events (Savage, 1997). In particular, as processes by which art intervenes in public spaces and is instilled with values and ideals of those making decisions about public space usage, public art influences the fabrication of social memory and fosters identity-making. Without a narrative—whether shared or not—a monument may lose in legibility. However, if approached through an understanding of a shared role between spectators and artists in producing art, monuments, when not plagued with reductionist constraints of assigned signification, can be seen as a sculptural genre communicating possibilities through metaphors.

In the city, monuments have been apprehended as specific expressions of place. In representing place, they also legitimize it and give it permanence. However, as systems of fixed and consensual symbolic representations and metaphoric demonstrations drawing on collective and individual imaginaries, monuments are challenged by the concept of anti-monument (or counter-monumental commemoration), which emerged to oppose the hegemonic memory discourses advanced by intentional monuments. Pushing Riegl's unintentional monument further into the subjective realm, the counter-monument "aligns much better than the traditionally static and non-dialogic monuments with the ongoing post-modern transformation and fluidity of contemporary urban spaces" (Krzyżanowska, 2016, p. 465). It is characterized by polysemic interpretations contingent on viewers' gazes. Also able to accept inconvenient narratives, it is an "open work" entertaining possibilities. As interactive artifact, it consists of a "form of a game in which the spectator is invited to interact in a novel way with identity of the place and its history (genius loci) and collective memory, all anchored in a specific space" (p. 471).

Counter-monuments commemorate people and places, create an "interpretive plain" leading to multiple and personal commemorations, and allow for the "de-politicisation and de-ideologisation of commemoration that should be at the foundation of remembrance in contemporary urban spaces of diversity and inclusion" (Krzyżanowska, 2016, p. 482)—akin to altercommemoration, which destabilizes an "elemental, intrinsic relationship between place and community" (Sherman, 2006, p. 139).

The demolished Cherche-Midi prison: historical and spatial context

Patrimonialization and demolition have often been observed concurrently. The destruction of buildings addresses the spatial dimension of place memory because it leaves an absence of relics. It constitutes symbolic violence and a denial of memory (Veschambre, 2008). Demolition removes elements needed for memory to anchor itself in visible and tangible remnants. When rebuilding does not occur, those traces left behind may be valorized or reinvented, so commemoration-monumentalization practices can be implemented along with other mnemonic productions. Place, memory, and absence converge in monuments and other memorial devices, allowing people to reappropriate place through marking (Veschambre, 2008). Examples may be the creation of "negative space" to express loss (Saltzman, 2006) or commemorative plaques described as "historical fragments" inserted into public space as "monuments of conciseness" contributing historical narratives and requiring precise geographical and spatial connections with place (Dutour, 2006). Plaques usually feature names. In France in particular, the names of the dead are central to commemorative practices (Sherman, 2006). Not only do plaques affirm the dead's worth, but they also legitimate the cause for which they died

The Cherche-Midi prison was one of three military prisons in Paris—in use from 1851 to the mid-1960s. Historians have unveiled that civilians too were incarcerated there. Interest in the prison has strengthened, but for a long time it was considered "a black hole in history" (Tronel, 2009). In particular, lesser-known aspects of the prison's history were revealed through recent research on penitentiary history and military justice (Tronel, web). During the Second World War, after being evacuated, the prison fell under German command. The Nazis would torture *résistants* there, including women. After Paris was liberated, German war prisoners were detained there, including high-ranking officers who hanged themselves while awaiting trial. Entirely razed between 1961 and 1966, the prison remains a relatively recent memory for long-term neighborhood residents. Despite historians' recent exploration into the diversity of prisoners in the Cherche-Midi Prison, collective memory has centered on Capitaine Alfred Dreyfus' incarceration during his trial, which took place across the street at the Conseil de Guerre (military tribunal) and ended in a wrongful treason conviction in 1894 (leading to Zola's *J'accuse*); the Capitaine de Corvette (warship) Honoré d'Estienne d'Orves, executed by the Nazis for organizing *Résistance* networks; and high school and university students, imprisoned in 1940 after a three-day insurrection against the Nazis following de Gaulle's 18 June call. Amidst this weighty context, the prison once shaped the identity of the neighborhood, which is otherwise further laden with Second World War memories. Across the street stands the Hôtel Lutétia, which during the war housed the headquarters for the German *Abwehr* (counter-espionage and intelligence) and *Geheime Feldpolizei* (secret military police). After the war the hotel became a reception center for Jews who had survived the camps, so they could possibly reunite with their family. That short stretch along the Boulevard with the old prison site and the hotel are part of a local memory that is readily accessible because it is recent and still painful. Beyond the local, both these sites are also etched in the history of the nation, the social trauma of German occupation in the city, and the liberation from Nazi Germany in August 1944. Today, although the only physical remnants of the legal-carceral military complex are located elsewhere (in Parisian suburbs), the site of the prison functions as a multi-layered *lieu de mémoire*, where local collective memory of place gets entangled with the history of Paris as a city marked by Nazi occupation, as well as with a broader national memory.

Haber's sculpture: intent and realization

The 25 blocks of blue granite from a Breton quarry (funders denied a more expensive granite from Labrador) were installed in August 1973, shortly after the completion of the MSH building. The inspiration for the building and the concept of the MSH—a center of research, documentation, and teaching—was found in the US where Depondt had studied architecture and where Fernand Braudel and other key founders of the center had traveled extensively in search of an institutional and organizational model to facilitate scientific advancement and intellectual exchanges across the disciplines. Early on, the building and the sculpture were envisaged jointly in the architects' design as part of the *1% décoratif* law, which prescribes that 1% of the total paid by the state for the construction of public buildings (at the time, it aimed specifically at buildings dedicated to education) must support the realization of a work of contemporary art that should be fully integrated in the architectural design. A review of the correspondence between architects Lods, Depondt, Malizard, and Beauclair and various institutional instances at the Ministry of Cultural Affairs (now the Ministry of Culture) indicates that they recommended Haber's work to the architectural and artistic board and the Paris Academy *Recteur*. Negotiations ensued for three years before the project was accepted, but architectural plans projected space allocation for the sculpture from the early conception and construction of the building. For example, written exchanges between the Service Constructeur de l'Académie de Paris and the Services de la Création Artistique reveal that the ground was reinforced because of the projected weight of the sculpture (at a substantial additional cost) and the building was set back from the street boundary. Widening the sidewalk was necessary to accommodate the reception of eight low blocks that are scattered in front and on the side of the building, leading to the main cluster where stones stand vertically to over 8 meters in height. Flowing water was planned, but the mechanism quickly malfunctioned. Additionally, three low blocks are dispersed in the area around the main cluster, echoed by three more blocks in the interior garden (Figure 37.2).

Figure 37.2 Shamaï Haber, Maison des Sciences de l'Homme, Paris. Dispersed blocks.

Already precisely established in the building plans, the sculpture's layout was not fortuitous. A rehearsal even took place before the blocks left the quarry. The dispersed sculpture precludes viewers from seeing it all in one glance and from one vantage point. Haber forces a mobile interaction with the piece: viewers must navigate around it and around the building, and in and out from the street. There are three potential levels of accessibility and exchange with the art piece. To the passersby, the eight low blocks on the street are not only visible but also accessible physically and tactically. People routinely use the blocks to sit on while eating lunch, talk on the phone, smoke a cigarette, or prop their bikes. Recently, homeless refugees have used the slabs closest to the building's overhang, where they set up night encampment, as tables to stash their things. The more manifest 11-block cluster (bases topped with five very tall blocks) is only accessible visually since it stands in an enclosed space, separated from the street. Since the building's renovations, a low metal separator was replaced by a tall transparent palisade, further disconnecting the sculpture from viewers. Finally, although the MSH is a public building, it is unlikely that many passersby would enter the building to get to the patio and experience the blocks that are there. Only MSH staff or people entering the building for a working purpose are likely to see that aspect of the sculpture. Over several years, vox pops (street interviews or conversing with passersby on the fly or with café customers across the street) revealed that many people literally do not see the 300-ton mass. They may have never noticed it even if they walk by it regularly. During the recent building face-lift, some showed complete misunderstanding in thinking that the low blocks on the street were not going to stay and that they were simply there as part of the renovation work—rubble. Eventually some of the blocks were boarded up for protection, while others were deliberately used as storage for construction material. Some, including workers involved in the renovation, were well aware of the sculpture. They sometimes made the explicit connection with the old prison, but offered other explanations as well.

"*L'artiste qui fait une œuvre ne sait pas ce qu'il fait*" explains Marcel Duchamp (1960) to express that artists have no control over the aesthetic result, meaning, and value of their creations. Correspondences show that Lods agreed with Haber that his purpose was not about beautifying, but about modifying place to contrast sharply with the surroundings—bringing mineral masses, "rich of hundreds of millions of years, radiating through their deep crystallization a creative force drawing from genesis and yet still a mystery for human sciences" (Motinot, 1974). Letters between the architects and the Ministère des Affaires Culturelles highlight that decoration was not part of artistic intent. Instead, the architects invoked a piece conceived as a landscape, a modification of the environment, and a hyphen between spaces, which would have to be experienced and have an impact on passersby as it allows for the convergence of the street, the building, the garden, and the idea behind the building itself, i.e., a pivot for intellectual thought. The artwork was presented as unbounded and affirming of the open spatiality of the city. The sculpture connects to the building spatially, technically, and aesthetically, as well as semiotically. However, conversations with a MSH officer cautioned that administrative justifications should be separated from the artistic process and the complex origins for the idea, intimating that even if commemorative intent had existed, it would have been silenced because incompatible with the mode of funding sought.

Reception: a nameless polysemic monument

Straightaway, the sculpture raised the question of abstract and non-figurative art. Even today contemporary art triggers vigorous debates in Paris, sometimes leading to oeuvres being defaced. Jeff Koons, Anish Kapoor, and Paul McCarthy can all testify to this. In the 1970s it was even more challenging. For instance, hindered by the lack of title and explicit intent, the *Recteur* first rejected Haber's project, finding the sculpture "not sculpted enough". He only came around once he had assimilated it as ancient structures symbolizing early humanity: menhirs (prehistoric megaliths found in Brittany). It made sense to him aesthetically and semiotically that the artist must have wanted to remind people of primitive societies, in contrast to the MSH as a center of knowledge.

The immediate reception from the neighborhood's residents was amply negative. Proving a conservative vision of the city, some elected representatives suggested the sculpture be moved and presented a *question écrite* to members of the government to decry the sculpture as "incoherent prisms", "model of Manhattan", "priapic erection", "pure geometrical forms" or, worse, an "American city". Readers' responses in the neighborhood's newsletter harshly rejected the piece, attacking both the sculpture and the modernist building. Additional elements of the complaint mention the blocks serving homeless people or even dogs urinating. People nevertheless tried to attribute logic to the untitled work by associating it, like the *Recteur*, with the symbolism of the menhirs. They engaged in reflections about the crudeness ("unfinished", "blocks left behind"), massiveness of the piece ("mountain of granite") in contrast to the refinement of the city and the stark opposition they felt it created between the natural (mineral) and the urban. In this respect, Haber was precisely successful, since his approach to urban art was to awaken consciousness about the place of nature in the city.

An unusual—even whimsical—complaint came in the form of a petition-poem by Mireille Dreisine (1921–2015, aka *La Bande à Caillou*), poet, researcher and habitual visitor at the MSH. It obtained 72 signatures. In her poem, Dreisine relays disappointed hopes and the reactions of the people who worked in the MSH when they discovered the sculpture. Like other commentators, she looked for intent in the sculpture by recognizing shapes and objects: giant cacti, menhir, obelisk, phallic symbol, dendrites, meteorite flowers, hearses, funerary slabs, altar, sanctuary. "We are left with interrogations"; there is "no legend"; it "leaves the spectator mute". But, it "offers a choice". She too speaks of gigantism, inanimate rock, dried-up stream (the fountain mechanism malfunctioned), and even death. While neighborhood commentaries were vehemently negative, the Parisian and regional press was less critical, remaining open to modernist architecture and the semiotic challenges posed. There, emphasis was placed on successful harmonies with the building and the use of granite as a medium operating a rapprochement between the heart of Paris and France's regions. The sculpture was seen as honoring the mineral riches of the nation and the savoir-faire of granite cutters. Other enthusiastic reactions discerned book stacks out of the grooved pillars, highlighting the symbolism of knowledge production and transmission. However, none of the documents consulted gave any account of the prison, aside from a fleeting allusion to "walls" in a few letters.

Contemporary interpretations are remarkable in how diverse and ungrounded in reality they are. When prompted for their reaction to the monument, passersby, café workers and patrons, construction workers (during the renovation) and people who use the MSH provide answers that are particularly revealing of their need to assign meaning to the sculpture. Reponses draw on two registers: the prison site (the past) and the site of knowledge (today's MSH). A selection follows:

- *It's there because there was a prison before. All the prisoners' names are inscribed on the stones.* (There are no names on the sculpture.)

- *This is what's left of the prison walls.* (Only a portal remains, moved elsewhere.)
- *The slabs are antique pieces, Roman or Egyptian.* (Untrue, although echoed by another passerby.)
- *The parallel grooves draw prison bars.* (They are due to the granite extraction technique.)
- *It's a human figure, it's Man.* (The sculpture is non-figurative.)
- *Here it's a research center and that represents an archeological dig.*
- *Book stacks, that's what they evoke.*

Key to the reception context is that, shortly after the installation of the sculpture, two plaques were appended by the city government on the ground in close vicinity to the sculpture, one on each side of the main cluster. They commemorate Dreyfus, d'Estienne d'Orves and the Second World War resisting students. These individuals continue to be celebrated on this site on pertinent national holidays or anniversaries meaningful to the city's history. I suggest that in the multiplicity of interpretations advanced about the sculpture, it is the integration of those names on the commemorative plaques as direct memorial signage that contributed to fixing an explicit narrative for the sculpture and to attaching its meaning directly to the old prison and the memory that still resides in the site itself, in effect making it legible so it could become "monument". The plaques and the sculpture make sense *together*; the sculpture and the building make sense *together*. Ultimately, the plaques conferred to the site the elements needed for people to make sense of the sculpture, not as evocative or representative of any particular object—it remains abstract—but of a place. It is the place itself that produced not the sculpture, but the monument. Eventually accepted, this memorial narrative became part of the neighborhood's storyline, with people today readily making the association between the sculpture and the old prison. The sculpture might still be illegible, and for some even invisible, but the monument is obvious, or, to use Haber's word: "evident" (Motinot, 1974).

A counter-monument appropriated through place

This case study reinforces that a monument is a process stemming from a multi-layered context (physical, spatial, social, historical, cultural, or financial) and deriving from a web of influences that are often rooted in place: purpose, location, functionality, perception, reception, and appropriation. Haber's "monument in the city" constitutes a spatio-temporal hyphen, which is facilitated by the artist not seeking to recall anything and instead leaving interpretation open. The absence of deliberate commemorative intent in design pushed me to interrogate how commemorative value had been attributed to the monument "from below". More than a work to be seen, it is a work to be contemplated. Without referencing the prison, Haber challenges the limitations of memory and the constraints of its representation in the form of memorial, linking history, memory, place, cityscape, all the while freeing viewers' perception. The sculptor did not seek to create legibility or propose bounded meanings, not even through a name. The untitled piece is monumentalized through the eyes of the seers based on place, past, and present coming together. Passed the initial negative reactions, commemorative value was shaped through public (mis) understanding. It is the near-by commemorative plaques that rendered place explicit and granted the sculpture a raison-d'être as *lieu de mémoire*, limiting—yet not muting—other interpretations.

Reading the sculpture through the lens of place has given it monumental value, in line with Riegl's idea of subjective reception creating the unintentional monument. Irrelevant in making the monument, original intent is eclipsed by perception in the present.

Today, the sculpture stands as an organic part of the collective memory of place in the neighborhood. Its commemorative value emerging out of its openness, the sculpture allows for multiple meaning-making and metaphors. Walls, primitive societies' menhirs, progress and knowledge transmission, human control over matter, tombs of the fallen, these interpretations all reveal people's need to justify the abstract design to give it memorial significance. Place is central to the process that grants the sculpture its memorial charge, whatever the reading made, while a feedback loop may in turn reaffirm place. In this respect the art piece has escaped its creator as viewers have appropriated it. Instead of the commemorative landscape making place, it is place that creates the commemorative landscape, transforming the sculpture into an unintentional monument based on the history of place and its function in the present. The unintentional monument is defined not by its maker, but by the public. As the "anti-sculptor" (Nouvelles Littéraires, 1974), Haber apprehends space and reshapes it with purpose but without precise evocation. In the end, it is place that yields meaning to the dispersed volumes. Haber's unnamed sculpture at the corner of the Rue du Cherche-Midi arises as a counter-monument, polyvocal, open to interpretations, and emanating from place in multiple intersecting ways.

Acknowledgments

Many thanks to the *Archives Nationales* (Fontainebleau), the MSH Director and staff, particularly Jacky Tronel and the late Mireille Dreisine, for their precious help.

References

Caves, R.W. (ed.) (2005) *Encyclopaedia of the city*, London: Routledge.

de Certeau, M. (1984) *The practice of everyday life*, trans. S. Rendall, Berkeley: University of California Press.

Drouain, M. (2006) *Patrimoine et patrimonialisation: du Québec et d'ailleurs*, Montréal: MultiMondes.

Duchamp, M. (1960, 1994) *Interview with Georges Charbonnier*, Marseille: Editions André Dimanche.

Dutour, J. (2006) 'Les plaques commémoratives: entre appropriation de l'espace et histoire publique dans la ville', *Socio-anthropologie*, 19 [Online] https://journals.openedition.org/socio-anthropologie/603

Halbwachs, M. (1950) *La mémoire collective*, Paris: Presses Universitaires de France.

Heinich, N. (2009) *La Fabrique du patrimoine: de la cathédrale à la petite cuillère*, Paris: MSH.

Howells W.D. (1866) 'Question of monuments', *Atlantic Monthly*, May: 646–649.

Jeudy, H. P. (2008) *La machine patrimoniale*, Paris: Circé.

Krzyżanowska, N. (2016) The discourse of counter-monuments: semiotics of material commemoration in contemporary urban spaces, *Social Semiotics*, 26(5): 465–485.

Lowenthal, D. (1985) The past is a foreign country, New York/Cambridge: Cambridge University Press.

Lowenthal, D. (2015) *The past is a foreign country revisited*, New York/Cambridge: Cambridge University Press.

Motinot, R. (1974) Le granit de la Maison des Sciences de l'Homme, *Le Mausolée: Technique Mensuelle*, 453: 904–912.

Nora, P. (1996) *Realms of memory: rethinking the French past*, New York: Columbia University Press.

Nora, P. (1997) *Les lieux de mémoire, Volumes I, II, III (1984, 1986, 1992)*, Paris: Gallimard.

Nouvelles Littéraires (1974) Haber, le déménageur de volumes, 20 May.

Post, C. (2018) Making place through the memorial landscape, in J. Smith (ed.) *Explorations in place attachment*, Oxon/New York: Routledge, pp. 83–96.

Poulot, D. (2006) *Une histoire du patrimoine en occident*, Paris: Presses Universitaires de France.

Riegl, A. (1982) 'The modern cult of monuments: its character and its origin', translated by K. Foster & D. Ghirardo, in *Oppositions*, 25: 21-51. Original text: 'Moderne Denkmalkultus: sein Wesen und seine Entstehung', (Wien: K. K. Zentral-Kommission für Kunst- und Historische Denkmale: Braumüller, 1903).

Saltzman L. (2006) *Making memory matter: Strategies of remembrance in contemporary art*, Chicago: University of Chicago Press.

Savage, K. (1997) *Standing soldiers, kneeling slaves: Race, war, and monument in nineteenth-century America*, Princeton: Princeton University Press.

Sherman D.J. (2006) Naming and the violence of place, In D.J. Sherman and T. Nardin (eds) *Terror, culture, politics: Rethinking 9/11*, Bloomington: Indiana University Press, pp 121–145.

Tronel J. (2009) La Prison du Cherche-Midi. un trou noir de l'histoire, Exhibition-Conference held at the Maison des Sciences de l'Homme 54, bld Raspail, Paris on June 12.

Tronel J. (web) https://prisons-cherche-midi-mauzac.com/accueil.

Veschambre V. (2008) *Traces et mémoires urbaines, enjeux sociaux de la patrimonialisation et de la démolition*, Rennes: Presses universitaires de Rennes.

Young J. (1994) *The texture of memory*, New Haven, CT: Yale University Press.

Young, J. (2016) Memorializing the holocaust, in C. Krause Knight, and H. F. Senie (eds) *A Companion to Public Art*, Hoboken, NJ: John Wiley & Sons, pp 13–50.

38

Place and heritage conservation

James Lesh

Introduction

Place has a distinctive meaning in the field of heritage conservation. Within conservation practice, understandings of place tend towards the pragmatic and rationalised. This contrasts with the diverse notions of place theorised and applied in other chapters in this volume, as well as within the academic field of heritage studies. Readers of this volume are engaged in a critical and dynamic dialogue around the idea of place. Conservation practitioners, however, employ specific renderings of place in support of their efforts to safeguard buildings, precincts and neighbourhoods. Furthermore, conservation operates as part of dominant architectural, planning and economic processes, which shape how conservationists apply concepts. There is a divide between critical conceptions of place and the practice of conservation.

This chapter explores the shifting and often difficult relationship between conservation and the concept of place. The 1979 Burra Charter – Australian practitioner guidelines for the safeguarding of cultural heritage significance – forcefully introduced 'place' as a formal unit of practice within conservation. Embodied within the Burra Charter and many other international heritage charters has been, however, a specific and problematic tension: an artificial distinction between that which is tangible and that which is intangible at places. Within conservation practice, the tangible relates directly to historic fabric and physical forms. The intangible embraces the social and human perceptions and constructed meanings of place, incorporating that which cannot be overtly seen, touched or observed.

In Western conceptions of heritage, and for historical reasons explained below, the tangible is perceived as more real and important than the intangible, and so, in attempts to achieve authentic or best-practice heritage outcomes, the former tends to be the focus of conservation at the expense of the latter. This conception of place, distinguishing between the tangible and intangible, moreover, conflicts with conceptions of place within human geography and other academic fields, which have long-avoided such clear-cut divisions. The continued emphasis on the tangible over the intangible means that people's diverse expectations and experiences of place are always at risk of being sanitised and erased when a place enters the remit of conservation. As Madgin et al. (2018: 587) argue, 'international [heritage] charters provide a recognition that whilst material fabric does have value we also need to be aware that this value is intimately connected to the feel, use, and experience of place'. Indeed, there is an urgency to reconcile the intangible and tangible in conservation, and to do so would make strides towards conserving places in ways that are more responsive to communities. Heritage scholars Schofield and Szymanski (2011) have suggested this is achievable: a 'sense of place can be understood, assessed and accommodated in the increasingly democratic process of managing change, at least as it is represented through heritage practices in the developed world'.

As this chapter explores, the notion of place in conservation has evolved in recent decades to reach this point. However, there has been little examination of this specific term. Often the moniker 'place' has simply been adopted in practice in ways that may appear to simply refer to the location, area, site or building subject to conservation, rather than engaging with the full realm of possibilities that the concept of place offers. Within what Smith (2006) calls the 'Authorised Heritage Discourse', particular ways of doing heritage come to dominate practice; limiting the prospects of conservation, and so concepts such as place. This chapter analyses the 'grey literature' of conservation since the 1960s: the specialist documents and international charters that inform the day-to-day activities of practitioners. In conservation, this literature holds considerable sway. The geographic emphasis of this chapter is Australia, because this is where the concept of place was formally introduced into global conservation practice via the Burra Charter. To understand the meaning of place within conservation, Silberman suggests focussing on the processes, rather than the sites, of heritage conservation (2015: 30).

Finding place

Conservation is an area of specialised professional practice that is concerned with the identification, assessment, rejuvenation, maintenance and interpretation of places (Aplin, 2002; Avrami et al., 2000; Delafons, 1997; Fairclough et al., 2008). Depending on intellectual and geographic context, places can also be called monuments, ensembles, areas, landscapes and historic environments (Sonkoly, 2017: 42); however, the focus of this chapter is specifically on the term 'place'. Heritage places range in scale from plaques, memorials and buildings to entire streets, neighbourhoods and towns. The key actors within this segment of heritage management, or the heritage industry, include conservation architects, urban planners, policymakers, generalist consultants, civil society and academics. Heritage governance regimes for places operate at various scales, ranging from the local municipal sphere, to the nation state, to UNESCO World Heritage (Sonkoly, 2017). The International Council on Monuments and Sites (ICOMOS) – the key advisor on historic environments to UNESCO – operates alongside various national and local practitioner and civil society and activist organisations involved in discussions about heritage conservation, including approaches to place. Authorities regulate conservation through urban planning, environmental and heritage legislation. Within this world of practice, it is standards, benchmarks, regulations, reports and charters – heritage management's grey literature – which constitute renderings for best-practice conservation and, for our purposes, the meaning of the term 'place'. At the same time, urban development and the property market exert tremendous sway on conservation outcomes (Logan, 2017).

The interdisciplinary academic field of heritage studies is concerned with the theory and practice of conservation. Scholars of heritage studies are drawn from urban studies, social and architectural history, sociology, geography, archaeology and elsewhere, and many explore heritage places. In particular, Smith (2006) has written about how conservation expertise and practice becomes distanced from communities. The emphasis in conservation practice on the tangible means concepts such as 'place' operate in ways that prioritise materiality and fabric and tend to de-emphasise communities and their sense of place. In order to overcome this, Harrison (2018) has suggested the need for refreshed ontological models (conceptual apparatus) for heritage. Recent efforts within heritage studies, necessarily involving allies in practice, are seeking fresh approaches to provide more nuanced conceptions of place (cf. Tait and While, 2009).

Since the 2000s, the notion of 'intangible heritage' has been intended as one such antidote (Craith and Kockel, 2015; Smith and Akagawa, 2009). Harrison (2013: 13), for instance, writes about the 'process of "dematerialising" heritage by introducing an ever-increasing emphasis on intangible aspects of heritage and tradition as part of the exponential growth in objects, places and practices that are considered to be defined as heritage'. Within and beyond the remit of the conservation of places, intangible heritage refers to 'practices, representations, expressions, knowledge, and skills' (UNESCO, Convention for the Safeguarding of the Intangible Cultural Heritage, 2003), and so can have both material and immaterial aspects. Scholars are also turning their attention to the social, affective, emotional and experiential aspects of place in order to overcome the enduring privileging of fabric and materiality (Garduño-Freeman, 2018; Madgin et al., 2018; Smith and Campbell, 2015).

An assumption of this research agenda, and an explicit contention of this chapter, is that the physical forms and social meanings of place cannot be neatly disentangled. Materiality first became privileged in heritage due to the modern field's origins in nineteenth-century Europe. From Eugene Emmanuel Viollet-le-Duc, John Ruskin and William Morris onwards, the idea was that the physicality of a building, its aesthetics and bricks and mortar, should be the focus of conservation. Buildings were to be restored in prescriptive ways, restoring or conserving them to a specific state that was imagined as genuine and thus authentic. Conservationists were also to resist processes of change, development and decay in order to safeguard heritage (Jokilehto, 1999; Pendlebury, 2013). Operating in a Western tradition indebted to Cartesian dualism, conservation favoured that which could be seen and touched (the rational) over what could be experienced and felt (the irrational) (Byrne, 2014; Harrison, 2013). The notion that there is only one (European) best-practice way for achieving authentic conservation outcomes also has overlapping genealogies (Winter, 2014).

In present-day practice, the privileging of fabric in conservation means the replication of outmoded conceptions for heritage. A key goal for critically engaged conservation, then, has been to overcome this tradition; to recognise that the tangible and intangible are just as socially and politically charged as each other, and that heritage and change are overlapping rather than opposed phenomena (Bandarin and van Oers, 2012). To conserve a place is always to physically and socially change it, and this happens through development, intersecting with property interests. Within heritage processes – themselves tied to social, economic, planning and architectural factors – historic fabric is not any more or less important than the social meanings of a place and requires equal effort to achieve positive heritage and social outcomes.

The crucial heritage concept in need of destabilisation is the longstanding principle of inherency. Ordinarily implied rather than made explicit, the principle of inherency refers to the idea that building fabric and physical forms seemingly have enduring and objective meaning beyond their historical or social context (Byrne et al., 2003; cf. Sullivan, 2015). In reality, place cultivates heritage significance only because people attach meaning to that said place in all its physical and social dimensions. Whether we are thinking about its tangible or intangible aspects, there is nothing naturally or inherently important about a place or its parts. Heritage places become valuable due to factors such as age and rarity, and for historical, aesthetic, architectural, social, archaeological, scientific and spiritual reasons (Avrami et al., 2000; Gibson and Pendlebury, 2009). A place develops meaning at once for and because of people. Geographer Mike Crang (2010, p. 103) writes that 'Spaces become places as they become "time-thickened". They have a past and a future that binds people together around them'. Even seemingly enduring values such as age – the notion that the older a place the more important it is – have varied perceptions of import attached to them. A patina of age can be just as crucial as actual longevity. Many of the distinctions and priorities of conservation practice prove themselves to be different to how people actually relate to places, which is through a diversity of social, memory and experiential practices (Sleight, 2018).

The principle of inherency is a legacy of historical conservation approaches and international heritage charters. The most influential source of this principle is the Venice Charter (1964). The founding document of ICOMOS, the Venice Charter was a European mid-twentieth-century modernist document containing a series of stipulations for the conservation of monuments (Hardy, 2008). The scholarship on the Venice Charter has identified that it assumed the intrinsic importance of the sites subject to its stipulations, and that it focussed almost exclusively on materiality and fabric (Glendinning, 2013; Jokilehto, 1999). Being a modernist endeavour, it assumed the possibility of wholly knowing, documenting and restoring heritage for and on behalf of the community by experts who adopted universal approaches. The UNESCO World Heritage Convention of 1973 had a similar ideological basis. The absolute privileging of expertise in conservation arguably continues to the present day (Schofield, 2014; Wells, 2010).

The subject of conservation in the Venice Charter was the 'monument'. Traditionally, monuments were buildings and objects deliberately designed and built to commemorate events and people, typically in service of city or state. Alongside antiquities, relics and ruins, physical monuments had been the emphasis of conservation back to Western Antiquity, though overtly so since the nineteenth century. The 'monument' also had its critics. In 1903, Riegl (1996) problematised this term, suggesting the need to distinguish between 'deliberate monuments' and 'artistic and historical monuments'. The former related to the ancient remit of conservation, while the latter referred to an expanding range of heritage sites bound to creative and human endeavours – classical ruins, striking architecture, pleasing boulevards and other physical signposts representing the progress and achievements of human civilisation – which modern conservationists sought to protect. The adoption of the term 'monument' in the Venice Charter – as in the 1933 Athens Charter – was the conventional and convenient European shorthand for the focus of conservation during the first half of the twentieth century (if not much of the second half). By the 1970s, however, even the 'main author' of the Venice Charter, Raymond M. Lemaire, took issue with the term. Lemaire proposed the idea of 'urban ensembles' to better engage with historic cities, including in his own country of Belgium (Houbart, 2014). Referring to heritage only as 'monuments' was becoming problematic.

From monument to place

As in Europe and North America (cf. Harwood and Powers, 2004; Page and Mason, 2004), the 1960s–1970s were a tumultuous period for conservation in Australia (Davison and McConville, 1991). The influential Australian heritage movement – incorporating environmentalists, resident action groups, construction unions, architects and planners, policymakers, academics and many others – brought about sizeable shifts in the management and regulation of heritage. Emboldened by their success at home, at the fifth general assembly of ICOMOS in Moscow in 1978, Australian conservationists, and specifically architectural historian Miles Lewis, sought to have the Venice Charter revised to incorporate a broader conception of 'monuments' so as to embrace a wider range of contexts including Australia. Lewis (1985, 1990, 2011) perceived the conservation philosophy embedded within the Venice Charter as serving a narrow range of heritage sites: 'great monuments of stone like the temples, cathedrals and palaces of Europe' (Lewis, 2011). The Canadian delegation was in general agreement with Australia that regional contexts beyond Europe were inadequately addressed within international conservation guidelines (ICOMOS, 1978). After a four-hour debate, however, the delegates carried a motion that the Venice Charter was itself a 'monument' and would not be revised (cf. Erder, 1977). The concerns of the heritage periphery had seemingly fallen by the wayside in Europe.

The events in Moscow symbolised a broader concern among Australian conservationists that European practices inadequately addressed their local contexts (Walker, 2014). Australian conservationists believed their recognised historic environments – effectively dating from 1788 onwards following British colonisation – required a distinctive approach. Similarly, the Australian landscape was already understood as just as ancient as Europe (Maynard, 1979), with Indigenous people having lived and constructed settlements on the land for millennia. The Australian reworking of the Venice Charter was completed in 1979 and called The Australia ICOMOS Guidelines for the Conservation of Places of Cultural Significance, or the Burra Charter for short, after the pastoral town in South Australia where it was ratified. Unlike the Venice Charter – '[which] takes it for granted that we know what our historic monuments are, what makes them historic, and how we want to preserve them' (Lewis, 1990) – the Burra Charter proposed a systematic and rationalised approach to heritage management. It provided the flexibility to identify, assess and safeguard a broad range of sites, which would now be called 'places'. For these reasons, the Burra Charter has been described as a 'relativist' or 'postmodern' heritage charter (Glendinning, 2013; Shua, 2018). The Burra Charter also suggested that conservation should be undertaken sequentially: a place needed to be understood before any decisions about its future could be made. To understand a place required that its significance be assessed against specific cultural heritage values: aesthetic, historical, social, scientific and, subsequently, spiritual (Avrami et al., 2000). More so than its adoption of the moniker 'place' as the unit of conservation, the scholarship on the Burra Charter has interrogated its value-based approach.

As part of the symbolic breaking away from traditional European heritage philosophy, the Burra Charter, indeed, substituted 'monument' for 'place'. The moniker 'place' had been used by the Australian Government to define the 'national estate' since 1973–1975 (Lesh, 2019a), however the 1979 Burra Charter was the crucial evolutionary moment leading to the term's wider Australian and then international adoption. Place assumed increased importance in conservation against the backdrop of the humanistic turn in geography. By employing 'place', the Burra Charter sought to expand the remit of heritage conservation to embrace urban areas, regional towns, industrial and vernacular buildings, Indigenous heritage, colonial-era construction technologies, and even objects such as shipwrecks and statues. By doing so, conservation could address more kinds of places to which people related. Many of the Australian principles – particularly around the systematic assessment and protection of cultural heritage significance – have been adopted worldwide because they provide a rationalised means for conserving a diverse range of places (Jokilehto, 1999; Sonkoly, 2017). Furthermore, the Burra Charter itself would not be a 'monument': it has been revised four times to date.

The Burra Charter has not been without its critics. Heritage scholars suggest it is constitutive of the 'Authorised Heritage Discourse' (Waterton et al., 2006). The Burra Charter, after all, is largely consistent with traditional Western conceptions of heritage – it was only ever intended as an Australian reworking of the Venice Charter. It has also provoked strong reactions in Europe. German architectural historian Michael Petzet (2009: 10) commented that Australians 'avoid the term monument just like the devil shuns the holy water'. For traditionalists such as Petzet, the Australian conception for the heritage place provided an ill-defined remit for the practice of conservation. Having done away with architectural and historical connoisseurship as conservation benchmark, the range, scope and scale of heritage places had become seemingly unlimited, making the determination of precisely where, when and to what extent practitioners should intervene more variable than in the past. Moreover, the European conception of the 'monument' not only expanded in the 1970s – for instance, through Lemaire's interventions (for urban ensembles) – but the Europeans would never have agreed the term was as narrow as the Australians imagined when they rejected it for 'place'.

Understanding place

Conservation practice engaged with ideas of place long before the Burra Charter, and it has continued to do so both within and separate from the terms of the Burra Charter. Importantly, conservation takes different forms depending on the kind of place being addressed, as well as the national and local context. Some efforts have been made within the academic field of heritage studies to examine the meanings of 'place' in order to introduce more critically engaged approaches into practice. Relationships between heritage, place and people appears within the longer history of conservation, dating back to Western Antiquity. Silberman (2015: 30) has traced 'the evolving social role of heritage places, from their initial roles as sites for pilgrimage and ritual to their formalization as national institutions [in] the early nineteenth century, to their multicultural context in the early twenty-first century'. Silberman surmises that societies' changing conceptions of themselves, and what they value for the future, becomes constitutive of the heritage place. Historically and theoretically informed approaches to place encourage us to re-think the idea of what a heritage place might be in the future as community priorities and aspirations for conservation continue to change.

After the Burra Charter, the changing priorities of and approaches to conservation have been reflected in more recent international heritage charters. Overlapping interactions between tangible and intangible heritage were made explicit in, for instance, the Nara Document on Authenticity (ICOMOS, 1994), the Declaration of San Antonio (ICOMOS, 1996), the Convention for the Safeguarding of the Intangible Cultural Heritage (UNESCO, 2003) and the Québec City Declaration on the Preservation of the Spirit of Place (ICOMOS, 2008). Using the example of shrines and temples in Asia, Byrne (2014: 97) has written that adopting the 'Venice Charter principle that authenticity resides in original fabric rather than in the traditional [Asian] practice of building and rebuilding' means erasing heritage significance. Across the world, the relationship between people and place as mediated through conservation becomes ruptured through a singular devotion to historic fabric. As with the Burra Charter in the Australian context, the Nara Document was intended to pluralise heritage authenticity for Asia, albeit not specifically for place.

In his examination of heritage charters, Wells (2010: 468) has suggested three main ways that 'place' has been conceived within the field: 'spirit of place', 'sense of place' and 'place attachment'. Each of these conceptions for place has its own associated academic literature and methodology for inquiry. Each also has its own historical and theoretical lineage. The 'spirit of place' – genius loci – has its origins in Roman beliefs and rituals, overlapping with Silberman's timeline of the heritage place. Place attachment has its basis in 1970s' quantitative social sciences. It still resonates within urban studies because it involves the definition and measured study of the interaction between people and place (Lewicka, 2011). The most influential conception for place within conservation has related to 'sense of place' (Schofield and Szymanski, 2011), which can be demonstrated in how the Burra Charter has been applied, for instance.

Place was originally defined in Article 1 of the 1979 Burra Charter as any 'site, area, building or other work'. This definition has been reworked to address the expanding remit of location-based heritage conservation. Article 1.1 of the 2013 Burra Charter states: 'Place means a geographically defined area. It may include elements, objects, spaces and views. Place may have tangible and intangible dimensions'. An explanatory note for Article 1.1 continues:

> Place has a broad scope and includes natural and cultural features. Place can be large or small: for example, a memorial, a tree, an individual building or group of buildings, the location of an historical event, an urban area or town, a cultural landscape, a garden, an industrial plant, a shipwreck, a site with in-situ remains, a stone arrangement, a road or travel route, a community meeting place, a site with spiritual or religious connections.

Under the terms of the Burra Charter, place engages with a variety of contexts and environments. Nonetheless, place appears to be defined in the Burra Charter only by precedent. Prior instances of conservation serve as a template for defining future heritage. The use of precedent makes heritage processes more predictable, but also has the potential to stifle innovative thinking, making heritage less responsive to community expectations. Moreover, the suggestion that a place may but does not necessarily have 'tangible and intangible dimensions' reflects the Burra Charter's historical ties to the Venice Charter and its fabric-focussed conception for heritage. As we know, place cannot be convincingly demarcated in this way.

To examine the extent to which conservation, and specifically the Burra Charter, tends towards the 'sense of place' definition, therefore, requires reference to the grey literature. Discussions in heritage have engaged with a variety of writers on place including Kevin Lynch, Jane Jacobs, Yi-Fu Tuan, Edward Relph and David Lowenthal. Focussing on the social-scientific dimensions of place, this generation of writers suggested the possibility of wholly knowing a place through detailed quantitative and qualitative study. Their ideas were circulating in conservation circles in the 1970s–1980s when the UNESCO World Heritage Convention and the Burra Charter were developed. The ways their ideas were introduced into practice did not necessarily lead to a sustained engagement with communities, however, and the emphasis during this period remained fabric as part of conscious efforts to produce systematic, predictable and expert-driven heritage processes (Davison and McConville, 1991).

More recent attempts have been made to introduce refreshed ideas of place into conservation. The 1990s–2000s writings of, for instance, John Agnew, John Tunbridge, Gregory Ashworth, Dolores Hayden, Setha Low and Doreen Massey have had some influence in practice. Practitioners again looked to this cutting-edge literature on place. In the Australian context, this happened as part of explorations of the role of social value and community perspectives in conservation (Lesh, 2019b; Garduño-Freeman, 2018). Drawing on similar ideas, the 2008 Québec City Declaration is the current exemplar for understandings of place within heritage because it states:

> the spirit of place is made up of tangible (sites, buildings, landscapes, routes, objects) as well as intangible elements (memories, narratives, written documents, festivals, commemorations, rituals, traditional knowledge, values, textures, colors, odors, etc.), which all significantly contribute to making place and to giving it spirit.
>
> *(ICOMOS, 2008)*

The Québec City Declaration seeks to overcome the tangible and intangible heritage divide. As with many of these charters, the longer-term impacts for understandings of conservation remain to be seen. After all, the 'spirit' of a place may well be perceived as not necessarily fundamental to conserving a place. Furthermore, to again demarcate place into its intangible and tangible aspects arguably re-produces this artificial division rather than overcoming it. Nonetheless, the Québec City Declaration (along with the recent scholarship highlighted in this chapter) opens the possibility for conservation to embrace richer conceptions of place.

Place prospects

Current heritage scholarship takes the interactions between heritage, place and people to be less fixed, measurable and predictable than in the past. Today, innovative practitioners strive to employ ideas that emphasise the dynamism, contingency and experience of places. While communities and civil society have played an important role in conservation since at least the 1960s–1970s, their relationship to place is again being examined. The socially constructed meanings of place have increasing importance in conservation (Wells, 2010). The scholarship also suggests that heritage places might be conceived of as a social assemblage of people, things and meanings (Harrison, 2018; Pendlebury, 2013). This results in the heritage place being understood as always in a process of becoming, continuously re-imagined and re-created through interactions between at once material and immaterial actors and elements. The affective, emotional and experiential aspects of historic places are also being brought to the fore in safeguarding heritage places and the urban studies place-making agenda is making inroads into conservation (Madgin and Lesh, forthcoming).

At the same time, conservation always operates as part of broader urban, social and economic processes. Architectural and urban planning paradigms exert a strong influence on practice. Conservation also happens at places that are both publicly and privately owned, sites impacted by the property market and the development industry, and so outcomes are always swayed by many stakeholders and interests. The inability of conservation to satisfactorily embrace the concept of place, to date, has not been caused by a lack of sophisticated articulations for the concept within academic and practitioner circles. Rather, the vagaries of place are difficult to reconcile within property boundaries, particularly when irreconcilable demands for the future of a place are circulating. Conservation is often contested and different groups – whether practitioners, policymakers, developers, civil society and so forth – have conflicting perspectives on how heritage and conservation should be done. An achievement of the Burra Charter was to empower practitioners to reflect on the spatial context of their activities, and to empower them to engage with a wider scope and scale of sites.

To integrate current social priorities and critical concerns within dominant rationalised conservation practice will prove challenging, especially while heritage is still dogged by the tangible and intangible tautology. As this chapter has demonstrated, conservation is historically contingent and takes considerable time to evolve. There will always be a lag between heritage thought and practice. The role of heritage studies becomes to, at once, question and strengthen the pragmatic practice of conservation. The global circulation of ideas around place and heritage, and attempts to integrate these ideas into practice, is evident. The conservation of place is also clearly locally specific as different societies and cultures wrestle with the most appropriate modes for meaningfully protecting places for people. The concept of place reminds us that heritage value can never truly be contained within the physical boundaries of a site. Place indeed proves itself to be an essential building block of heritage. Its provocations have the potential to enhance the relationship between conservation and people.

Acknowledgements

The author thanks Cameron Logan and Rebecca Madgin for their helpful comments on this chapter.

References

Aplin, G. (2002), *Heritage: Identification, Conservation, and Management*, Melbourne: Oxford University Press.

Australia ICOMOS (1979, 1981, 1988, 1999, 2013), "Australia ICOMOS Charter for the Conservation of Places of Cultural Significance ("Burra Charter")".

Avrami, E., Mason, R. and de la.Torre, M. (2000), *Values and Heritage Conservation*, Los Angeles: Getty Conservation Institute.

Bandarin, F. and van Oers, R. (2012), *The Historic Urban Landscape: Managing Heritage in an Urban Century*, Hoboken: Wiley.

Byrne, D. (2014), *Counterheritage: Critical Perspectives on Heritage Conservation in Asia*, New York: Routledge.

Byrne, D., Brayshaw, H. and Ireland, T. (2003), *Social Significance: A Discussion Paper*, Sydney: NSW National Parks and Wildlife Service.

Craith, M.N. and Kockel, U. (2015), "(Re-)Building heritage", in Logan, W., Craith, M.N. and Kockel, U. (Eds.), *A Companion to Heritage Studies*, Chichester: John Wiley & Sons, Inc, pp. 27–40.

Crang, P. (2010), "Cultural geography: after a fashion", *Cultural Geographies*, 17(2): 191–201.

Davison, G. and McConville, C. (Eds.). (1991) *A Heritage Handbook*, Sydney: Allen & Unwin.

Delafons, J. (1997), *Politics and Preservation: A Policy History of the Built Heritage, 1882-1996*, London: E. & F.N. Spon.

Erder, C. (1977), "The Venice Charter under Review", *Journal of Faculty of Architecture, METU, Ankara*, 25: 24–31.

Fairclough, G., Harrison, R., Jameson, J. and Schofield, J. (Eds.). (2008), *The Heritage Reader*, London: Routledge.

Garduño-Freeman, C. (2018), *Participatory Culture and the Social Value of an Architectural Icon: Sydney Opera House*, New York: Routledge.

Gibson, L. and Pendlebury, J. (Eds.). (2009), *Valuing Historic Environments*, Farnham: Ashgat.

Glendinning, M. (2013), *The Conservation Movement: A History of Architectural Preservation: Antiquity to Modernity*, London: Routledge.

Hardy, M. (Ed.). (2008) *The Venice Charter Revisited: Modernism, Conservation and Tradition in the 21st Century*, Newcastle upon Tyne: Cambridge Scholars.

Harrison, R. (2013), *Heritage: Critical Approaches*, London: Routledge.

Harrison, R. (2018), "On Heritage Ontologies: Rethinking the Material Worlds of Heritage", *Anthropological Quarterly*, 91: 1365–1383.

Harwood, E. and Powers, A. (Eds.). (2004) *The Heroic Period of Conservation*, London: Twentieth Century Society.

Houbart, C. (2014), "Deconsecrating a Doctrinal Monument: Raymond M. Lemaire (1921–1997) and the Revisions of the Venice Charter", *Change Over Time*, 4(2): 218–243.

ICOMOS. (1978). "Compte rendu resume de la Vième Assemblée Générale de l'ICOMOS", *Moscow-Souzdal: ICOMOS*, 22 May 1978, 10–17.

ICOMOS. (1994). *Nara Document on Authenticity*. Nara, Japan: ICOMOS.

ICOMOS. (1996). *Declaration of San Antonio*. San Antonio, USA: ICOMOS.

ICOMOS. (2008). *Québec City Declaration on the Preservation of the Spirit of Place*. Québec, Canada: ICOMOS.

Jokilehto, J. (1999), *History of Architectural Conservation*, Oxford: Butterworth-Heinemann.

Lesh, J. (2019a), "The National Estate (and the city), 1969–75: a significant Australian heritage phenomenon", *International Journal of Heritage Studies*, 25(2): 113–127.

Lesh, J. (2019b), "Social Value and the Conservation of Urban Heritage Places in Australia", *Historic Environment* 31(1): 42–62.

Lewicka, M. (2011), "Place attachment: How far have we come in the last 40 years?", *Journal of Environmental Psychology*, 31(3): 207–230.

Lewis, M. (1985), "A Regional Conservation Manifesto", *UNESCO (Australia) Review*, 10: 20–24.

Lewis, M. (1990), "Philosophy of Restoration", Australian Council of National Trusts (Ed.), *Heritage and Conservation: The Challenges in the Asia/Pacific Basin: Conference Papers*, Canberra, Australian Council of National Trusts.

Lewis, M. (2011), "Oral History (for Australian ICOMOS, National Library of Australia)". Interview by Bronwyn Hanna.

Logan, C. (2017), *Historic Capital: Preservation, Race, and Real Estate in Washington, D.C*, Minneapolis: University of Minnesota Press.

Madgin, R. and Lesh, J. (Eds) (forthcoming), *Methodologies for Exploring Emotional Attachments to Historic Urban Places*, London: Routledge.

Madgin, R., Webb, D., Ruiz, P. and Snelson, T. (2018), "Resisting relocation and reconceptualising authenticity: the experiential and emotional values of the Southbank Undercroft, London, UK", *International Journal of Heritage Studies*, 24(6): 585–598.

Maynard, L. (1979), "The archaeology of Australian Aboriginal art", Mead, S.M. (Ed.), *Exploring the Visual Art of Oceania: Australia, Melanesia, Micronesia, and Polynesia*, Honolulu: University Press of Hawaii, pp. 83–110.

Page, M. and Mason, R. (Eds.). (2004) *Giving Preservation a History: Histories of Historic Preservation in the United States*, New York: Routledge.

Pendlebury, J. (2013), "Conservation values, the authorised heritage discourse and the conservation-planning assemblage", *International Journal of Heritage Studies*, 19(7): 709–727.

Petzet, M. (2009), *International Principles of Preservation*, Berlin: ICOMOS and Bäßler.

Riegl, A. (1996), "The modern cult of monuments: its essence and its development (1903)", Nicholas Stanley Price, Mansfield Kirby Talley and Alessandra Melucco Vaccaro (Eds..), *Readings in Conversation: Historical and Philosophical Issues in the Conservation of Cultural Heritage*, Los Angeles: Getty Conservation Institute, pp. 69–83.

Schofield, J. (Ed.). (2014) *Who Needs Experts?: Counter-Mapping Cultural Heritage*, Farnham: Ashgate.

Schofield, J. and Szymanski, R. (2011), "Sense of Place in a Changing World", Schofield, J. and Szymanski, R. (Eds.), *Local Heritage, Global Context: Cultural Perspectives on Sense of Place*, Farnham: Ashgate, pp. 1–12.

Shua, Y. (2018), "Bridging Positivist and Relativist Approaches in Recent Community-Managed Architectural Conservation Projects in Singapore", *Sojourn: Journal of Social Issues in Southeast Asia*, 33 (3): 647–676.

Silberman, N. (2015) "Heritage Places: Evolving Conceptions and Changing Forms", In Logan, W., Craith, M. and Kockel, U. (Eds.), *A Companion to Heritage Studies*, Chichester: Wiley, pp. 27–40.

Sleight, S. (2018), "Memory and the city", Maerker, A., Sleight, S. and Sutcliffe, A. (Eds.), *History, Memory and Public Life: The Past in the Present*, London: Routledge, pp. 126–158.

Smith, L. (2006), *Uses of Heritage*, New York: Routledge.

Smith, L. and Akagawa, N. (Eds.). (2009) *Intangible Heritage*, London: Routledge.

Smith, L. and Campbell, G. (2015), "The elephant in the room", Logan, W., Craith, M. and Kockel, U. (Eds.), *A Companion to Heritage Studies*, London: Wiley, pp. 443–460.

Sonkoly, G. (2017), *Historical Urban Landscape*, Cambridge: Palgrave Macmillan.
Sullivan, S. (2015), "Does the practice of heritage as we know it have a future?", *Historic Environment*, 27 (2): 110–117.
Tait, M. and While, A. (2009), "Ontology and the conservation of built heritage", *Environment and Planning D: Society and Space*, 27(4): 721–737.
UNESCO. (2003). *Convention for the Safeguarding of the Intangible Cultural Heritage*. Paris: UNESCO.
Walker, M. (2014), "The development of the Australia ICOMOS Burra Charter", *APT Bulletin*, 45(2/3): 9–16.
Waterton, E., Smith, L. and Campbell, G. (2006), "The utility of discourse analysis to heritage studies: The Burra Charter and social inclusion", *International Journal of Heritage Studies*, 12(4): 339–355.
Wells, J. (2010), "Our history is not false: perspectives from the revitalisation culture", *International Journal of Heritage Studies*, 16(6): 464–485.
Winter, T. (2014), "Beyond Eurocentrism? Heritage conservation and the politics of difference", *International Journal of Heritage Studies*, 20(2): 123–137.

Part V

Displacement, loss and emplacement

Uma Kothari

Introduction: displacement, loss and emplacement

In considering displacement and emplacement, the notion that place is better understood as an event rather than a thing is reinforced (Howitt, 2002). More specifically, these processes conjure up ideas of movement and mobility, issues of especial pertinence today as we witness a humanitarian generational catastrophe, with millions of people being displaced and compelled to move because of conflict, violence and persecution in their places of origin. Strikingly, over half of Syria's pre-war population of 22 million have become displaced and, at a global level, nearly 60 million people have been forcibly moved from their places of origin. Regular news bulletins, television documentaries, newspaper articles, radio programmes and all forms of social media continuously depict and discuss these processes of displacement.

Loss and abandonment are central to displacement. As people are compelled to leave a place – a place that may subsequently become abandoned – they are likely to experience a sense of loss of community, as well as possessions, homes, histories and identities. Yet, displacement also plays a role in the creation of new identities such as that of refugee or of citizen (Ballinger, 2015).

The journeys of displaced people have attracted much attention, with large numbers of refugees dying along journeys as they seek safety. This focus on the journey has challenged characterisations of migration as constituting a move from one fixed place to another. When understanding motives for moving or attempting to map patterns of mobility there has been a tendency to think in terms of simplistic dichotomies, such as the push from a place characterised by conflict and the pull to another place that is envisaged as peaceful. Most movements, however, are more complicated and involve multiple migrations and journeys through various places often with no known destination. Thus, as Doreen Massey (1993) says, the emphasis should not only be on the places that people move from and to, but also on the inter-connectedness between places and people; for, as people are displaced, they not only create connections between places but are also a part of place-making processes.

Furthermore, as refugees pass through and arrive in different places after long, arduous and often traumatic journeys, issues of relocation, return, settlement and emplacement feature prominently. In this context, emplacement includes the social processes through which displaced individuals can build or rebuild networks of connection through which places become constituted (Glick Schiller and Çağlar, 2013). Accordingly, through displacement and emplacement, it becomes evident that a sense of place is not solely geographically bounded but is instead constructed out of a particular constellation of relations, through processes, encounters and connections.

It is not only refugees who are displaced and emplaced. Indeed, as Mimi Sheller and John Urry (2006) write, the world we live in can be best understood as being constituted by mobilities, not just of people but of finance, technology, ideas and media. Furthermore, as we see in this part, material objects are also accumulated, displaced and emplaced to create a sense of home and belonging, and such practices are also entangled in the sense of place created by non-humans. Where displacement is related to physical movement, studies largely focus on those who leave, where they leave and the places to which they go. Less emphasis has been placed on their emotional and virtual attachments to the places left behind (Malkki, 1995: 515). Additionally, displacement does not necessarily require physical movement from one place to another (Kalb, 2013). For example, people can become displaced as their social and economic environment changes such as when they are made unemployed or experience downward social mobility (Morell, 2014), and places can become abandoned as industries decline.

Given the particularity and almost everydayness of being-in-place, displacement and emplacement are significant for how place is conceptualised (Lems, 2016). Indeed, notions of displacement and emplacement can simultaneously de-essentialise notions of place and reify a bounded notion of place in implying movement from one particular place to another prescribed, fixed place. The notion of emplacement can signify spatial fixity; being in place. For Harri Englund (2002: 267), 'emplacement refers to a perspective in which the subject is inextricably situated in a historically and existentially specific condition, defined, for brevity, as a "place"'. Yet as Casey (1996: 39) writes, 'we are never anywhere, anywhen, but in place'. Bjarnesen and Vigh (2016: 13) suggest that the concept of emplacement emphasises a continuous process of embeddedness rather than the locus of such processes. In such a conception, emplacement implies a conceptual move away from place as location toward place as a process of socio-affective attachment, an engagement and entanglement with the lived environment and with processes of place-making.

Though they may appear opposites, whereby one implies being unsettled and compelled to move out of place while the other suggests being put in place, immobile and sedentary, there is a complex interplay between displacement and emplacement. They may also be conceived as part of a continuum, with displacement at one end (anomic, disempowering and disruptive) and emplacement at the other extreme (connoting rootedness and belonging). Yet, displacement and emplacement are mutually constitutive processes; they go hand in hand, since when that which is removed, it must be re-placed elsewhere at another time. For example, in some low-lying small island states, those who are displaced through the effects of sea-level rise must be resettled elsewhere, and yet these mobile connections between places are not uni-directional. As Arne (2015) shows, islanders displaced by coastal erosion experience multiple and repeated losses, forced movements and arrivals that result in 'circuits of displacement and emplacement'. Yet, while these 'circuits' can be repeated, they can also be ruptured and disrupted, each cycle being slightly different, playing a constitutive role in the ceaseless process of place-making. Thus, 'rather than propagating a free-floating placeless paradigm', these interconnected processes reflect how people 'come to be (dis)associated – and (dis)associate themselves – with or from place' (Jansen and Löfving, 2009: 6).

The chapters in this part collectively reflect these multiple dimensions of displacement and emplacement, underpin how people relate to notions of attachment, loss and abandonment, and how they are constituent of place-making processes. Chapters investigate how people, living non-humans and things can be displaced, moved, discarded and re-placed while continuously creating and transforming places. In exploring the subjectivities and spaces of those who are displaced, must keep moving, are regarded as out of place or considered threatening to the stability of place, these chapters discuss the homeless, refugees, migrants and non-humans as those caught up in such processes. Themes also focus on how places are reconfigured through abandonment and the re-ordering of place through processes of material accumulation and emplacement. Together, the chapters reveal the multi-scalar dimensions of displacements and emplacements from those who are exiled beyond national borders to things that are moved within a home and beyond.

The chapters reveal displacement and emplacement as simultaneously spatial and temporal processes. While the spatial aspects of place-making and of displacement and emplacement have received significant attention, the temporal dimensions have been less well detailed. Ali Madanipour addresses this lacuna in exploring the temporariness of place. He examines two aspects of temporary places, the movement of people and the change in places to illuminate how they are entwined in urban processes. He illustrates the particularities of time, space and meaning in the context of movement, and suggests that when people move to free themselves from various constraints they welcome the opportunity, whereas when they are forcibly moved and displaced they feel transient and vulnerable. Similarly, while change can bring flexibility and opportunities, when places are transformed beyond their control, a sense of displacement and disorientation may prevail.

Cathrine Brun's chapter also investigates the temporal dimensions of place, focussing on the experiences of refugees placed in camps. Some refugees experience prolonged, protracted and multiple displacements with repeated temporary emplacements or re-placements. As Brun shows, for refugees, theirs is a 'detached geography' caused by dislocation and social rupture from their homelands, from outside the camps, and from legal and social protection. Placed in camps, detention centres or guarded in holding places they are neither assimilated nor rejected but placed in a state of limbo. As such they become 'stuck' (Brun, 2016), fixed and stilled in place (Hyndman and Giles, 2011) so that they cannot escape from an undesirable place and future aspirations cannot be attained. Invoking the spatio-temporal dimensions of place, Brun explores the diverse and conflictual temporalities of place articulated by humanitarian aid workers and refugees, parties who envisage futures very differently. She concludes by reflecting on what kind of place the camp is; indeed, she considers whether it can be considered a place at all.

Despite being in 'limbo', refugees can also re-create viable, liveable places in camps, in the places in which they are resettled and while on the move. For those who are displaced from a familiar place, becoming emplaced is often associated with re-making a sense of home in order to mediate the emotional distance enforced by displacement (Ballinger, 2015). Bjarnesen and Vigh (2016) suggest that processes of place-making are linked to those of existential life-making. This is an issue addressed in Luis Eduardo Perez Murcia's chapter in which he illustrates how displaced people create and negotiate a sense of home on the move. Drawing on the experiences of internally displaced people in Colombia, he shows how home is not necessarily and only a physical place of shelter but encompasses embodied, affective and emotional attachments. He further suggests that while displacement often refers to a separation from a place that is known and described as 'homely', settling in a new place does not automatically invoke a sense of being at home. Thus, while the physical movement of displacement may come to an end, an existential condition of being displaced remains until people are able to remake a material, symbolic and affective place called home.

In their chapter on homelessness in Melbourne, Alison Young and James Petty examine how homeless individuals are prevented from investing places with the affective attachments that are often associated with houses. Yet they show that since those with no houses must find shelter and a place to store their possessions, they find, adapt and use various urban places in which to carry out the kinds of activities often associated with life in a place called home. In considering homeless people's relationship to place, they reveal these homes to be temporary and contingent. Specifically, they show how the arrangement and placing of their bodies as well as their possessions are significant in this process of home-making in urban places that are not houses.

As Alison Young and James Petty's chapter shows, things are displaced and emplaced in processes of home-making. For Bjarnesen and Vigh (2016: 12), emplacement includes the 'settling, mounting, situation, or location of an object'. These objects can be understood as matter-in-place and matter out-of-place. For Pikner and Jauhiainen (2014), matter-in-place refers to things that have a socially-defined 'correct' place and are in that place. In contrast, matter-out-of-place signifies disposed objects that have no place or are in an incorrect place or that are in an appropriate location but of which no use can be made. Tracey Potts' chapter examines the accumulation and ordering of stuff and the place-making qualities of accumulated things. While Young and Petty show how the possessions that homeless people carry with them are minimal, often lost, displaced or discarded, the stuff that is needed to make a home in a fixed place like a house can be overwhelming, overflowing and simply too much. Tracey Potts refers to this kind of stuff as clutter, 'wild things that disturb domestic order by straying beyond their bounds'. She considers the placing, displacing and replacing of things that she suggests are necessary to create a home but at the same time notes that they can overwhelm, challenging the order of place. She focusses on three senses of place, the orderly, homely and material to highlight the dynamics involved in place-making.

This straying in and out of bounds is also explored in the chapter by Catherine Phillips and Sarah Robertson in which they focus on human and non-human interactions. They examine these interactions in a city park, suggesting that urban realms may constitute key sites for discussions about how non-humans challenge human-centric ideas of place and boundaries. They argue that there has been very limited research on non-humans and how they relate to place and that we need to consider more seriously how they are co-creators, shaping the ways in which humans experience and understand places. Challenging human-centric ideas, they develop a conception of non-human place that questions and experiments with research methods, engages with Indigenous scholarship and entails ethical and political negotiations. They trouble human and non-human dualisms by showing how non-humans in the park are seen as 'out-of-place' yet are also at 'home' and in place, while humans might be conceived conversely as out of place or confined and controlled. They conclude by contemplating how cities might be built and organised in ways that can simultaneously accommodate the well-being of humans and non-humans, and how a notion of place can be wrought through examining the placings and places of living non-humans in urban spaces.

Taken together, these chapters illuminate how displacement and emplacement are founded upon notions of place-attachment and reflect a sense of loss and abandonment. Patrick Devine-Wright illuminates emotional, symbolic and affective dimensions of people's thinking about and feeling for places. Here, place attachment refers to the meanings and emotional bonds that are felt by individuals and groups towards specific places. In exploring scholarly research in this area, he considers place attachments that are characterised by negative feelings or ambivalence before examining how people form attachments to multiple types of places at diverse scales. He further explores the dynamics of place attachments, suggesting how a relatively static phenomenon, emphasising stability and continuity in relations with place, are continually unfolding over time just as places themselves are in a continual process of becoming. He also examines the temporality and intensity of place attachments whereby the strength of the attachment can vary between people and over time. Finally, through elaborating on the diversities of place attachment he shows how people can have attachments to multiple places at the same time.

While Devine-Wright explores the multiple dimensions of attachment to place, Justin Armstrong focusses on abandoned places. He explores long-lost villages in the Faroe Islands, forgotten settlements in Newfoundland and abandoned farms and factories on the Hornstrandir peninsula in Iceland to reveal how 'soon-to-be-invisible cultural landscapes get swallowed-up by a networked world gone global'. The chapter examines the cultural relevance and historical significance of the agency of ruins to reimagine these locations as dynamic, generative museums and sites of artistic, touristic and academic engagement. Through rethinking the social and cultural importance of ruins he provides a toolkit for developing an everyday ethnography of ruination, 'beyond the museum, beyond the curator, and beyond the edge of our known world'.

References

Arne, H., 2015, 'Leaving Lohāchāra: On circuits of emplacement and displacement in the Ganges Delta', *Global Environmental Change*, 8(1): 62–85.

Ballinger, P., 2015, 'Borders and the rhythms of displacement, emplacement and mobility', T. Wilson and H. Donnan (eds), *The Blackwell Companion to Border Studies*, Oxford: Blackwell, pp. 389–404.

Bjarnesen, J. and H. Vigh, 2016, 'The dialectics of displacement and emplacement', *Conflict and Society: Advances in Research*, 2: 9–15.

Brun, C., 2016, 'There is no future in humanitarianism: Emergency, temporality and protracted displacement', *History and Anthropology*, 27(4): 393–410.

Çağlar, A., and N. Glick Schiller, 2011, 'Introduction: Migrants and cities', In N. Glick Schiller and A. Çağlar, (eds), *Locating Migration: Rescaling Cities and Migrants*, Ithaca, NY: Cornell University Press, pp. 1–22.

Glick Schiller, N., and A. Çağlar, 2013, 'Locating migrant pathways of economic emplacement: Thinking beyond the ethnic lens', *Ethnicities*, 13(4): 494–514.

Howitt, R., 2002, 'Scale and the other: Levinas and geography'. *Geoforum*, 33(3): 299–313.

Lems, A., 2016, 'Placing displacement: Place-making in a world of movement', *Ethnos*, 81(2): 315–337.

Massey, D., 1993, 'Power-geometry and a progressive sense of place', In J. Bird, B. Curtis, T. Putnam and L.Tickner (eds), *Mapping the Futures: Local Cultures, Global Change*, London: Routledge, pp. 60–70.

Pikner, T. and J. Jauhiainen, 2014, 'Dis/appearing waste and afterwards', *Geoforum*, 54: 39–48.

Sheller, M. and J. Urry, 2006, 'The new mobilities paradigm'. *Environment and Planning A*, 38(2): 207–226.

39

Temporary places

Moving people and changing spaces

Ali Madanipour

This chapter explores the temporariness of places. First, the triangular relationship between time, space and meaning in temporary places is introduced, showing how their various combinations are characterized by particularity and contingency. The chapter then examines two forms of temporariness of places: when people move and when places are altered. Both of these forms of temporariness, of being in a place and the changing composition of a place, are short-lived arrangements and experiences. Temporary places are events in space and time, which characterize many urban experiences, in a wide range of circumstances and with completely different meanings and impacts (Madanipour, 2017a), which would depend on the social position and disposition of those involved (Bourdieu, 2000). These forms of temporariness are associated with conditions ranging from opportunity and emancipation to vulnerability and precarity.

Particularity and contingency

The increasing attention to place has been closely associated with the rise of a humanistic approach and a phenomenological perspective, approaches which have looked for meaning and value (Cloke and Johnston, 2005; Holloway et al., 2003). The modernist attitude, which prevailed in the mid-twentieth century, advocated a concept of space as an open, abstract expanse, which could be carved out for any required functions (Le Corbusier, 1987; Lefebvre, 1991). On this basis, vast urban development projects changed the cities around the world. In response to what appeared as cold, mechanistic functionalism and rapid change, displacing large numbers of people in the name of improvement, the humanistic notion of place indicated an emphasis on specific locations and personal and communal relations (Clark, 1985; Gans, 1968; Jacobs, 1961). Place was thus subsequently seen as a centre of 'felt value', offering security and stability, in contrast to the exposure and alienation of the undifferentiated space (Tuan, 1977: 3–6). Modernist space encouraged movement and change (Giedion, 1967), whereas humanistic place provided repose and continuity.

However, the temporariness of place tends to undermine this desire for stability, as it is not merely domesticating space into familiar places, but it also destabilizes these familiar places by making them a stopover in a journey, rather than the resting place of a home. If the meanings of the word place include home, status, location, and position, the temporariness of place undercuts the sense of stability and security that is attached to these notions.

A defining feature of 'place' is its spatial particularity, referring to a particular location, which distinguishes it from the generality of 'space'. This is evident in the everyday meaning of the word, as defined by the *Oxford English Dictionary*: 'A particular position, point, or area in space; a location'. It is also evident in the more specialist language of spatial fields, such as geography and planning (such as Cloke and Johnston, 2005; Healey, 2010; Holloway et al., 2003; Massey, 2005). While the space–place distinction is widely discussed, their relationship with time also requires attention, especially when their temporality fundamentally changes their character. By adding the adjective 'temporary', the characteristic spatial particularity of place is extended to its temporal dimension. Although spatiality is never without temporality, in a temporary place we have a dual process of spatial and temporal delineation, creating a short-lived spatial arrangement, which is not only a direct opposite of the idea of timeless, abstract space, but also at odds with the enduring and reassuring sense of place. A temporary place, therefore, becomes defined by its spatial and temporal particularity: a particular space at a particular time.

In addition to particularity in space and time, the notion of place also indicates particularity in meaning for a person or a group. While space is abstract and apparently devoid of emotional content, place is thought to be concrete and 'meaningful' (Tuan, 1977). Place is often linked with this particularity of meaning, which includes its spatial and temporal specificity, but also other forms of delineation of its attributes. When a place is thought to be temporary, these meanings may be unsettled, short-lived and open to frequent changes. Particularity of meaning, moreover, does not reduce the place to a single account, but can include a range of meanings for people who are related with the place. Each person or group may have a different idea and experience of the place, and so it may not be limited to a single 'genius loci' but would afford multiple interpretations from different perspectives. Temporariness of place would add an additional sense of contingency to the multiplicity of these meanings, with far-reaching consequences. A temporary place, therefore, signifies particularity in space, time and meanings, characterized by multiplicity and contingency.

The notion of place indicates a triangular relationship between time, space and meaning. The relationship between space and meaning creates a place. In other words, place is the outcome of attributing meaning to space, which gives it a particular character. This particular character, however, is not singular, as different perspectives generate different meanings for a place, which undermines the expectation that the emphasis on place can offer a sense of harmony and clarity. Moreover, the relationship between space and meaning is inevitably temporal. Temporariness, however, fragments and accelerates the dimension of time: it points to short-lived events and experiences. As such, it undermines the sense of endurance and stability that is assumed to inhere in the idea of place. A temporary place, therefore, displays multiple accounts of short-lived events and fragmented experiences, revealing its intrinsic contingency and plurality, and the potential tensions and incompatibilities between these three forms of particularity (space, time, meaning).

A paradox, therefore, lies inside the temporariness of place. On the one hand, it emphasizes the particular against the general and the universal, focussing on the concrete, affective and experiential as against the abstract, remote and rationalistic. It offers familiarity and continuity, evoking a sense of enduring calm, with a degree of permanence, as distinctive from the ever-changing face of the space, a place of identity in the space of flows (Castells, 1997). It indicates a sense of community (Tönnies, 1957) and a desire for rooted meaning in a sea of alienation and rootlessness (Heidegger, 1978; Norberg-Schultz, 1980).

On the other hand, temporariness of place reveals a sense of contingency and precarity, ignoring and negating the promises of continuity and identity. What was hoped to be achieved by putting the emphasis on place seems to evaporate as soon as it is exposed to the particularity of time, and challenged by the multiplicity of perspectives, relations and meanings. While the definition of temporary places shows an apparently harmonious combination of particularity in space, time and meaning, a closer look reveals a clash of spatiality, temporality and relationality. While space is tamed into place, temporariness challenges this domesticity, opening it to the turbulent tides of diversity and change.

How can this paradoxical triangle be analyzed and understood? What are the causes, conditions of possibility and consequences of these added dimensions of temporal contingency and relational multiplicity to that concrete spatial rootedness that was thought to offer security and endurance? These questions are addressed in the following section.

When people move

In the broadest sense, all experiences may be considered to be temporary, when seen in the context of long historical scales, whereby persons and places are no more than short-lived events. An entire human life may be considered to be a temporary existence, a short interval between birth and death. Even the life of cities, nations and civilizations may appear as brief moments in the long history of the earth, where nothing is ultimately permanent. This sense of transience has occupied much of human thought over centuries, as expressed in the sciences, arts, philosophy and religion. Observing the flux of the material world, philosophers such as Heraclitus have seen the world as continually changing like the water flowing in a river (Plato, *Cratylus*). In the face of death, poets have reflected on the ephemerality of life (Omar Khayyam); philosophers have articulated the sense of anxiety that this confrontation generates (Kierkegaard and Heidegger) or the opportunities that it offers (Sartre); and religions have promised a return at a higher status (for instance, Buddhism) or a happy eternal life after death (for example, Abrahamic religions) to compensate for this ephemeral existence. In the face of climate change, scientists have warned about the fragility of the planet as a whole and of the conditions that make life possible (IPCC, 2014). At this mega-scale, all places and experiences are existentially temporary, from the level of a single human's lifespan to the entire species, and even to the world as a whole.

The conditions of temporariness may also be existential at the social level, in the sense of being a way of life that is integral to the identity of a population. It may be a way of life shaped by the conditions over which individuals and groups may not have much control, conditions that are embedded in environmental circumstances, as well as in habits, traditions and social institutions. For most of human history the life of hunters and gatherers was lived in temporary places. With limited environmental resources, these foragers had to move from one place to another in search of food, as no alternative way of life was possible or conceivable (Bouquet-Appel, 2011). Even after the advent of agriculture and the settled life of farmers and artisans in towns and villages, moving in search of food was continued by nomadic tribes, who followed the seasons looking for new pastures.

A place can be temporary from the perspective of a person or group who, after a short stay, leaves it by choice or by force. In this case, the place is a temporary one for those who leave but may be more enduring for the others who remain. Temporariness applies to the person and not to the place but, as far as that person is concerned, the place has been experienced for a short period of time and therefore it has been a temporary place. The memory of the place may linger in the mind of the leaver and strong emotional attachments may have been developed, but the experience of being in the place is, nevertheless, a temporary one, an event in the past. An example of this case is a traveller for whom the journey is a series of brief experiences on the way, a sequence of temporary places. There is, however, a distinction to be made by those who have to move and those who want to move; for the former the temporariness of a place is a loss and an imposition, for the latter it is an opportunity and an emancipation.

Nomadism implies temporal and spatial flexibility, while in fact it is a different condition of stability, as nomads have repetitive patterns of movement according to the rhythms of seasons and within predetermined territories. As Deleuze and Guattari (1987: 482) advance in quoting Toynbee, nomads do not really move. Nomads are bound to temporal and spatial frameworks, which may be more flexible than settled farmers and town dwellers, but are nevertheless still fixed. After all, nomads have clear spatial and temporal routines, which they have repeated for thousands of years, moving along paths that are well trodden, which are only altered when they face major challenges. The nomad becomes the image of a desire for existential permanence that has adjusted itself to changing and precarious circumstances.

For travelling people, the meaning of place is fundamentally different from that of settled populations. Unlike the locals, who are characterized by their potentially deep emotional and material links to a particular location, the stranger is seen to be free from these connections, and therefore suspected to be without a commitment to the common values and causes. This is why the nomad is perceived by the villager as a threat and treated with hostility and suspicion, a tension which continues even in contemporary cosmopolitan urban life, where the Roma, the traveller, the refugee and the immigrant may be treated as unwelcome outsiders, disconnected from the place and unable to grasp its 'true' meanings. This is how the 'strangers' are identified and how they are treated as temporary people, socially excluded from the rest (Madanipour, 2016).

Modern urban life is itself a form of nomadism. Technologies of transportation, from the railways onwards, have expanded urban space, leading to a mobile lifestyle for many inhabitants of large cities, who may take for granted the necessity of travelling long distances and for long hours in search of their basic needs. Globalization, structural economic change and the technologies of information and communication have hastened the growth of larger cities (European Commission and UN-Habitat, 2016). The combination of movement and speed, which was so much an inspiration for the modernists (Giedion, 1967; Le Corbusier, 1987), has created a sense of perpetual transience in large urban areas. This transience is visible in central areas, in urban gateways such as railway stations, ports and airport, and in the business and leisure districts, where large numbers gather for work and entertainment. The accelerated tempo of life, meanwhile, has reduced the contact between people and places, many becoming no more than a detached gaze or a fleeting encounter. This transience is reflected in deeper levels still, in a growing trend of insecure, short-term forms of employment and housing, particularly for the young, in the process of casualization that is integral to what is termed the gig-economy. Nomadic urbanity shapes the experience of places, in which all urbanites are perpetual travellers, for whom most places are temporary places (Madanipour, 2017a).

Temporariness of place reflects opportunities and threats, bringing freedoms and precarities. When temporariness of place is thought to be an existential condition for a group, from which no escape is possible, they become accustomed to it, learning to live with these possibilities and limitations. However, when these limitations are thought to have been imposed by others, trapping people in unfavourable circumstances, the desire for emancipation arises. The notion of place can be an extremely conservative concept, allocating a social position and status to individuals and groups, who are then expected 'to know their place' and behave according to their station. On this basis, a hierarchical system of stratification would be built, whereby people were born and brought up within a social place, unable to leave it at any time. The ancient castes and Victorian social classes were some of the prime examples of this rigid structure. For those who rebelled against these systems, a place in the structure was not permanent and the accident of birth was not a determinant of life; it had to be a temporary place from which they could escape. When places are oppressive, temporariness of place is imagined as a window into the future, a hopeful promise of liberation and change, treating all current arrangements as temporary and contingent.

Even if there is no rebellion or revolution, there may be a quiet resistance, with the expectation to leave a place and move to a different, better one. This is common among many young people, who do not imagine that they will remain in the same place, even if it appears to be problem-free, as it would trap them in a particular circumstance, preventing them from experiencing and exploring the world. The young would then wish to move out and up, either through rebellion against the oppressive places that are allocated to them, or by working through the normal course of events while holding an expectation for change. For example, young people who live in a shared apartment at one stage of their life may aspire to move on to a bigger place, a place of their own. For previous generations in the Anglophone world this has been a place in the suburbs, which appears to be increasingly losing its attractions or be out of reach for many, forcing them to live in a sequence of temporary phases and temporary places.

Temporariness then, at one stage, may be a form of liberation, escaping from limited chances and oppressive arrangements and looking for a safer, better place. What appears to be permanent is imagined to be temporary, as a way of overcoming the obstacles. A hope for the future means treating the current situation as temporary, knowing that people are not bound to a place and can move to somewhere else. Even the dream of moving, and thinking that the current conditions are not fixed, would offer hope for the future. If such a hope is denied, and the prospect of change is not imaginable, the weight of rigid places may drown the hope for the future and all temporariness becomes a lost dream. Being stuck within a condition, caused by economic disability or political limitations and social conventions, or believing that the chances of change have passed, as in old age, all hope may be lost and a current place may be accepted and, consequently, adjustments are made to survive. For some, accepting this limitation is the beginning of happiness but it may also act against any hope for change.

There is no guarantee, however, that moving can be the answer to all dreams, as experienced by many migrants. Moving from one place to another is not a destination in itself but a desire for a better life. Rather than arriving at a better, more secure situation however, travellers and migrants may have to live through the uncertainty of continually living in temporary places. The life of the ancient foragers and nomads was an adjustment to the natural environment, with the possibility of establishing a stable social framework for survival. The life of the modern nomad, such as the metropolitan worker, the young and the migrant, however, may not offer them the possibility of constructing such a cohesive life experience. For them, life may be precarious, lived through temporary places, and the promise of a secure and stable future may not be within easy reach or equally distributed among them.

Nevertheless, the drive for change and a disposition that treats places as temporary may lead to innovation and creativity, remaking the world in new forms and conditions rather than accepting it as it is. The most common form of such creativity is travel, physically and/or mentally, which can enable individuals and groups to see the world from a new perspective, rather than remaining within a closed horizon and accepting the status quo. A key question, however, is who initiates a move: do people decide to move or are they forced to move? Liberation may be the aspiration of the rebellious, exploratory and creative, but the outcome for the deprived and disadvantaged may just be precarity and displacement. When people move, they create temporary places, which might be imbued with hope, innovation and transformation, or may be laden with uncertainty and insecurity (Madanipour, 2018). A similar dynamic may be observed in the second category of temporary places, when places change, as I now discuss.

When places change

Another form of temporariness of a place is when its characteristics are altered, going through a metamorphosis, physically or socially. If these changes are short-lived, the place is temporary in its makeup and relations before a new arrangement is made. In this case, temporariness applies to the place itself, as distinctive from the temporariness of people in a place. Examples of this category include festivals and street markets which are run for a short period of time in a particular location as a one-off event. The nomadic tent may be the iconic image of a temporary place, built of lightweight and mobile material, which may be removed and relocated as easily as it can be set up. The current trend of temporary urbanism, in which places are arranged on a short-term basis, built by, for example, pavilions, containers and stalls, appears to provide modern tents for the urban nomads of today (Madanipour, 2017a). The questions that emerge are about the causes and consequences of these changes and trends.

The economic drive behind the temporary use of space is partly related to the structural changes in global and urban economies, whereby fixed routines of manufacturing production are replaced with the more flexible patterns of work in services. In these flexible patterns of work, investment in physical capital, which was a hallmark of manufacturing, gives way to investment in human capital, where innovation is the keyword in knowledge-based economies. As the nodes of the global economy, cities undergo substantial transformation to accommodate these changes, which may metamorphose the city from a city of industrial working class to one of middle-class knowledge-workers (Madanipour, 2011). These changes may include the gentrification of former working-class areas through hesitant steps that might be manifest in the temporary use of space, experimenting and exploring new possibilities for the future shape of these areas.

Temporary places are also made and transformed as part of a process of city marketing and branding, accompanying iconic architecture and mega projects such as sports competitions, arts and cultural festivals and entertainment events. At the local level, they accompany the property market's search for new hotspots, helping to develop a new image which can attract financial and human capital, the investment of which would raise the prices of land and property. The aim is to adjust the city to the needs and demands of the new workers, maximize the monetary value of existing assets and put a city on the global map, making it more attractive to investors, visitors and mobile workers, partly through the images of vibrancy and conviviality.

The global economy's heavy reliance on accelerating the rate and increasing the volume of consuming goods and services triggers the rearrangement of the urban space. The festivalization of urban space has long been used to regenerate the abandoned sites of industrial cities into lively spaces of consumption. Along with the shops and restaurants that populate these areas, temporary activities such as street markets, artists and entertainers are used to create an attractive buzz. These temporary activities often take place in the public or semi-public spaces of these regeneration areas, a low-cost solution for changing the image of an area from an abandoned no-go area into a people-friendly and vibrant place. In the UK, there has been an existential threat to the high street, undermined by online and supermarket shopping. In response, the animation of public spaces through temporary installations is used to bring people out to these places; these retail spaces curate an urban experience so as to increase opportunities to survive and prosper.

In addition to the structural economic change in globalized cities from manufacturing to services, the creation of temporary places is also a response to the global economic crisis of 2007–2008, which created a large mismatch between the supply of and demand for space. Debt-fuelled economic expansion had over-produced shops, offices and residential units which were no longer needed, leaving ghost towns in various countries around the world. Retailers went bankrupt and many shops were left empty for years. In some British towns, up to a third of shops became vacant (Local Data Company, 2013). There were too many spaces with too few activities to fill them. In response, prices were cut, conditions were made more flexible and temporary use of space was encouraged. In the words of a retail guru, 'it is better to have something in them than left empty' (Portas, 2011: 40). Government regulations were relaxed and specialist organizations and websites were set up, all to facilitate new, temporary uses for empty spaces, waiting for the market to return to better conditions.

The rising attention to temporary places, therefore, has been a result of, and a response to, two generations of global crises, closely associated with technological changes which have reshaped the map of the world: the shift from manufacturing to services and, a generation later, the near-total collapse of the global financial system. The two crises marked the start and transformation of a neoliberal, globalizing phase in the world economy, reflected directly in urban spaces and lives. In both crises, temporary places have been used as an interim measure, a place-holder while the market recovers, a stopgap that paves the way for other, more enduring activities. The growing abundance of temporary places has formed a trend, with a variety of different names, reflecting the sense of contingency that these major structural changes have brought about. They generate new meanings for those who are involved. The public, private and social sectors may each have a different role and expectation in this process.

From the perspective of public authorities, the temporary construction and use of space serves a number of purposes. On the one hand, it may be a catalyst in the regeneration of an area, as exemplified by the garden festivals, which were an early form of temporary use of space in the UK. They were a way of preparing the ground for attracting private sector investors into an abandoned post-industrial area, an example of interim interventions towards long-term aims. Between 1984 and 1992 a number of garden festivals in cities such as Liverpool and Glasgow transformed the contaminated and abandoned industrial sites into gardens, not as permanent parks but as temporary places (Madanipour, 2017b). On the other hand, temporary use of space, such as street markets, has become a source of income for the dwindling budgets of public authorities.

Private sector developers may also see the temporary use of space as a catalyst towards higher returns on investment. It is a way of fighting the crisis of empty properties, filling the gap that is created by business cycles and large-scale crises. Large urban development projects take a long time to be planned and implemented, and may cross from one economic cycle to another. Long gaps may occur while waiting for the right economic conditions or for being granted planning permission or funds, gaps which are filled with temporary uses of space (Madanipour, 2018). These temporary uses can change the image of an area, or the image of the developer in the local area, by filling the space with temporary, popular activities which may range from a temporary park to street food stalls and low-rent workplaces for creative entrepreneurs. For private retailers on the high street meanwhile, temporary activities may generate an attractive, vibrant atmosphere through which they may compete with the digital giants that are increasingly dominating the retail sector.

For civil society actors, community groups and young entrepreneurs, access to temporary places may provide new opportunities at affordable prices (Madanipour, 2018). It may enable them to have access to space which they might not otherwise have. They may be able to fill the gap between supply and demand in the production and use of space. They may be seen to make creative use of empty space, experimenting and exploring new possibilities, and developing new cultural and economic pathways. At the same time, the notion that they do not have a more secure access to space reveals the precarious conditions that they face, and temporary access may merely normalize this precarity. Temporary places, therefore, may mean completely different things for different actors, depending on their position in urban processes of change. This ambiguity in function and meaning facilitates a shift of meaning from necessity to desirability.

One of the concepts of time in ancient Greece was Kairos, an opportune moment which seized 'all the possibilities contained within a given moment' (Hadot, 1995: 221). Temporary construction and use of space is an example of this concept, as it seems to offer opportunities to different stakeholders. It tends to transform the idea of time as consistent and predictable into a medley of precarious and ambiguous moments within incessantly recurrent storms and crises associated with globalization. With its normalization, opening up possibilities to seek opportunity and seize these moments, a new cultural trend emerges, increasingly disconnected from the necessities and vulnerabilities that formed it in the first place. It is in this context that major corporations embrace the idea of setting up pop-up shops and restaurants, to give the impression that they are offering something unique but fleeting, encouraging potential customers to hurry before it is too late. Pop-up shops test the market, re-image brands in new ways and participate in creating a fashionable trend (Calladine, 2012). This cultural trend generates a symbolic value for the sense of ephemerality, which is presented as fresh, innovative and opportune, even if it also means vulnerability and fragility.

Conclusion

The two aspects of temporary places, the movement of people and the change in places, are intertwined in urban processes. They show the particularity of time, space and meaning, which is in a continuous process of transformation and instability, with different meanings for those involved. If they wish to move, and see this as a form of liberation from constraints and a search for new horizons, some people may welcome the opportunity. If they are forced to move, however, they might become displaced, living through temporary places that carry a sense of transience and vulnerability. The same is true of the second category, when places change. Temporariness of spaces and activities within them offers a degree of flexibility to those who can benefit from the opportunity, filling the gaps and using temporary places as a catalyst for change. But when places are transformed beyond their control, a sense of displacement and disorientation prevails. It may provide new opportunities for creativity through a flexible relationship between people and places, but also normalizing precarity and uncertainty for the less privileged.

References

Bouquet-Appel, J.-P., 2011, 'When the world's population took off: The springboard of the Neolithic Demographic Transition', *Science*, 333: 560–561, 29 July 2011.
Bourdieu, P., 2000, *Pascalian Meditations*, Cambridge: Polity Press.
Calladine, D., 2012, 'Advice and resources', *London Pop-ups*, www.londonpopups.com/p/advice-resources.html, 17.11.2012.
Casey, E. S., 1996, 'How to get from space to place in a fairly short stretch of time: Phenomenological prolegomena', In S. Feld and K. H. Basso (eds), *Senses of Place*, Santa FE, NM: School of American Research Press, pp. 13–52.
Castells, M., 1997, *The Power of Identity*, Oxford: Blackwell.
Clark, A. N., 1985, *Longman Dictionary of Geography*, Harlow: Longman.
Cloke, P. and R. Johnston, eds, 2005, *Spaces of Geographical Thought*, London: Sage.
Deleuze, G and F. Guattari, 1987, *A Thousand Plateaus*, Minneapolis, MN: University of Minnesota Press.
European Commission & UN-Habitat, 2016, *The State of European Cities 2016*, Brussels: European Commission.
Gans, H., 1968, *People and Plans*, New York: Basic Books.
Giedion, S., 1967, *Space, Time and Architecture: The Growth of a New Tradition*, Cambridge, MA: Harvard University Press.
Hadot, P., 1995, *Philosophy as a Way of Life*, Oxford: Blackwell.
Healey, P., 2010, *Making Better Places*, Basingstoke: Palgrave Macmillan.
Heidegger, M., 1978, *Basic Writings*, David Farrell Krell ed., London: Routledge.
Holloway, S., S. Rice and G. Valentine, 2003, *Key Concepts in Geography*, London: Sage.
Hyndman, J. and W. Giles, 2011, 'Waiting for what? The feminization of asylum in protracted situations', *Gender, Place & Culture*, 18(3): 361–379.
IPCC, 2014, *Climate Change 2014: Synthesis Report*. Contribution of Working Groups I, II and III to the Fifth Assessment Report of the Intergovernmental Panel on Climate Change [Core Writing Team, R. K. Pachauri and L.A. Meyer (eds)]. Geneva: IPCC.
Jacobs, J., 1961, *The Death and Life of Great American Cities*, New York: Random House.
Jansen, S. and S. Löfving eds, 2009, *Struggles for Home: Violence, Hope and the Movement of People* (Vol. 3), New York: Berghahn Books.
Kalb, D., 2013, 'Regimes of value and worthlessness: Two stories I know, plus a Marxian reflection', *Max Planck Institute for Social Anthropology Working Papers*, 147.
Le Corbusier, 1987, *The City of Tomorrow and Its Planning*, London: The Architectural Press.
Lefebvre, H., 1991, *The Production of Space*, Oxford: Blackwell.
Lems, A., 2016, 'Placing displacement: Place-making in a world of movement', *Ethnos*, 81(2): 315–337.
Local Data Company, 2013, *The Knowledge Centre*, www.localdatacompany.com/knowledge, 23.5.2013.
Madanipour, A., 2011, *Knowledge Economy and the City*, London: Routledge.
Madanipour, A., 2016, 'Social exclusion and space', In R. LeGates and F.Stout (eds), *The City Reader*, Sixth Edition, London: Routledge, pp. 203–211.
Madanipour, A., 2017a, *Cities in Time: Temporary Urbanism and the Future of the city*, London: Bloomsbury.
Madanipour, A., 2017b, 'Ephemeral landscapes and urban shrinkage', *Landscape Research*, 42(7): 795–805.
Madanipour, A., 2018, 'Temporary use of space: Urban processes between flexibility, opportunity and precarity', *Urban Studies*, 55(5): 1093–1110.
Malkki, L.H., 1995, 'Refugees and exile: From "refugee studies" to the national order of things', *Annual Review of Anthropology*, 24(1): 495–523.
Massey, D., 2005, *For Space*, London: Sage.
Morell, M., 2014, 'When space draws the line on class', In D. Kalb and J. Carrier (eds), *Anthropologies of Class: Power, Practice, and Inequality*, Cambridge: Cambridge University Press, pp. 102–117.
Norberg-Schultz, C., 1980, *Meaning in Western Architecture*, London: Studio Vista.
Portas, M., 2011, *The Portas Review: An Independent Review into the Future of Our High Streets*, London: Department for Business, Innovation and Skills.
Tönnies, F., 1957, *Community and Society*, New York: Harper and Row.
Tuan, Yi-Fu, 1977, *Space and Place: The Perspective of Experience*, London: Edward Arnold.

40

The place of the camp in protracted displacement

Cathrine Brun

Introduction: the place of the camp

The camp has received significant attention over the last two decades in academic research and humanitarian policy and practice. Camps have been the most common mode of managing refugees, but have fallen out of fashion and become the least sought after way of assisting people on the move. While still a common practice, the attention has moved from camps towards the city with the realisation that between 60 and 80% of the world's forced migrants live outside camps and mainly in urban areas (Park, 2018). The presence and arrival of refugees in urban areas did not happen overnight and there are several intersecting processes that have contributed to the shift away from camps: a 'Southern urban turn' with historically unprecedented rates of urbanisation in Southern cities (Leitner and Sheppard, 2016); decreasing funding over time when a displacement crisis becomes protracted; and more emphasis on self-reliance rather than dependence on aid. Additionally, a critical body of research on the camp has emerged, questioning the purpose, role, usefulness and general logic of keeping refugees in camps. The temporary makeshift spaces of the camp are now commonly understood as sub-standard and undignified living spaces that deprive people of agency and control over their future.

Yet, the camp has far from lost its relevance. More than 2.6 million refugees live in camps (UNHCR, 2018) and many camps continue to expand and become more permanent features. Most camps are built for refugees and asylum seekers, people who have crossed an internationally recognised border. However, the number of refugees is smaller than the number of internally displaced persons (IDPs) dwelling in camps across the globe. IDPs are people who have been displaced, often for the same reasons as refugees, but who have not crossed an internationally recognised border. In Iraq, outside Mosul for example, humanitarian agencies in collaboration with the military are, at the time of writing (December 2018), erecting camps and administering humanitarian assistance for more than 500,000 people. In Syria, the Internal Displacement Monitoring Unit estimated that 750,000 people live in camps and informal settlements (IDMC, 2018).

These camps are driven by a mix of reasons, custody, care and control (McConnachie, 2016; Minca, 2015a). One prominent argument for camps has been that it eases coordination of assistance and helps in accessing people in need. The camps, however, can also be understood as mechanisms of control and monitoring of subjects as well as creating greater dependency, uncertainty and exposure of people to human rights violations or other violence. Still, humanitarian assistance to forced migrants outside camps tends to maintain and reproduce the logic of the camp; assistance and attitudes towards forced migrants continues to be based on care and control, dependency, limited freedom of movement and limited understanding of the agency that displaced persons can utilise to manage their lives.

In this chapter I ask *what is the place of the camp in forced migration?* I concentrate largely on situations of protracted displacement – enduring situations of displacement with no end in sight (Crisp, 2003). I will explore the camp in its multiple dimensions and over time. The camp is associated with place in numerous ways, and my starting point for considering the place of the camp is Doreen Massey's definition of a place as a local articulation:

> For me, places are articulations of 'natural' and social relations, relations that are not fully contained within the place itself. So, first, places are not closed or bounded – which, politically, lays the ground for critiques of exclusivity. Second, places are not 'given' – they are always in open-ended process. They are in that sense 'events'. Third, they and their identity will always be contested.
>
> *(Massey, 2010)*

I discuss the place of the camp in the context of this 'progressive sense of place', considering a place that is always in progress because it is created by the social interactions that are tied together, a place defined by the particular linkages to the outside and always with a contested identity and notion of community (Massey, 1991).

With this starting point, I first discuss the different meanings and practices of the camp. Second, I describe the changing position, status and reputation of the camp as a way of organising people and assistance in the context of a refugee crisis. I show how the status of the camp has changed, waxed and waned, over the past decades. Third, I address symbolic, political and practical meanings of the camp within the context of events and processes taking place outside the camp itself. Finally, I discuss the place of the camp by reflecting on what kind of place the camp is and whether, indeed, it can be considered as a place at all.

The blurred boundaries of the camp: shapes, governance and inhabitation

A 'camp' may have several meanings, not least because they come in different shapes, are located in different contexts, are run with varying degrees of control and are inhabited in particular ways. In order to understand the place of the camp in protracted displacement, I adopt an open, dynamic and flexible definition of the 'camp' to reflect the diversity of spaces that may be defined as a camp.

The shape of the camp

A predominant image of a camp is the tented city, supplied almost entirely from the outside (Black, 1998). We often see aerial photos of the large paradigmatic camps such as Dadaab in Kenya and Zataari in Jordan. However, these images represent only one kind of camp and contrast with other forms and shapes of camps such as 'small, open settlements where the refugee communities have been able to maintain a village atmosphere' (Bowles, 1998: 11). The camp may also look like a poor neighbourhood in a city – a space that blends in with its surroundings. Without local knowledge it might be difficult to distinguish the camp from the surrounding non-camp which may also be a marginalised environment. These urban camps constitute the main form of camp for the many Palestinian refugees in Lebanon, Jordan and in Syria.

There are other types of institutional and physical structures that constitute camps, such as the 'collective centres' in the former Soviet Union and the Balkans. The collective centre is often established in hotels, student dormitories, military barracks, kindergartens and hospitals and is a structure used to house forced migrants. Generally not intended for permanent living, collective centres nevertheless often become enduring dwellings for displaced people (Brun, 2012). Additionally, the boundaries between 'camps', 'settlements' and 'informal' neighbourhoods are often blurred. Informal settlements may be an unofficial group of temporary residential structures of varying sizes and numbers where displaced people have come together to seek shelter (OCHA, 2016).

The common trait for the different structures and shapes of the camp discussed here is that they are a place where displaced people have gathered. A camp is often characterised by containment and segregation and it has a specific temporality attached to it. The camp is intended to be a temporary space until such time as forced migrants return or move on. However, camp-spaces often become characterised by 'permanent impermanence' (Brun, 2003, 2012) and, while they may gradually blend in with the surrounding neighbourhoods, they often stand out from the wider environment, being more makeshift, looking poorer and in worse condition. Yet, these spaces are to varying degrees governed differently to their immediate neighbourhoods. Camps are subject to rules, statuses and management structures that separate and exclude them from the wider polity in which they are located.

Degrees of control and governance in the camp

The level of institutionalisation distinguishes different forms of camps from each other. Camps are often constructed and maintained by humanitarian organisations, sometimes with involvement from host governments, but camps may also be made, organised and managed by refugees and displaced people themselves. There are thus different degrees of control, institutionalisation, formality and informality attached to the camps. Some may start out as informal settlements and then become more formalised. Other camps, such as collective centres, may become permanent dwellings and subject to privatisation, but often continue to be marked by the stigma associated with displacement. In this sense, it may be important to differentiate between camps for refugees and camps for internally displaced people, but these boundaries are sometimes blurred, as in Jordan where Palestinians continue to live in camps even though they are Jordanian citizens.

Camps for forced migrants feature the biopolitical governance attached to humanitarian assistance. Camps can be understood as an attempt to control and institutionalise populations that are considered as 'matter out of place' (Malkki, 1992): spaces where governance techniques such as headcounts, situation reports, control over space and the movement of people dominate (Ramadan, 2013). Hanna Arendt (1973) understood camps as spaces where political life is extinguished: the camp represents a deprivation of agency, political rights and identities. Inspired by the Italian philosopher Giorgio Agamben, some commentators see camps as 'spaces of exception' which lie outside of the juridical order with a particular set of biopolitical governing technologies attached to them (Minca, 2006). Minca (2015b) sees the camp as an institution which is fundamentally based on a process of desubjectivation and mobility restriction. However, the heterogeneous nature of what a camp may be indicates that this is too narrow a definition of what constitutes a camp and how it is governed. It is a definition that deprives people residing in the camp of agency and at the same time overlooks the political significance and actual impact of the camp in its local, national and international context.

When conceptualised as a humanitarian space (Ramadan, 2013), the camp is seen as reflecting the attempts of the international community to institutionalise a state of protection and relief for refugees which leads in time to particular mechanisms of control and containment, often in close connection with the interest of host state and surrounding populations. Processes of institutionalisation may be introduced at different stages of a camp's life. For example, institutionalisation may be imposed on the space after its establishment in cases where the camp was not established by a humanitarian community or a state, but by forced migrants who organised themselves in camps as a way of staying together, protecting and supporting each other and finding shelter. Gradually, however, these spaces may attract assistance and governance structures may be established and exerted over them. What starts as a practical measure for staying together may end up in a different and less self-governed space. Consequently, the governance of the camp is also determined by the agency of its inhabitants.

Inhabiting the camp

Despite the temporariness of the camp, homemaking practices are noticeable from very early on in the formation of a camp. The material shelter – a tent, a caravan or another makeshift structure often provided by the international community or a host government – is almost always modified. Camps develop over time because, as people go about their daily lives in those spaces, they improve, expand, build and make temporary homes within those structures (Brun and Fabos, 2015). The first time I visited the camps for internally displaced Muslims in Sri Lanka in 1994, one of the camps had just burnt down. I witnessed how people came together and transformed over only a couple of days the burnt-out ruins into a dwelling place where everyday life continued in the newly erected huts made out of coconut leaves. In Ein El Hilweh, a Palestinian refugee camp in Southern Lebanon destroyed in the 1982 Lebanon War, women came together and rebuilt their concrete houses while their husbands were imprisoned and there was an embargo on building materials (Al-Jana, 2010). Similarly, in Zataari camp for Syrian refugees in Jordan, building materials cannot be brought into the camp, but residents have still, bit by bit, managed to extend the original caravan into a family compound with kitchen, reception areas and private bedrooms (Dorai with colleagues, 2018). In the same camp, residents have built a local economy and established a main shopping street, Shams-Elysées ('Sham' is the local term for Syria), which in addition to constituting livelihoods and a camp-economy, also offers public meeting spaces for residents (Ababsa, 2018). Residents invest in their lives and dwelling places in the camp, but at the same time imagine a life elsewhere and nurture networks and transnational relations with the outside (Horst, 2007). One important outcome of such relations is the remittances sent by relatives and friends elsewhere. At the same time, political mobilisation, militarisation and the making and recruitment of so-called 'refugee warriors' in camps represent common cross-border practices between country of origin and host country (Harpviken and Lischer, 2013). Over time, everyday practices and political action in the marginal and restricted space of the camp enable individual and collective formation, nurturing and expansion of relations within the camp and with the outside.

The changing place of the camp in managing forced migration

The displaced persons camp is a product of modernity and the contemporary state (McConnachie, 2016). McConnachie traces the camp back to the prisoners of war (PoW) camps in the late 1700s and the wars following the French Revolution, via internment camps from late 1800s, and up the present day refugee camps originating with the camps for Armenians fleeing genocide from 1915. Containment, temporality, the governing of people out of place, not belonging and hence not being trusted are common characteristics of the camp. Widespread use of camps to manage forced migrants became more established following World War II when Displaced Persons (DP) camps were created across Europe, first to house people fleeing the Nazi regime and later to accommodate people who had been freed from concentration camps or who were fleeing the Soviet Army (McConnachie, 2016). Additionally, setting up camps became a widespread practice with decolonisation and the establishment of new states. Perhaps the largest example was the partition of India and Pakistan in 1947 that created massive population displacements. As many as 10 million ended up in 'colonies' and camplike informal spaces which continue up to the present day to be marginal spaces (Sanyal, 2014b). Israel's declaration of independence in 1948 resulted in the mass displacement of Palestinians and refugee camps were established in Gaza and the West Bank and in the neighbouring countries of Lebanon, Jordan and Syria. Many of these camps continue to exist today with new generations being added to those already in exile. The United Nations Relief and Works Agency, established in 1949, took over the management of these camps from 1950 and, at different times since then, have established new camps (Katz, 2015). Globally, new camps continued to be established: for example, during the Cold War especially from 1977–1985 and across South East Asia, Central America, the Horn of Africa and Southern Africa (McConnachie, 2016). With the end of the Cold War, the humanitarian category of internally displaced persons (IDPs) was established and in civil wars across the globe containment spaces and more open camps were established for the internally displaced (Brun, 2010).

The end of the Cold War corresponded with a growing and maturing of the academic field of forced migration studies, a more consolidated policy response and renewed discussions of the place of the camp. A landmark publication, Barbara Harrell-Bond's *Imposing Aid* published in 1986 (Harrell-Bond, 1986), critically scrutinised the humanitarian community and the containment of people. The book was followed by a large body of work that examined the 'warehousing' of people in camps (Malkki, 1995; Smith, 2004). In 1998, a special section of *Forced Migration Review* (*FMR*) gathered some of the main arguments for and against camps and Jennifer Hyndman's (2000) *Managing Displacement* added a critical and geopolitical dimension to the understanding of the camp as a relational space. Increasing emphasis was placed on the question of why camps were so often the preferred means of managing refugees by host governments and international organisations (Black, 1998). In this context, Harrell-Bond (1998) argued that camps are driven by the demands of donors and humanitarian organisations, and that confinement in camps renders refugees dependent on assistance, deprives them of access to networks of social and economic support, controls their movements and contributes to a representation of refugees as helpless and dependent.

In the context of increasing critique against camps and more emphasis on urban refugees, the United Nations High Commissioner for Refugees (UNHCR) published a *Comprehensive Policy on Urban Refugees* in 1997. The policy was highly controversial and was criticised for trying to prevent urban displacement rather than assisting refugees in the city (Crisp, 2017;

Fàbos and Kibreab, 2007). The document promoted the view that a policy to assist the displaced and refugees in the city would increase urbanisation and urban poverty and thus implicitly suggested that camps were a better option. However, UNHCR was increasingly confronted with refugee populations who were not living in camps (Crisp, 2017), some because they did not want to live in them, others because the host country, such as Lebanon, did not promote a camp policy. There were narratives of people who left the camps for the city and who chose to live unregistered and more informal lives in order to gain more freedom and access to employment and to escape the containment in the camp. Their experiences led to an increase in academic work on urban refugees (Fàbos and Kibreab, 2007; Landau, 2006; Sommers, 2001).

The location of the camp

Despite ample evidence that camps are the least preferred way to assist and protect the displaced, they continue to exist to manage displacement. The camp literally takes place. It has a territorial reach which may vary from camp to camp, but which makes its material presence visible. The camp has boundaries and it may be fenced in, but sometimes these boundaries may not be visible and may be acted out at a symbolic level. Camp boundaries can thus be porous but are nonetheless exclusionary.

Despite its features of containment, segregation and exclusion, the camp only makes sense in relation to various social, political, legal and economic processes outside the camp. As such, the location of the camp can be identified as the meeting point between the humanitarian regime, those who they want to protect, and the state, whose aim may be to protect themselves against the refugees (Agier, 2002). The camp has become a normalised practice utilised to govern mobility and movement. In this way some camps constitute external national borders by serving as a proxy for a state's asylum practices. Border enforcement may be present in the camp by asylum application processing and is often deployed to enclose, contain and confine people away from aspirational destinations (Mountz, 2011). Hence, a camp represents for its inhabitants a not quite not yet presence in the nation state on whose territory they are present (Thorshaug, 2018). Camps may also then be understood to embody wider geopolitical regimes managing migration movements, such as those that handle the externalisation of asylum in centres on the island of Nauru, for example, or may more generally serve as a never-ending waiting space for people hoping for resettlement.

Camps have overlapping technologies of control from within and from outside. From within, organising principles formulated by humanitarian organisations and based on humanitarian labels, such as 'refugee', 'asylum seeker' and 'displaced persons', and levels of vulnerability, control access to assistance. At the same time there may be a camp leader, camp committee and organisations representing the displaced. These representatives are sometimes appointed by humanitarian organisations responsible for the camp or they may be elected by the residents of the camp. The camp leader and committee often represent the camp when dealing with assistance and host governments. From outside, the asylum regime and the rules of the country in which the camp is located also apply to the camp residents. Martin (2015) introduces the notion of 'campscape' to indicate how the camp effects, and is impacted by, its surroundings. The camp can also play a significant role in the local and national politics of the host states: the camp promises temporariness and a separate space to avoid competing with the needs of the host population. In this way the camp becomes a national compromise between pleasing a polity and assisting people in need.

What kind of place is the camp?

> Refugee camps and settlements are not, of course, 'normal' places, particularly in situations where the population has little or no access to land or wage labour, and must therefore rely on external assistance.
>
> *(UNHCR, 1995: 235)*

The camp is associated with place in numerous ways and in this final section I reflect on what kind of place the camp may be. I return to Doreen Massey's definition of place as a local articulation: as an open space, a process and always with a contested identity. Following Massey, I have shown that camps can be understood as an assemblage created from the connection of a range of heterogeneous components (Ramadan, 2013). The quote from UNHCR above emphasises that a camp is not a 'normal place'. Based on the definition from Massey, 'normal place' may be seen as an oxymoron. However, the formation, development and persistence of camps as an established space inspire two important discussions for understanding what the place of the camp may be: is the camp a community and is the camp a city?

Is the camp a community?

> A refugee community is an *institution* created specifically for the purpose of providing protection and assistance to a group of people who are not citizens of the country in which they are living.
>
> *(Hyndman, 2000: 145, original emphasis)*

As this quotation from Hyndman suggests, scholars have questioned the understanding that a camp can be a community. Places as communities are often based on a territorially bound group of people with formal place-based institutions such as local government, education, economy and religion (Bradshaw, 2008). Hyndman (2000) argued that camps are 'simulated communities', established by humanitarian agencies. Such camps cannot, in her view, operate as a village or a civil society, despite employing community-development principles such as self-governance and democratic decision making. Therefore, in Hyndman's view, refugee camps are not communities, they are institutions organised as temporary solutions to displacement and thus are more like enforced colonies than communities defined by voluntary association. Hyndman's description clearly has relevance for some camps and settlements. Understanding place as a local articulation, however, indicates that communities are not just place-based, but exist out of shared interests and commitments that unite a set of varied groups and activities (Selznick, 1992). With this definition of place, we may approach the question of camp as a community differently. Malkki (1997: 91) adopted the term 'accidental communities of memory' – 'a biographical, microhistorical, unevenly emerging sense of accidental sharings of memory and transitory experience'. Examples of such accidental communities of memory might be people who have experienced war together and people who live together in a refugee camp. Such accidental communities, according to Malkki, bring together people who might not otherwise have met in the ordinary course of their lives. Communities are always full of and constituted by contradictions (Staeheli, 2008). However, the collective identities and accompanying political agency, framed by Sigona (2015) as 'campzenship', that are produced in the camp show how the camp may represent a basis for collective forms of action and mobilisation (Pasquetti, 2015). The uniqueness of the camp is represented by how the camp residents, who are only temporarily present and whose legal status is contentious, have come together, built the camp and over time developed bonds and act as a place-based community due to their common history.

Is the camp a city?

The camp represents interconnections of shared histories and there is an increasing body of work that engages with the camp as a city (Agier, 2002; Knudsen, 2016; Martin, 2015; Sanyal, 2014a). Many authors have found a fascination in how the camp starts as a blank sheet and imprints itself in the landscape and gradually becomes a more established place. As it grows, it adopts a city-like structure, composition and functions. While those 'blank sheets' are of course not an empty space without a history, the persistence of the camp as a social and material space has contributed to the discussion of the camp as a city. Camps represent a population density similar to a city or an urban neighbourhood and develop economic functions and a political economy that may resemble a city economy. As I showed above, institutionalised camps may become integrated with a city as an urban neighbourhood, although often with stigma, processes of exclusion and marginalisation attached to them. As a result, some authors argue for unsettling the current distinction between the camp and the city to consider both as places where residents 'negotiate access to scarce material and symbolic resources in the context of powerful agencies of control' (Pasquetti, 2015: 704). Camps as cities, and following Massey's definition of place as a local articulation, may then be understood as products of their social, economic and material development, a political space with multiple engagements with the wider world. A fundamental consequence of considering the camp as a city – and as a community – is that it requires us to recognise its inhabitants and see the camp as a living place in which social relations are formed and where politics of connectivity are forged.

Understanding the overlapping spaces of camps and non-camps helps to conceptualise a forced migration setting that constitutes the inside and the outside of camps. Refugees, regardless of being in camps or out of camps, are subject to control and exclusion due to their legal status and right to be present. Informality and precarious life outside camps may not be a sought-after alternative and may even be a parallel space of marginalisation and exclusion to the camp (Knudsen, 2016; Pasquetti, 2015; Sanyal, 2014b).

Concluding reflections: the place of the camp

Following Massey's progressive sense of place, I suggest that the place of the camp is a dynamic and relational space where status and meaning change over time shaped by dwelling, homemaking practices and the potential for collective action. However, the camp also represents a specific set of technologies of governance shaped by the legal, social, economic and political status of its inhabitants; and in this way, camps are 'places out of place' (Knudsen, 2016: 444). On the one hand, camps are spaces of refuge, hospitality, identity formation and preservation (Ramadan, 2013). On the other hand, camps are spaces created by the absence of hospitality when a solution is not a possibility (Knudsen, 2016). Camps are almost always places on the margin, and the status of the camp will always constitute an exclusionary element even if it becomes a city and the residents identify with the camp as a community. However, currently camp dwellers seem to be stuck between a rock and a hard place: the international refugee regime has not come up with an alternative to camps that includes social and legal protection and even forced migrants in a non-camp solution cannot escape the camp-logic of care and control, temporary lives and exclusion.

References

Ababsa, M. 2018. An Urbanizing Camp? Zataari Refugee Camp in Jordan. *Conflits et migrations*, https://lajeh.hypotheses.org/1076 (last accessed 10 May 2018).

Agier, M. 2002. Between war and city. Towards an urban anthropology of refugee camps. *Ethnography* 3 (3): 317–341.

Al-Jana. 2010. *The Kingdom of Women: Ein El Hilweh.* (2010). [DVD] Directed by D. Abourahme. Lebanon: Arab Resource Center for Popular Arts, Al-Jana.

Arendt, H. 1973. *The Origins of Totalitarianism.* Orlando, Harcourt Inc.

Black, R. 1998. Putting refugees in camps. *Forced Migration Review* 2: 4–7.

Bowles, E. 1998. From village to camp: refugee camp life in transition on the Thailand-Burma border. *Forced Migration Review* 2: 11–14.

Bradshaw, T.K. 2008. The post-place community: contributions to the debate about the definition of community. *Community Development* 39(1): 5–16.

Brun, C. 2003. Local citizens or internally displaced persons? Dilemmas of long term displacement in Sri Lanka. *Journal of Refugee Studies* 16(4): 376–397.

Brun, C. 2010. Hospitality: becoming 'IDPs' and 'hosts' in protracted displacement. *Journal of Refugee Studies* 23(3): 337–355.

Brun, C. 2012. Home in temporary dwellings. In S.J. Smith, M. Elsinga, L. Fox O'Mahony, O. S. Eng, S. Wachter, R. Dowling (eds) *International Encyclopedia of Housing and Home.* Oxford: Elsevier. 424–433.

Brun, C. and A.H. Fabos. 2015. Homemaking in limbo? A conceptual framework. *Refuge* 31(1): 5–18.

Crisp, J. 2003. *No solution in sight: the problem of protracted refugee situations in Africa.* New Issues in Refugee Research. 75. Geneva: Evaluation and Policy Unit, UNHCR.

Crisp, J. 2017. Finding space for protection: an inside account of the evolution of UNHCR's Urban Refugee Policy. *Refuge* 33(1): 87–96.

Dorai, K. and colleagues. 2018. Architecture of Displacement. Exhibition and roundtable at Conflict and Migration in the Middle East. LAJEH Research Programme, Final Conference, Beirut, June 5-7, 2018.

Fàbos, A.H. and G. Kibreab. 2007. Urban refugees: Introduction. *Refuge* 24(1): 3–10.

Harpviken, K. and S. Lischer. 2013. Refugee militancy in exile and upon return in Afghanistan and Rwanda. In J. Checkel (ed.) *Transnational Dynamics of Civil War*, pp. 89–119. Cambridge: Cambridge University Press.

Harrell-Bond, B. 1986. *Imposing Aid.* Oxford, Oxford University Press.

Harrell-Bond, B. 1998. Camps: literature review. *Forced Migration Review* 2: 22–23.

Horst, C. 2007. *Transnational Nomads. How Somalis cope with refugee life in the Dadaab camps of Kenya.* New York: Berghahn Books.

Hyndman, J. 2000. *Managing Displacement. Refugees and the Politics of Humanitarianism.* Minneapolis, University of Minnesota Press.

IDMC. 2018. Syria. www.internal-displacement.org/countries/syria (last accessed 15 November 2018).

Katz, I. 2015. From spaces of thanapolitics to spaces of natality – A commentary on 'Geographies of the camp'. *Political Geography* 49: 84–86.

Knudsen, A. 2016. Camp, Ghetto, Zinco, Slum: Lebanon's transitional zones of emplacement. *Humanity* 7(3): 443–457.

Landau, L. 2006. Protection and dignity in Johannesburg: Shortcomings of South Africa's urban refugee policy. *Journal of Refugee Studies* 19(3): 308–327.

Leitner, H. and E. Sheppard. 2016. Provincializing critical urban theory. Extending the ecosystem of possibilities. *International Journal of Urban and Regional Research* 40(1): 228–235.

Malkki, L. 1992. National Geographic: The Rooting of peoples and the territorialisation of national identity among scholars and refugees. *Cultural Anthropology* 7(1): 24–44.

Malkki, L.H. 1997. News and culture: transitory phenomena and the fieldwork tradition. In A. Gupta and J. Ferguson (eds.) 1997. *Anthropological Locations. Boundaries and Grounds of a Field Science*, pp. 86–101. Berkeley: University of California Press.
Malkki, Liisa H. 1995. *Purity and Exile. Violence, Memory, and National Cosmology among Hutu Refugees in Tanzania*. Chicago, University of Chicago Press.
Martin, D. 2015. From spaces of exception to 'campscapes': Palestinian refugee camps and informal settlements in Beirut. *Political Geography* 44: 9–18.
Massey, D. 1991. A global sense of place. *Marxim Today* June 1991.
Massey, D. 2010. The future of landscape: Doreen Massey. 3:AM, www.3ammagazine.com/3am/the-future-of-landscape-doreen-massey (last accessed 01 June 2018).
McConnachie, K. 2016. Camps of containment: A genealogy of the refugee camp. *Humanity* 7(3): 397–412.
Minca, C. 2006. Giorgio Agamben and the new biopolitical nomos. *Geografiska Annaler* 88B(4): 387–403.
Minca, C. 2015a. Geographies of the camp. *Political Geography* 49: 74–83.
Minca, C. 2015b. Counter-camps and other spatialities. *Political Geography* 49: 90–92.
Mountz, A. 2011. Border politics: spatial provision and geographical precision. In Interventions on rethinking 'the border' in border studies. *Political Geography* 30: 61–69.
OCHA. 2016. Informal Refugees of Syrian Refugees in Lebanon. OCHA Services. https://data.humdata.org/dataset/syrian-refugeees-informal-settlements-in-lebanon (last accessed 04 December 2018).
Park, H. 2018. The Power of Cities. UNHCR Innovation Service, www.unhcr.org/innovation/the-power-of-cities (last accessed 11 December 2018).
Pasquetti, S. 2015. Negotiating control. *City* 19(5): 702–713.
Ramadan, A. 2013. Spatialising the refugee camp. *Transactions of the Institute of British Geographers* 38: 65–77.
Sanyal, R. 2014a. Urbanizing refuge: interrogating spaces of displacement. *International Journal of Urban and Regional Research* 38(2): 558–572.
Sanyal, R. 2014b. How refuge creates informality: shelter politics in refugee camps in Beirut. *Jerusalem Quarterly* 60: 31–41.
Selznick, P. 1992. *The Moral Commonwealth. Social Theory and the Promise of Community*. Berkeley, University of California Press.
Sigona, N. 2015. Campzenship. Reimagining the camp as a social and political space. *Citizenship Studies* 19(1): 1–15.
Smith, M. 2004. Warehousing Refugees: A Denial of Rights and a Waste of Humanity. In US Committee for Refugees and Immigrants, World Refugee Survey 2004, pp. 38–42. http://refugees.org/wp-content/uploads/2015/12/Warehousing-Refugees-Campaign-Materials.pdf (last accessed 07 April 2017).
Sommers, M. 2001. *Fear in Bongoland. Burundi Refugees in Urban Tanzania*. Oxford, Berghahn.
Staeheli, L.A. 2008. Citizenship and the problem of the community. *Political Geography* 27: 5–21.
Thorshaug, R. Ø. 2018. Arrival In-Between: Analyzing the Lived Experiences of Different Forms of Accommodation for Asylum Seekers in Norway. In Meeus, B., van Heur, B. and Arnaut, K. (eds.) *Arrival infrastructures*, pp. 207–227. London: Springer International Publishing.
UNHCR. 1995. *State of the World's Refugees. In Search of Solutions*. Oxford: Oxford University Press.
UNHCR. 2018. What is a refugee camp? UNHCR, www.unrefugees.org/refugee-facts/camps (last accessed 15 November 2018).

41

Remaking a place called home following displacement

Luis Eduardo Perez Murcia

Introduction

Narratives of internally displaced people (IDP) in Colombia reveal multiple and varied experiences of being uprooted from one place and forced to move to another. Many fled in a state of panic, confusion and anxiety when rebels and paramilitary groups arrived in their communities, threatening families, murdering innocent people, raping women and destroying personal and communal assets. Most profoundly, they felt that they had lost a place they called home, a place where they had felt materially, socially, politically, culturally, emotionally and existentially embedded. However, many attempt to remake a home in the places in which they settle and to recover their place in the world (Arendt, 1966).

Processes of losing and remaking a home have been widely explored for those who have been displaced across national borders (Hammond, 2004; Korac, 2009; Jansen and Löfving, 2011; Loizos, 2011; Taylor, 2015a). However, the experiences of those who have been internally displaced have largely been under-researched as it is often assumed that they have remained 'at home'. This chapter examines how those who are internally displaced within Colombia seek to remake a home. While a home can be remade following displacement, this entails more than simply finding a physical place of shelter as there are other material and symbolic dimensions to home-making. The chapter examines these, focussing on the materiality of a home, home as a social world and a familiar place, and the emotional and existential aspects of home. In other words, remaking home involves regaining a sense of being physically, socially, politically, culturally and emotionally and existentially immersed in a place of security, community and hope. The material and sociocultural dimensions of emplacement have been addressed in forced migration research (see Hammond, 2004; Korac, 2009; Jansen and Löfving, 2011; Den Boer, 2015; Taylor, 2015a) but the emotional and existential dimensions have been largely under-acknowledged and yet, as shown below, are of great significance.

The research findings are based on 72 semi-structured interviews with displaced individuals who fled from their places and communities of origin because of conflict between left-wing guerrillas, right-wing paramilitary groups and the state army forces between 1 January 1980 and 31 December 2010. The term 'conflict-induced displacement' (also referred to as 'displacement') refers to a process of human mobility compelled by the dynamics of conflict. Those who flee but remain within national borders are often referred to as 'internally displaced people' while the term 'refugees' tends to be used to refer to those who flee across national boundaries. Interviews were carried out with IDPs in three cities, Bogotá, Medellín and Cartagena, between December 2013 and August 2014. Seventy-two per cent of interviewees were female and 28% male. Participants were aged between 18 and 68 and 36% of participants identified themselves as members of the black communities, 15% as Indigenous and 49% as mestizos. Participants had received an average of 8.5 years of schooling and most are blue collar workers working as housekeepers, street peddlers, taxi drivers, assistant cooks, hairdressers, plumbers, waste pickers and bodyguards.

The following section reviews the theoretical and conceptual debates on emplacement and remaking a place called home and section three provides the empirical findings. The chapter concludes in section four by suggesting that while following conflict and displacement homes may be lost, they can be reconstructed over time.

Conceptualising emplacement and home-making

This chapter draws on the idea that one's sense of home varies over time, place and space (Blunt and Dowling, 2006). Physical locations may acquire profound meaning for individuals and communities (Taylor, 2013a) but one's relationship with those places may change over time. The term 'place' is used here to refer to 'any kind of space that people, through their everyday lives, use, appropriate, and reflect on, thereby generating meaning through practice and association' (Hammond, 2004, 82). The term 'space' denotes both the geographic location from which the displaced were expelled (or forced to flee) and the topographical area in which they are currently settled.

Drawing on these concepts, it is suggested that home is dynamic. It can be lost following conflict and displacement but can also be made and remade. Even in dire circumstances, for instance in a refugee camp or in shanty towns, a meaningless place can be transformed into a meaningful home through daily social practices (Hammond, 2004). Indeed, even when warfare destroys people's 'worlds' – which in Nordstrom's (1995) terms denotes the loss of comfort, family members, social kinships and cultural traditions – the displaced are often able to imagine and even create a new world. These dynamic notions of home suggest that a sense of home can be refashioned by individuals, either in their places of origin or elsewhere (Zetter, 1999).

Despite the material and emotional hardship that displaced people often experience, they are able to transform an unfamiliar and at times hostile place into a new home (Hammond, 2004). The experiences of refugees from Burundi (Malkki, 1995a), Ethiopia (Hammond, 2004), Cyprus (Loizos, 2011; Taylor, 2013a, 2015a) and the successor states of Yugoslavia (Korac, 2009; Jansen, 2011), for example, suggest that following displacement it is possible to rebuild a sense of community, identity and belonging and to regain a sense of home. Based on the experiences of repatriated Ethiopian refugees from Sudan, Hammond (2004) shows that home can be refashioned on the move through two conceptually different but complementary and simultaneous processes: *emplacement* and *community formation*. The term 'emplacement' often denotes 'a continuous process of making one's place in the world' (Hammond, 2004: 82), meaning the transformation of an unknown geographical place into a personalised, liveable and socialised one. Others refer to emplacement simply as the 'flipside' of displacement (Malkki, 1995b: 517) or the process of 'nesting' (Korac, 2009: 39–42). Korac's conceptualisation of emplacement, whose account draws on the experiences of refugees from former Yugoslavia who were granted asylum in Rome and Amsterdam, entails two analytically distinctive processes: taking control of one's life and reconstructing life. While the first denotes the process of becoming self-sufficient and independent, that is, regaining a 'sense of normality' (Korac, 2009: 61), the second refers to the process of becoming part of a new community. As Korac (2009: 42) stresses, 'emplacement does not take place in a social vacuum; rather it occurs within the context of intra- and inter-group relations'. The term 'community formation' refers to a process of creating social, cultural and political attachments within one's place of settlement and the process of developing a sense of identity and belonging through everyday social practices (Hammond, 2004). Hammond's notion of 'community formation' is closely related to what Korac refers to as the process of 'reconstructing life'. What is central for this discussion is that in Korac's approach, the process of remaking home is shaped by both the *need for continuity* and the *need for change* where,

> the need for continuity implies a search for links with social roles, meanings and identities embedded in one's life as it was before the flight. The need for change requires a flexibility and openness to reshaping (some) of these old roles, meanings and identities.
>
> *(Korac, 2009: 40)*

Remaking home is best understood as a process (see Hammond, 2004; Taylor, 2015b) and one that is more complex than merely going back to one's place of origin or resettling elsewhere. Hage's (1997: 102) account of Lebanese migrants in Sydney suggests that the process of home-building is an affective construct which entails the 'building of the feeling of being "at home"'. This involves four other interrelated feelings: security, familiarity, community and a sense of possibility. A sense of security is felt by being in a place in which individuals can satisfy their basic human needs, make their own rules and experience 'the absence of harmful threatening otherness'. Familiarity involves the creation of a space in which people possess a maximal practical knowhow and a maximal spatial knowledge, where one knows 'how it works'. Community refers to the sense of being in a place in which 'one recognises people as "one's own" and where one feels recognised by them as such' – in other words, where cultural practices and even moral values are shared. The sense of possibility derives from being in a place open to opportunities for a better life, where one can experience a sense of personal development and progression (Hage, 1997: 102–103). Obeid (2013: 369) adds a further aspect to being at home: the feeling of having a 'project of rootednesses' by which she means 'purposeful activities that demand a long-term commitment and produce attachments that root persons in a particular place'.

That some of the displaced can remake home in a new cultural setting does not mean that the loss of home is a minor bereavement. As Habib (1996) and Al-Ali and Koser (2002) suggest, the loss of home might be significant to such a degree that some displaced persons are unable to ever regain a sense of home, either in their communities of origin or elsewhere. For others, however, return is not the only way to remake home. Indeed, home can be simultaneously fixed and mobile (Ralph and Staeheli, 2011). As Ahmed et al. (2003: 1) state, 'being grounded is not necessarily about being fixed; being mobile is not necessarily about being detached'.

Remaking a place called home

For many displaced people, remaking a home involves reconstructing four interrelated dimensions: a material place, a social world, a familiar landscape and an emotional and existential place.

Receiving a regular income is critical to finding shelter and food. For many this was challenging as their rural labour skills were not suited to the types of jobs available in the urban areas in which they settled. However, even for those who were able to provide these material needs of shelter and food, satisfying them did not necessarily bring a feeling of being at home as many struggled to transform their physical shelter into a home.

For many, such as Marissa 'owning a house is a fundamental step in the process of making a home'. Indeed, as Jansen (2011: 59) suggests, 'home is not simply a shelter from unsafety, but also a base where insecurity and uncertainty can be reduced and confronted'. Yet, several participants said that they had worked so hard to buy a house but were disappointed when they realised that a house was not necessarily synonymous with a home. For Cesar buying his own house was important but at the same time he realised that the people who would make it a home were absent. He said 'my siblings are dead and my mother lives far away from this place. I already have a house, but I do not know when it will become home'.

Clearly, time is a central aspect in the process of remaking home. With few exceptions, narratives of displacement show that transforming a shelter into a home takes time. Mateo struggled for over 25 years to find a home. He said that displacement destroyed his world and he struggled to reinvent his life in a different place:

> Making a new home takes time and effort. You don't only need to satisfy basic needs of food and shelter but to create a new world for you and your family. In other words, you need to create a world in which you no longer feel like a foreigner in your own country.

Besides finding a physical place to call home, the process of remaking home requires the reconstruction of an individual's social world: in other words, the making of a place that embodies a sense of family, community and belonging. That is, membership of a social group based on shared traditions, values and a 'way of life' (Nash, 2002; Bennett and McDowell, 2012) as well as feeling recognised and acknowledged by others as belonging to a community (Christou, 2011).

Those whose relatives were killed or left behind struggled to remake home on the move. Piedad explained how the loss of his family led to feelings of solitude and of being in a foreign land in one's own country.

> When I see families sharing together on the streets of Bogotá, I feel like what I am: a displaced person. I feel like a foreigner in Bogotá … You can overcome the material impacts of displacement but not the sense of losing a family.

Others such as Mateo found it easier to remake a home by reuniting their family and adopting an attitude of determination to make Medellín their new home. He says,

> We lost our home, I mean everything that is essential for human life. We will never forget that but Medellín is our new world. We have new ideas and new questions in this city. It's like a new beginning. I personally feel in my world again, in my place. This is home.

Displaced people also said that regaining a sense of community, of shared cultural identity and belonging in the new place was vital to reconstructing a home. Those who felt that host communities perceived them to be outsiders or with mistrust said they struggled to feel at home in their new environment. As Michael explained, 'I do not feel at home in this place. I do not belong here. No one looks after me … Nobody trusts in me and I trust in nobody'. In contrast, those who built new friendships based on trust found it easier to settle. Nancy's narrative illustrates this connection between community and home:

> The first time I experienced the sense of being at home in this city was when I took part in a TV show. The audience and then people in my neighbourhood showed me support as if I were part of this place. After many years living in this city, I smiled with happiness. It wasn't a fake smile but a real one … For the very first time in years I experienced the sense of being part of this community.

For others, replicating social and cultural practices of the places left behind in a new setting helped the displaced to 'feel more at home'. Diana said that reproducing cultural traditions in new settings helped her overcome her feelings of being out of place: 'Keeping my own culture is not only a way to express myself but to find my place in this city'. Marina explained that inviting neighbours to enjoy traditional recipes from her hometown enabled her to strengthen social ties in her new community and was a way for her to express her identity: 'We share the food and have such a great time together in this house. It's what the feeling of being at home is all about'.

Remaking a sense of place entails the transformation of an unfamiliar milieu and its social and cultural practices into a familiar space (Massey, 1995). This often involves different scales of place, from the nation to a specific building, as evidenced through the experiences of Mateo and Jasbleidy. Mateo was able to build a shelter for him and his family but only started to see it as home when he was able to create a 'world' in which he no longer felt like a foreigner in his own country. His process of remaking home involved the house, the neighbourhood, the city and the country. Jasbleidy, by contrast, focussed on a micro scale of place to remake her sense of home. She began to feel at home when she started to earn money and buy things to make her flat more comfortable.

Becoming familiar with a place takes a long time. Even a decade after being displaced, people continued to notice and reflect on differences between their new settings and the places left behind. This was especially the case for those who moved from rural areas. They often compared their previous lives, where they had daily contact with animals and lived close to 'nature', with their current existence in the city amidst tall buildings. These recollections reminded them of being far from home. Luzmila explained that in the place left behind she used to see happiness and peacefulness, and in the new place only sadness and solitude. While displaced people often romanticise their place of origin, Bennett and McDowell (2012) confirm that, for some, the colour goes out of their lives when they are displaced.

Socio-cultural engagement with neighbours also plays a central role in creating a sense of home. The displaced adapt to the city through everyday social practices and engagement with neighbours. Jasbleidy explained how, when celebrating black festivals and carnivals in Bogotá, locals become curious and often asked if they could join in. Through everyday social practices, the displaced give new meaning to the places in which they have settled. Through cooking, working or communicating with neighbours, the new place begins to become familiar. To illustrate, the recreation of recipes connects the displaced with their places of origin and helps them to overcome the emotional impacts of displacement. Jasbleidy stressed this when she said 'What I like most in terms of food is a dish called tapado [plantain, fresh fish and a sauce made with onion, tomato and spices]. Eating tapado, I feel like I'm at home'. The longing for traditional food is not to imply that the displaced aspire to return to their abandoned place. Instead, as Obeid writes (2013: 374), 'what seems like a yearning for the past can contribute very much to the creation of the present and the future'.

The internal decoration in places of dwelling also plays a role in remaking home. Many replicated the decor of their homes left behind as a way of evoking the feeling of being at home in the new place. Rosemary said, 'I have a painting which recreates the landscape of my hometown. When I look at this painting, I feel myself to be home'. In several cases, when fleeing their home, they chose objects that often had little monetary value but were of great emotional value. The first time Anilpa was compelled to move she selected a photo album and a notebook of her stories. 'By looking at my photos and reading my tales I used to feel at home. After displacement, these were the things that evoked my home'. Before her second displacement, however, her possessions were burned by those forcing her to move. The loss of her personal belongings ruptured her connection with her childhood and her home.

> My photos and tales inform who I used to be and who I am, and the paths I want to walk in the future. You know, human beings develop strong attachments to material things. But it's not because of the things themselves but because of what these things signify for every human being.

Anilpa's reference to the emotional value of material things relates to what Appadurai (1996) refers to as the non-material value of material things. Objects from the place left behind help the displaced to develop a sense of home in new cultural settings primarily because of the personal connection between the 'thing' and the individual.

Some people referred to emotional-existential dimensions of home whereby they attempted to 'heal the soul' and 'refashion their life project'. The former requires leaving past experiences of conflict and displacement behind and the latter the ability to formulate new aspirations and purpose in life.

In terms of healing the soul, many of the displaced emphasised the importance of overcoming the emotional impacts of conflict and displacement in order to remake a home. Maria Esperanza's account illustrates this:

> Because of the injustices we've lived, we [the displaced] are often full of anger and resentment. When you're full of hate, you see a future nowhere. The first two years following displacement I just wanted to die. I didn't even look after my physical appearance. I felt completely alone, thinking only about what I have lost. My partner died, my older son was left behind, my older daughter was living in a different city, and I had all the responsibility for looking after my youngest children ... Then God helped to heal my soul. If you don't overcome the desire for revenge, you can't move forward.

The reconstruction of the emotional-existential feeling of being at home, however, is not a straightforward nor an easy process. Sorrow, pain, psychological distress and melancholy for the place and life left behind, and the material and emotional struggle of living in the present and planning for the future, were recurrent themes in the narratives of the displaced.

Those who fled after being raped and tortured, or after having witnessed the murder of relatives, said that they struggle to find an existential home. Piedad, who was raped and witnessed the murder of her parents, said 'I would not feel at home again if I cannot heal all my wounds'.

The ability to regain a sense of hope, to find a purpose in life and to use their voice and develop a capacity to aspire to transform their lives were all central to creating a sense of home. Agency and the capacity to aspire enabled the displaced to see themselves as purposive social actors with the ability to look after their own interests and those of others (Appadurai, 2004). Although conflict and displacement compromised their agency to contest and enquire, an overwhelming number of people said that they had become increasingly conscious of their political rights following displacement. For example, women whose role in the places left behind was often restricted to the domestic sphere became more aware of how they could use their voice to transform their social reality. For example, the experience of attending university helped María-Esperanza to understand her role in society. While her mother had taught her that 'women don't need education to look after their houses, children and husbands', following her studies on human rights she began to lobby for the rights of the displaced and to challenge dominant perceptions about women's role in society. She stated that she had spent many years following being displaced in searching for a place called home. But it was only when she was able to 'heal her soul' and use her voice that she was able to 'relaunch' her life and begin to see Bogotá as her home.

Conclusion

Following conflict and displacement, people forced to move from one place to another often lose their homes. The loss of a home is so profound and devastating that its remaking almost always involves a long process for which there is often no obvious start or end point. While those interviewed associated the loss of their home with time- and place-specific experiences of conflict and displacement, they struggle to identify if, when and how they will ever begin to feel 'at home' again. While it is often assumed that over time the longing for their place of origin will diminish and displaced people will start to remake a home, this is not always the case. The narratives of those who fled their homes in the early 1980s and 1990s reveal that, even after decades, their idea of home is only in their memories and making a new home remains an unfulfilled aspiration (Mallett, 2004; Arp Fallov, Jørgensen, and Knudsen, 2013; Taylor, 2015b).

The process of remaking a home involves much more than the physical structure of a house and finding a source of livelihood. It requires the reconstruction of a social world, of a sense of place and of emotional and existential feelings of being 'at home'.

Regaining a sense of family and of community were often described as some of the most critical aspects in the process of remaking a home. These were particularly challenging to achieve for those whose relatives were killed or forcibly disappeared and for those who had strong cultural attachments to a place such as Indigenous and black communities who tended to see home as a collective space. Those who fled with their families or were subsequently reunited with them were more readily able to remake a home.

A sense of place was also achieved through a process of familiarisation with the geographical features of a landscape and its social and cultural practices. Although the urban settings in which the displaced settle lack the geographical features familiar to them, creativity and imagination help many of them to transform an unfamiliar place into a home. Through the interior decoration of their houses, some have been successful in evoking the landscape, the mountains and rivers, that they were forced to abandon. Social and cultural practices such as sharing traditional ways of preparing food, rituals and religious ceremonies were also significant in transforming an unfamiliar place into a home. Ordinary objects such as photographs, drawings and ceramics recreating the landscape left behind constituted symbols of home-making (Zetter, 1999; Flanders, 2015).

Narratives of displacement indicate that people develop a sense of home in geographical and sociocultural terms but also through emotional-existential feelings. Many said that dealing with the trauma and sorrow associated with violence and death is central to the process of remaking home. For some, home could only be remade through 'healing the soul', in other words, when the experiences of violence and the desire for revenge could be consigned to the past. Because the emotional and existential impacts of conflict and displacement are often overwhelming, this process is replete with symbolism. These include rituals to remember murdered or forcibly disappeared relatives, writing or music therapy to exorcise violence, and reconciliation and forgiveness. In spite of their efforts to 'heal the soul' and find a new purpose in life, women who were raped and children who fled their communities after being forcibly recruited by illegal armed groups found the reconstruction of the emotional-existential home to be especially challenging.

Previous research on making a home following displacement has focussed on the material and sociocultural dimensions of emplacement (Hammond, 2004; Korac, 2009; Den Boer, 2015; Taylor, 2015a). What I have tried to show in this chapter is the centrality of the more symbolic, emotional and existential dimensions of the process of remaking home.

Acknowledgements

The findings presented in this chapter are part of a PhD research project carried out at the Global Development Institute at The University of Manchester with the financial support of the Administrative Department of Science, Technology and Innovation of Colombia, COLCIENCIAS. I am so grateful to Uma Kothari and Diana Mitlin for their constructive criticism, encouragement and invaluable support during my PhD. The fact that their engagement with my research did not end when I completed my PhD has contributed to my sense of belonging to the University of Manchester even more strongly. I also want to express my special gratitude to María del Pilar Bohada Rodríguez for her never-ending interest and support with my research and to all research participants whose experiences of conflict and displacement provided the empirical material and inspiration for this research. Last but not least, special thanks to the editors for having selected my chapter for this volume and to Uma Kothari for her insightful comments and editing my work.

References

Ahmed, S., Fortier, A.-M., Castañeda, C., Sheller, M., (2003). Introduction. Uprootings/Regroundings: Questions of Home and Migration, in: *Uprootings/Regroundings: Questions of Home and Migration*. Berg, Oxford and New York, pp. 1–19.

Al-Ali, N., Koser, K., (2002). Transnationalism, International Migration and Home, in: Al-Ali, N., Koser, K. (Eds.), *New Approaches to Migration?: Transnational Communities and the Transformation of Home*. Routledge, London and New York, pp. 1–14.

Appadurai, A., (1996). *Modernity at Large: Cultural Dimensions of Globalization*. University of Minnesota Press, Minneapolis, MN.

Appadurai, A., (2004). The Capacity to Aspire: Culture and the Terms of Recognition, in: Rao, V., Valton, M. (Eds), *Culture and Public Action*. Stanford University Press, Redwood City, CA, pp. 59–84.

Arendt, H., (1966). *The Origins of Totalitarianism*. New Ed/with added prefaces. Ed. New York. Harcourt Brace Jovanovich, New York.

Arp Fallov, M., Jørgensen, A., Knudsen, L.B., (2013). Mobile Forms of Belonging. *Mobilities* 8, 467–486.

Bennett, O., McDowell, C., (2012). *Displaced: The Human Cost of Development and Resettlement*. Palgrave Macmillan, New York.

Blunt, A., Dowling, R., (2006). *Home*. Routledge, London and New York.

Christou, A., (2011). Narrating Lives in (E)motion: Embodiment, Belongingness and Displacement in *Diasporic Spaces of Home and Return*. *Emotion, Space and Society* 4, 249–257.

Den Boer, R., (2015). Liminal Space in Protracted Exile: The Meaning of Place in Congolese Refugees' Narratives of Home and Belonging in Kampala. *Journal of Refugee Studies* 28(4), 486–504.

Flanders, J., (2015). *The Making of Home: The 500-Year Story of How Our Houses Became Homes*. Main edition. Ed. Atlantic Books, London.

Habib, N., (1996). The Search for Home. *Journal of Refugee Studies* 9, 96–102.

Hage, G., (1997). At Home in the Entrails of the West: Multiculturalism, "Ethnic Food," and Migrant Home-building, in: *Home/World: Space, Community and Marginality in Sydney's West*. Pluto, Sydney, pp. 99–153.

Hammond, L., (2004). *This Place Will Become Home: Refugee Repatriation to Ethiopia*. Cornell University Press, Ithaca and London.

Jansen, S., (2011). Troubled Locations: Return, the Life Course and Transformations of Home in Bosnia-Herzegovina, in S. Jansen and S. Löfving (eds) *Struggles for Home: Violence, Hope and the Movement of People*. Berghahn Dislocations, Berghahn Books, New York and Oxford, pp. 43–64.

Jansen, S., Löfving, S., (2011). Towards an Anthropology of Violence, Hope and the Movement of People, in: Jansen, S., Löfving, S. (Eds.), *Struggles for Home: Violence, Hope and the Movement of People*. Berghahn Dislocations. Berghahn Books, New York and Oxford, pp. 1–23.

Korac, M., (2009). *Remaking Home: Reconstructing Life, Place and Identity in Rome and Amsterdam*. Berghahn Books, New York and Oxford.

Loizos, P., (2011). The Loss of Home: From Passion to Pragmatism in Cyprus, in: Jansen, S., Löfving, S. (Eds.), *Struggles for Home: Violence, Hope and the Movement of People*. Berghahn Dislocations. Berghahn Books, New York and Oxford, pp. 65–84.

Malkki, L., (1995a). *Purity and Exile: Violence, Memory, and National Cosmology Among Hutu Refugees in Tanzania*. University of Chicago Press, Chicago, IL.

Malkki, L., (1995b). Refugees and Exile: From "Refugee Studies"; to the National Order of Things. *Annual Review of Anthropology* 24, 495–523.

Mallett, S., (2004). Understanding Home: a Critical Review of the Literature. *Sociological Review* 52, 62–89.

Massey, D., (1995). The Conceptualization of Place, in: Massey, D., Jess, P. (Eds.), *A Place in the World?: Places, Cultures and Globalization*. Oxford University Press, Oxford, pp. 45–85.

Nash, C., (2002). Genealogical identities. *Environment and Planning D: Society and Space* 20, 27–52.

Nordstrom, C., (1995). Creativity and Chaos: War on the Front Lines, in: Nordstrom, C., Robben, A. (Eds), *Fieldwork Under Fire: Contemporary Studies of Violence and Survival*. University of California Press, Berkeley, California, pp. 129–153.

Obeid, M., (2013). Home-Making in the Diaspora Bringing Palestine to London, in: Quayson, A., Daswani, G. (Eds.), *A Companion to Diaspora and Transnationalism*. John Wiley & Sons, Hoboken, NJ, pp. 366–380.

Ralph, D., Staeheli, L.A., (2011). Home and Migration: Mobilities, Belongings and Identities. *Geography Compass* 5, 517–530.

Taylor, H., (2013a). Refugees, the State and the Concept of Home. *Refugee Survey Quarterly* 32, 130–152.

Taylor, H., (2015a). *Refugees and the Meaning of Home, Migration, Diasporas and Citizenship*. Palgrave Macmillan UK, London.

Taylor, S., (2015b). "Home is Never Fully Achieved ... Even When We Are In It": Migration, Belonging and Social Exclusion within Punjabi Transnational Mobility. *Mobilities* 10(2), 193–210.

Zetter, R., (1999). Reconceptualizing the Myth of Return: Continuity and Transition Amongst the Greek-Cypriot Refugees of 1974. *Journal of Refugee Studies* 12, 1–22.

42
Homelessness and place

Alison Young and James Petty

Introduction: no place like home

A house is always more than a place in which we sleep, eat, and store our possessions. Housing creates a place that an individual can make into a home – a home being 'a particular location that has acquired a set of meanings and attachments' (Cresswell, 2009: 169) to do with privacy, refuge, family, and the self (see also Meier and Frank, 2016). The taken-for-grantedness of access to places categorised as homes can be seen in the fact that we do not commonly refer to people with housing as 'homed', whereas individuals who do not have access to housing are frequently referred to as 'homeless' rather than 'unhoused'. Thinking of this situation as constitutive of homelessness recognises that when an individual lacks access to housing, the place that is absent is far more than a location for sleeping, eating, and storage of goods.

Heidegger posited a conceptualisation of 'dwelling' as not merely occupation of a living space such as a house, but as *being-in-the-world* (1971); with dwelling in a single place seen as essential to the formation of the self in the world – a monadism criticised by Delueze and Guattari as 'the stable, the eternal, the identical, the always constant' (1987: 361), in contrast to the nomadic becomings through which a subject is always in process. Societal models of housing allocation whereby individuals or families hold tenure in singular houses or apartments tend towards the Heideggerian version of dwelling as the pre-eminent means by which individuals can discover 'a way of being, a way of doing, and a way of relating' (Brun, 2016: 426) that is approved by society.

Those without tenure over property that can provide them with shelter are compelled to create places other than houses in which to sleep, eat, and store goods. Homelessness disrupts the abilities of individuals to invest places with the affective attachments we commonly associate with homes; the need for shelter and storage, however, is an ongoing one, meaning that unhoused individuals must find places other than houses in which to sleep and eat. In considering the relationship to place of individuals experiencing homelessness in Melbourne, Australia, at the outset we would note the importance of housing to a sense of self, and the need for expansion of access to housing for people experiencing homelessness. In what follows, we seek to acknowledge and support the ways in which people experiencing homelessness utilise or adapt urban places for the activities associated with life in the place we call home (sleeping, eating, storing our belongings, and conducting relationships with family members; see Blunt and Dowling, 2006).

Homelessness is widely recognised to be one of the most serious social problems in Australia today. For several years running Melbourne was awarded the accolade of 'World's Most Liveable City' by *The Economist*, but its much-heralded liveability is inaccessible to many of its inhabitants. Like other Australian cities, Melbourne has a substantial population of individuals experiencing homelessness of varying kinds and duration (Mechkaroff et al., 2018) along with numerous individuals enduring social or economic precariousness (Standing, 2014). In recent years, homelessness has become a problem of national significance, experienced by as many as 1 in 200 individuals (Homelessness Australia, 2016). Around 90% of individuals annually are not assisted with their housing needs (Incerti, 2018), meaning most endure living conditions that are temporary, unsafe, unstable, or absent. Chamberlain and Johnson (2016) state that about 2.35 million Australians have experienced homelessness during their lifetimes, with 59% of them (about 1.4 million) sleeping rough as a result and Muir et al. (2018) note the increasing number of people experiencing homelessness is outpacing population growth.

Rough sleeping and seeking donations constitute the stereotypical activities associated with homelessness but involve a relatively small proportion of those without stable or permanent accommodation. People experiencing homelessness constitute a heterogeneous population (Jahiel, 1992). Murphy and Tobin refer to the 'housed continuum' (2011: 4), with those in stable accommodation at one end and rough sleepers at the other. Individuals experiencing homelessness might sleep rough but are more likely to couch surf or move from shared house to shared house with no secure tenancy, sleep in their car, move continually in and out of more or less stable accommodation, or sleep rough sometimes while accessing crisis accommodation at other times. Despite the diversity and complexity of this group and its practices, thanks to a dynamic through which the homed person attributes homelessness to a person who is begging or rough sleeping (Gerrard and Farrugia, 2014), individuals using public places for these activities tend to be categorised as 'homeless' even though, as Arnold (2004) points out, whether or not someone engages in practices such as begging does not reveal the nature of their current living conditions.

In this chapter, 'rough sleeping' describes the situation of sheltering within the locales of the city (rather than in a car, or on a sofa in a friend's house, and so on). 'Donation seeking' describes the act of asking for money, food or other items from strangers in public places – an act commonly, and often pejoratively, known as 'begging' in social policy and criminal law (Adams, 2014; Lynch, 2009; Walsh, 2011). In what follows, we discuss three aspects of place-making and place-regulation: first, some of the ways in which people experiencing public homelessness utilise elements of the urban environment in order to adapt places for the activities of donation seeking and rough sleeping; second, the process by which such uses of highly visible public places have been constructed as forms of urban disorder requiring enhanced regulatory responses; and, finally, some of the techniques deployed by local authorities to regulate public homelessness in urban places.

Homeless place-making

Individuals who are sleeping rough or seeking donations must find places within a city in which these activities can be carried out. These are distinct behaviours, although they may co-occur. However, a person may be begging in one area but sleeping elsewhere; further, begging does not in itself indicate that a person will be sleeping rough rather than couch-surfing, or sleeping in a car, or in temporary accommodation. The differences between these may appear minor but are important: the places chosen for each activity look different and are used differently.

What places are used by individuals seeking donations and what does such a place look like? A location must be visible enough to attract the attention of passers-by but, in order to minimise the chance of police intervention, donation seeking must not seem intrusive to pedestrians, allowing them to ignore it if they choose ('begging' continues to be an offence in many jurisdictions: Adams, 2014; Young and Petty, 2019 forthcoming). A location for donation seeking is likely to exhibit few features that would distinguish it from any other point on a street or pavement, but locations next to ATMs, small supermarkets, and convenience stores, venues that generate cash for pedestrians moving through the city, are the most pragmatic choices for donation seeking.

Once selected, a place is adapted for donation seeking through a combination of a small number of accoutrements that communicate the request for money or food to passers-by and through the presence of an individual whose body is arranged in a posture that combines supplication and minimal presence (Kawash, 1998). Individuals conventionally sit on the pavement, often with legs crossed or tucked under, back against the wall. Occasionally donation seekers sit with heads bowed, and occasionally kneel in a position of prostration, but most commonly donation seekers gaze straight ahead; their lowered sitting position means the line of their gaze intersects with the legs of passers-by.

Donation seekers might sit directly on the ground, but most sit on a folded blanket or sleeping bag, or on sheets of cardboard. Fabric and cardboard rarely protrude or extend from beneath, and individuals sit as though they aim to take up as little space on the footpath as possible. There may be an upturned cap or cardboard cup in which a few coins can be seen; a cardboard sign might also be propped against the wall or in front of the coin receptacle. Signs are hand-written; sometimes brief ('SPARE ANY CHANGE?', 'PLEASE HELP' and so on), or more discursive, setting out biographical details that led the individual to be seeking donations – a missed train, or lost job, seeking a new life after domestic violence, and so on ('GROWING UP, NEVER DID I DREAM I'D END UP LIVING ON THE STREETS … '). Occasionally signs engage the passer-by directly: 'WE ARE HOMELESS GRATEFULL FOR ANY & ALL HELP AFFORDED TO US HAVE A GREAT DAY AND GOD BLESS' and 'JUST BECAUSE I LOOK RESPECTFUL DOESN'T MEAN I'M NOT HOMELESS' read one sign displayed in Melbourne's CBD.

If there is no individual present – perhaps they have gone to use a public toilet, or to an appointment with a service provider – then the place of donation seeking may be identifiable only through the presence of items such as a few squares of cardboard, used for sitting on. At some sites, individuals would tidy away their unattended belongings in cubby-holes or alcoves in nearby laneways: cardboard might be left on the street but a blanket would be more likely to be removed. Places of donation seeking thus display minimal indicators that the place has been adapted from its imagined functions or qualities (thoroughfare, pavement, exterior wall).

Just as places for donation seeking are primarily chosen for proximity to the infrastructure of cash-generation, places for rough sleeping will provide comfort and shelter for the sleeper. These, however, are scarce; rough sleepers must adapt the meagre possibilities offered by existing buildings or use temporary materials to construct makeshift sleeping spaces. When adapting a place for sleeping, rough sleepers choose doorways, alcoves, recesses, archways, and other hollows that might be reconceptualised as frames for a bed, with sleeping bag or bedroll spread out within the recessed space. The narrowness of alcoves and doorways means the sleeper often lies on their side so as not to extend their body out onto the pavement, which could result in police attention (see Walsh (2008) on the policing of marginalised individuals such as those experiencing homelessness).

The activities of rough sleeping and donation seeking are often kept separate by individuals experiencing homelessness, but sometimes coincide. At such sites, sleeping bags remain unrolled, bags of belongings stacked alongside, while an individual sits or lies next to a coin receptacle and sign. The use of places for donation seeking or sleeping is often short-lived, with sites exhibiting the dynamic qualities of *emplacement* (adapted for sleeping or donation seeking) and *displacement* (when an individual leaves or is moved on by police) that Lems (2016) attributes to the place-making of individuals who are prevented from settling in place (see also Holmberg and Persson, 2016; note the contrasting metaphorical 'immobility' or fixity in place in Kerr's research recounted by individuals experiencing homelessness, 2016). However, when used both for sleeping and donation seeking, sites endure longer; and more substantial additions to the place are made.

One woman established a location for sleeping and donation seeking next to an ATM outside a convenience store, at the intersection of two main streets in Fitzroy, in the City of Yarra. She set up sections of two cardboard boxes, cut down on one side and positioned so that her head and feet would be contained within the walls of each box. In between, multiple layers of cardboard were placed. A bedroll, pillows and bed covers were laid inside this construction. Over the rear pieces of cardboard she stretched pieces of fabric, creating the effect of a tasselled bed head or canopy. A palm frond was inserted behind the pillows so that it arched up against the wall and extended over the bedding.

Plastic bags of belongings were initially placed beside her; after several days of living in this location, the woman obtained two plastic stick-on hooks which she attached to the wall next to the ATM; the plastic bags were then hung from them, keeping them off the ground and perhaps reducing the chances of them being removed or stolen. After a day or so more, a milk crate was added next to the cardboard boxes; the woman had covered it in black, yellow and cream-patterned fabric. Occasionally, when she was not at the site, a cardboard sign would be propped next to the crate, stating 'PLEASE DO NOT TAKE MY SEAT??? AS I'VE JUST GONE TO THE TOLIET THANK U'.

After occupying this location for around ten days, the woman moved her boxes and belongings around the corner. The boxes were sticky-taped to the wall to hold them in place, and the palm frond was re-installed. Two cardboard signs were now placed next to the bed: one stated that she had missed her train, the other that she had left a situation of domestic violence, lost custody of her children, and had hoped to stay with her niece but was unable to do so. Three plastic bags of belongings were lined up next to the cardboard bed; the fabric-covered milk crate had also been moved and now functioned like a bedside table, with donated food or drinks placed on it.

Despite having set up at this new location, after only a few days the woman had disappeared. The cardboard boxes remained, as did some personal items such as a notebook, a small towel, some drawings, and the hand-written cardboard signs. After a few days more, the boxes had been flattened, as though someone had jumped on them; the towel and palm frond could be glimpsed between the layers of cardboard.

Over the course of two and a half weeks, two 'homes' had been made by the woman, utilising a combination of donated items, personal belongings and materials such as cardboard and sticky tape. That these materials were vulnerable to being removed by others was something she was clearly aware of, writing a sign to explain temporary absence from the location in an effort to safeguard the milk crate, and storing her bags on hooks, but upon her relinquishing them, the materials utilised to adapt the locations for shelter and storage were devalued into waste. Her departure from the area made visible a moment 'in which things oscillate between value and worthlessness' (Boarder Giles, 2014: 94). Having been objects of utility with situational value in designating places to sleep and sit, the objects became worthless, compressed into layers and subsequently removed by the council (on the designation of materials as 'waste' see Edensor, 2005).

Removal of personal items as if they were unambiguously 'waste' was one of the proposals made by the City of Melbourne in 2017, generating months of debate around the impact of public homelessness in the city. Rough sleeping and donation seeking became the focus of media and policy controversy, as the permissible limits of homeless place-making were contested and reconstituted. In the midst of claims that people experiencing homelessness make places disorderly, we investigated public homelessness within two inner-urban municipalities, the City of Melbourne, and its neighbour, the City of Yarra.

The City of Melbourne claimed that publicly visible homelessness adversely impacted upon members of the public, by constituting an impediment for other users of public places and by generating excessive amounts of waste. To investigate these claims, we conducted discreet, non-intrusive observation of sites being used for these activities in the two municipalities (Young and Petty, forthcoming, 2019). Of the 29 places that we observed, many were situated in high-visibility commercial areas, such as Brunswick and Smith Streets in Fitzroy, Victoria Street in Richmond in the City of Yarra and Collins, Swanston, Bourke and Flinders Streets in Melbourne's CBD. These sites were the object of repeated observation for periods lasting from 10 minutes to approximately half an hour. In other places, data was collected in one-off periods of observation. The dynamic nature of the practices of individuals experiencing homelessness meant that several areas with high activity at the beginning of the study later became inactive, while additional sites were added, as places not initially slated for observation, but that became actively or frequently used.

The selected sites of visible homelessness, and their associated materials, varied dramatically. Some locations consisted of a single person sitting on the street against the wall of a building; others had multiple people, mattresses, milk crates, animals, blankets, food, and bags. Belongings were usually highly organised and neatly arranged, with minimal impact on other street users. Pedestrians had to avoid stepping on or running into the person seated on the pavement, although this would require only the same amount of care required for any non-homeless person encountered on the street.

Very occasionally, locations showed signs of more haphazard arrangements, or involved the occupation of more pavement space. In these sites, pedestrians were required to proactively navigate the materials to avoid stepping on them, or to step around the people using the site. All those observed managed this with ease, and the degree of impediment presented was no different to that caused by any group of people pausing mid-footpath to consult a map, make phone calls, or converse. The main difference worth noting is that a group of homeless individuals, or a cluster of their belongings, occupies a section of the footpath for what may be an extended period of time. However, sites with a lot of belongings were less likely to be in prime commercial spaces or on major pedestrian through-routes. The few locations that both had a considerable amount of materials and a location on busy pavements were generally maintained to a high degree of neatness.

Some sites were marked by low activity, with people either sleeping or huddled up. In others, people engaged actively with passers-by, chatting, calling out, or displaying a variety of signs. Researchers observed the public interacting with the people inhabiting observation sites many times, usually to donate. Donations usually appeared to be money, with food or drink observed as an alternative donation. No interactions were observed that appeared to be unsupportive or aggressive. Members of the public either continued their activities apparently unaffected, or engaged in an apparently supportive interaction with the homeless person. The relative frequency of donations, whether of money or food, and of conversations indicates that many hold a sympathetic perspective on homelessness and wish to engage in a helpful or positive manner. To that extent our research meshes with the findings generated in a survey of public attitudes conducted by the Victorian Department of Health and Human Services (2018), in which 28% of the 1,010 respondents indicated that they 'want to help'.

Making Melbourne's places of homelessness into places of disorder

Why did homeless place-making become contested in Melbourne in 2017? While no doubt in part due to the increased number of unhoused people engaging in rough sleeping, a number of high profile events intensified concerns about the place of homelessness in Melbourne. In 2015, many individuals experiencing homelessness had moved into central locations in the city, a fact that received extensive negative coverage in the local tabloid press (see, for example, Gillett, 2015; Masanauskas, 2015). The following year, individuals without housing set up a protest camp outside the Melbourne Town Hall; for three months, city officials and police dismantled the encampments and evicted homeless protesters repeatedly.

Six months later, in January 2017, an encampment of homeless people was established on Flinders Street, outside Melbourne's busiest railway station; a tourist attraction as well as a hub for thousands of commuters every day. The existence of the encampment was extensively, and negatively, covered in the news media. According to Victorian Police Commissioner Graeme Ashton, the camp was 'disgusting' and 'a very ugly sight' (Booker and Dow, 2017: unpaginated); one journalist called it a 'cesspit … like something you'd find in Delhi' (Panahi, 2017: unpaginated).

The intensity of the media coverage of the Flinders Street encampment prompted Melbourne's then Lord Mayor, Robert Doyle, to announce plans to amend Melbourne's Local Laws. Mayor Doyle claimed that homeless people's belongings were impeding free movement within public places and blocking access to city amenities for its other users (Doyle, 2017). Homelessness, said the mayor in a gesture consonant with the tendency to construct the homeless as a corrosive influence within contemporary societies (Parsell, 2011; Parsell and Phillips, 2013), was 'a blight on our city' (No Author, 2017).

Conceptualising homelessness as a 'blight' allowed the City of Melbourne to focus on the visual impact of the presence of people experiencing homelessness and their possessions within the streetscape. The changes would allow the council to excise unwanted goods or individuals from the streetscape just as a gardener or farmer would prune blight-affected parts of plants or crops. Council staff would target material items visible in public places – namely, temporary shelters and personal belongings. The former would be achieved by changing the nature of the municipality's ban on camping: council employees would be empowered to clear encampments even when established with materials such as cardboard.

The second proposed change would create powers to remove personal belongings:

> A person must not leave any item unattended in a public place. If an item is left unattended, an authorised officer may confiscate and impound the item, and can sell, destroy, or give away the item if a fee or charge is not paid within 14 days.
>
> *(Melbourne City Council, 2017)*

In addition to this 'retrieval fee', leaving belongings unattended would attract a $250 fine. In support of the proposed amendments, Mayor Doyle claimed that the belongings of homeless people constituted impediments to movement within public space and access to the amenities of the city for its other users (Doyle, 2017).

The public response was overwhelmingly opposed to the proposals: 84% of responses rejected the extended definition of camping and the power to confiscate possessions; 98% rejected the proposed fines (City of Melbourne, 2017a). In September 2017, the City of Melbourne decided not to adopt the amendments, opting instead for a formal Operating Protocol developed in conjunction with Victoria Police and homelessness service providers (City of Melbourne, 2017b).

But little was in fact changed from the proposed Local Law amendments to the Protocol. The appearance and material belongings of homeless individuals in the public places of the City of Melbourne are still judged according to notions of what is 'appropriate' and 'reasonable' in the public spaces of the city. The Protocol prohibits the gathering of groups of people sleeping rough in close proximity, specifies a 'reasonable' amount of possessions (namely, 'two bags which can be carried' and 'bedding like a sleeping bag, blanket or pillow'), and stipulates that ensuring unimpeded movement within and enjoyment of public space by members of the public is its primary aim. The Protocol's partners, Victoria Police, within one month of it coming into operation had used its guidelines to arrest 18 people in the CBD (Masanauskas, 2017). 2018 saw the election of a new Mayor in the City of Melbourne, Sally Capp, who campaigned around issues of support for homeless people in the municipality. After her election, the council announced that the 'Doyle-era crackdown' on homeless people was to be abandoned, because 'groups of people sleeping rough in close proximity have been positively managed by local laws officers and police, as has the amount and types of belongings left unattended in public spaces' (quoted in Lucas, 2018: unpaginated). Place-making by people experiencing homelessness is still considered to generate visual disorder and is still subject to judgement by council staff and police.

Conclusions: at home in the city?

People experiencing homelessness adapt places for shelter, storage of their possessions and the solicitation of donations to assist in their survival. Their place-making techniques are highly developed: from the ability to identify architectural forms which offer secluded spaces for sleeping to ways of maintaining the enforced tidiness demanded by city authorities of individuals who must transport their possessions with them in bags. Rather than being viewed as skilful, homeless place-making tends to be seen as a problem to be managed and constrained; to that extent, the City of Melbourne's Operating Protocol is typical of municipal approaches to public homelessness. However, thinking about homeless place-making should encourage us to sidestep tired debates about the visual impact of homeless encampments or the alleged impediments to pedestrian traffic created by people seeking donations from passers-by. Instead, homeless place-making should prompt consideration of two crucial issues. First, place is active and emplacement is a process: the place-making required of a person without access to housing is constant, relentless. A person who is publicly homeless must continually negotiate with others where they can sit, sleep or store their possessions – all activities that a homed person assumes as activities that can be accomplished every day. Second, sense of place is a question of positionality: the person experiencing homelessness will view and move through the city in a manner that is radically different to that of the homed person. Their sense of place is incommensurable with that of a homed person; further, it is as if urban places themselves are different. A doorway to a homed person is an entryway from the street into a building; a person experiencing homelessness is not meant to pass through it (if they do they may well be ejected); instead, a doorway is a recessed area that can offer shelter from the rain at night. Thinking through homeless place-making requires us to acknowledge that a city is more than a singular place; there are cities within the city, and their architectures and affordances are multiple and diverse.

References

Adams, L. (2014) 'Asking for change: tackling begging with enforcement in Melbourne', *Parity*, 27(9): 24–26.

Arnold, K.R. (2004) *Homelessness, Citizenship and Identity*, Albany: SUNY Press.

Blunt, A. and Dowling, R. (2006) *Home*, London: Routledge.

Boarder Giles, D. (2014) 'The anatomy of a dumpster: abject capital and the looking glass of value', *Social Text*, 32(1): 93–113.

Booker, C. and Dow, A. (2017) 'Melbourne CBD rough sleepers are pretending to be homeless: Victoria's top cop Graeme Ashton', *The Age*, 19 January, www.theage.com.au/victoria/melbourne-cbd-rough-sleepers-are-pretending-to-be-homeless-victorias-top-cop-graham-ashton-20170119-gtune7.html (accessed 23 November 2017).

Brun, C. (2016) 'Dwelling in the temporary', *Cultural Studies*, 30(3): 421–440.

Chamberlain, C. and Johnson, G. (2016) 'How many Australians have slept rough', *Australian Journal of Social Issues*, 50(4): 439–456.

City of Melbourne. (2017a) 'A new protocol to address rough sleeping in the city', 26 September, www.melbourne.vic.gov.au/sitecollectiondocuments/homelessness-operating-protocol.pdf (accessed 4 December 2017).

City of Melbourne. (2017b) 'Operating protocol/policy operating statement', www.melbourne.vic.gov.au/community/health-support-services/social-support/Pages/homelessness-protocol-local-laws.aspx (accessed 4 December 2017).

Cresswell, T. (2009) 'Place', in R. Kitchin and N. Thrift (eds) *International Encyclopedia of Human Geography*, pp. 169–177, London: Routledge.

Deleuze, G. and Guattari, F. (1987) *A Thousand Plateaux*, Minneapolis: University of Minnesota Press.

Department of Health and Human Services. (2018) *Hearts and Homes: Public Perceptions of Homelessness*, Melbourne: Victorian Government.

Doyle, R. (2017) 'We won't let the homeless dominate our city streets', *Herald Sun*, 24 October, www.heraldsun.com.au/news/opinion/robert-doyle-we-wont-let-the-homeless-dominate-our-city-streets/news-story/6719c9c04a3e21aaeb3d57ef48a6ba1a (accessed 23 November 2017).

Edensor, T. (2005) 'Waste matter – the debris of industrial ruins and the disordering of the material world', *Journal of Material Culture*, 10(3): 311–332.

Gerrard, J. and Farrugia, D. (2014) 'The "lamentable sight" of homelessness and the society of the spectacle', *Urban Studies*, 52(12): 2219–2233.

Gillett, C. (2015) 'Aggressive Melbourne CBD beggars earning up to $800 a day', *Herald Sun*, 17 October, www.heraldsun.com.au/news/victoria/aggressive-melbourne-cbd-beggars-earning-up-to-800-a-day/news-story/c90a1841a023f2d08be9add3d4c73fee (accessed 23 November 2017).

Heidegger, M. (1971) 'Building, dwelling, thinking', in *Poetry, Language, Thought*, New York: Harper Colophon Books.

Holmberg, I.M. and Persson, E. (2016) 'Ephemeral urban topographies of Swedish Roma: On dwelling at the mobile–immobile nexus', *Cultural Studies*, 30(3): 441–466.

Homelessness Australia. (2016) *Homelessness in Australia*, www.homelessnessaustralia.org.au/fact-sheets (accessed 15 November 2018).

Incerti, K. (2018) 'Responding to rough sleeping in the City of Port Phillip', *Parity*, 31(3): 44–45.

Jahiel, R. (ed.) (1992) *Homelessness: A Prevention-Oriented Approach*, Baltimore: Johns Hopkins University Press.

Kawash, S. (1998) 'The homeless body', *Public Culture*, 10(3): 319–339.

Kerr, D.R. (2016) '"Almost like I am in jail": Homelessness and the sense of immobility in Cleveland, Ohio', *Cultural Studies*, 30(3): 401–420.

Lems, A. (2016) 'Placing displacement: place-making in a world of movement', *Ethnos*, 81(2): 315–337.

Lucas, C. (2018) 'Melbourne City Council to formally drop Doyle-era crackdown on homeless', *The Age*, 21 June, www.theage.com.au/politics/victoria/melbourne-city-council-to-formally-drop-doyle-era-crackdown-on-homeless-20180621-p4zmys.html (accessed 23 June 2018).

Lynch, P. (2009) 'Begging for change: homelessness and the law', *Melbourne University Law Review*, 26(3): 609.

Masanauskas, J. (2015) 'Abusive Melbourne beggars put on notice for harassing, spitting on people', *Herald Sun*, 16 July, www.heraldsun.com.au/news/victoria/abusive-melbourne-beggars-targeted-for-harassing-spitting-on-people/news-story/7ea6c4853bd10ab9a4593e4b3ce5665d (accessed 23 November 2017).

Masanauskas, J. (2017) 'Police arrest 18 people in new policy to deal with professional beggars and homeless camps', *Herald Sun*, 2 November, www.heraldsun.com.au/news/victoria/police-arrest-18-people-in-new-policy-to-deal-woth-professional-beggars-and-homeless-camps/news-story/ac3b2357910e4386e951e8aeb754a5bf (accessed 19 February 2018).

Mechkaroff, N., Chapman, E. and Herbst, S. (2018) 'Tackling homelessness in Melbourne', *Parity*, 31(3): 6–7.

Meier, L. and Frank, S. (2016) 'Dwelling in mobile times: places, practices and contestations', *Cultural Studies*, 30(3): 362–375.

Melbourne City Council (2017) 'Activities (Public Amenity and Security) Local Law 2017', https://s3.ap-southeast-2.amazonaws.com/hdp.au.prod.app.com-participate.files/6714/8714/3841/Activities_Public_Amenity_and_Security_Local_Law_2017.pdf (accessed 4 December 2017).

Muir, K. et al. (2018) *Amplify Insights: Housing Affordability and Homelessness*, Sydney: UNSW Centre for Social Impact.

Murphy, J. and Tobin, K. (2011) *Homelessness Comes to School*, Thousand Oaks, CA: Corwin.

No Author. (2017) 'Melbourne lord mayor Robert Doyle to propose ban on sleeping rough in city', *The Age*, 20 January, www.theage.com.au/victoria/melbourne-lord-mayor-robert-doyle-to-propose-ban-on-sleeping-rough-in-city-20170119-gtv0fk.html (accessed 19 February 2018).

Panahi, R. (2017) 'Aggressive beggars turning Melbourne from the world's most liveable city into cesspit', *Herald Sun*, 18 January, www.heraldsun.com.au/news/opinion/rita-panahi/rita-panahi-aggressive-beggars-turning-melbourne-from-the-worlds-most-liveable-city-into-cesspit/news-story/e915040dcdb82e187ddd9c23afdea42c (accessed 23 November 2017).

Parsell, C. (2011) 'Homeless identities', *British Journal of Sociology*, 6(3): 442–463.

Parsell, C. and Phillips, R. (2013) 'Indigenous rough sleeping in Darwin', *Urban Studies*, 51(1): 185–202.

Standing, G. (2014) *A Precariat Charter: From Denizens to Citizens*, London: A and C Black.

Walsh, T. (2008) 'Policing disadvantage', *Alternative Law Journal*, 33(3): 160–164.

Walsh, T. (2011) *Homelessness and the Law*, Annandale: The Federation Press.

Young, A. and Petty, J. (2019) 'On visible homelessness and the micro-aesthetics of public space', *Australian and New Zealand Journal of Criminology*, 52(4): 444–461.

43
Clutter and place

Tracey Potts

A place for everything and everything in its place. In its received English usage, the word 'place' connotes order and organisation. To be in or out of place is to be more or less orderly or organised. To know one's place is to conform to the codes of the social order; while, in culinary French, *mis en place* translates as the putting into place of ingredients, tools and equipment prior to preparing a meal. The geographical framing of place – as distinct from space – adds home into the equation (Cresswell, 2015). If Yi-Fu Tuan (1977), in his foundational work on space and place, imagined planet Earth as *home* for its inhabitants, many thinkers have seen home as the exemplary place, even if, as Creswell notes, feminist geographers have questioned the gender politics of such thinking (Rose, 1993). As Dorothy in *The Wizard of Oz* tells us plaintively: '*There's no place like home*'. More, the production of homeliness, through practices of home-making, introduces material culture into considerations of place (Miller, 2001). Places are *made* via the material organisation of space (Woodward, 2007). Places matter (Hicks and Beaudry, 2010). Place thus comes more or less fully furnished: activated and enlivened with stuff (Miller, 2013a), made comfortable and intimate or, at least, materially significant (Attfield, 2007).

With these three senses of place in mind – as orderly, homely and material – the material element stands as a contradictory and stubborn presence. On the one hand, things are necessary to the production of place: for turning, as Dionne Warwick sings, *a house into a home*. On the other hand, the very same things pose a challenge to the order intrinsic of place: Mary Douglas' notion of '*matter out of place*' (2003: 44) distils the problem of dirt as distinctly spatial and organisational, while Judy Attfield considers the domestic interior to be quintessentially 'wild' (2000). Certainly, in recent years, the production of homely spaces has vied with a quest for order and control of the interior. The stuff so necessary to the creation of home is seen to be attacking its inhabitants, producing a situation of 'stuffocation' (Wallman, 2014), where 'stuff-a-lanches' (Brooker, 2012) threaten to engulf us and life itself is seen to exist somewhere *underneath* the things we own (Becker, 2016). So, *the stuff of place can be seen to attack the order of place destroying the home of place.*

The stuff that overwhelms place is commonly referred to as clutter. Clutter, as Attfield observes, can be seen to consist of 'wild things' (2000: 150): objects and items that disturb domestic order by straying beyond their bounds. The task of maintaining the place called home thus becomes one of taming its material culture, of disciplining the interior, above all by returning things back to their assigned drawers and cupboards. While this work of tidying things away has long been a feature of housekeeping (Beeton, 2008 (originally published1861)) and has a history longer than that of the domestic interior (Hicks and Beaudry, 2010), in recent years clutter has become a matter for professionals. Since the turn of this century, the world of professional organisation has become a growth industry complete with its own executive bodies and national associations. Accredited organisers, life coaches and storage gurus publish books, set up websites and blogs, offering consultancy and life coaching services dedicated to conquering the clutter that is deemed to blight contemporary existence. From space planning and organisation to time-management, the systems and skills advanced by professionals promise to help to cut through the demands of modern living, so that we might gain better control of everything from handbags to personal paperwork to kitchen cupboards.

The print publishing world alone is home to a thriving list devoted to the clutter 'crisis': in 2006 Amazon listed 139 separate titles, whereas now there are over 2,000. Lifestyle television, likewise, has seen an intensification of programming dealing with clutter, complete with TV tie-in publications, such as The Life Laundry (2002) and Tidying Up with Marie Kondo (2019). Each year, it seems, witnesses the reinvention of ways to control clutter: from the toothbrush principle (Chandra, 2010) to the KonMari method (Kondo, 2015, 2016) to Japanese ikigai (García and Miralles, 2017) to Swedish death cleaning (Magnusson, 2018), to the extent that journalists have begun quipping that decluttering books are themselves adding to the 'stuffocation' problem (Wiseman, 2015).

The promises of decluttering extend far beyond organised cupboards and homes, though. The clue is often in the title: *Clutter Busting Your Life: Clearing Physical and Emotional Clutter to Reconnect with Yourself and Others* (Palmer, 2012), and many others like it, pledge to enhance energy flow, cure illness, improve productivity and relationships, relieve stress, increase annual turnover, add value to property and reduce environmental impact. Storage solutions are offered, then, as vehicles to good living as well as containers for our possessions; decluttering is held to enhance psychological, ecological and spiritual wellbeing (Potts, 2007). As we are spurred on to organise our things, a minimalist refrain can be heard to resound through the hints, tips, hacks and guidelines: *living with less gives you more.*

As much as clutter appears to be the undoing of place, it, nevertheless, has a good deal to say to the notion of place; that is, *if listened to*, the things that constitute any given muddle and mess can be heard to *speak back* to place. Tuning into what clutter might have to say to place, though, requires admitting objects and other non-humans to be active agents, dynamic participants in human social life. Such an idea of object agency (Latour, 2005) figured in a range of approaches to the material world, from 'thing theory' (Brown, 2001) to 'new materialisms' (Dolphijn and van der Tuin, 2012) to 'object-oriented ontologies' (Bogost, 2012) to feminist materialisms (Alaimo and Hekman, 2008), thus helping to complicate the rather simple picture of stuff just messing up place by decentring humans in the drama of place making. The things corralled in the name of clutter, regularly found to be misbehaving in the place of home due to human fault or negligence, turn out, in new materialist thinking, to have ideas of their own. Granting agency to the things that comprise clutter thus means thinking of place otherwise. A more accommodating vision can be found in Doreen Massey's figuration: of place as a 'source of conflict' (2005: 140), which offers ample room for *all* of the actors, human and more-than-human, that congregate in any particular place. Before disturbing some of the claims made by professional declutterers, however, and introducing a more thingly consideration of place, it would be helpful to take them at their word, so as to appreciate more fully what is at stake in the battle between clutter and place.

How to banish clutter for life in five easy steps

Taking a lead from waste management strategy (i.e., *reduce, reuse, recycle*), a popular approach to clutter-control is one of radical space reduction. Phenomena such as the Tiny House Movement and what has been termed the New Minimalism are premised on the idea of minimising the accommodation offered to things. *Reducing space* thus *controls place* as a smaller architectural footprint can, so it seems, house less stuff. Rather than simply doing without, however, New Minimalists promise a life less burdensome, one free of twentieth-century values, where material goods operate as a measure of the good life. Graham Hill, founder of Treehugger.com, offers a typical story in his op-ed piece 'Living With Less. A Lot Less' (2013). From enjoying career success, a dot.com windfall, followed by frenzied status consumption (to the point of employing a personal shopper to spend his money), Hill finds himself anxious and overwhelmed with the complications of his supersized life. A period of drastic space reduction delivers a new design for living, together with a new metric of success, where less is considerably more: 'I have less – and enjoy more. My space is small. My life is big' (Ibid). Hill's narrative arc – affluence, conspicuous consumption, epiphany, radical downsizing – forms something of a hook in the New Minimalism. *Goodbye Things* (Sasaki, 2017), for instance, offers a variant on the same theme: its author's conversion from unhappy, messy maximalist to happy, tidy minimalist is one riven with similar paradoxes: smaller space, bigger life; emptier apartment, fuller soul; salary poor, time rich.

Aside from reducing the square footage of the house, the accommodation within the interior is further rationalised so as to banish all forms of *loitering*. Dawna Walter (2002) is, for instance, uncompromising in her advice to remove furniture, such as hallway or coffee tables, or indeed any surface where things might rest or hang about before being put away. Principles such as the 'one touch rule' dictate that keys, the mail, a reusable cup, must be marched directly to their assigned places. Marla Cilley, aka Flylady (Cilley, 2018), has a fantasy of designing 'hot' surfaces so that they repel anything that has the nerve to take rest, where imaginary sloping counters and coffee tables put a stop to unauthorised gatherings of things. Storage methods can also be operationalised to discourage items dawdling, nipping their potential to become clutter in the bud. Marie Kondo's folding principles in *The Life Changing Magic of Tidying* (2015) are designed along these lines to enable *vertical storage*. Stacking, piling and other horizontal methods are seen to squash the life out of things, creating an unconscious hierarchy (things at the bottom of the pile become neglected and overlooked) and making things recede from sight and thus hard to retrieve.

Technological developments help to lighten the load still further by shrinking or even evaporating our belongings. The evolution of smart objects, scaling from phones and watches to clothing and homes, promises to revolutionise everyday life and to tidy up the stuff-a-lanche into the bargain. One of the unique selling points of smart technology is its pledge, crudely, to reduce *volume* and *mass*. The history of computer storage offers a stark illustration as the bulk of punch cards, magnetic tape and hard drives the size of refrigerators have come to be replaced by flash drives, SD cards and amorphous cloud computing. New minimalist designs for living are, then, decisively underwritten by the digital revolution. In the words of the author of *Stuffocation*: 'we [will turn away from things largely] because we can. After all, what's the point in owning physical books and CDs when you can access them from the cloud?' (Wallman, 2014: 13). The excess baggage of all forms of hardware, from cameras to housekeys, is offloaded – or uploaded – as users are tempted toward lighter, smaller items and, ideally, away from material possessions entirely. Techno-fixes also help the move toward an experience economy: we are expedited by digital technology to *live more with less*, to opt for forms of consumption that, say, privilege travel adventures (complete with digital photo albums or Instagram accounts, which are crucially imagined to be immaterial) over the accumulation of what, revealingly, are sometimes referred to as the *trappings* of wealth.

Even in the most minimalist, technologically up-graded environment there remain stubborn leftovers that demand attention, however. Software upgrades, for instance, leave in their wake bits of digital clutter: redundant object code, installation files, string statements, what the tech industry refers to as 'cruft'. Likewise, our physical tabletops can turn on us in the blink of an eye. Clutter can, then, manifest under our very noses: a MacBook Air™, a clutch of index cards, a fountain pen, some notebooks, an iPad mini, two cups of cold coffee, a glasses case, journal articles, books, a book rest, a pack of tissues, two propelling pencils and a mobile phone sitting, *mis en place*, on a dining room table can *become* clutter at the point that a meal is about to be served. For the professional organiser, then, keeping such clutter under control is a matter of encouraging the right habits and making a solemn commitment to 'the on-going programme' (Walter, 2002: 106) of letting go and keeping constant control: 'require[ing] you to have discipline and never los[ing] focus of what is going on around you' (Ibid: 103). Rigorous maintenance regimes are, thus, the key to success and to remaining clutter-free for life.

The temporal flow of any place that aims to be clutter-free is thus oriented toward the present. The 'one in, one out' rule is revealing especially of the law establishing the movement of objects as being from front to back door. The wave of decluttering motion that propels material objects through the house relies on a regular action of *purging* the interior of its blockages. Book purchases, for instance, must be accompanied, like for like, by book disposal. Shelf space remains constant and stocks are culled to fit available space, resulting in a storage solution where books are pushed through space of the house as new titles are introduced. In order to keep things moving forward, decluttering practices range from annual events (the 'blitz' or what was once called spring cleaning) to monthly, weekly and daily routines to even '60 second sort-its'. The logic is clear at all scales. Decluttering is, then, a process of *throwing things away*, making even sanctioned items not safe from the periodic 'edits' that now feature as a permanent part of the practice of living: photographs need regularly to be reviewed and rehung; personal letters and cards need to be disposed of several weeks after their arrival, sentimental objects routinely inspected to ensure that they continue to *spark joy* (Kondo, 2016).

Decluttering and technologies of the self

The strategic plan of the clutter-free interior – reduced, rationalised, technologically upgraded, disciplined and edited – can, in spite of its claims, be critically re-described in terms of what Michel Foucault (1988) models as 'technologies of the self'. In Foucault's words:

> [technologies of the self] permit individuals to effect by their own means or with the help of others a certain number of operations on their own bodies and souls, thoughts, conduct, and way of being, so as to transform themselves in order to attain a certain state of happiness, purity, wisdom, perfection, or immortality.
>
> *(1988: 18)*

The skills and attitudes involved in decluttering and other forms of extreme tidying are advanced as guarantees: to deliver happiness, mindfulness and generalised wellbeing. The technical operations of organisation and place management, entailing the perpetual policing of persons and things, are thus designed to alter, permanently, the basis of the *relationships* between persons, materials and things, humans and non-humans. New rationalities and ways of being, i.e., forms of self-monitoring and behavioural modification, often draped in Buddhist robes (as Zen habits, for instance), are offered in exchange for a lifetime's peace and tranquillity.

The first clue that points toward decluttering as a complex technology of the self can be discerned in the level of policing of the interior. Just as beggars, vagrants and loiterers find themselves banished from urban space in the eighteenth century (Foucault, 2009), itinerant and wayfaring things become subject to forms of population control. So, items that cannot earn their keep or that have no homes to go to in the reduced accommodation of the minimalist apartment are corralled into categories of expulsion: donate, recycle, discard. Likewise, the smart habits that accompany the smart gadgetry of the new minimalism rely upon permanent practices and techniques of monitoring and self-monitoring, via one-touch methods and daily tidying regimes to ensure the smooth flow of things through the space of the house.

Moreover, elements of the task of policing become delegated to the things themselves in decluttering strategy. The belief that clutter can be eliminated by design: through spatial restriction or the removal of transitional furniture, such as hall tables, or the installation of sloping surfaces or the imposition of strict folding regulations, entrusts things, objects or artefacts with the burden of maintaining order. Robert Rosenberger's word for things that are inhospitable by design is 'callous objects' (2017) and his project highlights the way that devices such as bus stops and park benches are modelled to drive homeless people out of the city. The 'noninnocence' (Haraway, cited in Rosenburger, Ibid: xii) of such technologies and of the object world more broadly is pertinent, I would argue, to the question of clutter. Bringing the notion of the 'callous object' indoors, it becomes evident that domestic anti-loitering laws are in operation in the professional organiser's strategic plan. Flylady's dream of clutter-free fixtures and fittings operate very much like bus shelter benches constructed at an angle to deter rough sleepers. Equally, the frog-marching of objects to their assigned homes coupled with the removal of transitional resting places for things to foil potential loafers bears ready comparison with the logic of the city identified in Rosenberger's study of homelessness. Like a certain brand of refrigerator that attempts to control user taste by repelling fridge magnets, the blueprint of the clutter-free house is informed by principles found in hostile architecture. Designed to hinder and frustrate use, 'callous' features of the interior silently round up what become, in the process, delinquent things.

It is at this point that differentiating between things and objects becomes a helpful move. Without wishing to oversimplify what are complex discussions – usually involving protracted detours through Heidegger's tool analysis (Bogost, 2012; Harman, 2002) – Tim Ingold offers a more-than-good-enough distinction for thinking about clutter: *objects* present 'a fait accompli', they are 'already made' (2012: 435) and thus are functional, purposeful, determined and closed; in Latour's phrasing, '*matters of fact*' (cited in Ingold, 2012: 436). Whereas *things*, following Merleau-Ponty, are more 'stitched into the fabric of the world' (437), they constitute, following Heidegger, 'gathering[s] of materials in movement' (436) and, hence, *carry on being* in unforeseeable ways. While objects work in full-time occupations and are either in use, serving human needs, or else on stand-by, mis-en-placed or stored away in dedicated units, containers, drawers, cupboards and Hikidashi boxes, things exceed human intention and design.

With Ingold's explanation in mind, it becomes clear that a crucial trick in the clutter guru's repertoire is one not only of privileging *objects over things* but of turning objects *into* things in order to expedite their disposal. Marie Kondo's starting point, for instance, consists of dismantling the house by creating a vast heap of possessions. What were distinct objects, sitting on a bookshelf or in a wardrobe, are rendered as a pile of stuff, as things. What she sees as the poor treatment of possessions is, to all intents and purposes, *performative*, as even tidied-away objects (dresses hanging in a wardrobe, books sitting on a shelf) are wrenched from their respective homes and forced to reapply for their jobs, to earn their keep. Having created what is, to all intents and purposes, a hoard – 'it's very important to get an accurate grasp of the sheer volume for each category' (2016: 6) – a reverse trick is performed where each thing is pulled out of the pile and inspected to see if it 'sparks joy' before being consigned to its appropriate place. It is clear that what sparks joy for Kondo are objects: items with purpose, even if that purpose is simply to look decorative. Anything else is readied for disposal, sent on its way to the charity shop or the recycling plant.

If, as Bill Brown (2001) has it, things *get in the way* and, consequently, become conflated in the world of self-help with clutter, there is, nevertheless, no getting away from things. Ingold's framing of the *carrying on* of things as opposed to objects is especially instructive here:

> From an object-centred perspective, this carrying on is commonly rendered as recycling … From a materials-centred perspective, however, it is part of life.
>
> *(435)*

The declutterer's alibi of recycling is, further, shown to be short-sighted with its object-centred imaginary. It might well be objects that find themselves on the pavement in curb-side recycling schemes but it is *materials* that enter the waste stream, many of which are burned or buried in landfill sites. Samantha MacBride (2011) goes so far as to argue that contemporary recycling practices, especially with their focus on individual consumers, aggravate the environmental waste crisis by diverting attention away from grander scales of industrial waste production and by allowing us to freely dispose of objects by presuming that they stay somewhere in the consumption cycle and out of landfill.

Besides, things persist in other more subtle ways. Even if a given place is purged of things and its material culture is as disciplined as a surgeon's operating table, it is a short step from *mis-en-place* to mess, as Gregory Bateson demonstrates in his metalogue 'Why Do Things Get in a Muddle?' (2000). Staged as a playful conversation with his daughter, Bateson conducts a thought experiment around tidiness, which shows how even the most obedient, disciplined object contains the capacity to lapse into a state of disorderly thingness. Wagering that 'things will always go toward muddle and mixedness' (Ibid: 8), i.e., clutter, Bateson contrasts, by making fractional adjustments to his daughter's possessions, the 'very, very few places' (Ibid: 5) which are 'tidy' for any given object with the 'millions and millions and millions' of ways of constituting its untidy appearance (Ibid: 7). If there are millions and millions and millions of ways for things to be out of place then there are, correspondingly, equal numbers of potential ways of generating rubbish, especially if we follow the logic of stuffing a bin bag with 27 random items at high speed in an effort to keep the house in order (Cilley, 2018). The advice here is clear: clutter cannot be organised, only busted, reduced or binned. Kondo's trademarked method begins, tellingly, not with organisation but with disposal.

Technological objects are no less thingly nor are they resistant to muddles. The slow creep of software upgrades alone can render hardware obsolete and thinglike while we sleep (Chun, 2016). A fully functioning piece of technology becomes a thing not due to the machine itself wearing out but often by incompatibilities at the level of code, through scripts running silently in the background. If technologies effectively break without being broken, it also turns out that what appear to be the dematerialised spaces of the digital are thick with things. One element that is entirely missing from the imaginary of minimalists, who *lighten up* by swapping analogue for digital objects, is the vast infrastructure that supports the digital. If what is 'salient' about technology is that it 'is not salient, for most people, most of the time' (Edwards, 2003: 185), reducing and minimising one's possessions via technological means, say, to cloud storage, is, then, more an act of outsourcing, where clutter is sent packing to the complex architecture of the Internet: to its glass fibres, data warehouses and remote servers. Likewise, exchanging material possessions for travel adventures and consumer experiences disavows the entire infrastructure of the experience economy and, crucially, the elaborate scaffolding of leisure activity (Eide and Fuglsang, 2013), which is far from immaterial.

The obsolete technological object, whose hardware has outlived its software, like the laptop sitting on a table in a downsized apartment at lunchtime, is less matter out of place than *matter out of time*. In his work on waste, Will Viney calls into question the 'spatial bias of contemporary theories of waste' (2011, n.p.) in favour of an emphasis on time. If waste can be constituted as 'time's leftovers' (Ibid), then clutter is, all the more, time-bound. Clutter, as distinct from waste, can be glossed as *matter yet to find a place* and, in the homes of minimalists, as especially time-sensitive, whose *time is running out*. Place, then, in the imaginary of the declutterers, is not only squeezed spatially, it is on a clock. Equally, the temporality of the act of decluttering, with its 'one-touch' and 'one-in-one-out' rules is jittery, manic, not to say anxious, which is the very opposite of organisation. The time of place, in the dream of a clutter-free home, is now.

Doreen Massey (2005) offers an object lesson to the likes of Kondo in her altogether more messy configuration of place. Place is certainly less settled in Massey's view: 'you can't hold places still' (125), it involves ongoing dialogue and negotiation in dealing with the frictions and incompatibilities that surface in the effervescent space–time of place, what she refers to plainly as 'the here and the now' (139). Places are, thus, processual, heterogeneous, multiple, haphazard and, like clutter, marked by the 'throwntogetherness' (140) of people and things. This altogether more rowdy, *throwntogether* conception of place speaks back forcefully to the world of professional organisation and self-help. One thing is certain, the promise that resonates across the advice: of freeing oneself from clutter *permanently*, of banishing it for life, seriously underestimates the agency of the object world, whilst, simultaneously, overestimating the stability of place. The operations and techniques of decluttering rely upon a rather static notion of the items that are seen to constitute the mess and, equally, of the containers designed to organise and tidy them away. Indeed the very idea of a series of nesting containers housing, indeed disciplining things – scaling from drawers, to wardrobes, to rooms, to houses – proves incompatible with Massey's relational conception of place. Indeed, clutter is *more* rather than less likely to manifest in the reduced footprint of the tiny house or micro apartment, as tables and surfaces become multipurpose.

Conclusion

The ambition of controlling things as a means of controlling place radically underestimates the ontology of things. If the stuff of place is seen to attack the order of place threatening, in turn, the home of place, then controlling the stuffly element is by no means a straightforward task. What is clear is that the professional and self-help framing of the relationship between clutter and place certainly does not reckon for the wildness of things in its strategic plan. The five steps to permanent tidiness – reduce space, banish loitering, embrace the digital, adopt smart habits, ensure flow – unsettle place by installing a manic regime of control, where objects are constantly on the move or at best given temporary contracts. The smart habits of the clutter-free, in conforming to Foucauldian notions of technologies of the self, attempt to impose new relationships between people and things, potentially destroying the bond between humans and the material world. Decluttering rewrites the interrelation between order, home and material thus: *the order of place serves to attack the stuff of place destroying and dematerialising the home of place.*

Clutter, if listened to, has much to say to notions of place. Both the distinction between objects and things and Massey's figuration of place as *throwntogether* and irrepressible is helpful in considering how things consistently get in a muddle. More, the object-centred imaginary at the heart of decluttering serves to exacerbate rather than resolve the question of over-consumption by expediting clutter as a new species of waste. Decluttering is a new means of 'ridding' (Gregson, Metcalfe, and Crewe, 2007). The alibi of recycling and charity donation, which moves things along and out of the house (Gregson, Metcalfe, and Crewe, 2007a), turns out to be false: by far the biggest broken promise of the decluttering industry is the idea of getting away from stuff. The minimalist front stage thus masks a cluttered infrastructural backstage or even offstage (Goffman, 1990) and, when it comes to 'stuff', there certainly is no 'away' (Miller, 2013b). What we have, instead, is the far more awkward situation of facing our things, in all of their materiality and thinghood, as we try to work out how to live together. In Massey's schema, places are under constant negotiation and, rather than being 'locations of coherence', are sites of 'adventures and chance encounters' (2005: 180), not to say conflict and disagreement. It is, then, in the messy confrontation between all of the agents of place – human and non-human – that the challenge of configuring a more equitable and more ecological relationship between order, home and stuff must begin.

References

Alaimo, S. and S. J. Hekman. (2008) *Material Feminisms*. Bloomington, Indiana University Press.

Attfield, J. (2000) *Wild Things: The Material Culture of Everyday Life*. Oxford, Berg.

Attfield, J. (2007) *Bringing Modernity Home: Writings on Popular Design and Material Culture*. Manchester, Manchester University Press.

Bateson, G. (2000) *Steps to an Ecology of Mind*. Chicago, University of Chicago Press.

Becker, J. (2016) *The More of Less: Finding the Life You Want Under Everything You Own*. Colorado Springs, Colorado, Waterbrook Press.

Beeton, I. (2008) *Mrs Beeton's Book of Household Management: Abridged Edition*. Oxford, Oxford University Press.

Bogost, I. (2012) *Alien Phenomenology, Or, What It's Like to Be a Thing*. Minneapolis, University of Minnesota Press.

Brooker, C. (2012) *I Can Make You Hate*. London, Faber & Faber.

Brown, B. (2001). 'Thing Theory.' *Critical Inquiry* 28(1): 1–22.

Chandra, S. (2010) *Banish Clutter Forever: How the Toothbrush Principle Will Change Your Life*. New York, Random House.

Chun, W. (2016) *Updating to Remain the Same: Habitual New Media*. Cambridge, MA, MIT Press.
Cilley, M. (2018) *The CHAOS Cure: Clean Your House and Calm Your Soul in 15 Minutes*. London, Hachette UK.
Cresswell, T. (2015) *Place: An Introduction*. Oxford, John Wiley & Sons.
Dolphijn, R. and I. van der Tuin. (2012) *New Materialism: Interviews & Cartographies*. London, Open Humanities Press.
Douglas, M. (2003) *Purity and Danger: An Analysis of Concepts of Pollution and Taboo*. London, Routledge.
Edwards, P. (2003) 'Infrastructure and Modernity: Force, Time, and Social Organization in the History of Sociotechnical Systems.' in Thomas J. Misa, Philip Brey, Andrew Feenberg (eds.): *Modernity and Technology*. Cambridge, MA, MIT Press, pp. 185–225.
Eide, D. and L. Fuglsang. (2013) 'Networking in the Experience Economy: scaffolded networks between designed and emerging regional development', in J. Sundbo and F. Serensen (eds.): *Handbook on the Experience Economy*. Cheltenham, Edward Elgar, pp. 287–309.
Foucault, M. (1988) *Technologies of the Self: A Seminar with Michel Foucault*. Amherst, MA, University of Massachusetts Press.
Foucault, M. (2009) *Security, Territory, Population: Lectures at the Collège de France 1977–1978*. New York, St Martins Press.
García, H. and F. Miralles. (2017) *Ikigai: The Japanese Secret to a Long and Happy Life*. New York, Random House.
Goffman, E. (1990) *The Presentation of Self in Everyday Life*. London, Penguin.
Gregson, N., A. Metcalfe, and L. Crewe. (2007). 'Moving Things along: The Conduits and Practices of Divestment in Consumption.' *Transactions of the Institute of British Geographers* 32: 2.
Gregson, N., A. Metcalfe, and L. Crewe. (2007a). 'Identity, Mobility, and the Throwaway Society.' *Environment and Planning D: Society and Space* 25: 4: 682–700.
Harman, G. (2002) *Tool-Being: Heidegger and the Metaphysics of Objects*. Chicago, Open Court Publishing.
Hicks, D. and Mary C. Beaudry. (2010) *The Oxford Handbook of Material Culture Studies*. Oxford, Oxford University Press.
Hill, G., 'Living With Less. A Lot Less', *New York Times*, March 9, 2013, available at: www.nytimes.com/2013/03/10/opinion/sunday/living-with-less-a-lot-less.html?_r=0
Ingold, Tim. (2012). 'Toward an Ecology of Materials.' *Annual Review of Anthropology* 41: 1: 427–442.
Kondo, M. (2015) *The Life-Changing Magic of Tidying: The Japanese Art*. New York, Random House.
Kondo, M. (2016) *Spark Joy: An Illustrated Guide to the Japanese Art of Tidying*. New York, Random House.
Latour, B. (2005) *Reassembling the Social: An Introduction to Actor-Network-Theory*. Oxford, Oxford University Press.
MacBride, S. (2011) *Recycling Reconsidered: The Present Failure and Future Promise of Environmental Action in the United States*. Cambridge, MA, MIT Press.
Magnusson, M. (2017) *The Gentle Art of Swedish Death Cleaning*. Edinburgh, Canongate Books.
Massey, D. (2005) *For Space*. Thousand Oaks, CA, Sage Publishing.
Miller, Daniel. (2001) *Home Possessions: Material Culture Behind Closed Doors*. London, Bloomsbury.
Miller, Daniel. (2013a) *The Comfort of Things*. Oxford, John Wiley & Sons.
Miller, Daniel. (2013b) *Stuff*. Oxford, John Wiley & Sons.
Palmer, B. (2012) *Clutter Busting Your Life: Clearing Physical and Emotional Clutter to Reconnect with Yourself and Others*. Novato, CA, New World Library.
Potts, T. (2007). 'Organising Space: Clutter, Storage and Everyday Life.' *Key Words 5: A Journal of Cultural Materialism* 5: 88–105.
Rose, G. (1993). *Feminism & Geography: The Limits of Geographical Knowledge*. Minneapolis, MN, University of Minnesota Press.
Rose, G. (2013) *Feminism and Geography: The Limits of Geographical Knowledge*. Oxford, John Wiley & Sons.
Rosenberger, R. (2017) *Callous Objects: Designs Against the Homeless*. Cambridge, MA, University of Minnesota Press.
Sasaki, F. (2017) *Goodbye, Things: The New Japanese Minimalism*. New York, W. W. Norton.
Tuan, Y. (2001) *Space And Place: The Perspective of Experience*. Minneapolis, MN, University of Minnesota Press.
Tuan, Y. F. (1977). *Space and Place: The Perspective of Experience*. Minneapolis, MN, University of Minnesota Press.
Viney, W. (2011) 'Unproductive and Uninhabited': Wastes of Place and Time' https://narratingwaste.wordpress.com/tag/mary-douglas/. Accessed 4th March 2019.

Wallman, J. (2014) *Stuffocation: Living More with Less*. London, Penguin.
Walter, D. and M. Franks. (2002) *The Life Laundry: How to De-Junk Your Life*. London, BBC.
Wiseman E., 'Decluttering: a load of junk?', *Guardian*, 14 June 2015, available at: www.theguardian.com/books/2015/jun/14/decluttering-a-load-of-junk-the-life-changing-magic-of-tidying
Woodward, I. (2007) *Understanding Material Culture*. Thousand Oaks, CA, Sage Publications.

TV Programmes

The Life Laundry (2002). BBC 2, 30 January. Accessed 1 May 2019.
Tidying Up with Marie Kondo (2019). Netflix, 1 January. Accessed 1 May 2019.

44

Nonhuman place

Catherine Phillips and Sarah Robertson

In the inner suburbs of the sprawling and rapidly growing city of Melbourne in south-eastern Australia, the Merri Creek wends its way through a narrow corridor of revitalised parkland. Woodland and water birds, brush-tailed possums, geckos and blue-tongued lizards, butterflies and other insects, even the odd snake inhabit this place alongside diverse flora. Houses, powerlines, a series of paths, and rubbish crowd the banks of this parkland and mark this bushy place with signs of the city. About six kilometres along the path from Melbourne's central business district is CERES (the Centre for Education and Research in Environmental Strategies). This environmental park and community hub nestles among apartment developments, with the creek to its east. Visitors come to CERES to purchase organic produce or native and food-producing plants, to learn about renewable technologies, permaculture, or other sustainability initiatives, or just to walk through the urban farm or restored parkland. Once the site of a bluestone quarry, then a rubbish tip, now an urban sustainability demonstration park, CERES is a place thick with human influence. Yet, like the creek that runs along it, this is also a place of and for nonhumans.

What, if anything, has been said about nonhumans and how they relate to place? How do nonhumans condition, or even produce, the ways in which humans experience and understand places? These are questions that inform this chapter. Western conceptions of place have long been dominated by human-centric ideas focussed on meaning-making (Massey & Thrift, 2003). Yet geographic scholarship is significant in its contribution to expand thinking beyond anthropocentric accounts of the world, and we argue that such insights suggest it is time to rethink the meaning of place in such terms. Drawing on walks through CERES, we discuss insights gained through more-than-human scholarship and the implications of this for the idea of place. Though recent studies have enriched and expanded the conceptual boundaries of place, considering the import and implications of "more-than-human" (Whatmore, 2006) approaches for understanding place are still at the early stages.

This chapter deals with two expansive and troublesome concepts. First, as this Handbook suggests, "place" continues to be an elusive yet central concept for geography. To maintain clarity, we draw on Cresswell's consideration of place as focussed on meaning and experience: "Place is how we make the world meaningful and the way we experience the world" (2015, 19). Recent research highlights the dynamics involved in place-making. Massey (2005), for

Figure 44.1 Wandering CERES.
Source: Photos by the authors.

instance, insists that places are spatially and temporally extensive as collections of trajectories and "stories-so-far". In this way, the idea of place increasingly takes account of significant, if temporary, coalescing social relations. And yet, for the most part, place remains human-centred. There are, in other words, ongoing questions about how and why the "we" mentioned in Creswell's definition above might be rethought, and with what ramifications.

The term "nonhuman" is equally challenging. The idea covers a sweeping range of creatures and things from airplanes to microbes, trees to regulations, climate change to nanomaterials. Scholars continue to debate the concept, critiquing its perpetuation of problematic binaries by categorising all nonhumans as "other" rather than attending differentiations and diversity (Lulka, 2004). Though we acknowledge its limits we use the term here for, as Head has commented of Western scholarship, place is but one concept limited by a language that is "saddled with a separationist view of the human" (2016, 56).

Following the paths we walk at CERES (see Figure 44.1), our exploration focusses on living nonhumans such as birds, plants, and bees, as well as some abiotic elements including light. We consider: how geographical scholarship draws attention to inclusions and exclusions of nonhumans in places; and how it has troubled spatial and conceptual divisions through attention to nonhuman lives in cities. Each of these two sections highlights studies influencing a conception of nonhuman place and provides connections to our experiences at CERES. The final section outlines ways recent literature pushes discussions of nonhumans and place forward in relation to methods, Indigenous cosmologies, and ethico-politics.

Nonhumans in and out of place

In their influential work, Philo & Wilbert (2000) argued that geography tended to treat nonhuman animals as simply part of "nature", as backgrounds for human meaning-making. Instead, they suggested understanding animals as part of society and as creatures with their own worlds. Their conceptualisations of animal spaces and beastly places suggest how spatial and conceptual placements are inextricably tied. Since their intervention, though not always explicit, much research relating nonhumans to place is founded in questions of where nonhumans belong (or not) – weeds in the garden, animals (including pets) in the city, microplastics in the ocean. Power (2009), for instance, highlights how transgressing possums contribute to and challenge notions of home, while Barker (2008) explores the de/valuation of gorse as a non-native invasive species and the work involved in pursuing its exclusion. Such disputed belonging can create conflict, among humans as well as among humans and nonhumans, but they also provide opportunity to question boundary-making processes. Transgressing nonhumans can provoke thought, and action.

During our wanderings of CERES, we have experienced such boundary-making for and with nonhumans. For instance, we have had many encounters with chickens, meeting them in the nursery, outdoor market, café patio, and market gardens. Nonhuman animals, like chickens, are key to the ecological and social dynamics of CERES. Hovorka (2008) argues that urban livestock (including chickens) has tended to be ignored because people do not expect them to be in cities and because chickens tend to inhabit obscure spaces. They are, in other words, "out-of-place" and thus their fundamental roles in making cities are ignored. In contrast, at CERES chickens (or chooks) and the services they provide are highlighted rather than backgrounded. One sign above a gate leading off an open café area cautions visitors not to go any further: "Chooks only please". This may be read as containment and control; however, chook escapes and subsequent explorations are expected, even welcomed, by visitors and staff. Though the creatures are eventually caught and returned to their designated areas, steps are taken to prepare for and protect them during their regular excursions. In the nursery, signs warn that any four-pawed, furry companions must remain on short leashes and are not to approach the chooks. Human visitors are also alerted that the chooks do not generally appreciate being cuddled. In this way, we see that chickens here are "at home" while self-discipline is expected of usually favoured urban dwellers (humans and dogs) to ensure chook well-being.

The chooks at CERES serve as resources and labourers; they perform functions within a human-designed permaculture-inspired system by providing eggs, aerating and fertilising soil, and disposing of café "food waste". Simultaneously, their roosting, foraging, and mating shape this place. Moreover, chooks' inhabitations and antics alter how humans think, feel, and talk about this place. Chooks can enchant – making us pause in surprise or concern (see Bennett, 2010). At the end of the car park, a brood of chooks gathers on one side of a wire fence as Sarah crouches down on the other side. One chook stretches her short, rusted-orange-feathered neck through the wire and gives Sarah's extended hand a single peck. It seems an exploratory gesture. A quest for new food? A warning against further approach? A tentative offer of friendship (see Bingham, 2006). It is difficult to tell, much as it is challenging to interpret the messages shared in brief encounters between humans. The ways in which chickens experience and affect CERES as a place may be conditioned by the protections they are afforded through the bounds and warnings throughout the grounds; however, it seems clear that chooks continue to pattern this place in ways that suit them. In short, this is their place too.

Since calls in the 1990s for deeper attention to nonhuman worlds and human–nonhuman relations, geographical research has increasingly questioned human framings and categorisations of nonhumans in particular places, including cities (Hinchliffe *et al.*, 2005; Wolch, 1996, 2002), wilderness and remnants (Lorimer, 2015; Waitt *et al.*, 2009), fields and gardens (Atchison & Head, 2013; Ginn, 2014) zoos and laboratories (Anderson, 1995; Davies, 2012). Research has also focussed on transformative human encounters with nonhumans, from dogs (Haraway, 2008; Instone & Sweeney, 2014) to insects (Hatley, 2011;

Phillips, 2014), from plants (Head *et al.*, 2015; Jones & Cloke, 2002) to viruses (Greenhough, 2012), from waters (Neimanis, 2016; Thomas, 2015) to light (Edensor, 2017).

Much insight has been gained through these elaborations of human–nonhuman relations and the roles of nonhumans as co-producing agents within complex worlds. However, it is only recently that nonhumans have been considered as place-makers. Geographic scholarship has shifted to recognising that both humans and nonhumans are embedded in social relations with others and it is pursuing ways to further understand and communicate human–nonhuman and nonhuman place-making. For instance, Lorimer *et al.* (2019) consider how animals might be understood as having their own atmospheres, arguing that wolves and dogs influence the experience of spaces for others while making sense of spaces as individual subjects. van Dooren & Rose (2012) also explore nonhuman place-making through an investigation of how penguins and people make the Sydney coast their home, and the negotiations and points of contest that can arise. Extending such animal-oriented contemplations, Phillips & Atchison (2018) reiterate an understanding of urban places as collective, multinatural achievements and argue for engagement with the ways in which plants sense, belong, and negotiate places as urban residents.

In our own journey, beyond the chooks at CERES, word-of-mouth draws our attention to how nonhuman and human inhabitants make this place what it is. Other walkers alert us to the blue-tongued lizard's place where we must watch closely to witness its appearance without disrupting its routines, share their excitement and caution us to step carefully in areas where a brown snake has taken up residence near the path, and warn of the hordes of mosquitoes surrounding the pond. These and other nonhuman residents, and the multitudes to which we are not attuned, suggest a place rich with nonhuman diversity and agency that humans are only beginning to register.

Urban places: challenging spatial boundaries

Urban realms are key sites for discussions of how nonhumans challenge human-centric ideas of place and boundaries. Wolch (1996, 2002) has been prominent in calling for animal lives to be more central in urban geography and planning. In addition to attending to how particular animals live in cities, she emphasises the potential to breakdown framings of the city and nature as separate and opposed. There is important work on nonhuman places outside cities (see Bawaka *et al.*, 2016; Candea, 2010; Lulka, 2004); however, urban sites remain important to consider. More people than ever live in cities and, at the same time, cities are being recognised as hosting diverse flora and fauna, even representing hotspots for threatened species (Ives *et al.*, 2016). More to the point, nonhumans are "neither neatly bounded … nor insubstantial" entities (Hinchliffe *et al.*, 2005, 644), they are impacted by urbanisation but also disregard human-made rules and boundaries, calling attention to conceptual and political issues of urban places.

Urban political geographers have been central to problematising urban–nature dualisms and unravelling the apparent localisms of place (see Heynan *et al.*, 2006). As Braun asserts, "to construct the city as a social space we must continuously enroll nature but deny that we are doing so" (2005, 642). This paradox of understanding cities as human places is explored, for instance, by Kaika (2012) through attention to the flows of water that frame urban home-making, while Poe *et al.* (2014) and Robertson (2018) point to the importance of attention to both urbanisation of nature and nature in cities. Still, despite decades of academic critique and practical challenges, binaries of urban–nature, native–alien, domesticated–wild, among others, persist in everyday discourse (Castree, 2014) as does the image of the city as largely a human place (Houston *et al.*, 2017).

At CERES, the intertwinings of more-than-human experiences point to the challenges of troubling dualisms. Human visitors are drawn to this place for the "nature" experience it offers, particularly as a means of engaging with nature within the surrounding hardscaped urban areas of Melbourne. And, as we stroll through the nursery, displays encourage purchase and fostering of native plants. Research by Head (2012) points to how such categorisations of nativeness reinforce divisions between (some) humans and nonhumans as well as among plants. Despite this, Head finds potential to disrupt such logics in everyday practices that draw together plants of diverse origins. In the case of CERES, despite valorisation of native plants, there are plants from many places for sale and living throughout the site. Plants like lavender, rosemary, and citrus trees could have been contained to food-growing areas where, as non-natives, they might be welcomed based in the logic of dualisms; instead they are dispersed throughout the grounds. In this sense, dualisms become both reinforced and contested in complicated ways.

Inspired by these plants and their contested placements, we consider nonhumans that live within what property lines define as CERES as well as the connections beyond these bounds that make this place. Money changes hands, electricity flows through the overhead power lines, potting soil is bought in for propagation, school groups engaged in field trips. Buzzing bees also move our thinking in this direction. Aside from native bee habitats, Western honey bees have a designated apiary. Close to the water and flowering trees, a bit away from the most walked paths, we find signs marking this place. Similar to the signs about chickens, they warn us not enter the area without proper equipment and guidance. The area is also fenced and locked, reinforcing the directive meant to protect people as well as the hives. Of course, this is meaningless to bees. Honey bees forage throughout CERES, wherever there is nectar or pollen to be had. We also see and hear them zipping through the air beyond the grounds toward the creek or toward the city. Despite being non-native, these creatures thrive in Australia and their roles and places are contested across agriculture, conservation, and biosecurity arenas (Phillips, 2014). The activities of honey bees, among many other creatures at CERES, raise questions not only about human and nonhuman shapings of this place but also about how other places connect to here through flows of materials, ideas, and times.

Should, and how might cities be organised and built in ways that cater not only for humans but for nonhuman well-being? Wolch (1996) argued early on that renaturalising cities could help engender more ethical engagements with nonhumans, while Houston *et al.* (2017), inspired by Haraway (2016), suggest reorienting city-making through the idea of multi-species kinship. What might CERES contribute to such considerations? Mathews (2000) observes that CERES offers an opportunity for re-enchantment with nature and re-inhabitation of place, suggesting that experiences here might alter the sense of one's place in the world. Might CERES serve as a prompt for Plumwood's call to "work out new ways to live with the earth, to rework ourselves … [to] go onwards in a different mode of humanity" (2007, 1)? Van Dooren & Rose's (2012) discussion of multi-species ethics also details how penguins and flying foxes intimately know, use, and return to particular locations, which suggests that nonhuman as well as human places might reflect Massey's (2005) idea of place as "a simultaneity of stories so far". Might we, then, think beyond the human register to appreciate chickens, bees, and other nonhumans as place-makers, as inhabitants caught in and formulating the negotiations and trajectories of CERES? In light of such potential transformations wrought on visitors and inhabitants, might we even dare to consider this place to be acting itself as a nonhuman?

Developing pathways

Discussions about and affecting notions of nonhumans and their places continue to develop. In this section we outline three thematic areas enriching geographic ideas in the field: questioning and experimenting with research methods; engaging with Indigenous scholarship that challenges and/or supplements more-than-human approaches; and noticing and responding – in individual and collective terms – to involved ethical and political negotiations.

Questions of method

Consideration of more-than-human worlds has seen re-examination of the methods scholars use to investigate nonhumans and their relations with place (see Bastian *et al.*, 2017; Buller, 2015). Drawing attention to how different bodies learn (see Despret, 2004), these explorations of method tend to begin with an aim of enhancing human (and more pointedly researchers') capacities to sense and make sensible nonhuman relations and places. Tsing (2015), for instance, argues for developing "arts of noticing" to understand a multispecies eco-social world. In similar fashion, Bell *et al.*, (2018) draw upon Haraway (2008, 2016) and the Indigenous Australian concept of *Dadirri* to argue for an "engaged witnessing" to recognise the agencies of nonhumans and humans.

One way this has been explored is through walking-as-research. The practice of walking is a well-established means of exploring place and walking methods have been taken up across geography and anthropology to explore the places and placings of nonhumans (for a review of walking research, see Lorimer, 2016). In such studies, walking serves as a means to approach and account for the dynamism of place, its diverse human and nonhuman interactions, and its multiple temporalities (Edensor, 2017; Ingold & Vergunst, 2008). In this chapter, we took such an approach to reflect on encounters and contemplations experienced at CERES. If walking, or other similar practices, can be considered potentially sensitising, we can also consider sensitising devices such as journals detailing nonhuman traces (Hinchliffe *et al.*, 2005), videos highlighting gestures and paths (Lorimer, 2010), or sensors that detect and reveal more-than-human rhythms of places (Edensor, 2017; Gabrys, 2018). As the diversity of possible sensitising practices and devices suggests, an important part of strengthening analyses of nonhuman places will be multidisciplinary collaborations (see Buller, 2015). We will need, as Hinchliffe *et al.* (2005, 644) put it, to refuse "the old settlements of Science, Politics, Ethics, and Religion".

Indigenous cosmologies

Recent work on place in Indigenous scholarship challenges and extends conceptions of more-than-human and nonhuman place. Simultaneously, it draws attention to the ethical and political issues of knowledge appropriation and colonisation. Bawaka Country and colleagues (2016), among other work, articulates Indigenous cosmologies as positioning place as Country, as nonhuman or more-than-human.[1] Drawing a contrast between Euro-Western and Indigenous epistemological-ontological framings, Watts (2013) presents Haudenosaunee and Anishnaabe cosmologies of "Place-thought", the basis of which is that "land is alive and thinking and that humans and non-humans derive agency through the extensions of these thoughts" (Watts, 2013, 21). Building this notion of place, Larsen &

Johnson (2016) critique human-centric notions of Self and instead emphasise the agency of place that leads to a "*more-than-human* geographical self".

CERES sits within unceded Wurundjuri Country long inhabited by the Woi wurrung language group of the eastern Kulin Nation. Acknowledging the site's incarnations as a quarry, rubbish tip, and remediated land is only a small part of this place's history, only a hint of its living as "Country" (see Porter, 2018). Watts (2013) cautions that Euro-Western interpretations are often underpinned by an assumed distinction between epistemology/theory and ontology/praxis; an assumption, she argues, that "creates spaces for colonial practices to occur" (28). In addition to emphasising the livingness and agency of place, Indigenous scholarship has also demonstrated methods of listening, engaging, and story-telling with Country (Kovach, 2009; Wright *et al.*, 2012). As Thomas (2015, 976) writes, "further engagement with the transformative potential of Indigenous knowledges on environmental futures" is important if geographers are to pursue "opportunities to build a decolonised, political ethical project" and, we would add, to enrich understandings of nonhuman place.

Ethico-political mandates

From the beginning, ethical obligations and asymmetric power relations have been of interest to those concerned with human–nonhuman relations. Plumwood (2005, 2009) argues that anthropocentric approaches and innocent notions of place must be refused; instead, she invites us to rethink our respective places in shared worlds. Her suggestions include finding new ways of recognising "nature as powerful, agentic, and creative" (2009, 126) and taking responsibilities for complex ecosocial histories of places where we live and that support us. More recently, Haraway (2008, 2016) has been influential with her argument that everyday engagements with nonhumans foster new ways to build understanding and to facilitate less exploitative relations. For instance, Despret & Meuret (2016) explore the cosmoecology of shepherding as the work of recuperation for both humans and sheep, while Atchison *et al.*, (2017) and Gibbs & Warren (2014) explore how particular species, carp and sharks respectively, become killable in management strategies attempting to refuse nonhuman belonging in particular places. As such accounts suggest, more-than-human approaches tend to emphasise relational, everyday ethics, though this does not exclude considerations of justice or scales beyond the personal. It is not enough to learn nonhuman relations; rather, one must "become worldly and respond" (Haraway, 2008, 41). As this chapter suggests, places such as CERES are already engaging in this work though it is not simple or settled.

In addition to the challenge of developing skills to register nonhuman places then, there is a challenge in conveying what becomes learned and witnessed. Within and beyond geography, storytelling is increasingly employed in ways to attend to places as storied (see Massey, 2005), and how those stories are lived by humans and nonhumans, in relation or otherwise (Wright *et al.*, 2012; McKiernan & Instone, 2016; Kovach, 2009, ch. 5; Phillips & Atchison, 2018). For Tsing (2015), this requires critical description of complex, asymmetric, dynamic relations among species, which she demonstrates through "a riot of short chapters" that tell of the journeys, landscapes, and valuations of mushrooms. van Dooren & Rose (2012) explain that, for them, telling stories involves immersion in nonhuman worlds, multidisciplinary exploration, and ethico-political consequences; they suggest that stories draw researchers and audiences into new relations and obligations.

Conclusion

In relation to place, nonhumans have tended to be obscured as settings or backgrounds for human meaning-making; however, this is increasingly contested. A growing number of scholars are exploring the more-than-human dimensions of spatial and social relations, foregrounding practices and affects of nonhumans, and considering the challenges these insights pose. In this chapter, we have reflected upon (and raised questions about) nonhumans and their relations, and how admitting their importance into research involves reconsidering the key geographical concept of place.

Following the paths laid for us by CERES, we have explored some ways in which more-than-human research might inform a notion of nonhuman place. We have outlined the importance of the placings and places of living nonhumans, situating the discussion particularly within urban places. Considering the chickens, plants, and bees we encountered at CERES, we have contemplated possibilities of understanding nonhumans as place-makers for themselves and others. These nonhumans transgress human-imposed markers of place to follow their own paths and define other boundaries. And in following their activities, we can gain important insights into how places become made and contested, gaining and losing meaning in ways that make important differences.

We have argued that the concept of nonhuman places holds important potential for expanding geographical thought and for developing recognition of nonhumans. Human–nonhuman relations are infused with ethics and politics, as is how we, however that "we" is defined, live and become part of places. CERES is considered a place of demonstration, a manifest to some of the ways to live "urban sustainability". Ecological restoration and urban agriculture are certainly part of the agenda, but such a description seems to fall short. Exploring nonhuman place through CERES and more-than-human research has allowed us to draw attention to the limitations and potential of remaking cities and rethinking place in ways that acknowledge nonhumans as inhabitants.

The conversations and contestations involved in rethinking and respecting nonhuman place are far from over. We highlighted three areas that we find particularly encouraging for ongoing considerations: the methods researchers employ, insights from Indigenous studies, and the ethico-political mandate. We welcome these and other challenges in discussions about how to reorient lexicons and re/conceive how places become experienced and meaningful for humans and the diversity of nonhuman others with whom we share this planet.

Note

1 It is important to note the Western origins of these terms.

References

Anderson, K. (1995). Culture and nature at the adelaide zoo: at the frontiers of "human" geography. *Transactions of the Institute of British Geographers*, *20*(3), 275–294.

Atchison, J., Gibbs, L. & Taylor, E. (2017). Killing carp (Cyprinus carpio) as a volunteer practice: implications for community involvement in invasive species management and policy. *Australian Geographer*, *48*(3), 333–348.

Atchison, J. & Head, L. (2013). Eradicating bodies in invasive plant management. *Environment and Planning D*, *31*(6), 951–968.

Barker, K. (2008). Flexible boundaries in biosecurity: accommodating gorse in Aotearoa New Zealand. *Environment and Planning A*, *40*(7), 1598–1614.

Bastian, M., Jones, O., Moore, N. & Roe, E. (2017). Introduction. In M. Bastian, O. Jones, N. Moore & E. Roe (Eds.), *Participatory Research in More-than-human Worlds* (pp. 1–15). London: Routledge.
Bell, S. J., Instone, L. & Mee, K. J. (2018). Engaged witnessing: researching with the more-than-human. *Area, 50*(1), 136–144.
Bennett, J. (2010). *Vibrant Matter: A Political Ecology of Things.* Durham, NC: Duke University Press.
Bingham, N. (2006). Bees, butterflies, and bacteria: biotechnology and the politics of nonhuman friendship. *Environment and Planning A, 38*(3), 483–498.
Braun, B. (2005). Environmental issues: writing a more-than-human urban geography. *Progress in Human Geography, 29*(5), 635–650.
Buller, H. (2015). Animal geographies II. *Progress in Human Geography, 39*(3), 374–384.
Candea, M. (2010). "I fell in love with Carlos the meerkat": engagement and detachment in human-animal relations. *American Ethnologist, 37*(2), 241–258.
Castree, N. (2014). *Making Sense of Nature.* London: Routledge.
Cresswell, T. (2015). *Place: An Introduction*, second edition. Oxford: Wiley-Blackwell.
Country, B., Wright, S., Suchet-Pearson, S., Lloyd, K., Burarrwanga, L., Ganambarr, R., … Sweeney, J. (2016). Co-becoming Bawaka: towards a relational understanding of place/space. *Progress in Human Geography, 40*(4), 455–475.
Davies, G. (2012). What is a humanized mouse? Remaking the species and spaces of translational medicine. *Body & Society, 18*(3–4), 126–155.
Despret, V. (2004). The body we care for: figures of Anthropo-zoo-genesis. *Body & Society, 10*(2–3), 111–134.
Despret, V. & Meuret, M. (2016). Cosmoecological sheep and the arts of living on a damaged planet. *Environmental Humanities, 8*(1), 24–36.
Edensor, T. (2017). *From Light to Dark: Daylight, Illumination, and Loom.* Minnesota: University of Minnesota Press.
Gabrys, J. (2018). Sensing lichens. *Third Text, 32*(2–3), 350–367.
Gibbs, L. & Warren, A. (2014). Killing sharks: cultures and politics of encounter and the sea. *Australian Geographer, 45*(2), 101–107.
Ginn, F. (2014). Death, absence and afterlife in the garden. *Cultural Geographies, 21*(2), 229–245.
Greenhough, B. (2012). Where species meet and mingle: endemic human-virus relations, embodied communication and more-than-human agency at the Common Cold Unit 1946-90. *Cultural Geographies, 19*(3), 281–301.
Haraway, D. (2008). *When Species Meet.* Minneapolis, MI: University of Minnesota Press.
Haraway, D. (2016). *Staying with the Trouble.* Durham, NC: Duke University Press.
Hatley, J. (2011). Blood intimacies and biodicy: keeping faith with ticks. *Australian Humanities Review*, 50, 63–76.
Head, L. (2012). Decentring 1788: beyond biotic nativeness. *Geographical Research, 50*(2), 166–178.
Head, L. (2016). *Hope and Grief in the Anthropocene: Re-conceptualising Human-Nature Relations.* London: Routledge.
Head, L., Atchison, J. & Phillips, C. (2015). The distinctive capacities of plants: re-thinking difference via invasive species. *Transactions of the Institute of British Geographers, 40*(3), 399–413.
Heynan, N., Kaika, M. & Swyngedouw, E. (2006). Urban political ecology. In N. Heynan, M. Kaika & E. Swyngedouw (eds) *The Nature of Cities: Urban Political Ecology and the Politics of Urban Metabolism* (pp. 1–20). London: Routledge.
Hinchliffe, S., Kearnes, M. B., Degen, M. & Whatmore, S. (2005). Urban wild things: a cosmopolitical experiment. *Environment and Planning D, 23*(5), 643–658.
Houston, D., Hillier, J., MacCallum, D., Steele, W. & Byrne, J. (2017). Make kin, not cities! Multispecies entanglements and 'becoming-world' in planning theory. *Planning Theory, 17*(2): 190–212.
Hovorka, A. (2008). Transspecies urban theory: chickens in an African city. *Cultural Geographiies, 15*(1), 95–117.
Ingold, T. & Vergunst, J. (2008). *Ways of Walking: Ethnography and Practice on Foot.* Aldershot, UK: Ashgate Publishing.
Instone, L. & Sweeney, J. (2014). The trouble with dogs:'animaling'public space in the Australian city. *Continuum, 28*(6), 774–786.
Ives, C. D., Lentini, P. E., Threlfall, C. G., Ikin, K., Shanahan, D. F., Garrard, G. E., … Kendal, D. (2016). Cities are hotspots for threatened species. *Global Ecology and Biogeography, 25*(1), 117–126.

Jones, O. & Cloke, P. (2002). *Tree Cultures: The Place of Trees and Trees in Their Place.* Oxford: Berg.
Kaika, M. (2012). *City of Flows: Modernity, Nature, and the City.* St. Louis, USA: Routledge.
Kovach, M. (2009). *Indigenous Methodologies: Characteristics, Conversations and Contexts.* Toronto: CAN, University of Toronto Press.
Larsen, S. & Johnson, J. (2016). The agency of place: toward a more-than-human geographical self. *GeoHumanities, 2*(1), 149–166.
Lorimer, H. (2016). Walking. In T. Cresswell & P. Merriman (Eds.), *Geographies of Mobilities: Practices, Spaces, Subjects* (pp. 31–46). London: Routledge.
Lorimer, J. (2010). Moving image methodologies for more-than-human geographies. *Cultural Geographies, 17*(2), 237–258.
Lorimer, J. (2015). *Wildlife in the Anthropocene: Conservation After Nature.* Minneapolis, MI: University of Minnesota Press.
Lorimer, J., Hodgetts, T. & Barua, M. (2019). Animals' atmospheres. *Progress in Human Geography, 43*(1), 26–45.
Lulka, D. (2004). Stabilizing the herd: fixing the identity of nonhumans. *Environment and Planning D, 22* (3), 439–463.
Massey, D. (2005). *For Space.* London: Sage Publications.
Massey, D. & Thrift, N. (2003). The passion of place. In R. J. Johnston, M. Williams & British Academy (Eds.), *A Century of British Geography* (p. 674). Oxford, UK: Oxford University Press.
Mathews, F. (2000). CERES: singing up the city. *PAN: Philosophy Activism Nature, 1*, 5–15.
McKiernan, S. & Instone, L. (2016). From pest to partner: rethinking the Australian White Ibis in the more-than-human city. *Cultural Geographies, 23*(3), 475–494.
Neimanis, A. (2016). *Bodies of Water: Posthuman Feminist Phenomenology.* New York: Bloomsbury.
Phillips, C. (2014). Following beekeeping: more-than-human practice in agrifood. *Journal of Rural Studies, 36*, 149–159.
Phillips, C. & Atchison, J. (2018). Seeing the trees for the (urban) forest: more-than-human geographies and urban greening. *Australian Geographer.* doi: 10.1080/00049182.2018.1505285.
Philo, C. & Wilbert, C. (2000). *Animal Spaces, Beastly Places.* London: Routledge.
Plumwood, V. (2005). *Environmental Culture: The Ecological Crisis of Rreason.* London: Routledge.
Plumwood, V. (2007). A review of Deborah Bird Rose's 'Reports from a Wild Country: Ethics for Decolonisation'. *Australian Humanities Review, 42*, 1–4.
Plumwood, V. (2009). Nature in the active voice. *Australian Humanities Review, 46*, 32–47.
Poe, M., LeCompte, J., McLain, R. & Hurley, P. (2014). Urban foraging and the relational ecologies of belonging. *Social & Cultural Geography, 15*(8), 901–919.
Porter, L. (2018). From an urban country to urban *Country*: confronting the cult of denial in Australian cities. *Australian Geographer, 49*(2), 239–246.
Power, E. (2009). Border-processes and homemaking: encounters with possums in suburban Australian homes. *Cultural Geographies, 16*(1), 29–54.
Robertson, S. (2018). Rethinking relational ideas of place in more-than-human cities. *Geography Compass, 12*(4), e12367.
Thomas, A. (2015). Indigenous more-than-humanisms: relational ethics with the Hurunui River in Aotearoa New Zealand. *Social & Cultural Geography, 16*(8), 974–990.
Tsing, A. (2015). *The Mushroom at the End of the World: On the Possibility of Life in Capitalist Ruins.* Princeton, NJ: Princeton University Press.
van Dooren, T. & Rose, D. B. (2012). Storied-places in a multispecies city. *Humanimalia, 3*(2), 1–27.
Waitt, G., Gill, N. & Head, L. (2009). Walking practice and suburban nature-talk. *Social & Cultural Geography, 10*(1), 41–60.
Watts, V. (2013). Indigenous place-thought & agency amongst humans and non-humans (First Woman and Sky Woman go on a European world tour!). *Decolonization: Indigeneity, Education & Society, 2*(1), 20–34.
Whatmore, S. (2006). Materialist returns: practising cultural geography in and for a more-than-human world. *Cultural Geographies, 13*(4), 600–609.
Wolch, J. (1996). Zoöpolis. *Capitalism Nature Socialism*, 7(2), 21–47.
Wolch, J. (2002). Anima urbis. *Progress in Human Geography, 26*(2), 721–742.
Wright, S., Lloyd, K., Suchet-Pearson, S., Burarrwanga, L., Tofa, M. & Country, B. (2012). Telling stories in, through and with Country: engaging with Indigenous and more-than-human methodologies at Bawaka, NE Australia. *Journal of Cultural Geography, 29*(1), 39–60.

45

Place attachment

Patrick Devine-Wright

Introduction

This chapter focusses on the concept of place attachment, providing a summary of theoretical and methodological approaches, as well as a discussion of key debates in the literature. Place attachment, which refers to the emotional bonds that people form with particular places, has received considerable attention over the past three decades, leading to a flourishing body of research across disciplinary boundaries. Yet it also has some limitations, in part arising from the use of contrasting approaches that challenge the accumulation of a coherent body of knowledge. Here, I summarise different theoretical and methodological approaches to place attachment, as well as providing insights into ways that I have responded to strengths and weaknesses of the literature to develop a place-based and mixed method approach to understanding conflicts over the siting of energy technologies.

What is place attachment and why is it important?

The concept of place attachment became prominent following the publication of a seminal book with the same title edited by Altman and Low in 1992, with contributors from disciplines including sociology, anthropology, psychology and architecture. From that book emerged a classic definition: place attachment refers to 'positively experienced bonds, sometimes occurring without awareness, that are developed over time from the behavioural, affective and cognitive ties between individuals and/or groups and their sociophysical environment' (Brown and Perkins, 1992: 284).There are similarities between place attachment, which has become the prevalent term used within Environmental Psychology (Lewicka, 2011a), and related concepts such as topophilia and sense of place previously developed by Human Geographers (e.g., Tuan, 1974). Therefore, while emotional bonds that people hold with places have been a consistent topic of research over the past few decades, an array of terms has been used to describe these across several disciplines and there remains no single unified perspective. Instead, a plurality of theories and methods can be identified across disciplinary boundaries.

Place attachment is an important concept for several reasons. First, during the 1970s, a broad emphasis on the concept of 'place', encompassing people–place bonds, enabled geographers to challenge the conventional quantitative approach to understanding areas or regions in the world that was prevalent during previous decades. In contrast, it allowed scholars to foreground subjective experiences of places in the world, including emotional aspects, heavily influenced by phenomenological theory (Seamon, 2014). Second, place attachment provided scholars with a more useful term than residential satisfaction in order to better understand the feelings of grief and mourning that emerged amongst affected communities as a result of forced relocation in US cities during the 1960s (e.g., Fried, 2000), making visible emotional ties to place often taken for granted in the course of everyday lives (see Brown and Perkins, 1992). Given that disciplines such as Geography have evolved over recent decades to focus more on political and socio-cultural aspects of place, place attachment remains a useful way to understand the subjective experiences of places and their importance for self and identities (Antonsich, 2010). Third, psychological research suggests that place attachment is good for you. Empirical research has shown that individuals with strong attachments to place have greater life satisfaction, stronger ties with neighbours, a greater interest in family history and greater trust in others (Lewicka, 2011a), as well as an increased sense of belonging, self-esteem and meaning (Scannell and Gifford, 2017).

What makes place attachment especially interesting is that its aspects and consequences are not always as positive as the earlier literature would suggest. Manzo (2014) emphasised that place attachment has a 'shadow side' that has been persistently overlooked, pointing to the contribution of feminist scholars in making visible how residence places could represent both 'home as trap' as well as 'home as haven' (cf. Cooper Marcus, 1995). In contrast to the Brown and Perkins definition above (Brown and Perkins, 1992), Manzo viewed place attachments as complex, multivalenced bonds that may be positive, negative or ambivalent and evolve over time, distinct from the more normative view of place attachment in the literature as a whole (see Lewicka, 2011a).

Research has also shown that place attachments are implicated in social and territorial conflicts. For example, Dixon and Durrheim's research in South Africa showed that objections to the arrival of non-white incomers in white suburbs were embedded in place bonds and meanings linked to identity processes (Dixon and Durrheim, 2000). More broadly, Fried claimed that violent territorial conflicts were rooted in 'pathologies of community attachment' and argued that attenuated place bonds were therefore better suited to new conditions, transitions and wider opportunities (Fried, 2000: 193). Research on the consequences of environmental change supports this view. Marshall et al. (2012) studied the impacts of climate change on peanut farmers in Australia. The authors concluded that bonds with places that were rooted in occupational identities could be maladaptive if they served to prevent voluntary relocation and resettlement in other areas. Also linked to climate change, research has shown that community objections to the siting of renewable energy projects are rooted in attachments to place that spoil the character of sea or landscape contexts (Devine-Wright and Howes, 2010). Here, the immediate threat to place posed by the siting of energy technologies seems to outweigh the broader threats posed by global warming, lessening the potential for rapid and extensive changes to energy systems of provision called for by climate experts (IPCC [Intergovernmental Panel on Climate Change], 2018).

We can summarise these aspects of place attachment in four key dimensions. First, place attachments, while mainly positive, can also be characterised by negative feelings or ambivalence (Manzo, 2014). Second, people can form attachments to multiple types of places at diverse scales, from the intimacy of a bedroom to a neighbourhood, region or beyond. Third, place attachments are dynamic. While the term can suggest a relatively static phenomenon, in fact it is necessary to view place attachment as a dynamic process that is continually unfolding over time (Devine-Wright, 2014), just as places themselves are said to be in a continual process of becoming (Massey, 2005). Research on intensity of place attachment reflects the fact that the strength of the attachment bond can vary between people and over time, becoming more or less intense depending upon circumstances. Fourth, people can have attachments to multiple places at one and the same time, even if much of the research literature tends to focus upon relationships with the current residence place (Gustafson, 2014). These diversities outline why place attachment is both an interesting and challenging concept for research.

Theoretical developments have explored various dimensions of attachment, utilising varied methodological approaches and applying the concept to different problems, some of which are discussed below.

Theoretical approaches

Some scholars have defined place attachment in a holistic manner. From this perspective, place attachment is used to encompass multiple aspects of people–place relations, including emotions, meanings and actions (Altman and Low, 1992). By contrast, other scholars have used a more specific approach, confining place attachment to the emotional bond that people have with a place (Hernández, Hidalgo, and Ruiz, 2014). From this perspective, place attachments are often contrasted with two related yet distinct terms. Place identity refers to the ways that places are drawn upon by individuals and groups to convey a sense of identity at personal and social levels (Proshansky, Fabian, and Kaminoff, 1983). Place dependence refers to functional aspects of places – how well they provide the amenities or services to support our needs and intended uses (Williams, 2014).

Williams (2014) distinguished between two aspects: place as locus of meaning and place as locus of attachment. The former refers to the beliefs, memories and experiences associated with a given place by a person or group of people. It is these meanings that imbue place with its sense of uniqueness and particularity, and they are often stressed by scholars informed by a phenomenological approach. The latter refers to the emotional bond that people feel with a place. While not uniform, many scholars of place attachment within Environmental Psychology have tended to adopt an approach consistent with the social psychology literature on attitudes and behaviour. This perspective adopts a positivist epistemology and individualist ontology, assuming the attachment bond to be similar to an attitude (positive or negative) that leads to action (Stedman, 2002b). I have been critical of research adopting this perspective for its reductive tendency to overlook the social and political embeddedness of place bonds and related place meanings (see Batel et al., 2015; Devine-Wright, 2009).

Contrasting approaches are visible in ways that scholars have proposed conceptual frameworks. For example, Gustafson (2001) drew on qualitative research to argue that place attachments involve three distinct dimensions of meanings, using a triangle framework to represent inter-dependent personal, social and environmental aspects (see Figure 45.1).

By contrast, Scannell and Gifford (Scannell and Gifford, 2010a) identified a tripartite framework made up of person, place (social and physical) and process dimensions, with the latter said to involve cognitive, emotional and behavioural aspects (see Figure 45.2). The

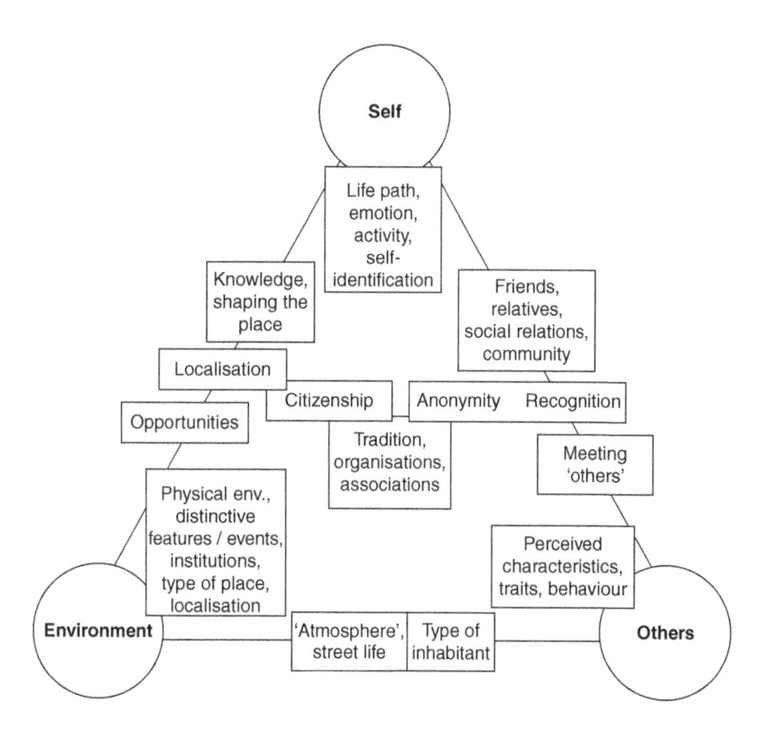

Figure 45.1 A multi-dimensional conceptualisation of place meanings.

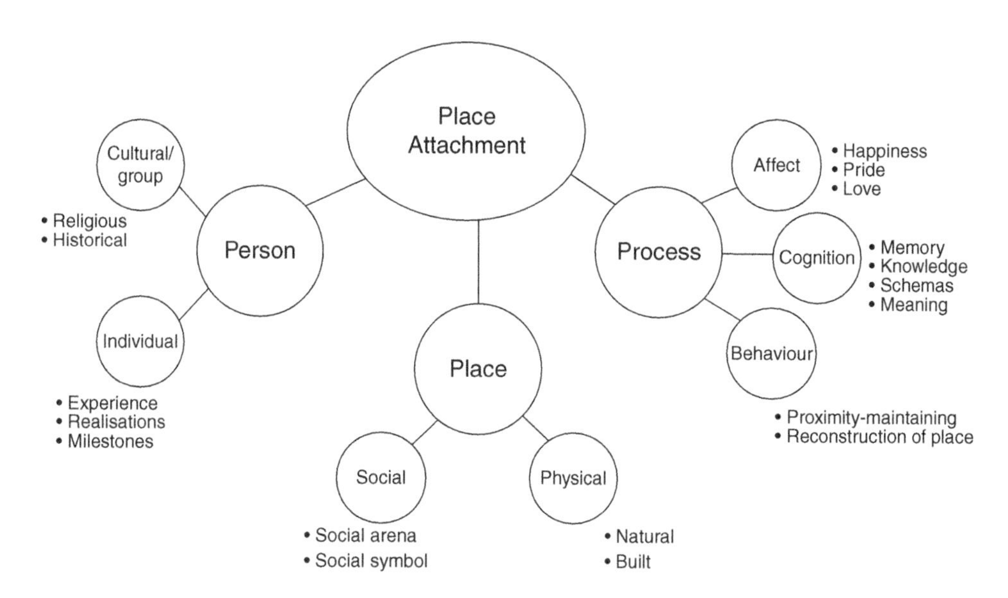

Figure 45.2 The tri-partite framework of place attachment.

authors also drew on quantitative survey data to distinguish between natural and civic dimensions of place attachment, reflecting different emphases upon environmental and social aspects of relationships with a local place (Scannell and Gifford, 2010b).

These dimensions and typologies reveal the complexities of attachments to place as well as the diverse ways that these have been theorised. Researchers informed by social psychology have been critical of the lack of clarity and empiricism in much research on place meanings that is informed by phenomenology (e.g., Stedman, 2002b). On the other hand, these same 'attitude'-based approaches have been criticised for neglecting broader socio-cultural and political dimensions (Devine-Wright, 2009; Di Masso, Dixon, and Durrheim, 2014; Manzo, 2014), and for presenting a rather static view of people–environment relations (such as the Scannell and Gifford framework) (Devine-Wright, 2014).

An important contribution has been the use of typologies to describe different ways that people relate to places. Work by Hummon (1992) and Lewicka (2011b, 2013) has drawn on qualitative and quantitative evidence to summarise different people–place relationships in terms of emotional valence (positive to negative), intensity (strong to weak) and activity (active to passive) (Lewicka, 2011b, 2013). According to this approach, there are multiple ways of feeling strong, positive attachments to a place (described as 'traditional' and 'active') and multiple ways of feeling weak or absent attachments (described as 'alienation', 'place relativity' and 'placelessness'; Lewicka, 2011b). Traditional attachment involves a taken-for-granted relationship with a place, while active attachment is defined by a high level of conscious attachment (Lewicka, 2011a). Alienation involves a negative feeling towards a place and a strong desire to leave; relativity refers to a fundamentally ambivalent relationship of moderate intensity, while placelessness refers to those for whom connections with place are not important or relevant to their lives.

Table 45.1 Summary table of life-place trajectory attributes (from Bailey, Devine-Wright, and Batel, 2016)

Life-place trajectory/variety of current place relation	*Length of local residence*	*Level of mobility*	*Level of settlement continuity*	*Feelings toward past residence place*	*Feelings towards current residence place*
Long-term residence in a single place/traditional attachment (7)	Long	Low	n/a	n/a	Strong, positive
Return to the home place/traditional-active attachment (5)	Long	Low – moderate	Moderate	Strong, negative	Strong, positive
Residential mobility with continuity in settlement type/active attachment (3)	Moderate – long	Moderate	High	Strong, positive	Strong, positive
Residential mobility with discontinuity in settlement type/place estrangement (7)	Low – moderate	Moderate	Low	Strong, positive	Strong, negative/ ambivalent
High residential mobility/placelessness (3)	Low – moderate	High	n/a	n/a	n/a

Note: The numbers in brackets show how many participants indicated each life-place trajectory.

Bailey, Devine-Wright, and Batel (2016) set these relationships with the residence place in the broader context of a person's life-course. Using qualitative, narrative interviews, five 'life-place trajectories' were identified (Table 45.1) including 'long term residence', characterised by decades of living in the same place and a sense of positive attachment and 'insideness' there; 'return to the home place' representing a person who grew up in a particular place, moved away for a short period before returning to the same place with the intentions of remaining there and retaining a positive attachment to that place; 'residential mobility with settlement continuity', representing a person who moved into a place in adulthood having previously lived in similar types of places, and holding a positive attachment to the current residence place; 'residential mobility with settlement discontinuity', representing a person who moved into a place in adulthood having previously lived in different types of places, and holding a negative or ambivalent attachment to the current residence place; and finally 'placelessness', representing a person for whom attachment to a place is unimportant by comparison with other ties or associations, such as to relatives or friends.

These trajectories take account of several of the dimensions of place attachment outlined above – their dynamics over time, the potential for attachment to multiple places and to multiple types of place, as well as both positive and negative bonds with the current place of residence. Moreover, they build on a neglected concept in the literature – settlement identity – that describes ways that people come to construct a sense of self embedded in a particular type of place (for instance, as 'a big city' or 'countryside' kind of person; Feldman, 1990). It is the continuity or discontinuity of settlement type as a consequence of residential mobility that was found by Bailey, Devine-Wright, and Batel (2016) to explain whether feelings towards the current residence place were broadly positive, negative or ambivalent.

These studies of place attachment are notable for revealing a facet of place attachment that has received considerably less focus in the literature: 'placelessness'. When interviewed, individuals with life trajectories characterised by high levels of residential mobility tended to find place attachment largely irrelevant to their personal lives (Bailey, Devine-Wright, and Batel, 2016). Given the extent of interest in hyper-mobilities associated with globalisation (Castells, 2000), it is necessary to examine whether the prevalence of these different trajectories is changing over time to make long-term residence more rare and placelessness more commonplace, something about which we have relatively little empirical evidence currently.

It is also important to note that frequent travel, such as for work purposes, has been shown to relate to place attachment in interesting and counter-intuitive ways. Gustafson (2009) showed that 'cosmopolitans' who travelled frequently for work also indicated strong attachments to their places of residence and were often more active in local clubs and associations than individuals who travelled less frequently for work. The attachment/mobility nexus is therefore complex, referring to residential mobilities and other forms, an issue discussed further below.

What these studies illustrate, apart from the complexity of the topic, is a proclivity amongst researchers to invent new terms and types that seem to fit with their approach or their findings. While this has stimulated novelty in the conceptual language brought to bear on the topic, it has led to some concern about an increasingly confusing literature that lacks coherence with scholars reacting to this concern in different ways. While some have claimed that certain conceptualisations are superior to others, and therefore should be more widely adopted, as with attitude-based social psychology (Stedman, 2002b), others have argued for

Table 45.2 Diverse approaches to place attachment research

Ontological focus	*Conceptual focus*	*Epistemology*	*Method*	*Examples*
Individual (processes of cognition and affect)	Place attachment as an attitude	Positivist	Quantitative methods (e.g., surveys)	Stedman (2002b) Hernández, Hidalgo, and Ruiz, 2014
Individual in context (life experiences or 'lifeworld')	Place meanings	Phenomenology	Qualitative (e.g., interviews and life narratives)	Seamon (2014) Manzo (2005)
Individual within a social context	Place meanings as social representations	Critical realist	Mixed methods (e.g., interviews, focus groups and surveys)	Devine-Wright (2009) Devine-Wright and Howes (2010) Batel and Devine-Wright (2015)
Socio-cultural	Discourse, narratives	Constructivist/ post-structuralist	Qualitative (e.g., interviews, documentary analyses)	Di Masso, Dixon, and Pol (2011)

tolerance of varied approaches, founded upon claims about the impossibility of unity or consensus in the field (Patterson and Williams, 2005).

Methods and approaches

The multiplicity of theoretical terms and ideas referred to above is mirrored by the varied ways that researchers have studied place attachment (Table 45.2).

Environmental Psychologists have often used questionnaire surveys as the primary method to research place attachment, preferring methods that offer the potential for quantitative analyses of numerical data that fit well with a positivist epistemology and potential for hypothesis testing. Specific items and scales have been developed as tools to measure the different dimensional aspects of place attachment referred to above. A notable example is Williams et al. (1992) who devised survey items that have subsequently been employed in numerous studies on a wide variety of topics and are much cited (Williams, 2014). An example is Stedman's work in the US state of Wisconsin (Stedman, 2002b). He used a survey to identify the views of lakeshore property owners towards a new housing development proposed nearby. His analysis showed that willingness to take action depended upon the kinds of meanings associated with the lakeshore area, as well as levels of current satisfaction with that place. Those people who viewed the area as somewhat wild and distant from urban areas, and were dissatisfied with its current condition, were least accepting of change and most intent on engaging in 'place protective behaviours' (Stedman, 2002b).

By contrast, other researchers have persisted with the phenomenological focus deployed by Humanist Geographers. This draws on personal experiences of place, using qualitative techniques such as in-depth interviews. Underlying this interest is a conceptual approach based on a continual dialectic in place experience between a sense of insideness and outsideness (Relph, 1976) or belongingness and alienation (Seamon, 1996). An example of phenomenology-informed research is Manzo's (2005) study that drew on interviews with residents of New York with the aim of learning more about the kinds of places that were meaningful to them, the roles that these places might play in their lives and the ways in which they develop meaning. The study revealed negative as well as positive and dramatic as well as mundane meanings associated with diverse places. All participants went beyond the current residence place when identifying important places in their lives, places that were said to provide opportunities for reflection, sanctuary or were associated with moments of life transitions. She concluded that the experiences of places which people find meaningful demonstrate the socio-political underpinnings of emotional relationships to places, particularly the impact of gender, race, class and sexuality.

Qualitative methods are also employed by research informed by post-structuralist approaches (e.g., Di Masso, Dixon, and Durrheim, 2014). Here, individual experiences of place are interpreted as reflections of broader socio-cultural discourses of dwelling, belonging and identity, inflected with power relations and asymmetries. Attention to the language used to convey place-related experiences and meanings is central to this approach, which is critical of previous studies that have tended to view language as a neutral container of meaning and to treat place attachment as an internal or mentalist process.

Building on a longstanding interest in mapping methods (Lynch, 1960), researchers have begun to use Geographical Information Systems (GIS) to digitally map place meanings and values (Brown, Raymond, and Corcoran, 2015). Informed by the 'mobility turn' in sociology, there has also been increasing interest in the use of situated or mobile methods, for example 'go along' or walking interviews (where the interview takes place outside and in motion and the interviewee may determine route choice and topics for discussion; Rishbeth, 2014) and photo elicitation (where the participant draws on their own or other's images of places to recount what is important to them; Stedman et al., 2014).

In my own work, I have attempted to overcome the limitations of solely qualitative or quantitative approaches by combining both in mixed-method case study research informed by a critical realist epistemology. This suits the dual emphasis upon place as locus of meaning and place as locus of attachment (Williams, 2014). When conducting a study of local acceptance of a proposed offshore wind farm in North Wales, I selected two coastal towns a similar distance from the project site (Devine-Wright and Howes, 2010). This was important due to assumptions about spatial proximity embedded in the 'NIMBY' (not in my back yard) concept. The research involved three stages. First, in-depth interviews were conducted with key actors, including local protestors, the developer and statutory bodies. Second, questionnaire surveys were completed by residents in each town, with items derived from the literature and from the interview analyses, which provided specific framings of the project and its impacts (e.g., that the wind turbines would 'fence in the bay'). Third, focus group discussions were held with residents in each location, extending some of the quantitative analyses from the surveys. Taken together, the methods provided insights into the meanings associated with each place and the wind farm that enhanced understanding of so-called 'NIMBY' reasons for objection, including opening up some new ideas.

To take one example, participants in the focus groups revealed unexplored narratives relating the wind farm with the place (Devine-Wright, 2011b). They suggested that if you imagined the wind farm as something that the Victorian-era founders of the coastal towns would have admired and indeed developed themselves, a potential narrative emerges for a local yet large-scale wind energy development that fits into the history of the area. This way of seeing the wind farm stands in contrast to the narrative of place threat proposed by the local action group, which emphasised how it would destroy the historical continuity and distinctiveness of the town. It also suggests ways that the energy project could be redesigned to achieve greater acceptance, not only with a different narrative but with a different structure involving greater local ownership in contrast to the wind farm that was proposed by a consortium of German companies.

A new project about shale gas takes this place-based, multi-method approach in new directions by combining national level, longitudinal public attitude surveys with social media analyses and local-level ethnography. This approach is better able to capture societal and local community responses as they develop over time in interconnected ways. Analysis of the content of social media postings by action groups and community organisations will provide an additional source of understanding about how meanings associated with places and energy technologies are interpreted to 'fit' or be out of place and, as a consequence, become integral to narratives of support or objection and taken up elsewhere. Geolocation of Twitter data will enable analyses of community responses to be underpinned by quantitative indicators of spatial proximity. Taken together, the aim is to provide robust, multi-level empirical analyses that take forward our understanding of how place attachments are implicated in societal and community responses to energy projects.

Current debates and new directions

While the 'mobilities turn' has had extensive influence across the social sciences, it has had less influence upon psychological research on place attachment, which has tended to remain focussed upon a rather static view of attachment to the current residence place. This was critiqued in a recent article that re-orients place attachment research by proposing an underlying fixity–flow continuum (Di Masso et al., 2019). Acknowledging the potential for stasis in dwelling and emotional bonds with place, the paper opened up more fluid understandings of place attachment, while making the case that increased mobilities associated with globalisation had not, despite many predictions, reduced the importance of place in the world. A novel conceptual framework outlined how place attachment could be viewed at either pole of a fixity–flow continuum, as well as multiple intermediate points in between, and traced the implications of each configuration. This framework has great potential for application to important contemporary challenges, including migration, remote working and resettlement after disaster.

The potential for attachment to distant places has been less commonly researched than relations to the (nearby) residence place. Gurney et al. (2017) proposed that changes to distant places can affect non-local people and could have important consequences for community-driven environmental stewardship. Empirical survey data was analysed with a focus upon the endangered Great Barrier Reef in Australia, drawing on a large-scale sample (n=5403) that combined coastal residents, tourist visitors from Australia and other countries, reef tourism operators and fishers. Four distinct communities were identified that indicated diverse forms of attachment to the reef: 'Armchair enthusiasts' (n=2314) represented individuals for whom the reef does not provide day-to-day functions, but instead is important for identity and the way of life it supports; 'Reef connected' (n=839) represented individuals for whom the reef was important across all facets of place attachment, from everyday use and livelihoods to more symbolic and emotional aspects; 'Reef users' (n=1096) represented individuals for whom the reef was more important for livelihoods than for identity; and 'Reef disconnected' (n=1154) represented individuals for whom the reef was not strongly important. The study is important in showing that a sense of attachment can lay a foundation for communal acts of environmental stewardship that transcend local or national boundaries.

Much of the existing literature on place attachment presumes the development of attachment to be a slow process, connected to length of dwelling or frequency of visits. Raymond, Kytta, and Stedman (2017) challenged this assumption by introducing ideas from affordance theory. The paper views the person as emplaced in a specific environmental context and argues that sensory engagement with the environment has been neglected in favour of the study of discourse or socially-constructed meanings. Instead of viewing these as opposite, a complementary approach is taken that presents two modalities, one slow (social construction) and the other fast (perception-action) whereby places can become meaningful and important. These ideas have implications for urban planning and management. City planners may not be able to design environments that increase place attachment directly, but by creating spaces that afford different opportunities for meaning-making by diverse user groups, attachment can arise as an emergent property of good design.

Conclusions

Place attachment refers to the meanings and emotional bonds that are felt by individuals and groups towards specific places. They have been subject to decades of sustained research by scholars in multiple disciplines including geography, psychology, sociology, anthropology and urban studies. A complex concept, fundamentals of place attachment include the following dimensions: multivalent emotional bonds, not just positive; varying in focus and scale of place, not just residential settings; dynamic over time, not static; and multiple, not singular. In part arising from its multidisciplinary relevance and importance, many concepts and frameworks have been developed to deepen our understanding of place attachment and, as a consequence, some debate has ensued about whether this is merely a source of confusion or necessitates a tolerance for varied approaches that are embedded in diverse methods allied to ontological and epistemological assumptions. In my own research, centred upon issues of acceptance of energy technologies, I have sought to make sense of these diversities by employing a conceptual and empirical approach that embraces the positive attributes of two disciplines: Environmental Psychology, with its focus upon robust methods and quantitative analyses, and Human Geography, which is open to qualitative and novel methods, as well as an emphasis upon the inherently social and political dimensions of place change. Such combinations produce new challenges, both conceptual and empirical, but when taken alongside the literature above, reflect the potential for research on place attachment to offer novel insights into contemporary challenges.

References

Altman, I. and Low, S. (eds) (1992). *Place attachment.* New York: Plenum Press.

Antonsich, M. (2010). 'Meanings of place and aspects of the self: an interdisciplinary and empirical account', *GeoJournal*, 75: 119–132.

Bailey, E., Devine-Wright, P. and Batel, S. (2016). 'Using a narrative approach to understand place attachments and responses to power line proposals: the importance of life-place trajectories', *Journal of Environmental Psychology*, 48: 200–211.

Batel, S., Devine-Wright, P., Wold, L., Egeland, H., Jacobsen, G. and Aas, O. (2015). 'The role of (de-)essentialisation within siting conflicts: an interdisciplinary approach', *Journal of Environmental Psychology*, 44: 149–159.

Brown, B. and Perkins, D. (1992) Disruptions to place attachment. In I. Altman and S. Low (eds) *Place attachment* (pp. 279–304). New York: Plenum Press.

Brown, G., Raymond, C.M. and Corcoran, J. (2015). 'Mapping and measuring place attachment', *Applied Geography*, 57: 42–53.

Castells, M. (2000) *The Rise of the Network Society*. Oxford: Blackwell.

Cooper Marcus, C. (1995) *House as a mirror of self: Exploring the deeper meaning of home.* Berkeley, CA: Conari Press.

Devine-Wright, P. (2009) 'Rethinking nimbyism: the role of place attachment and place identity in explaining place protective action', *Journal of Community and Applied Social Psychology*, 19(6): 426–441.

Devine-Wright, P. (2011a). 'Public engagement with large-scale renewable energy: breaking the NIMBY cycle', *Wiley Interdisciplinary Reviews: Climate Change*, 2: 19–26.

Devine-Wright, P. (2011b) From backyards to places: public engagement and the emplacement of renewable energy technologies. In P. Devine-Wright (ed) *Public engagement with renewable energy: From NIMBY to participation* (pp. 57–70). London: Earthscan.

Devine-Wright, P. (2014) Dynamics of place attachment in a climate changed world. In L. Manzo and P. Devine-Wright (eds) *Place attachment: advances in theory, method and applications* (pp. 165–177). Abingdon: Routledge.

Devine-Wright, P. and Howes, Y. (2010). 'Disruption to place attachment and the protection of restorative environments: a wind energy case study', *Journal of Environmental Psychology*, 30: 271–280.

Di Masso, A., Dixon, J. and Durrheim, K. (2014) Place attachment as discursive practice. In L.C. Manzo and P. Devine-Wright (eds) *Place attachment:Advances in theory, methods, and research* (pp. 75–86). Oxon: Routledge.

Di Masso, A., Dixon, J. and Pol, E. (2011) 'On the contested nature of 'Figuera's Well', 'The hole of shame' and the ideological struggle over space in Barcelona', *Journal of Enviromental Psychology*, 31(3): 27–44.

Di Masso, A., Williams, D. R., Raymond, C. M., Buchecker, M., Degenhardt, B., Devine-Wright, P., Hertzog, A., Lewicka, M., Manzo, L., Shahrad, A., Stedman, R., Verbrugge, L. and Von Wirth, T. (2019) 'Between fixities and flows: Navigating place attachments in an increasingly mobile world', *Journal of Environmental Psychology*, 61: 125–133. doi:10.1016/j.jenvp.2019.01.006

Dixon, J. and Durrheim, K. (2000). 'Displacing place identity: a discursive approach to locating self and other', *British Journal of Social Psychology*, 39: 27–44.

Feldman, R. (1990). 'Settlement identity: psychological bonds with home places in a mobile society', *Environment & Behavior*, 22: 183–229.

Fried, M. (2000). 'Continuities and discontinuities of place', *Journal of EnvironmentalPsychology*, 20: 193–205.

Gurney, G.G., Blythe, J., Adams, H., Adger, W.N., Curnock, M., Faulkner, L., James, T. and Marshall, N.A. (2017) 'Redefining community based on place attachment in a connected world', *PNAS*, 114(38): 10077–10082.

Gustafson, P. (2001). 'Roots and routes: exploring the relationship between place attachment and mobility', *Environment & Behavior*, 33: 667–686.

Gustafson, P. (2009). 'More cosmopolitan, no less local: the orientations of international travellers', *European Societies*, 11: 25–47.

Gustafson, P. (2014) Place attachment in an age of mobility. In L.C. Manzo and P. Devine-Wright (eds) *Place attachment: Advances in theory, methods and applications* (pp. 37–48). London: Routledge.

Hernández, B., Hidalgo, M.C., Ruiz, C. (2014) Theoretical and methodological aspects of research on place attachment. In L.C. Manzo and P. Devine-Wright (eds) *Place Attachment: Advances in Theory, Methods and Applications* (pp. 125–137). London: Routledge.

Hummon, D. (1992) Community attachment, local sentiment and sense of place. In I. Altman and S. Low (eds) *Place attachment* (pp. 253–278). New York: Plenum Press.

IPCC [Intergovernmental Panel on Climate Change] 2018. Global warming of 1.5°C: Summary for Policymakers. Available online: http://report.ipcc.ch/sr15/pdf/sr15_spm_final.pdf.

Lewicka, M. (2011a). 'Place attachment: How far have we come in the last 40 years?' *Journal of Environmental Psychology*, 31: 207–230.

Lewicka, M. (2011b). 'On the varieties of people's relationships with places: hummon's typology revisited', *Environment and Behavior*, 3: 676–709.

Lewicka, M. (2013). 'Localism and activity as two dimensions of people-place bonding: the role of cultural capital', *Journal of Environmental Psychology*, 36: 43–53.

Lynch, K. (1960) *The image of the city*. Cambridge, MA: MIT Press.

Manzo, L. (2014) Exploring the shadow side: Place attachment in the context of stigma, displacement and social housing. In L.C. Manzo and P. Devine-Wright (eds) *Place attachment: Advances in theory, methods and applications* (pp. 178–190). London: Routledge.

Manzo, L.C. (2005). 'For better or worse: exploring multiple dimensions of place Meaning', *Journal of Environmental Psychology*, 25: 67–86.

Marshall, N.A., Park, S.E., Adger, N.E., Brown, K. and Howden, S.M. (2012) 'Transformational capacity and the influence of place and identity', *Environmental Research Letter*, 7: 034032. (9 pages).

Massey, D. (2005) *For space*. London: Sage.

Patterson, M.E. and Williams, D. R. (2005). 'Maintaining research traditions on place: diversity of thought and scientific progress', *Journal of Environmental Psychology*, 25: 361–380.

Proshansky, H., Fabian, H. K. and Kaminoff, R. (1983). 'Place identity: physical world socialisation of the self', *Journal of Environmental Psychology*, 3: 57–83.

Raymond, C., Kytta, M. and Stedman, R. (2017). 'Sense of place, fast and slow: the potential contributions of affordance theory to sense of place', *Frontiers in Psychology*, 8: 1674.

Relph, E.C. (1976) *Place and placelessness*. Pion: London.

Rishbeth, C. (2014) Articulating transnational attachments through on-site narratives. In L.C. Manzo and P. Devine-Wright (eds) *Place attachment: Advances in theory, methods and applications* (pp. 100–111). London: Routledge.

Scannell, L. and Gifford, R. (2010a). 'Defining place attachment: a tripartite organising Framework', *Journal of Environmental Psychology*, 30: 1–10.

Scannell, L. and Gifford, R. (2010b). 'The relations between natural and civic place attachment and pro-environmental behavior', *Journal of Environmental Psychology*, 30: 289–297.

Scannell, L. and Gifford, R. (2017). 'The experienced psychological benefits of place attachment', *Journal of Environmental Psychology*, 51: 256–269.

Seamon, D. (1996) 'A singular impact', *Environmental and Architectural Phenomenology Newsletter*, 7(3): 5–8.

Seamon, D. (2014) Place attachment and phenomenology: The synergistic dynamism of place. In L. C. Manzo and P. Devine-Wright (eds) *Place attachment: Advances in theory, methods and applications* (pp. 11–22). London: Routledge.

Stedman, R. (2002b). 'Toward a social psychology of place: predicting behavior from place-based cognitions, attitude and identity', *Environment and Behavior*, 34: 561–581.

Stedman, R., Amsden, B.L., Beckley, T.M. and Tidball, K.G. (2014) Photo-based methods for understanding place meanings as foundations of attachment. In L.C. Manzo and P. Devine-Wright (eds) *Place Attachment: Advances in theory, methods and applications* (pp. 112–124). London: Routledge.

Tuan, Y.F. (1974) *Topophilia: A study of environmental perception, attitudes, and values*.Englewood Cliffs, NJ: Prentice Hall.

Williams, D. (2014) "Beyond the commodity metaphor," revisited: Some methodological reflections on place attachment research. In Lynne C. Manzo and Patrick Devine-Wright (eds) *Place attachment: Advances in theory, methods, and research* (pp. 89–99). Oxon: Routledge.

Williams, D.R., Patterson, M.E., Roggenbuck, J.W. and Watson, A.E. (1992). 'Beyond the commodity metaphor: Explaining emotional and symbolic attachment to place', *Leisure Sciences*, 14: 29–46.

46

In the presence of absence

Meditations on the cultural significance of abandoned places

Justin Armstrong

At some point, every place is a remainder, something left behind. In the wake of this leaving, a place begins its journey to abandonment as its signifiers untie themselves to float freely in a liminal cultural oblivion (Augé and Young 2004). In what follows, I cannot claim to answer the question of how and why places become abandoned, but I hope to develop a useful toolkit for considering and analyzing these sites as they emerge and fade in and out of our cultural and historical consciousness. Here, I present a series of inquiries and suggestions for how and why it can be useful to develop a critical framework around the processes and outcomes of abandonment in place. Drawing on over a decade of ethnographic research spent studying these types of places around the world, I offer a potential outline for thinking about, and engaging with, abandonment in an era increasingly dominated by the virtual and global. In this way abandonment becomes a lived inquiry, a site of material interaction, and a translatable cultural text that can illuminate forces much more intricate than the simple absence of people. I cannot (nor do I want to) answer the question of abandonment here, rather I offer a series of inquiries and reflections that can be useful in forming an ontology or taxonomy, an ontological taxonomy, of these places.

In 1982 Uranium City, Saskatchewan, was home to almost 5,000 people. That same year the mine closed and the majority of the residents left as soon as Lake Athabasca froze over and they could get moving trucks in on the ice road. As of 2016, 73 people live in this mostly forgotten corner of the province. This place is so far north that most provincial maps exclude it, appearing only on a tiny inset map at best. Abandoned in the real world and abandoned by road atlas cartographers, Uranium City is the quintessential abandoned place. Still, it is not empty, it is not a place without people, but it has been abandoned and displaced from the rest of the world. Most people who live in Saskatchewan have never heard of it. My grandparents, who live in the southern part of the province, knew its name and general location but nothing more, a kind of myth, an *ultima Thule* to the rest of Saskatchewan, something north beyond the known world.

Down a one-time street at the western edge of town, a lost suburb, a place where fire hydrants hide amongst the spindly sub-arctic trees gouging their way through sidewalks now covered in leaf litter and moss. An apartment block built but never occupied rests slack-jawed and mute, wires hanging like tendons in a late summer already getting cool. As I wander the remainders of these people, places and things, I'm building narratives and assembling meaning, partially from my own positionality, but also from the collective fragmented nostalgia absorbed through cultural osmosis. Here, part of me feels like I am breathing in a new kind of haunting, one that has been patched together around something I never knew. Memories becoming untethered, slippery and amorphous.

Without a doubt this place is being abandoned, erasing itself quietly in solitude. I wonder if I should mourn this place. Maybe I am simply fetishizing this encounter and offering misplaced sympathy to a place I have never known. In many ways, charting a place's life cycle requires thinking of abandonment as a necessary component of placemaking, that human engagement and inscription on landscapes that frames our interactions. Place is shifting, dynamic and ultimately plastic, and perhaps the most productive way for us to understand its significance is to see the processes and outcomes of abandonment as natural and inevitable, and incomplete. Here, Derrida's (2006) concept of *hauntology* is particularly relevant. For Derrida, hauntology is the way in which we come to know and understand haunting, how we deconstruct and reconcile an encounter with an unrealized future. Using this lens to view Uranium City helped to bring into focus the psychological and cultural importance of deeply considering abandoned place. A hauntology of place forms a contemporary archeology of the could-have-been-but-never-was (see Buchli and Lucas 2001).

In a corner of the dead liquor store along one of Uranium City's main streets, I am holding a smeared and mouldy letter from a middle school girl to her best friend—more markers of abandonment left in place, then displaced and now being reassembled. A moment of time in miniature buried just below the invisible surface, something plainly hidden. This letter and its place form a tangible abandonment, something anchored and fading that marks this place as abandoned—once occupied and now floating between existence and non-existence, a true ghost.

Defining abandoned place is at once simple and complicated. Perhaps it is only remnants, ideas and things left behind in the wake of departures. Maybe it is just *waste*, something Brian Thill (2015: 8) describes as "every object, plus time". From a more spectral perspective, it could be the direct and observable presence of absence, human occupation under slow-motion erasure. Here it is important to consider why we might care, as anthropologists, geographers, archeologists, or any other sort of person moving through the environment during the course of our everyday lives. In beginning to examine this world of here and nowhere as a location for critical inquiry, it becomes necessary to locate the betwixt and between (Turner et al. 1969) of abandoned place.

In the final line of Marc Augé's seminal *Non-Places: An Introduction to Supermodernity*, the author suggests the need for "an ethnology of solitude" (2009: 98), a challenge that I enthusiastically take up in this chapter. Here, it is important to define ethnology a bit more specifically before moving on to offer my own framework for what a practice of hauntological anthropology of place might look like and how it can be applied to discussions of abandoned places and their translation.

Ethnology examines a given set of cultural data from a variety of sources in order to draw a conclusion about a particular phenomenon (incest taboos, myth, religion, polygamy, etc.) (Lévi-Strauss 1974), whereas ethnography examines a specific cultural context in order to ask and answer questions about the people, places and things of the anthropologist's chosen fieldsite. With this delineation in mind, I suggest something more akin to an *ethnography* of solitude, wherein understanding something about the nature of abandonment comes from direct experience of it, from the first-hand ethnographic contact with the observable ghosts/remainders of abandoned place.

As an anthropologist, I have been studying abandoned and becoming-abandoned places for the better part of fifteen years. Here and there, a person or two, a curtain hesitantly parted in a nearly evaporated town in South Dakota, a roadside conversation with one of the four residents of Tway, Saskatchewan. There has been relatively little discussion of the roles and outcomes of abandonment in the process of placemaking (or what might be called re-placemaking), but here I hope to illuminate some of the ways in which we can begin to read the *presence in absence* embedded in abandonment as a specific genre of place. The notion of a *genre of place* is important here, and builds on Perec's (2008) view of species of spaces, offering a taxonomy of locations and locationality. This framework allows for mode for understanding the defining characteristics of certain kinds of spaces and places. Moving forward, it is useful here to outline the ways in which I conceptualize space and place (and non-place). For Tuan (2001), the relationship between space and place is akin to that of container and contents. Human beings draw place (significance) from space (context) by embedding affect, history, and culture in a location. In outlining this concept for my students, I often use the example of their dorm rooms. As they experience them, these rooms serve as blank slates where, over the course of the year, they undertake the process of placemaking to fill these containers with meaning, thus changing the *space* into a lived and affective *place*. Of course this explanation very often leads to the question of whether placeness remains in the absence of placemakers, an inquiry I answer with reference to Jon Anderson's (2015) discussion of the *trace*. For Anderson, traces "are marks, residues or remnants left in place by cultural life" (2015: 4), a notion that dovetails effectively into the current discussion around how we begin to translate an apparent emptiness of abandoned places into a dynamic cultural text. The work of assembling meaning in the presence of absence has been addressed by a number of scholars (Armstrong 2010; DeSilvey 2006, 2007, 2017; DeSilvey and Edensor 2013; Edensor 2005; Pétursdóttir 2013), forming several valuable departure points for presenting a discussion of how and why abandoned place deserves our ethnographic and everyday consideration.

In reflecting on the life cycle of place and the process of placemaking, it can be said that a location begins as a place (in the moment that space is injected with affect), becomes misplaced in the time of pre-abandonment, moves to displacement when permanent occupation ceases, and gets re-placed when former inhabitants and new visitors (re)form their own narratives. This precarity is exactly what makes abandoned place so compelling—it is always at the edge of dissolution, always leaking presence toward complete absence. Abandoned places offer a window into inevitable emptiness.

But what is the usefulness of this ability to notice, observe and translate abandoned places? What does this awareness *do*? Simply put, recognizing abandoned place as more than a basic aesthetic experience presents the opportunity for translation of embedded dialogs between people, places and things. In an abandoned place the narratives are much more fluid in the absence of human occupation. The observer can assemble their own meaning in search of those moments of illumination and insight. To carefully and consciously notice abandonment and seek out these sites is to add a valuable counterpoint to our overpopulated, oversaturated places of inhabitation. To know abandoned place is to know inhabited place. To apprehend these places as lived experience allows for the (re)centering of our peripheries.

Of course there is a phenomenological aspect to this form of engagement with place, a practice that involves a careful notice of the almost-unseen, the just-peripheral and the becoming-buried. In a sense, access to these locations and their associated resonance requires a new form of attention—an attention to the palimpsests of place. In this way, abandonment sees itself as both new and old, a liminal location between being present and absent, it is at once a new ruin and an archaic monument. This sentiment is echoed in Huyssen's (2006: 11) statement that "(r)eal ruins of different kinds function as projective screens for modernity's articulation of asynchronous temporalities and for its fear of and obsession with the passing of time". If we assume that there is always some ruination inherent in abandoned place, then Huyssen's reflection can easily be transposed onto our relationship to place. Abandonment is the immediate precursor (and companion) to ruin, setting the stage for this type of resonance. The abandoned place offers a perfect cultural echo chamber in which to wrestle with our fraught relationship to shifting, contracting and expanding time. Here, we can also consider the cultural relativity of abandoned place and ask ourselves if Western conceptions of abandonment and history can and should be applied to places. It only seems logical to assume that a Global North experience of abandonment is not universal (see Fabian and Bunzl 2014 for an in-depth discussion of the cultural relativity of time).

Part of this phenomenological and historical interaction with the presence of absence lies in the aesthetic and ideological allure of abandoned place, the pull of the unknown (unknowable) history, the hope for a sparkling instance of Benjamin's (1978) profane illumination, that moment when, as the result of careful notice, the ordinary becomes extraordinary for an instant and unveils its hidden meaning. Our engagement with these places seeks meaning in both the past and the future, looking for clues to what has happened and what will happen as remainder, reminder and remnant.

In this way, our contemporary fascination with abandonment appears to emerge—at least in part—from a kind of fetishization of the condition of displacement and ruination. The aesthetics of the ruined or abandoned place as a modern cultural trope may offer clues about the slide into what Augé and Young (2004) have termed *oblivion*, the necessity of forgetting. Perhaps abandoned places serve as a last stand against the inevitable dissolution of time and memory in our collective imagination. Again, this fascination becomes a privileged perspective wherein the late-capitalist West maintains the luxury of reflecting on abandonment from a comfortable distance. Rarely do we consider the significance of abandonment as the result of forced migration, military and political upheaval, or economic displacement. The monied West often sees these places as aesthetic curiosities or quaint historical reverberations. Once more, in moving toward a culturally relative and comprehensive view of abandoned place, it is important to consider the polyvocality of these locations and their associated interpretations and translations.

A critical engagement with the place of abandonment also requires a rethinking of how we understand the historical trajectory of place as an accumulation/stratification (DeSilvey 2006) of meaning. Increasingly, as an anthropologist studying abandonment, I find myself moving toward post-processual archeological methods (Earle et al. 1987; Miller 1997; Trigger 2006) for reading and translating these sites and adopting what Victor Buchli and Gavin Lucas (2001) have termed "archaeologies of the contemporary past". From this vantage point it becomes possible to consider abandoned place as a location where past, present and future intertwine and form a dialog, making these sites resonate on multiple planes. Here, the abandoned place forms a kind of immanent becoming, a rhizomatic (Deleuze and Guattari 1987) location without beginning or end, a place that is all places at once, depending on the positionality of the observer. And while this meditation may appear abstract, it is beneficial to consider the subjectivity and cultural/ontological impositions we readily apply to abandoned place, as well as how we define these locations in various contexts.

Within the scope of these discussions it is important to consider when a place has reached its tipping point from inhabitation to dissolution, to recognize when it has become truly abandoned. From my perspective, an abandoned place is a culturally resonant location without direct and continued human occupation. It is this resonance that we can read as a text, translating the echoes of lives once lived in the human inscriptions on place. These locations are not placeless, nor have they reverted back to pre-placed spaces, rather they have become misplaced in the contemporary geographic and cultural consciousness. Following Caitlin DeSilvey's claim that "the disintegration of structural integrity does not necessarily lead to the evacuation of meaning" (2017: 5), it is important to consider the significance of ruins and ruination in the production and analysis of abandoned place. In many ways, ruins are an essential element of these sites.

In presenting a potential intersection between ruins and abandonment, it might be helpful to think of the ruin as the remnant of placemaking, the visible marker of abandonment, and therefore the placeholder for certain types of temporality and cultural engagements with historicity. Therefore, it seems unlikely that abandoned place could exist without material ruin. On the other hand, abandoned place might also be formed by what could be called *psychic ruins*, or imagined ruins. In this way, abandoned space is populated by mental projections, even in the absence of material ruins. Here, abandoned place becomes a frame for projections both individual and collective.

Below, using three key examples from my ethnographic fieldwork experiences, I draw out the importance of recognition, analysis and translation, forming a dialog between people, places and things, as well as past, present and future, making an argument for the multiplicity of cultural significance present in an abandoned place. The Hornstrandir region of northern Iceland offers a view to long-scale abandonment; my time spent living in the ghost town of Múli in the Faroe Islands provides an illustration of life lived in abandoned place; and the tiny outport community of Grand Bruit, Newfoundland, sets the stage for unpacking the sense of pre-abandonment. These accounts aim to reify abandonment in place and describe the processes and outcomes of this form of place(re)making.

Late June in Iceland's remote Hornstrandir Nature Reserve—the sun never sets, slowly moving across the sky in a lazy, bent loop. It's 1 a.m., and I am sitting in the hollow of an old farmhouse foundation in Barðsvik, the rock walls and caved-in roof made out of driftwood that floated over from Siberia are being devoured by nature around me. I am putting myself in the world of my subjects, the abandoned world of lives lived at the edge of the world, lives nobody has lived since the 1950s, but for hundreds of years before that.

In an attempt to understand something about this place, I undertook a week-long hike retracing the footpaths and mountain passes once used by the residents of this place to travel between the isolated farms that dotted the rugged coast only a few miles from the Arctic Circle. As a cultural anthropologist, one of my primary methodologies is participant-observation, something that is quite difficult in the relative absence of human agents. In order to gain a better understanding of the lives these people once lived, I endeavored to briefly embody their outlook and daily understanding of place. In essence, I wanted to become the ghost of my vanished subjects to gain a tiny window into the past of this abandoned present.

And while this approach to ethnographic reasoning is admittedly experimental and impressionistic, it furnished me with a more complete view of this place, allowing me to assemble fragments of the history of this now abandoned place. During the course of my trek, I interrogated this place by attuning myself to its *aura*, what Benjamin (2010) saw as the appearance of nearness at a distance. By walking these paths and sleeping in these

Figure 46.1 Abandoned church, Barðsvik, Iceland.

emptied fjords among the ruined farms, I found several instances of revelation about what it was to experience abandoned place from a more *emic* (inside) and less *etic* (outside) perspective. I gained a new dimension to my framing and was given access to a sliver of the lives once lived out here at the edge of everything. These places became less purely abandoned and formed new and dynamic structures of feeling (Williams 1983), in a strange way re-populating these abandoned and displaced places.

At the tip of the Faroese island of Borðoy a ten-mile single-lane dirt road hugs the side of a primordial mountain, my borrowed hatchback bumps its way along the deteriorating track to Múli, a pinprick of a village that after years of steady depopulation was finally abandoned in 2002. This place had been occupied for hundreds of years, with records of its existence reaching back to the 1400s, and it is widely understood to have been one of the oldest settlements in the Faroes. When the government eventually built a road connecting the village to the rest of the country in the 1990s (Múli was the last settlement in the Faroes to get electricity in 1970) the villagers seized the opportunity and promptly left. Now I was here, about to spend three weeks living as the lone resident of an abandoned village at the edge of the North Atlantic. I hoped to be able to answer some of the questions that had been following me through my years of research on abandoned places. Here, I wanted to get a glimpse (perhaps what James Clifford (2010) calls a "partial truth") of these places at the cusp of abandonment.

As a sort of precursor/companion piece to my project in Hornstrandir, and building on the work I'd done in Newfoundland (see below), my aim was to experience an abandoned

Figure 46.2 Múli, Faroe Islands.

place as the sole resident, to occupy an emptied microcosm to develop a better understanding of the various ways that we recognize and categorize abandoned places. During my time in the village I developed everyday patterns of being that gave me new insight into the true presence in absence. I became the presence that was often absent in my other interactions with these sorts of places. I was a ghost of my own design and I was performing a sort of self-ethnography of haunted place with myself as primary subject. Again, I had been offered a view into the prehistory of abandonment through my brief occupation of this place.

The provincial ferry rocks itself back and forth along Newfoundland's southern coast en route to Grand Bruit, a tiny dot of occupation amongst the seemingly endless coastal inlets. Sparsely populated and desolate, this landscape holds some of the few remaining outport communities in the province. These tiny villages were once the predominant form of settlement throughout rural Newfoundland until the advent of the Resettlement Act of 1965 which sought to centralize these isolated communities into "growth centers". As part of the Resettlement Act, communities had to vote on relocation and were given a one-time payment to simply walk away from their homes and, in the process, creating dozens of abandoned communities around the island. Now known as the Community Relocation Policy, resettlement requires at least 90% of residents to volunteer to leave. Since 2013, each household in a given outport is offered $270,000 in compensation for relocating.

By the time I arrived in Grand Bruit in June of 2008, only 15 permanent residents remained. This phase of my fieldwork marked the beginning of my experiments with a lived experience of abandoned place. Over the next several weeks I hoped to understand something about lives lived in a state of pre-abandonment. This state of becoming-abandoned offered a moment of ethnographic encounter with the almost-already form of abandoned place, a site of potential and inevitable ruin. The summer following my time in Grand Bruit, the community voted to resettle, making concrete its future abandonment. As the last ferry pulled out of the harbor, the electricity was cut and Grand Bruit went dark forever. Yet, to this day, some of the fishermen still return to the village for the summer lobster season. Does this blip of occupation render the place un-abandoned, or is a place abandoned in the moment that the last permanent resident leaves? From my perspective temporary occupation does not constitute reinhabitation. The original form of placemaking has been overwritten, and returning to these places becomes more of a pilgrimage to a holy ruin. Many of the outport communities hold regular "come home years" when past residents briefly return to the abandoned outports where they once lived. This ritual becomes a mode of affirmation and memorialization of displaced place, similar to what DeSilvey (2006) has described as the constellation of meaning in place. In many ways, this idea has always been central to my work—the constant *reassemblage* (Minh-ha 1982) of place in the absence of its primary authors, the formation of a *bricolage* (Lévi-Strauss 1966) of abandoned fragments of place as a form of cultural translation (Clifford 1997). And perhaps this view is where the true essence of abandoned place comes into focus; it is that unwritten and always written location that hovers at the edge of the everyday. And again, it is the notice of these displaced, misplaced and replaced locations that form the sites of Benjamin's (1978) profane illumination. It is in this very moment of notice that our place-based imaginations take shape. That summer in Grand Bruit, as I wandered along the pathways and chatted with the few remaining residents, I began to see the flickers of these illuminations, the specters of becoming-abandoned haunting the edges of my vision. Now, once again, I occupied the present future of an almost-abandoned place, watching its threads unravel slowly in the lives of its people. Then and now, Grand Bruit is everything at once, it is anchored and drifting, present and absent, and then, now and forever.

An ethnographic account of the processes and outcomes of abandonment presents a novel way of understanding the creation, maintenance and dissolution of place. Here, abandoned place forms a site of indefinite liminality that is not bound by materiality, a place that resonates and translates its embedded significance through careful attention and an attuned awareness to its everyday ruptures (Stewart 2007) and illuminations. Here, archeology, cultural anthropology, cultural geography and philosophy offer us a readymade toolkit for crafting valuable inquiries and gaining new perspectives. An abandoned place becomes, by turn, placed, displaced, misplaced and replaced. It is useful to view no place as truly abandoned, it is never a finished project, it is always in the process of becoming abandoned. It is forever liminal, always evaporating and dissolving without ever disappearing. Abandoned place is everywhere and nowhere, never and always. Our duty as both cultural researchers and global citizens is to heed its call and work to find and translate its polyvocal cultural significance. It is not a location of mourning, nostalgia or fetishization, rather it is a place where history, the present and the future collide in a messy, beautiful dialog between people, places and things.

And then, again, in one of these collisions, I find myself in a place in-between, laying on my back, on a dense mat of moss, in the tumbled down stone foundation of a century-old Icelandic farmhouse. Here, in a ruining world at the edge of the world, the slow-fade of culture reveals itself in a constellation of dimly lit fragments. Just as in Grand Bruit, Uranium City and Múli, I find myself in silent dialog with the place and its things. For a moment, I am its people. Briefly, I can

Figure 46.3 Grand Bruit, Newfoundland.

sense its agency. I believe that the toolkit that I have described above offers a new way of knowing the world through the cultivation of ethnographic feedback loops whereby places without people can offer complex encounters to anthropologists and critical thinkers of all types. In attempting to understand and document this world that rests just below our everyday experience, it is necessary to reconfigure and re-attune our awareness in order to sense the interactions in places without people. Everything and every place speaks its truth in the absence of people.

I look up at the nightless sky bordered by the rounded earthen walls: a cradle, a home, a coffin. An artwork, a story, a beautifully profane dreamworld. The remainder of a farmhouse in Hornstrandir that is unlived, but not unloved.

References

Anderson, Jon. *Understanding Cultural Geography: Places and Traces*. 2 edition. Abingdon: Routledge, 2015.

Armstrong, Justin. "On the Possibility of Spectral Ethnography." *Cultural Studies ↔ Critical Methodologies*, vol. 10, no. 3, June 2010, pp. 243–250. *SAGE Journals*, doi: 10.1177/1532708609359510.

Augé, Marc. *Non-Places: An Introduction to Supermodernity*. Translated by John Howe, 2 edition. London: Verso, 2009.

Augé, Marc, and James E. Young. *Oblivion*. Translated by Marjolijn De Jager, 1 edition. Minneapolis, MN: University of Minnesota Press, 2004.

Benjamin, Walter, Edmund Jephcott and Peter Demetz. *Reflections: Essays, Aphorisms, Autobiographical Writings*. New York: Harcourt Brace Jovanovich. p. 10, 1978.

Benjamin, Walter. *The Work of Art in the Age of Mechanical Reproduction.* Scotts Valley, CA: CreateSpace Independent Publishing Platform, 2010.
Buchli, Victor. *Archaeologies of the Contemporary Past.* 1 edition. Abingdon: Routledge, 2001.
Buchli, Victor and Gavin Lucas. *Archaeologies of the Contemporary Past.* Abington: Psychology Press, 2001.
Clifford, James. *Routes: Travel and Translation in the Late Twentieth Century.* Cambridge, MA: Harvard University Press, 1997.
Deleuze, Gilles, and Felix Guattari. *A Thousand Plateaus: Capitalism and Schizophrenia.* Translated by Brian Massumi, 2 edition. Minneapolis, MN: University of Minnesota Press, 1987.
Derrida, Jacques. *Specters of Marx: The State of the Debt, The Work of Mourning & the New International.* 1 edition. Abingdon: Routledge, 2006.
DeSilvey, Caitlin. "Observed Decay: Telling Stories with Mutable Things." *Journal of Material Culture,* vol. 11, no. 3, Nov. 2006, pp. 318–338. *SAGE Journals,* doi: 10.1177/1359183506068808.
DeSilvey, Caitlin. "Salvage memory: constellating material histories on a hardscrabble homestead." *Cultural Geographies,* vol. 14, no. 3. 2007, pp. 401–424. doi:10.1177/1474474007078206
DeSilvey, Caitlin. *Curated Decay.* 1 edition. Minneapolis, MN: University of Minnesota Press, 2017.
DeSilvey, Caitlin, and Tim Edensor. "Reckoning with Ruins." *Progress in Human Geography,* vol. 37, no. 4, Aug. 2013, pp. 465–485. *SAGE Journals,* doi: 10.1177/0309132512462271.
Earle, T., R. Preucel, E. Brumfiel, C. Carr, W. Limp, C. Chippindale, ... R. Zeitlin. "Processual archaeology and the radical critique [and comments and reply]." *Current Anthropology,* vol. 28, no. 4, 1987, pp. 501–538. Retrieved March 4, 2020, from www.jstor.org/stable/2743487.
Edensor, Tim. "The Ghosts of Industrial Ruins: Ordering and Disordering Memory in Excessive Space." *Environment and Planning D: Society and Space,* vol. 23, no. 6, Dec. 2005, pp. 829–849. *SAGE Journals,* doi: 10.1068/d58j.
Fabian, Johannes, and Matti Bunzl. *Time and the Other: How Anthropology Makes Its Object.* With a New Postscript by the Author edition. New York: Columbia University Press, 2014.
Fortun, Mike, and Kim Fortun. *Writing Culture: The Poetics and Politics of Ethnography.* Edited by James Clifford and George E. Marcus, 2 edition, 25th Anniversary edition. Berkeley, CA: University of California Press, 2010.
Huyssen, Andreas. "Nostalgia for Ruins." *Grey Room,* vol. 23, Apr. 2006, pp. 6–21. *mitpressjournals.org (Atypon),* doi: 10.1162/grey.2006.1.23.6.
Lévi-Strauss, Claude. *Structural Anthropology.* Revised edition. New York: Basic Books, 1974.
Lévi-Strauss, Claude. *The Savage Mind.* Chicago, IL: The University of Chicago Press, 1966.
Miller, Daniel. *Material Culture and Mass Consumption.* 1 edition. Chichester: Wiley-Blackwell, 1997.
Minh-ha, Trinh T. *Reassemblage* [film]. Produced by Jean-Paul Bourdier; directed, photographed, written and edited by Trinh T. Minh-ha. (198220002009). New York: Women Make Movies, Inc., 1982.
Perec, Georges. *Species of Spaces and other Pieces.* Edited by John Sturrock, New edition. London: Penguin Classics, 2008.
Pétursdóttir, Þóra. "Concrete Matters: Ruins of Modernity and the Things Called Heritage." *Journal of Social Archaeology,* vol. 13, no. 1, Feb. 2013, pp. 31–53. *SAGE Journals,* doi: 10.1177/1469605312456342.
Stewart, Kathleen. *Ordinary Affects.* Durham, NC: Duke University Press Books, 2007.
Thill, Brian, et al. *Waste.* London: Bloomsbury Academic, 2015.
Trigger, Bruce G. *A History of Archaeological Thought: Second Edition.* Cambridge: Cambridge University Press, 2006.
Tuan, Yi-Fu. *Space and Place: The Perspective of Experience.* Reprint edition. Minneapolis, MN: University of Minnesota Press, 2001.
Turner, Victor W., et al. *The Ritual Process: Structure and Anti-Structure.* 1 edition, Chicago, IL: Aldine Transaction, 1969.
Williams, Raymond. *Keywords: A Vocabulary of Culture and Society.* Oxford: Oxford University Press, 1983.

Part VI
Economic geographies of place

Ares Kalandides

Introduction: economies of place

Examining and writing about the economies *of* places necessarily involves writing about the economy *in* places, but, as this part will show, the two are far from being identical. Indeed, for many people any memory of the geography classes at school would include learning about places, both nationally and globally: their physical features, their socio-political structures – and their economies. This idiographic approach, where each place – be it a country, a region or a city – would be considered individually and described in detail, was for a long time one of the key features in geographical education (Livingstone, 1992; Dunbar, 2001; Withers and Mayhew, 2002).

There is no doubt that economies 'happen' *in* places, but, as geographers would argue today, also *across* places. This however has implications for geographical theory and practice. We can identify very diverging interpretations of the 'economies of place', each one linked to particular schools of thought, but also with important geographical variations. We can broadly distinguish between several approaches: ideographic-descriptive, phenomenological, social-constructionist and political-activist. Such approaches can and do overlap since most understandings of place today tend to blend some or all of the above.

In the ideographic-descriptive approach, economies of places are examined as closed systems that operate according to their own internal place-based logic. The job of academics, practitioners and policymakers would then be to find and influence those local factors that make the economy what it is (Lorimer, 2003). Theories of endogenous growth, concepts of local innovation systems, etc., are examples of this particular understanding of economies *in* place (Aghion et al., 1998; De la Mothe and Paquet, 2012). More recently, place marketing, the practice of positioning places in a competitive world market for investment, visitors or talent, is based on the same premise: that there is a need to find and emphasise the local characteristics that will make the particular location more attractive (Kavaratzis and Ashworth, 2005; Warnaby and Medway, 2013).

The phenomenological approach is one where what is examined is the relationship between place and the human: the mental images that we form about places, how place enters into the constitution of human (individual or group) identity and, vice versa, how human identities form places (Tuan, 1977). The discussion in economic geography about place-based trust relationships and the importance of face-to-face relationships in economic development are examples of this understanding of place (Granovetter, 1985; Storper and Venables, 2004). The recent surge in co-working spaces, innovation labs, etc., where apparently unconnected individuals come together *in place* to form new economic relationships is another (Gandini, 2015). Finally, a recent interest in the notion of the 'sense of place' in place marketing and branding practices is very much concerned with questions of authenticity and how individuals form mental images about places (Melewar and Skinner, 2018).

The social-constructionist approach is concerned on the one hand with how places are constituted through social relations (including, but not limited to economic ones) and, vice versa, how that (socially produced constitution of) place then reproduces, challenges or changes social relations – also those of production, distribution and consumption (Harvey, 1982; Massey, 1984). As social relations are imbued with power relations (of class, gender, sexuality, race, etc.) there has also been a growing interest in how such diverse 'power fault lines', in particular gender, structure the unequal economies of place (McRobbie, 2016; McDowell, 2018). The social-constructionist approach is also about the interrelations *between* places, those of interdependency, dominance and subordination, competition or cooperation – as well as the interrelated uneven development of places (Smith, 1984). The economies of places are not given but are constructed through both local and global forces that intersect in a location (Massey, 1991). Borders and institutional boundaries play a very important role in the way that economies work, but they are considered socio-political constructs that serve particular economic needs. Although still concerned with the local, here the openness of systems is stressed: local innovation systems are linked to global innovation networks (Coe and Bunnell, 2003); global economies are embedded in places (Dicken, 2014); global production chains connect places with each other in complex relations of production, distribution and consumption (Gereffi, 2014). The constitution of place is paramount to the way that a globalised and interdependent economic system takes form on the ground and, vice versa, the former will influence the latter's way of functioning. In other words, although globalisation *influences* places it is also produced *in* places (Massey, 2004).

Finally, place can be a call for political action: a turn towards a different type of economic system, conscious of global connections, but where local production, distribution and consumption become demands for political change (De Neve et al., 2008; Kalfagianni and Skordili, 2018). The rise of concerns for social justice, economic exploitation and environmental degradation have created new political movements that challenge the sustainability of our global economic system. They are returning to the local, to *place*, and are designing new economies of place to deal with them. From fabrication labs, makers' spaces, urban gardening to local value chains – or even the well-known slogan of 'think globally–act locally' bear testimony to such claims (Certomà, 2001; Carrier, 2008; Schmidt et al., 2014).

What still seems to be a contentious and unresolvable issue here is the relationship between economic and non-economic processes in the production of places. While the compromise today seems to be an understanding of places as 'assemblages', produced both by economic and non-economic relations (including those between humans and nonhumans or even with inanimate physical objects), the contributions to this part show the wide variety – and sometimes polarisation – of approaches on the matter.

Costis Hadjimichalis, in his introduction to the concept of place in economic geography, traces its pre-eminence in academic debate in the dominance of Anglophone theory creation and policy. He argues that the rise of place to prominence occurs with the European industrial crisis of the 1970s and 1980s and the rise of neoliberalism. With the end of the so-called 'welfare regionalism' that aimed at social and spatial redistribution, as well as the realisation that places with similar industrial structures reacted differently to the effects of de-industrialisation, academics and policy-makers turned their attention to the differences between places. Hadjimichalis examines in turn several strands of this turn: the British industrial restructuring perspective, the Italian school of 'Third Italy's' Industrial Districts, the French School of 'Milieu Innovateur', the US school of 'agglomeration economies' and 'transaction cost analysis', the 'New Regionalism' debate of the 1980s, etc. Hadjimichalis closes his chapter with a very critical assessment of European territorial policies.

Ray Hudson, in a straightforward Marxian analysis, examines places within capitalism and, in particular, the relationship between places, capital accumulation and its constituent processes of uneven and combined development. He argues that although non-capitalist social relationships may help shape places, it is capitalist social relationships that routinely have a decisive shaping influence within capitalist socio-spatial formations. Within the social relations of capital, places can therefore be thought of as complex assemblages of people and things which come to have varying meanings for different people. For the owners and managers of capital, places are above all *work*places, sites for the production of profit; for workers and their families places also have a different and more complicated meaning as work*places*. Consequently, especially in historical phases of place-specific devalorisation, a common element of an on-going process of creative destruction, we observe a variety of attempts to defend places and to seek to ensure their successful reproduction, involving complex processes that are material, social and discursive.

Dawn Lyon is concerned with interconnections of work, place and rhythm, as well as processes of place-making at and through work, in particular with the role of rhythm in place-making at work. She examines how certain workspaces come into being as places, how work and workplaces take shape in space and time, and considers the character of workplaces brought about by the rhythms that permeate them. Her conceptual point of departure is the work of thinkers such as Henri Lefebvre ('rhythm' and 'rhythmanalysis') as well as Edward Casey's formulation of places as 'thick' and 'thin'. By bringing temporality to the fore, this chapter draws on empirical material from the Billingsgate fish market, a site of display, interaction, movement, negotiation and exchange, where place is performed through the multiple overlapping trajectories and rhythms of people and things moving in the market. Lyon explores ways that place encompasses affects and practices that 'thicken' or 'thin' the embodied experience and meaning of being-in-place.

Bastian Lange examines the format of 'open worklabs' to describe the constitutive features of the broader phenomenon of new types of self-organised collaborative workplaces and the associated practices of social and grassroots innovation. He argues that open worklabs, with all their complexity and open-endedness, need to be understood as emerging alternatives in a post-growth economic paradigm. He examines how actors involved in open worklabs integrate their work-related approach into their everyday lives, and conversely incorporate their vernacular knowledge, their 'private' ideation, their habitus as urbanites into the supposedly closed small world of open worklabs. Places, Lange contends, especially the density and heterogeneity of urban social contexts, have a strong triggering function as they re-configure older and create new combinations of economic agency, bringing about a temporary sticking together of heterogeneous social and material elements. He concludes that open worklabs are provisional socio-spatial arrangements that are absorbed into the urban field as a constitutive element of its emergence and where the separation between private and work space blurs.

Mark Jayne examines recent developments in our understanding of the relationship between consumption, identity and place. Through empirical research in Petržalka, a high-rise housing estate in Bratislava, Slovakia, and an examination of what it means to be wearing 'comfy' clothes in public space, he shows the connections between the consumption of fashion and identity. After a review of 'archetypal' depictions of the ways in which consumption is moulded by place, and how places are moulded, he argues for the need to move beyond the theorisation of place and consumption only based on studies from a handful of large metropolises in Europe and North America. Jayne also points out how recent theoretical and empirical work on assemblages, materialities, mobilities, bodies, emotions or affect may offer significant opportunities to re-invigorate study of consumption and place for the twenty-first century.

Gary Warnaby and Dominic Medway examine not only the role of 'place' but also that of 'time' in the place marketing context. They show how the interplay of the past, present and future can be an important aspect of urban place marketing and how this influences its practice. They are particularly interested in the concept of 'retroscapes' as deliberate evocations of times past and how such retroscapes are used to generate *genius loci*, or a 'sense of place'. Genius loci, and the multiplicity of temporal layers (past, present and future) that constitute it, can then be appropriated and capitalised upon in place marketing practice, in order to convey authenticity and the essence of 'being in place'. However, place marketing practice will face the same challenge as history as it is inevitably selective – what is to be narrated and how is it to be interpreted. They finally argue that this marketing/branding activity is likely to be most effective when temporally-bound concepts such as memory work and forgetting, and the role they might play in the representation work relating to places, are fully embraced and better understood by place marketing stakeholders and institutions.

Judith Mair examines Master Planned Communities in Australia, i.e., large-scale private housing developments, in relation to place, placemaking and place marketing. In particular, creative placemaking, the strategic use of arts and cultural activities, including festivals and events (eventification), are used by developers to shape the physical and social character of places, with the aim of bringing together and inspiring diverse individuals and groups of people. Places and the communities which thrive in them, argues Mair, are multifaceted, complex, constantly evolving and in a permanent process of being negotiated by those who live in them. Imposing particular narratives of place and community on a development does not automatically result in any single fixed state of place or community, particularly not an idealised or 'imagined' one. However, events and festivals offer an opportunity for those unscripted encounters which are vital for the organic development of sense of place and perhaps in that way they can play a role in the ongoing development of place and community in these places-to-be.

Finally, Jaime Hernández-Garcia and Beau B. Beza critically discuss the contribution of informal settlements to tourism and place, with examples from Bogotá and Medellín in Colombia. They argue that informal settlements are social and collective constructions which may be best understood from the *inside looking out* (rather than from the *outside looking in*) and that place is a major component in this understanding. Place in these settings has become a commodity that tourists wish to visit. In Bogotá and Medellín, tourism to informal settlements is a relatively recently activity and has so far been predominantly organised through the initiatives of *barrio* residents. The authors discuss how informal settlements contribute to both the physical and the socio-cultural experience of tourists, suggesting place may act as a socio-spatial unit and, in effect, is the attractor. They also look at how informal settlements are gradually included in city branding strategies, even though such areas are still predominantly considered problematic and unsafe.

References

Aghion, P., Ljungqvist, L., Howitt, P., Howitt, P.W., Brant-Collett, M., and García-Peñalosa, C. (1998) *Endogenous Growth Theory*, Cambridge: MIT press.

Certomà, C. (2011) 'Critical urban gardening as a post-environmentalist practice'. *Local Environment*, 16 (10): 977–987.

Carrier, J. (2008) 'Think locally, act globally: the political economy of ethical consumption', in: G. De Neve, L. Peter, J. Pratt, and D.C.Wood (eds) *Hidden Hands in the Market: Ethnographies of Fair Trade, Ethical Consumption, and Corporate Social Responsibility*, Bingley: Emerald, pp. 31–51.

Coe, N., and Bunnell, T. (2003) '"Spatializing" knowledge communities: towards a conceptualization of transnational innovation networks'. *Global Networks*, 3(4): 437–456.

De la Mothe, J., and Paquet, G. (eds) (2012) *Local and Regional Systems of Innovation*, Berlin: Springer.

De Neve, G., Peter, L., Pratt, J., and Wood, D. (eds) (2008) *Hidden Hands in the Market: Ethnographies of Fair Trade, Ethical Consumption, and Corporate Social Responsibility*, Bingley: Emerald.

Dicken, P. (2017) *Global shift: Mapping the Changing Contours of the World Economy*, Thousand Oaks: SAGE Publications.

Dunbar, G. (ed) (2001) *Geography: Discipline, Profession and Subject since 1870: an International Survey*, Berlin: Springer.

Gandini, A. (2015) 'The rise of coworking spaces: a literature review'. *Ephemera*, 15(1): 193.

Gereffi, G. (2014) 'Global value chains in a post-Washington consensus world'. *Review of International Political Economy*, 21(1): 9–37.

Granovetter, M. (1985) 'Economic action and social structure: the problem of embeddedness'. *American Journal of Sociology*, 91(3): 481–510.

Harvey, D. (1982) *Limits to Capital*, Oxford: Basil Blackwell.

Kalfagianni, A., and Skordili, S. (eds) (2018) *Localizing Global Food: Short Food Supply Chains as Responses to Agri-food System Challenges*, London: Routledge.

Kavaratzis, M., and Ashworth, G. (2005) 'City branding: an effective assertion of identity or a transitory marketing trick?'. *Tijdschrift voor Economische en Sociale Geografie*, 96(5): 506–514.

Livingstone, D. (1992) *The Geographical Tradition: Episodes in the History of a Contested Enterprise*, Oxford: Blackwell.

Lorimer, H. (2003) 'Telling small stories: spaces of knowledge and the practice of geography'. *Transactions of the Institute of British Geographers*, 28(2): 197–217.

Massey, D. (1984) *Spatial Divisions of Labour: Social Structures and the Geography of Production*, London: Macmillan.

Massey, D. (1991) 'A global sense of place'. *Marxism Today*, (June): 24–29.

Massey, D. (2004) 'Geographies of responsibility'. *Geografiska Annaler: Series B, Human Geography*, 86(1): 5–18.

McDowell, L. (2018) *Gender, Identity and Place: Understanding Feminist Geographies*, Hoboken: Wiley.

McRobbie, A. (2016) *Be Creative: Making a Living in the New Culture Industries*, Cambridge: The Polity Press.

Melewar, T., and Skinner, H. (2018) 'Territorial brand management: beer, authenticity, and sense of place'. *Journal of Business Research: Online First*, https://doi.org/10.1016/j.jbusres.2018.03.038.

Schmidt, S., Brinks, V., and Brinkhoff, S. (2014) 'Innovation and creativity labs in Berlin'. *Zeitschrift für Wirtschaftsgeographie*, 58(1): 232–247.

Smith, N. (1984) *Uneven Development: Nature, Capital and the Production of Space*, Oxford: Blackwell.

Storper, M., and Venables, A. (2004) 'Buzz: face-to-face contact and the urban economy'. *Journal of Economic Geography*, 4(4): 351–370.

Tuan, Y.F. (1977) *Space and Place: The Perspective of Experience*, Minneapolis: University of Minnesota Press.

Warnaby, G., and Medway, D. (2013) 'What about the "place" in place marketing?'. *Marketing Theory*, 13 (3): 345–363.

Withers, C., and Mayhew, R. (2002) 'Rethinking "disciplinary" history: geography in British universities, c. 1580–1887'. *Transactions of the Institute of British Geographers*, 27(1): 11–29.

47

Place and economic development

Costis Hadjimichalis

Places have been for centuries the loci par excellence for economic production and exchange. Their growth and decline has depended on the dominant economic activity of the times and on their particular physical and human resources. Long-distance trade, piracy, collection of agricultural surplus, finance, mineral extraction, industry, tourism and more have produced particular places, embedded in the capitalist landscape. Historical geographers, such as Lucien Febvre and Fernand Braudel; radical geographers such as Doreen Massey, Franscesco Indovina, David Harvey and Ray Hudson; radical anthropologists and sociologists such as Michael Blim, Susana Narostsky, Arnaldo Bagnasco, Huw Beynon and Enzo Mingione, among many others, have famously explained why and how particular places flourish or die.

Before moving on, two important acknowledgements should be made. First, my approach is Eurocentric, i.e., it does not take into account experiences from other parts of the world. This is mainly due to my sporadic knowledge, insufficient for a rigorous exposé. I am sure others will close this gap, providing also a critique of the usually problematic 'transfer' of European-made theories and policies to other socio-spatial contexts. Second, the meaning of 'place' in English is quite different in other languages and geographical traditions. In Latin languages it is often carried by the words for territory (territorio, território, territoire, etc.). This can lead to confusion since the English term 'territory' is much stronger in the sense of a coherent, bounded space. In Italian the term for place is '*luogo*', but this rarely enters geographical discourses. Instead the word '*locale*' (locality) is associated with local economic transactions and business, an important parameter in Italian theories of local development (*sviluppo locale*), as we will see below. Finally, the Greek term for place '*topos*' and its adjective '*topikos*' is closer to the English 'local' and the Italian '*locale*', rather than 'place'.

Bearing this in mind, Anglophone economic geography and planning usually use place as location, a specific point or area, as a sense of place – the subjective feelings of people – and as locale, a setting for people's daily interactions and routines (Castree, 2003). My own understanding is that place has connotations of attachment and meaning, whereas local is more just a reference to a spatial scale. Capital and labour often have conflicting attitudes towards place, although they could also build histories of consensus politics as to what is 'best for the place'. Beynon et al. (1994) illustrate this in their excellent study of Teesside in the North East of the UK, one of the early heavy industrial cores, dating from the mid-nineteenth century. In their words:

> locations that, for capital, are a (temporary) space for profitable production are for workers, their families and friends places in which to live; places in which they have considerable individual and collective cultural investment; places to which they are often deeply attached, and which may hold powerful emotional ties and socially endowed symbolic meaning for them.
>
> *(p. 5)*

Beynon et al. (1994) continue by describing how patriarchy in a male-dominated labour market and paternalism via company unionism control social relations in such a place:

> Teesside was built around distinct gender relations and profound class divisions, but these did not necessarily erupt into open conflict because of the way in which social relationships were constituted and understood there. From an early stage, gender relations were molded into patriarchal forms and class relations into strongly paternalistic ones that extended beyond the workplace and gave form to an emergent local society.
>
> *(p. 185)*

Finally, Beynon et al. (1994) give particular attention to external factors regulating Teesside's future, namely government planning regulations and party politics and, most importantly, the changing international markets that affected Teesside's de-industrialisation in the late 1970s to the early 1980s.

I use the Teesside example for two main reasons. First, because it is one of the best critical accounts in Europe of the rise and decline of a heavy industrial pole. The writers analyse in depth the production and social reproduction of place from a radical perspective, taking into account capital, labour, gender, politics, state involvement and globalised external forces. And, second, because it illustrates a major shift in Anglophonic geography and planning. Until the late 1970s there was no or very little discussion about place. This was due to, on the one hand, the negative history of the term as a unique and bounded area and, on the other, to relatively stable economic conditions and strong state intervention via macro regional development projects – so-called 'welfare regionalism' – aiming at social and spatial redistribution. Since then, however, the dual effect of de-industrialisation and the withdrawal of government support has opened questions such as how and why places with similar industrial structures, and subject to the same external forces, react differently and how they cope socially with the negative effects of unemployment. Similar effects, but with different causes, were visible in other European countries with different socio-spatial formations (see, for Spain, Charnock et al., 2014; for Italy, Paci, 1972 and Martinelli, 2009; for Greece, Labrianidis, 2008 and Skordili, 1999). Language, however, affects the production, consumption and circulation of geographical knowledge, as per my short discussion on the different meanings of place above. At present, the dominant discourse originates from the Anglophone literature which determines the themes and terms of geographical debate, assuming that it produces 'theory'. Theory thus constituted is based on local knowledge which constructs and sees itself as general and universal. By this token, theoretical formulations from non-dominant languages and areas are not considered as 'theory' but rather as lagging examples or 'local' illustrations (Hadjimichalis and Vaiou, 2004). This is not only a theoretical problem but enters the field of practice via 'one-fits-all' policies.

In what follows I will look at how places were used, ignored and rediscovered in economic geography and planning and in different linguistic milieux, unpacking some misreading and fallacies and looking at how theories of and policies for place followed the current general depoliticisation of academia and politics. I give special attention to policies for the economic development of places, known as 'local economic development', and how the content of the latter changed after the institutionalisation of neoliberalism.

Rediscovering place

As stated above, the importance of place in economic geography and planning was rediscovered during the industrial crisis of the 1970s and 1980s and the gradual shift towards neoliberalism. It was the period in which major old industrial centres in the birthplaces of capitalism in Europe and the USA started falling apart due to capital's falling rates of profits and the political attack against the strongholds of unionised labour (Dicken, 2015). The turn to neoliberal policies by the end of the 1980s introduced free-market capitalism, conservative macroeconomic policy and less state intervention, all resulting in diminishing interest in social and spatial redistribution. Empirical investigations in the early 1980s discovered, first, that places with similar economic and social structures responded differently to globalised forces and to the neoliberal political attack. And, second, researchers discovered cases of supposedly 'spontaneous' place-based growth, in places far from old industrial cores and subsidised backward agricultural regions; growth that occurred without direct assistance from the central state or inward investment, where small firms with strong entrepreneurial spirit initiated 'bottom-up' local growth.

A good example of the first investigation strand was the British *industrial restructuring perspective*, following Doreen Massey's original ideas (1984), applied in the mid-1980s in the research programme 'The Changing Urban and Regional System' (CURS). The programme, known also as 'locality studies', focussed on seven localities which were chosen and studied by geographers, sociologists, planners and a few economists for their different experiences of decline or growth.[1] The research was criticised by some on the left for overemphasising the local and the empirical at the expense of wider processes and general theory. Among the protagonists of the 'locality studies' research was P. Cooke, later joined in this perspective by K. Morgan, A. Amin and N. Thrift, while other CURS participants, such as R. Hudson and H. Beynon, developed a much more critical view, as in the Teesside case. Major findings of CURS include the negative effect of Thatcherism on regional policy, the need to link local to global processes for understanding restructuring, the importance of local proactivity and the propensity for networking by local actors. The European Union, following a similar theoretical framework, launched several cross-country research projects such as the following, among many others: 'The New International Division of Labour and Regional De-Industrialisation in the European Community' (1984–1987), 'Regional and Urban Restructuring in Europe' (in collaboration with the European Science Foundation, 1991–1994) and 'Local Development Strategies in Economically Disintegrated Areas' (1992–1995).

Following the second strand of place rediscovery we find pioneering research on place-based development coming from different directions. First, there is the Italian School of Third Italy's Industrial Districts (IDs) which, since the 1970s, has provided brilliant analyses, mainly by economists, of networked small and medium-sized industries in various parts of north-east and central Italy (Becattini et al., 2003). The Italian school has made several ground-breaking contributions such as the 'local flexible production system', the notion of 'diffused industrialisation', the importance of small and medium enterprises and their networking, the role of local culture, local social relations and the knowledge and re-use of

Alfred Marshall's notion of localised external economies. Key figures include G. Becattini, R. Camagni, A. Sforzi, G. Garofolli, G. Dei Ottati and many more.[2]

Second, there is the French School of *Milieu Innovateur* (innovative local environment, although '*milieu*' in French has a wider meaning than local) developed by the economic research group GREMI (Groupement de Recherche Européen sur les Milieux Innovateurs) in the early 1980s (Aydalot, 1986). Their hypothesis was based on Philippe Aydalot's idea that innovational behaviour is based on local or regional factors rather than national or global factors. Extending Schumpeter's ideas on innovation, GREMI combined two French traditions: the older milieu concept from sociology and regional geography and that of regulation regimes. Apart from Aydalot himself, other contributors include M. Quévit, D. Maillat, O.Crevoiser and, later, C. Courlet and B. Pecqueur.

Third, in California, again in the 1980s, economic geographers developed innovative research on agglomeration economies and transaction cost analysis. Inspired by Italianate industrial districts and regulation theory, agglomeration and transaction are analysed for both their quantitative and qualitative dimensions (Scott and Storper, 1986). A. Scott, A. Saxenian, M. Storper and D. Walker were among this group of researchers. Although their focus was on particular industrial localities, they underline the importance of competitive regional economies in general at times of globalisation and they stress technological learning, based on traded (input–output relations) and untraded interdependencies (labour markets, regional conventions, norms and values, public or semi-public institutions).

Small firms in these places, according to the above-mentioned researchers, avoided the problems of Fordism and successfully took advantage of market opportunities by mobilising locally existing resources (particularly knowledge and learning) combined with new forms of production organisation such as flexibility and networking. These new industrial places highlighted the role of competition and co-operation at the local level with the assistance of strong place-based cultural traditions, local institutions and associations. Researchers and policy makers agree that a distinctive feature of these places and localities is the embeddedness of certain noneconomic factors such as social capital, trust and reciprocity based on familiarity, face-to-face exchange, co-operation, embedded routines, habits and norms, and local conventions of communication and interaction, all of which have a key role to play in successful local endogenous development (Garofoli, 1983; Vàzquez-Barquero, 1992; for a critique see Blim, 1989; Hudson, 1999; Hadjimichalis, 2006).

These pioneering analyses, despite important differences between the places and localities studied, were used as paradigmatic examples for the introduction of a new development paradigm named 'local development' and later 'local sustainable development' to incorporate environmental issues. Local development was based on synthetic and policy-oriented concepts such as 'endogenous growth', 'networked firms and regions', 'industrial clusters', 'learning firms and regions', 'innovative firms and regions' and 'local social capital', forming what has been called by others 'New Regionalism' (McLeod, 2001). According to both theorists and policy makers, while in the past capitalist development took place in a spontaneous manner, it can now be designed to implement a policy 'from below' (Becattini, 1998; Becattini et al., 2003).

Misreadings, omissions and fallacies

In the 1980s and early 1990s local development became the new catch phrase and 'New Regionalism' was a new kind of analytical and development doctrine during this period of major political changes in the global North after the institutionalisation of neoliberalism as the dominant economic, political and cultural regulation. In the euphoria that followed the 'discovery' of dynamic small firms in these paradigmatic industrial places, several misreadings, omissions and fallacies became apparent.

First, there was a simplistic binary opposition between mass production and flexible specialisation; Fordism was never dominant in the regions and places in which growth was directed by flexible small industrial production ensembles. Very few Anglophone researchers, however, took into account the many sweatshops or understood the highly exploitative working conditions in small Southern European firms and beyond, where the gender and age division of family labour is the cornerstone of their flexible operation (Paci, 1972; Vaiou and Hadjimichalis, 1997)

Second, there was an emphasis on endogenous growth, ignoring exogenous factors. In reality capitalist growth and development is always the outcome of the dialectical tension between the two, while the negative social and environmental consequences of this tension are experienced mainly locally. There is a shifting importance between endogenous and exogenous factors, which are always uneven and combined in particular places and times. But uneven capitalist development is never exclusively endogenous.

Third, there was a major confusion of scale, when local development was analysed in some cases with reference to traditional agglomerations of small industrial firms in areas with a few thousand inhabitants, to whole regions with multiple sectors, or to large urban conurbations with populations of several hundred thousand. Finally, some cases are identified as 'local' simply due to their coincidence with lower scale administrative boundaries. Furthermore, mainstream views take account of local productive and institutional structures, over-emphasising the supply side. In this respect, they give scant, if any, attention to the empirical dynamic of the demand side and the global capitalist competition in the sphere of circulation.

Fourth, in the relevant literature a very selective appropriation of the complexity and richness of local productive systems in Southern Europe has taken place, in which only certain general economic, organisational and institutional issues have been taken on board. Others remain in the shadows such as class and gender, power and inequalities among small firms, the limitations of networking and learning, what co-operation, reciprocity and 'social capital' really mean for firms and labour, and conditions of work in the informal economy, to mention but a few. Some Anglophone researchers ignored these issues, while others who took account of them after field work provided a more complete critical picture (see Blim, 1990; Dunford, 1991; Amin, 2003; Dunford and Greco, 2006; Pickles and Smith, 2011).

Fifth, there is a lack of attention to the role of the state and other supranational entities in introducing various protectionist and assistance measures. From labour legislation and the devaluation of national currency to particular incentives for small firms and international agreements (such as the Multi-Fibre agreement) there was never such a thing as 'without assistance from the central state', as argued by Piore and Sabel (1984).

Sixth, by looking only at success in the context of competitive capitalism, there has been interest only in a few paradigmatic industrial sectors and in a few advanced service providers in a limited number of European and North American places and regions. As result of the above, there has been a remarkable silence of other important sectors in Southern Europe that provide multiple employment, such as tourism, local trade, construction and agriculture. These sectors rarely attract interest in Anglophone economic geography and planning (Hadjimichalis and Melissourgos, 2013).

Finally, seventh, there is the fallacy of 'good practice', as in the promotion by international agencies, such as the World Bank, OECD and others, and supranational governing bodies such as the EU, of a few 'success stories' as good practice transferable to other places. Although any student of economic geography knows that what happens in a particular place cannot be transferred elsewhere, this a-theoretical and a-spatial framework and empirically unjustifiable process became the standard illustration in the glossy manuals of local development.

Perhaps the most important omission in this research framework and in much of the 'New Regionalism' literature, what really remains unspoken, is the *uneven relations among places*, successful and unsuccessful alike, in other words the *hard question of uneven geographical development under capitalism* (Harvey, 1982; Hadjimichalis and Hudson, 2007; Hadjimichalis, 2018; Hudson, Chapter 48, this volume). This was evident in how little attention was paid to the capitalist crisis that was already visible from the mid-1990s in some industrial districts of Third Italy and in other emblematic localities and regions, not to speak of the 2009–2018 general crisis in Southern Europe. These developments directly challenge grandiose claims about flex-spec small firms in industrial districts as models for the future and about the social and cultural continuity of the paradigm based on trust, reciprocity and social capital. Just at the very moment that policy prescriptions based upon the assumed bases of success in these places were becoming generalised within regional and urban policies across the globe, the conditions on which success was based in these exemplar places were being eroded. This was all too evident in the 2008–2018 general crisis in Southern Europe and the Eurozone (Hadjimichalis, 2018).

Ignoring these problems since the 1990s, many countries and major global institutions such as the OECD, the World Bank and the European Union (EU) launched programmes dealing explicitly with places and local development using the aforementioned success stories as 'best practice'. In 1991 the EU LEADER initiative was announced for the period 1991–1994 as a Europe-wide experiment. It involved the first 217 Local Action Groups (LAGs), representatives from small and medium-sized enterprises, local authorities, non-profit organisations and other local players from various sectors, who agreed on a common approach to the interests involved in developing their local rural areas. Three other initiatives also embodied similar principles of local development, especially that of partnership: URBAN for urban areas; INTERREG for cross-border and interregional collaboration; EQUAL for partnership in social integration, combating discrimination in whatever guise, inequality in the labour market and continued vocational training. In Europe another actor was LEDA, an independent association which brings together organisations, practitioners, research centres and others who are active in the creation of local employment in EU countries. The regional employment pacts launched in 1977 provided opportunities to places with high unemployment, but attracted limited attention and were dropped from 2000 onwards. Application of these programmes succeeded in generating jobs and reducing unemployment in some places, but the latter was dependent on local leadership quality and knowledge of the EU's highly complicated and bureaucratic procedures that made possible a productive and on time absorption of EU funds.

The OECD's involvement with place and local development dates from the 1970s, with various programmes and recommendations, but perhaps the most advanced framework was initiated in early 1980 with LEED (Local Economic and Employment Development) which had three main objectives: to improve the quality of public policy through continuous monitoring and assessment of current practices; to promote innovation in local economic and employment development across the globe; to support the design, implementation and evaluation of development strategies to help grow local economies. The OECD, overcoming some limitations of 'New Regionalism' theories, gives particular attention to tourism and culture as important pillars for local development. It finances particular projects in many countries and, since 2003, has operated the LEED Trento Centre for Local Development in the Autonomous Province of Trento in Northern Italy.

The World Bank also promotes local economic development projects (LED) worldwide, particularly in Latin America and SE Asia, without any particular spatial reference. For the Bank, local development means creating a favourable environment for business success and job creation in urban areas. To do this optimally, LED is undertaken through partnerships between local government, business and community interests. Finally, the UN in the 1992 World Earth Summit in Rio adapted the Local Agenda 21, which provides guidance to local governments in the preparation of local development plans, integrating the principles of sustainable development. Local Agenda 21 introduces explicitly the environmental dimension in LED, such as the concern with the per capita ecological footprint. In this respect, it is a major innovation and a critique of previous models, but unfortunately with limited applications.

The de-politicisation of local economic development

Despite problems and fallacies, the original conceptualisation of local development was an explicitly social and territorial approach to development including not only economic growth but also employment creation, poverty reduction, quality of life and environmental sustainability. In its original formulation, local development aimed to empower local communities to shape, in a participatory and democratic way, the future of the place where they live. In this respect and despite the inherent contradictions previously described, this original formulation of local development was nevertheless a politicised project.

Application of local economic development projects that incorporated imitations of 'good practice' did not deliver on all their promises, although without these programmes the situation in marginal European places could have been much worse. Spending for these programmes became 'media unfriendly' at the turn of the century, as argued in an EU policy brief in 2011. In the context of neoliberal budgetary cuts for social programmes, local economic development became somewhat forgotten. Attention now turns to policies not requiring public funding such as competitiveness among places to attract investment or to strengthen local comparative advantages in the new knowledge economy, entering a straightforward neoliberal and de-politicised phase in the relation between place and economic development.

Academia and mainstream theory-building played its role in this shift. Despite its progressive intentions, the 'New Regionalism' treatment of place and local economic development is often highly compatible with a neoliberal view. I don't argue that theories of local development or the New Regionalism models are neoliberal *stricto sensu*, or that their exponents are neoliberals. Neither is my aim here to take issue with colleagues, with whom I agree on many other counts. What I am saying is that in the crucial decade of the 1990s the way they formulated, directly or indirectly, the original question posed by D. Massey back in the late 1970s, '*in what sense a regional problem?*' (in our case *local problem?*), de-politicised it at the same time as neoliberalism was making a frontal attack in economic geography and planning. Thus, the highly needed resistance and fighting back never materialised from 'New Regionalism' researchers. This has made their views easy to absorb into neoliberal policies, making it sometimes difficult to differentiate progressive from regressive applications. In that sense there was a gradual sliding towards the dominant neoliberal discourse. Let me clarify this with four points.

First, there was the treatment of places and localities as quasi-individuals, obliged to find their own way to economic prosperity, competing and fighting with each other. What places need, according to New Regionalism, is less politics, more competition, more innovation and tolerance to business demands. Second, the use of a few super-star places as best practice ignores those millions of 'ordinary' places that form the majority everywhere and are in real need of development and prosperity. This elitist approach fits perfectly with the neoliberal promotion of elitism everywhere as efficient (Dorling, 2010). Third, the current lack of interest in social issues beyond success and competitiveness is also evident in inadequate analyses of working conditions, poverty and informal employment in New Regionalism theories, not to mention class and gender. And fourth, while initial formulations focussed on small industrial firms, in other words on production and employment creation, planning now shifts to consumption, leisure and digitalisation, in particular in tourism and culture. The rationale behind this shift is that Europe has plenty of history and cultural heritage, both appropriate to promote post-industrial European places.

In the 2000s European economic development strategy for places seems to have invented a new 'magical' recipe: culture-led development. It is the practice of using a community's public cultural amenities to make economic progress. This approach focusses on the unique cultural features of particular places, building on existing assets and using them through place-marketing to attract new investment and visitors and to strengthen existing businesses. The idea in itself is not new or inappropriate. But its association with the neoliberal competitive dogma may produce negative socio-spatial effects, such as those already visible in long-established historical urban centres or island and coastal mass tourist destinations (Melissourgos, 2008).

Despite its popularity, however, the culture-led development in neoliberal Europe (EU, 2010; KEA European Affairs, 2006) seems far from beneficial to the majority of people (Sacco et al., 2013). Many economic geographers love the idea and showcase the profit-making of innovative cultural industries and the leading role of the so-called 'creative class' in various case studies (among the many see Scott, 2000; Florida, 2002; Landry, 2008 and others). Few, however, investigate the negative socio-spatial consequences of a development path previously self-generated and now introduced as a desirable planning alternative.

Critics of culture-led development focussed on the poor theoretical background and insufficient empirical evidence; on the socio-spatial exclusion of local marginal population segments; on the rise of land rent, serving vested interests of real estate developers and high-income professionals; on the limited provision of well-paid and secure jobs, while the majority of cultural workers work seasonally on low-paid contacts or in the informal economy; and, finally, on a mono-causal development scheme based only on culture and particularly visible in historical urban centres and tourist destinations (Peck, 2005). Furthermore, the culture of a place is a collective product and belongs to the commons. Its appropriation by private firms is equivalent to public assets' dispossession, dominant in the era of neoliberalism. Private firms on their own or in partnership with some municipality, capture commonly produced use value from place-embedded cultural assets and turn it to privately appropriated exchange value.

Thus, I have major reservations about the European Community's (2010) grandiose claim that culture-led development as an economic strategy is particularly relevant in today's age of globalism 'to create growth and jobs ... in the framework of the Lisbon Strategy' (pp. 1–3). It is worth remembering that both the Lisbon Strategy until 2014 and its successor 'Europe 2020' promoting 'smart growth' reproduce the contradiction between introducing macro-economic austerity policies while pursuing employment and social cohesion. The situation is getting worse when we take on board the current trend that expects employment and social cohesion as the result of the promotion of cultural activities and not as a prime target, as in the 1990s programmes. That said, I don't underestimate the potential of culture-led development, provided that it takes into account the previous critical points and in particular avoids becoming a mono-causal strategy.

Concluding comment

Planning for local and regional economic development in Europe, from the 1970s to the 1990s, succeeded in providing a substantial number of jobs in peripheral areas and in reducing inequalities among places and regions. Although deliverables were behind promises and despite inherent contradictions, attempts for social and spatial redistribution during this period belong to what is known as 'welfare regionalism'. The picture has changed dramatically since the 1990s with the neoliberal victory. Now competitiveness replaces redistribution and each place must find its way in the globalised framework. Using the existing cultural assets of a place or creating new ones could help some communities but this remains inadequate as a general economic development strategy. Turning cultural assets into traded commodities is self-destructive for local residences, as the paradigmatic cases of Venice, Barcelona, Santorini and many others clearly show. A much more comprehensive vision is required in which ending austerity and dealing with the uneven development among places are prerequisites.

Notes

1 Ray Hudson, in a personal conversation, believes that locality as a term was just a product of the 1980s' debates and a desire to say something different about the local in the context of the emphasis on uneven economic development in the era of neoliberal globalisation.
2 Anglophone researchers discovered Third Italy via mainly the book by Piore and Sabel (1984) *Third Industrial Divide*. Descriptions and explanations of Italian small firms and workshops plus their industrial organisation – termed 'flexible specialisation' – was very idealistic and based on only five interviews with small mechanical firms in Emilia Romagna. From this thin evidence, Piore and Sable became fierce advocates of 'flexible specialisation' that guided others in wrong directions. Furthermore, they argue that this was the path that Proudhon advocated back in 1848, which was the association of small producers who could democratically control their work and local life.

References

Amin, A. (2003) 'Industrial districts', in: T.J. Barnes, and E. Sheppard (eds) *A Companion to Economic Geography*, Oxford: Blackwell, pp. 149–168.

Aydalot, Ph. (1986) *Millieux innovateurs en Europe*, Paris: GREMI.

Becattini, G. (1998) *Distretti industrially e made in Italy*, Torino: Bonigheri.

Becattini, G., Bellandi, M., Dei Ottati, G., and Sforzi, F. (2003) (eds) *From Idustrial Districts to Local Development*, Cheltenham: E. Elgar.

Beynon, H., Hudson, R., and Saddler, D. (1994) *A Place Called Teesside*, Edinburgh: Edinburgh University Press.

Blim, M. (1990) *Made in Italy: Small Scale Industrialization and Its Consequences*, Νέα Υόρκη: Praeger.

Castree, N. (2003) 'Place: connections and boundaries in an interdependent word', in: S. Holloway, St. Rice, and G. Valentine (eds) *Key Concepts in Geography*, London: Sage, pp. 165–186.

Charnock, G., Purcell, Th., and Ribera-Fumaz, R. (2014) *The Limits to Capital in Spain: Crisis and Revolt in the European South*, London: Palgrave MacMillan.

Dicken, P. (2015) *Global Shift. Mapping the Changing Contours of the World Economy*, London: Sage (seventh edition).

Dorling, D. (2010) *Injustice. Why Social Inequality Persists*, Bristol: The Polity Press.
Dunford, M. (1991) 'Industrial trajectories and social relations in areas of new industrial growth', in: G. Benco, and M. Dunford (eds) *Industrial Change and Regional Development*, London: Belhaven, pp. 51–82.
Dunford, M., and Greco, L. (2006) *After the Three Italies. Wealth, Inequality and Industrial Change*, Oxford: Blackwell.
EU. (2010) *Unlocking the Potential of Cultural and Creative Industries*, Brussels: Green Paper.
Florida, R. (2002) *The Rise of the Creative Class, Leisure, Community and Everyday Life*, New York: Harper.
Garofoli, G. (1983) *Industrializazione diffuza in Lombardia*, Milano: F Angeli.
Hadjimichalis, C. (2006) 'The end of third Italy as we knew it?', *Antipode*, 38(1): 82–106.
Hadjimichalis, C. (2018) *Crises Spaces. Structures, Struggles and Solidarity in Southern Europe*, London: Routledge.
Hadjimichalis, C., and Hudson, R. (2007) 'Rethinking local and regional development: implications for radical political practice in Europe', *European Urban and Regional Studies*, 14(2): 99–113.
Hadjimichalis, C., and Melissourgos, Y. (2013) 'Tourism and economic geography: a comment on the (still) missing link', in: O. Rullan (ed) *En l'espai-temps: Homenatage a Alicia Bauzà van Slingerlandt*, Palma: Universitat de Illes Baleares, pp. 71–84.
Hadjimichalis, C., and Vaiou, D. (2004) 'Local' illustrations for 'International' geographical theory', in: K. Simonsen, and J.O. Bœrenholdt (eds) *Space Odysseys*, Alkdershot: Ashgate, pp. 171–182.
Harvey, D. (1982) *The Limits to Capital*, Oxford: Basil Blackwell.
Hudson, R. (1999) 'The learning economy, the learning firm and the learning region: a sympathetic critique of the limits to learning', *EURS*, 6(1): 59–72.
KEA European Affairs. (2006) *The Economy of Culture in Europe*. Report for the EC, Brussels.
Labrianidis, L. (ed) (2008) *The Moving Frontier. The Changing Geography of Production in Labour-intensive Industries*, Ashgate: Aldershot.
Landry, C. (2008) *The Creative City. A Toolkit for Urban Innovators*, London: Earthscan.
Martinelli, F. (2009) 'The Mezzogiorno', in: C. Hadjimichalis (ed.) *International Encyclopedia of Human Geography*, Regional Development Section, Amsterdam: Elsevier.
McLeod, G. (2001) 'New Regionalism reconsidered: globalization and the remaking of political economic space', *IJURR*, 25(4): 804–829.
Melissourgos, G. (2008) *Local/regional development and social conflicts: two cases of large scale tourist development in Spain and Greece*, PhD Dissertation, Department of Geography, Harokopio University Athens (in Greek).
Paci, M. (1972) *Mercato del lavoro e classi sociali in Italia*, Bologna: Il Mulino.
Peck, J. (2005) 'Struggling with the creative class', *International Journal of Urban and Regional Research*, 29: 740–770.
Pickles, J., and Smith, A. (2011) 'Delocalization and the persistence in the European clothing industry', *Regional Studies*, 45: 167–185.
Piore, M, and Sabel, C. (1984) *The Second Industrial Divide: Possibilities for Prosperity*, New York: Basic Books.
Sacco, P., Ferilli, G., and Tavano Blessi, G. (2014) 'Understanding culture-led local development: a critique of alternative theoretical explanations', *Urban Studies*, 51: 2806–2822.
Scott, A.J. (2000) *The Culture Economy of Cities: Essays on the Geography of Image-producing Industries*, London: Sage.
Scott, A., and Storper, M. (1986) *Production, Work, Territory*, Boston, MA: Allen and Unwin.
Skordili, S. (1999) *Geographical Restructuring of Manufacturing: the case of the food sector in Greece*, PhD Dissertation, Department of Planning and Regional Development, AUTH (in Greek).
Vaiou, D., and Hadjimichalis, C. (1997) *With the Sewing Machine in the Kitchen and the Poles in the Fields: Cities, Regions and Informal Work*, Athens: Exandas (in Greek).
Vàzquez-Barquero, A. (1992) 'Local development and flexible accumulation: learning from history and policy', in: G. Garofoli (ed) *Endogenous Development and Southern Europe*, Aldershot: Avebury, pp. 31–48.

48

Place and uneven development

Ray Hudson

Conceptualising places within capitalist socio-spatial formations

Places were a familiar presence in the landscape long before the rise of capitalism and this pre-capitalist history and legacy has often had a prominent impact on their subsequent evolution. However, in this chapter I want to focus specifically on places within capitalism, and in particular the relationship between places, capital accumulation and its constituent processes of uneven and combined development. For as Merrifield (1993, 520) observes: "the global capitalist system ... has to ground itself and be acted out in specific places".There has, however, been a long-running debate within geography and cognate disciplines as to how most appropriately to conceptualise place. For many years there was an emphasis upon places as locally-based, bounded, settled and closed communities. More recently, there has been growing emphasis on places as open, with permeable boundaries, as sites through which all sorts of flows of capital, commodities and people take place, leaving their traces selectively sedimented in the cultures of places and in their social and economic structures as a result of their articulation with these structures. While many places may be now less bounded, more open and more interconnected than they once were, to some degree they have always had these characteristics and so "maybe those notions of a coherent settled place were always inaccurate. This means that *in principle* the conception of places as bounded and undisturbed is incorrect" (Massey, 1995, 64, emphasis in original) and that all places are to a degree open (Harvey, 1996, 310).

Harvey's and Massey's comments emphasise that there have always been connections in various ways to a wider world, as places have *never* been entirely closed, especially since the onset of industrial capitalism. Indeed, the growing density and complexity of linkages as capitalist social relationships expanded had a crucial formative influence on places. People and a variety of commodities such as raw materials required in production flowed in, finished commodities and money capital flowed out. Conversely, however, places have never been entirely open. For people thrown together in the new industrial sites and spaces that were integral to the establishment of industrial capitalism, migrants from a diverse set of origins, surviving and coming to terms with life as proletarian wage labourers required developing place-based institutions and so, to a degree, creating closed and bounded places as part of their strategies for them, their families and friends to survive and create a degree of predictability in their lives within the emergent processes of combined and uneven capitalist development.

Furthermore, the frequency, intensity and spatial reach of connections linking places have altered over time as the social relationships of capitalism have become more stretched and re-defined social spaces in new ways, especially with the intensification of time–space compression in recent years. This has enabled some places to become economically dominant, key nodes and command and control centres within a globalising economy. At the same time, socio-spatial inequality within these places has deepened. In contrast, while at one time increasing, in many places the density and geography of linkages with the wider world subsequently declined – for example, as a result of disinvestment decisions by transnational companies and national governments, undermining their coherence as places and provoking various forms of resistance to the threat that these changes posed.

Capitalist development and the growth and decline of places

Places are socially produced complex assemblages made up of material objects, things and buildings including factories, offices, hospitals and houses, transport routes and terminals, and of workers and their families, capitalist enterprises and representatives of capital. They are bound together in systems of social relations in which those of capital are dominant but which also encompass distinct cultures, multiple meanings, identities and practices that are produced, modified and destroyed in the course of capitalist development. As Harvey (1996, 295) puts it, "places arise, constituted as fixed capital in the land and configurations of organisations, social relations, institutions etc. on the land". Non-capitalist social relationships may help shape these places – and often do – but capitalist social relationships routinely have a decisive, although not inevitable, shaping influence within capitalist socio-spatial formations. Places thus constitute both a product of and a setting in which the production, exchange and sale of commodities occurs. They provide the material and social settings through which the totality of production systems and the circulation of capital can be organised. Capital and wage labour co-exist in a dialectical relationship, each requiring the presence of the other: capital needs workers to be in their places in the social, technical and spatial divisions of labour while workers and their dependents need capital to be in *their* places to provide waged labour.

In the specific context of the production of profits through the production of things, the growth and decline of places is linked to their relationship to the dynamics of circuits of capital, capital accumulation and economic growth. Places must therefore be envisioned in the context of capitalist relations of production "stretched out" in various ways over space and coming together as a spatially and temporally specific condensation of intersecting social relationships as those of capital interact with other social forces. Peck (1995) draws a useful distinction between *work*place and work*place* and there is a crucial and often contested relationship between the production, exchange and sale of commodities in *work*places and social reproduction in the wider community beyond the *work*place in work*places*. As Polanyi (1944) emphasised, capitalist economies are always embedded in wider systems of social relationships that extend beyond the *work*place, many of which involve non-commodified forms of production in work*places*. The places that grow economically as *work*places do so as a consequence of their relationship to various growth mechanisms that characterise particular phases of capitalist development. Many places relate only marginally – or not at all – to the dominant growth mechanism(s) at a particular time but may nonetheless be successfully reproduced while others have remained beyond the reach of circuits of capital but have nonetheless been reproduced as places as a result of the presence and effectiveness of non-capitalist social relationships. Other places, on the other hand, which once were central to processes of capitalist development no longer relate directly to processes of growth but to processes of decline as capital there is devalorised and in some cases physically destroyed and as a result their coherence as *work*places is eroded and their viability as work*places* is compromised, sometimes fatally.

Capitalist economies are thus characterised by a tension between processes of fixity and fluidity of flows of capital and by processes of the uneven and combined development of places in the economic landscape. Any settlement of social relations into a spatial form within the parameters of the capitalist mode of production will, however, only ever be temporary. Nevertheless, some of these settlements last longer than others, forming relative permanences in the landscape and in a world of dynamic and fluid processes of flows of capital in its various forms. Some places have therefore become major centres of capital accumulation and economic growth and continued to expand as a result of various forms of agglomeration economies while others have emerged and then declined quite quickly as a result of the place-specific devalorisation of capital (Harvey, 1982). The production and destruction of places has therefore always been an integral aspect of the process of uneven and combined development that is inherent to the capitalist mode of production; production and realisation of profits through the circuit of industrial capital requires that the various component parts of capital (fixed, constant and variable) be brought together in places of production, in *work*places – factories, mines, offices, shops and so on. Crucially, capital needs a supply of suitably qualified labour, the source of surplus-value on which its capacity to produce and realise profits depend. But people need to live, to survive outside the *work*place and this in turn requires the building of houses, the provision of schools, medical and social facilities, and various leisure and recreational facilities provided either via private capital seeking profit, via the state at national or local level providing public services and facilities, or via some other process of provision – for example, co-operatively produced social housing. People in their places make demands for the provision of facilities and services that have no direct bearing upon, or are not materially necessary for, profitable production but are necessary for these places to become and be socially reproduced as viable work*places*. The processes of production of places are thus contested, as different social groups seek to shape the geographies and landscapes of capitalism to reflect and further their particular interests in production in *work*places and social reproduction in the wider community beyond the *work*place in work*places*.

Differential meanings and attachments to place

Within the social relations of capital, places can therefore be thought of as complex assemblages of people and things which come to have varying meanings for different people. For the owners and managers of capital, places are above all *work*places, sites for the production of profit; for workers and their families, recognising that the working class is characterised by an "infinite fragmentation of interest and rank" (Beynon, 1999, 38) as a result of the growth of social and technical divisions of labour, places also have a different and more complicated meaning as work*places*. Furthermore, because economic activities have to take place somewhere, but cannot take place everywhere, workers can also be divided from one another, or may divide themselves, by place as they compete to attract or retain investment and jobs and preserve both their *work*place and their work*place*. A *work*place is one in which to earn a wage and *work*places are critical within capitalist economies, both to capital as the places in which surplus-value is produced and profits realised and to workers and their families who rely upon wage incomes in order to live. However, the wider places of reproduction, the work*places*, the neighbourhoods, villages, towns and cities, the regions and national territories, in which *work*places are embedded, are much more than just places in which to work for a wage, as they are places in which to enjoy life as well as work. As a result, people develop emotional attachments to these places and so may act in various ways to defend them when they are threatened by the decisions of capital or national states.

With one critical exception, capital and labour thus have very different and largely mutually exclusive interests in and commitments to places. That exception reflects the need for capital to be able to purchase and employ labour-power in the production and realisation of profits and, reciprocally, for workers to be able to find waged work. For many people places must yield a money income to pay for the costs of living there and this entails successfully selling their labour power to capitals or, perhaps, the state. In this sense, workers, their families, and capitals – both those who buy their labour-power and those with whom they spend the resultant wage income – share an interest in that place as a *work*place, as a location for the successful production and realisation of profits. While not all capitals are equally mobile, and not all labourers are equally immobile, occupationally, industrially or geographically, in general capital is more mobile than labour and increasingly, as a result of technological and organisational innovations, has been able scour the globe for *work*places in which to produce more profitably. Places that are, for capital, merely a temporary location for profitable production, albeit one to which some capitals are more closely tied than others, become work*places* in which workers, their families and friends live and to which they come to have longer-term and more nuanced attachments and commitments. For people do not simply exist or create their identities and define their lives in terms of the social relationships of commodity production and consumption. To borrow a phrase made famous by the striking steel workers of Longwy in north eastern France in the late 1970s as they fought to preserve their communities, people live and learn, as well as work, in places (and not all work is waged work). This highlights the importance of social relationships other than those of capitalist production. Relations of ethnicity, gender, household, family and friendship, and indeed non-capitalist social relations of production and work, may be central to the constitution of place as both *work*place and work*place*. For these reasons places become imbued with meanings in a variety of ways (Beynon and Hudson, 1993). Places created "from scratch" as *work*places in "new industrial spaces", with capitalist relations of production established there "for the first time", thus become translated into "meaningful places", work*places* in which people can live their lives as socialised human beings beyond the *work*place. People make considerable individual and collective cultural and emotional investments in such work*places* which consequently acquire deeply symbolic meanings. As a result, people living there may become deeply attached to them, tied to these work*places* by powerful emotional forces, although with the important qualification that attachments and meanings may vary significantly between different people resident in the same location.

This intra-place variation was present even in the "mono-industrial" or "single occupational communities" (Dennis et al., 1956) once found in many parts of the industrial capitalist landscape. Even in these very particular cases, claims about social homogeneity, shared consensus viewpoints and unanimous attachment to bounded closed places need to be treated with caution. There was often marked social differentiation within such work*places*, reflected in distinctive intra-place social geographies. Steel towns were marked by social distinction which flowed from the occupational hierarchy within the *work*place, for example (Bell, 1985). Even in the arch-typical coal-mining villages of northern England, Scotland and Wales, social differences translated from the division of labour in the *work*places of the mines into the social geography of the work*place*. "Office Street", the housing for mine managers adjacent to the colliery, typically contained houses that were larger than those in the densely packed back-to-back terraces for the miners. While the terraces for the miners may have had a back yard, the houses for the managers would often have individual gardens. The coal owner typically lived in a large residence beyond, or on the fringes of, the colliery village

(Austrin and Beynon, 1994). For each of these social groups, the village as both *work*place and work*place* had significantly different meanings.

While such mono-industrial places can be thought of as a relatively simple limit case, more generally there will always be multiple, co-existing characterisations and understanding of particular places. As places become larger and more complex, so too do these understandings and meanings. Different social groups within a place will have different, even opposed, and contested, readings of its character and different stakes or interests in it as both *work*place and work*place*. For example, many of the former mono-industrial towns and villages of north east England have been partially and selectively transformed into commuter settlements as a result of an influx of speculative capital into private sector housing developments inhabited by new in-migrants. While these new residents live in the place – part of the meaningful work*place* of others, although many residents remain unemployed and without waged work, at least in the formal economy – their own *work*place lies elsewhere and residence in the work*place* of the former mining village is simply a temporary staging post prior to their next move in the housing and labour markets. Consequently, and more generally, attachments to and feelings about place become caught up in "the power relations which structure all our lives" (Rose, 1995, 89). Furthermore, as a consequence of these asymmetrical power relations, different individuals and groups also have differential powers and resources to promote and materialise their vision of the place. Nonetheless, despite these differences, such attachments are in general strongly indicative of the ways in which many people live their lives in work*places* as socialised human beings with ties of community, friends and family and not merely as the abstract commodity labour-power in *work*places or as the long-term unemployed who no longer are of interest to capital as a source of labour-power to deploy in those *work*places.

Defending places: tangled relationships between attachments to place and class

As a product of a sense of shared identity and interest in the place by a range of social groups and forces, places may develop a "structured coherence", which is expressed via a particular "structure of feeling" (Williams, 1989a), with their existence as work*places* typically predicated upon capital having an interest in continuing to produce there and as a result providing waged work in them as *work*places. This particular convergence of interests between capital and workers and their families and residents in place can however be ruptured and the "coherence" of places, the symbiotic relationship between *work*place and work*place*, pulled apart if the economic rationale for production there is eroded. This is one aspect of the destructive character of capitalism, place-specific devalorisation and possibly physical destruction of fixed capital as an element of an on-going process of creative destruction. Consequently, especially in situations in which the main or sole source of waged incomes is a single employer or industry, such places and their constitutive cultures may often lead a precarious existence, vulnerable to the disinvestment decisions of companies and the policies of national (and emergent supranational) states. This can lead to a variety of attempts to "defend places" as both *work*places and work*places* and to seek to ensure their successful reproduction, involving complex processes that are material, social and discursive. These can range from the development of a politically progressive place-based "militant particularism" (Williams, 1989b), while recognising the difficulties of generalising these to more broadly based political programmes, to blatantly regressive campaigns that pit place against place in a divisive and often desperate struggle to secure capital investment and waged employment. For example, an instrumental attachment to a location as a *work*place, a source of wage income, may lead workers and their families who reside there to engage in competition with other workers and their families in their places to keep existing jobs or gain new ones. This was exemplified in the 1980s by the competition between steel plants and towns in Britain and France and, most forcefully, coal mining communities in Britain, as between those in Nottingham dominated by the Union of Democratic Mineworkers and those in the remaining coalfields in which the National Union of Mineworkers was dominant, as coal miners and steel workers fought to keep jobs in their *work*place and secure the future of their work*place* as attachments to place took precedence over class solidarity (for example, see Beynon, 1985; Beynon and Hudson, 2020; Hudson and Sadler, 1983, 1986).

In other times and places, rather than place-based intra-class conflict, campaigns to retain existing investment and jobs in place or to compete for new investment and fresh employment in new industries have involved cross-class alliances as a shared interest in place trumps class solidarity. The defence of place is pursued via the emergence of place-based cross-class alliances that seek to preserve both *work*places and work*places*. Workers, their families and friends, either personally or via their institutions such as trades unions, may form alliances with small companies, such as those involved in retailing and the provision of personal services that are tied to that place and see their future threatened by the loss of jobs and working class incomes. There are also situations in which capitalists or their representatives, such as plant managers, become involved in the social relations of the work*place* as a necessary requirement of reproducing the conditions needed for the production of profits in *work*places there. Once this requirement no longer holds, however, the bonds of the social relations of place for them may dissolve, though for some forms of capital and some capitalists and their representatives coming to terms with this dissolution is more difficult than for others. The actual or potential threat to a place may thus provide the catalyst that stimulates campaigns that can encompass some capitalist interests within more broadly based social movements to defend work*places* against the dangers posed by job loss and *work*place closure. For example, in the 1930s in north east England the owners and managers of major capitalist enterprises took the leading role in lobbying central government for the emergence of new forms of regional policy to protect their place-based assets and interests in *work*places while incorporating trades unionists concerned to attract new *work*places and preserve existing work*places* (Hudson, 1989).

Such campaigns to defend and promote places as new *work*places involve modes of place re-presentation and processes of identity re-formation as well as material changes there to sell the place to new capitalist interests. This involves representations of shared interests in place, centred around particular social groups that discursively construct these place-specific interests. In this way, echoing an old idea from Marxian class analysis, places are transformed from being "in themselves" to being "for themselves" (Lipietz, 1993). For some, this involves seeing places themselves becoming agents (Cooke, 1989). Such views run a risk of embracing spatial fetishism, however, with considerable political and theoretical dangers, as places as such do not become "pro-active". Rather it is the case that some social groups foster the construction of a shared place-based interest which becomes a basis for action in campaigns to defend places or promote their interests in other ways. Drawing on the legacy of Gramsci, such a territorial social bloc, a place-sensitive alliance of social forces, acquires a legitimate capacity to act and speak on behalf of a place via political and social struggle. While places are not coherent, integrated wholes but are internally fragmented as well as being open to a variety of external influences and flows (Allen et al., 1998), nevertheless the figure of the place as such, as a coherent and bounded whole with unified interests, can be mobilised rhetorically and politically. Places as apparently coherent entities can be (re) produced discursively, as when local political leaders "speak for" the place "as a whole" and claim to represent "its interests". But such discursive constructions are always contestable and are often contested, although inequalities in power may result in one such construction becoming dominant or even hegemonic.

Campaigns to defend places as *work*places can therefore either involve seeking to preserve their existing economic basis or to create a new one in competition with other places. Often, decline of established industries is grudgingly accepted, without challenge, as the focus becomes the attraction of new jobs and activities in new *work*places. Sometimes, however, it is vigorously contested. One option is to mount a political campaign rooted within democratic processes, to allocate more state resources to these places via central government policies or restructure the territorial organisation of the state, devolving more powers for economic development to local and regional levels. As the experiences of north east England, a region that experienced a severe deindustrialisation with many examples of local place-specific economic crises but in which an offer of limited devolution to a regional assembly was decisively rejected in 2003, such devolution is not always seen as the answer to the problems of economic decline (Hudson, 2006). However, such demands for devolution can become particularly powerful when issues of economic development become interlinked with those of culture and identity and the preservation of the wider work*place* at varying spatial scales (Anderson, 1995). Such pressures can also coincide with national states wishing, or needing, to draw in the scope of their activities as they seek to become more "hollowed out" institutions, decentralising responsibility if not commensurate resources to sub-national levels of governance in the face of fiscal crises at national level.

Sometimes, however, the possibilities of democratic channels seem to be exhausted and to have failed. Another option therefore is to move beyond those forms of protest that are generally regarded as legitimate within democracies and adopt tactics of direct action, challenging the power of capital and the state head-on as people in threatened places turn to more direct methods of defending their *work*places and work*place*. This threat to place can be especially stark in mono-industrial settlements. As a result, it can lead to campaigns to defend place grounded in a resurgence of place-based class conflict. This is particularly so in circumstances in which the state is seen to be directly involved in the processes of economic decline via sectoral or industrial policies, such as those of nationalisation and public ownership, especially in mono-industrial places (Beynon, 1985; Hudson, 1989; Hudson and Sadler, 1983, 1986). This can involve strikes and industrial disputes and workers occupying their *work*places (for example, see Foster, 2013). It can also involve workers, their friends and families taking their concerns beyond the boundaries of the *work*place and onto the streets of their work*place* and beyond. For example, in the late 1970s and 1980s, steel workers in north east France engaged in a range of direct actions, including protesting violently on the streets of Paris, the national capital city, strongly opposed by the French state (Hudson and Sadler, 1983). But the processes of contested closures were perhaps most dramatically, and most forcefully, seen in Britain in those coalfields in which the National Union of Mineworkers was dominant, as coal miners there fought to keep jobs in their *work*places through a year-long strike in 1984/5 in opposition to the Thatcher government, which was determined to close their mines and crush the NUM while encouraging the Union of Democratic Mineworkers to keep working in their mines in Nottingham (for example, see Beynon, 1985). Central government deployed the power of the state in unprecedented ways to break the power of the NUM and destroy the coal mining communities as both *work*places and work*places* in the areas in which it was dominant, revealed in the physical violence and brute force displayed in Easington in the summer of 1984 and at Orgreave cokeworks later in that year (Beynon and Hudson, 2020).

Such campaigns to defend places, as both *work*place and work*places*, however prosecuted, both draw on and reproduce, maybe in modified form, the cultures in which they are grounded and the understandings of the world and what is possible within it that these provide. In this sense, it can be argued that "militant particularisms" are always in some sense profoundly conservative because they seek to perpetuate patterns of social relations and community solidarities – loyalties – achieved under a certain sort of oppressive and uncaring industrial order (Harvey, 1996, 40). But, partly because attachment to place is grounded in more than just access to profits and wages in *work*places, it does not necessarily follow that such a "conservative" defence of work*place* is divisive or socially regressive, setting places in competition with one another. Place-based campaigns to defend *work*places need not necessarily be place-bound and can become a basis from which to try progressively to overcome, rather than reinforce, the negative social impacts of spatial differentiation whilst respecting the specificities of people in their work*places*. They can be grounded in a radical politics that seeks to forge wider connections and affiliations in the best traditions of socialism. The 1984/5 British coal miners' strike is one, albeit in the end unsuccessful, example of this process (Beynon, 1985; Beynon and Hudson, 1993, 2020). Indeed, this strike in defence of mining jobs in *work*places, albeit that these were typically dirty and dangerous jobs, and communities in work*places*, albeit that these were often marked by internal division and tensions, can only be understood in terms of its grounding in a deep, place-based communal solidarity that embraced both *work*place and work*place* and sought to challenge the political-economic project of Thatcherism, albeit in the end in vain.

The bitter residue of failure

The failure of the 1984/5 miners' strike powerfully illustrates the immanent dangers involved in seeking to move political activity from one level of abstraction grounded in a place-based militant particularism to broader campaigns that seek to link a variety of issues and places, encompassing both *work*places and work*places*. For "in the act of translating, something important gets lost, leaving behind a bitter residue of always unresolved tension" (Harvey, 1996, 359). A combination of an unprecedented (ab)use of state power, brutality and violence, the extent of which only became revealed in subsequent court cases, a bitter inter-union dispute between the National Union of Mineworkers and the nascent Union of Democratic Mineworkers, which was also expressed as a territorial conflict between places in Nottingham and the other coalfields, and a particular, partial and selective representation of the struggle in most of the media, led to the failure of the attempt to defend *work*places and work*place*-based communities via strike action. It generated powerful intra-class conflicts as workers in one place competed with workers in other places to preserve their jobs. Class solidarity was subordinated to more immediate concerns with life and work in territorially delimited communities as the dialectic of class and place became worked out in particular ways (see also Mandel, 1963). The subsequent post-strike privatisation and then closure of the remaining deep mines in the territories of both the UDM and NUM, the legacies of widespread worklessness, poverty and ill-health in the former coalfield communities as they struggled to survive as work*places* in the absence of *work*places, the destruction of the NUM as a powerful industrial union and the continuing competition between the former coalfields for new jobs in new *work*places do indeed constitute a "bitter residue of unresolved tension".

References

Allen J., Cochrane A. and Massey D. (1998). *Re-thinking the Region*. London, Routledge.
Anderson J. (1995). "The exaggerated death of the nation state". In Anderson J., Brook C. and Cochrane A. (Eds.). *A Global World*. Oxford, Oxford University Press: 65–112.
Austrin T. and Beynon H. (1994). *Masters and Servants*. London, Rivers Oram.
Bell, L. (1985). *At the Works*. London, Virago.
Beynon H. (Ed.) (1985). *Digging Deeper*. London, Verso.
Beynon H. (1999). "A classless society?" In Beynon H. and Glavanis P. (Eds). *Patterns of Social Inequality*. Harlow, Longman: 36–53.
Beynon H. and Hudson R. (1993). "Place and space in contemporary Europe: some lessons and reflections". *Antipode*, 25: 177–190.
Beynon H. and Hudson R. (2020). *The Shadow of the Mine: The Decline of Industrial Britain*. London, Verso (forthcoming).
Cooke P. (1989). *Localities: The Changing Face of Urban Britain*. London, Unwin Hyman.
Dennis N., Henriques F. and Slaughter C. (1956). *Coal is Our Life: An Analysis of a Yorkshire Mining Village*. London, Eyre and Spottiswoode.
Foster J. (2013). "The 1971-72 UCS work-in revisited". *Our History, 9(New Series)*: 1–36.
Harvey D. (1982). *The Limits to Capital*. London, Arnold.
Harvey D. (1996). *Justice, Nature and the Geography of Difference*. Oxford, Blackwell.
Hudson R. (1989). *Wrecking a Region: State Policies, Party Politics and Regional Change*. London, Pion.
Hudson R. (2006). "Regional devolution and regional economic success: myths and illusions about power". *Geografiska Annaler B*, 88: 159–171.
Hudson R. and Sadler D. (1983). "Region, class and the politics of steel closures in the European community". *Society and Space*, 1: 405–428.
Hudson R. and Sadler D. (1986). "Contesting works closures in Western Europe's old industrial regions; defending place or betraying class". In Scott A. J. and Storper M. (Eds.). *Production, Territory, Work*. London, Allen and Unwin: 172–193.
Lipietz A. (1993). "The local and the global: regional individuality or interregionalism?" *Transactions of the Institute of British Geographers NS*, 18: 6–18.
Mandel E. (1963). "The dialectic of class and region in Belgium". *New Left Review*, 20: 5–31.
Massey D. (1995). "The conceptualization of place". In Massey D. and Jess P. (Eds.). *A Place in the World? Place, Culture and Globalization*. Oxford, Oxford University Press: 45–86.
Merrifield A. (1993). "Place and space: a Lebebvrian reconciliation". *Transactions of the Institute of British Geographers NS*, 18: 516–531.
Peck J. (1995). *Workplace: The Social Regulation of Labour Markets*. New York, Guildford.
Polanyi K. (1944). *The Great Transformation: The Political and Economic Origins of Our Time*. Boston, Beacon Press.
Rose G. (1995). "Place and identity: a sense of place". In Massey D. and Jess P. (Eds.). *A Place in the World? Place, Cultures and Globalization*. Oxford, Oxford University Press: 87–132.
Williams R. (1989a). *The Politics of Modernism*. London, Verso.
Williams R. (1989b). *Resources of Hope*. London, Verso.

49

Place-making at work

The role of rhythm in the production of 'thick' places

Dawn Lyon

Introduction

Recently, a friend and colleague from another institution came to give a talk at the University of Kent where I work. The context was our regular departmental seminar series which forms an important part of the rhythm of a term-time week for many academic staff and postgraduate students. We had a little time to spare so I decided to show him around. The School is based in a relatively new building configured so that workspaces – from individual offices to open plan suites – are interconnected via social areas and meeting rooms. Whatever the internal spatial politics, it looks good and he was duly impressed. I felt a keen sense of belonging as I introduced him to other colleagues and led the way through the building (my key card with out-of-hours access making itself felt in my pocket). The thing is though, I don't actually have a permanent office on this site. There is a separate, smaller campus where my books and papers sit behind a locked door with my name on it. Perhaps this is why I notice when I feel such a territorial claim. My friend – I'll call him Michael – is not long retired. He is experiencing some of the unease of such a transition and he too is sensitive to the connections of work, place and belonging. A colleague asks him how retirement is going in a tone that anticipates a positive response. 'I feel disembedded!' comes his reply (Michael is well known for his work on social, political and economic ties). It's a powerful statement about work, time and place across the institutional, normative, regulatory, affective and material dimensions of place and it helps me figure out what I want to say here.

This chapter is concerned with place-making at work, with particular attention to the role of rhythm. First, I want to say something about *how* certain workspaces come into being as places; and second, I want to comment on *what* kinds of workplaces are produced. To do this, I make use of French philosopher, sociologist and urban scholar Henri Lefebvre's (2004) ideas of rhythm and rhythmanalysis as a conceptual and methodological means for tracing how work and workplaces take shape in space and time. In addition, the discussion draws on philosopher Edward Casey's (2001) formulation of places as 'thick' and 'thin' to think about the character of workplaces brought about by the 'ensemble of rhythms' that permeate them (Edensor, 2010b: 69). The chapter draws on empirical material to explore processes of place-making at/through work if temporality is brought to the fore.

Billingsgate fish market, where I conducted an audio-visual ethnography (Lyon, 2016), is a useful setting for thinking about interconnections of work, place and rhythm and a point of departure to question and imagine their reach beyond the marketplace itself.

Work, place and the 'thickness' of rhythm

The concept of place is of course central to geography (Creswell, 2004; Lefebvre, 1991; Massey, 1994; Tuan, 1977), with a strong emphasis on the production of urban space. In sociology (the disciplinary 'place' I write from), if the language of place and place-making has long been present across different fields, place – and space – has largely been treated as a container for social life or a backdrop to concerns about social divisions and power relations of class, gender and race. In the sociology of work in particular, a 'spatial turn' has led to the recognition of interconnections in working lives at a global macro scale (Herod et al., 2007; Ward, 2007). At a more micro level, ethnographies of work and communities often convey a rich sense of place and recognise the spatial (and temporal) interconnections of different spheres of life (Orr, 1996; Westwood, 1984). However, while place permeates such accounts, it doesn't necessarily figure as an explicit analytical concept (Halford, 2004, 2008). In organisation and critical management studies, analyses of the everyday practices of place-making have been more prominent. Studies have documented architectural, interactional, sensory and affective dimensions of organisational life which analyse power, control, meaning and the everyday lived experience of organisational life (e.g., Best and Hindmarsh, 2019; Gherardi, 2019; Riach and Warren, 2015).

This chapter explores processes of place-making at work with particular attention to rhythm. Rhythm has long informed norms, practices and understandings of work (Roy, 1959). The development of clock-time as a form of work-discipline was central to the Industrial Revolution (Thompson, 1967), as were time and motion studies as a means of regulation of workers' bodies in the factory system in the early twentieth century. In the twenty-first century, the intensification of labour is accompanied by new (and some not-so-new) forms of 'calibration' (Sharma, 2014) of workers' bodies in time and space. In the world of finance for instance, traders must align themselves to global patterns of exchange and keep watch over the operation of their deals and the algorithms that underlie them (Borch et al., 2015; Snyder, 2016). In the platform or gig economies, despite discretion over the place of work (most often the home), workers must work long hours and make themselves available at unsocial times or for irregular hours to meet the needs of clients in different time zones (Wood et al., 2019). These forms of work have implications for everyday rhythms of living as well as for any sense of embeddedness in place – material or digital – through work.

My starting point for working with rhythm is Henri Lefebvre's (2004) *Rhythmanalysis: Space, Time and Everyday Life*, published in English in 2004. Rhythmanalysis has been taken up in geography and used in particular to study mobility and place (Edensor, 2010a; Stratford, 2015). It shows how places are made: through mobile flows of capital, people, objects, energy or matter 'which course through and around them'. It 'emphasizes the dynamic and processual qualities of place' identifying the 'rhythms through which spatial order is sustained'. Places, then, are sites in which multiple temporalities come together, in harmony or at odds with one another to produce an 'ever changing polyrhythmic constellation' (Edensor, 2011: 190–191). Rhythmanalysis has been explicitly used to study work, including the performance of work (Simpson, 2012; Snyder, 2016), the procrastinating body's refusal of work (Potts, 2010), the rhythmic ordering of organisational life and the socio-economic relations of work (Borch et al., 2015; Lyon, 2016; Nash, 2020). And there is renewed interest in Lefebvre's thinking more generally for the study of work and organisations (Chari and Gidwani, 2018; Dale et al., 2018). Since rhythmanalysis always combines a mutual attention to space and time, place – or 'lived space' in Lefebvre's terms – is part of the story from the start. The conceptual repertoire of rhythmanalysis offers a rich and 'thick' (in Clifford Geertz' sense) description of everyday life and a methodological 'orientation' (Highmore, 2002: 175) for identifying and analysing particular rhythms, their constellations and their effects (Lyon, 2018). However, where the character of place is the focus of attention – as in this chapter – the discussion might be enhanced, bringing other conceptualisations into dialogue with rhythmanalysis.

Edward Casey's (2001, 2009a, 2009b) discussions of place and characterisations of place as 'thick' or 'thin' are fruitful here. Thick place refers to places characterised by a 'greater density of meaning, affect, relations, habits, memories' (De Backer and Pavoni, 2018: 11). Such resonances are absent in 'thin' places, a formulation which has much in common with Marc Augé's (1995) 'non-places'. However, this kind of binary opposition is problematic. De Becker and Pavoni (2018) argue that Casey's distinction between thick and thin rests on the assumption that otherwise inert space is 'heated up' by personal experience and activities. This leads to nostalgic views of thick and warm community space contrasted with thinner and colder public spaces. De Backer and Pavoni nevertheless contend that using a notion of thickness with a more phenomenological understanding of the fundamentally embodied character of space is more promising.

Whilst I recognise the dangers of an oversimplified opposition between thick and thin, I am drawn to Casey's ideas for the scope they offer for thinking about the effects – and affects – of rhythm in the production of place in processual terms. My focus on Billingsgate Market, with its vibrant atmosphere and lively interactions, makes it easy to hold onto a dynamic sense of the 'unfolding' and layering of the place and allows me to attend to overlapping processes of the 'thickening' and 'thinning' of place in accordance with different constellations of rhythm. 'Thickening' refers to the affective charge that arises through combinations of materials, bodies, things (not least the fish), interactions and practices – as well as through the application of the regulations, inspections and surveillance in the market place. Conversely, the dilution of these intensities corresponds to a 'thinning' of place. Both processes may gather or lose momentum over time or come about suddenly, triggered by the presence of key figures, fish or interactions.

Thick places – following Casey (2001) – are dense places, the result of encounters between bodies and things in time. The character of place can also be discussed in terms of atmosphere (Bohme, 1993; Sumartojo, Chapter 16, this volume) and Edensor (2012) has made explicit connections between Casey's thinking on place and the production of atmosphere. In this chapter, I give prominence instead to the relationship between place and rhythm and processes of thickening and thinning in time. At the start of the night before the market gets going, an observer familiar with the rhythms of Billingsgate might have a sense that the place is poised for activity. But at midnight or 1 a.m., the market hall is more or less empty and still, a far cry from the mood that will later take hold here. So what moves it? How does it take shape and become the lively site and workplace that it is widely known for?

I seek to address these questions in several ways. First, I consider the role of memory and rhythm in the consolidation of 'thick' space. Second, I discuss the performance of place through the different trajectories of fish, customer, merchants and porters moving in rhythm and their everyday working practices as they do. This allows me to think literally about how work 'takes place' and contributes to place-making in the market. The material presented here is taken from the original ethnographic research I conducted at Billingsgate, primarily in autumn 2012. I visited the market several times a week, arriving between 1 and 4 a.m., and carried out informal and formal interviews and observations with fish merchants, inspectors, salespeople, porters/fish handlers and customers. I spent hours each day wandering around the market hall, sensing the polyrhythmic complexity of the place, interspersed with time trying to make sense of the life of the market and disentangle its elements and configurations, usually from a spot in one of the cafes. In addition, I made a short film of the market with Kevin Reynolds from set-up to close as a means of documenting, perceiving and analysing the rhythms of the market. (See Lyon, 2016 for a full discussion of the project and the film.)

Memory and rhythm in the making of 'thick' space

Billingsgate is the UK's largest inland fish market and has been located in East London since 1982. It moved there from its former site in the City of London where it had operated for several hundred years. Fresh and some frozen fish and seafood is sold to catering firms, hoteliers, processors and fishmongers as well as to the public (mostly on a Saturday); supermarkets in contrast have no presence at the market, operating through separate supply chains. Billingsgate's traders are mostly older (over fifty), white, working-class men alongside more recently established south Asian and Indian sellers. It's a self-contained site of exchange and distribution, tightly defined in time and space in 'a flow of dispersion-concentration-dispersion' (Harvey et al., 2002: 205). Fish are legally permitted to leave the market from four o'clock in the morning although trade continues until at least eight.

Many of the vendors are long-serving fish merchants or employees. In our conversations, they often lament the character of the present-day market and the closure and loss of the original site in the City. Talk of the differences between the old market and the new one suggests that the latter is a paler, 'thinner' version of the so-called original, even though the broad rhythms of market work persist (and despite my direct experience of the current market's vibrancy and humour which is at odds with claims that the market is not lively). But the story is more complicated. I am told that the new market opens earlier than the old one by a couple of hours, which deepens the temporal dissonance some market workers experience in relation to friends and family, although the old market used to operate six days a week instead of the current five. The new market is no longer strictly wholesale which means it is a less bounded, more porous place leading to an erosion of the insider community that had an exclusive claim to be there. In addition, the old market was reportedly 'friendlier' with more scope to enjoy socialising (with the presence of a higher volume of pubs and tea shops). Then again, the new market is described as a 'light-hearted place' based on close ties. There are also differences in the material environment. The porters (becoming fish-handlers in an explicit instance of deskilling and loss of status in 2012) use trolleys in the market hall rather than carrying their loads (mostly atop their 'bobbin' hats) and casual workers no longer wait 'under the clock' each day for work. This changes the rhythms and atmosphere of the market as bodies, wheels, feet and fish negotiate their own paths through the space. Finally, in the new market, the volume of fish is significantly less. It now takes two weeks to move the amount of fish that passed through the market in a single day, reports Jerry who tells me that he started out at Billingsgate in 1952.

These tales weave connections and continuities between places and generations. Despite the felt experience of rupture – or perhaps because of it – the longevity of the market and the rhythms and relationships of former ways of working are transmitted and brought into the new space through these reckonings. In effect, memory 'thickens' the sense of place of the new market, contributing to the attachment that current merchants and workers feel. Casey talks about this in terms of 'place memory' (2009a). In Tim Cresswell's words, this refers to 'the ability of place to make the past come to life in the present and thus contribute to the production and reproduction of social memory' (2004: 87). So even if the present site is 'lacking' in relation to the former market and carries little in the way of objects or sites of memorialisation, its very existence stimulates memory and narration about place, and the everyday work of the market replicates and sustains long-established rhythms.

Enhanced by the living presence of figures in the market with experience of both sites, a sense of place is swiftly transmitted to newcomers through the inculcation of the rhythms of work and belonging. Furthermore, according to workers, Billingsgate is an institution that 'gets under your skin', so much so that a number claim to have tried to leave, even experimented with different lines of work, but find themselves drawn back to the market. Place affects them and 'insinuate[s] itself into the very heart of personal identity' (Casey, 2001: 684).

> I qualified to do something totally different and I'd got a few months to spare, and he said, well you're not going to sit on your arse for three months, he said you get down to Billingsgate, this is all those years ago in the old market. And in that three months' period, I was going to be a draftsman, architect and all that with Shell-Mex. The way of life so got under my skin I didn't want to do anything else. The way of life is unique in Billingsgate Market, yeah, yeah.
>
> *(Terry Daniels, fish merchant)*

In some cases, there is a certain 'readiness' on the part of new recruits to 'take on' the meaning, practices and rhythms of the market so they too learn to anticipate their attunement to place (Edensor, 2012: 1114). Prior relationships act as 'thickeners' (De Backer and Pavoni, 2018: 11), tying place and self together. In other words, thick places quite literally 'hold the self in place' (Duff, 2010: 882).

> It's mostly family handed down. Business or friends or people who worked ... in that company that would take over the business ... when people do pass on or leave, they hand it over or they pay some money for it and they continue the business.
>
> *(Market administrator)*

Even where there are economic and social ties which underpin these transitions and a sense of belonging that has already been cultivated, starting out can be a mixed experience. Jimmy, a fish merchant's son, recalls his first day at 15 or 16, coming to work with his Dad and being 'scared of the live fish, lobsters, crabs and things like that'. He continues: 'And one, the guy who was working for us at the time picked two great big lobsters up and chased me round the market with them and I was like, I was going mad, I was'. This suggests a kind of rite of passage to initiate the younger man into the fold rather than a gesture of exclusion. It also points to the specificity of this kind of workplace which involves dealing with live seafood as well as fish securely iced and packed in polystyrene boxes. And although he found the experience 'really hard', 'too much' and said 'I'm never going down there again', literally in the next breath Jimmy goes on to show his affection for and attachment to market culture: 'It's good, it's good, you get used to it. Straightaway you get used to it, silly things, a lot of banter and joking about. But it makes it, it makes a great atmosphere and that's, that's what I like'.

This section has demonstrated the importance of memory and narration in contributing to the production of thick place and how rhythm inflects place-making. Even when stories highlight differences in the rhythms of the old and 'new' markets, there is a sense in which connections are made between them in the telling as well as in the enactment of similar working practices across time. The next section focusses on how place is performed and produced in everyday embodied practices of work on the move.

Moving in rhythm: the everyday practice of work and the performance of place

The market hall at Billingsgate is the central site of display, interaction, movement, negotiation and exchange of the fish market and it is where trade can be seen and heard. Its architecture is important for the production of a sense of place and with more than one hundred shops or stands back to back in three 'corridors' with several cross-cutting paths. The central aisle structures and organises mobility and stimulates encounters as customers walk the routes made for them (Harvey et al., 2002: 206 on Covent Garden), the verges lined with fish. This physical layout of the market also contributes to everyday sociality as people 'rub along' across class and cultural differences (Watson, 2006).

In this section I discuss the performance of place through the multiple overlapping trajectories and rhythms of people and things moving in the market. Lefebvre advocated rhythmanalysis primarily as an immersive phenomenological practice: 'The rhythmanalyst calls on all his [*sic*] senses. […] He thinks with his body, not in the abstract, but in lived temporality' (2004: 21). Indeed, 'to grasp a rhythm, it is necessary to have been ***grasped*** by it; one must let oneself go, give oneself over, abandon oneself to its duration' (27, emphasis in original). I started my own immersion in the market by wandering around trying to take it in. My movements were staccato at first as I quite literally did not know where to place myself. There is no discrete position from which to stand and observe as the market is constantly on the move (the cafes do offer some respite – as does the first floor gallery). I was out of step and out of place, lacking the 'spatial confidence' (Nash, 2020: 10) of the insider.

This experience of 'arrhythmia' made me aware of the fluidity of the rhythms of the key figures in the market – the fish merchants, sales people and regular buyers who know how to move and 'bend' to its activity at an embodied level (Lefebvre, 2004: 39–40). Lefebvre calls this *dressage* to refer to the process of bodily entrainment and repetition through which rhythm is learnt and makes itself felt – and amenable to observation – in the body over time. I started to learn by matching my pace to the fish inspectors and porters I shadowed. The porters are busy from the moment the fish can leave the market. They must decipher items on the order sheet (hand-written by the fish merchant), identify them in the chiller, load their trolleys and take them to a waiting van. Speed matters as the buyers usually have somewhere else to be. At the other end of the spectrum, the fish inspectors operate at a different pace making regular visits to the market floor during the course of the night. They walk slowly and deliberately, a white hat bobbing in the crowd, stopping to look at produce they suspect is 'unfit for the food of man's body' (according to the Fishmongers' Company Royal Charter, 1604), ready to 'condemn' it with their regulatory gaze. The mood is intense, and people still and quieten at these moments. When the inspector moves on, there is often a burst of chatter and a sense of relief expressed in greetings, complaints, jokes and banter (cf. Porcu, 2005).

Louise Nash's (2020) analysis of organisation and place-making in the City of London shows how the performances and rhythms of City workers effectively 'bring into being' the lived space of the City itself, as they do at Billingsgate. The space of the City exceeds the offices and formal meeting places which mark the territory. People interact and engage directly in the street, making plans and doing deals in person and through mobile connections. It's a fast-paced world and in her walking interviews Nash takes on the (mostly frenetic) rhythms which characterise different spatial and temporal zones, remembering her own experiences of the relentless rhythms of City work and noticing 'the way in which rhythms shape how place is performed' (2020: 1).

Back at Billingsgate, the fish merchants and salespeople call attention to their work in different ways. Their range of movement is more restricted as they display and maintain the fish on sale and show it off to customers. If their patch is smaller at the peak of the market, earlier in the night, they tend to walk around the site to check one another's stock and prices. Buyers, on the other hand, continue to be mobile, forming, renewing or eschewing attachments and making judgements all the while stimulating interactions. In the meantime, the fish wait – and work their own magic as 'thickening agents' (De Backer and Pavoni, 2018: 14) of trade. Their colour, shape and texture draw buyers in and, combined with the right sales pitch and price, make for a fast exit from the market. That said, different fish have different rhythms. Sometimes speed is key and 'you've just got to get the fish into the system', explains Jim Dillon, a long-established salesman. However, the process can be slowed 'if the fish is very good to begin with' as it may have 'another three, four, five days' life in it'. Each species has its own rhythms of decay which the fish merchant attempts to slow down through careful attention and an icy environment. Indeed, the fish themselves are enlivened by the performance of the work of seller and buyer as well as the inspector's gaze which keeps them all alert while the market is happening. Then the lids are closed on the remaining boxes and they are returned to the thin air of the chiller.

Casey (following Heidegger) discusses 'the micro-practices that tie the geographical subject to his or her place-world' and which are 'continually *put into action*' such that 'place and self are intimately interlocked in the world of concrete work' (2001: 684–687). This section has shown how the presence, movement and rhythms of bodies, materials and fish generate the thickening or thinning of the space of the market. Places do not have enduring 'single, essential identities' (Massey, 1994) but there is a phase in the night when the market comes together as a polyrhythmic and eurhythmic whole. This collective choreography arises from these different but interconnected work practices performed by bodies in place.

Conclusions

The appeal of Casey's characterisations of place as thick or thin lies in their affective, social and material resonances. They make sense in terms of the felt sense of place and offer a vocabulary to make distinctions between different places – or within different places at different times (despite the dangers of the binary opposition on which they rest). Place encompasses affects and practices that 'thicken' or 'thin' the embodied experience and meaning of being-in-place. However, Casey does not provide a sense of how thick or thin places might be identified, or indeed how they emerge and fade (Duff, 2010). In the example discussed in this chapter, my aim has been to show how, drawing on Lefebvre's (2004) rhythmanalysis, the thickening/thinning of place can be recognised through attention to rhythm; and accounts of thickened or thinned space effectively capture what rhythms do. Indeed, rhythmanalysis offers an opportunity to trace 'the relationship between rhythms and the performances of place' (Nash, 2020: 1). I have sought to show what rhythm does in place – at Billingsgate – and what kinds of places arise from different constellations of rhythm, as they make themselves felt through memory and narration, and mobility and interaction.

References

Augé, M (1995) *Non-places: Introduction to an Anthropology of Supermodernity. London*: Verso.

Best, K and J Hindmarsh (2019) 'Embodied spatial practices and everyday organization: The work of tour guides and their audiences', *Human Relations*, 72(2) 248–271.

Bohme, G (1993) 'Atmosphere as the fundamental concept of a new aesthetics', *Thesis Eleven*, 36 113–126.

Borch, C, K Bondo Hansen and AC Lange (2015) 'Markets, bodies, and rhythms: A rhythmanalysis of financial markets from open-outcry trading to high-frequency trading', *Environment and Planning D*, 33(6) 1080–1097.

Casey, E (2001) 'Between geography and philosophy: What does it mean to be in the place-world?' *Annals of the Association of American Geographers*, 91(4) 683–693.

Casey, E (2009a, second edition) *Getting Back into Place: Toward a Renewed Understanding of the Place-World*. Bloomington: Indiana University Press.

Casey, E (2009b, second edition) *Remembering: A Phenomenological Study*. Bloomington: Indiana University Press.

Chari, S and V Gidwani (2018) 'Introduction: Grounds for a spatial ethnography of labor', *Ethnography*, 6 (3) 267–281.

Creswell, T (2004) *Place: An Introduction*. Oxford: Wiley-Blackwell.

Dale, K, S F Kingma and V Wasserman (eds) (2018) *Organisational Space and Beyond: The Significance of Henri Lefebvre for Organisation Studies*. London: Routledge.

De Backer, M and A Pavoni (2018) 'Through thick and thin: Young people's affective geographies in Brussels', *Emotion, Space and Society*, 27 9–15.

Duff, C (2010) 'On the role of affect and practice in the production of place', *Environment and Planning D: Society and Space*, 28 881–895.

Edensor, T (ed) (2010a) *Geographies of Rhythm: Nature, Place, Mobilities and Bodies*. Farnham: Ashgate.

Edensor, T (2010b) 'Walking in rhythms: Place, regulation, style and the flow of experience', *Visual Studies*, 25(1) 69–79.

Edensor, T (2011) 'Commuter: Mobility, rhythm and commuting', in T Cresswell and P Merriman (eds) *Geographies of Mobilities: Practices, Spaces, Subjects*. Farnham: Ashgate, pp. 189–203.

Edensor, T (2012) 'Illuminated atmospheres: Anticipating and reproducing the flow od affective experience in Blackpool', *Environment and Planning D*, 30 1103–1122.

Gherardi, S (2019) 'Theorizing affective ethnography for organization studies', *Organization*, 26(6) 741–760.

Halford, S (2004) 'Towards a sociology of organizational space', *Sociological Research Online*, 9(1) 13–28.

Halford, S (2008) 'Sociologies of Space, work and organisation: From fragments to spatial theory', *Sociology Compass*, 2(3) 925–943.

Harvey, M, S Quilley and H Benyon (2002) *Exploring the Tomato, Transformations of Nature, Society and Economy*. Cheltenham: Edward Elgar.

Herod, A. *et al* (2007) 'Working space: Why incorporating the geographical is central to theorizing work and employment practices', *Work, Employment and Society*, 21(2) 247–264.

Highmore, B (2002) '*Street Life in London*: Towards a rhythmanalysis of London in the late nineteenth century', *New Formations*, 47 171–193.

Lefebvre, H (1991) *The Production of Space*. Oxford: Wiley-Blackwell.

Lefebvre, H (2004) *Rhythmanalysis: Space, Time and Everyday Life*. London: Continuum International Publishing Group Ltd.

Lyon, D (2016) 'Doing audio-visual montage to explore time and space: The everyday rhythms of Billingsgate Fish Market', *Sociological Research Online*, 21(3) 12. www.socresonline.org.uk/21/3/12.html.

Lyon, D (2018) *What is Rhythmanalysis?* London: Bloomsbury.

Massey, D (1994) *Space, Place and Gender*. Minneapolis: University of Minnesota Press.

Nash, L (2020) 'Performing place: A rhythmanalysis of the city of London', *Organization Studies*, 41(3) 301–321.

Orr, JE (1996) *Talking about Machines: An Ethnography of a Modern Job*. Ithaca and New York: Cornell University Press.

Porcu, L (2005) 'Fishy business: Humour in a Sardinian fish market', *Humour*, 18(1) 69–102.

Potts, T (2010) 'Life hacking and everyday rhythm', In Edensor, T (ed) *Geographies of Rhythm: Nature, Place, Mobilities and Bodies*. Farnham: Ashgate, pp. 33–44.

Riach, K and T Warren (2015) 'Smell organization: Bodies and corporeal porosity in office work', *Human Relations*, 68(5) 789–809.
Roy, DF (1959) '"Banana Time", Job satisfaction and informal interaction', *Human Organization*, 18(4) 158–168.
Sharma, S (2014) *In the Meantime, Temporality and Cultural Politics*. Durham and London: Duke University Press.
Simpson, P (2012) 'Apprehending everyday rhythms: Rhythmanalysis, time-lapse photography, and the space-time of everyday street performance', *Cultural Geographies*, 19(4) 423–445.
Snyder, B (2016) *The Disrupted Workplace, Time and the Moral Order of Flexible Capitalism*. Oxford: Oxford University Press.
Stratford, E (2015) *Geographies, Mobilities, and Rhythms over the Life-Course*. Abingdon: Routledge.
Thompson, EP (1967) 'Time, work-discipline, and industrial capitalism', *Past & Present*, 38 56–97.
Tuan, YF (1977) *Space and Place: The Perspective of Experience*. Minneapolis: University of Minnesota Press.
Ward, K. (2007) 'Thinking geographically about work, employment and society', *Work, Employment and Society*, 21(2) 265–276.
Watson, S (2006) *City Publics: The (Dis)enchantments of Urban Encounters*. London: Routledge.
Westwood, S (1984) *All Day, Every Day*. London: Pluto Press.
Wood, AJ, M Graham, V Lehdonvirta and I Hjorth (2019) 'Good gig, bad gig: Autonomy and algorithmic control in the global gig economy', *Work, Employment and Society*, 33(1) 56–75.

50

Alternative economies and places

Bastian Lange

Contextualizing alternative economies and places

Current debates on knowledge-based and creative locational development in the Global North have come to recently deal with small urban places of novelty that formerly remained unnoticed (Gibson-Graham, 2008; Krueger, Schulz, & Gibbs, 2018). Needless to say, bottom-up places such as urban gardening, open worklabs and others have always had a more existential dimension for the people in the Global South. Paradoxically, due to severe crises in Western Europe, these places increased in numbers and gained enormous interest for everyday problem solving. A plethora of new spatial expression recently emerged in unplanned and uncoordinated ways, bearing odd names such as FabLabs, Coworking Spaces, Open Worklabs, RealLabs, Open Design Cities, Techshops, Repair Cafés, and more (Smith et al., 2017). They are pointing toward new modes of producing and working in spatial context and shed light on post-capitalistic modes of degrowth and postgrowth.

Established political bodies have been surprised by the recent emergence of these initiatives; at the same time, standard epistemic tools of the social sciences and economics have been rendered unfit to understand these socio-spatial formations. More concise analytical reconstructions are needed to adequately capture the variety and complexity of these place-specific 'labs', their heterogeneous causation, their contingent proceedings, their surplus of latency, their peculiar power relations and their local embeddings.

Urban social contexts have a strong triggering function as they help to re-configure older, and create new, combinations of heterogeneous social and economic agency. Meanwhile strong elements of grassroots innovation (Smith et al., 2017) have informed the formation of various models of alternative work and production. Taking the format of open worklabs as a revealing example, I will describe the constitutive features of these new types of self-organized collaborative workplaces and the associated practices of social innovation. A fresh gaze on the complexity and open-endedness of socio-material formations may help to better understand the nature of emerging alternative and post-growth economies.

The makers movement as a generator of new places

The recent wave of urban grassroots innovation left even experts of alternative scenes baffled. Virtually out of the blue uncoordinated small-scale experimentation with local work and production proliferated where well-organized SME production, craft and the cultural business formerly dominated urban scenes. Meanwhile a tremendous variety of heterogeneous approaches, work models and types of localization have emerged. There are new names galore, such as makerspaces (Anderson, 2012), fab labs (Dickel, Ferdinand, & Petschow, 2014; Fleischmann, Hielscher, & Merritt, 2016; Hielscher & Smith, 2014; Lange, 2015; Schneidewind & Scheck, 2013), open worklabs (Smith et al., 2016), open garages, real labs (Liedtke et al., 2015), urban laboratories (Evans & Karvonen, 2014), living labs (Liedtke et al., 2015), real laboratories (Karvonen & van Heur, 2014), open worklabs (Lange, 2017), tech shops, repair cafés (Baier et al., 2016) and many more.

Following Anderson (2012) and other scholars, the obvious euphoria of the 'makers turn' is associated with expectations for the freedom to choose the objects, routines and social forms of work. Furthermore, there are claims for self-determined experimentation, learning and producing beyond just 'doing economy' in a different, socially meaningful way.Several scholars have taken a closer look at the maker movement and have thus been criticized due to its affirmative reading of the phenomenon (Lee, 2015), due to its overwhelming local optimism (Catney, 2014), or its positivistic governance interpretation (Liedtke et al., 2015), so it is obvious now that innovation processes have left their branch-specific and exclusive spatial confines and entered different terrains.

In particular, they are driven by new civic collectives (Moulaert et al., 2013; Smith et al., 2017). While innovation had formerly often been restricted to narrower techno-economic fields such as SME networks or industrial clusters, it is currently involved in varied social and collective practices which partly involve economic activities and spaces and partly, however, exclude them. Such 'open innovation' has not only come to permeate alternative or green economies; it also points at a needs-based and user-centred social logic.Makers labs and open worklabs spontaneously get involved in resource-saving, environmentally friendly, socially balanced work by establishing the principles of open access, mutual learning, knowledge sharing and do-it-yourself. My central research interest is to discuss if I am facing a radically new phenomenon which emphasizes social justice while establishing its socio-economic practices. Moreover, I seek to clarify the question of whether it has the potential to substantially transform the organization of urban work and radiate into broader areas of production.

Focussing on the case of open worklabs, I take up this fundamental question and seek to carve out the particular features that set apart these alternative types of labour. I assume that open worklabs initially create a marginal phenomenon in the sense that self-organized innovative work can neither be assigned to an economy organized in sectors and branches nor to social practices which lack any reference to the economy. It might not simply add another 'industrial revolution' (Anderson, 2012) to a series of preceding technological and economic shifts, as well as it might not offer reliable bridges between social needs and economic activities. Initial observations indicate that it blurs the traditional dividing lines between producers, consumers, and groups which have been excluded from formalized work, production, and consumption. It clandestinely reassigns the tools of production to those who are engaged in basic social practices, and in so doing it points more to the presumed beginnings of post-capitalism (Mason, 2016) and post-growth satisfaction of basic needs (Paech, 2015) than to sophisticated, 'creative' or innovative urban capitalism.

While the possible outcomes of such experiments, involving partial shifts in the mode of production, are still hardly tangible, the social movement of the so-called makers has also more practical, no less important, political implications for the local level. It touches upon urging questions of urban justice and civic participation in the aftermath of neoliberal capitalism. Small actors in their everyday lives are obviously leaving behind the limits of functional systems, branches, neighbourhoods or whatever political concepts had been invented to frame social and economic practices. Their search for a new meaning of 'good' work (not necessarily a purely social, economic or cultural one) has an emancipatory potential similar to those of social movements during their initial stages of evolution (cf. Smith et al., 2017).

But since the scope of action is limited to small worklabs, educational projects, sharing initiatives, etc., the scope and the types of emancipation are indeterminate, and it is unclear which approach will gain political momentum. It is also unclear which kind of local environment will be created by their activities. I assume that the new search for meaning, and the open innovation it entails, draws upon '*Resonanz*' (German for 'resonance') (Rosa, 2010). According to Rosa (Ibid.), resonance is a capacity arising from the endeavour of social actors to acquire what surrounds them, thereby changing the meaning and the socio-spatial positions of the elements they address. Movements of various elements within a field equipped with resonance may get temporarily 'frozen' in the shape of socio-material formation which lend themselves to empirical analysis. This is the perspective from which I approach alternative economies and their place-based scene practices.

I will have a closer look at the aggregations which are produced within a local framework in open worklabs so that the starting points and potentials of local emancipatory practices become visible. To get hold of the implications of concrete types of work for questions of emancipation, social inclusion, autonomy and community-based self-determination I focus here on very down-to-earth practices such as tinkering, repairing and experimenting with everyday materials. They are best represented in open worklabs. They combine crafts, tool-sharing, open communication and community building into something that has local and also more general effects. I will ask how the users of open worklabs develop their particular attitudes towards work within a favourable context provided by their peers, and how they position themselves within the socio-material context they are part of, as well as how they are positioned by their surroundings. Do users make use of the open workshop in innovative ways so that it becomes a resource of new emancipation? Do they explicitly claim for authenticity and the right to bottom-up practices, knowing that they might defy market constraints and social coercion? Do they make use of particular options for value creation that emerge from the socially induced ways of approaching work and managing the everyday necessities of sharing, repairing and acquiring technological knowledge?

Open worklabs as new place: first approximation and manifestations of alternative economies

Open worklabs are multi-faceted phenomena. Some of them are indebted to micro-economic initiative and SME start-up activities, others to social empowerment and everyday practices (Simons, Petschow, & Peuckert, 2016). Recent attempts to define the term 'open workshop' roughly relate to alternative ways of life and informal modes of production (Petschow, 2016; Rigi, 2012), occasionally addressing them as a home for tinkerers or a pastime for the like-minded. Being a collective term for various open projects and initiatives, the only consistent definition so far has been delivered by the Verbund Offener Werkstätten (VOW) (German for 'Association of Open Worklabs'):

> Open worklabs are at the disposal of all those who want to be active in self-organized crafts or arts. Frequently, open worklabs emerge out of private initiative, sometimes

> they are part of cultural, citizens' or youth centres, more rarely of companies. While some command experience of several decades, others are still under construction.
>
> *(Verbund_Offener Werkstätten, 2016)*

Accordingly, open worklabs are engaged in the open-ended development of self-organized work. This nevertheless affords particular knowledge, tools, materials, machinery and spaces. Open worklabs are therefore 'places of opportunity for many, not of business for few. They offer the necessary space and a productive infrastructure for self-initiative and independent work' (Verbund_Offener_Werkstätten, 2016). The VOW emphasizes maximum openness (for all and everyone) and collective non-profit attitudes (i.e., no material profit-orientation), which initially leads to limited possibilities for incorporating entrepreneurial initiatives or options for commercial activities.

Characteristically, the concept of open worklabs first appeared in experts' and practitioners' discourses before it became a point of academic interest. Basically it refers to practical forms of collective work, repair, testing and producing, practiced at a concrete physical-material location (Simons, Petschow, & Peuckert, 2016). The objects processed vary between low-tech and high-tech apparatus, ranging from bikes and trousers to toasters and computers. The practical topics and activities of open worklabs span from traditional craftsmanship to technologically advanced, innovative fabrication techniques such as 3D printing. From a comparative perspective, the different manifestations of open worklabs seem to point more to differences than similarities. In addition to various technological specializations (3D printing, laser cutting, wood processing, metal processing, repair of bicycles, production of clothing, etc.), there are important differences in 'business' models and organizational forms. The spectrum comprises informally organized neighbourhood groups, non-profit organizations and commercial companies. Other important differences exist concerning the individual objectives pursued by its protagonists.

Relational articulations of practices that form and structure alternative places

Analog repair – flexible prototyping

A key component of these alternative places in the Global North is the increased interest in repairing everyday goods. It is an essential feature of open worklabs. At a given time, participants can bring malfunctioning everyday technical objects (hair dryers, remote controls, coffee machines, bicycles, video recorders, etc.) to an open worklab for fixing, as far as this is possible without the help of professionals. In spontaneous groups, exemplary solutions are developed for technical problems that are specific of the individual objects. Over time the growing trial-and-error competencies, which are necessary for these tasks, may assume a semi-professional quality:

> Our network and its worklabs, working places and offices together create spaces for physically realizing ideas. Professionalism is important, people can learn that with us.
>
> *(Anonymous respondent of a worklab)*

Users are gradually enabled to take over the roles of the seeker and finder on their own responsibility. They are obviously driven by the curiosity to 'chop' or 'hack' (i.e., open up) the switching and control plans of electrical and electronic everyday objects, which can be fairly complex.

The practice of disclosing the interior of technical objects, and the subsequent process of reassembling, also encourage protagonists to become creative, for example in the area of rapid prototyping. This means that they not only fix non-functioning objects but, in some cases, also develop similar objects anew. For example, two-dimensional templates are used to produce full-plastic device parts using 3D printers. In this case the technical procedures require special social practices. Just as the building plans are technically translated (2D to 3D), the processual sequences have to be clarified through social interaction and intense communication. It is only under this condition that necessary technical understanding can be successfully transferred from one to another participant. This transfer does not follow professional conventions but rather happens by free interaction within a synchronized social framing. The corresponding relational arrangement of technical-material and social elements of the concrete assemblage forms the basis for the unfolding of hoped-for effects.

Many of them emerge on the fly, without even having been intended, e.g., in the course of context-oriented, spontaneous problem solving. Yet there is one important restriction: alternative arrangements (e.g., without the inclusion of 3D computer technology) are not conceivable for the participants since their development would leave the core area of the open workshop. On the whole, users reduce the incapacitation generated by the inaccessibility of technology, and they crisscross the role assignment of passive consumers which has been made by the capitalist economy. The emancipatory potential involved is twofold: i) the actors here serve as local pathfinders who inspire their peers and neighbours to take initiative on their own account; ii) they explore individual and collective ways of acquiring ownership of working-producing-repairing that might show the way to post-growth practices.

Amateurs (socially motivated) vs. professionals (economically motivated)

Each worklab hosts evolving individual competences. They develop according to two contradictory logics which have a decisive influence on the workshop's stability and dynamics of development. On the one hand, there is a basic attitude which can be characterized as amateurish. The actors prefer experiments and spontaneous exploration but are at the same time competent in the everyday handling of things. On the other hand, they gradually develop biographical, social and sometimes also commercial commitments. Together they test a wide variety of topics and working forms. They act as communicating peers who organize open thematic alliances to achieve small self-imposed goals. These alliances give them a special social impact and credibility, in spite of their status as laymen and amateurs. Often their social image does not range below that of professionals.

Taking a closer look at the development dynamics of open worklabs it becomes clear that the core organizers assume a guiding role, mainly in the formation of thematic-technological interfaces to professional freelancer and SME networks. Under the curation by their organizers the worklabs act as a kind of social laboratory, i.e., as a place where technical resources are communitized and explored. Former amateur competencies thus turn into semi-professional skills and learning routines. The resulting programmatic design does not rely on the exclusivity of knowledge and the regulated access to private spaces, as is the case in many industrial enterprises and R&D labs. Rather, an open competence-based approach is being pursued from the outset: learning offers, courses, coaching and tutoring potentially prepare those for money making and income generation who had formerly been mainly socially motivated, thereby creating an integrated approach. As one of the interviewees put it: '[w]hile founding the repair café it occurred that many craftsmen, technicians and engineers got in touch so that we jointly decided to pass on our knowledge and our experience especially to youths and migrants' (Anonymous respondent of a worklab).

In addition to direct communication, the social reaches of open worklabs are also indirectly expanded, mainly through increased reputation building. In the course of time, the professionalization of the actors and the growing diversification of activities are often carried out with the aim of providing a high standard of reliable service provision for a local demand.

The formation in question thus turns from freely shifting encounters between social and material elements to more systematically arranged structures and purposeful action. This clearly reveals the permutation of amateur practice into professional practice, including the occasional chance to generate income. This process is accompanied by the development of situated configurations of value creation. The option to switch to a 'real' economic area, including the docking to professional and SME networks, varies with the competencies of the stakeholders and the institutionalization of these options (as realized, for example, with the help of initiators).

In this respect, the places perform a constant swing between two antagonistic drivers, i.e., 'autonomy of the amateurs' and 'professional search for returns'. It creates a slightly subversive momentum that undermines market relations and a fixed system of production and consumption. Paradoxically, at the same time these formations drift towards stabilization and adaptability to this system. It is this 'constructive' approach to alternative practices, and, as a consequence, local participation, that characterizes open worklabs.

This approach involves several internal tensions. One of them is expressed through the so-called 'paradox of professionalization' (Lange, 2014: 190): this second-order concept helps to understand that in the first-order concepts there is a contradiction between an actor's commitment to a community motivated by social issues and the economization of formerly purpose-free activities. Such economic reframing does not only affect the infrastructures of open worklabs, which on principle are thematically open, but also socially more distant areas of action. Some of the business-oriented worklabs (e.g., semi-commercial fablabs) resolve this paradox by more closely approaching private-sector networks, professionalizing the competencies established by their community and focussing their activities more strongly on business.

Another important tension is constituted by two competing understandings of community. On the one hand, there is an understanding oriented towards peer-groups. It involves post-growth values (e.g., the priority of social needs). On the other hand, a network-related understanding has been established which entails a sense of community created by soft ties between workers who pursue a thematic or even economic interest. Both formations, i.e., the one which results from 'social commitment/autonomy' and the one characterized by 'economization/professionalization', hardly interact. They have a life of their own. Although they seem to emerge according to distinct rules their evolution is fundamentally unpredictable, provided that they are not affected by interventions by core initiators and other powerful agents. Only through such interventions are the number of potentially existing options restricted – or, the variability of potential 'push and pull' between heterogeneous elements is reduced. Emancipation thus is limited by the degree of community-specific search for adaptability to a given economic system or a 'business environment'.

Flexible value creation

As social formations, the individual open worklabs bind together alternative ideas concerning products and the processing of materials, as well as related action patterns, experimental practices, search routines and the sharing of knowledge, often in a very unpredictable manner. This has immediate consequences for the economic side of these socio-material arrangements.Particularly the occasions and places of value creation are often not determined from the outset, nor are the characteristics of the later products and services. They are gradually developed and defined by contingent learning processes and the trial and error activities of interacting peers. This practice is similar to the emergence of configurations of flexible value creation as can be observed in arts and the creative industries, but it differs in the degree of contingency which governs open worklabs.

While the creative industries mostly operate within pre-fixed organizational schemes or pre-set networks of producers, open worklabs happen to explore value creation by chance. The step into economically relevant activities is often carried out almost casually, as a by-product of hedonistic or interactively initiated attitudes. 'Knowledge, education and fun for all!' – this is the corresponding basic orientation. This, however, does not manifest itself as a universal 'manual' for all and everything is transformed into a variety of different intentions and micro-strategies.Value creation is therefore always guided by a social logic that oscillates between purposeful experimentation, already preformulated social objectives (e.g., the realization of self-determined, exploitation-free labour), the chance of realizing economic returns and the necessity to generate income. Concrete definitions of places and objects of value creation, e.g., products and services developed on the fly, are found according to the shifting shapes and circumstances of the particular assemblage.

Conclusion

This chapter sought to conceptualize open worklabs as places of alternative, self-organized work in the Global North. They have been described as formations, namely as the result of random coincidences and the temporary sticking together of heterogeneous social and material elements. Within this conceptual framing, the users, organizers, tools, technical equipment, social practices and spaces appeared as being arranged according to the principle of relational exteriority, i.e., by the open interaction of these elements without a previously established, system-immanent logic.

The key results reveal that open worklabs are driven by curiosity, self-organized learning, experiments and trial-and-error activities, based on a collective focus on social needs. What might initially appear as a spontaneous and practical staging of the dichotomy between 'the social' and 'the economic' has, nevertheless, wider implications for society as a whole as well as for local environments.

The concepts I identified indicate a tentative return to the very basics of societal production, including a playful reappropriation of the means of production. This might evoke connotations of Marxian aspirations to reduce the alienation of workers and eliminate the extraction of value added by capitalists. But there is no clear political ideology involved in the evolution of open worklabs, and it would certainly be a gross overinterpretation of 'normalized' everyday formation to say that all these practically minded actors have been on their way to realize the communist dream.

Nevertheless, the will to push back economic constraints in favour of 'doing something different' cannot be overlooked. The marked turning away from profit-seeking links open worklabs to the current post-growth debate. Without necessarily being fully aware of it, the protagonists put into practice post-growth principles, such as open access to resources and tools, sustainable usage of resources, needs-based production, or the loading of work with new social meaning (Paech, 2015).

This is a different approach from what theorists of post-capitalism have imagined as an advancement of the current mode of production. Authors such as Mason (2016) conceived the way out of exploitative capitalism as an incremental permutation of production and work by means of digitization, the internet and advanced communication technologies. Crowd funding, crowd sourcing and the distribution of work through virtual networks – such attributes are only marginally represented in open worklabs.

If at all, they appear in technologically advanced work arrangements like 3D printing. Moreover, what sets open worklabs even more apart from the technological bias of post-capitalism is their strong local attachment. The formations I identified have been working on the basis of the local co-presence of actors and the direct handling of materials 'on the ground'. If they ever point at future system change, they probably do so on the basis of an approach which radically differs from technologically informed revolutions. Rather, this approach valuates small everyday activities, self-empowerment and emancipation from below. It does so by making salient the principle of all-embracing openness which can be directly experienced by everyone. In its different manifestations this principle is recognizable as the basic feature of spatial formations.

The clearly anti-commercial, often even anti-monetary, approach of open worklabs often combines with the incalculable changeability of these formations. The habit of doing something different all the time and avoiding money-making turns them into unreliable partners of some projects of political reform. Recent political attempts to embrace open worklabs as experimental harbingers of future economies (e.g., as conceived by the German Ministry of Education and Research, BMBF) have largely misunderstood the anti-economic core of open worklabs and maker spaces. For the moment it appears obsolete to try and integrate these experimental models into mainstream economies or apply a neoliberal political approach which seeks to make them productive for urban development.

What does this mean for emancipatory and participatory urban policies? First of all, there is more detailed knowledge required to fully realize the significance of open worklabs: they can be understood as cross-sectional, heterotopic phenomena by nature and thus overcome a clear and undisputed rule for the traditional economy. It is indispensable to gain more in-depth insights into their innovative workings and discuss the emancipatory potentials they generate. This means that the study of open worklabs, labs, etc., must refrain from a priori classification and categorization.

My description of the socio-material constellations that emerge as distinct formation may be a first step in this direction. However, after having reconstructed the inductive logic of first-order concepts, another analytical step must follow which makes formations intelligible within a wider discursive context. Second-order concepts, preferably those based on common understandings of alternative ways of life or post-growth grassroots, must be consulted to clarify the social and economic significance of open worklabs.

My findings let me conclude that studies supporting this objective must emphasize the plurality, heterogeneity, and diversity of the phenomenon. This is a necessary, but not sufficient, starting point. Further theoretical descriptions are necessary which deepen our understanding of work-related 'grassroots innovation movements' (Smith et al., 2017).For example, if one looks at 'open worklabs as assemblage', it becomes obvious that the usual analytical differentiation into micro- and macro-phenomena is obsolete. 'Micro' and 'macro' have already been inscribed into the genesis of the assemblages from the very beginning, albeit without any fixed order. They are subject to flexible scaling and rapid changes in their social, economic and political meanings. Unintended intertwining, rather than purposeful interrelation, is their effect. For instance, the digitization and virtualization of individual communication adds a global dimension to the small activities of local protagonists. Conversely, globally negotiated beliefs, political attitudes and social interests are drawn into the smallest micro-social ramifications of community building and the social valuation of work.

Open worklabs must therefore be conceptualized as emergent formations which are fundamentally variable in their material composition, character, coherence and social significance. Only on the basis of their shifting relationality can these formations be plausibly explained. For possible future strands of theoretical and empirical reconstruction this means that tentative concepts must be continually revised, together with the short-term changes in relations that characterize assemblages.

Open worklabs thus present themselves as elusive places of social phenomena. Their detachment from economic rationality and their latent opposition to neoliberal political ideology make it difficult to calculate the possible effects that they might have on the orientation of local environments. At present it might be helpful to carefully observe the urban contexts in which open worklabs are positioned. These contexts cannot be understood as categorically separated phenomena to which the worklabs would have to communicate, for example, in the form of 'external relations'.

On the contrary, according to their inherent logic of development, open worklabs are provisional socio-spatial arrangements, which are absorbed into the urban field as a constitutive element of its emergence. Specifically, actors integrate their work-related approach into their everyday lives, be it the intention to understand the blueprints of things or becoming self-reliant through repairing their belongings. They incorporate their vernacular knowledge, their 'private' ideation, their habitus as urbanites into the supposedly closed small world of open worklabs. The social and economic significance of such emancipatory habit-in-development still awaits further exploration.

References

Anderson, C. (2012). *Makers: The New Industrial Revolution*. New York: Crown Business.

Baier, A., Hansing, T., Müller, C., & Werner, K. (2016). *Die Welt reparieren - Open Source und Selbermachen als postkapitalistische Praxis*. Bielefeld: Transcript.

Catney, P. (2014). Big society, little justice? Community renewable energy and the politics of localism. *Local Environment, 19*, 715–730.

Dickel, S., Ferdinand, J.-P., & Petschow, U. (2014). Shared machine shops as real-life La-boratories. *Journal of Peer Production, 5* http://peerproduction.net/issues/issue-5-shared-machine-shops/peer-reviewed-articles/shared-machine-shops-as-real-life-laboratories/ (accessed 13 June, 2016).

Evans, J., & Karvonen, A. (2014). Give me a laboratory and I will lower your carbon footprint!' — Urban laboratories and the governance of low-carbon futures. *International Journal of Urban and Regional Research, 38*(2), 413–430. 10.1111/1468-2427.12077.

Fleischmann, K., Hielscher, S., & Merritt, T. (2016). Making things in fab labs: a case study on sustainability and co-creation. *Digital Creativity, 27*(2), 113–131.

Gibson-Graham, J. K. (2008). Diverse economies: performative practices for 'other worlds'. *Progress in Human Geography, 32*(5), 613–632.

Hielscher, S., & Smith, A. (2014). Community-based digital fabrication workshops: a review of the research literature. *SWPS 2014-08. Available at SSRN: https://ssrn.com/abstract=2742121 or http://dx.doi.org/10.2139/ssrn.2742121 May 21*(2014).

Karvonen, A., & van Heur, B. (2014). Urban laboratories: experiments in reworking cities. *International Journal of Urban and Regional Research, 38*(2), 379–392.

Krueger, R., Schulz, C., & Gibbs, D. (2018). Institutionalizing alternative economic spaces? An interpretivist perspective on diverse economies. *Progress in Human Geography, 42*(4), 569–589.

Lange, B. (2014). Entrepreneurship in creative industries: the paradox between individual professionalization and dependence from social contexts and professional scenes. In R. Sternberg &

G. Krauss (Eds.), *Handbook on Research on Creativity and Entrepreneurship* (pp. 177–208). Abingdon: Routledge.

Lange, B. (2015). FabLabs und Hackerspaces. *Ökologisches Wirtschaften, 30*(1), 8–9.

Lange, B. (2017). Offene Werkstätten und Postwachstumsökonomien: kollaborative Orte als Wegbereiter transformativer Wirtschaftsentwicklungen? *Zeitschrift für Wirtschaftsgeographie, 61*(1), 38–55.

Lee, M. (2015). The promise of the maker movement for education. *Journal of Pre-College Engineering Education Research* (*J-PEER*), *5*(1), Article 4, https://doi.org/10.7771/2157-9288.1099.

Liedtke, C., Baedeker, C., Hasselkuß, M., Rohn, H., & Grinewitschus, V. (2015). User-integrated innovation in Sustainable LivingLabs: an experimental infrastructure for researching and developing sustainable product service systems. *Journal of Cleaner Production, 97*, 106–116. 10.1016/j.jclepro.2014.04.070.

Mason, P. (2016). *Postkapitalismus. Grundrisse einer kommenden Ökonomie. Unter Mitarbeit von Stephan Gebauer.* Berlin: Suhrkamp.

Moulaert, F., MacCallum, D., Mehmood, A., & Hamdouch, A. (2013). *The International Handbook on Social Innovation. Collective Action, Social Learning and Transdisciplinary Research.* Cheltenham: Edward Elgar.

Paech, N. (2015). Die Sharing Economy – ein Konzept zur Überwindung von Wachstumsgrenzen? *Wirtschaftsdienst, 95*(2), 101–105.

Petschow, U. (2016). How Decentralized Technologies Can Enable Commons-Based and Sustainable Futures for Value Creation. In J.-P. Ferdinand, U. Petschow, & S. Dickel (Eds.), *The Decentralized and Networked Future of Value Creation* (pp. 237–255). Wiesbade: Springer.

Rigi, J. (2012). Peer production as an alternative to capitalism: a new communist Horizon. *Journal of Peer Production, 1*(1), http://publications.ceu.hu/node/33879 download am 33813.33807.32016.

Rosa, H. (2010). *Alienation and Acceleration. Towards a Critical Theory of Late-Modern Temporality.* Malmö/Arhus: NSU Press.

Schneidewind, U., & Scheck, H. (2013). Die Stadt als „Reallabor"für Systeminnovationen. In J. Rückert-John (Ed.), *Soziale Innovation und Nachhaltigkeit. Perspektiven sozialen Wandels* (pp. 229–248). Wiesbaden: VS-Verlag für Sozialwissenschaften.

Simons, A., Petschow, U., & Peuckert, J. (2016). *Offene Werkstätten – nachhaltig innovativ? Potenziale gemeinsamen Arbeitens und Produzierens in der gesellschaftlichen Transformation.* Berlin (IÖW): Schriftenreihe des IÖW. 212/16.

Smith, A., Fressoli, M., Abrol, D., Arond, E., & Ely, A. (2017). *Grassroots Innovation Movements.* London: Routledge.

Smith, A., Hargreaves, T., Hielscher, S., Martiskainen, M., & Seyfang, G. (2016). Making the most of community energies: three perspectives on grassroots innovation. *Environment and Planning A, 48*(2), 407–432.

Verbund_Offener_Werkstätten. (2016). Was sind Offene Werkstätten?.

51
Consuming places

Mark Jayne

Introduction

Consumption has been at the vanguard of advances in research across the social sciences over the past 30 years (see Jayne 2005; Jayne and Ward 2016). Study of consumption has been important in advancing our understanding of global interdependence, drawing connections between politics and governance, economic restructuring and changes in employment; hand-in-hand with consideration of our everyday social and cultural lives – where and how we spend our leisure time; where and what we eat and drink; where and when we do our shopping, what we wear, how we decorate our homes, etc. In these terms consumption has been theorized as standing at the intersection of different realms; the public and private, the political and personal, the social and the individual. Indeed, as Miles (1998) suggests, consumption can be understood as a means and motor of political, economic and social change; playing a vital role in contracting our identities and lifestyles; and as an active constituent in the construction and experience of place.

Despite the centrality of consumption to advances in understanding of historic and contemporary worlds, in this chapter I explore how sustained attention to 'placing' consumption can add value to theoretical and empirical scholarship. To that end, subsequent sections begin with a review of strengths and weaknesses of 'archetypal' depictions of the ways in which consumption is moulded by place, and how places are moulded. I then highlight recent arguments regarding a need to move beyond theorization of place and consumption dominated by studies from a handful of large metropolises in Europe and North America, and in doing so point to theoretical and empirical opportunities offered by the emergence of thinking on assemblages, materialities, mobilities, bodies, emotions and affect, etc. I elaborate this argument through a focus on consumption, identity and fashion, specifically 'placing' consumption through considering the wearing of 'comfy' clothes in public space. In doing so I respond to Doreen Massey's (1991: 29) contention that:

> instead of thinking of places as areas with boundaries around, they can be imagined as articulated moments in networks of social relation [and as such we should focus on] ... a sense of place, an understanding of 'its character', which can only be constructed by linking that place to places beyond.

Consuming archetypes

In most places in the world, human beings are aware that they are consumers – that our lives, the places we live, the places we go, the things we do are differentially and discursively structured around consumption. Alongside such commonsense understanding, significant academic attention has focussed on defining consumption. For example, consumption has been described as being about the 'selection, purchase, use, reuse and disposal of goods and services' (Campbell 1995: 104); as 'comprising a set of practices which permit people to express self-identity, to mark attachment to social groups, to accumulate resources, to exhibit social distinctions, to ensure participation in social activities' (Warde 1997: 304); and also as constituting the ways we construct, interpret and experience place (Urry 1995). Other theorists have also shown that consumption is not just about goods that are manufactured, sold, (re)used and so on, but that it increasingly relates to ideas, services and knowledges and, moreover, that places, sights, smells, sounds and atmospheres can all be consumed (see Miles 1998; Jayne 2005). Such work has been vital in exposing ideological dimensions bound up in the growth of consumer society, and has enabled fascinating theoretical insights and rich and detailed empirical research into how the 'ubiquitous nature of consumption is reconstructed on a day-to-day basis' – and how consumption has become 'a way of life' (Miles 1998: 4).

Indeed, one of the most significant contributions that studies of consumption have made to understanding of both historic and contemporary worlds has been to focus on urban place-based identities, lifestyles and forms of sociability. Theorists have famously reflected on the development of the modern city (1880–1930) by investigating the proliferation of commercial culture that shaped public life and which served the consumption identities and practices of the urban bourgeoisie in Europe and North America. This work shows us that consumer culture was not simply a product of industrialization and intellectual successes of modern thought, but rather 'the consumer' and experiences of consumption were integral to physical and infrastructural developments, political and economic structural and institutional landscapes, and everyday sociability and social relations. The remarkable changes that took place in cities from the end of the eighteenth century were mapped through the emergence of key topographical landmarks such as shopping arcades and department stores, theatres, boulevards, cafes, public squares and plazas, parks and so on (Zukin 1989).

For example, Walter Benjamin (1982) famously described the emergence of shopping arcades in 1920s' and 1930s' Paris as 'magical' new worlds enabled through architectural design and materials that included large plate glass windows and steel structures. Benjamin argues that shopping arcades created 'artificial behaviour' in a 'world in miniature'. Music shops, wine merchants, hosiers, haberdashers, tailors, boot makers, bookshops and restaurants were part of a 'dream world', a utopian and imagined place, constructed as better than the real world – an ideological celebration of capitalism. In a similar vein, Sennett et al. (2002) highlight how department stores grew from industrialization and increasing international trading led to the availability of goods from around the world, with inventive displays generating all manner of new pleasures, sights, sounds, smells and so on. Frisby (2001) describes the dramatic expansion of metropolitan life in Berlin in the 1930s, noting the importance of cafes and street-side consumption alongside the 'mass ornamentation' of public and commercial buildings through the abundant use of marble, carpets, furnishings and decoration that inextricably linked femininity to consumption in the imaginations and practices of middle-class consumers. Zukin (1989) reminds us, nonetheless, that from an initial air of exclusivity for upper-class patrons, consumption became accessible to both working- and middle-class consumers. Social mixing took place in shopping districts, bars and pubs, at racetracks, in commercial and public space with liminal, illicit and illegal activities often crossing class, ethnic and sexual lines.

Studies of later 'archetypal' historical periods have also explored consumption, identity and place. In the USA theorists have focussed on spatialities of consumer culture through the emergence of inner city ghettos and suburban shopping malls in the 1950s; 'post-modern' shopping districts and loft living in the 1980s; out-of-town retail parks in the 1990s (Zukin 1989) as well as 'hyper-real' simulacrum of 'Disneyfied' consumption spaces in Las Vegas (Baudrillard 1993). When read together these studies offer groundbreaking insights into the relationships between consumption and place pursued through consideration of complex interactions of political, economic, social, cultural and spatial practices and processes. Indeed, this work importantly highlights how place should not be considered a passive backdrop to consumption, but is an active ingredient in political and economic formations, social relations, subjectivity and social selfhood.

Re-invigorating study of consumption and place

Despite the valuable advances offered by generalizable, 'ideal' or archetypal studies of consumption, place and identity, it has been argued that such an approach nonetheless fails to fully explore the difference that place makes (Crewe and Lowe 1995). This assertion has important implications in two ways: first, pointing to the need to move away from the 'tyranny' of studying single sites such as malls, department stories and urban villages to consider the complex 'tissue of sites' that constitute and connect localities, neighbourhoods, cities, regions, nations and global spatialities. Second, and implicated in such arguments, is a need to better account for political, economic and cultural practices and processes 'embedded' in particular localized social systems (Jackson and Thrift 1995). As such, consumption, lifestyles, identities and social relations must be understood as being differentially and discursively constructed and negotiated in specific ways that give places their 'unique' characteristics through relational connections with places elsewhere. While consumption theorists have arguably been relatively slow to fully embrace and respond to these theoretical and empirical challenges, this chapter now turns to highlight rich and fruitful theoretical and empirical avenues that have responded to the challenges set out by Massey (1991) that opened this chapter, and in doing so point to significant opportunities to reinvigorate study of consumption and place.

First, there is a growing call to re-balance research agendas overly dominated by studies of places in a small handful of large metropolises in Europe and North America (Bell and Jayne 2006, 2009; Jayne et al. 2010; Jayne 2018). Second, numerous theorists have been working to address both territorial and relational comparative approaches to place to focus on networks that connect different places (see Robinson 2006; Ward 2010; Jayne et al. 2011, 2013; Edensor and Jayne 2012). This work has advanced understanding of the overlapping networks that come together to produce distinctive places in order, as Robinson (2006: 544) highlights, to enable creative 'consideration of multiple social networks of varying intensity, associated with many different kinds of economic and social processes, and with different kinds of locales, or places'. Such overlapping networks of relations bring together people, resources and ideas in diverse combinations, making cultural, political, design, planning, informal trading, religious, financial, institutional and intergovernmental connections. As Ward (2010: 476) stresses, interconnected trajectories show how places are dynamic aggregations of social relations and interactions that are often entangled with processes in other places at varying scales whilst recognizing grounded, local dynamics 'implicated in each other's past, present and future which moves us away from searching for similarities and differences between two mutually exclusive contexts and instead towards relational comparisons that uses different cities to pose questions of one another'.

Such insights offer new theoretical inspiration to address consumption and place through consideration of images, fantasies, desires and pleasures, memories, sensations and manifold rhythms that emerge through daily encounters and experiences, in all manner of cultures, strategies and contingencies. Third, recently emerging theoretical and empirical advances in understandings of mobilities (Urry 1995), materialities (Latham and McCormack 2004), rhythm (Edensor 2010), assemblages (McFarlane 2011), emotions and embodiment (Davidson et al. 2005) and affect (Thrift 2004) also offer valuable resources that offer opportunities to re-invigorate understanding of consumption and place. The remainder of this chapter engages with these rich and fruitful overlapping theoretical and empirical avenues by discussing fashion and consumer culture.[1]

Theorising the 'place' of comfort: fashion and consumption

It will be of no surprise to anyone with a popular interest in fashion, nor theorists interested in consumption, that wearing 'comfortable' clothing in public spaces – pyjamas, sportswear and other clothing most often thought of as only being acceptable to wear 'at home' – is a feature of urban life around the world. For example, in Liverpool it is popular for women, in preparing for the weekly 'big night out' on Fridays and Saturdays, to wear comfortable clothes *and* walk around the city with 'curlers in their hair' (see Figure 51.1). In Middlesbrough (BBC 2011) parents who wear pyjamas when taking children to school and who attend school parent/teacher meetings in their nightwear have been criticised, with it being noted that they 'drop them [their children] off in the morning and are collecting them wearing the same pyjamas' suggesting that this practice is creating a 'bad impression'. In Belfast, jobseekers have been banned from wearing pyjamas at the local social security office, and in Cardiff a supermarket has outlawed 'night wear' in its aisles (Irish Central 2012). In Shanghai, China, Iossifova (2012: 202–203) describes how local customs such as 'wearing pyjamas in public, drying blankets on the street and spitting, for instance [are] being portrayed as backward or rural ... by the Government and the media'. In

Figure 51.1 Scouse women 'at home' in Liverpool city centre.

Source: Jenny Poole – www.flickr.com/photos/16873035@N00/4893357883 accessed 16 June 2013.

contrast, the trend of wearing comfy clothing in public spaces is prevalent in the affluent upper east side of New York City with fashion conscious women and teenagers parading their Louis Vuitton and other designer pyjamas on the streets of Manhattan. This fashion is not, however, new, with a headline in the *New York Times* in 1929 highlighting 'Court sanctions pyjamas in the street'. The article discusses how a man was arrested only to be released by a judge who warned the police that 'Neither you nor I are censors of modern fashion' (Manjoo 2012).

Writers such as Veblen (1898), Simmel (1903), Benjamin (1982), Hebdige (1979), Wilson (1985) and Maffesoli (1994), to name but a few, have long articulated the spatial, social and psychological processes of consuming fashion. Recent writing has nonetheless contrasted such 'generalised' or 'archetypal' depictions with focus on fashion consumption in specific places (Entwistle 2000). Studies have considered a diverse array of contexts and topics, for example fashion in junior schools in the UK (Swain 2002), fashion in the late Qing period in Shanghai with reference to sexuality, desire and mixing of 'western' and Chinese styles (Li 1998) and 'postmodern' style in Milan (Bovone 2006).

Seeking to advance this approach here I focus on research that unpacks consumption of fashion and identity in Petržalka, a high-rise housing estate in Bratislava, Slovakia. The wearing of 'comfortable' clothing by diverse socio-economic groups and across generations in Petržalka is a way that citizens make public spaces more 'homely', an expression of collective local identity and 'belonging', formulated in opposition to individualized 'petit-bourgeois' consumption (for more detail see Ferenčuhová and Jayne 2013; Jayne and Ferencuhova 2015). Informality and appreciation of personal 'comfort' was noted by respondents as being related to physical separation from the rest of the city (by the river Danube and a motorway) and because of the 'estate's lack of architectural beauty' (see Figures 51.2, 51.3 and 51.4).

For example, one respondent in the ethnographic research project called Babeta (female, 60, professor) described her surprise when, while moving into an apartment in Petržalka, she witnessed a conversation between two neighbours dressed in negligees 'shouting' across the street from the window of one high-rise building to another, and František (male, 65, retired) who recounts how residents claimed the newly built quarter 'as their own':

> people present themselves as 'being in Petržalka' and they behave accordingly [edit] ... They have to try and feel at home in Petržalka and they often behave as if in their own kitchen. [edit] ... because everyone knows each other and lives so close together in the flats that all look the same ... [edit] they have no scruples to go out in tracksuits, or in shorts, to the street, or to the shop [edit] ... Petržalka, its public spaces, are part of their home [edit] ... If they went downtown [to the city centre], they wouldn't dress like that, but here, they do [edit] ... Here whether it is school kids, middle-class, or loafers it doesn't matter [edit] ... They think this is their place and they set the rules.

It is possible to theorise the relationship between place and consumption in Petržalka through Latham and McCormack's (2004: 703) interest in assemblages of human and non-human actors that 'multiply the pathways along which the complex materialities of the urban might be apprehended'. Drawing on the post-structuralist writing of Latour (2005) and Deleuze and Guattari (1988), work on urban networks, practices and places has sought to speak

Figure 51.2 The motorway that connects/divides Petržalka with/from the rest of Bratislava.
Source: Mark Jayne.

Figure 51.3 The River Danube, a boundary between Petržalka and the rest of Bratislava.
Source: Mark Jayne.

Figure 51.4 Public space in Petržalka.
Source: Mark Jayne.

> not to static arrangement or a set of parts, whether organized under some logic or collected randomly, but to processes of arranging, organizing, fitting together ... where assemblage is a whole of some sort that expresses some identity and claims to a territory.
>
> *(Ward 2006: 56)*

Focussing on non-human relations of comfy clothes in Petržalka helps to better understand

> the multiple spatial networks that any city is embroiled in, and to ... [allowing consideration of relationships between place and consumption to be explored through] the full force of those networks and their juxtaposition in a given city upon local dynamics.
>
> *(Amin 2002, 112)*

For example, respondents suggested that elsewhere in the city such informal dressing in public spaces was frowned on as a signifier of 'lack of culture' or 'rural-like character', but in Petržalka familiarity and informality expressed via 'comfy' clothing was celebrated as a marker of identity:

> Well, yes! That relaxed attitude. Everyone knows each other. I find it fantastic [edit] ... if one guy is on drugs and the other has a university degree and a good job they still know each other. They might have different social backgrounds but people have

> something in common [edit] … everyone is so close to each other … [edit] living together in the flats everyone lives on top of each other amongst the concrete Lukáš.
>
> *(Male, 28, project manager)*

Respondents also pointed to the 'placed' human and non-human relations of physical isolation from the rest of the city, socio-economic mixing, the 'newness' and concentration of the blocks of flats, as being different to other parts of the city:

> Of course, when you are at the estate and you go to the grove or by the lake, you wear sporty clothes [edit] … You don't go in high-heels [edit] … from the perspective of the inhabitants, with everyone living near each other in flats, meeting on the stairs and all the concrete that surrounds, you don't need to dress up and wear those high-heels [edit] … You wear the two-piece, or skirt when you go elsewhere, because it is part of the game in the city … Anna.
>
> *(Female, 50, secretary)*

> Here in Petržalka, no one takes notice of me, of course, why should they [edit] … look at the buildings that surround you, it is not a fashionable place … [elsewhere in the city] they pay attention to what they are wearing. [edit] … Here, when you find someone dressed up decently, you would think he is going downtown. František.
>
> *(Male, 65, retired)*

The 'game' as noted by Anna acknowledges how local residents individually and collectively rejected bourgeois 'distinction' practices – 'dressing to impress' was not part of everyday life in Petržalka.

Reflecting on geographies of everyday life in the socialist era, respondents also talked about the ways in which clothes were too expensive and less fashionable than western clothes, that there was a lack of choice and, moreover, that clothes were often 'uncomfortable' to wear:

> in the 1960s and 1970s [edit] … When I bought Italian shoes for 400 crowns, unfortunately I couldn't afford to buy trousers at the same price, jacket at the same price, shirt at the same price, tie at the same price, raincoat at the same price so I only could buy some of these. Very few people at this time could afford to buy a whole outfit [edit] … Then, more things became available in seventies and early eighties but were even more expensive … [edit] to afford fashionable clothing meant not being able to afford something else. If you had a good salary you could afford eating and some clothes, or, if poorly paid, some people bought clothing only, and didn't eat. František.
>
> *(Male, 65, retired)*

> Women did not have boutiques here, but we always had fashion magazines. I used to make clothes for my kids and for myself [edit] … when my sister went to England [in the 1960s] they said – yes, Slovak girls are very beautiful, but they wear very ugly skirts. Babeta.
>
> *(Female, 60, professor)*

Babeta also commented on the texture of clothing fabrics, suggesting 'the dress used to bite', and that being able to dress in 'comfy' clothing when back 'at home' in Petržalka was 'something of a relief' from having to wear 'biting' clothing when at work.

Exploring these empirical findings in order to highlight the relationship between place and consumption via McFarlane's (2011: 209) interest in 'the intensity and excessiveness of the moment', it is possible to argue that the wearing of comfortable clothing in Petržalka represents a

> disruption of pattern … [where consumption of comfy clothes] generate new encounters with people and objects, and invents new connections and ways of inhabiting everyday urban life … [and in doing so represents] the potential of urban histories and everyday life to be imagined and put to work differently.

In these terms our understanding of consumption and place in Petržalka is advanced through a focus on co-functioning heterogeneous human and non-human actors, networks, practices, ideas and learning – with the research findings pointing to the importance of a 'parliament of things', including materials, technological artifacts, bodies, texts, concepts and symbols (Latour 2005), and in and beyond Petržalka to residents' explicit knowledge and engagement with qualities, intensities, speeds and topologies of territorial, proximate and relationally distant connections and flows (Deleuze and Guattari 1988).

As such, the consumption of comfortable clothing in Petržalka can be understood not only as a 'local' response to the assemblages that constituted the symbolic built infrastructure of nationalistic socialism and the mobilities and materialities of everyday life in a modernist housing estate, but to what McFarlane (2011: 219) calls cosmopolitanism as a kind of 'worldliness' which takes four relational forms:

> as a *knowledge*, of how difference might be negotiated or how mutuality across differences might operate; as *a disposition*, either as progressive orientation to urban cultural diversity or as regressive exclusionary sensibility replied in relation to other cultures; a *resource* as means of coping and getting by, surviving and managing uncertainty in the city; and finally as, *an ideal*, openness to and celebration of urban diversity and togetherness to be worked towards.

These relational forms are clearly present in experience of living in Petržalka, acknowledged both by locals and non-residents and performed through materialities/mobilities of 'comfortable' clothing.

Following the revolution in 1989, it is important to note that housing estates such as Petržalka came under a spotlight in new ways, no longer celebrated as ideological and infrastructural success stories of state socialism, but rather through critical definition as places of chaos, disharmony and discomfort. Unemployment, increasing poverty and criminal gangs in Petržalka and elsewhere in Bratislava ensured that, for a short while after 1989, for some residents the wearing of comfy clothes became a way to avoid social distinction, rather than as a marker of belonging. More broadly, the extremes of wealth and poverty that emerged as capitalist accumulation took hold were made visible through the growth of spectacular buildings in the city centre, an influx of tourists to Bratislava's historical city centre and the rise of gated communities and affluent suburbs. Such spaces and places contrasted to Petržalka and other quarters, now increasingly blighted by poor maintenance of buildings (due to withdrawal of state funding for infrastructural maintenance) and 'unkempt' and decaying public spaces (see Figure 51.4). While everyday life and the concrete materialities of Petržalka had been long derided by its residents, the growing characterization of the estate as representative of the worst kind of state-socialist urban planning from 'outside' challenged and problematized local pride in comfort and informality.

However, it is in these changing structural contexts that the continued importance of comfortable clothes for individual and collective identities in Petržalka since 1989 can be understood with reference not only to historic associations of 'homeliness' and 'belonging', but also to changes in the geographies of consumption landscape beyond the quarter. As the following quotes show, the discursive construction of spatial isolation and socio-economic mixing in Petržalka was re-imagined with reference to new consumption spaces that had appeared elsewhere in the city. Lukáš, for example, talked about the proliferation of consumer culture with reference to city centre shops and suburban shopping malls, which represented the growing presence of international chains and global brands and increased 'choice' in the clothing available. In a similar vein, Anna pointed to the increases in the opportunities to promenade in the city and an intensification of social differences expressed through fashion elsewhere in the city:

> After 1989, *Korzo* became even more popular [edit] ... more and more people started to go and get dressed up, and not just on the weekends [edit] ... now you see so many events in the city centre ... Christmas markets, concerts, open air cinema, New Year's Eve, or so ... [edit] but now it is much easier to see differences in social background. Anna.
>
> *(Female, 50, secretary)*

> I left Bratislava when I was 18 [edit] ... When I came back I felt that Bratislava was dressing very chic and that people go out dressed up, girls wear make-up, men dress in fashionable shirts and suits even to go out to a pub, where it is not really necessary to dress that well [edit].
>
> *(Lukáš, 28, project manager)*

Responding to a question as to whether Petržalka had changed in similar ways, Lukáš suggested that:

> [Petržalka is] everything behind the Danube, everything south of the Danube. There is a clear division, us and them, and we are separated by the river. Petržalka has always been different [edit] ... We always understood it as different and they always understood it so, too [edit] ... we are still known for wearing comfortable clothing.

Such responses show that while increasing social and spatial differentiation expressed through fashion had proliferated throughout Bratislava, collective identity across socio-economic groups and generations in the wearing of comfortable clothing nonetheless remained an important marker of collective identity in Petržalka. The collective identity of 'comfy clothing', initially promoted as a response to economic and material conditions of socialism, was now re-articulated because of increased social divisions associated with geographies of the emergence of consumer capitalism.

Haraway's (1990) ethnographies of political possibilities of (non)human hybridities through a focus on 'attachment sites' where assemblages are formed offers a useful insight into the 'stubborn' and enduring nature of the relationship between 'comfy clothes', consumption and place in Petržalka. Such critical perspectives highlight complex ways in which people make/are able to make decisions (and define boundaries) regarding 'dwelling' and material conditions, social relations, emotions, embodiment, affective atmospheres as relational comparison/ experiences through comfort and consumption. The 'improvisations in dwelling' bound up with wearing comfy clothes are not merely impromptu but are the product of bringing together materials obtained through different connections with places and people as part of an ongoing assemblage. Deleuze and Guattari's (1988) depiction of matter and energy of becoming through encounters with (non)humans was clearly present in the emotional and embodied work undertaken through consumption to make Petržalka 'comfy' and 'homely' in relation to multiple connections and imaginaries with other places/times (see Jayne and Hall 2019).

Conclusion

At the heart of study of consumption over the past few centuries has been a focus on the 'place' of identity, lifestyle and forms of sociability. Archetypal, 'ideal' and generalizable accounts have offered rich and detailed insights into the political, economic, social, cultural and spatial emergence and proliferation of consumer culture around the world. In order to advance that vital work, in this chapter I have pointed to the opportunities afforded by recent theoretical and empirical work focussed on assemblages, materialities, mobilities, bodies, emotions and affect, etc. The theoretical and empirical resources enabled by such critical territorial/relational perspectives offer significant opportunities to re-invigorate study of consumption and place for the twenty-first century.

Acknowledgements

I would like to thank the Chinese National Social Science Foundation funding, entitled 'Mobilities theory and practice in contemporary western criticism' for allowing time to write up the final draft of this chapter- 当代西方批评中的'移动性'理论与实践研究（西部项目）18XZW004.

Note

1 This chapter draws on material already published in Jayne, M. and Ferenčuhová, S. (2015) 'Comfort, identity and fashion in the post-socialist city: assemblages, materialities and context', *Journal of Consumer Culture*, 15(3): 329–350.

References

Amin, A. (2002) 'Spatialities of globalization', *Environment and Planning A*, 34, 385–399.
Baudrillard, J. (1993) *Symbolic Exchange and Death*, London: Sage.
BBC News. (2011) *Head teacher appeal to school run 'pyjama parents'*, http://www.bbc.co.uk/news/uk-engalnd-tees-13511668
Bell, D. and Jayne, M. (eds) (2006) *Small Cities: Urban Life Beyond the Metropolis*, London: Routledge.
Bell, D. and Jayne, M. (2009) '*Small* cities? towards a research agenda', *International Journal of Urban and Regional Research*, 33 (3), 683–699.
Benjamin, W. (1982) *The Arcades Project*, Harvard University Press: Harvard.
Bovone, L. (2006) 'Urban style cultures and urban cultural production in Milan: postmodern identity and the transformation of fashion', *Poetics*, 34, 370–382.
Campbell, C. (1995) 'The sociology of consumption', In Miller, D. (ed) *Acknowledging Consumption: A Review of New Studies*, London: Routledge, 96–126.
Crewe, L. and Lowe, M. (1995) 'Gap on the map? towards a geography of consumption and identity', *Environment and Planning A*, 27, 1877–1898.
Davidson, J., Smith, M. and Bondi, L. (eds) (2005) *Emotional Geographies*, Aldershot: Ashgate.
Deleuze, G. and Guattari, F. (1988) *A Thousand Plateaus*, trans. B. Massumi. Atlhone Press: London.
Edensor, T. (2010) 'Introduction: thinking about rhythm and space', In Edensor T. (ed) *Geographies of Rhythm: Nature, Place, Mobilities and Bodsse*, Aldershot: Ashgate, 1–20.

Edensor, T. and Jayne, M. (2012) *Urban Theory beyond 'the West': A World of Cities*, London: Routledge.
Entwistle, J. (2000) *The Fashioned Body: Fashion, Dress and Modern Social Theory*, Cambridge: Polity Press.
Ferenčuhová, S. and Jayne, M. (2013) 'Zvyknúť si na Petržalku: každodenný život, běžná spotreba a vzťah k socialistickému sídlisku', *Český lid*, 1, 303–318.
Frisby, D. (2001) *Cityscapes of Modernity*, Cambridge: Polity Press.
Haraway, D. (1990) 'A manifesto for cyborgs: science, technology, and socialist feminism in the 1980s', In Nicholson, L. (ed) *Feminism/Postmodernism*, London: Routledge, 190–233.
Hebdige, D. (1979) *Subculture: The Meaning of Style*, London: Methuen.
Iossifova, D. (2012) 'Shanghai borderlands: the rise of a new urbanity?', In Edensor, T. and Jayne, M. (eds) *Urban Theory beyond the West: A World of Cities*, London: Routledge, 195–208.
Irish Central. (2012) *Jobseekers Banned from Wearing Pyjamas at Dublin Dole Office*, www.irishcentral.com accessed 17th March 2012.
Jackson, P. and Thrift, N. (1995) 'Geographies of consumption', In Miller, D. (ed) *Acknowledginf Consumption: A Review of New Studies*, London: Routledge, 204–237.
Jayne, M. (2005) *Cities and Consumption*, Abingdon: Routledge.
Jayne, M. (2018) *Chinese Urbanism: Critical Perspectives*, London: Routledge.
Jayne, M. and Ferencuhova, S. (2015) 'Comfort, identity and fashion in the post- socialist city': assemblages, materialities and context', *Journal of Consumer Culture*, 15 (3), 329–350.
Jayne, M., Gibson, C., Waitt, G. and Bell, D. (2010) 'The cultural economy of small cities', *Geography Compass*, 4 (9), 1409–1417.
Jayne, M. and Hall, S. M. (2019) 'Urban assemblages, (in)formality and housing in the Global North', *Annals of the American Association of Geographers*, 109 (3), 685–704.
Jayne, M., Hubbard, P. and Bell, D. (2011) 'Worlding a city: twinning and urban theory', *City: Analysis of Urban Trends, Culture, Theory, Policy and Action*, 15 (1), 25–41.
Jayne, M., Hubbard, P. and Bell, D. (2013) 'Twin cities: territorial and relational geographies of worldly Manchester', *Urban Studies*, 50 (2), 239–254.
Jayne, M. and Ward, K. (2016) *Urban Theory: Critical Perspectives*, London: Routledge.
Latham, D. and McCormack, D. (2004) 'Moving cities: rethinking he materialities of urban geographies', *Progress in Human Geography*, 28 (6), 701–724.
Latour, B. (2005) *Reassembling the Social: An Introduction to Actor–network Theory*, Oxford: Oxford University Press.
Li, X. (1998) 'Fashioning the body in Post-Mao China', In Brydon, A. and Niessen, S. (eds) *Consuming Fashion: Adorning the Transnational Body*, Oxford: Berg, 71–95.
Maffesoli, M. (1994) *The Time of Tribes: The Decline of Individualism in Mass Society*, London: Sage.
Manjoo, F. (2012) *The pyjama manifesto*, www.slate.com accessed 28th March 2012.
Massey, D. (1991) 'A global sense of place', *Marxism Today* June, 24–29.
McFarlane, C. (2011) 'On context': assemblage, political economy and structure', *City*, 15 (3-4), 375–388.
Miles, S. (1998) 'The consuming paradox: a new research agenda for urban consumption', *Urban Studies*, 35, 1001–1008.
Robinson, J. (2006) *The Ordinary City: Between Modernity and Development*, London: Routledge.
Sennett, R., Lipovetsky, G. and Porter, C. (2002) *The Empire of Fashion: Dressing Modern Democracy*, Princeton: Princeton University Press.
Simmel, G. (1903) *Die Großstädte und das Geistesleben [The Metropolis and Mental Life]*, Dresden: Peetermann.
Swain, J. (2002) 'The right stuff: fashioning an identity through clothing in a junior school', *Gender and Education*, 14 (1), 53–69.
Urry, J. (1995) *Consuming Places*, London: Routledge.
Veblen, T. (1898) *The Theory of the Leisure Classes*, London: Penguin.
Ward, K. (2006) 'Policies in motion, urban management and state restructuring: the trans-local expansion of Business Improvement Districts', *International Journal of Urban and Regional Research*, 30, 54–70.
Ward, K. (2010) 'Towards a relational comparative approach to the study of cities', *Progress in Human Geography*, 34, 471–487.
Warde, A. (1997) *Consumption, Food and Taste*, London: Sage.
Wilson, E. (1985) *Adorned in Dreams: Fashion and Modernity*, London: Virago.
Zukin, S. (1989) 'Urban lifestyles: diversity and standardization in spaces of consumption', *Urban Studies*, 35 (5–6), 825–839.

52

Memory and forgetting in city marketing

(Re)writing the history of urban place?

Gary Warnaby and Dominic Medway

Introduction

Huyssen (2003: 11) has emphasised the importance of what he terms a 'culture of memory'; namely, 'a turning towards the past that stands in stark contrast to the privileging of the future so characteristic of earlier decades of twentieth-century modernity'. This culture of memory – manifest in 'memory discourses' – has, he argues, pervaded contemporary cultural and political spheres. He identifies various 'subplots' of this phenomenon, including, in particular, obsessive self-musealisation by 'video recorder, memoir writing, and confessional literature' (Ibid: 14). In the intervening period since Huyssen's work, such trends have become ever more pervasive with the rise of social media and blogging where people are able to record their own remembered versions and understandings of events for others to see – potentially in perpetuity. Similarly, others have analysed how the consumption of artefacts, souvenirs, clothing, family narratives and bodily modifications can stimulate and facilitate 'memory work' by individuals (see Buse and Twigg, 2015; Epp and Price, 2010; Marcoux, 2017; Steadman et al., 2018). Moving such notions from the level of the individual consumer to broader marketing Brown (1999, 2001) employs the term 'retro-marketing', which he argues is an amorphous concept with no consensus definition, but could best be described in terms of 'yesterday's tomorrow's, today!' (1999: 365).

In this chapter we consider this aphorism in relation to the marketing of towns and cities, arguing that the interplay of the past, present and future is an important aspect of urban place marketing, which, in turn, has significant implications for its practice. There is – perhaps inevitably – a place-related aspect to such a phenomenon. This is manifest, for example, in the notion of 'retroscapes' (see Brown and Sherry, 2003), reflecting the fact that '[m]arketing is *inherently* spatial, *relentlessly* geographical, *unavoidably* locational' (Brown, 2001: 142, original emphasis). Retroscapes – as deliberate 'evocations of times past' (Brown, 2003: 3) – can be regarded as attempts to generate *genius loci*, or a sense/spirit of place. In a marketing context this spatial focus on the past is evident at various scales (see Brown, 2003; Brown and Sherry, 2003). The importance of *genius loci* is stressed by Sherry (2000: 275) in terms of conveying authenticity – and the essence of 'being in place'. This importance is highlighted in the context of place marketing by Skinner (2011), who describes *genius loci* as the 'essence' of a place brand.

Huyssen similarly emphasises that an important aspect of a 'culture of memory' is place-specificity, noting that although 'it is important to recognize that although memory discourses appear to be global in one register, at their core they remain tied to the histories of specific nations and states' (2003: 16). Place-based histories, and associated 'memory discourses', can also exist at other spatial scales, such as the *urban*, which is our particular focus in this chapter. For example, Italo Calvino, in *Invisible Cities*, states that the city consists 'of relationships between the measurements of its space and the events of its past' (1997 [1972]: 10). However, our aim in this chapter is to move beyond an exclusive focus on the past, to consider temporality more broadly; specifically, the *interweaving* of time and place in the context of the current and future practice of urban place marketing. We begin by discussing the complex relationship between temporality and the nature of the spatial entity being marketed, and the extent to which '[a]s fundamentally contingent categories of historically rooted perception, time and space are always bound up with each other in complex ways' (Huyssen, 2003: 12). This contention draws closely on the concepts of 'memory work' and forgetting, and we analyse these aspects in the context of the 'representation work' undertaken by marketers to promote places.

The ticking of urban time

In his theoretical contextualisation of time in an urban context, Madanipour (2017) makes a basic distinction between what he terms 'instrumental' and 'existential' temporality. *Instrumental* temporality relates to 'how time has become treated as an instrument and an asset, how it has been subject to the pressures of acceleration in the process of globalization and how transience and ephemerality are the outcome of these pressures' (Ibid: 5). Thus, time can be seen as a social institution, to which a numerical value can be ascribed. Madanipour notes that this has resulted in various historical trends, including: a quest for precision (where timekeeping is ever more detailed and precise); the systematisation and standardisation of time and the production of a universal time across the globe; the materialisation of timekeeping through the objects used to measure time; and 'the externalization and collectivization of timekeeping through public display and personal ownership of these material objects and the organization of space'. He notes that the effect of these trends together 'constitute a public infrastructure that regulates social life' (2017: 23), and time thus becomes 'a social institution for understanding and managing change' (Ibid: 26).

Applying this notion to the context of place marketing, various aspects of temporality become – implicitly or explicitly – part of marketing messages for towns and cities. For example, accessibility and ease of inter-place and intra-place mobility by various modes of transport is an oft-repeated trope of many place marketing messages (Warnaby, 2009), which has obvious temporal implications resonant with the adage that 'time is money'. Thus, any aspect of place that facilitates or enables the efficient use – and saving – of 'instrumental' time will provide some form of place 'competitive advantage'. In addition, the marketing strapline for the northern English town of Middlesbrough – 'Middlesbrough – Moving Forward' – is a reminder that the orientation of much place marketing activity is about capitalising upon potential *future* economic development opportunities. Furthermore, moving in the opposite direction on the temporal continuum, many places will inevitably draw upon their past for the purposes of place marketing, especially where the materiality of the past (in the form of archaeological heritage, for example) is an attraction for tourist visitors. However, Ross et al. (2017: 37–38) also highlight the 'socio-cultural and historical value' of more intangible heritage which can communicate the 'incorporeal significance and subtle meanings' inherent in place. The immaterial means by which history suffuses through the urban experience is described by Madanipour as follows:

> even when a city changes beyond recognition, it still keeps some of its old characters in a kind of unconscious realm, through small traces that are left here and there, and in habits and routines, concepts and beliefs that people carry with them and share in their social life. Without being visible, this unconscious realm maintains a degree of continuity in society, even when the city's fabric and social institutions have significantly changed.
>
> *(2017: 90)*

This links to the second of Madanipour's (2017) fundamental time-related distinctions in an urban context, namely *existential* temporality, which relates to the notion of a personal sense of time, which is more subjective, and linked to memory and identity. This more phenomenological perspective relates to what Cresswell and Hoskins (2008: 394) term the 'realm of meaning' about a place that is linked to the lived experience of being in 'place'. There are evident parallels with Huyssen's (2003) concept of a culture of memory, which could incorporate notions of nostalgia – defined by Baker and Kennedy (1994: 169) in terms of a 'sentimental or bittersweet yearning for an experience, product or service from the past'. This runs counter to the future-oriented 'moving-forward' narratives of much place marketing activity. However, it could be argued that precisely by focussing on this more phenomenological concept of urban time, through the development and articulation of place-related 'memory discourses' (Huyssen, 2003), urban place marketers might actually foster feelings of place attachment and more effectively communicate the *genius loci* that constitutes the 'essence' of a place brand (Skinner, 2011). In other words, it is in this way that the 'place' featured in the 'representation work' of place marketers is actually created, and we now move to discuss this 'representation work' below.

Place marketing 'representation work'

One of the defining characteristics of city marketing involves the commodification of particular urban attributes to promote a positive image of the place as a holistic entity (Warnaby et al., 2002). Notwithstanding the importance of communicating context-specific *genius loci* in order to differentiate a place from competing destinations, a longstanding criticism of place marketing campaigns highlights their *lack* of distinctiveness, regardless of the intrinsic nature of the places that are their subjects (e.g., Barke and Harrop, 1994; Burgess, 1982; Harvey, 1987; Holcomb, 1994). Thus, Eisenschitz (2010: 27) argues that many of the same marketing techniques are commonly used across different places. Similar critiques exist in relation to place *branding*, with Eisenschitz again highlighting a lack of differentiation of many place brands, where difference 'is often bolted-on rather than built-in' (2010: 28). Supporting this, Medway and Warnaby (2014) criticise the repeated and unimaginative use of particular toponymic tropes and anodyne sloganising in place branding. Clearly linked to the production of the messages that constitute place-marketing-oriented 'representation work' is their inherently selective nature, where perceived positives are explicitly accentuated (Short, 1999). The temporal implications of creating positive 'memory discourses' (Huyssen, 2003), to create an attractive 'realm of meaning' (Cresswell and Hoskins, 2008), are highlighted by Griffiths:

> Place marketing works by creating a selective relationship between (projected) image and (real) identity: in the process of reimaging a city, some aspects of its identity are ignored, denied or marginalised. For example, attention may be drawn to a city's industrial or mercantile heritage, while the practices of class exploitation and slavery that may have made this possible remain under a veil of silence. Strong local loyalties and civic pride may be highlighted, but not the traditions of trade union militancy or revolutionary politics. Great play may be made of a city's cultural diversity but not the systematic racial discrimination that in all probability accompanied it.
>
> *(1998: 53)*

The use of such selective 'memory discourses' in place marketing is indicative of an inherent paradox in the wider 'turn toward memory and the past'; namely, that the 'boom in memory' is 'inevitably accompanied by a boom in forgetting' (Huyssen, 2003: 16–17). Huyssen suggests that the resulting 'amnesia' can be couched in a critique of the media, whereby although communications media 'make ever more memory available to us day by day', its veracity may be questioned because 'many of the mass-marketed memories we consume are "imagined memories" to begin with' (2003: 17). Ideas about forgetting also have resonance with Marcoux's (2017) work on souvenirs of the World Trade Centre attacks on 11 September 2001. He discusses the role of *forgetting* in 'memory work', highlighting that although it might be portrayed as 'delinquent, disrespectful and threatening', and possibly (after Ricoeur, 2004) a 'betrayal', there is a more positive view to offer. Thus (after Connerton, 2008), forgetting can be regarded as 'the process of allowing certain things or details from the past to slip from memory in order to move forward' and '[s]elective forgetting is a way to come to terms with the past' (Marcoux, 2017: 952).

As highlighted in Griffiths' (1998) quote above, such fears about forgetting resonate in a place marketing/branding context. Linking to notions of selectivity in place marketing representation work, where only the positive is accentuated (and the negative is downplayed or, indeed, air-brushed out of these officially sanctioned place narratives), Short uses a metaphor of light and shadow in describing the resulting discourses:

> The first [discourse] is the positive portrayal of a city; the city is presented in a flattering light to attract investors, promote 'development', and influence local politics. But every bright light casts a shadow. The second discourse involves the identification of the shadow, the dark side that has to be contained controlled or ignored.
>
> *(1999: 40)*

This 'representation work' means that place marketing strategies are 'both a social and political construction', consistent with the agendas of hegemonic groups (Sadler, 1993: 191). Consequently, the perceived need for a place brand 'to be positioned as elemental, definitive and different' (Clegg and Kornberger, 2010: 8), coupled with the associated imperative to put the most positive (selective) *gloss* on the place being promoted, means that the act of 'forgetting' by place marketers entails a deliberate *glossing over* of undesirable historical aspects – a case of forgetting about the past in order to 'move forward' (Marcoux, 2017). This could have potential moral and ethical overtones, given the possibility for contestation over both the content and process(es) of representation work, arising where there are diverse groups of stakeholders who may be adversely affected by conflicting interpretations of what is worth remembering and/or forgetting about a place (Clegg and Kornberger, 2010). Paraphrasing one of the well-known military history aphorisms about history being written by the victors, perhaps in this particular context we need to consider the extent to which place history is (or, indeed, should be) written by the marketers?

Cities of future past?

Thus, in the 'representation work' of place marketing we suggest that there is a case for incorporating aspects of both the past and the present in order to develop distinctive and differentiated communications messages about towns and cities in the pursuit of future economic and socio-political opportunities. In their discussion of the evaluation of historical significance in the context of the particular qualities of place, Cresswell and Hoskins identify three facets of place that are 'prevalent in geographers' writings on memory, where the geographical functions both as a realm for outsourcing memories and as a location in which reminiscence occurs' (2008: 394). These facets (which all incorporate an inherent temporality) are:

1 *Materiality* – in that any place has a tangible material form, which can act as a 'stabilizing persistence' over time;
2 *Meaning* – relating to how the place is perceived and interpreted, which can be built up over time, and which may be contested; and,
3 *Practice* – relating to the lived and performative experience of place (Ibid: 394–395).

Implicitly highlighting temporality, Cresswell and Hoskins (2008: 395) note that places 'are a complicated mixture of fixity and flow, stability and change'. This is evident in Madanipour's description of the city of Newcastle upon Tyne, in the north of England:

> If we stand on the riverside ... we can look around and see the remnants of two thousand years of history. What is now called the Swing Bridge has replaced a Roman bridge over the river Tyne, a node on the wall that the emperor Hadrian built to protect his northernmost territories from the ancestors of the Scots. The straight and long streets such as Westgate are built along the wall, which ended in Wallsend further east. From this vantage point, we can see the Castle Keep, which was built a millennium ago after William conquered Britain and built many castles to dominate the country. We can see the remains of the medieval walls and streets of the city, and buildings from medieval Georgian and Victorian periods, as well as the twentieth and twenty-first centuries. From our vantage point, they are all episodes of the past, all traces of ideas and practices that we may no longer recognize.
>
> *(2017: 91–92)*

This description emphasises a multi-layering of time, where 'different origins and durations' are all present at once (Madanipour, 2017: 80). It resonates with Crinson and Tyrer's (2005: 67) discussion of philosopher Michel Serres' notion of time as percolating rather than flowing:

> urban time is not like a line, as architectural historians would have it, a continuous sequence of monuments and events. Rather, in urban time some elements are filtered out, jumped over, left behind or forgotten. Serres uses metaphors like crumpling,

> folding, and liquid turbulence to capture the complex diversity of time – a multi-temporality that can better be studied by its topology than measured by its metrical geometry or chronology.

So how, and to what extent, can the history of the city be (re)written to capitalise upon the past for the purposes of place marketing, particularly in terms of developing some form of distinctiveness in an ever-more competitive environment?

'Place' marketing implications?

As mentioned above, a key characteristic of place marketing is the commodification of selected attributes of the place (Warnaby et al., 2002) and their subsequent incorporation into the representation work of marketers. Thus, the urban place as a 'product' to be marketed has been conceptualised as consisting of a holistic 'nuclear' entity, comprising a variety of 'contributory elements' (see Ashworth and Voogd, 1990a; Sleipen, 1988). Developing this notion, Van den Berg and Braun (1999) identify 'three levels' of (urban) place marketing, comprising: (1) *individual goods and services* in a location; (2) *clusters of related services*; and (3) the place as a whole. This third level (unlike the previous categories) may not in itself be a well-defined 'product' and is, consequently, open to various interpretations, as different *combinations* of individual goods/services and clusters therein may be promoted to distinct market segments (Ashworth and Voogd, 1990b). This process is further complicated by the fact that individual places are nested in spatial hierarchies. This has a number of implications for their marketing (see Boisen et al., 2011), not least the potential existence of scale discrepancies in the definition of the product between those responsible for shaping, marketing and managing it, and those who consume it (Ashworth and Voogd, 1990b).

Such conceptualisations of places as 'products' fit with the notion of assemblage (Deleuze and Guattari, 1987), which regards the properties of a particular entity as emerging from 'the interactions between parts' (DeLanda, 2006: 5). In the context of places, Anderson (2012: 579) states that '[i]t is how these component parts relate, connect and interact that forms particular places'. Furthermore, he notes that:

> These parts do not come together necessarily by intention or design or have an essential permanence that makes their connection insoluble; rather their aggregation keeps their coherence as an individual unit intact but, nevertheless, forms a larger whole through their connection with others.
>
> *(Ibid: 578)*

Implied in the above discussion is an emphasis on the *material* 'contributory elements' of a place, which could include such things as infrastructure, built environment, etc. Some of these may have an overt temporal dimension, consistent with Ross et al. (2017) who highlight the emphasis on the tangible in conventional classifications of heritage and its tourism potential and historical significance. However, Anderson also stresses the importance of the *immaterial* in assemblages:

> Places as assemblages are constituted not simply by 'things' but also by practices (e.g., the building of houses, the movement of people, the 'flows of life'). As a consequence these places are also constituted by the experiences of those involved in these practices,

> and the meanings they bring to these places, and the intensities of affect produced by their interactions with the other connecting parts.
>
> *(2012: 579)*

This focus on the immaterial links to Cresswell and Hoskins' (2008: 394) notion of place as incorporating both 'a certain materiality' and a 'less concrete realm of meaning', and highlighting place as 'a lived concept'. Thus, if in their 'representation work' place marketers are seeking to create a distinctive narrative about a particular town or city, we suggest that they should focus on both the material and immaterial aspects of the place, and incorporate a more overt temporal perspective. This does not necessarily imply a focus on heritage *per se* (although this might be appropriate for some historic places with extensive material remains that are attractive to tourist visitors). However, acknowledgement of the interweaving of the past and present and the 'layering' of time discussed above in the development of a marketing-oriented place narrative that effectively incorporates and communicates the *genius loci*, is, we argue, more likely to foster feelings of place attachment and the positive associations that are fundamental to successful place brands (Zenker and Braun, 2017).

However, crafting such a place narrative (as with all place marketing representation work) is inevitably selective. This also applies to the integration of a more overt temporal dimension. In his discussion of the writing of history, Marwick notes that the historian 'inevitably has a problem of *selection* [because] of the intense richness and complexity of historical experience' (1970: 143, original emphasis). This has implications for 'historical writing' in that 'not only must the historian represent the complexity of past experience, he must represent it in movement through time' (Ibid: 143–144). This is analogous to the 'representation work' of place marketers, who must incorporate (perceived positive) aspects of the city's past – whether in the more usual terms of material heritage (see Ross et al., 2017), or a more immaterial 'realm of meaning' (Cresswell and Hoskins, 2008) – into future-oriented narratives which seek to influence and facilitate economic imperatives in an increasingly competitive environment.

An important question relates to how this could be achieved, and we now briefly outline possible historical manifestations (both material and more immaterial) of the 'layering' of time (Madanipour, 2017) in an urban context, and how they could be used in place marketing activity. A material example of what Warnaby (2019) has termed historical 'urban fragments' are 'ghost signs', referring to outdoor advertising from the period 1930–1955, where advertising on walls/buildings was hand-painted directly onto the surface itself. Many such signs – unlike many of the brands and businesses they advertise – have survived to the present day (see Schutt et al., 2017 for a detailed examination). Whilst they are throwbacks to a bygone time, Schutt (2017) argues that they have found new lives through books of collected images and, more importantly, through online social media, such as Flickr and Facebook. This has made it increasingly easy to record and share images through dedicated websites (e.g., www.hatads.org.uk/catalogue/ghostsigns.aspx), some of which are specific to particular places and which act as a focus for discussion about the places in question. They are also the subject of commercial tours through urban space (see www.ghostsigns.co.uk/tours). Schutt argues that such activity can serve to document small, but evocative, historic place details (which would otherwise potentially be missed) that could contribute to *genius loci*.

Another more immaterial manifestation of this is technical applications that use a geographic information system capable of overlapping place-based data (i.e., relating to the history of the user's current location) onto maps in layers. This offers the ability to visualise new relationships folded into a representation of space (e.g., maps), enabling the user to move virtually through space *and* time (Koeck and Warnaby, 2015). Such applications emphasise an overtly experiential dimension to urban places, in part by incorporating kinaesthetic aspects of movement through space. Like the ghostsign tours mentioned above, they also illustrate de Certeau's (1984: 93) contention that walking is 'an elementary form of this experience of the city'. Koeck and Warnaby (2015) provide a range of examples of such applications, which serve to highlight the lived history of place.

Concluding comments

Previous authors have suggested that the practice of place marketing needs to better understand the role of 'place' (Warnaby and Medway, 2013). In this chapter we have demonstrated the need to also recognise the importance of 'time' in a place marketing context, which may have particular traction when focussing on the urban experience. This reflects Madanipour's (2017) contention that the multiplicity of temporal layers (past, present *and* future) that are evident in a city will contribute to its *genius loci*, and that these layerings will be brought into sharp(er) focus through the mechanisms of both temporal continuity *and* change. It is a phenomenon that, as we have indicated above, can be appropriated and capitalised upon for marketing/branding purposes – resonating with Brown's description of retro-marketing as 'yesterday's tomorrow's, today!' (1999; 365). However, this marketing/branding activity is likely to be most effective when temporally bound concepts such as memory work and forgetting, and the role they might play in the representation work relating to places, are fully embraced and better understood by place marketing stakeholders and institutions. Such temporal sensitivity unlocks the potential to produce representational constructions that highlight the uniqueness of the place in question. This, in turn, may counter critiques that place marketing strategies only serve to homogenise what may be very distinct – and unique – spatial entities.

References

Anderson, J. (2012) 'Relational places: The surfed wave as assemblage and convergence', *Environment and Planning D: Society and Space*, 30: 570–587.

Ashworth, G. and Voogd, H. (1990a) *Selling the City*. London: Belhaven.

Ashworth, G. J. and Voogd, H. (1990b) 'Can places be sold for tourism?', in Ashworth, G. and Goodall, B. (eds), *Marketing Tourism Places*, pp. 1–16. London: Routledge.

Baker, S. M. and Kennedy, P. F. (1994) 'Death by nostalgia: A diagnosis of context-specific cases', *Advances in Consumer Research*, 21: 169–174.

Barke, M. and Harrop, K. (1994) 'Selling the industrial town: Identity, image and illusion', in J. R. Gold and S. V. Ward (eds), *Place Promotion: The Use of Publicity and Marketing to Sell Towns and Regions*, pp. 93–114. Chichester: John Wiley & Sons Ltd.

Boisen, M., Terlouw, K. and van Gorp, B. (2011) 'The selective nature of place branding and the layering of spatial identities', *Journal of Place Management and Development*, 4 (2): 135–147.

Brown, S. (1999) 'Retro-marketing: Yesterday's tomorrows, today!', *Marketing Intelligence & Planning*, 17 (7): 363–376.

Brown, S. (2001) *Marketing – The Retro Revolution*. London, Thousand Oaks and New Delhi: Sage Publications.

Brown, S. (2003) 'No then there – Of time, space and the market', in S. Brown and J. F. Sherry, Jr. (eds), *Time, Space and the Market – Retroscapes Rising*, pp. 3–16. Armonk and London: M. E. Sharp.

Brown, S. and Sherry, J. F. Jr. (2003) *Time, Space and the Market – Retroscapes Rising*. Armonk and London: M. E. Sharp.

Burgess, J. (1982) 'Selling places: Environmental images for the executive', *Regional Studies*, 16 (1): 1–17.

Buse, C. and Twigg, J. (2015) 'Materialising memories: Exploring the stories of people with dementia through dress', *Ageing & Society*, 36 (6): 1115–1135.

Calvino, I. (1997 [1972]) *Invisible Cities*. (trans. W. Weaver). London: Vintage.

Clegg, S. R. and Kornberger, M. (2010) 'An organizational perspective on space and place branding', in F. M. Go and R. Govers (eds), *International Place branding Yearbook 2010: Place Branding in the New Age of Innovation*, pp. 3–11. Houndmills: Palgrave Macmillan.

Connerton, P. (2008) 'Seven types of forgetting', *Memory Studies*, 1 (1): 59–71.

Cresswell, T. and Hoskins, G. (2008) 'Place, persistence, and practice: Evaluating historical significance at Angel Island, San Francisco, and Maxwell Street, Chicago', *Annals of the Association of American Geographers*, 98 (2): 392–413.

Crinson, M. and Tyrer, P. (2005) 'Clocking off in Ancoats: Time and remembrance in the post-industrial city', in M. Crinson (ed), *Urban Memory: History and Amnesia in the Modern City*, pp. 49–71. London and New York: Routledge.

De Certeau, M. (1984) *The Practice of Everyday Life*. (trans. S. Rendall). Berkeley, Los Angeles and London: University of California Press.

DeLanda, M. (2006) *A New Philosophy of Society: Assemblage Theory and Social Complexity*. London: Continuum Books.

Deleuze, G. and Guattari, F. (1987) *A Thousand Plateaus: Capitalism and Schizophrenia*. (trans. B. Massumi). London: The Athlone Press.

Eisenschitz, A. (2010) 'Place marketing as politics: The limits of neoliberalism', in F. M. Go and F. Govers (eds), *International Place Branding Yearbook 2010: Place Branding in the New Age of Innovation*, pp. 21–30. Houndmills: Palgrave Macmillan.

Epp, A. and Price, L. (2010) 'The storied life of singularized objects: Forces of agency and network transformation', *Journal of Consumer Research*, 36 (5): 820–837.

Griffiths, R. (1998) 'Making sameness: Place marketing and the new urban entrepreneurialism', in N. Oatley (ed), *Cities, Economic Competition and Urban Policy*, pp. 41–57. London: Paul Chapman Publishing.

Harvey, D. (1987) 'Flexible accumulation through urbanisation: Reflections on postmodernism in the American city', *Antipode*, 19 (3): 260–286.

Holcomb, B. (1994) 'City make-overs: Marketing the post-industrial city', in J. R. Gold and S. V. Ward (eds), *Place Promotion: The Use of Publicity and Marketing to Sell Towns and Regions*, pp. 115–132. Chichester: John Wiley & Sons Ltd.

Huyssen, A. (2003) *Present Past: Urban Palimpsests and the Politics of Memory*. Stanford: Stanford University Press.

Koeck, R. and Warnaby, G. (2015) 'Digital chorographies: Conceptualising experiential representation and marketing of urban/architectural geographies', *Architectural Research Quarterly*, *19* (2): 183–191.

Madanipour, A. (2017) *Cities in Time: Temporary Urbanism and the Future of the City*. London: Bloomsbury Academic.

Marcoux, J-S. (2017) 'Souvenirs to forget', *Journal of Consumer Research*, 43 (6): 950–969.

Marwick, A. (1970) *The Nature of History*. London and Basingstoke: The Macmillan Press.

Medway, D. and Warnaby, G. (2014) 'What's in a name? Place branding and toponymic commodification', *Environment and Planning A*, 46 (1): 153–167.

Ricœur, P. (2004) *Memory, History, Forgetting*. (trans. D. Pellauer). Chicago: University of Chicago Press.

Ross, D., Saxena, G., Correia, F. and Deutz, P. (2017) 'Archaeological tourism: A creative approach', *Annals of Tourism Research*, 67: 37–47.

Sadler, D. (1993) 'Place marketing, competitive places and the construction of hegemony in Britain in the1980s', in G. Kearns and C. Philo (eds), *Selling Places: The City as Cultural Capital Past and Present*, pp. 175–192. Oxford: Pergamon Press.

Schutt, S. (2017) 'Rewriting the book of the city: On old signs, new technologies, and reinventing adelaide'', *Urban Geography*, *38* (1): 47–65.

Schutt, S., Roberts, S. and White, L. (2017) *Advertising and Public Memory: Social, Cultural and Historical Perspectives on Ghost Signs*. New York and London: Routledge.

Sherry, J. F. Jr. (2000) 'Place, technology, and representation', *Journal of Consumer Research*, 27 ((Sept)): 273–278.

Short, J. R. (1999) 'Urban imagineers: Boosterism and the representation of cities', in A. E. G. Jonas and D. Wilson (eds), *The Urban Growth Machine: Critical Perspectives Two Decades Later*, pp. 37–54. New York: State University of New York Press.

Skinner, H. (2011) 'In search of the *genius loci*: The essence of the place brand', *The Marketing Review*, *11* (3): 281–292.

Sleipen, W. (1988) *Marketing van de Historische Omgeving*. Breda: Netherlands Research Institute for Tourism. Cited in Ashworth and Voogd (1990a).

Steadman, C., Banister, E. and Medway, D. (2018) 'Ma(r)king memories: Exploring embodied processes of remembering and forgetting temporal experiences', *Consumption Markets & Culture*, 1–17. doi:10.1080/10253866.2018.1474107.

Van den Berg, L. and Braun, E. (1999) 'Urban competitiveness, marketing and the need for organising capacity', *Urban Studies*, 36 (5–6): 987–999.

Warnaby, G. (2009) 'Non-place marketing: Transport hubs as gateways, flagships and symbols?', *Journal of Place Management and Development*, 2 (3): 211–219.

Warnaby, G. (2019) 'Of time and the city: Curating urban fragments for the purposes of place marketing?', *Journal of Place Management and Development*. doi:10.1108/JPMD-08-2018-0063.

Warnaby, G., Bennison, D., Davies, B. J. and Hughes, H. (2002) 'Marketing UK towns and cities as shopping destinations', *Journal of Marketing Management*, 18 (9/10): 877–904.

Warnaby, G. and Medway, D. (2013) 'What about the "place" in place marketing?', *Marketing Theory*, 13 (3): 345–363.

Zenker, S. and Braun, E. (2017) 'Questioning a "one size fits all" city brand: Developing a branding house strategy', *Journal of Place Management and Development*, 10 (3): 270–287.

53

Making new places

The role of events in Master-Planned Communities

Judith Mair

Introduction

In Australia, as in many developed Western countries, the nature of urban development, housing and community configurations is undergoing significant changes (Williams and Pocock, 2010). This has resulted not only in changes to where and how people live, but also how they experience their living environment (Lloyd, Fullagar, and Reid, 2016). The Master-Planned Community (MPC), sometimes referred to as master-planned estate, has become an established form of placemaking in Australia, particularly, although not exclusively, among families with young children (Cheshire, Walters, and Wickes, 2010). While there is no exact definition of a MPC as such, they are generally taken to refer to large-scale private housing developments, which incorporate both physical and social infrastructure, generally located on the urban fringes of major cities (Gwyther, 2005). These MPCs continue to dominate both housing development and population growth in Australian cities, despite criticisms relating to a growing divide between inner and outer suburbs in terms of service provision, health outcomes, transport and social disadvantage (Andrews, Johnson, and Warner, 2018). The creation of new suburbs or towns is naturally not new – new settlements sprang up in Europe following industrialisation; the Garden City movement in the UK aimed to provide a healthier and more peaceful place to live than heavily polluted cities; and strong post-war demand in the United States, the United Kingdom and parts of Europe saw the creation of many large-scale housing estates (Cheshire et al., 2010). However, as Cheshire et al. (2010) point out, MPCs represent a new kind of residential development, as their appeal is not necessarily simply spatial. Instead, MPCs are marketed on the basis of their ability to appeal to those seeking a change in lifestyle, as well as their capacity to provide a 'ready-made' or a priori community (Walters and Rosenblatt, 2008). Events form one key part of the MPC developer's 'community building' arsenal.

'Community' is naturally a highly contested term and this chapter doesn't allow for a full investigation of the multifaceted nature of community nor how the concept has been understood and used (see Schultze, Chapter 25, this volume, for a more detailed exploration of the concept). According to Voydanoff (2001), community can be defined in locational and/or relational terms. This differentiation has resonance with the ideas of *Gemeinschaft* (used to refer to a place-based community) and *Gesellschaft* (social relations characterised by mutual interest as initially proposed by Tonnies (1887/1963). In relation to the MPC, it appears that the developers are referencing communities of place, and indeed local place (Walters and Rosenblatt, 2008). However, it is crucial to note that what is in fact being created is a particular notion of community and place – a set of narratives and ideals created by developers in order to sell houses. As Cheshire et al. (2010, p. 284) suggest, developers are attempting 'to infuse these new estates with a sense of community through symbolic and material practices of placemaking'. Despite Massey's (1994) observation that place and community have only rarely been coterminous, and the contention from Dixon and Dupuis (2003) that the role of physical space in the creation of community is overplayed, nonetheless place is the assumed basis for the creation of community in MPCs.

Place and MPCs

Agnew (1987) famously noted that place can be conceptualised in three key ways – as location (a spot on the map); locale (somewhere material where social relations can play out, such as public spaces, shops, halls, etc.); and sense of place (the relationship between humans and places, referred to as the subjective and emotional attachment that people have to places). Kalandides (2011) similarly considers place to be formed of three elements, referred to as materiality, institutions and practices, and further notes that only a combination of these three elements allows us to begin to understand place. He also highlights the importance of meaning, noting that all spaces and places have some meanings for people, but that these meanings may not be shared; indeed, 'sometimes, the ways that people give meaning to places may even stand in stark contrast to each other' (Kalandides, 2018, p. 150). How then might we conceptualise MPCs in relation to place? Location is not a concern – these are definitely on the map, although it depends on the extent to which construction has been completed as to whether there is much to see in the way of development, or whether the location remains a green field with an optimistic advertising hoarding promising 'a new community coming soon'. Assuming that the MPC is at least in the process of construction, there is a locus that offers the potential for social relations. In some MPC developments, initial public spaces are constructed along with the early subdivisions of construction but these may simply be small parks or squares without any additional amenities. Therefore, depending on the MPC, there are likely to be locales, identifiable forms of place where social relations can occur, thus fulfilling the second criteria of place according to Agnew (1987). Nonetheless, it is worth noting that some of these may be in their infancy, as it is more common for the construction and development of larger public facilities (such as shopping malls, schools, community halls and recreational facilities) to form part of later development stages.

However, it is the third criterion, the development of a sense of place, or a subjective or emotional attachment to place, that is likely to provide the biggest challenge for developers constructing these large new estates. Narratives and identities are developed about places through human interaction, as people begin to develop attachments over time. Further, places are constantly being reinforced by people doing things (Thrift, 2008); yet in newly built MPCs, due to the time lag between initial house building and the development of a fully-fledged MPC complete with public spaces and infrastructure, people often lack the ability to do things in that location, focussing instead on commuting to work, driving to the shops and travelling to visit friends and relatives. As Tuan (1977, p. 183) notes, 'abstract knowledge about a place can be acquired in short order [...] but the feel of a place takes longer to acquire It is made up of experiences, mostly fleeting and undramatic, repeated day after day over the span of years'. Therefore, should we begin to argue that MPCs, rather than being 'new places', are actually an exemplar of what Relph (1976) considered to be inauthentic places? Are residents moving into MPCs destined to become 'surrounded by creeping placelessness, marked by an inability to have an authentic relationship to place' (Relph, 1976, p. 90)? Perhaps more generously, we might consider them as 'places-to-be', waiting in the wings for the social relations and narratives to build meaning over time.

Cresswell (2014) suggests that we are surrounded by spaces which could be anywhere, with the same sights, smells and sounds, as well as the same brands as everywhere else. Whilst this notion may be somewhat overstated, nonetheless the miles and miles of new developments on the fringes of major Australian cities do present a certain level of homogeneity. The housing styles may vary slightly, and the architecture of the public spaces may be somewhat different, but visually, many of these new developments appear interchangeable. Sometimes only a slight change in the format of the street signage advises you that you have left one MPC and moved into another. New shopping malls, plazas and town squares, too, are all filled with the same supermarket brands and, while some of the coffee shops may be independently owned, they nonetheless mostly seem to conform to our accepted notions of what they should look like. However, as Massey (2007) argues in her conceptualisation of a progressive sense of place, the central construct lies within the ongoing negotiations and contestations around the place, rather than simply in the location itself. Given the physical resemblance between many of these MPCs, the competitive nature of the housing marketplace and the assertion that facilitating social relations is key to building communities, it is not surprising that each developer is trying to make their product (their MPC) distinctive. As well as offering social hubs and meeting places, leisure and recreation facilities and the formation of residents' associations, clubs and societies, the provision of community festivals and events by these developers is a key part of their marketing efforts and thus is it valuable to examine place marketing in a little more depth.

Place marketing and branding

Place marketing has increased and become more competitive over recent years as efforts are made to distinguish places from each other through branding (Warnaby and Medway, 2013). This has further exacerbated the commodification of place (Barke and Harrop, 1994). The aim of place marketing, very broadly speaking, is to create a marketable product out of a particular location, that is targeted at a specific audience, often to attract tourism, inward investment or other economic development. One of the key tools available to place marketers is place branding, which involves the construction of narratives about a place. However, these narratives are what Kalandides (2011) calls officially sanctioned – they are carefully constructed by urban managers to emphasise the positive aspects of a place, while minimising or glossing over any negatives. This can lead to a lack of acknowledgement of the material-structural inequalities which may exist there (Colomb and Kalandides, 2010). Such selective narratives may marginalise particular groups or individuals, along with their ways of life, sometimes to such an extent that the dominant narrative may be scarcely recognisable to those who have an alternative experience of the place in question.

As an illustration of the type of place marketing and branding that is being used in the promotion of MPCs, Table 53.1 identifies the USP (unique selling proposition) as revealed

Table 53.1 USPs used by property developers in their advertising messages

Developer advertising message/s
'When we create a community, the aim is to create lifestyles, employment opportunities, leisure spaces, relationships and opportunities.'
'Huntlee [MPC] is a brand new community designed to grow as a traditional vibrant neighbourhood.'
'Springfield Lakes [MPC] is a community where you can really feel you belong – a place where you can be part of the community.'
'Springfield Lakes [MPC] was created as a place for people to "live" their own way.'
'Edgebrook [MPC] is where community spirit thrives.'
'Edgebrook [MPC] is designed to become an iconic meeting place ... a place where you will know your neighbours ... it's not just any place, it's a special place.'
'Community living is all about feeling a sense of belonging, knowing your neighbours, everyone feeling welcome, being connected to all life's essentials and being part of something bigger.'
'By joining the community at Avon Ridge [MPC], you'll be entering into an estate with a variety of residents ... community living is more than finding an address for your new home, it's choosing the lifestyle you want to live.'
'We create the perfect conditions for a community to flourish with carefully designed spaces that give people a chance to connect with each other.'
'At The Orchards [MPC], we believe you're truly at home when you love your community.'

in the advertising messages on the websites of five of the main property developers involved in building MPCs in Australia. As can be seen, the choice of words used strongly reflects the ideals that are being used to sell houses in these developments. Continual repetition of the word '*community*', along with emotive phrases like '*create lifestyles*', '*a special place*' and '*truly at home*' emphasises the point that developers are working hard to sell not just a house, but almost membership of a community. The notion of membership is reinforced by the use of phrases such as: '*a community where you can really feel you belong*'; '*where community spirit thrives*'; '*feeling a sense of belonging*'; and '*spaces that give people a chance to connect with each other*'. This serves to underscore the importance placed on the notion of community and place by these property developers.

There are a range of criticisms of the use of place marketing and place branding, not least of which is the contention by Pred (1984, p. 279) that the overuse of visual description can lead to places being seen as 'little more than frozen scenes for human activity'. Additionally, Colomb and Kalandides (2010) highlight other criticisms, including the fact that place branding, while aiming for distinctiveness in a crowded marketplace, has instead led to homogeneity, both in physical resemblance of places and in marketing slogans. Finally, place marketing and branding is often intended to have a dual purpose – to make a place attractive externally in order to attract investment and to build civic pride and social cohesion internally. However, the choice of narrow official narratives, which are unlikely to represent the diversity of any place, and the fact that places are continually in a state of flux, makes achieving internal social cohesion and civic pride a tall order. Additionally, research to date seems to suggest that while place marketing messages around community are clearly successful in attracting home buyers, and MPC residents appear to benefit from a sense of community, nonetheless they do not consider themselves necessarily responsible for the co-production of sense of community (Cheshire et al., 2010).

Placemaking

As early as the 1960s, the notion of placemaking was beginning to find favour in the US, aligned with calls to pay more attention to the human scale of place (see, for example, Jacobs, 1961). Additionally, notions of mobility and 'liquid modernity' (Bauman, 2001) have drawn attention to the fluid nature of places, thus necessitating a different conceptualisation of place. Coming from an urban planning background, Wyckoff (2014, p. 2) defines placemaking as 'the process of creating quality places that people want to live, work, play and learn in'. For Wyckoff, quality places are those where people and business want to be – 'active, unique, interesting, visually attractive with public art and creative activities' (p. 2). It is questionable whether this kind of placemaking is concerned with making places that are meaningful and constantly evolving; rather, placemaking as it is used here appears to be in search of a 'finished' place, appealing to a range of people, co-constructed but nonetheless presenting the selected narrative as being singly emblematic of the place in question. De Brito and Richards (2017, p. 2) consider placemaking quite simply to be 'the art of making places better for people'. Although this may sound conceptually rather weak, the notion is nonetheless an attractive one. Wyckoff (2014) additionally identifies some specific types of placemaking, each with different aims and activities associated with them. *Strategic placemaking* usually refers to specific efforts to create places that are attractive to talented workers who will want to live there. *Tactical* placemaking involves small and incremental activities as a way to prepare for more significant changes over time, such as pop-ups or one-off events. More relevant to this chapter is *creative* placemaking, the strategic use of arts and cultural activities, including festivals and events, to shape the physical and social character of places, with the aim of bringing together and inspiring diverse individuals and groups of people. As Fountain and Mackay (2017) point out, placemaking is an inherently political practice as it requires choices and value judgements about how a place should be recognised at any given time. Strydom, Puren, and Drewes (2018) document the history and theoretical development of the placemaking concept and also draw attention to the move away from creating a physical end-product and towards a process, ideally democratic, which allows for the active involvement of all interested parties in the negotiation of place with an emphasis on tolerance of diversity. Thus, there are certain components of placemaking, including its collaborative nature and the use of projects and activities such as events and festivals, that make it of some relevance to the discussion in this chapter, which will now move on to consider how events and placemaking may be linked in the context of MPCs.

Events and place

The introduction of festivals into city planning has become a prominent planning tool to advance local urban and economic development, consumer experiences and city images and have become an established part of the repertoire of contemporary urban planning (Fincher and Iveson, 2008; Jakob, 2013). The notion of 'festivalisation', once limited to the context of mega-event spectacles such as the Olympic Games, is now a common part of local, regional and urban economic development and has been broadened out to encompass many types of events. This is often referred to as 'eventification' – a process in which the consumption of both products and space is turned into an event (Jakob, 2013). Many of the studies which have investigated events and their social impacts have taken a community focus rather than a place focus, but, given that often the community under study is considered to be a geographically bound concept, it is likely that much of this research has significant implications for place (Coghlan et al., 2017). In particular, a strand of research has investigated the relationship between festivals and community, whether that be community identity and pride (Rogers and Anastasiadou, 2011), sense of community (Derrett, 2003), group and place identity (De Bres and Davis, 2001), regional development (Moscardo, 2007) and power and hegemony (Clarke and Jepson, 2011). Additionally, events constitute 'a complex experience where expectations, experience and perception interact in the mind of attendees to create a place, either permanent or transient' (Barrera-Fernández and Hernández-Escampa, 2017, p. 25). Coghlan et al. (2017) argue that it is essential to establish events and festivals which link to the lived identity of a place, but highlight that in order to do this, the event organisers need to value the community outcomes of the events they are organising, understand the diversity of meanings that local residents attach to the place and integrate at least some of that meaning into the event design. Arts and cultural events in particular have been shown to play a role in making spaces unique and contributing towards the negotiation of place meaning (Rota and Salone, 2014).

Events and festivals offer tangible and intangible experiences than connect people to place (Derrett, 2003) and have been a component of placemaking for some time. As Getz (1991) pointed out, events are opportunities to express shared, collective meanings and values. They can play an important role in placemaking efforts, largely because they serve as catalysts to bring people together (De Brito and Richards, 2017). Most of the work that has been done on events and placemaking has taken major and mega events as its focus, including work on the Olympics and Commonwealth Games (De Brito and Richards, 2017). Additionally, a significant amount of the research that has investigated events and placemaking has taken an urban/city focus (Barrera-Fernández and Hernández-Escampa, 2017). However, smaller events can have similar placemaking effects, albeit coming from a more grassroots perspective. For example, Gibson (2010), in his examination of rural Australia, noted the importance of events and festivals in small-scale placemaking activities.It has long been established (e.g., Agnew, 1997; Cresswell, 2014; Massey, 1994) that in order for places to be meaningful, social relations between humans must take place: indeed, places are continually constituted by social relations. Locale is needed for such relations to take place, yet research suggests that even when material spaces are provided, there are no guidelines as to how to bring about relations in the social realm (Andrews et al., 2018). It is here that the concept of 'unscripted encounter', which offers opportunities to experience the diversity of life around us (Fincher and Iveson, 2008), appears to offer us some useful ideas. Here, encounter is simply taken to mean the various ways in which different groups of people are able to cross paths and meet each other in urban contexts and, through these unplanned meetings, are able to see and potentially appreciate the different ways of living in a city or place. Despite some criticism as to the way that encounter has been conceptualised (see, for example, Wilson, 2017), the notion of unscripted encounter allows us to interrogate the way that social relations can be facilitated. While Peattie (1998) suggests that conviviality is strongly connected to sociability, and this most often occurs in the cafes and bars where we eat and drink, supermarkets, post offices, community or drop-in centres where individuals need to engage in some form of business exchange in order to inhabit that space, others (e.g., Amin, 2002) argue that any space can act as a site of conviviality, including street corners, parks and city squares. However, Duffy and Mair (2018) highlight the potential for festivals and events to act as sites of encounter. Festivals are meeting points between local and extra-local actors and forces (Quinn, 2006) and thus allow for significant exchanges. As Fincher and Iveson (2008: pp. 176, 177) point out, festivals 'have become an established part of the repertoire of contemporary urban planning', and one of the main drivers of this is the effectiveness of festivals in offering both scripted and unscripted activities, formal and informal sites of encounter and a space of conviviality. Spaces of convivial encounter, such as those at festivals, 'enable participants to step out of the conventional stances they hold towards one another and enter instead into a shared status of participation' (Duffy and Mair, 2018, p. 57).

Whilst there are arguments as to the extent to which events and festivals can facilitate long-term positive changes, Richards (2015) points to the fact that events are spatial phenomena that have lasting effects beyond the duration of the event and are increasingly being designed with the aim of enacting place transformation (whether this is urban regeneration, rural revitalisation or other transformations). Such permanent changes can include new infrastructure specific to the events industry including building or upgrading venues and sporting facilities, but can also refer to long-term public infrastructure such as improved roads, public transport services and leisure facilities. Additionally, these lasting changes can relate to intangible outcomes, such as increased levels of civic pride or improved destination image. Richards (2017) further argues that events contribute materials, meanings and creativity to places – the physical spaces necessary for the production of the event, the social and cultural context of the event and the lived experience of those attending which arises from the creative use of the space. However, he goes on to caution that not all events have the ability to encourage positive changes; indeed, some events simply reinforce the status quo rather than challenging it. This has resonance with the narratives that are constructed around events, as part of place marketing and branding.

In the context of MPCs, there appears to be little research to date on the impacts of running events to enhance community building efforts by the developers. Nonetheless, each of the five major property developers building MPCs in Australia are using community events as part of their community building strategies. For example, at the Huntlee MPC development in New South Wales, the developer notes on their website that '*Huntlee will create a sense of vitality and fun through a well-planned community events program*' (LPC, 2019), while at Springfield Lakes, Queensland, '*fun runs, community picnics, festivals and more*' are promised (Delfin Lendlease, 2019). At Edgebrook near Melbourne in Victoria, Stockland's community development program will offer '*free events and activities to encourage residents to connect with like-minded people and help create a supportive and safe environment*' (Stockland, 2019). Peet offers community markets and events at its Avon Ridge development in Western Australia (Peet, 2019) and Sekisui House (2019) emphasises that they provide '*options for fitness, recreation, events, relaxation, inspiration and entertainment in your new community* [The Orchards in New South Wales]'. This, in addition to prior research by Gwyther (2005), Andrews et al. (2018), Cheshire, Wickes, and White (2013) and Cheshire et al. (2010), shows how vital MPC developers feel events to be in their creation of place and community. However, research by Andrews et al. (2018) suggests that while residents felt place-based relationships to be important, nevertheless most place-based relationships (e.g., with neighbours) were found to be very superficial. The strongest relationships that had developed in the MPC under study in the Andrews et al. (2018) case were instead built on communities of interest. Further, Walters and Rosenblatt (2008) found that while residents are happy to attend entertainment and events provided by the developer, and that these events contribute to a sense of 'imagined' community, residents are not as interested in active participation in this imagined community, preferring to leave the establishment of community to the developer. This suggests that while events can and do play a role in the way that place meanings are developed over time, the ubiquitous use of events by MPCs as an intentional strategy designed to 'magically' create community seems likely to be doomed to failure.

Conclusions

Master-Planned Communities are undoubtedly an important part of the current and future landscape of urban Australia. As Australia continues to receive significant numbers of international migrants, and as long as some of those living in the inner cities seek to fulfil their Australian dream of suburban home ownership, it is likely that MPCs will thrive for some time to come. There seems little question as to the viability of the physical infrastructure that is being developed – residential homes are needed to meet demand and where there are houses, soon schools, shops, public transport and leisure facilities will follow. However, what is less clear is how successful the MPC developers will be in providing ready-made communities for new residents to integrate seamlessly into. Places (and the communities which thrive in them) are multifaceted, complex, constantly evolving and in a permanent process of being negotiated by those who live in them. Imposing particular narratives of place and community on a development does not automatically result in any particular fixed state of place or community, particularly not an idealised or 'imagined' community. However, events and festivals offer an opportunity for those unscripted encounters which are vital for the organic development of sense of place and perhaps in that way (rather than as a top-down instrument) they can play a role in the ongoing development of place and community in these places-to-be.

References

Agnew, J. (1987). *Place and politics: the geographical mediation of state and society.* Boston, MA: Allen and Unwin.

Amin, A. (2002). 'Ethnicity and the multicultural city: living with diversity'. *Environment and Planning A, 34*(6), pp. 959–980.

Andrews, F.J., Johnson, L. and Warner, E. (2018). 'A tapestry without instructions. Lived experiences of community in an outer suburb of Melbourne, Australia'. *Journal of Urbanism: International Research on Placemaking and Urban Sustainability, 11*(3), pp. 257–276.

Barke, M. and Harrop, K. (1994), 'Selling the industrial town: identity image and illusion', in J.R. Gold and S.V. Ward (Eds), *Place promotion: the use of publicity and marketing to sell towns and regions.* Chichester: Wiley, pp. 93–114.

Barrera-Fernández, D. and Hernández-Escampa, M. (2017). 'Events and placemaking: the case of the Festival Internacional Cervantino in Guanajuato, Mexico'. *International Journal of Event and Festival Management, 8*(1), pp. 24–38.

Bauman, Z. (2001). *Community: seeking safety in an insecure world.* Cambridge: Polity Press.

Cheshire, L., Walters, P. and Wickes, R. (2010). 'Privatisation, security and community: how master planned estates are changing suburban Australia'. *Urban Policy and Research, 28*(4), pp. 359–373.

Cheshire, L., Wickes, R. and White, G. (2013). 'New suburbs in the making? locating master planned estates in a comparative analysis of suburbs in South-East Queensland'. *Urban Policy and Research, 31* (3), pp. 281–299.

Clarke, A. and Jepson, A. (2011). 'Power and hegemony within a community festival'. *International Journal of Event and Festival Management, 2*(1), pp. 7–19.

Coghlan, A., Sparks, B., Liu, W. and Winlaw, M. (2017). 'Reconnecting with place through events: collaborating with precinct managers in the placemaking agenda'. *International Journal of Event and Festival Management, 8*(1), pp. 66–83.

Colomb, C. and Kalandides, A. (2010). 'The "be Berlin" campaign: old wine in new bottles or innovative form of participatory place branding?', in G. Ashworth and M. Kavaratzis (Eds), *Towards effective place brand management.* Cheltenham: Edward Elgar Publishing, pp. 173–190.

Cresswell, T. (2014). *Place: an introduction.* Chichester: John Wiley & Sons.

De Bres, K. and Davis, J. (2001). 'Celebrating group and place identity: a case study of a new regional festival'. *Tourism Geographies, 3*(3), pp. 326–337.

De Brito, M.P. and Richards, G.W. (2017). 'Events and placemaking'. *International Journal of Event and Festival Management, 8* (1), pp. 8–23.
Delfin Lendlease. (2019). 'Springfield Lakes'. Available at: https://communities.lendlease.com/queensland/springfield-lakes/living-in-springfield-lakes/quick-facts/ (accessed January 2019).
Derrett, R. (2003). 'Making sense of how festivals demonstrate a community's sense of place'. *Event Management, 8*(1), pp. 49–58.
Dixon, J. and Dupuis, A.N.N. (2003). 'Urban intensification in Auckland, New Zealand: a challenge for new urbanism'. *Housing Studies, 18*(3), pp. 353–368.
Duffy, M., and Mair, J. (2018). *Festival encounters: theoretical perspectives on festival events.* Abingdon: Routledge.
Fincher, R. and Iveson, K. (2008). *Planning and diversity in the city: redistribution, recognition and encounter.* London: Macmillan International Higher Education.
Fountain, J. and Mackay, M. (2017). 'Creating an eventful rural place: Akaroa's French Festival'. *International Journal of Event and Festival Management, 8*(1), pp. 84–98.
Getz, D. (1991). *Festival events and tourism.* New York, NY: Van Nostrand Reinhold.
Gibson, C.R. (2010). 'Place making: mapping culture, creating places: collisions of science and art'. *Local-global: Identity, Security, Community,* 7, pp. 66–83.
Gwyther, G. (2005). 'Paradise planned: community formation and the master planned estate'. *Urban Policy and Research, 23*(1), pp. 57–72.
Jacobs, J. (1961). *The death and life of great American cities*'. New York, NY: Random House.
Jakob, D. (2013). 'The eventification of place: Urban development and experience consumption in Berlin and New York City'. *European Urban and Regional Studies, 20*(4), pp. 447–459.
Kalandides, A. (2011). 'The problem with spatial identity: revisiting the "sense of place"'. *Journal of Place Management and Development, 4*(1), pp. 28–39.
Kalandides, A. (2018). 'Editorial'. *Journal of Place Management and Development, 11*(2), pp. 150–151.
Lloyd, K., Fullagar, S. and Reid, S. (2016). 'Where is the 'social' in constructions of 'liveability'? exploring community, social interaction and social cohesion in changing urban environments'. *Urban Policy and Research, 34*(4), pp. 343–355.
LPC. (2019) 'Huntlee'. Available at: https://huntlee.com.au/community/ (accessed January 2019).
Massey, D. (1994), 'A global sense of place', in D. Massey (Ed.), *Space, place and gender.* Cambridge: Polity Press, pp. 146–156.
Massey, D. (2007). *World city.* Cambridge: Polity Press.
Moscardo, G. (2007). 'Analyzing the role of festivals and events in regional development'. *Event Management, 11*(1–2), pp. 23–32.
Peattie, L. (1998). 'Convivial cities', in M. Douglas and J. Friedmann (Eds), *Cities for citizens.* Chichester: John Wiley & Sons, pp. 247–253.
Peet. (2019) 'Allura'. Available at: www.peet.com.au/about-us/life-in-our-communities/communitylife (accessed January 2019).
Pred, A. (1984). 'Place as historically contingent process: structuration and the time-geography of becoming places'. *Annals of the Association of American Geographers, 74*(2), pp. 279–297.
Quinn, B. (2006). 'Problematising 'festival tourism': arts festivals and sustainable development in Ireland'. *Journal of Sustainable Tourism, 14*(3), pp. 288–306.
Relph, E. (1976). *Place and placelessness.* London: Pion.
Richards, G. (2015). 'Events in the network society: the role of pulsar and iterative events'. *Event Management, 19*(4), pp. 553–566.
Richards, G. (2017). 'From place branding to placemaking: the role of events'. *International Journal of Event and Festival Management, 8*(1), pp. 8–23.
Rogers, P. and Anastasiadou, C. (2011). 'Community involvement in festivals: exploring ways of increasing local participation'. *Event Management, 15*(4), pp. 387–399.
Rota, F.S. and Salone, C. (2014). 'Place-making processes in unconventional cultural practices. The case of Turin's contemporary art festival Paratissima'. *Cities, 40,* pp. 90–98.
Sekisui House. (2019) 'The Orchards'. Available at: www.sekisuihouse.com.au/the-orchards/masterplan (accessed January 2019).
Stockland. (2019) 'Edgebrook'. Available at: www.stockland.com.au/residential/vic/edgebrook/life-at-edgebrook (accessed January 2019).

Strydom, W., Puren, K. and Drewes, E. (2018). 'Exploring theoretical trends in placemaking: towards new perspectives in spatial planning'. *Journal of Place Management and Development, 11*(2), pp. 165–180.

Thrift, N. (2008). *Non-representational theory: space, politics, affect*. Abingdon: Routledge.

Tonnies, F. (1887/1963). *Community and society: gemeinschaft and gesellschaft*. Translated and edited by Charles P. Loomis. New York, NY: Harper and Row.

Tuan, Y-F. (1977). *Space and place: the perspective of experience*. Minneapolis, MN: University of Minnesota Press.

Voydanoff, P. (2001). 'Incorporating community into work and family research: a review of basic relationships'. *Human relations*, 54, pp. 1609–1637.

Walters, P. and Rosenblatt, T.E.D. (2008). 'Co-operation or co-presence? the comforting ideal of community in a master planned estate'. *Urban Policy and Research, 26*(4), pp. 397–413.

Warnaby, G. and Medway, D. (2013). 'What about the 'place' in place marketing?'. *Marketing Theory, 13* (3), pp. 345–363.

Williams, P. and Pocock, B. (2010). 'Building 'community' for different stages of life: physical and social infrastructure in master planned communities'. *Community, Work & Family, 13*(1), pp. 71–87.

Wilson, H.F. (2017). 'On geography and encounter: bodies, borders, and difference'. *Progress in Human Geography, 41*(4), pp. 451–471.

Wyckoff, M.A. (2014), 'Definition of placemaking: four different types'. Planning & Zoning News, January, Available at: www.canr.msu.edu/uploads/375/65814/4typesplacemaking_pzn_wyckoff_january2014.pdf (accessed January 2019).

54

Place as commodity

Informal settlements' contribution to tourism in Bogotá and Medellin

Jaime Hernández-Garcia and Beau B. Beza

Introduction

Informal settlements are an overwhelming reality for most Latin American cities and the developing world, with an estimated 1 billion people living in these areas (UN-HABITAT, 2006). In Latin America more than 80% of the population live in cities, and 23.5% or 113 million people of the continent's urban population are estimated to be living in conditions defined by United Nations Human Settlements as slums [informal settlements] (ONU-HABITAT, 2012). These settlements are a considerable part of Latin American cities and in Bogotá, for example, nearly 50% of the city has grown from informal patterns of development (Martin-Molano, 2000: 66). Settings such as these have been for many the only way to gain access to housing and urban facilities (Hernández-García, 2013a) and are met with mixed reactions. That is, some city residents look upon these informal settlements with fear and trepidation, which are perceptions of these settings developed usually from *the outside looking in*. Frenzel et al. (2015) explain that "[t]he 'slum' then symbolizes the 'dark', the 'low', the 'unknown' side of the city; slums are 'places of the unknown Other'" and "[t]he imagined geography of 'the slum' is that of another world – chaotic, uncivilised, and horrifying" (p. 240).

Largely, these perceptions of informal settlements are developed by *outsiders* making comparisons between their settings (or their own expectations and desires) and the place they are visiting (or avoid visiting). However, these *informal* settings and the places within them are regarded by some people to exhibit exceptional social capital and self-organisation practices (Brugmann, 2010), apart from the spatial and built environment characteristics that rival the realisation of planned places in the more "developed" settings around the world. Consequently, informal settlements are more than just spatial units; they display distinctive socio-spatial qualities and dynamics that result from, and reflect, particular bottom-up urban design processes characterised by conflicting and collaborative situations (Hernández-García, 2013a). Perhaps it is not everybody's choice, but, increasingly, tourists from a range of local and international places wish to experience "the more tangible experiences [of informal settlements such as]: the smells, sounds, maze-like streets, and atmospheres" (Frenzel et al., 2015: 242). In effect, these informal areas now participate globally in the tourist sector. Attracting a variety of tourists to experience, at some level, the qualities of the setting that led it to become a place of some acclaim. This chapter argues that these places have become a commodity exhibiting social, urban and/or architectonic qualities that are desired, by tourists, to be experienced. To examine this experience Hernandez-Garcia and Lopez-Mozo (2011) asked if there is a role for informal settlements in the branding of cities. With evidence from the *barrios* of Bogotá, these authors concluded that informal areas do in fact have something to offer, and, what is more, they are interesting places for tourists. Although considerations now have to be made to avoid the risk of gentrification of those areas (due to increasing interest) and particularly to assure that the benefit of tourism can reach the communities being visited and not just the tourist operators.

The aim of this chapter is to critically discuss the contribution of informal settlements to tourism and place, arguing, as mentioned above, how place in these settings has become a commodity that is wished to be experienced by tourists. Following this introduction, a brief theoretical framework is presented in terms of tourism, place and informal settlements. The findings and discussion are then presented to draw one's attention to the commodification of place in informal settlements and their direct links to tourism, and, to a lesser extent, with branding. This chapter draws from a 15-year longitudinal study of the *barrios* in Bogotá and Medellin, and a re-visiting of many of them in 2017 and 2018. A critical review of the longitudinal studies and their re-examination took place as part of the data gathering stage of this project, which allowed for a broadening and deepening of the analysis of data. In this regard, the built-form and social analysis was conducted through in-field observations and interviews with residents of the *barrios* in these cities along with interviews with public servants and residents in what may be considered the more affluent parts of Bogotá and Medellin.

Tourism and place in informal settlements

Tourism and place, when it comes to informal settlements, is something that is not fully understood and may be controversial. Especially when place is generally conceived and described with positive leanings (e.g., Porter, 2016), even in its contribution to city branding. Informal settlements are not usually considered appropriate for touristic activities and even less for complementing a city brand. However, the official brands of Bogotá and Medellin, which include a global imagery of informal settlements, use their diverse city features as examples to support investment, tourism and possessing cultural qualities worth experiencing. But, in terms of specific public programmes that may support an informal settlement's tourism plans these are virtually non-existent (with a few exceptions in Medellin), and many initiatives are those of the community based on private tourist operator's desires. Once a city's diversity is shown (with general images of informal settlements), more detailed clips associated with a place's "re-branding" largely depict wealthy parts of the city, including its historic centre. Yet, the informal settlements in these cities, and in Colombia in general, are consistent parts of the social and urban fabric that the world has come to know of and, arguably, that continue to exist due to people's individual and collective efforts.

Tourism in informal settlements has been labelled in literature as slum tourism and that literature has covered tourism and tourists (e.g., Frenzel, 2017), tourist operators (e.g., Frenzel et al., 2015), the perspective of the residents (e.g., Lombard, 2014) and, to some extent, their contribution to place branding/iconic developments (e.g., Frenzel et al., 2015; Hernández-García, 2013b). Yet, little has been said about the place (i.e., the informal settlement) and how it has developed, and the contribution, with pros and cons, to tourism and branding it has/can make. The lens from which we "view" informal settlements in this chapter is from the perspective of critically appraising social self-inventiveness and ingenuity, to argue that place in informal settlements can be seen as a commodity, generating interesting outcomes, but that also accompanying it are big challenges. Bogotá and Medellin are taken as case studies because they represent different stages of tourism interest and are differing examples of places becoming commodity and illustrations of informal place-making.

In terms of informal settlements, they are defined by the United Nations (ONU-HABITAT, 2012) as built settings that do not comply with planning and building regulations (i.e., that usually lack basic services and infrastructures). However, this view has been criticised for being deprecatory and for not recognising the positive characteristics of such neighbourhoods (Gilbert, 2007). An alternative view is to then consider informal settlements in terms of the patterns or processes of development that they followed or continue to follow in their realisation. This chapter adopts this alternative view and recognises informal settlements as socio-spatial urban developments in which local residents have a central role in the production and transformation of their built environment (Hernández-García, 2013a).

The origin of informal settlements in Colombia can usually be related to one of the following factors or a combination of them (see Beza and Hernández-García, 2018): "Pirate" urbanisation (illegal developers who sell plots to poor people at low rural prices but with no individual property ownership and no access to public services); or land invasion (squatters invading public or private land). Indistinct as to their origins, many of these settlements gradually move towards consolidation, owing their development to self-build and self-help practices. From these early beginnings, these places start to develop a particular significance to people, because the informal settlement has struggled to come to be, and, later on, is transformed by them (Hernández-García, 2013a). Informal settlements are, then, unfinished projects in which the agency and creativity of the occupant-builders are central, in contrast to architect-produced architecture, which emphasises the physical form of the building(s) often at the expense of users (Kellett, 2008). In an urban-design-related sense, informality is associated with informal settlements (informal housing, self-built/developed housing) while, in urban/economic terms, it is associated with street vendors particularly and, according to Santos (2000) in relation to culture, it is linked to subaltern practices or non-mainstream urban elements, such as rap music, graffiti arts, etc. Informal settlements constitute not only a spatial practice but also a social response to the challenges of living in cities. Hence, the production of place in informal settlements is an on-going process in Bogotá and Medellin, with many older *barrios* now fully legalised through municipal appropriation while, at the same time, new *barrios* are continuing to be created.

People living outside of informal settlements may consider these settings to be undesirable urban places to live in and this perspective can be positioned in a number of ways: theoretically and perceptually. In the first instance, literature tends to articulate this link through a range of arguments and/or positions, including being violent and unsafe (e.g., Koonings and Venestra, 2007). Perceptually from the perspective of the outsider looking in, for example, if one lives in a *favela* (i.e., an informal settlement) of Brazil, that person may be considered by those who do not live in this same setting as a criminal (Johnson, 2012). Yet, from the perspective of the people that have worked to realise the *favela*, their setting is considered a place where life occurs and that is, essentially, home. The process used to realise settings like this is complex and a variety of approaches are used to examine the socio-physical fabric that makes up these places. The concept of place-making is one tool, among potentially many, commonly used to provide a theoretical structure to help explain how these settings came to be. In itself, place-making is considered "a socio-political process where value and meaning are assigned to settings" (Beza and Hernández-García, 2018: 192) by people that experience the resulting spatial and social outcome(s).

Interestingly, place-making (and place) is a construct developed and refined in settings such as the USA and Australia where a highly regulatory planning system supports community engagement with urban decision makers to realise public space outcomes. It has a relatively short history, having evolved from discussions on place in which Relph (1976) and Tuan (1977) worked to "establish a thematic or experience-based categorisation [...] to explain one's 'connection' (or not) with an existing setting" (Beza and Hernández-García, 2018: 193). The issue when applying place-making as a tool to examine informal settlements is that, fundamentally, the people that realise these settings work outside of formal planning structures and with little to no engagement with city officials. This consideration is one alternative view presented by Lombard (2014) in which she argues place-making and place are key conceptual elements of Mexico's *colonias populares*.

However, arguing as to whether or not place-making should or should not be used or that an alternative conceptual tool is better suited for use in informal settlement research is beyond the scope of this chapter. Nonetheless, we suggest caution when using place-making as a reference point in the examination of informal settlements. Hence, issues/outcomes in the informal settlements of Bogotá and Medellin may best be addressed by considering the resulting (re)development process and outcomes along a socio-spatial "spectrum"; where, for example, social innovation may be bookended by social capital and inventiveness, both of which support community action and can be best understood from the *inside looking out*. Judgement of a setting by scholars and/or tourists may then be based on the innovation expressed in an outcome rather than the displayed visual aesthetic or common negative leaning stereotype(s).

Commodifying place in informal settlements of Bogotá and Medellin

The above discussion outlines perceptions of informal settlements derived mainly from people living outside of these settings and a tool commonly used to examine people's realisation of these places. This section presents views people have of these settings, which are developed from living within the informal settlement. The section then uses these perceptions as a mechanism to identify the informal settlements' contribution to tourism and to argue the commodification of place.

When residents positively described the informal settlement of San Luis, on the north-east periphery of Bogotá, they responded with comments revolving around the beauty of the neighbourhood, its safety, and that it is set within the mountains and allows for views of the city. In particular they commented that "it has the best view of Bogotá" or "friends visiting us get surprised how beautiful the *barrio* is", "our *barrio* is very safe, people which do not know it think the opposite, but once they come, they change their minds". When describing the person/place connection with the *barrio* they also explain that "I feel very connected here" and "that this is my home". Interestingly, when referring to how people from outside the *barrio* perceive this setting, residents suggest "people in Bogotá don't know where it is".

When talking to the residents about how tourism might be viewed, their response was generally positive. However, when talking to residents in the *barrio* of Tierra Roja, an informal settlement on the city's south periphery, they describe the setting with mixed emotions. They do describe the neighbourhood as beautiful but talk about how the young people do not study or work and become a target for drug dealers. Although they generally say the *barrio* is safe, many people from *the outside looking in* think Tierra Roja is unsafe. Note that even newcomers to this area, such as Miguel and his family, report that "this is a good place to live and raise the family".

The production of space in Bogotá's informal settlements is closely linked to the people that live in these settings, making these places distinctive, in terms of the realisation process and resulting end product, from the rest of the city where professionals (architects, planners) are more involved. The process referred to here is resident-oriented; where through a great deal of struggle and achievement the outcomes are in a constant state of evolution. This makes these places distinctive, and different to the other. Hence, informal settlements are not only a significant part of Bogotá in terms of size, but also because they display distinctive physical and social characteristics. In this regard, they are differentiated places (Kavaratzis and Ashworth, 2005), and to some extent this difference is what is attractive to tourists. Different from Medellin's place development, which focusses on social urbanism, the informal settlements of Bogotá have not been the setting of large infrastructure projects but rather are a conglomerate of regular upgrading initiatives in which some come from the residents and some from the municipality. The characteristics these upgrades display come from a range of sources and the settlements' developed areas display cultural expressions that can be of interest beyond their own boundaries, showing the creativity and richness found in *barrios*, and which individually work together to contribute to the branding of the informal settlement and generally reinforce tourist interest.

For example, in terms of tourists experiencing life in the informal settlement, the *barrios* of Bogotá are full of activity: people can be seen in the settlements' streets and playing in their parks, chatting to each other, buying things in the *tiendas* (local stores), and, in essence, transforming space constantly to accommodate these interactions, which provide the base elements that Frenzel et al. (2015) describe as important to tourists. There are also several particular cultural expressions found in Bogotá's *barrios*, and, for the purposes of helping to achieve the aim of this chapter, they provide distinctive experiences for tourists. The following are examples of the cultural expressions that can be found in Bogotá's *barrios*: traditional games (e.g., '*tejo*'), religious expressions (e.g., shrines), celebrations and events (e.g., Christmas, community meetings), and special provincial and traditional dishes (e.g., '*tamales*').

These are among the services and amenities that are not only part of the everyday life found in the *barrios* but they are also the life of the place that generates interest to tourists and to local residents of wider Bogotá. For example, Margarita, a university lecturer living in a more affluent part of Bogotá, says that "I always go to la Andrea [an informal area on the south-west periphery of the city] to get the best *tamales* [a maize pudding filled with chicken and/or other meat(s)] and *lechona* [a suckling pig stuffed with rice and vegetables]".

There is also a range of informal meetings, celebrations and events that can be observed and expressed in the *barrios*, involving chatting, drinking, eating, music, and sometimes dancing. In Bogotá's *barrios* one of the most notable of the above phenomena is the music and food festival in Ciudad Bolivar (the biggest informal settlement of Bogotá, with more than 600,000 inhabitants) called: *Ojo al Sancocho* (eye to the soup). This event is rapidly escalating to be an important cultural element of the city and it is also gathering attention from tourists.

More major celebrations are usually held during Christmas and Easter, connecting social practices with religious traditions. As Rojas and Guerrero (1997) found in their research, communities organise themselves to collect money for painting houses, decorating the pavement and streets with Christmas decorations and organising parties. People may close off one street (placing barriers at each side to control the entrance), decorate it (with colourful banners) and hold *novenas* (nativity prayers) over the nine days before Christmas. Rosa, who lives in the *barrio* Santa Cecilia, suggests that in this place "we decorate our street to be the most beautiful of the *barrio* to make the best *novenas*".

Connected with celebrations such as this and a major cultural manifestation in the *barrios* are the activities around eating and drinking. In some settlements the *olla comunitaria* (community picnic) is a monthly (or as required) community event, where those living near the park get together to cook, eat and drink. The food, most of the time, is *sancocho* (a soup, with reference to the above name *olla*, meaning pot) and, of course, *tamales* and *lechona* are popular in some areas. However, there are many other specialties which are usually connected to the different provincial origins of the people.

Arguably these cultural expressions, including the built environment, found in the *barrios* of Bogotá are very distinctive compared with other areas of the city, and display in a similar way to what Porter (2016) describes as the cultural commodification of place. It is not only particular in the way it may be considered as an end product, but also in the way in which it is realised (the process). In this sense, the physical and social aesthetics (Hernández-García, 2013a) of informal settlements are distinctive and interesting.

What the above discussion begins to suggest is that these settings possess qualities that contrast with the generally negative perceptions of these *barrios* by outsiders. The result is that these places can be branded as positive place examples of informal settlements. Additionally, in terms of the beauty they are described to possess, they provide an alternative to the general aesthetic that people from the outside may originally consider the setting to exhibit. For example, Hernández-García (2013a) found three design language themes displayed in these areas. The first one is hybridisation (Garcia Canclini, 1989), which involves the use of different design elements that correspond to different building styles. That is, a vocabulary or design language is taken from different geographical, temporal and social contexts and used to produce something new. The second theme is permanent transformation, which implies potential to expand, fragment and support rhizomatic growth (Berenstein Jacques, 2001) as well as a permanent state of flux (Brillembourg Tamayo et al., 2005). The third theme is *engalle* (over-decoration) (Carvajalino, 2004), which can be described as 'the more the better', including a strong relationship with popular expression and exhibiting an aspirational language. These themes provide access to concepts heard commonly in the literature on the subject of the aesthetic of informal settlements, such as diversity and complexity, and others that may be not so common, such as order or a different kind of order, a never-ending product, and fragment vs wholeness.

Collectively these aesthetic considerations suggest a distinct language that conveys also distinct meanings and appropriation of places. For example, a park or a house is a finite product in formal terms. Each is designed and built, and although they may be upgraded/improved in years to come, they are designed as final products representing a specific place/time commitment. While, on the contrary, an informal spatial-product is to design a never-ending product/place relationship that, as Porter (2016) suggests (in terms of landscape branding), is not based upon a "fixed set of relations" (p. 50). That is, this relationship, in terms of the informal settlement's product(s), represents a transformation of ideas where residents translate abstraction (i.e., ideas of social, urban and/or architectural qualities) into commodities (i.e., evolving built-form outcomes) that tourists wish to experience.

As an illustration of this evolving built-form outcome, since the 1990s Medellin has undertaken programmes and projects in informal settlements to physically upgrade them and to integrate them both physically and socially into the urban fabric found within the *barrios* of the city (Hernández-García, 2013b). Possibly the first, and suggested here as the more influential, project was the over-ground metro system introduced in 1995. But it was in 2004, when "the city implemented the world's first modern urban aerial cable-car public transport system" (Brand and Davila, 2011: 648) to reach the city's hillside informal settlements, that Medellin became "known" to the world as a positive example of slum development. Additionally, public space upgrade(s) and community services such as schools and libraries designed by prestigious architects have enriched the *barrios*' atmosphere, and to some extent the quality of life, in these impoverished areas. To the point that the *barrios* of Medellin are commonly visited not only by Colombians but also by international visitors (especially from Latin America) who want to see firsthand the projects and how the settlements and the city have changed. The municipality that realised these facilities has called this initiative social urbanism and the term and concept is now used as a mechanism to help brand the city (Brand, 2010; Echeverry and Orsini, 2010).

Additionally, Medellin has been engaged with several social, economic and urban projects over the years, resulting in an urban transformation that now is seen as a positive example in the country and in the continent. The city in less than a decade has changed its face, from an insecure and violent city to a place regarded as exhibiting hope and interest. The transformation is observable in the city with the landmarks they have produced (e.g., the cable car), but also in the social improvements they instigated. However, much still needs to be done, especially in addressing poverty, reducing social inequalities and providing mechanisms for full urban and social inclusion.

Interestingly, this path has become the subject of attention by authorities, academics and common people in the country (and beyond) who desire to visit Medellin to see first-hand the transformations and learn from the experience (Alvarado-Renner, 2012: 17). This path has also been called Medellin's model or Medellin's integral development model, but, again, it is commonly known as social urbanism. Yet, social urbanism is seen as highly controversial because it can be argued that urbanism, at its core, is social.

Nonetheless, to some extent the term social urbanism has been validated by the several papers produced on the subject (e.g., Brand, 2010; Echeverry and Orsini, 2010) and the official documents in which it has been used. Social urbanism has been directly referred to in the city's *Proyectos Urbanos Integrales* (Integral Urban Projects) and, in this regard, to the social provision (i.e., community participation) of space in the *barrios*. Hence, high quality architecture and public space, when upgrading occurs in informal settlements, were driven by ideas of interventions in the *barrios*. These ideas were intended to not only provide urban facilities but also to provide pride and a sense of belonging to the residents (Hernández-García, 2013b).

What the above discussion suggests is that in Bogotá and Medellin residents themselves help to develop and transform their own places (with sometimes little help from the outside), including housing, urban facilities and public space, which can lead to and contributes to a city brand. In that process, they also build a particular social and spatial space that is part of the identity of the city. In this sense, for example, this suggestion supports Hernandez-Garcia and Lopez-Mozo's (2011) argument that these areas contribute to the branding of the city, not only benefitting the city itself, but also as an economic and a social tool that can enhance life and the wellbeing of people in the *barrios*. As explained in the Bogotá example, cultural expressions found in these settlements can be of interest beyond its own boundaries, showing the creativity and richness found in these areas and contributing to an enrichment of tourism and branding initiatives. It can also exhibit the unique built environment found between formal interventions (as highlighted in Medellin), vernacular developments and the everyday appropriation and transformation of space in Bogotá. Which is arguably another distinctive characteristic that can contribute greatly to place development and consideration of place as commodity. Among those issues, the idea of branding the place, or branding the social occurrences (events and socially related activities) that may take place in it, is interesting to explore, particularly in terms of tourism. On the one hand, tourism and branding can refer to a place as a whole, itself a distinct entity formed from physical structures, functions, activities, differing atmospheres and even symbolic values that somehow become encapsulated in the name of a particular city. And, on the other hand, tourism can be attracted "to quite specific services, facilities or attributes that occur at such a place" (Ashworth and Voogd, 1990: 66). In the case of tourism and branding in informal settlements, it can be argued that both the physical space and the activities displayed within it are a subject of interest, in other words, and again, these places can be attractive to tourists and are brandable.

Conclusion

In Bogotá and Medellin, tourism to informal settlements is a relatively recent activity (especially in Bogotá), and so far it has been predominantly organised through the initiative(s) of *barrio* residents. In a similar sense, informal settlements are discussed as contributing to both a spatial experience (built environment, infrastructures and nature) and a socio-cultural experience which is able to attract tourism, suggesting place may act as a socio-spatial unit and, in effect, is the attractor. Also, as inferred in this chapter, informal settlements' contribution to city branding gradually takes place and these cities are becoming aware and municipal staff are beginning to understand that these areas possess value (although they are still predominantly considered problematic areas that are unsafe, poverty stricken and without hope). Yet, these "problematic" perceptions of the informal settlement suggest a cursory understanding of these settings. The value these places possess revolves around complex relationships involving the people, activities and physical features displayed in the *barrios* of Bogotá and Medellin. Essentially, it is the informal settlement's unique socio-cultural expression(s) and design language (exhibited through hybridization, permanent transformation and *engalle*) that work to generate tourist interest and compose an identity, which may work to change the marginal image of these areas. In this sense, place is increasingly being used as a commodity in informal settlements and although benefits may result from this relationship, more research into identifying the potential risks of this association is needed.

What this latter discussion highlights is that the informal settlement is a social and collective construction which may be best understood from the *inside looking out* (rather than from the *outside looking in*), and that place is a major component in this understanding. Interest in, experience of and, importantly, judgement regarding the residents' achievements in the informal settlement may then be based on their social self-inventiveness and ingenuity, which a municipality can potentially integrate into its tourism strategies and/or city branding. However, a key ingredient in this integration is that the municipality works with the people that live within the informal settlement to develop plans and strategies together, and, above all, for any socio-economic benefits generated from tourism/branding to flow back to the people of the *barrios*.

References

Alvarado-Renner, N. (2012). Medellin, Experiencia de Transformación Urbana y Ciudadana, in A. de Medellin. (Ed.), *Laboratorio de 004Dedellín, Catalogo de Diez Practicas Vivas*, Mesa Editores, Medellin, pp. 16–17.

Ashworth, G. and Voogd, H. (1990). *Selling the City: Marketing Approaches in Public Sector Urban Planning*, Belhaven Press, London.

Berenstein Jacques, P. (2001). The Aesthetics of the Favela: The Case of an Extreme, in Fiori, J. and Hinsley, H. (Eds.), *Transforming Cities: Design in the Favelas of Rio de Janeiro*, AA Publications, London, pp. 28–30.

Beza, B.B. and Hernández-García, J. (2018). From placemaking to sustainability citizenship: an evolution in the understanding of community realised public spaces in Bogotá's informal settlements. *Journal of Place Management and Development*, Vol. 11, No. 2, pp. 192–207.

Brand, P. (2010). El Urbanismo Social de Medellin, Colombia. *Revista de Arquitectura COAM*, (Colegio Oficial de Arquitectos de Madrid), Vol. 359, pp. 99–103.

Brand, P. and Davila, J. (2011). Mobility innovation at the urban margins, medellin`s metrocables. *City*, Vol. 15, No. 6, pp. 647–661.

Brillembourg Tamayo, A., Feireiss, K. and Klumpner, H. (2005). *Informal City: Caracas Case*, Prestel, Munich and London.

Brugmann, J. (2010). *Welcome to the Urban Revolution: How Cities are Changing the World*, Bloomsbury Press, New York, NY.

Carvajalino, H. (2004). *Estética de lo Popular: Los Engalles de la Casa*, Serie Ciudad y Hábitat, Ediciones Barrio Taller, Bogotá.

Echeverry, A and Orsini, F. (2010). Informalidad y Urbanismo Social en Medellin, in Hermelin, M., Echeverry, A. and Giraldo, J. (Eds.), *Medio Ambiente, Urbanismo y Sociedad*, Universidad EAFIT, Medellin, pp. 130–152.

Frenzel, F. (2017). Tourist agency as valorisation: Making Dharavi into a tourist attraction. *Annals of Tourism Research*, Vol. 66, pp. 159–169.

Frenzel, F., Koens, K., Steinbrink, M. and Rogerson, C. M. (2015). Slum tourism state of the art. *Tourism Review International*, Vol. 18, pp. 237–252.

Garcia Canclini, N. (1989). *Culturas Hibridas*, Editorial Grijalbo, México.

Gilbert, A. (2007). The return of the slum: Does language matter? *International Journal of Urban and Regional Research*, Vol. 31, No. 4, pp. 697–713.

Hernández-García, J. (2013a). *Public Space in Informal Settlements, the Barrios of Bogotá*, Cambridge Scholars Publishing, UK.

Hernández-García, J. (2013b). Slum tourism, city branding, and social urbanism: The case of Medellin, Colombia. *Journal of Place Management and Development*, Vol. 6, No. 1, pp. 43–51.

Hernandez-Garcia, J. and Lopez-Mozo, C. (2011). Is there a role for informal settlements in branding cities? *The Journal of Place Management and Development*, Vol. 4, No. 1, pp. 93–109.

Johnson, C. (2012). *We're from the Favela but We're Not Favelados: The Intersection of Race, Space, and Violence in Northeastern Brazil*, The London School of Economics and Political Science, London, UK.

Kavaratzis, M. and Ashworth, G. (2005). city branding: An effective assertion of identity or a transitory marketing trick? *Tijdschrift voor Economishe en Sociale Geografie*, Vol. 96, pp. 506–514.

Kellett, P. (2008). *Constructive Journeys: Dwelling Consolidation and Social Practices in a Squatter Settlement*, Department of Antrophology, Durham University, Durham, UK.

Koonings, K. and Venestra, S. (2007). Exclusion social, actores armados y violencia urbana en Rio de Janeiro. *Foro internacional*, Vol. 47, No. 3, pp. 616–636.

Lombard, M. (2014). Constructing ordinary places: Place-making in urban informal settlements in Mexico. *Progress in Planning*, Vol. 94, pp. 1–53.

Martin-Molano, J. (2000). *Formación y Consolidación de la Ciudad Espontanea en Santafe de Bogotá: el caso de Altos de la Estancia en Ciudad Bolívar*, CIDER (Centro Interdisciplinario de Estudios regionales), Universidad de Los Andes, Bogotá.

ONU-HABITAT. (2012). *Estado de las Ciudades en America Latina y el Caribe. Rumbo a una Nueva Transición Urbana*, Programa de las Naciones Unidas para los Asentamientos Humanos, ONU-Habitat, Nairobi.

Porter, N. (2016). *Landscape and Branding: The Promotion and Production of Place*, Routledge, Abingdon, Oxan and New York, NY.
Relph, E. (1976). *Place and Placelessness*, Pinion Limited, London.
Rojas, E. and Guerrero, M. (1997). *La Calle del Barrio Popular: Fragmento de una Ciudad Fragmentada*, Series Ciudad y Hábitat, Ediciones Barrio Taller, Bogotá.
Santos, M. (2000). *La Naturaleza del Espacio*, Ariel Geografia, Barcelona.
Tuan, Y. F. (1977). *Space and Place: The Perspective of Experience*, University of Minnesota Press, Minneapolis, USA.
UN-HABITAT. (2006). *State of the World's Cities 2006/7*, Earthscan, London.

Part VII

Creative engagements with place

David Cooper

Introduction: creativity and place

In 2018, Dirk Hoffmann and his colleagues published an article, in the journal *Science*, attributing cave art found at three sites in Spain to Neanderthals. The authors of the paper acknowledge that they are not the first researchers to make this claim; but, at the same time, they stress that recent technical and technological developments 'enable the possibility of obtaining age constraints for cave art by U-Th dating of associated carbonate precipitates' (Hoffmann et al. 2018a: 912). For Hoffmann et al., this methodological innovation leads to the conclusion that the art is older than 64.8 thousand years. This research has not been uncontroversial and, a few months later, Ludovic Slimak et al. published an article, also in the pages of *Science*, questioning the approach that led to the dating of the cave art (2018): a critique which the authors of the original paper subsequently, and robustly, rebutted (Hoffmann et al. 2018b). At the time of writing this Introduction, therefore, the rigorous scientific debate continues to unfold. If Hoffmann et al. are ultimately proved to be correct, however, then their findings are staggering. That is to say, the art that they researched at La Pasiega (Cantabria), Maltravieso (Extremadura), and Ardales (Andalucía) – images that include geometric patterns, painted and engraved figures, and figurative representations of animals – was created some 20 thousand years before the first modern humans arrived in Europe. In their paper, Hoffmann et al. also focus on the art itself. More specifically, they consider the geographical location of the hand stencils and argue that, because they 'seem to have been deliberately placed in relation to natural features in caves rather than randomly created on accessible surfaces, it is difficult to see them as anything but meaningful symbols placed in meaningful places' (2018a: 914). For the Neanderthal artist – according to Hoffmann et al. – place mattered; and this 'painting activity constitutes a symbolic behaviour by definition, and one that is deeply rooted' (2018a: 914). This ground-breaking paper, therefore, sets up the proposition that the long, deep relationship between creativity and place is even longer and deeper than might previously have been assumed.

If we fast forward to the last millennium, then the Neanderthal impulse to situate art *in* place has similarly informed the site-specific, creative place-making practices of artists working with three-dimensional forms: think – to cite two radically different examples – of the gardens at Versailles designed by André Le Nôtre or Rachel Whiteread's sculptural intervention, *House*, in London's East End in the 1990s. Moreover, the place-attentiveness and rootedness that Hoffmann et al. identify in the Iberian cave art is evident in the work of visual artists who have repeatedly revisited particular circumscribed geographies over extended periods. Katsushika Hokusai is celebrated all over the world for his series of woodblock prints, *Thirty-six Views of Mount Fuji*, and Paul Cézanne famously produced a sequence of paintings documenting Mont Sainte-Victoire from his home in Aix-en-Provence. Literary history is similarly rich with writers who have demonstrated an artistic commitment to particular places and, as a result, have served to create those places within the geographical imagination. This process – emerging out of an entanglement of writers and writings, material places and representational spaces – is evident in the practices and products of contemporary literary tourism: from guided walks around Dante's Florence to driving tours of Alice Walker's rural Georgia; from the production of digital maps of William Wordsworth's Lake District to public access to the St Petersburg homes of Fyodor Dostoevsky.

The history of the relationship between creativity and place is littered, then, with examples of practitioners who have become synonymous with specific sites through the imaginative interrogation – across different media – of being-*in*-place. There are also artists, however, who have been drawn to the creative expression of what it means to move-*through*-place. Homer's *Odyssey*, of course, is predicated on the experience of being-out-of-place; whilst journeying from place to place is similarly integral to the poetry of Matsuo Bashō. Movement, displacement, and deracination, have figured prominently, too, in the more recent history of the moving image. Charlie Chaplin's silent classic *The Immigrant* finds humour in a penniless European moving to New York, whilst Ousmane Sembène's seminal film *Black Girl* documents the experiences of a Senegalese nanny who is hired to work for a wealthy family in Antibes, south-east France. The struggle to be emplaced has also been explored in music. *Lieder eines fahrenden Gesellen* ('Songs of a Wayfarer') is a song cycle by Gustav Mahler in which a jilted lover ultimately endeavours to find some consolation in the natural world. In musical contrast, 'I've Been Everywhere' is a song written by the country singer Geoff Mack, in which the nomadic narrator breathlessly recounts the dozens of Australian places through which he has passed.

It would be erroneous, however, to suggest that creative responses to place have focussed exclusively on the mapping – textual and pictorial, filmic and sonic, and so on – of real-world geographies. On the big screen, superheroes stalk Gotham City and Dorothy visits the Wizard of Oz in the Emerald City. On the printed page, readers enter the imaginative spaces of the entirely fictionalised worlds of Jonathan Swift's Brobdingnag, Tolkien's Middle-earth, and Mervyn Peake's Castle Gormenghast. Cultural history is also populated by creative works in which the boundaries between real-world and imagined places are consistently, and often confusingly, blurred. L S Lowry's paintings of industrial Manchester and Salford, for example, were created through the unavoidably unreliable remembering of his encounters with local places. Margaret Atwood's dystopian novel *The Handmaid's Tale* is set in what was once – within the futuristic imaginative space of the fiction – Cambridge, Massachusetts, whilst R K Narayan's town of Malgudi may be fictitious but it also partly emerges – as with Hardy's Wessex and Faulkner's Yoknapatawpha – from the author's own embodied geographical experiences of growing up in south India. In such cases, the reader may be aware that the geographical setting is a fiction; but, at the same time, there frequently remains a readerly compulsion to place that imagined location onto a real-world map. The knottiness of the relationships between actual and imagined geographies – and, by extension, creative texts and cultural tourism – can be illustrated by turning to the example provided by Peter Jackson's cinematic adaptations of *The Lord of the Rings*. During the filming of this epic trilogy, a stage set was built – in the dairy farming landscape around Matamata in New Zealand – to portray the Shire; but, upon cessation of filming, the homes of the fictional hobbits were dismantled. Remarkably, though, the set has since been reconstructed to satisfy the demands of the scores of tourists that arrived in search of the fictional setting (Peaslee 2011). Today, then, the visitor is invited to 'experience the real Middle-earth ™ Movie Set where, in the heart of the Waikoto region, you can step into the lush pastures of the Shire ™, as seen in *The Lord of the Rings* and *The Hobbit* trilogies' ('Hobbiton Movie Set').

Over the past decade or so, there has been an energetic and energising emphasis, within the academy, on how geographical thought might advance the understanding of our imbricated experiences as both emplaced and creative beings. In 2007, the Association of American Geographers, in partnership with the University of Virginia and the American Council of Learned Societies, organised 'Geography and the Humanities', a symposium gathering together researchers 'working at the fertile intersections where geography and the humanities meet' (Richardson et al. 2007). Responding to the increasing flow of ideas, vocabularies, and methodologies across disciplinary boundaries, the organisers sought 'to examine these reciprocal effects and investigate new pathways for potential future collaborations' (Richardson 2007). The Symposium, then, was an attempt both to celebrate and facilitate the creation of interdisciplinary – perhaps even transdisciplinary – scholarly spaces. At the same time, the organisers aimed to blur the perceived boundaries between conventional critical research and exploratory creative practices: they were keen to interrogate how 'creative pursuits' might be harnessed to challenge 'geographic concepts' (Richardson 2007); and they were keen to examine how the braiding of geographical thought and creative methodologies might dismantle traditional borders between academic research and wider publics.

Two landmark collections emerged from this symposium. The first volume, *Envisioning Landscapes, Making Worlds: Geography and the Humanities* – edited by Stephen Daniels, Dydia DeLyser, J. Nicholas Entrikin, and Douglas Richardson – endeavours to 'capture the rich diversity of geographical engagements with the humanities' (Daniels et al. 2011: xxvi). Over the course of its 29 chapters, scholars drawn from a heterogeneous range of disciplinary backgrounds think geographically about a heterogeneous range of cultural objects including 'maps, photographs, paintings, films, novels, poems, performances, monuments, buildings, traveler's tales and geography texts' (Daniels et al. 2011: xxvi). Organised into four thematic clusters ('Mapping', 'Reflecting', 'Representing', and 'Performing'), the contributions thereby set up new critical dialogues between geographical concepts and extant creative 'texts'. In the process, these contributions collectively invite, as Robert Summerby-Murray has put it, 'seasoned members of the academy to rethink the artificial barriers we have constructed between and within our disciplines' (2014: e21).

The second collection was *GeoHumanities: Art, History, Text at the Edge of Place* edited by Michael Dear, Jim Ketchum, Sarah Luria and, once again, Douglas Richardson. In line with the structure of its sister publication, the 30 contributions to *GeoHumanities* are organised into four main parts: 'Creative Places', 'Spatial Literacies', 'Visual Geographies', and 'Spatial Histories'. Some of the contributions are characterised by the commitment to interdisciplinary critical interpretation that is threaded through *Envisioning Landscapes, Making Worlds*. So, to give just two examples, Caren Kaplan examines the aerial photographs of Sophie Ristelhueber (Kaplan 2011) and Stuart C. Aitken and Deborah P. Dixon explore the 'cinematic landscapes of the American West' (Aitken and Dixon 2011). Yet, in 'Introducing the GeoHumanities', the co-editors patently shift the methodological emphasis from critical analysis to creative production: 'The term "geohumanities" refers to the rapidly growing zone of *creative* [my italics] interaction between geography and the humanities' (Richardson et al. 2011: 3). *GeoHumanities*, therefore, is a collection that is defined by methodological innovation and formal experimentation. More specifically, it is a volume that is defined by presentations of, and reflections on, creative processes: acts, for instance, of making and/or curating 'wordmaps' and digital geovisualisations, satellite imagery and counter-cartographies, urban photographs and children's artwork. In contrast to *Envisioning Landscapes, Making Worlds*, then, the contributions in *GeoHumanities* are primarily preoccupied with the ways that creative practices can produce geographical knowledge. Moreover, the collection's titular compound gestures towards the radical transdisciplinarity of such difficult-to-categorise creative-critical practices.

Unsurprisingly, theorists and theories of place feature in both collections. In *Envisioning Landscapes*, to cite three prominent examples, Edward S Casey offers a 'geo-philosophical inquiry' into places as 'the pre-eminent bearer of edges' (2011: 65), Tim Cresswell proposes 'a geosophical approach' to 'Race, Mobility and the Humanities' (2011: 74), and Yi-Fu Tuan argues that the 'notion that the good inherit the earth [...] is axiomatic' (2011: 139). As the book's title indicates, however, it is 'landscape', rather than 'place', that is the geographical keyword in many of the essays within the collection (see, for instance, Dubow 2011; Livingstone 2011; Olwig 2011). Saliently, place is foregrounded in the sub-title of *GeoHumanities* and, in his Afterword, Dear asserts that: 'The geohumanities that emerges in this book is a transdisciplinary and multimethodological inquiry that begins with the human meanings of place' (2011: 312). In between, Dear examines 'Creativity and Place' (2011: 9–18), Sarah Luria explores 'the role of place in the life of literature, and the role of literature in the production of place' (2011: 67), and Trevor M Harris et al. consider how GIS might be used to record the subjective 'experience or perception of place' (2011: 230). According to John Agnew, however: 'Substantively, it is the "spaces of cities" that tend to dominate the empirical stories told in the book' (2012: 514). For Agnew: 'although the language of place appears episodically in the volume, it is the modernist concern with space as expressed in fields such as architecture, photography, map art, and geographic information systems (GIS)/cartography that prevails overall' (2012: 514).

Since the publication of these twin collections – and as we have moved further away from the difficult-to-pinpoint moment of the so-called 'spatial turn' across the humanities (see, for instance, Warf and Arias 2008) – the theoretical balance has demonstrably shifted towards place. This growing conceptual preoccupation can be traced within the critical research spaces that have continued to aggravate traditional disciplinary boundaries over the past decade. A direct methodological line can be drawn, for example, between *Envisioning Landscapes* and *Literary Geographies*: an international journal launched in 2015. In opting for a pluralistic title, the co-editors of this journal attempted 'to accommodate and encourage' (Alexander 2015: 5) diverse ways of conceiving and practicing literary geography; yet, as with the 2011 collection, the journal's methodological interest clearly resides in acts of 'geographically-attuned' (Hones et al. 2015: 2) critical interpretation. Since the inaugural issue, concepts of place have emerged as central to such hermeneutic processes. More particularly, the work of such placial thinkers as Edward Relph and Tuan, Doreen Massey and Tim Cresswell, Caitlin Desilvey and Dydia DeLyser, has proved to be of fundamental theoretical importance to scholars exploring the representation, across different literary forms, of named places. As Jon Anderson succinctly puts it in an article that moves 'towards an assemblage approach to literary geography': 'To both geography and literature, place matters' (2015: 121).

Place has come to occupy a similarly prominent position in the ongoing evolution of creative geographical practices. In the same year that *Literary Geographies* appeared for the first time, the inaugural issue of *GeoHumanities* was published: a new journal from the American Association of Geographers. In their inaugural editorial, Cresswell and Deborah P. Dixon indicate how their vision for the new journal is 'attentive to past endeavors over the course of the twentieth century – endeavors that saw cultural geography become both a mainstay of the discipline [geography] and an arena where dialogue with other disciplines was encouraged and facilitated' (2015: 2). In charting such disciplinary porosity, Cresswell and Dixon acknowledge

> a shared concern in history, literary studies, philosophy, arts and performance, feminist studies, postcolonial and cultural studies, science and technology studies, and the digital humanities and geographic information systems (GIS) to pose and rethink spatial imaginaries, from topologies to third spaces.
>
> *(2015: 2)*

Cresswell and Dixon, therefore, make it clear that their journal remains resolutely committed to work that offers critical analyses of 'spatial imaginaries'. Crucially, however, the co-editors also afford textual space to more exploratory work within a section entitled 'Practices and Curations': 'Here, we can showcase a wealth of creative and experimental work being undertaken by geographers, academics from other disciplines, and creative artists, ranging from photo-essays and innovative geovisualiations to artists' pages and poetry' (2015: 3). The inaugural issue of the *GeoHumanities* journal foregrounded the importance of place in such creative geographical practice by including, among other contributions, a creative non-fiction account of reading poetry in place (Olstad 2015) and a discussion of creative collaboration as a tool for excavating the emotional archaeologies embedded within the palimpsestic nature of place (Bolland 2015). By extension, place, since that first issue, has emerged – as in the pages of *Literary Geographies* – as the key geographical concept in work that *creates* 'spatial imaginaries'.

The possible reasons for this conceptual realignment are both multifarious and concatenated. The reassertion of place can be read as a riposte, at least in part, to the spatial abstractions generated by late-capitalist, neo-liberal globalisation. Connected with this, a focus on the local and near-at-hand might also be understood as an attempted re-thickening of the geographical experience of being-in-the-world within an increasingly digital age. Above and beyond all of this, of course, the return to place has been shaped by the desperately urgent challenges generated by Anthropogenic change. Yet, at the same time, the seismic ruptures in so many political landscapes have meant that the concept of place has become invariably and inevitably entangled within debates about belongingness, national identity, and even autochthony. More than ever, therefore, place is a complicatedly contested term that is both absolutely essential and essentially problematic; and more than ever, it is crucial that academic researchers and creative practitioners (and those whose work braids the two modes) think carefully and critically about place and its possible meanings.

The contributions gathered together in this part of the *Handbook* tackle this imperative and, in the process, illustrate the breadth and depth of contemporary methodological approaches to the imbrications of creativity and place. The part opens with a scene-setting chapter in which Cara Courage maps out a critical typology of art practices and processes in twenty-first-century urban placemaking. In drawing this map, Courage highlights and unpicks a fundamental tension. In one sense, according to Courage, the arts have become routinely absorbed into the 'symbolic and fiscal economy' of the culturised city; yet, at the same time, the work of creative practitioners can still radically resist such top-down narratives of place. Scene-setting also informs the second chapter in this part: David Cooper's attempt to define place writing as a distinct literary genre. In a literary geographical reading of contemporary British writing, Cooper argues that there is a need for a flexible and expansive definition of place writing that incorporates both textual art and works of fiction as well as the first-person accounts to be found in much creative non-fiction and lyrical poetry. The discussion of the relationship between place and literary form and language continues in the following chapter by Helen Mort. Whereas Cooper offers a critical reading of extant literary writing, however, Mort – an award-winning poet, novelist, editor and writer of creative non-fiction – reflects on the role that place plays in her own creative practice. More particularly, Mort articulates a concept of 'ghost-rhetoric' – a belief in the value of fictional accounts – in reflecting on her own imaginative engagements with the urban, rural, and edgeland places of South Yorkshire and the Peak District.

Autoethnography similarly informs the next two chapters in this part. In his creative-critical chapter, 'Navigating cinematic geographies', Les Roberts offers a deep, and deeply personal, reflection on one film: *The Navigator: A Medieval Odyssey* (1988) by the New Zealand director Vincent Ward. Roberts recalls how familiarity with the locations in Ward's film shaped his engagement with the material places of Auckland during a backpacking trip in the early 1990s. In foregrounding film as a spatial practice, Roberts opens up wider theoretical thinking about cinematic geographies and the situatedness in place of the individual cinephile. Preoccupations with both place and visual culture also underpin the chapter by the artist and cultural geographer Veronica Vickery. Focussing on three of her own creative projects, Vickery reflects on the tensions between place attachment and detachment, house and home: reflections that are articulated against the backdrop of a personal relocation to the city of Bristol as well as the wider politics of place in the United Kingdom in the second half of the 2010s.

The following chapter is by another visual artist, Mike Collier, who also directs the WALK (Walking Art, Landskip and Knowledge) research centre at the University of Sunderland. In 'Place-walking', Collier reflects on his creative responses to the experience of moving through four different landscapes; and the experimentally 'meandering' form of the chapter is an attempt to encapsulate Tim Ingold's concept of place-as-meshwork. Saliently, Collier also moves beyond visual experiences and practices by exploring the creative geographies of sound. An energetic movement between different creative forms is also evident in the work of Marita Dyson and Stuart Flanagan: multimedia artists, based in Melbourne, whose responses to place, history, and memory take on the forms of songs, recordings, performances, and visual art. In her chapter, 'Walking west', Dyson reflects on *Newer Volcanics*: a body of work – including poems, songs, and visual art – mapping the waterways in the north and west of the city of Narrm/Melbourne. Dyson's chapter leads into 'Place and music', a chapter in which the artist and cultural geographer Rob St John reflects on the production of a place-based sound installation Concrete Antenna in Newhaven, Edinburgh. Drawing upon relational and dynamic understandings of place,

St John sets out three types of place-attentiveness in reflecting upon creative work that blurs the perceived boundaries between sound and music.

This part of the *Handbook* ends with an authoritatively wide-ranging chapter in which Mike Pearson offers a history of the places of dramatic performance. Here, Pearson 'traces the theatre's spatial and architectural evolution and elaboration': a survey that takes in – via the spatial thought of Michel Foucault, Deleuze and Guattari, and Ingold – Japanese *noh theatre*, the proscenium arch, and the use of a twentieth-century car plant to restage the medieval Welsh poem, *Y Gododdin*. 'Of all places … ', then, opens up conceptual thinking about creative places of performance. At the same time, the chapter itself can be read – through both its imaginative points of connection and the lyrically elliptical nature of its prose – as a creative act. It is fitting, then, that the part ends with a chapter that brings together some of the theoretical and methodological threads running through so much recent work on the relationship between creativity and place. It is similarly fitting that the chapter is written by an artist-academic whose influential work, over a sustained period of time, has dissolved both traditional boundaries between scholarly disciplines and historical borders between creative and critical approaches to place.

References

Agnew, J. (2012). A review of envisioning landscapes, making worlds: geography and the humanities and geohumanities: art, history, text at the edge of place. *Annals of the Association of American Geographers*, 102 (2), 514–516.

Aitken, S. and Dixon, D. (2011). Avarice and tenderness in cinematic landscapes of the American West. In M. Dear et al. (eds.), *GeoHumanities: Art, History, Text at the Edge of Place*, pp. 196–205. London: Routledge.

Alexander, N. (2015). On literary geography. *Literary Geographies*, 1 (1), 3–6.

Anderson, J. (2015). Towards an assemblage approach to literary geography. *Literary Geographies*, 1 (2), 120–137.

Bolland, E. (2015). Every place a palimpsest: creative practice, emotional archaeology, and the post-traumatic landscape. *GeoHumanities*, 1 (1), 198–206.

Casey, E. (2011). Do places have edges? a geo-philosophical inquiry. In S. Daniels (ed.), *Envisioning Landscapes, Making Worlds: Geography and the Humanities*, pp. 65–73. London: Routledge.

Cresswell, T. (2011). Race, mobility and the humanities: a geosophical approach. In S. Daniels (ed.), *Envisioning Landscapes, Making Worlds: Geography and the Humanities*, pp. 74–83. London: Routledge.

Cresswell, T. and Dixon, D. (2015). Editorial – imagining and practicing the geohumanities: past, present, future. *GeoHumanities*, 1 (1), 1–3.

Daniels, S. et al. (2011). Introduction: envisioning landscapes, making worlds. In S. Daniels (eds,), *Envisioning Landscapes, Making Worlds: Geography and the Humanities*, pp. xxvi–xxxii. London: Routledge.

Dear, M. (2011). Afterword: historical moments in the rise of the geohumanities. In M. Dear et al. (eds.), *GeoHumanities: Art, History, Text at the Edge of Place*, pp. 309–314. London: Routledge.

Dubow, J. (2011). Still-life, after-life, *Nature Morte*: W. G. Sebald and the demands of landscape. In S. Daniels (ed.), *Envisioning Landscapes, Making Worlds: Geography and the Humanities*, pp. 188–197. London: Routledge.

Harris, T. et al. (2011). Humanities GIS: place, spatial storytelling, and immersive visualization in the humanities. In M. Dear et al. (eds.), *GeoHumanities: Art, History, Text at the Edge of Place*, pp. 226–240. London: Routledge.

Hobbiton Movie Set. www.hobbitontours.com/en/ (date accessed 18 September 2019).

Hoffmann, D. (2018a). U-Th dating of carbonate crusts reveals Neandertal origin of Iberian cave art. *Science*, 359, 912–915.

Hoffmann, D. (2018b). Response to comment on 'U-Th Dating of carbonate crusts reveals Neandertal origin of Iberian cave art'. *Science*, 362, 1736.

Hones, S. et al. (2015). Editorial. *Literary Geographies*, 1 (1), 1–2.

Kaplan, C. (2011). The space of ambiguity: Sophie Ristelhueber's aerial perspective. In M. Dear et al. (eds.), *GeoHumanities: Art, History, Text at the Edge of Place*. London: Routledge.
Livingstone, D. (2011). Darwinian landscapes. In S. Daniels et al. (eds.), *Envisioning Landscapes, Making Worlds: Geography and the Humanities*, pp. 106–117. London: Routledge.
Luria, S. (2011). Geotexts. In M. Dear et al. (eds.), *GeoHumanities: Art, History, Text at the Edge of Place*, pp. 67–70. London: Routledge.
Olstad, T. (2015). An island, a heron, Jim Harrison, and I: Reading poetry in place. *GeoHumanities*, 1 (1), 171–176.
Olwig, K. (2011). *Choros, Chora* and the question of landscape. In S. Daniels et al. (eds.), *Envisioning Landscapes, Making Worlds: Geography and the Humanities*, pp. 44–54. London: Routledge.
Peaslee, R. (2011). One ring, many circles: The Hobbiton tour experience and a spatial approach to media power. *Tourist Studies*, 11 (1), 37–53.
Richardson, D. (2007). Geography and the humanities symposium program. University of Virginia. www.aag.org/galleries/project-programs-files/Geography_and_the_Humanities_Symposium_20007_program.pdf (date accessed 18 September 2019).
Richardson, D. et al. (2011). Introducing the geohumanities. In M. Dear et al. (eds.), *GeoHumanities: Art, History, Text at the Edge of Place*, pp. 3–4. London: Routledge.
Slimak, L. (2018). Comment on response to comment on 'U-Th Dating of carbonate crusts reveals Neandertal origin of Iberian *Cave Art*'. *Science*, 361, eaau1371.
Summerby-Murray, R. (2014). Review: envisioning landscapes, making worlds: geography and the humanities. *The Canadian Geographer/Le Géographe canadien*, 58 (1), e21–e22.
The Navigator: A Medieval Odyssey (1988). Directed by V. Ward. New Zealand: Arenafilm.
Tuan, Y-F. (2011). The good inherit the Earth. In S. Daniels et al. (eds.), *Envisioning Landscapes, Making Worlds: Geography and the Humanities*, pp. 127–140. London: Routledge.
Warf, B. and Arias, S. (eds.) (2008). *The Spatial Turn: Interdisciplinary Perspectives*. London: Routledge.

55

The art of placemaking

A typology of art practices in placemaking

Cara Courage

Introducing current placemaking practices

There is no one definition of placemaking but, rather, a set of 'understandings' of a field of practice. These 'understandings' differ, depending on the subject position of the person doing the defining, and they converge and diverge in various ways, thereby illustrating the contestation around issues of place management and analysis. Normatively, placemaking is used to describe the process of creating the material and social spaces of place so that they are desirable for the public to visit and spend time in. Placemaking holds an assumption of a strong mutually constitutive and positive affect between the built environment and behaviour: public spaces are positioned in a value-framework, viewed as essential for urban civility and for creating rich relations between strangers (Sennett, in Watson 2006: 14), where people shape the place around them according to their needs and desires. This practice is concerned with the creation or improvement of (predominantly) urban environments that reflect community values and their socio-economic and environmental connect (Legge 2012: 34) and has an improvement function to better the material quality of public space and – ergo, according to placemaking logic – the quality of lives within it. Placemaking recognises that being in a place is an affective and social phenomenon; that sights, sound, environmental factors, ambiance, and imagination all play a role – historically and contemporaneously – in the making of place and how it is used. The term 'sense of place', therefore, is a common utterance in placemaking discourse and practice. Global placemaking agency Project for Public Spaces positions placemaking as an 'overarching idea and a hands-on tool' (Project for Public Spaces 2014) for urban improvement. Thus, placemaking is the 'set of social, political and material processes by which people iteratively create and recreate the experienced geographies in which they live' and is a networked process 'constituted by the socio-spatial relationships that link individuals together through a common place-frame' (Pierce et al. 2011: 54). It involves participation in both the production of meaning and the means of production of a locale (Lepofsky and Fraser 2003: 128) to deliberately shape space to improve a community's quality of life in place (Silberberg 2013: 1–2).

An arts practice, process or practitioner is often found in placemaking, and placemaking that has an explicit arts element to it, in the vernacular, is grouped under the term 'creative placemaking'. Unlike the umbrella placemaking term, creative placemaking has a boundaried definition as proposed by its creators, from a White Paper for the USA's National Endowment for the Arts (NEA):

> In creative placemaking, partners from public, private, non-profit, and community sectors strategically shape the physical and social character of a neighborhood, town, city, or region around arts and cultural activities. Creative placemaking animates public and private spaces, rejuvenates structures and streetscapes, improves local businesses viability and public safety, and brings diverse people together to celebrate, inspire and be inspired.
>
> *(Markusen and Gadwa 2010b: 3)*

Explicitly, then, creative placemaking utilises arts in public–private partnerships in the proactive making of place, with an economic imperative to precipitate localised economic development: creating jobs; fostering entrepreneurs and cultural industries, new products and services; and attracting and retaining businesses and skilled workers (Gallant 2013; Markusen and Gadwa 2010b: 3). The arts are a tool for the 'revitalisation of the city to encourage its businesses and citizens to undertake their own making of place' (Markusen and Gadwa 2010a, 3). Arts in this context have been further instrumentalised to drive a number of key 'liveability goals', including, for example, 'public safety, community identity, environmental quality, affordable housing, workplace options for creative workers, collaboration between civic, non-profit and for-profit partners', as well as aims to instil 'more beautiful and reliable public transport' (Lilliendahl Larsen 2014: 5). The NEA creative placemaking definition is important as benchmarking practice and funding in the USA and, as a sector thought-leader, globally. It has directly informed the NEA's ArtPlace funding scheme, an Obama administration funding initiative of thirteen foundations, eight federal agencies and six banks that positioned creative placemaking as an investment in art and culture for strategic cultural and fiscal ends.

Arguably, this is the culturisation (Zukin 2010: 3) that assigns the arts an entrepreneurial role in city-making and in purporting casual links between cultural activity and neighbourhood improvement (SIAP 2007) and, consequently, market improvement (Knight Soul of the Community 2010; Tonkiss 2013: 165). Normative placemakings, as top-down urban design, can be non-contextual and productive of generalist outcomes, and generative of further fixed notions of community and public space based on a 'pseudo-participation' (Petrescu 2006: 83) model that is organised and manipulated, idealised, uncritical and concerned with reaching consensus, the process effectively silencing the voices it is meant to articulate. Cohen (2009: 145) attributes the term 'place-faking' to the process whereby artists are placed into a project and where placemaking is done to a community, not emergent from it. The authors of the NEA White Paper themselves critique creative placemaking. Markusen states that there is confusion as to what the outcomes of creative placemaking should be, ranging widely from job creation and tourism, for example, to an increase in property values and the provision of residential services. There is also a questioning of its power imbalances (Markusen and Gadwa Nicodemus 2017: 16–19): a culturised flattening that is ignorant of art's socially transformative values (Stern, in Lowe and Stern, in Finkelpearl 2000: 150). Others have a more expansive understanding of (creative) placemaking, viewing it more as a grassroots design tool used by and for the community in question, the tool kit comprising arts-based field research, social organising, community rituals, public interventions, educational workshops, audio documentation and performance art (Rochielle n.d.: 40–41). This creative placemaking is a mode of cultural work in the social sphere (Thompson 2012: 86) which encompasses urban planning, sociology and pedagogy. It is also part of the democratisation of creativity – of the move from elitist to populist and exclusive to inclusive; not so much 'social acts' as 'social creativity' – and the depth of the work is determined by the degree of meaningful change that work has made in the social group (Hermansen 2011: 2–3).

The breadth and depth of arts practices across placemaking activities has not been captured in the understandings of placemaking nor the definition of creative placemaking offered so far. The terms placemaking and creative placemaking can be found inadequate, singularly or used in reference to each other, to describe what is a multiscalar practice field. Further, there is an adverse stratification of placemaking vis-à-vis the formal planning sector and a creative and citizen-led approach (Legge 2013). It was community arts' failure to construct its own theoretical framework that was the reason for its relative devaluing in the art sector (Kelly 1984: 29). If the placemaking sector does not create its own theoretical framework, therefore, it risks a similar reduction of a 'naïve romanticism' of its claims to outcomes and a side-lining in urban design and planning as a creative, worthy 'welfare arts' (Kelly 1984: 29) adjunct to be deployed for city culturisation, marketing and regeneration, rather than as a meaningful strategy for urban living (Schneekloth and Shibley 2000, 130). This demands a more nuanced understanding of the arts, creativity and the agency of arts practice in placemaking, as is the focus of this chapter.

A placemaking typology

With generalisations and the blurring of subject positions and intents in placemaking, the spectrum of activity across the field needs to be articulated whilst, at the same time, not drawing fixed subject boundaries. The placemaking typology offered here provides an analytical framework for the placemaking sector and those that interface with it. It marries modes of placemaking – strategic, tactical and opportunistic (Legge 2013) – into a matrix designed to be used to offer insight into practice, to celebrate the breadth and depth of art-based placemaking, and to better communicate the spectrum of arts and participatory practices used in placemaking. In doing so, the typology functions to advance the understanding of the arts in placemaking and takes into account, and also celebrates, the sector's multiple standpoints. It acts as a promulgation of the differentiation of placemaking practices as well as a mode of critique and sector reflexivity. Whilst the boundaries between the placemaking modes are fluid as symptomatic of participatory, creative and co-produced practices – where the tactics of placemakings deployed may vary from site to site and temporality within a placemaking project – the operational relation is the degree to which a placemaking practice is engaged with people-in-place: as one reads it left to right, the placemaking practice in question increasingly engages the users of a place with an increasingly co-produced arts practice. As a matrix, it illustrates operational fluidity: for example, a placemaking practice may be operating at a strategic level, and have public realm or participatory outputs. The figures below (Figures 55.1–55.3) first present the placemaking typology and then its utilisation to identify two example forms of placemaking: strategic public realm and opportunistic participatory placemaking. The subsequent section goes on to detail the classifications and modalities of placemaking therein, with a particular focus on social practice placemaking as a site of emerging and innovative placemaking practice.

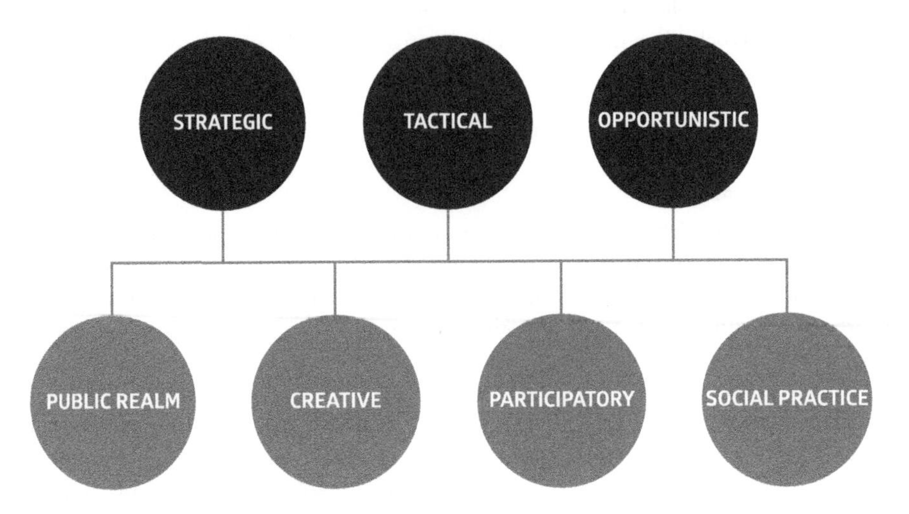

Figure 55.1 Placemaking typology.

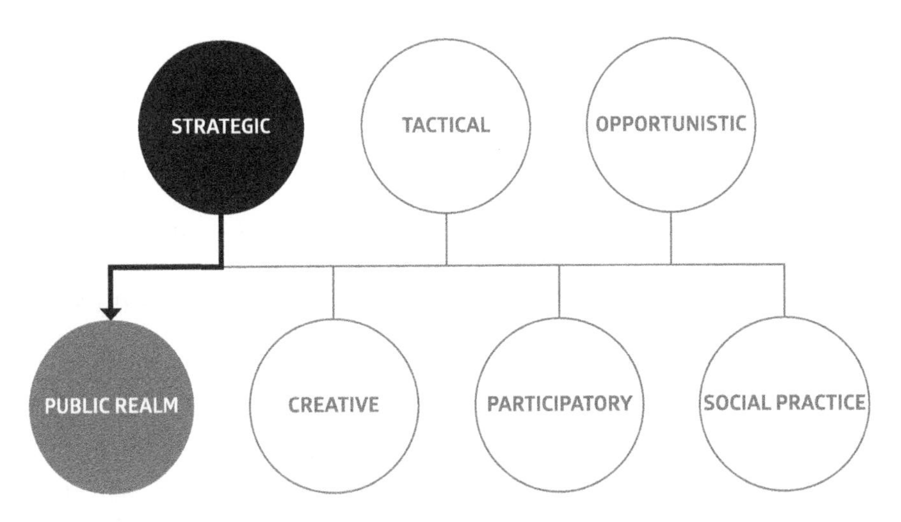

Figure 55.2 Strategic public realm placemaking matrix illustration.

Classifications and modalities of placemaking

The placemaking typology draws on Legge's (2013) three classifications of placemaking: *strategic*, undertaking and engaging with in-depth research into the local social, political, economic, physical and cultural context to define its placemaking strategy and its implementation; *tactical*, referring to the collaborative, citizen-led interventions that focus on place improvement, community capacity building and economic development, which can in turn feed into a larger

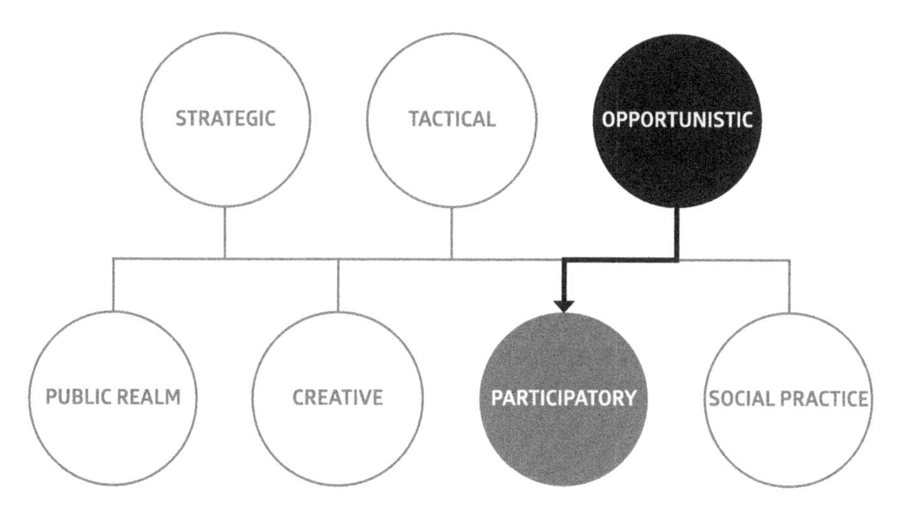

Figure 55.3 Opportunistic participatory placemaking matrix illustration.

strategy or objective (read 'tactical urbanism', Lydon and Garcia 2015); and *opportunistic*, ad hoc and unprogrammed interventions on the micro-scale enacted by small groups of people or lone citizens in response to an immediate need.

The typology further augments and extends this categorisation by including a portfolio of common modalities of placemaking practice. First, public realm. When Fleming (2007) talks of the 'art of placemaking', the projects offered as examples are ones of a masterplanned strategy analogous to the practice of monumental Public Art and regeneration. The practice restricts public engagement and utilises a managerial and visual language of urban development that creates homogeneous 'spaces' that lack a 'psychological fit between people and their physical surroundings' (Sime 1986: 49) and are 'safe for, and accommodating of, transnational investment flows and that class of economic actors who attempt to ride them' (Tonkiss 2013: 11). Second, creative, as pertaining to that practice recognised by Markusen and Gadwa (2010a) which has deeper public and private connections and relationship with the arts in practice and process. Third, participatory. This practice brings the public into an active and dynamic decision-making process, moving placemaking towards 'user-generated urbanism' or 'collaborative city-making' (Marker, in Kuskins 2013). Participatory placemaking has emerged in response to an increased demand from developers and governments for all project stakeholders to share responsibility and decision-making (Legge 2012: 34), as well as from a body of placemakers that view citizen participation in placemaking as a moral imperative and the start of a process that is based on Lefebvre's (1968) right to the city of local citizenship. Participatory placemaking operates with an inclusive design approach (Lehmann 2009; Newman 2001; Sorensen 2009) that begins to align professional and non-professional constituents in dialogue and holds that the placemaking professional should have a level of personal involvement in a place and recognise a place's diverse range of users and patterns of behaviour and experience which give it meaning. This practice is not without its own challenges, however, centring, as with participatory art, around issues of inclusivity, gentrification, capacity and standardisation. Participatory placemaking may not be the inclusive process it appears to be on the surface but another 'pseudo-participation' (Petrescu 2006: 83) that in turn manifests 'pseudopublic spaces' (Sorkin, in Crawford 1999: 22). Furthermore, there is tension in some participatory placemaking practice claiming both an economic and a community benefit, which may be mutually exclusive. There are also issues of capacity and practice standardisation: there is a danger that the practice becomes the sector's go-to one-size-fits-all solution to placemaking, attractive to city authorities in a time of fiscal austerity, appealing to impoverished administrations as an attractive box-ticking and 'cheap' solution that acts as a salve to urban realm problems without any structural and meaningful change.

Social practice placemaking

This chapter now moves on to consider social practice placemaking (Courage 2017), a practice that recognises the active social engagement of a new wave of placemaking practice, that both represents a progression in art practice within the placemaking sector and a deepening of the work of socially engaged art as a situated emplaced art form.

Social practice placemaking represents a new performative aesthetic lens for placemaking, as encountered, relational and dialogic. Its art practice and process is both revealed and inserted into everyday situations and practices (Froggett et al. 2011: 101), a 'spatio-temporal event' that locates place in the social and cultural context, over and above the built environment manifestation of space (Massey 2005: 130–131; Sen and Silverman 2014: 4). These heterogeneous encounters form an immediate 'urban-aesthetic discourse' (Deutsche 1996, in Beyes 2010: 231), required to emancipate new urban collective experiences and strengthen socially-connective tissue (Crawford, in Bishop and Williams 2012: 89).

The art heritage from which socially engaged art emanates, and which, in turn, informs social practice placemaking, emerged out of the situational turn in the arts out of the gallery, predominantly from the 1960s onwards. This had an ancillary move away from monologically expressive art where art functioned as a 'repository of values' (Kester 2004: 87), with the artist abstracted from the audience via an intermediary art object (Finkelpearl 2000: 27; Kester 2004: 89; Reed 2005: 28–29). Artists such as Beuys and Kaprow, and collectives such as Fluxus and The Diggers, removed the artist ego position and created space for subsequent anti-individualistic art practice. They desired to act as 'radically-related', with a focus on the means and ways of interaction and intersubjectivity (Gablik 1992: 2–6), inviting the audience closer to the artwork (Kwon 2004: 66), reflective of the culture of the community in which the 'viewer's physical and cognitive interaction is integral to the work itself' (Kester 2004: 51). Such art operates Freire's (1972) 'problem-posing pedagogy'; it does not replicate the dominant ideologies or modes of production but operates in a third space between art and its own critique, 'revealing temporality and renewed possibility of society as horizontal and dialogic' (Sherlock 1998: 219). The situational turn expresses a site-specific tendency, art work responding to, reflecting and exploring 'the temporal and circumstantial context in which it inhabits' (Klanten et al. 2012: 131). In the public realm specifically, such artists operate to question the prescribed city function and engage local people in a social criticism of this, involving strategies of micro-communities of human interaction (Kester 2011: 29). Artists turn to a place-based social practice: from the push factors of wanting to work 'under their own enquiry' (Cornford 2008) with the everyday and to be disruptive to macro politics; to the pull factors of frustrations with the regeneration process and the desire to materialise a less bureaucratically bound public realm strategy by working immersed in the community locale and taking artistic cues from the site (Brown 2012: 10).

As a socially engaged arts practice, social practice placemaking involves a deeper level of engagement in the process of art making than is to be found in other placemakings. Artists in this practice are moving beyond participatory art to co-production, working to the side of cultural gatekeepers and with a 'critical materialism' (Whybrow 2011: 28) where art does not equal an outcome of labour but is the labour itself. Using socially engaged arts processes, social practice placemaking works to 'jolt cultural assumptions', using the performative everyday to engender an 'engagement through alienation' (Klanten and Hübner 2010: 3; Rendell 2006) and a 'disruption of the sensible' expected urban norm (from Rancière, Beyes 2010: 231) in the urban event space (Hannah 2009: 117). As performative, it is 'enmeshed unselfconsciously within social life, rather than a conscious decision to participate in capital "C" culture' (Messham-Muir 2009: 120–123) of public realm placemaking or creative placemaking for example. Social practice placemaking, then, is concerned with the social aesthetic encounter – an 'encounter art' (Rancière 2006) – found in both relational (Bourriaud 1998/2006: 161–165) and dialogic (Kester 2011) aesthetics and the 'relational specificity' of the interactions between objects, people and spaces (Kwon, in Rendell 2006: 33). This is an art process of meetings, encounters, events and various types of collaboration between people that works around intersubjective exchange (Kester 2011: 29) of seven forms: cooperation, interaction, participation, sustainability, responsibility, authorship and feedback (Bosch and Theis 2012: 14). It works from a collaborative arts epistemology in which all participants have legitimate claims to knowledge construction and devising that aligns to a people-in-place-based placemaking and the democratisation of urban design as a social concern. Its process is one of dissolved categories of artist and non-artist, participant and audience, where the artist is moved from the participatory stance of 'elevated outsider' to co-producing 'engaged partner' (Adamek and Lorenz 2008: 57). This has a consequential deconstruction of relationships to, and roles in, urban public space. As a co-produced endeavour, knowledge creation is not singular but relational, occurring between community members that possess a holistic view and knowledge of their place. Relative expert skills will be deployed strategically and tactically at different stages of a project, with the locus of power with the community. This collaboration results in diverse outcomes, personal growth and knowledge exchange (Lehmann 2009: 18) where residents become co-designers in the process of urban regeneration, acting as a connect between the proscribed urban designers and the users of the city (Klanten et al. 2012: 209). The social practice locale is materially localised as site-specific (Klanten and Hübner 2010: 63) and concerned with creating an urban built environment that enfranchises urban dwellers through an active appropriation of space (Sherlock 1998: 219–220).

Such a place-framing approach harnesses socio-spatial relations and networks and 'socio-spatial positionality' (Martin 2013: 85–86) holding a transformative agency to motivate a place-based activism towards an aim of effecting change. Social practice placemaking can be viewed as indicative of a street-level activism, a type of politics where people are involved at the hyperlocal in issues that affect them, from a point of disaffection with formal politics (Bishop and Williams 2012: 138). Functions and aesthetics of the city are questioned, as is private ownership of the public realm, its access and use rights (Sherlock 1998: 220). Community conscientisation, via co-production in social practice placemaking, is a process of developing a community-level critical awareness of lived experience (Colombo et al. 2001: 457; Madyaningrum and Sonn 2011: 360; Sorensen 2009: 207) and the affirmation of the community's own narrative (Colombo et al. 2001: 460; Grodach 2010: 489; Madyaningrum and Sonn 2011): the process is one of place attachment and leads to individual and community empowerment (Bishop and Williams 2012: 23; Hall and Smith 2005: 175). This conscientisation leads to a subconscious desire to be involved in culture at a deeper level (Messham-Muir 2009: 123) and can (re)form new spatial, cultural and social identities (Franck and Stevens 2007; Hall and Smith 2005: 176). This process has facilitated a nuanced and applied understanding of community in placemaking discourse and marked a turn from '"what makes a good place" to "what – and who – make a good placemaking process?"' (Silberberg 2013: 51): a sector shift, in some practices, away from 'place making' to 'place shaping', a 'making-focused paradigm for the practice' (Silberberg 2013: 11). This position accepts that places are already existent before a placemaking intervention – and that placemaking is a subsequent, additional mode of intervention – and signals the community loci in placemaking practice. It also aligns with new urbanism (the notion that 'well-designed public space, centrally located within an urban village, will foster or create community by bringing people closer together' (Iveson 2014, in Gieseking and Mangold 2014: 188)) through its horizontal and self-determining participation models. A 'virtuous cycle of placemaking' is created of mutual and consequent community and place transformation (Silberberg 2013: 3). Taking these cues this chapter closes with a consideration of the implications of the placemaking typology for the sector.

Implications of the typology for placemaking

Across placemaking, stakeholders enter the common field with diverse agendas and knowledges and have inherently varied and numerous aims and objectives. When working at its optimum, no single discipline or site of knowledge takes precedence over another in placemaking, but may be called into primary use at different stages in the process. This should be the norm of placemaking and the placemaking typology aims to both elucidate this condition as well as to aid the sector to locate its practices in any given context. It is hoped that the typology will work to counter placemaking as an 'ill-defined buzzword' (Fleming 2007: 13) subject to contradictory overuse by planners and developers as a branding tool – a 'place wash' (Legge 2013) of homogenised and surface rendering of place work. The purpose of the typology is to share knowledge across types of placemaking and to redress exclusory power practices by uncovering the many different types of placemaking undertaken by different ecologies of practice and people and result in the opening up of a continually negotiated border position. With this border negotiation, the placemaking sector too can engage those 'outside' of it through a clear articulation of the variety of practice and the value of these practices. Appreciating that no single lens can be adopted in placemaking – given that it is a multiscalar field of varied actors – the typology aims to be of developmental, conceptual and pragmatic use to a rapidly evolving sector. It is offered as a classification of observed placemaking practices and is designed to be operationally tested against and in practice to provide a 'new analytical utility' (Pierce et al. 2011: 54) for the sector. The collaborative, co-produced aspect of placemaking also aids Roberts' (2009: 440) call for 'intra-professional action learning sets'. These sets include the community in a wide network of actors and could extend the placemaking sector further towards a practice field where it can operate as an open process, filled with democratic possibility. This could draw together, on an equal basis, amateurs and professionals (artists, planners, policy makers, architects et al.): people with a range of skills and experience who congregate and collaborate, shifting their previously fixed positions, becoming alternately producers and spectators, viewers and evaluators (Hope 2010: 69).

By drawing attention to the increasing volume of work by critically and socially engaged artists in place-based, co-produced work, the typology – and the further new classification of social practice placemaking – is also part of a critical discourse around the role of arts in the wider placemaking endeavour, spotlighting the power relations of how place is made in the city, who is making it and the use of aesthetic power in the urban discourse. Any expert appropriation of placemaking renders placemaking effectively redundant as it denies the community/ies in question the opportunity to take control over their lives. This is useful to inform the thinking of different types of placemaking and to consider where these practices could be located and their relative power relations: for example, is a strategic position concerned with defending power in imposing a view of urban form and function, whereas a tactical position is defined by the absence of power and in iterative manoeuvrings within place? Placing art into the Lefebvre (1968/1984) triad of space and representation activates its agency as a relational practice. It is of networks, as a place of exchange and interaction; of manifold political, social and cultural borders; and of denoting difference in lived experiences. In the culturised city, 'the arts' become part of the city's symbolic and fiscal economy. However, creativity – and by extension, arts in placemaking – is also a site of resistance to culturisation in the city, a 'call and response among different social groups' (Zukin 1995: 2) to find, create and maintain sites of different cultural value through new city visualisations, moving placemaking towards being an intersectional practice of space and place.

References

Adamek, M. and Lorenz, K. (2008). Be a crossroads: public art practice and the cultural hybrid. In Cartiere, C. and Willis, S. (eds.), *The Practice of Public Art*, pp. 70–79. New York: Routledge.

Beyes, T. (2010). Uncontained: the art and politics of reconfiguring urban space. *Culture and Organisation*, 16 (2), 229–246.

Bishop, P. and Williams, L. (2012). *The Temporary City*. Abingdon: Routledge.

Bosch, S. and Theis, A. (eds.) (2012). *Connection: Artists in Communication*. Belfast: Interface.

Bourriaud, N. (1998/2006). Relational aesthetics. In Bishop C. (ed.), *Participation*, pp. 161–165. London: Whitechapel Gallery.

Brown, A. (2012). All the world's a stage: venues, settings and the role they play in shaping patterns of arts participation. *GIA Reader*, 23 (2). Available at: www.giarts.org/article/all-worlds-stage (date accessed 22 August 2019).

Cohen, M. (2009). Place making and place faking: the continuing presence of cultural ephemera. In Lehmann, S. (ed.), *Back to the City: Strategies for Informal Intervention*. Ostfildern: Hatje Cantz.

Colombo, M., Mosso, C. and de Piccoli, N. (2001). Sense of community and participation in urban contexts. *Journal of Community and Applied Social Psychology*, 11 (6), 457–464.

Cornford, M. (2008). Takin' it to the streets. In *ixia*. Available at http://ixia-info.com/new-writing/mat thewcornford/ [Accessed: 10th March 2014].

Courage, C. (2017). *Arts in Place: The Arts, the Urban and Social Practice*. Abingdon: Routledge.

Crawford, M. (1999). Blurring the boundaries: public space and private life. In Chase, J., Crawford, M. and Kalishi, J. (eds.), *Everyday Urbanism*. New York: The Monacelli Press.

Deutsche, R. (1996). *Evictions: Art and Spatial Politics*. Cambridge, MA: MIT Press.

Finkelpearl, T. (2000). Interview: Paulo Freire: discussing dialogue. In Finkelpearl, T. (ed.), *Dialogues in Public Art*, pp. 276–293. Cambridge, MA: MIT Press.

Fleming, R. L. (2007). *The Art of Placemaking: Interpreting Community through Public Art and Urban Design*. London: Merrell.

Franck, K. A. and Stevens, Q. (2007). Patterns of the unplanned: urban catalyst. In Franck, K. A. and Stevens, Q. (eds.), *Loose Space: Possibility and Diversity in Urban Life*, pp. 271–288. Abingdon: Routledge.

Freire, P. (1972). *Cultural Action for Freedom*. Harmondsworth: Penguin.

Froggett, L., Little, R., Roy, A. and Whitaker, L. (2011). *New Model of Visual Arts Organisations and Social Engagement University of Central Lancashire Psychosocial Research Unit*. [Online]. Available at: http://clok. uclan.ac.uk/3024/1/WzW-NMI_Report%5B1%5D.pdf. (Accessed: 12th August 2015).

Gablik, S. (1992). Connective aesthetics. *American Art*, 6 (2), 2–7.

Gallant, M. (2013). A vibrant transformation: cities and states take creative placemaking to new heights. In NEA Arts *Arts and Culture at the Core*, 3rd November 2013. Available at http://arts.gov/ NEARTS/2012v3-arts-and-culture-core/vibrant-transformation. [Accessed on 4th December 2013].

Gieseking, J. and Mangold, W. (eds.) (2014). *The People, Place, and Space Reader*. New York: Routledge.

Grodach, C. (2010). Art spaces, public space and the link to community development. *Community Development Journal*, 45 (4), 474–493.

Hall, T. and Smith, C. (2005). Public art in the city: meanings, values, attitudes and roles. In Miles, M. and Hall, T. (eds.), *Interventions: Advances in Art and Urban Futures (Vol 4)*, pp. 175–179. Bristol: Intellect Books.

Hannah, D. (2009). City as event space: defying all calculation. In Lehmann, S. (ed.), *Back to the City: Strategies for Informal Intervention*, pp. 114–119. Ostfildern: Hatje Cantz.

Hermansen, C. (2011). Social creativity. In *SCRIBE: Scarcity and Creativity in the Built Environment working paper no3* (January 2011). Available at www.scibe.eu/wp-content/uploads/2010/11/03-CH.pdf. [Accessed on 2nd July 2013].

Hope, S. (2010) Who speaks? who listens? Het Reservaat and critical friends. In Walwin, J. (ed.), *Searching for Art's New Publics*, pp. 64–78. Bristol: Intellect.

Iveson, K. (2014) 'Putting the public back into public space' (1998). In Gieseking, J., Mangold, W., Katz, C., Low, S. and Saegert, S. (eds.), *The People, Place, and Space Reader*, pp. 221–225. London: Routledge.

Kelly, O. (1984). *Community, Art and the State*. London: Comedia.

Kester, G. H. (2004). *Conversation Pieces: Community and Communication in Modern Art*. Berkeley, CA: University of California Press.

Kester, G. H. (2011) *The One and the Many: Contemporary Collaborative Art in a Global Context*. Durham: Duke University Press.

Klanten, R., Ehmann, S., Borges, S, Hübner, M. and Feireiss, L. (2012). *Going Public: Public Architecture, Urbanism and Interventions*. Berlin: Gestalten.

Klanten, R. and Hübner, M. (2010). *Urban Interventions: Personal Projects in Public Spaces*. Berlin: Gestalten.

Kuskins, J. (2013). Love or hate it, user-generated urbanism may be the future of cities. In *Urbanism*. Available at: http://gizmodo.com/love-it-or-hate-it-user-generated-urbanism-may-be-the-1344794381. [Accessed on 4th December 2013].

Kwon, M. (2004). *One Place after Another: Site-Specific Art and Located Identity*. Cambridge, MA: The MIT Press.

Lefebvre, H. (1968/1984). *The Production of Space*. (translated, Nicholson-Smith, D.). Malden, MA: Blackwell Publishing.

Legge, K. (2012). *Doing It Differently*. Sydney: Place Partners.

Legge, K. (2013). *Future City Solutions*. Sydney: Place Partners.

Lehmann, S. (2009). Hidden in the urban fabric: art and architecture, a case study of collaboration in interdisciplinary contexts. In Lehmann, S. (ed.), *Back to the City: Strategies for Informal Interventions*. Ostfildern: Hatje Cantz.

Lepofsky, J. and Fraser, J. C. (2003). Building community citizens: claiming the right to place-making in the city. *Urban Studies*, 40 (1), 127–142.

Lilliendahl Larsen, J. (2014). Lefebvrean vagueness: going beyond diversion in the production of new spaces. Inn Stanek, Ł., Schmid, C. and Moravânszky, Á. (eds.), *Urban Revolution Now: Henri Lefebvre in Social Research and Architecture*, pp. 319–339. Farnham: Ashgate.

Lydon, M. and Garcia, A. (2015). *Tactical Urbanism: Short-term Action for Long-term Change*. Washington: Island Press.

Madyaningrum, M.E. and Sonn, C. (2011). Exploring the meaning of participation in a community art project: a case study on the seeming project. *Journal of Community and Applied Social Psychology*, 21 (4), 358–370.

Markusen, A. and Gadwa, A. (2010a). *Creative Placemaking White Paper*. Available at: www.nea.gov/pub/CreativePlacemaking-Paper.pdf. [Accessed: 5th October 2013].

Markusen, A. and Gadwa, A. (2010b). *Creative Placemaking White Paper Executive Summary*. Available at: www.nea.gov/pub/CreativePlacemaking-Paper.pdf. [Accessed: 5th October 2013].

Markusen, A. and Gadwa Nicodemus, A. (2017). Creative placemaking: reflections on a 21st-century American arts policy initiative. In Courage, C. and McKeown, A. (eds.), *Creative Placemaking: Research, Theory and Practice*, pp. 11–27. Abingdon: Routledge.

Martin, D. G. (2013). Place frames: analysing practice and production of place in contentious politics. In Nicholls, W., Miller, B. and Beaumont, J. (eds.), *Spaces of Contention: Spatialities and Social Movements*, pp. 85–102. Farnham: Ashgate.

Massey, D. (2005). *For Space*. London: Sage.

Messham-Muir, K. (2009). Beneath the pavement. In Lehmann, S. (ed.), *Back to the City: Strategies for Informal Urban Interventions*, pp. 120–123. Ostfildern: Hatje Cantz.

Mitchell, D. (2014) *The Right to the City*. New York: The Guildford Press.
Newman, J. (2001). *Modernising Governance: New Labour, Policy and Society*. London: Sage.
Petrescu, D. (2006). Working with uncertainty towards a real public space. In *If You Can't Find It, Give Us A Ring: public works*.
Pierce, J., Martin, D. G. and Murphy, J. (2011). Relational placemaking: the networked politics of place. *Transactions of the Institute of British Geographers*, 36 (1), 54–70.
Project for Public Spaces (2014). *All Placemaking is Creative: How a Shared Focus on Place Builds Vibrant Destinations*. Available at www.pps.org/blog/placemaking-as-community-creativity-how-a-shared-focus-on-place-builds-vibrant-destinations/. [Accessed on 4th January 2014].
Rancière, J. (2006). Problems and transformations in critical art. In Bishop, C. (ed.), *Participation*. London: Whitechapel Gallery and The MIT Press.
Reed, S. (2005). Art and citizenship. In Charity, R. (ed.), *ReViews: Artists and Public Space*. London: Black Dog.
Rendell, J. (2006). *Art and Architecture: A Place Between*. London: I. B. Tauris.
Roberts, P. (2009). Shaping, making and managing places: creating and maintaining sustainable communities through the delivery of enhanced skills and knowledge. *Town Planning Review*, 80 (4–5).
Rochielle, J. (n.d.). Taking it to the streets; artists hit the road using creativity, communication and food to address social issues. *Public Art Review*, 13 (2), 40–42.
Schneekloth, L. H. and Shibley, R. G. (2000). Implacing architecture into the practice of placemaking. *Journal of Architectural Education*, 53 (3), 130–140.
Sen, A. and Silverman, L. (2014). Introduction – embodied placemaking: an important category of critical analysis. In Sen, A. and Silverman, L. (eds.), *Making Place: Space and Embodiment in the City*, pp. 1–18. Bloomington: Indiana University Press.
Sherlock, M. (1998). Postscript - no loitering: art as social practice. In Harper, G. (ed.), *Interventions and Provocations: Conversations on Art, Culture and Resistance*, pp. 219–225. Albany: State University of New York Press.
Silberberg, S. (2013). *Places in the Making: How Placemaking Builds Places and Communities*. MIT Department of Urban Studies and Planning [Online]. Available at: http://dusp.mit.edu/cdd/project/placemaking. [Accessed: 13th August 2015].
Sime, J. D. (1986). Creating places or designing spaces? *Journal of Environmental Psychology*, 6 (1), 49–63.
Social Impact of the Arts Project (SIAP) of University of Pennsylvania (2007). *The Power of Place-Making: A Summary of Creativity and Neighborhood Development: Strategies for Community Investment*. Available at www.sp2.upenn.edu/siap/docs/cultural_and_community_revitalization/power_of_placemaking.pdf. [Accessed: 2nd July 2013].
Sorensen, A. (2009). Neighbourhood streets as meaningful spaces: claiming rights to shared spaces in Tokyo. *City and Society*, 21 (2), 207–229.
The Knight Foundation (2010). *Soul of the Community*. Available at: www.soulofthecommunity.org/. [Accessed on 7th September 2013].
Thompson, N. (2012). Socially engaged art is a mess worth making. *Architect*, (August 2012), 86–87.
Tonkiss, F. (2013). *Cities by Design: The Social Life of Urban Form*. Cambridge: Polity.
Watson, S. (2006). *City Publics: The (Dis)enchantments of Urban Encounters*. London: Routledge.
Whybrow, N. (2011). *Art and the City*. London: I B Tauris.
Zukin, S. (1995). *The Cultures of Cities*. Malden, MA: Blackwell Publishers.
Zukin, S. (2010). *Naked City: The Death and Life of Authentic Urban Place*. Oxford: Oxford University Press.

56

Contemporary British place writing

Towards a definition

David Cooper

Introduction: contemporary British writing and place

Place, in contemporary British literature, is everywhere. It is evident in the emergence of independent publishers such as Little Toller, Longbarrow Press, Penned in the Margins and Uniform Books: presses with radically differing creative agendas and aesthetics that share a commitment to new writing that explores the topographies and texturalities of particular locations. It is similarly evident in the proliferation of online fora and magazines, including *Caught by the River* and *Elsewhere: A Journal of Place*, that act as digital meeting-points for writers and readers interested in the literary articulation of geographical experiences and imaginaries. The preoccupation is also manifest in the cultural pages of the national newspapers as, during an age in which the journalistic space afforded to literary writing has been ever shrinking, there has been an exponential growth in the number of articles and reviews examining the literature of place. Central to much of this activity has been the work of Robert Macfarlane: a writer whose creative-critical practices are threaded through with the knowledge that 'placeless events are inconceivable, in that everything that happens must happen somewhere' (2010: 113). Perhaps more than anyone, Macfarlane, through a long-standing imaginative excavation of named landscapes and their cultural representation, has both captured and created the sense that, in British literary culture, we are living in a period of place.

Feeding off this wider cultural context, this chapter's principal interest is the term 'place writing' itself: a collocation that has increasingly entered creative and critical literary discourse (Smith, 2013); but a label that has yet to receive substantial scholarly scrutiny. Immediately, critical questions proliferate. Some relate to the reader's identification of place writing as a mode of literary expression. What are the cardinal characteristics of such work? Is place writing associated with a particular literary form or is it applied as a generic category? Further questions inevitably relate to content. What are the dominant thematic tropes in contemporary place writing? How do writers use literary forms and language to reconfigure and/or reinscribe extant understandings of place and placelessness? The exploration of these questions can be particularised by turning to *Towards Re-Enchantment: Place and its Meanings* (2010) and *Ground Work: Writings on Places and People* (2018): two field-defining collections that feature writers as prominent and various as Kathleen Jamie, Richard Mabey, Helen Macdonald, Macfarlane, Alice Oswald, Iain Sinclair and Marina Warner. This chapter proposes that these twin publications provide influential snapshots of contemporary British place writing. By extension, in focussing on the forms and themes that emerge from these collections, this chapter moves towards a critical definition of a term that has crept into the cultural consciousness, and literary critical lexicon, in Britain over the first two decades of this century.

Collecting place writing: two landmark publications

In 2010–2011, Artevents co-ordinated 'The Re-Enchantment': a national project that sought 'to interrogate the various meanings of "place" in the twenty-first century' (2010). As part of this project, Artevents published *Towards Re-Enchantment: Place and Its Meanings*: a collection of new writing, edited by Gareth Evans and Di Robson, in which eleven contributors were invited to reflect 'on specific locations from across the diverse landscapes of the British Isles, and on the potential for "re-enchantment"' (2010). In publishing work by a heterogeneous range of voices – including Jay Griffiths, Jane Rendell and Ken Worpole, as well as Jamie, Macfarlane and Sinclair – the co-editors implicitly challenge the borders between extant strands of cultural categorisation as the disparate work of psychogeographers, nature writers, architectural historians and landscape poets are gathered together in one textual space. In framing this heterogeneity, the co-editors eschew a conventional introduction in favour of a three-sentence note, on the front cover, that proffers the creative content as phenomenological panacea: 'Here are paths, offered like an open hand, towards a new way of being in the world' for 'the multiple alienations of modern society' (2010). The result is the creation of a literary map of England, Scotland and Wales – stretching from the Isle of Lewis to the port city of Aberdeen, from Ystrad Fflur in Cardiganshire to Upper Clapton in London – in which the contributions are unified by a collective concern with 'the importance of "place" to creative possibility in life and art' (2010). In turning to 'place' as their cardinal term, Evans and Robson also destabilise imaginatively entrenched distinctions between the urban and the rural to demonstrate – both intratextually across the collection and within individual contributions – the complex messiness of contemporary geographies and their 'meanings'.

Eight years later, another anthology, *Ground Work: Writings on Places and People*, edited by Tim Dee, featured the work of 31 writers and artists. There is some contributor crossover as writings by Mabey and Worpole feature in both books. There's also a shared spatial breadth as Dee's collection 'is a book of writing about places' with 'the personal geographies' coming 'from as many acres as people' (2018: 1). As with *Towards Re-Enchantment*, it is explicitly place – in all of its knotty and contradictory 'meanings' – that provides the experiential and imaginative foundations for Dee's volume. In contrast to *Towards Re-Enchantment*, however, *Ground Work* is prefaced by a relatively expansive and poetically polemical Introduction in which Dee examines how places are 'anthropogenic creations called into being by the meeting of humans and their environment' (2018: 1). Saliently, Dee uses the compound 'place-writing' to refer to both earlier work and the new work that he has brought together. Dee doesn't trace the lineage of this term nor does he self-reflexively interrogate his own use of the label. Yet, through this absence of editorial contextualisation, he implicitly assumes that the reader will share his understanding of what place writing (hyphenated or not) might be and do. Given that this particular collocation does not feature in the brief prefatory note to *Towards Re-Enchantment*, Dee's confident use of the term suggests that, by 2018, the label place writing – a label that can be retrospectively applied to the collection edited by Evans and Robson – had flowed into the literary mainstream.

Dee uses his Introduction to identify some of the 'cultural, ecological or spiritual' (Evans and Robson, 2010) preoccupations of the contemporary British place writer. Throughout, Dee's language is distinctly Heideggerian as he contends that place writing carries the capacity to remind the reader of 'the *place-ness* of place' in an age in which 'most of the time most of us are *unplaced*' (2018: 1–3). Dee perceives place writing as a creative counter to the spatial abstractions of late-capitalist, neo-liberal globalisation: a world in which 'specificities have been dulled, local habitations and names globalised, the instress or haecceity of every street or field driven from common memory' (2018: 3). In other words, place writing can highlight and celebrate what the environmental charity Common Ground refers to as '*local* distinctiveness': 'the diverse, local and intimate connections that people have had, and might yet have, with the landscape that surrounds them' (2018: 11). Moreover, such literary work can offer an imaginative antidote to how the practice of everyday life increasingly unfolds within 'untextured places': 'the unmuddy world of the depthless screen and the sealed space' (2018: 3). Place writing, therefore, reminds the reader of the embodied situatedness of what it means to-be-in-the-world; by extension, it opens up 'the potential', for both writer and reader, of quotidian 're-enchantment' (Evans and Robson, 2010).

A series of issues and questions, however, are raised by the framings in both books, some overtly relating to the politics of place. Back in 2010, Evans and Robson's prefatory reference to 'our sense of belonging' might have seemed urgently apposite for the Anthropocene; but, almost a decade and a schismatic referendum later, that 'our' also feels problematically exclusionary. Dee's persistent emphasis on the re-thickening of the phenomenological experience of place may be founded upon the admirable environmental principles that underpin the work of Common Ground. Yet he is also vulnerable to the critique that he is offering an intrinsically nostalgic vision of place – and conception of place writing – that is predicated on both a reactionary rejection of the affordances of digital technologies and a reductive and elitist critique of 'non-places' (2018: 2). Dee appears closed to the possibility that digital technologies might potentially enrich an understanding of place; and he remains wedded to an analysis that denies that 'airports … offices, hospitals, supermarkets' (2018: 2) might also be experienced as places of attachment and meaning (see Chapter 8 by Bissell, this volume). Other issues and questions are of a more literary nature. Most significantly, within the context of this chapter, it remains unclear – from reading both the prefatory note to *Towards Re-Enchantment* and Dee's Introduction – as to what literary form(s) place writing takes.

Creative non-fiction: place, prose and the authorial 'I'

Creative non-fiction, as a literary form, remains surprisingly under-theorised. According to Lee Gutkind, creative non-fictional writing is necessarily predicated on the telling of truths and, as a result, ought to be 'as accurate as the most meticulous reportage': 'names, dates, places, descriptions, quotations may not be created or altered for any reason, at any time' (1997: 10). Crucially, for Gutkind, what elevates creative non-fiction above documentary journalism is the imaginative space that it allows for the articulation of subjective thought: 'More often than not, writers turn to the creative nonfiction genre because they feel passionately about a person, *place* [my italics], subject, or issue and have no interest or intention of maintaining a balanced or objective tone or viewpoint' (1997: 12). In short, creative non-fiction is characterised by the unapologetic presence of the authorial self.

Drawing upon these definitions, it becomes evident that, in situating *Towards Re-Enchantment* and *Ground Work* as field-shaping collections, creative non-fictional prose lies at the centre of contemporary British place writing. In bringing together writers from a suite of cultural and disciplinary backgrounds, the collections present the reader with both a disparate range of topographies and of disciplinary approaches. Worpole, for example, weaves personal reflection, literary criticism and a professional knowledge of the European Landscape Convention, to think about the aesthetics of 'the Essex coastal landscape' (2010). Tim Ingold, on the other hand, implicitly draws upon anthropological thought to reflect upon 'somewhere in northern Karelia' (2018: 132). Many of the creative non-fictional contributions, however, share formal characteristics in that the dominant mode is the prose essay that draws upon the author's personal experience of how 'places work on us' as well as 'what places might look like to themselves' (Dee, 2018: 1). They are writings, to apply Gutkind's terms, that present the reader with a subjective account of the truth of place.

The overarching effect is an emphasis on authenticity in terms of both the sincerity of the writerly voice and the veracity of the 'placial' (Casey, 2002: 351) experiences that are subjected to literary narration. When Griffiths writes about the grave of Dafydd dap Gwilym, therefore, the reader is not encouraged to disbelieve that the author felt the rain on her skin as she entered the Ystrad Fflur graveyard in which the poet is buried (Griffiths, 2010). Perhaps inevitably, in articulating first-hand 'meetings of people and world' (Dee, 2018: 7), many of the contributions in both books offer neo-Romantic reflections – almost invariably made through the prism of a named place – on the entanglements of memory and the geographies of the everyday. In 'Tekels Park', for instance, Macdonald attempts to catch sight of her childhood home – 'a place that draws me because it exists neither wholly in the past, nor in the present, but is caught in a space in between' (2018: 155) – as she drives along the M3 motorway in Surrey. In exploring the imbrications of self and location, such creative non-fiction writers address twin site-specific questions posed by Macfarlane: 'firstly, what do I know when I am in this place that I can know nowhere else? And then, vainly, what does this place know of me that I cannot know of myself?' (2012: 27).

Many of the autobiographical writers included in both collections aren't demonstrably preoccupied with expanding the formal properties and possibilities of place writing. Connected with this, there is a tendency for these writers, in privileging the documentation of actual geographical experiences, to refrain from straying too far into the worlds of the imagined and the unmappable. Clearly, they are invariably concerned with how literary language can be pushed and pulled to encapsulate encounters with place; but, at the same time, the preponderance of autobiographical accounts raises questions regarding the extent to which radical creative experimentation underpins such writing. It's important to note that there are exceptions to these dominant tropes. Peter Davidson, for example, begins by highlighting the imagined territories opened up by archival objects as he contemplates 'a place which I have never seen and which I know by heart': a coloured drawing that 'John Aubrey (1626–97) made of his family house, grounds and farmlands at Easton Piers, near Kington St Michael in Wiltshire' (2018: 71). Alternatively, David Matless eschews the first-person in creating a difficult-to-define prose essay, set on the Norfolk coast, that thinks about the 'Anthroposcenic': the landscape 'emblematic of processes marking the Anthroposcene' (2018: 185). Even here, though, Matless's fragmented landscape descriptions are interlaced with gnomic memories of being-in-place: 'Holidays at East Runton; forty years ago, with predictions of a new ice age' (2018: 187). Ultimately, then, both *Towards Re-Enchantment* and *Ground Work* project the overriding sense that the prose exploration of the authorial self-in-place is *the* dominant and defining mode of contemporary place writing.

Poetry: place, quiet lyrics and radical landscapes

In spite of the dominance of prose non-fiction, the editors of both collections also afford space to the poetry of place. Most of the poems they include are lyrics. The term 'lyric' is notoriously slippery but there 'are certain consistent features in definitions': 'it is characterised by brevity, deploys a first-person speaker or persona, involves performance, and is an outlet for personal emotion' (Brewster, 2009: 1). *Towards Re-Enchantment* opens with 'Tillydrone Motte': a fifty-line lyric in which Robin Robertson revisits a childhood landscape of Seaton Park, Aberdeen. On the surface, 'Tillydrone Motte' is a poem of nostalgia: an imaginative return to the 'fifteen years' that the poet spent 'here on this highest edge,/this hill, in this park; my garden/spread out for me two hundred feet below,/the Don coursing through it, out towards the sea' (2010: 9). It is a poem that laments the loss of the deep, embodied knowledge of place generated through childhood play. This Romantic nostalgia, though, is undercut by a self-awareness forged out of the experience and knowledge – both geographical and personal – that the poet has accrued during the intervening years. Other poems within the two collections similarly situate the authorial self-in-place. In 'Hevenyssh', Lavinia Greenlaw's first-person speaker meditates on the large sky and liquid landscape at Holkham on the north Norfolk coast: 'There is no place as airy and dilute/as level or simple' (2010: 105). *Ground Work*, on the other hand, contains 'Waders': a long poem by the former Poet Laureate, Andrew Motion. Consisting of ten sonnets, the poem's place is the speaker's childhood home and the text's overriding tone is elegiac as, through a series of objects, the poet-speaker imaginatively reconstructs 'the former glories of the house' (2018: 193).

There are clear correspondences between creative non-fictional prose and such loco-specific lyrical poetries. As with much contemporary creative non-fiction, the readers of such poems of place aren't dissuaded from eliding the authorial self with the first-person speaker; and in both types of writing there's an inclination to root the literary text within an unambiguously verifiable, mappable world. There are thematic cross-fertilisations too in terms of the shared – and largely topophilic – concerns with memory, nostalgia and the intimate spaces of childhood. It's not coincidental that a number of contemporary writers of place – including Kathleen Jamie, Jean Sprackland, Paul Farley and Michael Symmons Roberts – have oscillated between poetry and non-fictional prose when writing about personal geographies. Through a reading of both *Towards Re-Enchantment* and *Ground Work*, therefore, it appears uncontentious to assert that geographically focussed contemporary lyric poetry can also be categorised as place writing.

There's a need, however, to look beyond the lyric. In spite of the diversity of their chosen places, Robertson, Greenlaw and Motion share a commitment to traditional poetic forms: a commitment that's ordinarily associated with the poetic 'mainstream' in Britain and Ireland (Alexander and Cooper, 2013: 2). In contrast, Elisabeth Bletsoe – whose 'Votives to St Wite' is included in *Towards Re-Enchantment* – is a poet whose work 'flouts the categorisations of contemporary British literature, particularly those related to nature writing, ecopoetics, ecofeminism, L=A=N=G-U=A=G=E poetry, experimentalism and the avant-garde' (Ryan, 2018: 82). 'Votives to St Wite' encapsulates the 'palpable eclecticism' (Ibid: 82) of Bletsoe's practice: a long poem, written in neo-Modernist free verse, that ventriloquises the spirit of the saint of Whitchurch Canonicorum in Dorset. 'The result', according to Jeremy Hooker, 'is a wonderful poetic embodiment, in which St Wite is "wedded" to the place, an incarnate spirit present in the total ecology, natural and geological, and in language – words of her own time, and words of now' (2010).

This psychogeographic 'verbal palimpsest' (Hooker, 2011) offers an alternative version and vision of the contemporary place poem by eschewing the convention of the stable autobiographical 'I'. Bletsoe's poem draws attention to how the term place writing needs to afford space for the polyphonic poetic geographies of the imagination as well as the embodied geographies of authentic experience. At the same time, the shifting temporalities of the poem indicate how contemporary place writing can think about a site's histories that stretch far beyond the 'imprisoned' (Bletsoe, 2010: 87) temporality of an author's own lifetime. By extension, through the imagining of a deeper past, Bletsoe circumnavigates the problem of nostalgia evident in Dee's later editorial framing of place writing. In short, the inclusion of 'Votives to St Wite' in *Towards Re-Enchantment* underlines that the term place writing needs to incorporate formally experimental 'radical landscape poetry' (Tarlo, 2011) as well as the traditional loco-lyric.

Images and texts: place, writing and the visual arts

The definition of place writing is further elasticised through visual artistic practices. Each contribution to *Towards Re-Enchantment* is prefaced by a black and white image. Some are reproductions of paintings: the opening lines of 'Votive to St Wite', for example, appear opposite the reproduction of a painting, 'CLOUD PIERCING: Charmouth', by Frances Hatch (2010: 82). Primarily, though, these visual paratexts are grainy black and white photographs – some taken by the authors themselves – of the place that provides the geographical focus for the literary writing that follows. On the surface, the photographic images visually signify the truthfulness, to return to Gutkind's cardinal term, of the creative non-fictional text; the photographs reassure the reader that the writer has documented a material, experienced place.

Crucially, however, the practice also carries visual echoes of the work of W. G. Sebald: a writer whose spectral influence hovers over much contemporary creative non-fictional place writing; and who simultaneously reinforced and subverted the way that 'photographs have generally been regarded as a mode of documentation and continue to carry this denotation' (Furst, 2006: 220). There's a clear distinction to be made: whereas Sebald – in *The Rings of Saturn*, for instance – 'intersperses' (Furst, 2006: 220) photographs within the main body of his work, the images included within *Towards Re-Enchantment* uniformly sit outside the textual frames. Yet, by sharing Sebald's preoccupation with the difficult-to-discern, caption-free image, the editors of *Towards Re-Enchantment* implicitly raise readerly doubts about the 'realism' of place writing (Furst, 2006: 220). The image (Shawyer, 2010: 28) that prefaces Mabey's 'On the virtues of dis-enchantment', for instance, seemingly has three constituent parts – a cloudscape; a thin strip of land; and seawater – and, as the reader moves into the prose essay, it becomes apparent that the facing image is of the flat, littoral landscape of East Anglia (Mabey, 2010). The photograph, however, doesn't securely anchor the reader in this liminal lowland. Instead, the image, as in Sebald's work, both creates a sense of documentary realism and unsettles the reader: the dark horizontal line across the middle of the photograph appears to be land, but the absence of any discernible topographical features leads to hermeneutic uncertainty; the reflection of the clouds in the water ambiguates the distinction between sky and sea. The photograph serves more than a merely illustrative function by opening up the possibilities of uncertainty and the uncanny experience of place. By extension, its presence destabilises and undermines the implied veracity of contemporary place writing through the visualisation of the strange and difficult-to-define. The image prompts the reader to remain alert to uncertainties and instabilities within the main text that follows: qualities that emerge as key to Mabey's autobiographical reflections on East Anglia as a landscape that '*floats* on water' (2010: 29).

Dee, in *Ground Work*, displays even greater editorial openness towards visual practices. The contributions by Dexter Petley and Greg Poole both feature pencil sketches of natural phenomena – birds and beetles, flora and fauna – encountered in their respective home topographies of the Northumberland fishing village of Craster (Petley, 2018) and the city of Bristol (Poole, 2018). *Ground Work* also contains 'Childhood ground abiding places': a bringing together of text and a photographic image by the artist Richard Long. On the left-hand page appears a fourteen-line poem, reproduced in Long's signature font of Gill Sans, remembering key sites from the artist's past: 'The cliff ledge den/The look-out tree/ The bicycle racing track in the wood' (2018: 138). Imaginatively, the reader is situated in Robin Robertson territory as Long recalls the days and public places of childhood play. Opposite this poetic text, however, is a difficult-to-discern photographic image that consumes the entire page. Instead of providing a photographic representation of, say, the 'look-out tree', Long adopts his characteristic method of looking down and photographing the textured landscape at his feet (2018: 139). The resultant image is of a cracked, scarred pavement of rock. It's possible that this image is of the 'place where we dug some quartz with hammers' (2018: 138). Long deliberately resists a literalist yoking of text and photograph, however, and through this visual abstraction, invites the reader to think about the childhood experience of place as a pre-toponymised site of near-at-hand phenomenological encounter. If Long's early work challenges conventional understandings of what ought to be displayed within the environmentally controlled spaces of the gallery, Dee, by including Long's practice in *Ground Work* – alongside work by Petley and Poole – implicitly questions the cultural assumption that the term place writing should only be applied to texts published in conventional literary forms.

Fiction: place, narrative strategies and imagined geographies

Although this chapter is edging towards a definition of place writing, there remains a need to address a key issue: the relationship between place writing and narrative fiction. Since the publication of his first book of creative non-fiction, *Mountains of the Mind* (2003), Macfarlane has repeatedly acknowledged his imaginative indebtedness to writers of fiction: 'I have learned much myself as a writer – at the levels of the image, sentence and chapter – from the techniques of novelists' (2015). Crucially, the convergences extend to textual content as well as literary stylistics. That's to say, in spite of Gutkind's insistence that non-fiction ought to be rooted in fact, the suggestion that the writing of creative non-fiction allows for 'a *different* [my italics] level of truth' (1997: 10) also opens up the potential for textual elements that swerve away from real-world geographies or the authenticable personal experience of place.

The possibilities of the fictional, embedded within the frame of non-fiction, can be illustrated by moving beyond the two collections and towards *Common Ground*: a book in which Rob Cowen documents his obsession with an edgeland on the outskirts of the Yorkshire town of Harrogate following his relocation from London. *Common Ground* offers a deep mapping of a bounded plot in the 'no man's land between town and country' (2015: 3): a 'patch of earth' that Cowen comes to recognise as 'a place of transformation' (Ibid: 8). Unsurprisingly, Cowen interweaves this excavation of place with autobiographical reflections and, more particularly, recounts the efforts of home-making ahead of the arrival of his first child. Unusually, though, Cowen's text – in exploring the 'fusions of human and place' (Ibid: 205) – clearly oscillates between creative non-fictional and fictional modes. Most memorably, Cowen's use of the first-person becomes destabilised, in the chapter 'The union of opposites', as the narration is taken over by 'John Joseph Longthorne, born right here in Harrogate in 1945' (Ibid: 99). Cowen concocts a personal history for this character – a man who habitually sits in the corner of Café Nero in the centre of town – and imagines Longthorne's own deep engagement with the edgeland that provides the geographical focus for the text. As a result, *Common Ground* is a book that posits that a sense of place is constituted through a knotty synthesis of the real and the fictional, the experienced and the imagined. By extension, the act of fiction-making 'augments' the author's own embodied and 'instinctive' (Cowen, 2015: 205) sense of place.

Although *Common Ground* merges non-fictional and fictional modes and strategies, the book ostensibly remains rooted in the literary articulation of first-hand placial experience. However, could texts that are explicitly presented to the reader as works of fiction also be labelled as place writing? Neither *Towards Re-Enchantment* nor *Ground Work* contain any overtly fictionalised contributions; and as a result, the editors of both collections implicitly exclude fiction from their otherwise catholic presentations of the contemporary literary landscape. This impulse to distinguish between non-fictional and fictional responses to place is understandable in going some way to ensuring the autobiographical and material groundedness of place writing. Ultimately, though, the division is problematically artificial. Bletsoe's experimental poetics highlight the imaginative possibilities of giving voice to people from a place's deep past; and the Sebaldian preoccupation with difficult-to-make-out photographic images similarly contributes to the erosion of the unpoliceable border between fact and fiction. In addition, *Common Ground* exemplifies the potential for the place writer to embed fictional writing within the textual framework of the non-fictional. Given these textual uncertainties and instabilities, it seems reasonable that the term place writing could be applied to *any* kind of writing – including texts that are unambiguously packaged as works of fiction – that places place at its centre.

Scores of contemporary British novels could be used as illustrative models of fiction-as-place-writing: texts in which the narrative events could only ever unfold in the named locations in which they are set; literary works in which place itself emerges as a key agent. An example is provided by Zadie Smith's *NW* (2012): a novel whose title foregrounds the imaginative primacy of the suburban geographies of Willesden, London. *NW* maps out, through a range of narrative experimentations, the complicatedly coalescing lives of four locals. In giving a voice to each of these characters, Smith's polyphonic text explores how creative strategies can be used to encapsulate the dizzying multiplicity of contemporary (sub) urban experience. At the same time, Smith maps out place itself as *NW* is a novel that's as concerned with the particularities of Willesden – a suburb that is simultaneously singular and unremarkable – as it is with the characters whose lives are shaped by its streets and estates. The literary effect is the fictionalised reimagining of a real place that reinforces Doreen Massey's influential assertion – generated through her own autoethnographic account of nearby Kilburn High Road – that 'the specificity of place [...] derives from the fact that each place is the focus of a distinct mixture of wider and more local social relations' (1994: 156). Smith's novel captures the 'throwntogetherness' (Massey, 2005: 151) of contemporary

Willesden; and, as a result, the literary work offers a counter to Dee's anxiety (2018: 3) that globalisation necessarily leads to a diminishing of the particularities of place. Ultimately, *NW* calls attention to the ways that fiction can explore how places 'remain stubbornly *there*, itchy, palpable, determining' (Dee, 2018: 7).

Conclusion: towards a definition of place writing

The consideration of fiction leads back to a question posed at the beginning of this chapter: does place writing refer to particular literary forms or does it denote a literary genre? In revisiting this question, it's instructive to turn to the analogous label of 'life writing': a term, like place writing, that's variously used both with and without hyphenation. As Zachary Leader puts it: '"Life-writing" is a generic term used to describe a range of writings about lives or parts of lives, or which provide materials out of which lives or parts of lives are composed' (2015: 1). According to Leader, these writings straightforwardly include 'memoir, autobiography, biography, diaries'; but, saliently, he suggests that life writing can also refer to both 'autobiographical' and 'biographical fiction' (Ibid). Leader further complicates the genre by proposing that life writing includes a range of non-literary texts: 'letters, writs, wills, written anecdotes, depositions, court proceedings (*narratio* first existed not as a literary but as a legal term), marginalia, nonce writings, lyric poems, scientific and historical writings, and digital forms (including blogs, tweets, Facebook entries)' (Ibid). Scholars of place writing, in attempting to map out the emerging field of critical study, should adopt a similarly pluralistic approach when defining the *genre*. Place writing, as with life writing, ought to allow for creative non-fictional texts, overtly fictional texts, and texts in which fact and fiction are imbricated. It should also allow for a heterogeneous range of non-literary texts in which place provides a central point of interest. Contemporary place writing, then, includes the varied forms of poetry, fiction, memoirs, text-based art, tweets and digital notebooks, to name a few. Ultimately, place writing is a helpfully broad critical label that can capture the rich heterogeneity of contemporary texts – across a range of forms – that think deeply, and complicatedly, about place and its meanings.

References

Alexander, N. and Cooper, D. (2013) Introduction: Poetry & Geography, in N. Alexander and D. Cooper (eds), *Poetry & Geography: Space & Place in Post-War Poetry*. Liverpool: Liverpool University Press, pp. 1–18.

Artevents (2010). *The Re-Enchantment.*

Bletsoe, E. (2010) Votives to St Wite, in G. Evans and D. Robson (eds), *Towards Re-Enchantment: Place and Its Meanings*. London: Artevents, pp. 83–88.

Brewster, S. (2009) *Lyric*. London: Routledge.

Casey, E. S. (2002) *Representing Place: Landscape Painting and Maps*. Minneapolis, MN: University of Minnesota Press.

Cowen, R. (2015) *Common Ground*. London: Hutchinson.

Davidson, P. (2018) Bodleian Library, Oxford; Aubrey Manuscript 17, folio 12e, in T. Dee (ed.), *Ground Work: Writings on Places and People*. London: Jonathan Cape, pp. 71–76.

Dee, T. (2018) Introduction, in T. Dee (ed.), *Ground Work: Writings on Places and People*. London: Jonathan Cape, pp. 1–14.

Evans, G. and Robson, D. (eds) (2010) *Towards Re-Enchantment: Place and its Meanings*. London: Artevents.

Furst, L. R. (2006) Realism, Photography, and Degrees of Uncertainty, in S. Denham and M. McCulloh (eds), *W. G. Sebald: History-Memory-Trauma*. Berlin: Walter de Gruyter, pp. 219–229.

Greenlaw, L. (2010) Hevenyssh, in G. Evans and D. Robson (eds), *Towards Re-Enchantment: Place and Its Meanings*. London: Artevents, pp. 105.
Griffiths, J. (2010) The Grave of Dafydd, in G. Evans and D. Robson (eds), *Towards Re-Enchantment: Place and Its Meanings*. London: Artevents, pp. 91–103.
Gutkind, L. (1997) *The Art of Creative Nonfiction: Writing and Selling the Literature of Reality*. New York: John Wiley & Sons.
Hatch, F. (2010) CLOUD PIERCING: Charmouth, in G. Evans and D. Robson (eds), *Towards Re-Enchantment: Place and Its Meanings*. London: Artevents, p. 82.
Hooker, J. (2011) Spirit of the Place. *Resurgence & Ecologist*. 267. www.resurgence.org/magazine/article3437-spirit-of-the-place.html
Ingold, T. (2018) Somewhere, in Northern Karelia, in T. Dee (ed.), *Ground Work: Writings on People and Places*. London: Jonathan Cape, pp. 132–137.
Leader, Z. (2015) Introduction, in Z. Leader (ed.), *On Life-Writing*. Oxford: Oxford University Press, pp. 1–6.
Long, R. (2018) Childhood Ground Abiding Places, in T. Dee (ed.), *Ground Work: Writings on People and Places*. London: Jonathan Cape, pp. 138–139.
Mabey, R. (2010) On the Virtues of Dis-Enchantment, in G. Evans and D. Robson (eds), *Towards Re-Enchantment: Place and Its Meanings*. London: Artevents, pp. 29–38.
Macdonald, H. (2018) Tekels Park, in T. Dee (ed.), *Ground Work: Writings on People and Places*. London: Jonathan Cape, pp. 148–155.
Macfarlane, R. (2003) *Mountains of the Mind: A History of a Fascination*. London: Granta.
Macfarlane, R. (2010) A Counter-Desecration Phrasebook, in G. Evans and D. Robson (eds), *Towards Re-Enchantment: Place and Its Meanings*. London: Artevents, pp. 107–130.
Macfarlane, R. (2012) *The Old Ways: A Journey on Foot*. London: Hamish Hamilton.
Macfarlane, R. (2015) Question and Answer with Robert Macfarlane. *The Baillie Gifford Prize for Non-Fiction*. https://thebailliegiffordprize.co.uk/bailliegiffordprize/news/qa-robert-macfarlane
Massey, D. (1994) *Space, Place and Gender*. Oxford: Polity Press.
Massey, D. (2005) *For space*. London: Sage.
Matless, D. (2018) Seavoew: the Anthroposcenic, in T. Dee (ed.), *Ground Work: Writings on Places and People*. London: Jonathan Cape, pp. 185–188.
Motion, A. (2018) Waders, in T. Dee (ed.), *Ground Work: Writings on People and Places*. London: Jonathan Cape, pp. 189–193.
Petley, D. (2018) The Four Wents on Craster, in T. Dee (ed.), *Ground Work: Writings on People and Places*. London: Jonathan Cape, pp. 210–219.
Poole, G. (2018) Redland, Bristol, in T. Dee (ed.), *Ground Work: Writings on People and Places*. London: Jonathan Cape, pp. 220–227.
Robertson, R. (2010) Tillydrone Motte, in G. Evans and D. Robson (eds), *Towards Re-Enchantment: Place and Its Meanings*. London: Artevents, pp. 9–10.
Ryan, J. (2018) *Plants in Contemporary Poetry: Ecocriticism and the Botanical Imagination*. New York: Routledge.
Smith, J. (2013) An Archipelagic Literature: Re-framing 'The New Nature Writing'. *Green Letters*. 17.1. pp. 5–15.
Smith, Z. (2012) *NW*. London: Hamish Hamilton.
Shawyer, A. (2010) Norfolk, The Broads. View South West from Breydon Water towards Halvergate Marshes, in G. Evans and D. Robson (eds), *Towards Re-Enchantment: Place and Its Meanings*. London: Artevents, p. 28.
Tarlo, H. (2011) (ed.), *The Ground Aslant: An Anthology of Radical Landscape Poetry*. Exeter: Shearsman Books.
Worpole, K. (2010) East of Eden, in G. Evans and D. Robson (eds), *Towards Re-Enchantment: Place and Its Meanings*. London: Artevents, pp. 61–81.

57

Writing a place

Poetry and 'ghost rhetoric'

Helen Mort

This chapter is concerned with the subjective experience of 'writing a place' from the perspective of a creative writer. I consider what I mean by my 'imaginative landscape' and identify this as South Yorkshire and the Peak District. Examining my own poetry and prose, I explore two different engagements with place. The first is an 'independent' engagement through two poetry collections and a novel. The second is the creative work produced during my tenure as Derbyshire Poet Laureate from 2013–2015: a role that required a very specific – and often collective – way of responding to the Peak District. I outline the difference between these two approaches; but also how they might overlap or feed into one another. I argue that my creative engagements with place are unified by 'ghost-rhetoric': a belief in the value of fictional accounts – particularly ghost stories – in the constitution of place.

Ghost stories as narratives of place

There's a story at the front of David Bell's volume of collected *Derbyshire Ghosts and Legends* flippantly titled 'Beware of the Dog!' Every time I see the book on my shelf, I feel compelled to read it. I've been doing this for so long now I almost have the words by heart. The tale is set in Bradwell in the late 1700s and concerns two brothers, Will and Sam, who both worked in the local mines. Returning home late one night after playing cards, Sam was stopped dead in his tracks by the chilling sight of a giant black dog. He pointed, shaking, while his brother could only implore 'What yer staring at?':

> Sam could not believe that his brother was unable to see what was plainly in front of them […] it was so near that he could feel its hot breath on his face but as he stood in petrified silence the dog vanished before his very eyes. When he recovered his voice, he explained what he'd seen but his brother would have none of it. 'Naw, Sam,' he laughed, 'there wiz nowt there.' The two argued for the rest of their journey home.
>
> *(Bell, 1993: 21)*

Next day, Sam tried to convince his brother that neither of them should go to work down the mines. But while he stayed at home, Will dismissed his superstition and set off for the shift as usual. Later that day, a roof collapsed in the mines and Will was killed beneath the weight. The black dog that Sam saw had been a warning, fatal to ignore.

As well as being visually and metaphorically compelling – the visionary black dog evoking Churchill's famous image for depression – I'm drawn to 'Beware of the Dog!' because it seems an allegory for creative writing about place. This local legend has become part of the fabric of Bradwell for me – held in mind every time I run over the limestone there, or walk the lanes – but also an apt motif for my sense of what it is to write creatively about place. What matters in the story is not the evidence of one's eyes, but the constructed meaning, the warning that Sam takes from the omen. Stories about place intersect with everyday life, with the daily grind of industry. These stories would have been part of the imaginative life of miners across Derbyshire and South Yorkshire; and omens and folklore are often connected closely to human tragedy. But these legends also become artefacts in their own right, inviting a response from readers and writers.

My first serious engagement with writing about place was a chapbook of poems called *a pint for the ghost* (tall-lighthouse, 2010) which took as its premise the idea that myths, superstitions and stories we tell about place and space are central to their construction and identity. Set in a pub taproom late at night, each poem in the book re-imagined or re-told a local ghost story. Most of these were drawn from other sources (written texts and oral testimony); others were ghost stories I invented myself based on historical figures such as 'Stainless Stephen', a Sheffield music hall comedian who notoriously took to the stage in an outfit made from local steel. My guiding assertion was that all stories about place have equal validity in place writing and that we might understand a location as much through myth and legend as through measurable data or the more easily verifiable narratives offered by history.

This premise is founded on an approach to writing that I'm calling 'ghost-rhetoric': a belief that the act of storytelling is of primary importance to the formation of a sense of place identity and that the stories in question can sit alongside – or, in some cases, even overwrite – historically accurate or quantitatively provable narratives. In turn, a belief in the value of 'ghost-rhetoric' is founded on an assumption that the tendency to tell stories about place is a universal one. As Edward Boaden Thomas states at the beginning of *The Twelve Parts of Derbyshire*, an odyssey of the county in verse:

> Derbyshire may be as remote as China
> Where you are concerned, or like an old glove.
> In either case I have something to say
> To you about the shire as I see it:
> Whether you listen may matter or not;
> May matter or not, I mean, to yourself:
> Naturally I prefer to be heard
> Not being content to talk to the wind …
> (Thomas, 1988: 5)

There's a strange persuasiveness in this opening: though the narrator may 'prefer to be heard', we're made aware that he will tell his story even if it does turn out to only be the elements that listen to him. All the same, this approach (half-nonchalant, half-imploring) has the effect of making the reader feel they might be missing out by inattentiveness: it 'may matter or not' to ourselves, but we will be disappointing the eager narrator. The opening of *The Twelve Parts of Derbyshire* also frames storytelling as something that is of equal importance in establishing a sense of place ('Derbyshire may be as remote as China …') and overlaying it with new meaning: even if the landscape is 'like an old glove' to the reader, the poet has 'something to say' about his own subjective impressions.

'Strange interludes': the importance of subjective impressions

In his introduction to *Real Barnsley*, poet Ian McMillan also begins with an assumption about the parity of fictional and verifiable narratives about place. Outlining the context for his 'walking guide' to Barnsley town, the aftermath of the 1984–1985 miners' strike and the economic upheaval in its wake, he suggests that South Yorkshire is becoming a site of multiple stories again after the decline of its primary industry:

> We are looking backwards and forwards at the same time: we're reclaiming history and we're trying to reinvent ourselves as a place and as a population. Perhaps the same old tale, with slight variations, is being told in all the parts of the borough where the mines were. Perhaps my job as a writer is to note and celebrate nuance. Perhaps your job as a reader is to note and celebrate it alongside me.
>
> *(McMillan, 2017: 13)*

Again, an implicit contract between author-commentator and reader is subtly established at the start of the text as the nature and terms of the kind of 'storytelling' involved are explored. McMillan also states his intention to include creative interludes throughout the book alongside his 'walking and talking map' of Barnsley:

> as well as the topographical wanderings in the book I'll be writing what Eugene O'Neill might have called Strange Interludes: they'll take me across the borough and they'll take me back in time and back inside my head. Think of them as poemy things.
>
> *(2017: 13)*

This ends the introductory section: McMillan does not even need to defend the inclusion of his 'Strange Interludes', proceeding on the assumption that subjective impressions of a place ('poemy things') should be afforded equal precedence in his historical and topographical account of Barnsley. The 'Strange Interludes' McMillan describes form part of 'ghost-rhetoric', offering the reader a parallel view of Barnsley, one which does not need to be constrained by time or conventional logic, one that seems equally illuminating as a portrait of place.

My own belief in the primacy of this ghost-rhetoric was established as a child growing up in Chesterfield, a town that features prominently in Frank Rodgers' *Curiosities of the Peak District* and that is famed for its distinctive twisted church spire, a landmark that looms over the town like a warped ice cream cone. The spire is seventy metres high, has a twist of forty five degrees and a lean of nine feet and six inches. It is the only twisted spire in the UK, though other towns in Europe boast them. The crooked spire in Chesterfield, however, leans more than any other. I tried to articulate how my writing life has been influenced by the stories surrounding the spire in an essay first broadcast on BBC Radio 3:

> I grew up with crooked stories. One autumn day, the devil went to Bolsover to get new shoes. But the shoes he got from the blacksmith were badly fitting, painful, pinching at his toes. He leapt over Chesterfield in agony, kicking the spire as he went and

> booting it out of shape. Or else, the devil decided to rest on the spire and wrapped his tail around it, rooting himself there. When the people of Chesterfield rang the church bells that day, the devil was startled and tried to take flight with his tail still wound around the spire, twisting it as he moved. Or else, on a Spring morning, a beautiful virgin was to marry at St Mary's, she picked her way along the stone path to the church door. The spire was so surprised to see a virgin bride in Chesterfield, it turned around to look at her. Next time a virgin is married in the church, legend has it, the spire will twist back to true. These were the myths of my childhood. Mischief, vengeance, promiscuity [...] However problematic, I almost prefer the image of the incredulous, personified spire turning to ogle a local virgin to the truth of its likely construction, the effect of lead and unseasoned timbers, material too heavy for the spire's bracing. I'd rather picture a listless devil than a host of local craftsmen, labouring on the windlass that now rests in Chesterfield museum.
>
> *(Mort, 2016)*

I published my first pamphlet of poetry in 2007 and my most recent book in 2019 but, throughout this period, my engagement with writing place has taken two distinct forms. Establishing – and challenging – the differences between them is of central significance to what I understand by 'writing a place'. First, I've been engaging with place creatively as an 'independent' creative writer for the past decade. By 'independent', I mean as someone free to choose the subjects of my poems, stories and novels, as someone collecting material together and publishing work with a press I approached myself. The term 'independent', however, is perhaps best defined by opposition. I also engaged with writing about place as Derbyshire Poet Laureate between 2013 and 2015. This was a very specific role in which I was employed by Derbyshire County Council to respond to aspects of Derbyshire culture, geography and history and was given commissions and public functions to fulfil, culminating in a publication. Though I am distinguishing between my life as an 'independent' place writer and my life as Derbyshire Poet Laureate, it's also important to avoid an artificial separation between the different roles which have, of course, overlapped and influenced each other. While I was writing commissioned poems as Derbyshire Laureate and poems that emerged from local workshops, I was also working on my own individual writing projects. Equally, I brought some of the assumptions and experiences drawn from my existing life as a creative writer to the role of Derbyshire Poet Laureate, and my writing life since 2015 has been influenced by carrying out the Laureate role.

Charting the imaginative landscape

It might be helpful at this point to define the parameters of what I consider to be my imaginative landscape (at least in terms of the work considered in this chapter). I'm interested in 'The Peak District' in general; but – more specifically – I'm interested in both the rural and urban parts of the county of Derbyshire as well as the large South Yorkshire conurbation of Sheffield. In this sense, the landscape my writing is concerned with might seem messy, spanning two geographical counties. Though separated by a county boundary, however, Sheffield and Derbyshire are inextricably linked: driving towards the city from Hathersage, you pass alternating 'Welcome to Derbyshire' and 'Welcome to Sheffield' signs, repeatedly and seamlessly crossing the border as you travel. Along with Manchester, Sheffield is one of the major conurbations on the edge of the Peak District National Park, and the Park receives over ten million visitors each year, its proximity to urban places making it one of the most popular National Parks in the UK. David Cooper describes the 'heterogeneity' of the Peak District and how this has led to it being represented less extensively by writers than the Lake District. This, he notes, is also a source of freedom:

> whereas post-Romantic Lake District writers have been frequently burdened by 'the problem of precedent', contemporary poets have been able to perceive the heterogeneous landscapes of the Peak District as up for literary grabs.
>
> *(Cooper, 2017: 680)*

My sense of Derbyshire includes the pastoral landscapes and wild places of the Dark Peak and the White Peak; but – crucially – it also encompasses the 'edgelands' environments of North East Derbyshire, Dennis Skinner's constituency in and around Bolsover where I grew up. In their collaborative creative non-fiction book *Edgelands: Journeys Into England's True Wilderness*, Paul Farley and Michael Symmons Roberts define these landscapes through their own childhoods, citing geographer Marion Shoard who invented the 'edgelands' term to denote areas like 'the fringes of English towns and cities, where urban and rural negotiate and renegotiate their borders' (2011: 5). At the same time, their exploration of edgelands is predicated on assumptions of the fleeting and of invisibility:

> edgelands, by and large, are not meant to be seen, except perhaps as a blur from a car window, or as a backdrop to our most routine and mundane activities. Edgelands are part of the gravitational field of all our larger urban areas, a texture we build up speed to escape as we hurry towards the countryside, the distant wilderness.
>
> *(2011: 5)*

For Farley and Symmons Roberts, imagining and writing about edgelands becomes a crucial part of establishing their existence and significance. My first full-length poetry collection *Division Street* (2013) explored urban, rural and edgelands locations and took its name from a street which cuts through central Sheffield, a thoroughfare populated by cafes, shops and bars. Titling a book after a particular road might seem to signify a commitment to an approach to writing the 'parochial' which Irish poet Patrick Kavanagh expounded and which Robert Macfarlane mentioned in an article for the *Guardian* in 2005:

> For Kavanagh, the parish was not a perimeter, but an aperture: a space through which the world could be seen. 'Parochialism is universal,' he wrote. 'It deals with the fundamentals.' [...] Again and again in his writing, Kavanagh returned to this connection between the universal and the parochial, and to the idea that we learn by scrutiny of the close-at-hand. 'All great civilisations are based on parochialism,' he wrote, beautifully: 'To know fully even one field or one land is a lifetime's experience. In the world of poetic experience it is depth that counts, not width. A gap in a hedge, a smooth rock surfacing a narrow lane, a view of a woody meadow, the stream at the junction of four small fields – these are as much as a man can fully experience.'
>
> *(Macfarlane, 2005)*

I also chose *Division Street* as a title because of the different resonances I felt this place-name held within it: the idea of conflict and separation; and the notion of trying to demarcate landscape into different sections or 'divisions'. Both of these were themes in my approach to writing about the particulars or the parochial aspects of place. Inspired by the Sheffield musician Richard Hawley, who titles his albums after resonant city place-names, many of the poems in the collection drew on the names of places ('Lowedges') or the names of pubs ('Fagans') and their implications. In 'Lowedges', an area of South Sheffield is at once ordinary – populated by bichon frises, kids playing on the astroturf and bored smokers – and imbued with the significance of escapism:

> And if those doors to other worlds exist
> You'll find them here: Lowedges, where the city
> smooths its skirt down in the name of modesty ...
> (2013: 58)

I chose to end the book with 'Lowedges' because I felt the poem embodied the approach I had taken to place writing in the collection as a whole, that it was suffused with ghost-rhetoric. In Philip Pullman's 'His Dark Materials' trilogy, portals to other worlds can be found by tracing a knife through the air somewhere unremarkable on an ordinary day. In 'Lowedges', our imaginative life can serve the same function in a familiar place. I wanted to juxtapose the everyday and the possibility of the remarkable, the sense of escapism provided by our different and overlapping narratives of place. The poem concludes:

> If you're to leave this world, you'll leave it here:
> this salvaged Friday, shop lights dimmed. Look up –
> how easily the rain bisects the sky.
> (2013: 58)

These closing lines suggest that ordinary things can take on magical significance for the observer: even the regularity of Sheffield rain might be a marker of the boundary between worlds. It is the act of noticing which imbues places with transcendental qualities: the imperative to 'look up' that poet Kathleen Jamie explores at length in her book of essays *Findings*.

Other poems in the collection ('Litton Mill', 'Stainless Stephen') overlay familiar Sheffield and Derbyshire landscapes with strange visions, ghost stories, events that might be memory or invention. The book's central sequence, 'Scab', evokes the clash between picketing miners and police at Orgreave in 1984, setting historical events against biblical imagery and descriptions of conceptual artist Jeremy Deller's 2001 re-enactment of the Battle of Orgreave featuring 800 people, many of whom were ex-miners or police involved in the original encounter. The authorial intention here was to frame the act of writing about place as a kind of re-enactment in itself, as ludicrous and artificial (but also as convincing) as Deller's staged battle. Interestingly, the re-enactment that Deller set up – and documented in his film *The Battle of Orgreave* – took on its own elements of hostility and danger as the process stirred up emotions in the participants. Such is the case with writing about a place: the new stories created can be both artificial and authentic at the same time. The creative act is still an enactment, generating new stories and reactions.

Maps and undocumented stories

My second poetry collection, *No Map Could Show Them* (2016), framed the idea of ghost-rhetoric in landscape even more directly, setting out to uncover undocumented stories about place. My approach in this book was informed by Eavan Boland's poem 'That the Science of Cartography is Limited'. In Boland's poem, the narrator and a partner go to the borders of Connacht in Ireland, enter a wood and look down at the place where a famine road (built by the starving Irish in 1847, the road running out where the workers died of hunger) runs. The poem explores the significance of this absence, this line that maps cannot represent:

> [...] when I take down
> the map of this island, it is never so
> I can say here
> is the masterful, the apt rendering of
> the spherical as flat, nor
> an ingenious design which persuades a curve
> into a plane,
> but to tell myself again that
> the line which says woodland and cries hunger
> and gives out among sweet pine and cypress,
> and finds no horizon
> will not be there.
>
> (Boland, 2005: 205)

The double negative of the poem's ending (the road is incomplete, 'gives out' before it finds a horizon, but it is also utterly unmapped) has the paradoxical effect of giving the famine road a looming presence in the piece: it is a loud absence. The idea of unwritten or overwritten place narratives is also explored in depth in Kei Miller's collection *The Cartographer Tries to Map a Way to Zion*, where place-names and place histories are similarly contested. In this book-length sequence, a Jamaican rastaman and a cartographer enter into conversation and trade different ways of understanding Jamaica. Initially, the cartographer assumes that he can approach his work without bias, but the rastaman expounds the inextricability of Jamaican history, place and people to mapping the land. In *Edgelands*, Farley and Symmons Roberts cite the poet Ciaran Carson's suggestion that poets are fascinated by maps because of the way 'a map has to use shorthand, or symbols, or metaphor in this it resembles poetry' (2011: 16). Yet, in work like Boland's and Miller's which engages directly with the process of mapping, it seems apparent that poetry can also do the opposite, refusing the shorthand of the map, looking behind its partial representations.

Some of the neglected narratives of place I chose to respond to in my own collection of poems included: the experiences of Tibshelf long-distance runner Tom Hulatt; the life of Hull fishwife and activist Lillian Bilocca who campaigned for better safety provisions on trawlers; and the stories of early female rock climbers and mountaineers. Though the collection was informed by my physical appreciation of Derbyshire landscape through my life as an avid rock climber, representing these experiences on Peak gritstone and limestone directly or autobiographically proved challenging. Once again, I found the notion of ghost-rhetoric more helpful in my process. I found it easier to frame my own experiences of Derbyshire landscapes through the lens of the life of Alison Hargreaves, the Belper-born climber who died descending from K2 in 1995. This was certainly not my primary intention when writing about Alison Hargreaves' remarkable life and mountaineering achievements, but when writing the sequence (which uses the second person throughout, a direct address to Alison's memory) I was also imagining her in parts of the Peak District which I had climbed in, places which I was re-imagining through her life. Alison herself recognised the imaginative overlaying of places: in her diary on her last trip to K2, she wrote that she was dreaming of Derbyshire, of places like Black Rocks near Cromford where – appropriately enough – the rocks are overwritten with graffiti, visitors transcribing their own names and brief stories onto the stones.

If (inadvertently) re-imagining a landscape through the lens of another person's life seems like an audacious act, the approach I took to writing about place in my first novel, *Black Car Burning* (2019), might seem even more problematic. The novel explores trust and trauma through narratives of rock climbing and through the contested activities of South Yorkshire Police in East Sheffield. The book is set in 2015, but draws heavily on the Hillsborough disaster and the place of this collective trauma in landscape and community, looking at the aftermath of the tragic events of 1989. As such, stories about place and their enduring significance were once again foregrounded in the novel and the notion of contested stories in particular – police officers on duty at the Hillsborough stadium subsequently had their testimonies altered by senior officers. At the same time, the Sheffield of my novel is overwritten with competing stories: for an ex-policeman revisiting his past, the city is a site of trauma; for the rock climbers, it is a training ground, an access point for the Peak District where all their ambitions are channelled. To emphasise the way narratives of place intersect, I decided to intersperse 'place interludes' throughout the novel, short sections where places within Sheffield and Derbyshire 'narrate' and have a voice. Sometimes, the characters in the novel are viewed by something inanimate such as a quarry near Buxton, or the Millstones near Stanage Edge. Sometimes, the places which I chose to write 'from' try to articulate their world view. Such an approach might be an attempt to imaginatively inhabit a landscape further or it might be an act of extreme anthropocentrism. In this example extract, the village of Hathersage (a key setting for many of the chapters in the novel) narrates:

> Before the weddings and the slanting tea shops and the vintage cars, I was unbalanced, teetering underneath the moor, looking up at Stanage, waiting for the gritstone to topple and wind me. I was full of millstones and hard labour, men who drank their wages, farmers who worked me properly. I'm trying to be polite. I don't mind the coach parties and ramblers, the tourists weighed down by cameras larger than their heads, the day-trippers who come without walking boots and trudge up to the Plantation, tiptoeing over the muddy parts, the bogs where runners lose their shoes. But some days I watch the would-be climbers queuing up for gear in the outdoor shop or shovelling bacon butties down their necks in the upstairs café, comparing lightweight coats and brand-new guidebooks, and I want someone to hurt me, rattle me, pick me up and shake me, so they all topple out like coins from a pocket.
>
> *(Mort, 2019: 274)*

I wanted the village to say something that might counteract the image of it previously established by descriptions in the novel – picturesque, a playground for climbers and hikers. I hoped that some of these 'place interludes' might destabilise the reader's sense of the novel's various settings.

Writing to a brief: the role of Derbyshire Poet Laureate

In contrast to these independent engagements with the Peak District and Sheffield, my collection *Made in Derbyshire* (2015) was the product of a more specific and constructed series of encounters with place, space and community. In 2013, I was appointed in a two-year role as Derbyshire Poet Laureate, a scheme operated by Derbyshire County Council. Overall, the scheme ran from 2005–2015 and there were five different Laureates, each given a brief that involved responding to the varied landscape, culture and heritage of the Peak District in structured and unstructured ways, traversing the county from New Mills to Swadlincote and working with individuals and groups. My role as Laureate was diverse, but my main duties included giving public workshops and readings, writing commissioned poems for occasions and organisations, and crafting personal responses to Derbyshire landscape. My two-year tenure culminated in a publication, *Made in Derbyshire*, that aimed to bring this disparate work together. Attempting to sum up my varied role in a foreword for the book, I wrote:

> A day reading poems in every cafe the length of Chesterfield's Chatsworth Road, ambushing the shoppers in Morrisons. A workshop with patients at Newholme Hospital, Bakewell, who taught me that Sheffield was 'a dirty picture in a golden frame'. A visit to my old secondary school. A poem for Toyota, translated into Japanese. A poem for a football match. A week exploring the history of Eckington. A reading in a beautiful garden in Ashbourne, surrounded by rare flowers. A poem for a film about Shirebrook. A poem for a tea towel. These are some of the things I've done, but they aren't the whole story either.
>
> *(Mort, 2015: ii)*

The commissioned poems I wrote about Derbyshire mark a very specific kind of engagement with place because the parameters of the poem (whether thematic or structural) were often externally set. The restrictions imposed were sometimes fairly loose but at other times quite tight. For instance, when approached by the charity Hathersage Careline (a telephone support group providing regular contact for elderly residents of Hathersage village) to write a poem for their anniversary, I was asked to produce a piece of writing that responded to the theme of 'listening'. When asked to write a poem for the Made in Derbyshire festival in 2015 (celebrating local products and businesses), I was given both a theme and a formal restriction: as well as reflecting regional crafts, the piece had to fit on a tea towel – the finished piece would be a physical artefact as well as a poem. Often, I was involved in discussions with the commissioning body around the thematic content of the piece and so the process was dialogic. Again, then, I brought my own preoccupations and interests to bear on even these more restricted pieces. Equally, when I was working independently on poems that I thought might feature in the *Made in Derbyshire* book, I sometimes thought of these as self-imposed commissions: for example, aiming to write a small triptych of poems which reflected different activities (swimming, cycling, running) people might enjoy in Derbyshire. Perhaps even the act of working with form or rhyme is a kind of self-imposed restriction and the notion of separating commissioned poems from freely written pieces is an artificial one. That's to say, even when I choose the topic of a poem myself, I create my own constraints to help me construct the piece: a set number of syllables per line, a particular stanza formation.

One way of responding to the Peak District that was undeniably distinct, however, was the 'group poem'. I was asked to run a series of events in local libraries and schools where I would work with participants to produce a co-authored poem. This tended to take one of two forms. In my work with schools, the students were often given an exercise to work on individually ('imagine you are flying over your town: describe what you can see as vividly as you can') and I then worked with the children to identify a distinctive or interesting line from each poem, putting these individual lines together as a 'group' piece. In the library workshops – often working with older people who sometimes lacked the confidence to write – I took an even more collective approach. At the outset, we would discuss the kind of poem we wanted to create as a group: rhymed or unrhymed, long lines or short? I would then prompt a themed group discussion with a series of questions, write down snippets of the conversation and attempt to shape these fragments into a poem. The group pieces were published alongside my individual commissions and other poems. Cooper summarises the book's progression thus:

> Symbolically, Mort's sensitivity to the way the 'cultural landscape is shaped by a colloquium of voices' (David Matless) is reflected in the playful organisation of the collection's textual space. *Made in Derbyshire* opens with a series of first-person poems in which the reader is not discouraged to identify the lyrical 'I' with the authorial voice. On page 14, however, this seemingly stable pattern is disrupted by a sequence of 'Seven Group Poems' which have been brought-into-being via creative writing workshops – with both school groups and older writers – held within the Peak District and other locations in Derbyshire. The stable 'I' is dissolved, in the first of these poems, with the inclusion of an authorial footnote which explains the poem's genesis: 'By Linda, Mary, Jo, Margaret, Helen, and Michelle, with a bit of help from me'. Clearly, this self-deprecatory note contains the residual trace of a creative hierarchy. The poet's voice, however, is democratically subsumed within the collective; hers is simply one voice amongst many.
>
> *(2017: 682)*

The 'residual trace of creative hierarchy' is something that might have been problematic about my approach to 'group poems'. Whilst the pieces were constructed from words chosen by the participants, the way the words were ordered and arranged on the page was my decision, perhaps my imposition. In this sense, my role as Derbyshire Poet Laureate often involved the illusion of 'speaking for' others; but, ultimately, the creative control remained with me. To successfully implement a more democratic approach, more radical methods of co-authorship would need to be used which might conflict with group wishes: in many cases, the groups I interacted with indicated strongly that they wanted their group poem to rhyme. In order to do this, I had to impose some aspects of form onto the sentences which were spoken. As such, I was having to impose my own language on the group in order to satisfy one of their articulated desires.

This paradox is one of many encountered by a writer seeking to creatively engage with 'writing a place'. Focussing on the universal aspects of a locality might engage a wider audience, but it might alienate a local readership. Writing a subjective account might privilege the author's narrative over others. Attempting to co-author poems of place might mask underlying and pervasive creative hierarchies in a way that seems disingenuous. But poems – capable as they are of standing up to differing interpretations and containing even contradictory arguments – are at least well-equipped to embody paradox. And for me, the concept of 'ghost-rhetoric' has remained a helpful means of navigation through the challenges of writing a place: places are narrative palimpsests and one way of writing them does not preclude others. If anything, it might encourage writers (and readers) to expand on, subvert or refute texts, the way ghost stories continue to re-shape our rational appreciations of place and space.

References

Bell, D. (1993) *Derbyshire Ghosts and Legends*. Newbury: Countryside Books.
Boland, E. (2005) *New Collected Poems*. Manchester: Carcanet.
Cooper, D. (2017) A poetic playground: collaborative practices in the Peak District. *Landscape Research*, 42 (6), 677–689.
Farley, P. and Symmons Roberts, M. (2011) *Edgelands: Journeys into England's True Wilderness*. London: Jonathan Cape.
Macfarlane, R. (2005) Where the wild things were. *The Guardian*: www.theguardian.com/books/2005/jul/30/featuresreviews.guardianreview22 (date accessed 1 May 2019).
McMillan, I. (2017) *Real Barnsley*. Bridgend: Seren.
Mort, H. (2010) *A Pint for the Ghost*. London: tall-lighthouse.
Mort, H (2013) *Division Street*. London: Chatto and Windus.
Mort, H. (2015) *Made In Derbyshire*. Matlock: Derbyshire County Council.
Mort, H. (2016) *No Map Could Show Them*. London: Chatto and Windus.
Mort, H. (2019) *Black Car Burning*. London: Chatto and Windus.
Thomas, E. B. (1988) *The Twelve Parts of Derbyshire*. Matlock: Derbyshire County Council.

58

Navigating cinematic geographies

Reflections on film as spatial practice

Les Roberts

Introduction: a cinematic odyssey

In the early 1990s, on the final leg of a backpacking trip around the world, I found myself in Auckland. A couple of years earlier I had seen New Zealand director Vincent Ward's extraordinary time-travelling feature, *The Navigator: A Medieval Odyssey* (1988). The film had obviously left something of an impression as, while distractedly moving around the city and beyond, I had purposely kept one eye open for a key landmark from the film, a centrepiece in an on-screen odyssey which, for the medieval travellers whose journey we follow, had become a form of pilgrimage. Although my rather half-hearted efforts to locate the landmark in question (a tall church steeple) did not match the devotional intent of the Cumbrian pilgrims, burrowing, as they did, through the core of the earth to emerge on the other side six hundred years in the future, they nevertheless had the air of a quest about them. I now know the church to be St Patrick's Cathedral in Auckland, a fact that, in a pre-Internet age of information scarcity, I had not been able to reliably determine on location.

I relay this anecdote, not for reasons of nostalgia or to contrive some idea of a sacred journey inspired by an arbitrary landmark in a film, but for reasons that speak to the broader objectives of this chapter. That is, to offer some reflections on the relationship between film, place and location (loosely aggregated under the term 'cinematic geography') and, in particular, to consider the ways this relationship is the product or instigator of creative forms of engagement with ideas and affects of place. For a start, although I most probably would have baulked at the suggestion at the time, what I had perhaps been engaged with as part of my New Zealand excursion was a rudimentary form of screen tourism. Although my Auckland adventures could hardly be conceived of as film-*induced* tourism (Beeton, 2016), the film had conjured a certain imaginary of place that I had carried with me on my travels, even if, in the end, it was destined to remain unconsummated.

In this respect, *The Navigator* serves as a ready-to-hand autoethnographic example that allows us to approach the relationship between film, space and place from a number of different vantage points. For a start, what we might think of as a creative practice in relation to the production and consumption of *cinematic geographies* (Aitken and Zonn, 1994; Lukinbeal and Zonn, 2004; Roberts, 2005, 2012) is certainly not limited to that invested in the creation of the film as a text (the creativity of the director/filmmaker as auteur, for example). As with a growing number of case studies that have explored screen tourism through the prism of fandom and selfhood (Cateridge, 2015; Reijnders, 2011), the film text often provides the raw material from which fans and those consuming geographies of film can put into practice their own creative and performative responses to a given film or, more crucially, to the *spatiality* of the film that their situatedness renders present.

Cinematic geography and autoethnography

When I look back through some of the photographs my backpacking partner and I had taken when travelling around New Zealand in 1993 I am struck by the number of churches (many with tall steeples) that feature in the collection. In their locatedness, these images are thus a significant visual marker of memory and place as expressive of my own shared experience as a young backpacker with a keen interest in world cinema and a liminal disposition towards life and the transformational possibilities of travel (Turner, 1969; Winnicott, 1971). A purely visual analysis of these travel images, perhaps taking its lead from John Urry's notion of a 'tourist gaze' (2002), might conceivably throw up the following proposition: discursive constructions of otherness as filtered through the tourist gaze of a British/European traveller are focussed around sites/sights that bear the semiotic imprint of a British and European colonial heritage. In other words, they reinforce an idea of New Zealand as an antipodean landmass with a singularly European (and thus 'closer to home') cultural and historical identity. Perhaps, but a wider methodological net would need to be thrown before reaching this particular conclusion. Which is not to say this analysis does not have merit, it merely points to the many limitations attached to 'readings' of cultural landscapes that may be playing host to a wider, less *legible* (and hence less *representational*) range of intentions on the part of those engaging with these landscapes. Another thread of analysis, one that proceeds from my declared interest in *The Navigator* and its association with the places I was visiting as a tourist/traveller in New Zealand, might venture the argument that my gaze was constructed through my consumption of the film and that, as a consequence, my visual engagement with the landscape was indexically linked to the images I had carried with me. This line of argument builds on the tourist gaze thesis but fashions it more discreetly around a medium-specific point of entry: my suitcase of 'imaginaries' (Reijnders, 2011; Salazar and Graburn, 2014; for a critical take on the idea of 'tourism imaginaries' see Andrews, 2017), already burdensome, is unpacked and rendered portable by just focussing on the cinematic geographies that I have brought to my tourist experience. By this reckoning, I have become a 'film tourist' and it is that which colours my vision and directs my sightseeing gaze.

While, again, this may well hold up as a theoretically informed reflection on a very particular transaction between film and place, it does not fully account for the plurality and open-endedness of spatial practices that may well lack an obvious instrumental intention. The elements of chance, serendipity, or of a disaggregated sense of performative *doingness* (a phenomenology of 'practised place' (De Certeau, 1984: 117) in which the body and the senses play a more pivotal role) are no less likely to be important factors that inform the correspondence between film and place. There is, perhaps, a risk of over-determining the power of the 'gaze' as to render it a blunt instrument whereby any creative agency on the part of the traveller/tourist is significantly downplayed. For example, as best as I can now recall, my engagement with the natural and cultural landscapes of New Zealand was the product of a myriad of complex mediations and dispositions and can in no way be narrowed down to a single set of representational motives. Reflecting on my own embodied experience and memory, the image of the church steeple from *The Navigator* clearly had some phenomenological bearing on the wayfinding practices (Ingold, 2000; Lynch, 1960) that shaped my movements through, and immersion in, the places and landscapes I encountered. But it would be misleading and certainly reductionist to draw a straight line of causality between the cinematic image (the diegetic site of Christian pilgrimage and iconic marker of redemptive place) and my photographic 'gazing' at churches. By no stretch of the imagination could it be claimed that my travel itinerary had been filmically 'induced'. Although I am an admirer of Ward's film, it is not a text that would have motivated me in the sense of a fan seeking out a screen location as if a site of pilgrimage (Couldry, 2003: 75–94). If anything, the pilgrimage connotations were more likely to have found resonance with the wider liminal circumstances that underwrote my backpacking adventure as a youthful rite of passage. It was the idea of consummation that was important; of marking some sense of a symbolically significant waypoint. But the symbolism was not in itself specific to the film. In many respects it was quite arbitrary. As an icon, the image of the church steeple offered a creative rationale whereby I was able to *work myself in* to the landscapes through which I moved. In Ingoldian terms, it was not so much a landscape as a 'taskscape' (Ingold, 2000: 197) that defined my relationship with place. I was not merely sightseeing, if by this is meant a largely passive engagement with landscapes that we 'stand back' from (Bender, 2001: 3) and which we are thus not intrinsically entangled and enmeshed within. I was certainly seeing 'sites'; but there was a degree of openness and play in the way *The Navigator* church steeple had become woven into the spatial practices that defined my immersion in place. In the performative and in many respects accidental collision of film (*The Navigator*) and place (New Zealand) I had found and forged a connection that was as much embodied and material as it was visual. I was engaged with the production of what cultural geographer David Crouch refers to as 'lay geographies' (2010: 14–15), the small-scale modalities of dwelling and wayfinding that shape aesthetic entanglements of place and everyday creativity.

Glimpsed through the autoethnographic lens of a narrative of place that is, in part, a salvaged fragment of memory, the cinematic geographies that are marshalled around the film text spill out beyond the representational frame through which place is projected to lend their performative weight to what I have elsewhere referred to as a 'spatial anthropology' of film, place and memory (Roberts, 2012, 2018a). As with Marc Augé's (2009) discussion of the 1941 film *Casablanca*, proceeding on these terms is to render explicit the subjective and intersubjective tangles of affect, memory and everyday practices that are interlaced with the film but which, importantly, are not determinatively bound to it. As cartographies of place and memory (Bruno, 2002; Conley, 2007), the mappings that are set in motion through the kind of memory-work that Augé is transacting in *Casablanca: Movies and Memory* are not intrinsic to the places being projected on screen (whether these are filmed on location or in a studio). Memories *from* the film are braided with those that are remembered *of* the film: '[a film's] fictional scenes', as Augé writes in his earlier book *Oblivion* (2004: 73), 'dive into our real life, slip in like remembrances in the same capacity as those we have lived'. For Augé, *Casablanca* performs an autobiogeographical function insofar as reflection on the film transports him back to places he remembers as a young boy travelling with his mother in occupied France in the early 1940s (Augé, 2009; see also Roberts, 2018a: 162–164). Approached from the vantage point of spatial anthropology and autoethnography, and of interdisciplinary-infused scholarship in spatial humanities more generally, 'cinematic geography' can thus be seen to encompass a more expansive and less rigidly representational ontology of space, place and memory.

In the case of *The Navigator*, as with Augé's affective entanglements with *Casablanca*, my own reading of place has been coloured by: the memories of who I was at the time I first watched the film; how the landscapes and journeys depicted on screen had caught my imagination as a would-be traveller/pilgrim myself; the way the film 'travelled' with me as I negotiated the material and symbolic landscapes I encountered in situ/on location in 1993; who I was sharing (and subsequently remembering) these cinematic geographies with as fellow travellers; the oblique perspective the film offered in the touristic framing and consumption of the built cultural heritage (the particularities of the tourist gaze into which I was discursively inducted). No less important is a reflexive awareness that the place of home to which I would eventually return (London) was, by dint of its open and relational geographies of global connection (Massey, 1991, 2005), somewhere that had furnished me with the requisite cultural capital that exposure to international cinema had initially helped nurture (it is unlikely that the film would have been screened in many, if any, places outside of the capital and in pre-DVD days would have had, at best, a limited VHS release). In short, what such considerations point to is the extent to which a dialogue between film and place can be topologically mapped out in a number of different directions and spatio-temporal dimensions. Indeed, straddling the nexus of a plurality of voices, places and practices, cinematic geography (if by this we mean a constellation of approaches that in some way address the relationship between film and place) is arguably better put to work if thought about not in terms of what it tells us about *place*, but rather how it might inform understandings of the everyday performativity and practice of *space*. Cinematic geography, viewed thus, is concerned with the production and consumption of spatial stories (De Certeau, 1984; Roberts, 2012, 2018a). Analytically, this does not just amount to a different way of framing the correspondence between film texts and the places to which they are indexically linked. It speaks to a different, more dialectical understanding of place and space in which the moving image is but one facet in a complex and contingent process of space-in-the-making (Lefebvre, 1991). From such a vantage point, cinematic geography becomes part of a broader project of spatial anthropology.

The Navigator: a spatial story

Returning to *The Navigator*, what, then, are some of the other ways we might critically unpack the spatial practices that feed into and from the textuality of the film? Moreover, how does it creatively rework and remap the temporal geographies by which semiotic understandings of place are stitched into or from the material fabric of time? Time that is at once architectural, archaeological and geological in its spatial composition; time that is constellated and rendered spatial through the strategic deployment of mise-en-scène; time that is linear only insofar as it is aligned with the runtime of the film as experienced from the point of view of the spectator. To answer these and other questions it is necessary to take a closer look at the narrative geographies of *The Navigator* and to ruminate on the film's central spatio-temporal conceit: the simultaneously embodied movement through deep historical time (across six centuries) and the 'earthen matter' and 'primordial humus' (Bachelard, 2002: 60) through which the pilgrims navigate their pathway to redemption.

Geographically the film begins and ends in the plague-ridden landscapes of what is more famously known today as the Lake District, a mountainous region in the county of Cumbria in the north west of England. The spectre of the Black Death haunts fourteenth-century England and it is just a matter of time before the Cumbrian village that is home to nine-year-old Griffin falls victim to its curse. Griffin is a visionary who, in his dreams, glimpses fragmentary images of a 'celestial city' of colour and lights, an unknown destination that he intuitively knows has sacred and symbolic significance, offering hope of deliverance from the villagers' plight, but which he is unable to articulate with any great precision or clarity. It is a *feeling* that possesses young Griffin. Whether the vision (the image), or the emotional insight (the embodied affect) linked to that vision, came first is impossible to determine; one cannot be neatly prised apart from the other. Griffin simply *knows* and it is this embodied knowledge that propels the spatial story that unfolds in the film. If Griffin 'travels' through the power of the imagination and intuitive invocation his more worldly brother Connor is an explorer and adventurer whose homecoming brings with it grim tales of plague and death from across the land (he also, as we discover later in the film, brings home with him the plague itself). It is the combined will and faith of the two brothers – the plucky heroism of the traveller and the fervent conviction of the seer – that sets in motion the pilgrimage to the celestial city. The culmination of their sacred journey, the principal objective that will seal the fate of the villagers, involves the mounting of a cross on the steeple of the 'biggest Church in all of Christendom'. As a travel film and journey through time this vision of an iconic site of pilgrimage and the spatially performative response it triggers (the act of pilgrimage and the final scaling of the church steeple) provides *The Navigator*'s central narrative structure.

As well as geographically establishing the mountainous Cumbrian setting as the place of departure and eventual return, these bookending scenes are shot in grainy black and white. There is nothing Romantic or picturesque about the landscapes within which the medieval villagers dwell. The sense of place encapsulated by the name 'Lake District' could not be further removed from the mud-strewn, pockmarked and distinctly uninviting vistas we are confronted with in *The Navigator*. The places that Griffin and his denizens call home starkly convey an experiential space of dwelling that is unforgiving, playing host to toil, misery and earthly suffering. Perceived not as an enchanted locus of the sublime but as a bleak, twilit and hellish wasteland, the Cumbrian mise-en-scène speaks of the need for departure and deliverance, not of transcendence. In this respect, as with Andrew Kötting's 2001 film *This Filthy Earth*, or Ben Wheatley's *A Field in England* (2013), the primal earthiness and rugged materiality of the landscapes works against any cloying or idealised aesthetics of English rurality, emphasising instead a natural world that is pitted against those whose fate it is to claw and eke out some kind of brutalised existence. It is the shadow of Hobbes rather than Rousseau that casts its pall over these particular cinematic landscapes.

When the pilgrims finally embark on their journey (via a deep chasm in the earth and then by tunnelling their way vertically through layers of rock) we, as viewers, witness a transition that is in part chromatic (the film is now moving towards colour) and in part passage through a wormhole. Having left behind the grime and squalor of fourteenth-century Cumbria, six intrepid pilgrims (Griffin, Connor and four other villagers) emerge in modern-day Auckland, the celestial city of Griffin's dreams. A set of structural oppositions are drawn decisively around the boundary line that separates the two worlds (the crust and core of the Earth): medieval/modern; black and white/colour; darkness/illumination; rural/urban; primitive/civilised; religion/science; faith/rationality; body/mind; northern/southern (hemisphere); artisanal/industrial-mechanistic; slowness/speed, and so on. Despite these binary distinctions, there are at the same time certain allusions to coevality in the way the pilgrims and those they encounter interact. This is particularly the case with a group of soon-to-be-unemployed foundry workers who help the pilgrims to forge the copper ore they have brought with them from Cumbria into the cross that is to be mounted on the church steeple. Although the Aucklanders look upon the visitors with an air of curiosity and wonderment, a degree of acceptance of their respective differences informs the development of what quickly becomes a convivial working relationship that is marked by a sense of shared values and craftsmanship. Other themes that draw a line of connection between the two worlds are also hinted at: for example, the analogy between the Black Death and the HIV/AIDS epidemic that was at its peak in the late 1980s and early 1990s; also, in one scene, an allusion to nuclear conflagration, in the form of a Leviathan-like US submarine that surfaces while the pilgrims are crossing Auckland harbour, linking the threat of the plague to that oriented towards New Zealand's status as a nuclear-free zone. But these textual points of reference aside, when we again map out beyond the diegetic spaces that are encountered in the film, and begin to anchor its cinematic geographies in more diffusely contextual spaces of representation, then different and potentially multifarious readings can be extracted. As I have already intimated, such readings are the product of critical reflections on *The Navigator* as an intersectional locus of spatial practice, not as a text that merely serves as a means to frame a discussion on a given place by drawing a closed hermeneutic circle both around that place and the text with which it is brought into correspondence.

Myth, creativity, bricolage

Picking up the autoethnographic threads of reflection that precipitated this discussion, as a backpacker embarked on a liminal rite of passage, some of the binary distinctions I would doubtless have internalised with the help of this and countless other travel narratives consumed in the preceding years map seamlessly on to those that are transacted in *The Navigator*. I too went in search of otherness and authenticity (Graburn, 1989; MacCannell, 1976), but to the extent it could be looked upon as a sacred journey, my travels, taken in the round (i.e., beyond New Zealand as merely a standalone destination), followed a reverse spatio-temporal trajectory: from sprawling urban metropolis to rural or gemeinschaft idyll; from the disenchantments of modernity to an embrace of the pre-modern (a perceived locus of enchantment, an exotic counterpoint to the stultifyingly familiar); from the dromological intensity of a culture addicted to speed to a habitus of place in which the embodied rhythms of slow living are more keenly felt (Edensor, 2010; Roberts, 2018a: 71–76; Virilio, 2006). However essentialising or orientalist these imagined geographies may strike us critically, the fact remains that these sentiments and affective dispositions have long underwritten many travel practices, not least those 'journeys to the east' (Gemie and Ireland, 2017; Hesse, 1989) that characterised the mobilities of backpackers and travellers following the well-trodden routes of the hippie trail lain down in the 1970s. As such, at a mythological level of signification (Barthes, 1993; Lévi-Strauss, 1966, 1967), classic travellers' tales such as *The Navigator* bolt on very readily and very instructively to forms of material spatial practice in which the merits of transformation, enculturation (and deculturation), or perhaps even redemption are being earnestly sought. In hindsight it is without doubt these mythic structures of symbolic meaning that best account for the particular engagements with place that *The Navigator* brought to my experience as a young backpacker travelling around New Zealand in the early 1990s. Quite simply, the text works extremely effectively as a myth, a semiotic device with which to embellish and inject greater structural integrity to a liminal space and time where flux or 'anti-structure' (Turner, 1969) can at times erode the coherence and rationale afforded to the journey itself. The myth helps bind together a constellation of practices and motivations that, over time, has come to function as a map of memory.

Creatively, the use of a film text (or, for that matter, any other text) as a mythic device allows for interactions with place that play on ready-to-hand cultural resources in ways that potentially shape the experiential fabric of that place. This I have demonstrated in the case of

The Navigator in terms of my own creative engagements with landscapes and taskscapes of travel as a backpacking neophyte. But as a creative practice, what needs further refinement in our theoretical exposition of the *production* of cinematic geographies is the role played by mythopoeic narratives as tactical forms of 'spatial bricolage' (Roberts, 2018b): the art of poetically 'making do' (De Certeau, 1984: xv) as applied to the sociocultural production of place and space. The idea of the traveller-as-bricoleur takes its lead from the work of anthropologist Claude Lévi-Strauss who describes bricolage as '[the making] do with "whatever is at hand" ... [to address oneself] to a collection of oddments left over from human endeavours' (1966: 17, 19). Importantly, as Lévi-Strauss goes on to note, '[the bricoleur] may never complete his [sic] purpose but he always puts something of himself into it' (1966: 21; Roberts, 2018b: 2). What this more explicitly translates to, in the case of cinematic geography, is a mode of creative practice that is necessarily provisional, adaptive to environmental circumstances and strategically open to whatever intentionalities work themselves through those lived, happenstance moments when text, embodied subject and material geography interact and collide. From this collision something can potentially be fashioned, but it needn't be anything more than a shift in disposition; a landscape glimpsed from a new perspective; a deepening understanding or aesthetic appreciation of the film or films that are in some way tied to that landscape; an affective 're-booting' in the sense of drawing emotional energy from an experience that is at once situated, haptic, immersive, but also immaterially lodged in the imagined geographies that the traveller-bricoleur subjectively brings to 'the field' (and then puts *into practice*). In other words, the creative intentionalities that are invested in both film and place do not presuppose an obvious concrete output to which the bricoleur-as-producer can subsequently point (whether this be a film, poem, painting, photograph, for example). The elemental creativity of everyday spatial practices – the spatial anthropology that binds place, people and culture together in a holistic and often tangled meshwork of meanings, orientations and affectivities – becomes slavishly instrumental only insofar as its value is limited to that which can be profitably utilised by the culture and creative industries. Which is not to say, of course, that everyday creativity and spatiality isn't routinely harnessed for such ends.

The difference between, say, a film such as the Beatrix Potter biopic *Miss Potter* (2006) and Ward's *The Navigator* – both of which are engaged in the creative re-fashioning of the Cumbrian landscape – is that the former does little more than re-entrench existing imaginaries of place, drawing on authorised heritage discourses (Smith, 2006) in ways that allow for the symbolic capital invested in the film's Lakeland setting (the established brand and place-myth of which Beatrix Potter is a part) to service both a lucrative global entertainment product (the film) and equally lucrative global heritage and tourism industries (the film as a tool of place-marketing). *The Navigator*, by comparison, can be seen as part of a tradition of filmmaking that first and foremost draws on the materiality of place as a locus of atmospheres (Crouch, 2017) and creative affectivity. In the words of writer and self-professed psychogeographer Iain Sinclair: 'Landscape is the story, memory and meaning. You begin there' (2002: 34; Roberts, 2016: 375). You begin from an immersive and embodied experience of place and take your creative bearings from whatever it is that that experience throws up. As a spatial bricoleur (Roberts, 2018b), psychogeographer (Richardson, 2015), or whatever other label best fits, the cultural resources (myths, memories, narratives, imaginaries) you have in your possession are tools to be deployed in the flux and situatedness of the moment as it is lived. The intention is not to allow these resources to determine or engineer the experience that one might otherwise expect to 'consume' as a product of that place (as with, for example, a tourist visiting Cumbria expecting to find there a promised Beatrix Potter fantasyland). The tradition of filmmaking and visual arts practice that *The Navigator* can arguably be aligned with, therefore, is one in which the constitutive elements adumbrated above – spatial bricolage, making do, happenstance, serendipity, atmosphere, affectivity, wayfinding, immersion, corporeality, sensory geography, spatial anthropology, film as a spatial practice, psychogeography – feed into a creatively performative and locative process whereby the filmmaker is as much in the business of making him- or her*self* as s/he is an aesthetically crafted and realised visual output. The filmmaker-as-traveller-as-bricoleur – as inventively showcased in Agnès Varda's 2000 documentary *The Gleaners and I* (see Croft, 2018; Roberts, 2018b: 5–6) – is the embodiment of a creative practitioner who *always puts something of herself into it*, to borrow again from Lévi-Strauss. Put simply, the relationship between film and place is always mediated by social actors and agents but it is the anthropological and creatively reflexive make-up of these cultural brokers – in all their difference and diversity – that warrants closer analytical attention.

Conclusion: film as spatial practice

This chapter has not been *about* the 1988 film *The Navigator: A Medieval Odyssey*. The film has played a performative role in an autoethnographic and autobiogeographical narrative from which I have attempted to address broader questions surrounding the creative engagement with place as mobilised through different filmmaking practices and as expressive of different forms of spatial habitus (Roberts, 2018a: 23–24). Foregrounding film as a spatial practice is to provide a more expansive analytical canvas within which to explore and expound what we understand cinematic geographies to be from a range of perspectives. As we have seen, these may encompass developments in screen tourism and intangible cultural heritage; equally they may allow space for creative interventions that are reflective of a psychogeographic and bricolage sensibility that draws its affective charge from space and place as it is directly lived and experienced. However we approach it, the creative interface of film and place is by definition interdisciplinary in its methodological and theoretical spacing, and it is with this ambulant spirit of boundary-crossing, whether discursively or as part of our everyday spatial stories, that cinematic geographies are most propitiously navigated.

References

Aitken, S. C. and L. E. Zonn (eds.) (1994) *Place, Power, Situation and Spectacle: A Geography of Film*. Lantham, MD: Rowan & Littlefield.

Andrews, H. (2017) Becoming through tourism: imagination in practice. *Suomen Antropologi*, 42 (1), 31–44.

Augé, M. (2004) *Oblivion*. Minneapolis, MN: University of Minnesota Press.

Augé, M. (2009) *Casablanca: Movies and Memory*. Minneapolis, MN: University of Minnesota Press.

Bachelard, G. (2002) *Earth and the Reveries of Will: An Essay on the Imagination of Matter*. Dallas, TX: The Dallas Institute Publications.

Barthes, R. (1993) *Mythologies*. London: Vintage Books.

Beeton, S. (2016) *Film-induced Tourism*. 2nd edition. Clevedon: Channel View Publications.

Bender, B. (2001) Introduction. In B. Bender and M. Winer (eds.), *Contested Landscapes: Movement, Exile and Place*. Oxford: Berg, pp. 1–18.

Bruno, G. (2002) *Atlas of Emotion: Journeys in Art, Architecture and Film*. New York: Verso.

Cateridge, J. (2015) Deep mapping and screen tourism: the Oxford of Harry Potter. *Humanities*, 4 (3), 320–333. www.mdpi.com/journal/humanities/special_issues/DeepMapping.

Conley, T. (2007) *Cartographic Cinema*. Minneapolis, MN: University of Minnesota Press.

Couldry, N. (2003) *Media Rituals: A Critical Approach*. London: Routledge.
Croft, J. (2018) Gleaning and dreaming on Car Park Beach. *Humanities*, 7 (2). www.mdpi.com/journal/humanities/special_issues/spatial_bricolage.
Crouch, D. (2010) *Flirting with Space: Journeys and Creativity*. Farnham: Ashgate.
Crouch, D. (2017) Space, living, atmospheres, affectivities. In M. Nieuwenhuis and D. Crouch (eds.), *The Question of Space: Interrogating the Spatial Turn between Disciplines*. London: Rowman & Littlefield, pp. 1–21.
De Certeau, M. (1984) *The Practice of Everyday Life*. Translated by S. Rendall. London: University of California Press.
Edensor, T. (ed.) (2010) *Geographies of Rhythm: Nature, Place, Mobilities and Bodies*. Farnham: Ashgate.
Gemie, S. and B. Ireland. (2017) *The Hippie Trail: A History, 1957–78*. Manchester: Manchester University Press.
Graburn, N. H. H. (1989) Tourism: the sacred journey. In V. L. Smith (ed.), *Hosts and Guests: The Anthropology of Tourism*. 2nd Edition. Oxford: Blackwell, pp. 17–32.
Hesse, H. (1989) *The Journey to the East*. London: Paladin.
Ingold, T. (2000) *The Perception of the Environment: Essays in Livelihood: Dwelling and Skill*. London: Routledge.
Lefebvre, H. (1991) *The Production of Space*. Translated by D. Nicolson-Smith. Oxford: Blackwell.
Lévi-Strauss, C. (1966) *The Savage Mind*. London: Weidenfeld and Nicolson.
Lévi-Strauss, C. (1967) *Structural Anthropology*. Translated by C. Jacobson and B. Grundfest Schoepf. New York: Doubleday Anchor Books.
Lukinbeal, C. and L. Zonn (eds.) (2004) Cinematic geographies. Special issue of *GeoJournal*, 59 (4), 247–337.
Lynch, K. (1960) *The Image of the City*. Cambridge, MA: The MIT Press.
MacCannell, D. (1976) *The Tourist: A New Theory of the Leisure Class*. London: Macmillan.
Massey, D. (1991) A global sense of place. *Marxism Today*, 38, 24–29.
Massey, D. (2005) *For Space*. London: SAGE.
Reijnders, S. (2011) *Places of the Imagination: Media, Tourism, Culture*. Farnham: Ashgate.
Richardson, T. (ed.) (2015) *Walking Inside Out: Contemporary British Psychogeography*. London: Rowman & Littlefield.
Roberts, L. (2005) *Utopic Horizons: Cinematic Geographies of Travel and Migration*. PhD thesis, Middlesex University, 6 July 2005: http://liminoids.com/LesRoberts_PhD_2005.pdf.
Roberts, L. (2012) *Film, Mobility and Urban Space: A Cinematic Geography of Liverpool*. Liverpool: Liverpool University Press.
Roberts, L. (2016) Landscapes in the frame: exploring the hinterlands of the British procedural drama. *New Review of Film and Television*, 14 (3), 364–385.
Roberts, L. (2018a) *Spatial Anthropology: Excursions in Liminal Space*. London: Rowman & Littlefield.
Roberts, L. (2018b) Spatial bricolage: the art of poetically making do. *Humanities*, 7 (2), www.mdpi.com/journal/humanities/special_issues/spatial_bricolage.
Salazar, N. B. and N. H. H. Graburn (eds.) (2014) *Tourism Imaginaries: Anthropological Approaches*. Oxford: Berghahn.
Sinclair, I. (2002) Heartsnatch hotel. *Sight and Sound*, 12 (12), 32–34.
Smith, L. (2006) *Uses of Heritage*. Abingdon: Routledge.
Turner, V. (1969) *The Ritual Process: Structure and Anti-Structure*. New York: Cornell University Press.
Urry, J. (2002) *The Tourist Gaze*. 2nd edition. London: Sage.
Virilio, P. (2006) *Speed and Politics*. Los Angeles, CA: Semiotext(e).
Winnicott, D. W. (1971) *Playing and Reality*. London: Tavistock Publications.

59

Practices of home beyond place attachment

Veronica Vickery

Prelude

For me, place has meant home, perhaps even a yearning for home—frequent house and school moves as a child had left me feeling un-homed, out of place. During times of life change, my work as an artist and geographer always seems to come back to this feeling of displacement as a starting point from which to work. Now is one such transitional moment, a moment marked by an upcoming move from Cornwall where, for the first time in my life, I have lived in the same house for more than three years. This house in which I have brought up my family has been our home since 2001; a settling in one place that has been life-affirming. Those who know me well are therefore surprised to hear that I am leaving West Cornwall and moving to a city. Bristol is very different from these remote cliff tops near Land's End, four hours further west. A *house* became a *home*, only to now, once again, become a *house*.

Introduction

In this chapter I juxtapose *home* and *house* using the practice(s) of artist, geographer, and then would-be builder to: reflect on the tensions between place attachment and detachment, between being in place and then out of place; explore some of the complex personal politics involved in the pragmatic choices and compromises I am making in order to facilitate this move from cliff top to city; and then reflect more widely on issues of translocation amidst a geopolitical environment dominated by the othering of people deemed to be out of place. I approach this through discussion of a series of art projects: *BOShomes*, an 'estate agency' (2009); *Springs Farm*, a series of paintings worked in response to a derelict farmhouse in West Cornwall (2008–2011); and current work-in-progress *From home to house: VV Renovations* (working title). These art projects and the underlying politics are very much place or site-specific, deeply embedded amidst the scattered moorland settlements between Land's End and St Ives. It is to the rich but complex cultural heritage that entwines this holidayed place that I first turn, before considering the impact of this tourist economy on access to housing—the subject of *BOShomes*. The next project, *Springs Farm*, marks a shift in affective register with a more subjective exploration of place as homed, lived and material. Then, as artist-builder-geographer, I use *VV Renovations* to work with the tensions between house as commodity and home as dwelling place, between being emplaced and displaced in the earlier projects, to move towards something more complex and nuanced.

The face of West Penwith

> From Wicca to Levant the coastline emerges out of cairns and bracken and cultivated greenland, revealing on its varied faces a sea history and a land history of men within and without and a commerce of man with the weather. Here, in a small stretch of headland, cove and Atlantic adventure the most distant histories are near the surface as if the final convulsion of rock upheaval and cold incision setting in a violent sandwich of strata had directed the hide and seek of celtic pattern.
>
> *(Lanyon, 1950: 43)*

At the time of the project *BOShomes*, I was artist-in-residence with the National Trust at Bosigran in West Penwith, based on an Atlantic-facing farm nestled under Carn Galva. This wild, weathered, granite, post-industrial landscape, with its long heritage of mineral extraction, is indelibly woven into the texture of place in the settlements of the far west of Cornwall. Now one of Britain's top tourist destinations, Cornwall boasts not only beaches and moorland walks and a temperate climate, but also a range of tourist attractions cumulatively promoted as 'Cornwall's Heritage and Cultural Offer' (Hale, 2001; Haydu, 2017: 4) which currently constitute the primary economic drivers in this remote location.

Somewhat bucking the trend of the 'internalist and essentialist construction' of place critiqued by Massey (1995: 183), place-attachment in Cornwall has a long history of interconnectedness with elsewhere (Orange, 2012: 267). Triggering the Industrial Revolution, Cornish mining became 'a powerhouse of technical innovation and a nursery for the world's mine captains and engineers' who, taking 'those skills overseas [...] imagined themselves a global elite' (Kennedy and Kingcome, 1998: 47). Recognising this global impact, from Australia to the Americas, 'and all the influences which that brought' (Massey, 1995: 183), UNESCO listed the Cornwall and West Devon Mining Landscape as a World Heritage Site (WHS) in 2006. This pride in 'Cornwall's former position as an industrial world leader remains embedded within a modern sense of Cornish identity' (Orange, 2010: 101; Payton, 2005) and extends to Cornish diasporic communities overseas. Where I live, much of this cultural pride is manifested through the miners' institutes, pubs, the mining museum at Geevor and the annual St Just Feast, when descendants of the nineteenth-century Cornish diaspora—known as *Cousin Jacks*—return from all around the world (Payton, 1984, 2005). The return *home* therefore becomes an essential part of the imaging of diaspora and the spatialisation of memory (Basu, 2007).

Contrasting with the picture-postcard image, West Cornwall is widely recognised as having deceptively high levels of deprivation. The European Union continues (for now) to allocate Cornwall more substantial regional development structural funds than any other part of the UK (Brien, 2018). In West Cornwall, the distortion of the local housing market is severe; wages in the St Ives parliamentary constituency are only 68% of the UK weekly median (Haydu, 2017: 2, 8). Cornwall has the third highest number of rough sleepers in the UK and a significant problem with hidden homelessness (Ryan, 2017; Seria-Walker, 2018; Vergnault, 2017). This deprivation is driven by in-work poverty sitting 'alongside areas of considerable wealth and affluence', with work in the tourist economy particularly low paid, seasonal and subject to the vagaries of the weather (Haydu, 2017: 8).

This ambiguity of appropriation and identification, exemplified by a tourist-focusesd economy—associated in the Cornish imaginary with working-class labour—brings into stark relief longstanding issues of ethnicity, ownership and representation (Hale, 2001: 186), a tension summed up by Lippard with reference to another deprived area:

> if the tourist's landscape is perceived as the past, then present concerns need not interfere with superficial pleasure. Maine is a poor state, and tourists learn to avert their eyes from the less ingratiating sights, those incompatible with the invented past and our modern expectations thereof, as though only the nice old houses, shady lanes, and harbor vistas were visible; as though there were no shabby side streets, edge-of-town strips, tall old multi-family houses cowering in their tarpaper coats against the elements, no yards full of rusting cars, no toothless elders, and pale, overweight teenagers pushing strollers, no convenience stores on the corners.
>
> *(1999: 163)*

A few miles up the coast road lies the town of St Ives, 'a salt-encrusted barnacle of a town with a picture-book harbour, bathed in the luminous Atlantic light that has bewitched generations of painters' (Jackman, 2011). The former fishermen's cottages in the higgledy-piggledy, cobbled alleyways of *Downlong* are chock-a-block with visitors during the summer; only a couple survive as all-year-round homes. Echoing the slum clearances of the 1930s when, despite considerable local opposition, many residents were forcibly rehoused in newly built council houses isolated on the outskirts of the town, local people are again pushed to the margins. Priced out of the housing market by the inflationary impact of second homes, year-round residents are 'dispossessed from their immediate surroundings in a localised diaspora and removed from their livelihood and connection to place' (Laviolette and Baird, 2011: 67). As a National Trust ranger once told me, the only way he could return to nearby Zennor, the village in which he was raised and which he still considers *home*, would be 'in a box' on his death.

Cornwall Council undertook a survey of St Ives residents in 2016 to assess support for measures mitigating the impact of the second home market on access to housing. The results, supported by 80% of voters—in contrast to academic research on attitudes of local residents to second home ownership (Paris, 2009; Perkins and Thorns, 2006; Quinn, 2004) —demonstrated a clear mandate for action. The Council, supported by a High Court judgment, now refuses planning permission for any housing development not restricted to full-time residents (Morris, 2016).

Projects

BOShomes, house as commodity

With this context in mind, I wanted to focus attention on the heritage dissonance (Tunbridge and Ashworth, 1996) embodied in the contrast between house as 'home' and house as 'commodity'; so, whilst at Bosigran, I developed *BOShomes* (2009). The suffix *bos* is common in Cornish place-names, particularly in small settlements (MAGA, undated). In Kernewek (modern-day Cornish), *bos* relates to *being, becoming, or existing*; within the context of place-naming, therefore, it has overtones of belonging and, given the size of these settlements—often just a few cottages around a farm—embodies a sense of dwelling-place or home-making.

I traipsed the moor, searching for and photographing derelict eighteenth- and nineteenth-century cottages. The resulting project consisted of several elements: estate agency signs placed in the curtilage of the cottages (Figure 59.1); accompanying house particulars; a website boshomes.co.uk; and a board-game *BOSopoly* (floor-sized variation on the well-known property marketing game *Monopoly*). In *BOSopoly*, players purchase a derelict cottage

Figure 59.1 BOShomes estate agency sign at Mill Farm, Bosigran, 2008.

on the moors of West Cornwall, then install bespoke bathrooms, a designer kitchen and so on, to become the owner of a 'Grand Design' second home. Thus, as the game progresses, former homes are 'done up' to become houses commodified as investments or 'somewhere to escape' (for project details, see Vickery, 2019).

Springs Farm: a derelict home, an empty and overloaded place

I was particularly drawn to Springs Farm, its traces leaving an enduring and 'indelible mark' on my mind (Anderson, 2015: 177). Long deserted and non-existent in Land Registry records, it is situated amidst the iconic engine houses of Wheal Hearle on the edges of the West Penwith moor near the former mining village of Pendeen. As its name suggests, apart from dereliction and tumble-down stacks, its other notable feature is cattle-tramped bog.

The farmhouse was full of detritus left where it had fallen, abandoned by its former human residents. There were barn owl pellets, dead mice and rats, plastic flowers, ivy coming in through the widows, broken chairs and an old dilapidated range, bits of farm machinery, endless other dilapidated and piled up odds and ends, bric-a-brac and an old 'chef' utensil pot just like the one I remember sitting in the corner of my grandparents' kitchen. It was just about possible to climb the crumbling and far-from-safe stairs and negotiate the disintegrating floorboards to the main front bedroom where more personal belongings were strewn across the floor: old photos, fallen pictures, chairs waiting for someone to sit, toddler toys asking to be picked up and scooted across the boards, NatWest bank statements from the early 1980s, a school photograph of two children in St Just school uniform (the local school) and, potently, a Barbie doll lying on the floor nestled in Christmas tinsel amidst jumbled, just-left clothes. Feeling like an interloper, I was constantly on edge. With the building creaking precariously, the former and current residents, human and more-than-human, were still homing this eerie, disarming place. The decaying corpses reiterating the ongoing liveliness of vermin and the cattle outside baying, reminding me that the farmer might suddenly appear—and that here I was, out-of-bounds.

Figure 59.2a–d Photographs taken at Springs Farm, 2009.

Initially, I approached the site through photography, intending it to be part of the estate agency project (Figure 59.2a–d). In an early show, I exhibited the photographs as taxonomic groups, already anticipating a refusal on my part to see images as individual objects by exploring the 'grey borderland between remembering and forgetting' (DeSilvey, 2007: 893). People found them beautiful, alluring, and commented on my skill with a camera, clearly fascinated by the images as nostalgic objects. We have an overwhelming impulse for the old and the forgotten; here, this seemed to be the sole point of entry for the viewer and the photographs seemed unable to operate on any other terms. This photography of dereliction has been described as *ruin porn* in which, through the circulation of 'seductive' images online, empty ruins provide a compelling, immersive spectacle which 'reproduce[s] the viewing subject as a consumer of dereliction, the images

Figure 59.2a–d (Cont.)

mediating the ruin as a theme park to be drifted through' (Cunningham, 2011: 18). Thus, through the romanticising function of the pictorial frame, these depopulated photographic images of ruin ignore the social devastation of localised diaspora described by Laviolette and Baird (2011), and the 'contingent constitution of place', to produce instead an image of ruination 'fetishized as a sublime fossil, deprived of history and motion' (Lavery and Hassall, 2015: 113, 114–117).

DeSilvey and Edensor use writing to emphasise the accelerated and highly visible, processual, vital materiality of ruins and therefore moderate 'the overriding focus on the visual in ruin scholarship and focus attention on the ways in which the material qualities of ruins afford particular sensual and affective experiences' (2013: 16). This might be read as a critique of image-making processes such as painting. I would argue, however, that there are other avenues for countering the framing power of the static image that might still embrace image-making practices. Images, despite their frame, do not necessarily operate at a distance from either the pictorial or wider performative context of their reception. Joselit (2009), with reference to painting and later digital media (2013), argues that the dynamic transformation of matter through the transitive process of images (their making and reception) can operate to render images [and ruin] contingent [performative] in their representations, something I have discussed elsewhere with reference to images and landscape (Vickery, 2015). Similarly, Lavery, reflecting on Lee Hassall's film *Return to Battleship Island* (Lavery and Hassall, 2015) set amidst the dereliction of the Japanese island Hashima, makes a case for an alternative understanding 'of representation that problematize[s] the tendency of images to deny the destructive power of Time'. Lavery suggests that Hassall is not interested in the immediate representation of ruins or the production of the past in the present but, instead, uses film to 'purposefully set out to contest the tendency of ruin porn to blind us to the possibility of a future' and by so doing disrupts the gaze of ruin photography and draws attention to the politics of depopulated places (2015: 112).

I was heading in a similar direction by turning from photography to painting to 'disrupt' this photographic gaze (Vickery, 2015: 326). I share DeSilvey's sense that, 'in order to describe what is happening in these perforated places, I draw on new ways of storying matter—surfacing meaning that extends beyond cultural frames of reference' (2017: 6); but, whereas DeSilvey concentrates on the act of writing, I invited in other visual forms. As a result, I used the processes intrinsic to painting to explore the temporalities, traces and material residues of this depopulated place. Painting became an exploration of the shifting relationships between time and place (beyond the photographic gaze), of those allusive memories that one moment we can catch and the next are gone, only to re-emerge in some other shape at some other moment. '[T]he house shelters day-dreaming, the house protects the dreamer, the house allows one to dream in peace' (Bachelard, 1994: 6).

Around that time, I came across the interiors of the German painter Matthias Weischer, in which the world outside appears shut off. Whilst Weischer's paintings construct a convincing sense of architectural space, he simultaneously uses strategies to disrupt conventional perspective and expected spatial scales to create a sense of 'detachedness', a 'being-for-themselves' that heightens the intimacy of his rooms, so that the 'viewer is alone with himself (sic) and without the world' (Pfeffer, 2004: unpaginated). Weischer describes his paintings as an 'empty and overloaded space' (2007: 94); rather than attempting to represent a space through a photographic eye, he uses paint to open up images/interiors in a way that acknowledges the potential for spatial conflict. Through their architectural complexity and pictorial structure laid bare on the canvas surface, and the quietness and emptiness of direct human presence, they function as a place of imagining projected with a kind of unsettling intimacy.

In the *Springs Farm* series of paintings (Figure 59.3), I realised I was working in similar territory, the difference being that my work was situated and immersed in a very particular place. Using some of the photographs from the house as a starting point, I started to paint directly on the triptych of canvases: flat areas against perspectival depth and dense areas of paint against thin layers of glaze, all worked at larger-than-body-scale to demand a more performative interaction from the viewer. I worked on each component of the triptych separately, and then as a whole, breaking down edges, borders, boundaries and spatial logic. This enabled me to avoid closure and to remain 'open to inconsistencies in their systematic ordering, and to displacements that trouble the phantom of a coherent, bounded site [/frame]' (DeSilvey, 2007: 899); it allowed me to break the authority of the visual logic of interior representations and the logics of the ruin-as-image, house-as-object.

Figure 59.3 A shelter for daydreaming (oil on canvas, triptych, each panel 100 x 120 cm), 2010.

The title references Bachelard's metaphorical description of home, with its potential to hold the darkness of life (cellar) through the integrating comfort of its rooms for living, and the possibility of a space (garret) for imagining, hopes and dreams (Bachelard, 1994: 3–37). This metaphor of house as home is a phenomenological space, one of becoming, of potential, of dwellingness. It is 'a place of intimate and nurturing experiences' (Tuan, 1977: 138), a place that is past, present and future 'where every day is multiplied by all the days before it' (Stark, 1956 in Tuan, 1977: 144), acting as an intimate reservoir of day-dreaming in which our lives co-penetrate with place. However, Bachelard also writes that, when this sense of continuity is lost, 'man' (sic) then becomes 'a dispersed being' (1994: 7). To unsettle home further, I turn to this theme of dispersal and displacement.

VV Renovation: how home becomes house-as-holiday-let

Perhaps I should at this point tell you our plan. As I mentioned at the start, we are now leaving this place called home. We have a small narrowboat, *Mona*, moored on the River Avon between Bristol and Bath which my husband, working away from home, has stayed on midweek for the last four years. We will live on her, not forever I tell myself—and now you. Bristol is a wonderfully vibrant city, with friends, work connections and opportunities, the chance to experiment with something different, to spend more time together. On many levels moving is an exciting change; but still we are leaving the place we call home and renting out the house.

Originally, we intended to let it as a home to a local family. I am afraid we will now be Airbnb-ing. I excuse myself with

> 'we need to use the house as home for gatherings such as Christmas (after all we can't fit an adult family the size of ours on a 30′ narrowboat) … we still only have the one house … we will be able to return one day … the house needs to pay for itself … it's a means to an end … .

So, for much of the last year, my day job has been that of demolition worker, builder, carpenter, plasterer, transforming our somewhat neglected but very-much-loved home into a *house-as-holiday-let.* The line of former miners' cottages to which it belongs is now 50% holiday lets and second homes; and, for a time, ours will be yet another.

I almost needed my home to go through a period of dereliction (Figure 59.4a and b)—of de-homing, a process of adjustment and of letting go—'to understand the story of how self comes to be [and then undone] through a continual process of reexperiencing and redescribing the fragmented narratives encoded with objects', the material culture associated with home-making enacted in the dismantling of home (Crewe, 2011: 44; see also Gregson et al., 2007). I tried to be organised, it didn't last long. Going through cupboards, drawers and boxes under beds, sorting into containers variously labelled 'charity shop', 'free cycle', 'car boot', 'recycling/dump', 'keepsakes'. It all ended up in the front room while I demolished the tatty, twenty-year-old chain-store kitchen, knocked off Artex, plaster and decades of layered 'bodging', demolished the porch and then bit-by-bespoke-bit rebuilt, with the organised boxes of packed up things disrupted as more and more surplus, or stuff we just didn't quite know what to do with, ending up in the now inaccessible front room. At the time of writing we are emptying that room in time for Christmas and transforming the *house* to a returning, newly multi-generational family *home-in-transition*—quite an undertaking.

This period of de-homing became the start of a new art/research project. Still very much embryonic, I had a sense that this could lead somewhere. During the incubation phase of new projects, I tend to use an archival, documentary, gathering approach to keep the project open and exploratory until a practice/research focus starts to emerge out of the process itself.

Figure 59.4a and b VV Renovation, 2018.

Figure 59.4a and b (Cont.)

So, now as artist-builder-geographer, I attempted to document/photograph every item going out, being kept and coming in—things, fragments, traces—until it became overwhelming. I now have a hard-drive full of photographs and boxes full of scribbled lists, measurements, calculations, designs and wood cutting layouts; invoices, receipts, catalogues and all manner of marketing material from builders' merchants and home improvement firms. All of these things will become raw material from which to progress work in the studio.

At an affective level, these changes have created a feeling of 'severance' echoing my previous experience of housing precarity as a mother with young children (Brickell, 2011: 233). In the years since that experience, however, I became personally embedded in Cornwall: a place-attachment that has run through my personal life, jobs, practice and research. This leads me to the question 'Does being-in-place necessitate this locational presence?' or is there mileage in thinking location in terms of vulnerability and difference (Haraway, 1988: 590; Rose, 1993: 146)? Broadly, geographical thought has shifted from emphasising place attachment and identity—the coproduction of embodied space and self through the intimacies of localised place-presence (see Tuan, 1977)—to a networked and fluid sense of place and identity underpinned by 'power-geometries of time-space compression' (Massey, 1991: 25). Whilst the former emphasises a causal link between dwelling-in or rootedness to a particular place 'across lifelines and generations of families' and the resilience of locally based customs and practices—'those local walls that created the particularity of one place and distinguished it from others' (McDowell, 1999: 3)—the latter is concerned with the relationship between place and mobilities in which:

> all identities are a fluid amalgam of memories of places and origins, constructed by and through fragments and nuances, journeys and rests, of movements between. Thus, the 'in-between' is itself a process or a dynamic, not just a stage on the way to a more final identity.
>
> *(McDowell, 1999: 215)*

Indeed, the historical picture demonstrated by the globalised relation between Cornish diaspora, the collapse of nineteenth-century world tin-prices and the export of technology and labour, is just one example to demonstrate that place-identities have always been far more complex, politically poly-scalar and replete with 'unequal distribution' (Massey, 1991: 26) than our often nostalgic nationalistic imaginations might insist. People always have been of here, and of there—often subject to forces beyond their control.

Final thoughts

This chapter has discussed two art projects, *BOShomes* and *Springs Farm*, and the emerging project *VV Renovation*, encompassing painting, digital and located practice to juxtapose house as commodified object with the intimacies and discontinuities of house as home. In so doing, I have storied a personal journey from house to homing, and then a 'doing up' or transition from home into house again. In a few final reflections, I return to the original aim in writing this chapter: to think-through-practice, as an artist and geographer, about my shifting relationship to a particular place, to house and homing and to the anxiety involved in leaving. This writing, however, has also led me somewhere slightly different: first, to consider my positionality within this situated, intimate geography of change (Simandan, 2018); and, second, how this practice might resonate with wider geographies and politics of place and dis-placement.

Renovating, documenting, thinking, writing over the past months has helped bring more nuance to how I think *home* and, by extension, my shifting relationship to Cornwall. Whilst mindful of the trap of romanticising place-attachment in terms of a somewhat privileged being out of place—in contrast to the destitution experienced by so many—it helps me to have arrived at a place in which my way of living home becomes multiple, fluid, translocational: place-attachment does not necessarily involve place-presence. Accordingly, I have avoided any sense of profound loss of the materialised memories that are so much part of an identity that has become enmeshed with this home. I might be moving to Bristol but I have not left Cornwall.

Still work-in-progress, the archival gathering process of moving and its enactment, necessarily threaded through with inconsistencies of practice, auto-ethnographic uncertainties and partialities of positioning, provides material for developing and thinking-with studio practice in the months ahead. How I will creatively work with this material within the context of my move to Bristol, and the city's undercurrent of activism, is a challenge for the future.

> Responding to the historical and contemporary violence associated with mass destructions of home, Brickell reports that a recent concern of geographers has been to show how the intimate and personal spaces of home—and their loss—are closely bound up with, rather than separate from, wider power relations. The two clearest expressions of this are, first, the entangled relationships between home and poly-scalar politics and, second, experiences of homelessness.
>
> *(2011: 229)*

I am still not sure if it is possible or even ethical to try to reconcile my own complicity in the power geometries of holiday let and homelessness within the context of Airbnb-ing our house. Perhaps my task is to do something else with it, to work through it. Not long ago, I was moored on the narrowboat under the former railway bridge that is now a cycle path when a couple of cyclists yelled down at me, 'pikey': a UK term of abuse directed at members of the travelling community, usually those of Irish Traveller ethnicity. Despite years of working in community development challenging discrimination, such overt, personally directed abuse took me aback. Perhaps this work constitutes a small 'moment of resistance' (Brickell, 2011: 238), a refusal (now via writing and later through the next stages of practice) of a romanticised representation of home as necessarily singular, intimate, place-bound, secure. As Bolt writes:

> the disorientation and dislocation that are part of arriving in a new place are precisely the kind of (dis-) orientation that is required for feeling and seeing the world differently and for sensing patterns that become invisible in the familiarity of the everyday.
>
> *(in Boyd and Edwardes, 2019: 361)*

For me, this might be a redefining of my sense of self and enactment of home beyond place-bound attachment, and an integration/acceptance of 'hybridity' (McDowell, 1999: 215) or the vulnerabilities of place-disorientation within a necessarily more internalised sense of fluidly placed self and 'an insistence on difference' (222); a 'process of becoming' through change that is productive of and produced by geographical space (Simandan, 2018) worked through practice. My 'house' is going to be minimal and afloat, with tenuous rights, but it is still going to be at least part of my experience of self-homing and embodied futures.

I offer these thoughts and work-in-progress very much aware of the dangers of discussing the politics of location and translocation, of place and displacement, from the particularity of my own situation. However, I argue, in line with other feminist approaches (Brickell, 2011: 238), that it is imperative in today's climate of open hostility towards those perceived to be out of place (as notably signified by current British Government hostility towards migrants and refugees) that we reflect on what attachment to place and related practices of homing and un-homing mean to us, as individuals and as a basis to inform our research (for example, see Wylie, 2016). I contend that thinking with our own intimate, autobiographical stories can be a powerful way to hold up a mirror to the bigger picture without minimising the dehumanising experiences of others at the sharp end of globalised politics of place. As Hanson and Pratt, (2003: 19–20) write: 'by understanding that we as individuals move between/across margins and centers, we can destabilize unexamined dualisms and boundaries as we begin to see the inherent connections between inside/outside, center/margins, same/other'. This is a politically productive positioning that art can bring to the fore and contribute to critical geographies of the home and place (Brickell, 2011: 238). Without this reflection on our own experiences of place and practices of homing, how can we be in a position to contribute narratives with the potential to contribute any incremental adjustment to wider societal relations? How, with regard to those who experience destitution, displacement, disenfranchisement, or who are held to be in the wrong place, might we challenge the insidious implication of politicised calls for people to be returned to (put in) 'their place'?

References

Anderson J. (2015) *Understanding Cultural Geography: Places and Traces*. London: Routledge.

Bachelard G. (1994) *The Poetics of Space*. Boston: Beacon Press.

Basu P. (2007) *Highland Homecomings: Genealogy and Heritage Tourism in the Scottesh Diaspora*. London: Routledge.

Boyd C.P. and Edwardes C. (eds.) (2019) *Non-Representational Theory and the Creative Arts*. Basingstoke: Palgrave Macmillan.

Brickell K. (2011) 'Mapping' and 'doing' critical geographies of home. *Progress in Human Geography*, 36 (2), 225–244.

Brien P. (2018) *UK Funding from the EU*, London: UK Parliament. Available at: https://researchbriefings.parliament.uk/ResearchBriefing/Summary/CBP-7847, (accessed 01/ 01/2019).

Crewe L. (2011) Life itemised: lists, loss, unexpected significance, and the enduring geographies of discard. *Environment and Planning D: Society and Space*, 29 (1), 27–46.

Cunningham J. (2011) Boredom in the charnel house: theses on post-industrial ruins. *Variant Magazine*, 42 18–21.

DeSilvey C. (2007) Art and archive: memory-work on a Montana homestead. *Journal of Historical Geography*, 33 (4), 878–900.

DeSilvey C. (2017) *Curated Decay: Heritage Beyond Saving*. Minneapolis, MI: University of Minnesota Press.

DeSilvey C. and Edensor T. (2013) Reckoning with ruins. *Progress in Human Geography*, 37 (4), 465–485.

Gregson N., Metcalfe A. and Crewe L. (2007) Identity, mobility, and the throwaway society. *Environment and Planning D: Society and Space*, 25 (4), 682–700.

Hale A. (2001) Representing the Cornish: contesting heritage interpretation in Cornwall. *Tourist Studies*, 1 (2), 185–196.

Hanson S. and Pratt G. (2003) *Gender, Work and Space*. London: Routledge.

Haraway D. (1988) Situated knowledges: the science question in feminism and the privilege of partial perspective. *Feminist Studies*, 14 (3), 575–599.

Haydu T. (2017) *Cornwall's Vital Signs 2017*. Cornwall Community Foundation, UK Community Foundations. Available at: www.cornwallcommunityfoundation.com/wp-content/uploads/2018/04/CCF-Vital-Signs-2017-LowRes-002.pdf.

Jackman B. (2011) *Cornish Riviera Express: On a slow train to yesterday*. available at: www.telegraph.co.uk/travel/journeysbyrail/8516133/Cornish-Riviera-Express-On-a-slow-train-to-yesterday.html, (accessed 24/06/19).

Joselit D. (2009) Painting beside itself. *October*, 130 (Fall), 125–134.

Joselit D. (2013) *After Art*. Princeton: Princeton University Press.

Kennedy N. and Kingcome N. (1998) Disneyfication of Cornwall—developing a Poldark heritage complex. *International Journal of Heritage Studies*, 4 (1), 45–59.

Lanyon P. (1950) The face of Penwith. *Cornish Review*, 4 ((Spring 1950)), 42–125.

Lavery C. and Hassall L. (2015) A future for Hashima. *Performance Research*, 20 (3), 112–125.

Laviolette P. and Baird K. (2011) Lost innocence and land matters: community regeneration and memory mining. *European Journal of English Studies*, 15 (1), 57–71.

Lippard L.R. (1999) *On the Beaten Track: Tourism, Art, and Place*. New York: New Press.

MAGA (undated) *Cornish Dictionary/Gerlyver Kernewek*, Cornish Language Partnership, available at: www.cornishdictionary.org.uk, (accessed 03/01/19).

Massey D. (1991) A global sense of place. *Marxism Today*, 38 24–29.

Massey D. (1995) *Places and Their Pasts*. Oxford: Oxford University Press.

McDowell L. (1999) *Gender, Identity and Place: Understanding Feminist Geographies*. Cambridge: Polity Press.

Morris S. (2016) *St Ives Moves Step Closer to Restricting Second Home Ownership*, available at: www.theguardian.com/money/2016/nov/10/st-ives-moves-step-closer-to-restricting-second-home-ownership-cornwall, (accessed 04/01/2019).

Orange H. (2010) Exploring sense of place: an ethnography of the Cornish Mining World Heritage Site. In Schofield J. and Szymanski, R. (eds), *Local Heritage, Global Context: Cultural Perspectives on Sense of Place*. London: Routledge, pp. 99–117.

Orange H. (2012) *Cornish Mining Landscapes: Public Perceptions of Industrial Archaeology in a Post-Industrial Society*, University College London, Thesis.

Paris C. (2009) Re-positioning second homes within housing studies: household investment, gentrification, multiple residence, mobility and hyper-consumption. *Housing, Theory and Society*, 26 (4), 292–310.
Payton P. (1984) *The Cornish Miner in Australia: Cousin Jack Down Under*. Redruth: Dyllansow Turan.
Payton P. (2005) *The Cornish Overseas: A History of Cornwall's 'Great Emigration'*. Fowey: Cornwall Editions.
Perkins H.C. and Thorns D.C. (2006) Home away from home: the primary/second-home relationship. In McIntyre N., Williams D. and McHugh K. (eds), *Multiple Dwelling and Tourism: Negotiating Place, Home and Identity*. Wallingford: CABI, pp. 67–81.
Pfeffer S. (2004) 'Everyone Carries a Home inside Them' (Franz Kafka). In Pfeffer S. (ed), *Matthias Weischer: Simultan*. Ostfildern: Hanje Cantz, unpaginated.
Quinn B. (2004) Dwelling through multiple places: a case study of second home ownership in Ireland. In Hall C. M. and Muller D. K. (eds), *Tourism Mobility and Second Homes: Between Elite Landscapes and Common Ground*. Clevedon: Channel View Publications, pp. 45–59.
Rose G. (1993) *Feminism & Geography: The Limits of Geographical Knowledge*. Cambridge: Polity Press.
Ryan D. (2017) *Rough Sleeping Statistics Autumn 2017, England (Revised)*. London: Ministry of Housing, Communities & Local Government. Available at: https://assets.publishing.service.gov.uk/government/uploads/system/uploads/attachment_data/file/682001/Rough_Sleeping_Autumn_2017_Statistical_Release_-_revised.pdf.
Seria-Walker E. (2018) *Evidence Review: Adults with Complex Needs (with a Particular Focus on Street Begging and Street Sleeping)*. London: Public Health England. Available at: www.gov.uk/government/publications/homeless-adults-with-complex-needs-evidence-review.
Simandan D. (2018) Being surprised and surprising ourselves: a geography of personal and social change. *Progress in Human Geography*, 10.1177/0309132518810431 (accessed 1 June 2019).
Stark F. (1956) *Perseus in the Wind*. Boston: Beacon Press.
Tuan Y.-F. (1977) *Space and Place: The Perspective of Experience*. Minneapolis: University of Minnesota Press.
Tunbridge J.E. and Ashworth G.J. (1996) *Dissonant Heritage: The Management of the Past as a Resource in Conflict*. Chichester: Wiley.
Vergnault O. (2017) *This is the Shocking Truth About Homelessness in Cornwall*, available at: www.cornwalllive.com/news/cornwall-news/shocking-truth-homelessness-cornwall-463439 (accessed 9 May 2019).
Vickery V. (2009) *BOShomes* [online art project] www.boshomes.org.uk
Vickery V. (2015) Beyond painting, beyond landscape: working beyond the frame to unsettle representations of landscape. *GeoHumanities*, 1 (2), 321–344.
Vickery V. (2019) *Artist Website*, available at: http://veronicavickery.co.uk/artwork.html, (accessed 02/01/2019).
Weischer M. (2007) *Matthias Weischer: Malerie/Painting*. Ostfildern: Hatje Cantz.
Wylie J. (2016) A landscape cannot be a homeland. *Landscape Research*, 41 (4), 408–416.

60

Place-walking

The Umwelt explored through our creative imagination

Mike Collier

Introduction

In this chapter I will write about my direct, embodied experiences of walking through four different land/soundscapes (places) and my cultural/artistic responses to, and reflections on, such experiences. I will describe a journey of personal revelation and enrichment; a journey sometimes undertaken on my own, but more often shared with others on walks that I call *conversive meanders* or *pilgrimages* (Collier 2019). The chapter is a meander – rather like the walks themselves – in which thoughts, memories, feelings and sensuous reflections combine with folklore and scientific knowledge to create what Tim Ingold describes as *a meshwork* 'when everything tangles with everything else' (2015: 3). It is a journey that describes a gradual awakening and opening up of the body to a sensual engagement with the world, and a realisation that there may be expressive ways of re-presenting this embodied experience in visual form. Along the way, I will explore ways of resensitising ourselves with the dynamic presence of place: to think of place as a living, animated presence; the ground that implaces everything we do but provides more than the mere backdrop to our everyday lives (Casey 1993).

Methodologies of naming and meandering creatively

Every journey has a starting point. First, I want to focus on the activity of walking itself, looking at how and why it might begin to explain our phenomenological, embodied sense of being-in-the-world and how this experience creates a sense of Umwelt in relation to emplacement. *Umwelt* is 'a German word that means literally "the environment" or "the world around." Scientists studying animal behaviour, however, use it to evoke something more specific. For these biologists, the umwelt signifies the *perceived world*, the world sensed by an animal', human or non-human (Kaesuk Yoon 2009: 15). To begin, then, I want to consider ways of creatively re-presenting my embodied, perceptual experiences of place-walking through the world, moving towards a clearer understanding of how our polysensory engagement with atmosphere, weather and air can enfold and give vision to enmeshed experiences (Ingold 2015). As my walking and art practice has developed, I have become aware of the importance of meandering – the freedom of idleness (O'Connor 2018) – as a creative research methodology: an approach I am increasingly developing in my experimental writing too.

The aim of WALK (Walking, Art, Landscape and Knowledge), a research centre I run at the University of Sunderland, is to explore the world *creatively* as we walk through it. The key word here for me is *creatively*. The psychologist Liane Gaora asserts that: 'Walking and creativity are distinctly connected'. When walking, she argues, we can:

> engage in a stream of thoughts [...] people get ideas when they are out walking. You are away from things that are demanding your attention. You can stray farther and farther from the present by deconstructing or making sense of what happened in the past, but also by figuring out what could happen in the future.
>
> *(in Rubinstein 2015: 189–90)*

This creatively reflexive layering of thought is directly linked to the places we walk through, allowing us to feel that we are a part of, and not separate from, the world. It is this sense of 'emplacement', or Umwelt, that allows us to sense and to feel the creative ecologies that are in balance; to realise that in our more-than-human world, there is a rich array of different cultures – birds, plants, humans, and so on – constantly interacting. As Kaesuk Yoon puts it:

> the Umwelt sees a clear vision of life, of a natural order, but it is a vision that knows nothing of objectivity [...] and it cares even less. In fact, the Umwelt's vision of the natural order often stands in direct conflict with a scientific and evolutionary ordering of life. The Umwelt is instead thoroughly sensuous and wildly subjective.
>
> *(2009: 303–4)*

Walking creatively, and actively engaging our imagination in the process, allows us to build a sensuously layered experience of place; an experience that recognises and shares the lives of 'others' in a particular place.

As a hill walker, I am aware of *being* within space – I am spatially situated. My immersion in the environment moves beyond a purely visual perceptual response to place and extends to other senses – smell, touch, taste and hearing – that are all heightened and sharpened. When walking in the hills, I invariably need to *use* a combination of these senses to move about: touch becomes important as I scramble over rock; smell or taste combine to give me a sense of impending weather; sound allows me to pick up distances or particular features such as rivers, streams or waterfalls. Often, these senses are used intuitively and in combination as I make my way forward. It is an experience that is memorably articulated by Nan Shepherd in 'Being', the final chapter of *The Living Mountain*:

> Walking [...] hour after hour, the senses keyed, one walks the flesh transparent. But no metaphor, transparent, or light as air, is adequate. The body is not made negligible, but paramount. Flesh is not annihilated but fulfilled. One is not bodiless, but essential body.
>
> *(Shepherd 1977: 93)*

I will now take you on four different walks or meanders that provide the backbone to this chapter. Each individual part or walk, each diversion, describes a different experience of place or emplacement, of Umwelt. En route, I will explore what Deleuze and Guattari call *smooth space*.

According to Edward S. Casey, Deleuze and Guattari argue that '*striated space* is a space that is counted in order to be occupied; whereas *smooth space* is a space that is occupied without being counted' (1998: 303). Here, Casey uses the word 'counted' in the Cartesian sense that space is to be measured rather than sensuously experienced. *Smooth spaces*, on the other hand, are:

> wedded to a very particular type of multiplicity: nometric, acentred, rhizomatic, multiplicities that occupy space without counting it and can be explored only by legwork. They do not meet the visual condition of being observable from a point in space external to them: an example of this is the system of sounds or even of colours, as opposed to Euclidean space.
>
> *(Casey 1998: 303–4)*

A walk up Windy Gyle and along the border ridge

In 'Upper Coquetdale/Border Ridge' (2004; Figure 60.1a and b), I re-imagine a walk from Barrow Burn to Windy Gyle in North Northumberland. In this work, constructed of bold pastel colours and text, I refer to birds and plants specific to the walk. The anticipation is that the sounds, smells, textures and tastes I experienced might suggest themselves to the viewer – especially through the combined use of colour and text – both a sensual engagement with, and an observation of, place. I hope, for example, that the viewer might intuit that the inclusion of ***Sandpiper*** (a visitor from April–October), ***Eyebright*** (which flowers from June–October), ***Buttercup*** (which flowers from April–October) and ***Crowberry*** (which flowers in April and May) would set the date of this walk towards the end of May/early June. This gives an indication to the type of weather and temperature I might have experienced. Regarding the type of location, rocky streams are amongst the ***Sandpiper***'s favourite habitats. ***Eyebright*** can be

Figure 60.1a and b Upper Coquetdale/Border Ridge, Unison Pastel on hand-made paper, 70 x 70 cms, nine x nine-letter indicative species of flora and fauna encountered on a nine-mile walk in the Cheviots (2004).

Figure 60.1a and b (Cont.)

found on low or high grassland of a type familiar to anyone who has walked in the Cheviot Hills in Northumberland. ***Buttercup*** is commonly found on damper grasslands: for instance, on land adjacent to a stream. ***Crowberry***, on the other hand, only grows on moors and mountains. The ***Stonechat*** often perches on top of a gorse bush in scrubland on the hillsides of river valleys. ***Goldfinches*** can regularly be sighted or heard in neglected farmland, gardens and open areas with scattered trees. They will often be found carefully picking at the seeds of thistles in such habitats; whereas the ***Peregrine*** is a bird of crag, moor and mountain.

This list of nine flora and fauna (Figure 60.1a) also alludes to certain sounds to be heard in this particular part of Northumberland. The first time one usually spots a ***Sandpiper*** is when it is startled into flight by one's presence, rising with a loud, musical, high-pitched 'twi – wi – wi – wee', flying low just above the stream with alternate spells of gliding and flickering shallow wing beats, to settle down again some distance away. The call of the ***Stonechat*** resembles the sound of two pebbles being banged together and, if disturbed by a walker, its alarm call sounds a little like 'wee – tac – tac'. ***Goldfinches*** have a tinkling, bell-like call as they move periodically from plant to plant. Their song is a pretty, liquid sound – an elaborate version of the flight notes. The ***Peregrine***, on the other hand, is unlikely to be heard. Finally, what about tastes? ***Crowberry*** has a rather bland, berry-like, succulent taste, but it turns the mouth blue. ***Eyebright***, on the other hand, has a bitter and cool taste: it is useful as a remedy for conjunctivitis or styes – hence its name.

Reading 'Upper Coquetdale/Border Ridge' should be a slow process of discovery. In an analogy to the use of perspectival space in painting, I seek to use words in a way that strips away the cultural references that, through grammatical structure, attach themselves to ideas of power and hierarchy. I am especially interested in using words that signify ideas of place and our embodied relationship to our environment; and I will explore this further in the next walk as I describe the experience from the summit of a hill rather than the walk up it.

Helvellyn

I have long been fascinated by the poetry of place and mountain names, as well as the local names for plants and birds. One of the many pleasures of hill walking is to reach a summit and to see a whole range of hills – many of which I have climbed, some of which I intend to climb – and to be able to name them. One mountain, in particular – Helvellyn in the Lake District – has become a place of pilgrimage for me and I often make two or three visits each year to its airy summit. I have vivid memories of each of these journeys. My most recent memory is of a cloud inversion seen at Helvellyn's summit as the peaks of Lakeland's mountains appeared to 'breach' through the early morning mists: 'a thousand ridges [...] Gleaming like a silver shield!' as William Wordsworth put it (1936a: 173). I could see:

High Street (the street of the Britons),
Harter Fell (the mountain of the hart or stag),
Thornthwaite Crag (the hill of the clearing by the hawthorn trees),
Kentmere Pike (the peak from which Kentmere's springing fountain came),
Ill Bell (the treacherous bell-shaped hill),
Red Screes (named from its massive wall of reddish rock and loose stones),
Fairfield (beautiful field or open country),
Silver Howe (the silver hill),
Helm Crag (the Lion and the Lamb),
Wetherlam (weather-helmet),
Coniston Old Man (small Scandinavian mountain kingdom),
Grey Friar (hooded hill),
Harrison Stickle (the peak associated with the Harrison family),
Bowfell (the bowed mountain),
Scafell (the bald, stony, mountain),
Glaramara (the shieling at the ravines),
Yewbarrow (the hill grazed by ewes),
Great Gable (the enclosure or hunting ground of Gavel),
Green Gable (the grassy, gable-shaped hill),
Red Pike (the reddish peak),
Pillar (named from its vertical climbing wall),
High Stile (the high, steep place),
Hindscarth (the pass of the hind or female deer),
Robinson (associated with the Robinson family who in the 16th century owned land around Buttermere),
Grasmoor (the grassy upland waste),
Eel Crag (treacherous crags or rocky heights),
Causey Pike (the peak by the causeway or paved track),
Grisedale Pike (the hill above the valley where young pigs graze),
Skiddaw (the mountain with the jutting crag),
Blencathra (Saddleback – the summit of the seat-like mountain) and
Catsysam (the ridge by the wild cat's steep path).

(The descriptions are taken from Diana Whaley, *A Dictionary of Lake District Place Names*, 2006.)

I have climbed all of these hills and no longer need a guidebook to recognise their distinctive outlines on the horizon when seen from the grassy slopes of Helvellyn's summit. Indeed, I have stood beside many hilltop cairns and noticed that I am not alone in pointing to, and sometimes saying out loud, the names of these distant hills: an activity that is an important part of the embodied experience, I *feel*, of being up and out there in the open, breathing in the air and atmosphere.

For a sense as to why this might be so, I will turn, at length, to Keith H. Basso:

> Several years ago, when I was stringing a barbed wire fence with two Apache cowboys from Cibecue, I noticed that one of them was talking quietly to himself. When I listened carefully, I discovered that he was reciting a list of place-names – a long list that went on for ten minutes. Later, when I asked him about it, he said, 'I ride that way in my mind'. And on the dozens of other occasions when I have been working or travelling with Apaches, they have taken satisfaction in pointing out particular locations and pronouncing their names – once, twice, three times or more. Why? 'Because we like to' or, 'because the names are good to say'. More often, however, Apaches account for their enthusiastic use of place names by commenting on the precision with which the names depict their referents. 'That place looks just like its name', someone will explain or, 'that name makes me see that place like it really is' or, 'its name is like a picture'.
>
> *(Basso 2001: 90)*

In my work *The Ancient Brotherhood of Mountains: There Was a Loud Uproar in the Hills* (Figure 60.2), I develop this focus on place names and mountain names, referring specifically to a passage from 'To Joanna' by William Wordsworth: a text that appeared in the series, 'Poems on the Naming of Places', and which was addressed to the poet's future sister-in-law. In this complex and layered poem, Wordsworth contrasts a contemporary, urbane, pastoral view of the landscape with that of a more rooted, sensuous and pantheistic engagement with place. 'To Joanna' is also a poem that could only have been written, I think, by someone who has climbed these Lake District fells and who has stood on their summits. (Indeed, it is said that Wordsworth walked a distance of 175,000 to 180,000 miles in his lifetime: walking becoming a major part of his practice around which the content, structure and rhythm of his writing evolved (Collier 2014).) The lines below are pivotal to the poem:

> Joanna, looking in my eyes, beheld
> That ravishment of mine, and laughed aloud.
> The Rock, like something starting from a sleep,
> Took up the Lady's voice, and laughed again;
> That ancient Woman seated on Helm-crag
> Was ready with her cavern; Hammar-scar,
> And the tall Steep of Silver-how, sent forth
> A noise of laughter; southern Loughrigg heard,
> And Fairfield answered with a mountain tone;
> Helvellyn far into the clear blue sky
> Carried the Lady's voice, – old Skiddaw blew
> His speaking-trumpet; – back out of the clouds
> Of Glaramara southward came the voice;
> And Kirkstone tossed it from his misty head.

– Now whether (said I to our cordial Friend,
Who in the hey-day of astonishment
Smiled in my face) this were in simple truth
A work accomplished by the brotherhood
Of ancient mountains, or my ear was touched
With dreams and visionary impulses
To me alone imparted, sure I am
That there was a loud uproar in the hills.
And while we both were listening, to my side
The fair Joanna drew, as if she wished
To shelter from some object of her fear.

(1936b: 117)

Standing atop Helvellyn, as I have many times, I can readily imagine this ancient brotherhood of mountains calling to each other; and the circular form of the texts I developed echoed the sounds of what Wordsworth describes as '*a loud uproar in the hills*'. Here, Wordsworth – who was himself interested in both etymology and the use of colloquial words – also refers to these fells as 'ancient mountains'. Picking up on this phrase,

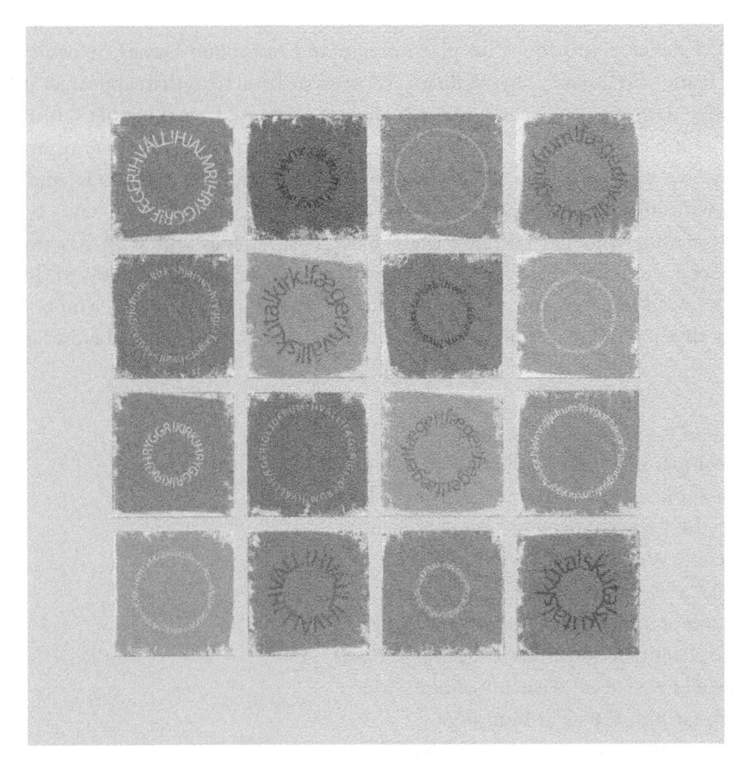

Figure 60.2 The Ancient Brotherhood of Mountains: there was a loud uproar in the hills 2012, (From 'To Joanna', 'Poems on the Naming of Places'), digital print 150 x 150 cms.

I used the oldest derivations of the names of the mountains I could find, once more using Whaley's *A Dictionary of Lake District Place Names* as my source:

Derivations

Helm Cragg	hjalmr
Loughrigg	laugrh
Fairfield	fæger
Helvellyn	hváll
Skiddaw	skúta
Glaramara	gljúfrum
Kirkstone	kirk
Silver How	haugr

A walk along the Sefton Coastal Footpath

My third walk, along the Sefton Coastal Footpath, takes me back to my childhood. It is also the first of the four walks, discussed in this chapter, that I would describe as both collaborative and conversive. In 2015, I was supported by Arts Council England to undertake this walk with: members of the public (who shared with me their experiences and personal memories of this special area); a natural historian (John Dempsey); a wildlife photographer (my brother Tim Collier); a poet (Jake Campbell); and, for the first time on my conversive walks, with a sound artist (Robert Strachan).

I was born in Sefton and lived there for eighteen years; and I regularly return to continue exploring its flora, fauna and land/soundscape. The Sefton Coast runs from the mouth of the River Mersey to the Ribble Estuary and holds one of England's largest undeveloped dune systems. It contains a National Nature Reserve and many Sites of Special Scientific Interest; and it is a European Special Protection Area because of the importance of its natural heritage. I have examined the ecology of this coastal region in an attempt to grasp what gives the place its identity. I have experienced its visceral and emotional character: the sand, the light, the weather. I have walked through this liminal landscape, researching the derivation of place names and the flora and fauna of its unique bio-region, examining its ecological footprint, feeling its Umwelt.

In *The Spell of the Sensuous*, David Abram suggests that expressive language came from our engaged and embodied relationship to the environment: 'Our task', he contends, 'is that of taking up the written word, with all of its potency, and patiently, carefully, writing language back into the land' (1995: 273). We once had a wonderfully rich and revelatory range of words for our local landscapes/soundscapes and the flora and fauna within it. However, increasingly, we now 'make do with an impoverished vocabulary for nature and landscape': a 'blandscape' as Robert Macfarlane memorably puts it (2010: 115). In *British Birds: Their Folklore, Names and Literature*, Francesca Greenoak points out that 'one of the most enjoyable aspects of investigating the names of birds is that you find an unpredictable and haphazard richness' (1997: 8); colloquial names for birds often being rooted in a sense of place. As a result, in *Six Birds of the Sefton Coast* (Figure 60.3), I have deliberately sought out colloquial and Anglo Saxon derivations of bird names; and the order in which these names are presented is based on a sense of the poetic quality of the words irrespective of a scientifically accepted structure of bird classification or any other hierarchical knowledge

Figure 60.3 Six Birds of the Sefton Coast, 2015, digital print on paper, 112 x 112 cm, produced in collaboration with EYELEVEL.

system. So: WASHTAIL; DEVILING; YARWHELP; GELVINAK; SPARLING; LAVEROCK.

These colloquial names are often a poetic reminder of a closer understanding and feeling for the natural environment we once had, and they frequently refer to the look, behaviour or sound of the bird. For instance, one of the many colloquial names for a Swift is DEVILING, perhaps because of its elusive speed in flight. The name WASHTAIL (Pied Wagtail) arises from the similarity between the constant up-and-down movement of the bird's tail and the action of dipping and lifting made by a person washing or scrubbing clothes (or dishes) by the waterside. Avocets utter loud yelping cries when disturbed, hence YARWHELP. GELVINAK is an old Cornish name for the Curlew: the name makes reference to its long bill (= gelvin). SPARLING makes reference to the harsh call of the Common Tern; and LAVEROCK (Skylark) is from Middle English *laverok* and Old English *lāwerce* lark (Collier 2015: 95). There are often many different colloquial names for the same bird, specific regions or places choosing to name each one distinctively.

In the next piece illustrated here, I begin to reference more directly what Ingold calls the 'weather world'. In *The Life of Lines*, Ingold writes:

> As the philosopher Maurice Merleau-Ponty wrote, in his essay 'Eye and Mind', 'There really is inspiration and expiration of being' [...] This, Merleau-Ponty insisted, is not to speak metaphorically [...] breathing the air we also perceive in the air. Normally we cannot see the air, though sometimes we can – as in the mist, or in rising smoke from chimneys, or in light snow when flakes, in their feathery decent, pick out the delicate

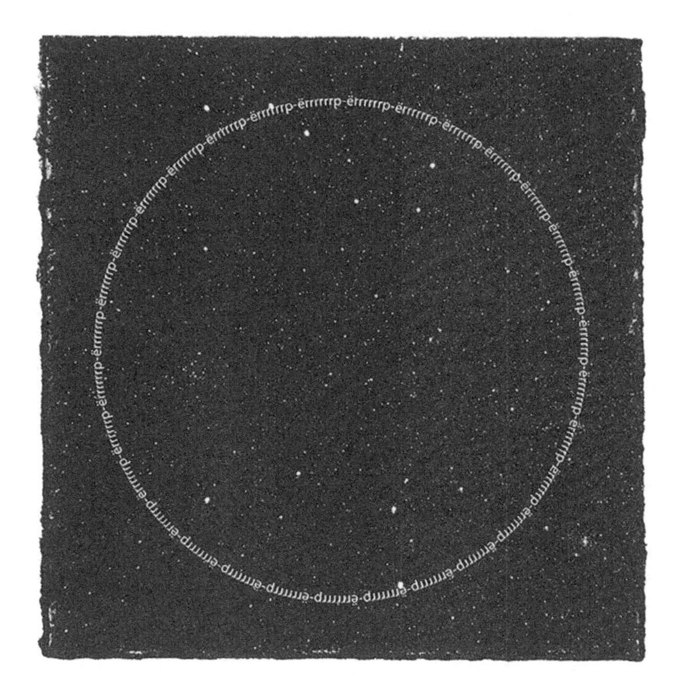

Figure 60.4 The Birkdale Nightingale, 2016, digital print on paper, 96 x 96 cm, produced in collaboration with EYELEVEL.

> tracery of aerial currents. Yet it is precisely because of the transparency of this life-sustaining medium that we can see. **Moreover, in its vibrations, air transmits sound waves, so that we can hear**.
>
> *(Ingold 2015: 68)*

I have highlighted this last sentence because it has significantly affected my own work, reinforcing the idea that it may be possible to visually represent embodied perception – especially soundscapes – as an example of the world made flesh through sound and movement. For example, the Sefton Coast is one of the most important breeding grounds in the UK for its rarest amphibian: the Natterjack Toad. The Natterjack is smaller than the common toad; but what it lacks in size, it more than makes up for with a loud rasping croak that echoes around the dunes on spring nights as the males go in search of a mate. It is the noisiest amphibian in Europe and its ratcheting call has brought it two local nicknames: the Birkdale Nightingale (Figure 60.4); and the Bootle Organ. It is one of the most haunting sounds of the Sefton Coast. The poet Jean Sprackland talks of

> the cosmic sound of [the toads] clamouring all around me. I knew it was the males calling the females to the mating pools, but it seemed, as I stood alone in that vertiginous darkness, that they were throwing their voices into the sky, a sound as timeless as the stars themselves.
>
> *(2015: 5)*

As a child, I can vividly remember hearing this throaty, guttural sound as I walked through the dunes.

Cheeseburn (singing the world) – a dawn (and evening) chorus

'The Birkdale Nightingale' was the first time, I think, that I had successfully managed to re-present and re-imagine a soundscape visually. This brings me to the final walk in this chapter, 'Singing the World: A Dawn Chorus – Mimesis and Birdsong': a dawn chorus walk with a natural historian (Keith Bowey of Glead Ecological and Environmental Services) around the grounds of Cheeseburn in north Northumberland. Edward Grey, in *The Charm of Birds*, describes the dawn chorus as 'a tapestry translated into sound' (1937: 89). This description accurately describes the embodied, woven, shimmering sonic experience I had whilst walking from 3.30 a.m. to 6.30 p.m. through the grounds of Cheeseburn Grange – a woodland garden – in mid-May: the time of the year at which the dawn chorus, in terms of species involvement, is at its most diverse. My particular chorus began in that liminal space *before* dawn – with a Redstart in song, followed by Blackbirds and Robins fifteen minutes later – rising to a full chorus of sixteen birds between 5.30 a.m. and 6.00 a.m.

I approached the re-imagining of my dawn chorus walk in a number of different ways, collaborating with printmaker Alex Charrington (Charrington Editions), composer and musician Bennett Hogg, and natural history sound recordist Geoff Sample. Together, my dawn choir of sixteen 'songsters' was re-enacted variously as digitally manipulated sonograms and musical transcriptions, and formed the basis of a show of screen prints, music and digital prints: a process-based sequence of music, sound and layered 'neumatically derived' notations. The gardens and original house at Cheeseburn date back several hundred years: originally the Grange, or farm, of the Augustinian Priory in Hexham, it was granted to the Priory by John de Normanville in 1297. I wanted to re-present this sense of layered history; a visual and sonic palimpsest that gave expression to a fully embodied sense of place.

As I walked *through* my Cheeseburn Dawn Chorus, I became aware of the way that each bird species interacted with one another: calling, answering, competing, announcing and establishing territories … as well as simply revelling in the sheer joy of singing. This work, therefore, examined how individual bird species interact through song in the dawn chorus: looking at ways of visually and musically re-invoking these patterns of cultural interaction in a more-than-human world; exploring the relationship between the natural world, its specific cultures and cultural ecologies, and our own sense of culture/s in relation to place. My aim

Figure 60.5 An example of early medieval neumes from 'Investigating Italian Gradual leaves'; a project by Tessa Cernik.

was to avoid anthropomorphising the birds and, instead, to democratically explore a sense of emplacement in both a human and more-than-human Umwelt.

In developing our work I looked at visualisations of birdsong in W.H. Thorpe's *Bird-Song: The Biology of Vocal Communication and Expression in Birds* (1961); and the rough, printed symbols taken from a 1950s oscilloscope bore a superficial resemblance to handwritten neumes, a medieval form of musical notation (Figure 60.5). Working from Geoff's sonograms of individual bird recordings from Northumberland, I drew the notations for the birds recorded. The music composed by Bennett consisted of two separate but interconnected pieces, both of which were made from transcriptions of the individual birds of the Cheeseburn dawn chorus. These were not straight transcriptions, however, after the manner of the French composer Olivier Messaien; they don't directly transcribe the sound of the birds. Instead, my visual rendering of Sample's sonograms formed the basis for melodies in medieval neumatic notation; and Bennett then transcribed these into modern musical notation. The musical notes, therefore, are 'twice-translated': from sonograms into medieval neumes; and from neumes into modern notation.

As a result, although the music is based on birdsong, it doesn't mimic birdsong; and in this it closely parallels my own approach. For a visitor to our exhibition it is probably not even apparent which piece is being heard, let alone which birds are 'present'. But this is not the point. Just as we can listen to a dawn chorus, and not necessarily recognise each and every species, so the combination of the music and the images is intended to 'stage', for want of a better word, an 'experience', inside of which we can make our own connections

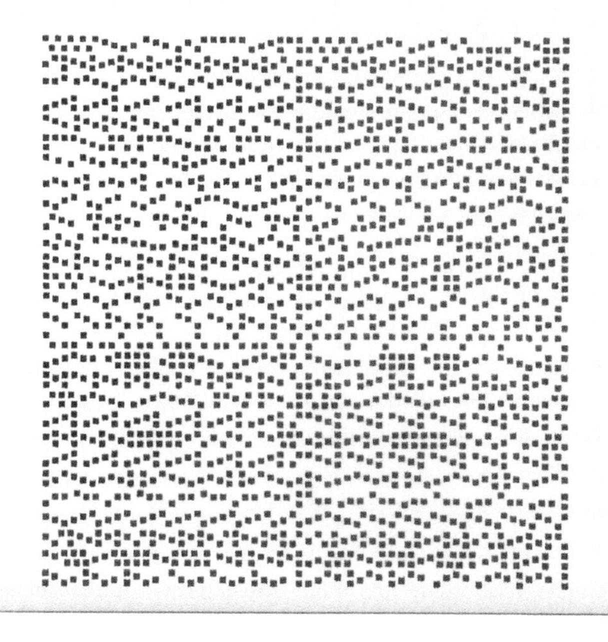

Figure 60.6 The Dawn Chorus Neumatic Notation # 1; 2018, produced in collaboration with Charrington Editions, Bennett Hogg and Geoff Sample, digital print, 50 x 50 cms.

(Hogg 2018). Crucially, a viewer has to walk around the gallery to see my dawn chorus narrative sequence of prints and, whilst doing so, the musical experience shifts sonically, just as my experience of the dawn chorus did as I walked through the woodland garden. This creates a parallel and re-imagined sense of emplacement (Figure 60.6).

Reflections

In this chapter I have described four very different walks and explored various ways of creatively re-presenting my embodied experiences of each 'phenomenological walk' that involved the 'gathering of synesthetic, material and sensory experiences as they unfolded [...] in the duration of the walk' (Tilley 2012).

I have explored how our experience of colour – as with the experiences of movement and depth that are key to a phenomenological understanding of being-in-the-world – effects a 'momentary crystallisation [...] awakening a transformative view of the world' (Wiskus 2017). Ingold explores sensation and atmosphere in the 'weather world' (2015); Merleau-Ponty discusses vibration and space. I would like to suggest that somewhere between these four words (or perhaps in the echo of these words) lies the sense of what both Ingold and Merleau-Ponty mean when they talk about the role of colour, sound and space in picture-making – a reimagining of an immersive, knowledge-rich walk.

In relation to text, my use of the list form can be seen both as a meditative monument to a particular place and an embodiment of things in their *sing*ularity. I like the possibility that this word singularity also references *sing*ing with respect to the spoken or sung word (Collier 2019). These lists can be read or sung out aloud, each word 'rooted in the felt experience induced by specific sounds and sound-shapes as they echo and contrast with one another [...] a particular way of "singing the world"' (Abram 1995: 76). Names can draw us 'into a world where art meets science, often producing names which scintillate in the mind and engage our feelings' (Marren 2015: 130).

Furthermore, these individual words, spoken together, map a very specific landscape. They allow us the intellectual and emotional space to fill in the gaps between the words – to *re-imagine* what these landscapes/soundscapes might be like – and so we can become participants in the imagined world summoned up. These place-specific lists are indicative of particular habitats: 'lists that anyone with a little knowledge of plant-life and ecological habitats might identify with' (Collier 2015: 95). They couple art and science: feeling and knowledge intertwined. I focus on embodied, phenomenological *smooth space* because I suggest that we need to re-engage our heart with the mind when exploring place. As Casey puts it:

> Deleuze and Guattari insist that their own unabashed preference for *smooth space* [...] does not entail an unconditional endorsement of such space [...] "never believe", they admonish, "that a *smooth space* will suffice to save us" [...] it is a matter of "not better, just different".
>
> *(1998: 308)*

My aim in producing this work is to explore ways of showing how we might better understand our complex relationship to a more-than-human world, enabling us to value the whole world (birds, plants, animals and people) as a living ecology of cultural differences. We need to release ourselves 'from the dichotomy of regarding nature either as a combination of *processes* or *things* [...] to recognize that nature is a communion of subjective, collaborative beings that organize and experience their own lives' (Hall 2011:

169). I would suggest that the idea and experience of walking creatively can enhance and expand the ecological imaginary in such a way that we might be open to considering more-than-human agency and subjectivity.

References

Abram, D. (1995) *The Spell of the Sensuous*. New York: Pantheon Books.
Basso, K.H. (2001) Stalking with stories: names, places and moral narratives among the Western Apaches. In Halpern, D. (ed.) *The Nature Reader*. London, Picador.
Casey, E.S. (1993) *Getting Back into Place: Toward a Renewed Understanding of the Place-world*. Bloomington: Indiana University Press.
Casey, E.S. (1998) *The Fate of Place: A Philosophical History*. Los Angeles: University of California Press.
Collier, M. (2014) Introduction. In Collier, M. (ed.) *Wordsworth and Bashō: Walking Poets*. Grasmere, Art Editions North and the Wordsworth Trust, 10–11.
Collier, M. (2015) Picturing language back into the land. In Collier, M. (ed.) *Ghosts of the Restless Shore: Space, Place and Memory of the Sefton Coast*. Sunderland, Art Editions North, 84–107.
Collier, M. (2019) Conversive pilgrimage. In Marland, P., Borthwick, D. and Stenning, A. (eds) *Walking, Landscape and Environment*. Abingdon, Routledge, 51–66.
Greenoak, F. (1997) *British Birds: Their Folklore, Names and Literature*. London: A & C Black.
Grey, E. (1937) *The Charm of Birds*. London: Hodder and Stoughton.
Hall, M. (2011) *Plants as Persons: A Philosophical Botany*. Albany: SUNY.
Hogg, B. (2018) Handout for the exhibition *A Dawn Chorus*, Drawing Projects UK.
Ingold, T. (2015) *The Life of Lines*. Abingdon: Routledge.
Kaesuk Yoon, C. (2009) *Naming Nature: The Clash Between Instinct and Science*. New York: W.W. Norton and Co.
Macfarlane, R. (2010) A counter desecration phrasebook. In Evans, G. and Robson, D. (eds.) *Towards Re-Enchantment: Place and its Meanings*. London, Artevents, 107–130.
Marren, P. (2015) *Rainbow Dust, Three Centuries of Delight in British Butterflies*. London: Square Peg.
Merleau-Ponty, M. (1968) *The Visible and the Invisible*. Evanston: Northwestern University Press.
O'Connor, B. (2018) *Idleness*. Princeton: Princeton University Press.
Rubinstein, D. (2015) *Born to Walk: The Transformative Power of a Pedestrian Act*. Ontario: ECW Press.
Shepherd, N. (1977) *The Living Mountain*. Aberdeen: Aberdeen University Press.
Sprackland, J. (2015) Foreword. In Collier, M. (ed.) *Ghosts of the Restless Shore: Space, Place and Memory of the Sefton Coast*. Sefton and Sunderland, Art Editions North and Sefton Council, 5.
Thorpe, W.H. (1961) *Bird-Song: The Biology of Vocal Communication and Expression in Birds*. Cambridge: Cambridge University Press.
Tilley, C. (2012) Walking the past in the present. In Árnason, A. et al. (eds) *Landscapes Beyond Land*. Oxford, Berghahn Books, p. 29.
Whaley, D. (2006) *A Dictionary of Lake District Place Names*. Nottingham: English Place-Name Society.
Wiskus, J. (2017) Cohesion and Expression: Merleau-Ponty on Cezanne. In Davcis H.D. and Hamrick W.S. (eds) *Merleau-Ponty and the Art of Perception*. New York: State University of New York Press, p. 77.
Wordsworth, W. (1936a) To - [Miss Blackett], on her first ascent to the summit of Helvellyn. In De Selincourt, E. (ed.) *Wordsworth: Complete Poetical Works*. Oxford, Oxford University Press, 173.
Wordsworth, W. (1936b) To Joanna. In De Selincourt, E. (ed.) *Wordsworth: Complete Poetical Works*. Oxford, Oxford University Press, 116–117.

61

Walking west

Newer Volcanics song project

Marita Dyson

We acknowledge the Traditional Owners and Custodians of the land and waterways upon which our work is written and produced, the Wathaurong, Wurundjeri (Woi Wurrung) and Boon Wurrung people of the greater Kulin Nation, and acknowledge their continued culture and connection to land and waters. We pay our respects to Elders past, present, future and emerging. We acknowledge we are writing on stolen land where sovereignty has never been ceded.

Stuart Flanagan and I are multidisciplinary artists interested in place, history and memory. We produce work as The Orbweavers, in the form of songs, recordings, performances and visual art. Over the last three years we have been researching Narrm/Melbourne waterways through the late-nineteenth and twentieth centuries, to write and produce a suite of creative works which explore industrial history and environmental change over time, and the lives of people who lived and worked alongside these waterways. This has resulted in *Newer Volcanics*, a suite of works which includes poems, songs and accompanying visual works that we describe as Song Maps. The project focusses on waterways in the north and west of Narrm/Melbourne: Merri Creek, Moonee Ponds Creek, the former wetland of West Melbourne Swamp, Maribyrnong and Birrarung Rivers, and Stony Creek in Yarraville.

Newer Volcanics is a geological province in south-west Victoria, extending into South Australia (Thomas, 1967) and includes the country south west of Narrm/Melbourne – the unceded sovereign lands of the Wathaurong, Wurundjeri (Woi Wurrung) and Boon Wurrung people of the greater Kulin Nation. It is a landscape characterised by water courses carved through ancient volcanic plains, silty clays, estuarine sediments and low-lying tidal salt marshes of abundant birdlife, where creeks and rivers meet the sea (Lack, 1991). Much of the land in the inner-west of Melbourne has been paved and overlaid by post-settlement industry and transport infrastructure but the waterways, and sometimes their immediate surrounds, persist. They are a living thread of the past and the present; a continuing flow and force through time.

We have learned about their materiality – the stone, water, vegetation, clay and mud that is particular to this place. We have become aware of patterns of industrial use where the value of waterways and their surrounds has been seen in resource terms, and repeatedly exploited as commodity, drain or waste receptacle. Through contemplating these histories, we reflect on our relationship to, and implication in actions of harm against, waterways and their ecosystems.

Why this place?

I grew up in Yarraville, a western suburb of Melbourne named for its location by the River Yarra (Birrarung). I never saw the riverbank growing up – it was hidden by fertiliser sheds, factories and shipping containers. My family still live in Yarraville and I also often work here, at Museums Victoria, in the nearby suburb of Spotswood. The Orbweavers' *Newer Volcanics* project has thus become an extension of the Narrm/Melbourne river crossings I have been making throughout my life.

My family's collective history in this country is short. Ancestors arrived in Perth in the nineteenth century, from Ireland, Scotland and England. My mother's family were farmers in central Victoria; she moved to Melbourne in 1972. As the first generation of my family born and living in Narrm/Melbourne, I grew up with limited knowledge of the city's social, cultural and geological histories. Through creative work, I am seeking to understand and more deeply connect with the place in which I live.

I acknowledge I cannot begin to understand the connection First Peoples have to place, and what that long connection to Country, through thousands of generations, feels like. Nayuka Gorrie, Kurnai/Gunai, Gunditjmara, Wiradjuri and Yorta Yorta writer, on Djap Wurrung culturally significant trees under threat from a proposed freeway extension, writes:

> It is important to situate myself when I write about these trees. My existence would perhaps not be possible without them. I am a Djap Wurrung person through my grandmother Sandra Onus (you can call her Aunty). These trees are Djap Wurrung people's inheritance. These trees are my inheritance, our inheritance. Their survival and our fight to keep them alive and safe are a cultural obligation and an assertion of our sovereignty. This sovereignty is a threat to the state. These words are my attempt to keep them alive. These words are too an assertion of sovereignty.
>
> No settler in Australia can look at any place in this country and know they have a blood connection spanning more than a few generations. There is probably a word for it in my language that I haven't learned yet but the closest I can come to describing it in English is an immense and utterly overwhelming sense of connection.
>
> The limitations and lack of sophistication in the English language, and that there is no word for this feeling, means the Anglo settler doesn't get to experience this and cannot possibly know this feeling. This was long traded by their ancestors. They can't understand what it means to be able to connect the blood coursing through your body to ancestors' blood soaked in ancient soil and ancient trees. To sit in a tree that saw your people birthed, your people massacred, and now your people's resistance is a feeling that the English language will never be able to capture.
>
> *(Gorrie, 2019)*

I am a descendant of white settlers who benefited from genocide, dispossession and theft of land and waters from people of the greater Kulin Nation. My ancestors contributed to the destruction of ecosystems through introduced crops and agricultural practices which required the draining of wetlands and re-engineering of waterways into irrigation channels. My creative work acknowledges and responds to these histories.

Psychogeography, *dérive* and Song Maps

When we began walking along the waterways, our knowledge of their natural and cultural histories was patchy. We needed to physically experience the places we wished to sing about and combine this with archival reference material. Psychogeography, and the practice of the *dérive* (Debord, 1956), describes how we drifted, letting our minds wander over layers of history visible as accretions, remnants and traces, contemplating personal connections and reactions to landscape, our *dérives* sought to build up a sense of place. We paid attention to smells: fresh and saltwater, diesel, anaerobic mud, wattle; physical textures: splinters of red gum from remnant piers, clay-caked shoes, prickly burred seeds, soft casuarina needles; and sounds: bird calls, bridge reverberance, surface reflections, traffic. Paying attention in this way forges new sensory memories and emotional associations with a place.

Dudley Flats: A Psychogeography, David Sornig's Creative Fellowship project at the State Library of Victoria (2015), first introduced us to the term *psychogeography*. David took us on a walk into what he describes as 'the Zone', the site of the former West Melbourne Swamp. Stuart and I were interested in this area for its lost saltwater Blue Lake, which we had first come across in writing by Robyn Annear (2009). David had researched the area for his book, *Blue Lake* (2018), and generously shared his archival finds of maps and early photographs. On our first conscious *dérive*, we walked 17 kilometres under freeways, by drainage canals, through rubbish tips and restricted port areas. David writes of this walk in *Blue Lake*:

> Entering the Zone, plotting sometimes unauthorised paths through its segments, challenged me to see through the veil of tightly construed designators that both sectioned it into areas of formal activity – port, road, rail, warehouse, leisure – and refused the natural connection of its parts. While I built from this a general impression of the Zone, it wasn't until I took a final circumnavigatory expedition around it in November 2016 that I came to see it whole. I saw connections where none were intended, that followed the logic of the land as it had once been: the erased river, the drained lagoon. Ultimately, entering the Zone required me to imagine it into existence.
>
> *(Sornig, 2018: 22)*

This led us to read Guy Debord and the Situationists and we subsequently sought works of local literary psychogeography by Sophie Cunningham and Nick Gadd. Both echo threads and moods we explore. Gadd (2016) describes the contemporary logistics landscape which covers the former West Melbourne Swamp as 'the edgelands [...] parts of the city that no one occupies, that no one cares for, that no one thinks about'; while Cunningham reminds us that:

> So much engineering has gone into shifting the city's waterways and the boundaries they represent. Into excavating canals, into rerouting rivers and into building freeways that float above drains that were once chains of ponds. There is no going back to the landscape that was once here, but – despite concerted effort – it has not been totally erased.
>
> *(2016: 30)*

For us, the *dérive* concerns walking in urban environments and being open to layers of meaning visible or sensed through their materiality and atmosphere, while also contemplating how these layers affect us psychologically. We refer to historical accounts and maps of a location before and after our *dérives*. We produce work through this contemplation of past and present.

When we return from *dérives*, I lay out large sheets of paper and words surface from the lake of memory. I recall impressions of the physicality of a place: geological formations, plants, architecture, building materials. Intangible and lost elements are also contemplated: former industries, industrial processes, histories, personal memories, folklore, emotions, residual ambiences. I list these as words – sometimes forming fragments of sentences. Then I draw around and between words, cross hatching and texturing different elements (the basalt outcrops, silty clay), plants (invasive weeds, remnant stands), architectural features (chimney stacks, iron girder bridge spans).

This process creates a Song Map, a way to collapse historical time and space, and generate new associations. An image that holds words and lyrics, Song Maps are also a way of expressing place absorbed and remembered as feel and resonance. Hidden stratigraphy and fragments of social and industrial histories are brought into the same space through marks, shade, shape and text. A nineteenth-century quarry can occupy the same visual plane as a lost ecosystem and a contemporary road. This process of conflation enables incongruous systems and histories to sit simultaneously with personal, emotional responses. Though the impetus to create Song Maps seemed to arrive intuitively, they have strong parallels to the *dérive* practices and map making described by Debord in 1956:

> With the aid of old maps, aerial photographs and experimental dérives, one can draw up hitherto lacking maps of influences, maps whose inevitable imprecision at this early stage is no worse than that of the first navigational charts. The only difference is that it is no longer a matter of precisely delineating stable continents, but of changing architecture and urbanism.

Our Song Maps were originally intended as a drafting tool for songwriting – literally to map place, record research notes and the lyric ideas they inspired. They soon became an integral part of our practice alongside audio recordings and live performances. Sometimes we place large format printed Song Maps on the floor, in front of our performance area, pointing out regions and details to the audience as we sing.

Early photographs, published histories and archaeological reports also prompt us to reimagine lost environments, both in ambience and physicality. In addition, historical maps from the State Library of Victoria have been an important reference for our work. Estuarine environments around the West Melbourne Swamp, where much of our work is situated, have been obliterated. Surveyor Charles Grimes' 1803 map documents the meandering Birrarung and basalt rises around waterways, prior to settler intervention. Robert Russell's 1837 map shows extensive sheoak (Casuarina) forests which covered the western lava ridge of the early Melbourne settlement, which Footscray historian John Lack laments were 'the first casualties of European settlement, mainly as fuel' (Lack, 1991: 3). A later Harbour Trust map indicates Coode's proposed dredging and river realignment project, with subsequent Victorian Railways maps evidencing the completed works.

Yarraville and Spotswood

The historic 1890s Sewerage Pumping Station, in Spotswood, exerts multiple influences: as my workplace (the site houses Museums Victoria's west Melbourne collection store), as a historical marker of place, and as the destination for waters that flow through me, and other Melburnians, as bodily waste or sewage. The pumping station is purposely situated at the lowest geographical point, allowing sewage to flow underground, by gravity from across the city, before being pumped uphill to the treatment plant (Lang and Vaughan, 2013). The Pumping Station sits alongside the Stony Creek estuary and Coode's realigned confluence of the Maribyrnong and Birrarung rivers, just south of the erased West Melbourne Swamp. With Melbourne's main underground sewer flowing beneath the Birrarung and Maribyrnong, there is water above and water below, the city's collective bodily waste waters converging in this sewer under the rivers, just as the rivers form a confluence above. The West Gate Bridge mirrors this east–west passage, a stream of traffic through the sky, from one side of the river to the other. There is a strong flow of people, water and gravity in this location. It is the city's concentrated *sink*.

I learned about Lucey Alford, the first woman employed at the Pumping Station in the late 1930s (Museums Victoria, 2019). Lucey was employed as a bacteriologist, to study bacteria which were corroding the concrete sewer linings. Her laboratory was situated in one of the station's impressive towers. Being the first woman employed on site, there was no existing women's toilet, so one was installed in the tower, next to her laboratory. I see Lucey as a sentinel of hydrocommons. From her tower laboratory, overlooking the re-engineered arterial riverine confluence, she worked to safeguard the cement-lined communal bowels of the city. Arteries above and civic bowels below, carrying our collective bodily waters, her tower toilet a direct line into the subterranean flows.

A little further upstream, arsenic slowly leaches from the infilled banks of the two rivers (Fyfe, 2005). Yarraville was the centre of Victoria's fertiliser industry from early settlement until the 1980s. Its riverside location was ideal for incoming bulk supplies by ship and barge, and from which to transport fertiliser to rural regions via railways. Production of superphosphate at the riverside Mount Lyell fertiliser works was a noxious process, involving the mixing of bone ash, obtained from upstream abattoirs and bone mills in Footscray and Kensington, with sulphur dioxide, extracted on site by roasting iron pyrites, a copper-mining waste product shipped from Mount Lyell, Tasmania. After burning, the iron pyrite cinder waste was used as infill along the Birrarung and Maribyrnong rivers (Chancellor, 2011), polluting the bank and waters with arsenic (Fyfe, 2005).

During a collection inventory at the museum one day, I found myself holding a jar of superphosphate produced at these fertiliser works, acquired by the museum in 1918 as an example of local innovation in agricultural technology (Museums Victoria, 1918). I considered its journey: from mining waste products and animal bones borne along the river, through furnaces and hoppers, to granules in sacks. Then carted from warehouses to railyards, delivered to farm sheds and into my grandparents' hands. I contemplated the white granules held in the museum's glass jar in *my hand*, and the direct relationship between this material and the infilled and polluted southern banks of the Birrarung, the river of Yarraville, my childhood suburb.

I considered how my settler ancestors grew introduced wheat in an unsuitable environment through the application of superphosphates. Their crops were irrigated by waters channelled from the Waranga basin at the expense of naturally occurring wetlands. A confluence of past and present eddied and swirled in my mind. My grandmother, a farmer, had recently died, and my mother gave each grandchild one hundred dollars – the dividends from my grandmother's shareholdings in Pivot, formerly Commonwealth Fertilisers, the superphosphate company who eventually subsumed Mount Lyell and the river sites in Yarraville (Barnard et al., 2000; Heritage Council Victoria, 2019). I bought a coat with my inheritance. Profits from the river system's contamination had flowed on, paid forward to me, and I was wearing them. I am implicated in the draining, channelling, dredging, the burning bones and pyrite cinder infill along the Birrarung and Maribyrnong. Sweating in my new coat, I felt porous; I was metaphorically leaching iron pyrite cinder infill.

This story is retold in the lyrics of *River River*, along with histories of munitions production, which occurred further upstream in Footscray and Maribyrnong, which similarly resulted in industrial contamination of riverside land (Figure 61.1).

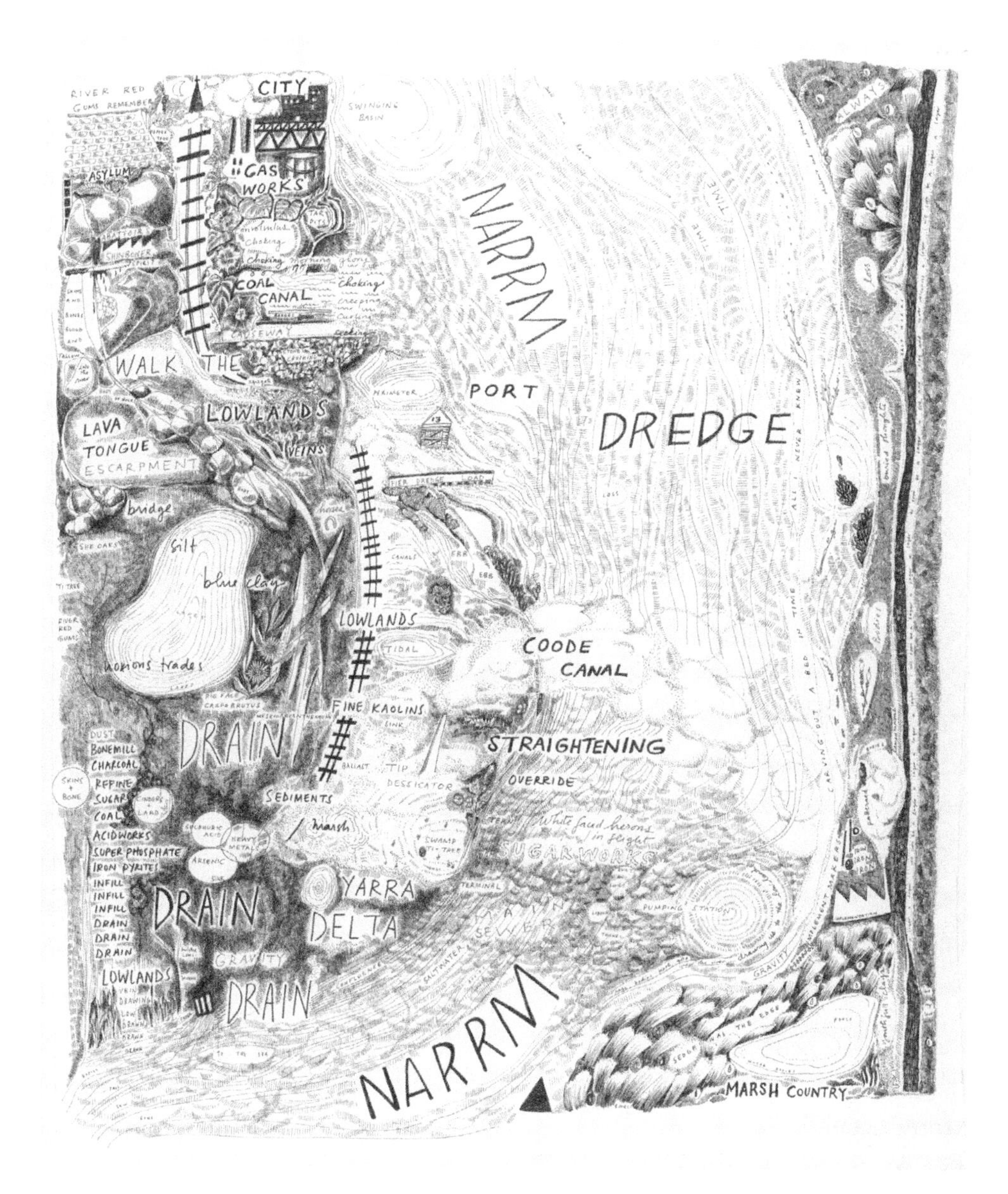

Figure 61.1 Confluence Song Map, Marita Dyson 2018.

SONG – *RIVER RIVER (MARIBYRNONG)*

Marita Dyson and Stuart Flanagan 2017

Cartridge maker by the River
Pouring powder into thimbles
Metal rolling bloody foil
Her body laid in flour and oil

Reaper, binder, boiler, packer
Superphosphate chimney stacks are
Multiplying down the River
Burning copper pyrites to a cinder

Take the cinders to the dredges
Fill the saltmarsh up to edges
This is how a city grows:
Border caution fencing forward

Woollen mills and cordite tresses
Nitroglycerine confessors
Now I dream and float there with you
Under Calder, under bridges

Ammunition safely stored
Warships downstream justly moored?
Moonlight over tidal song
There I find you
Maribyrnong

River, River, all your people
River, River, all your people
All the souls we never grieved
Untold bodies lost and given

Untold bodies lost and given
There I find you
Maribyrnong

Stony Creek

Stony Creek flows from St Albans in the west, through the suburbs of Sunshine, Braybrook and Yarraville. Quarry holes riddle its banks. Abundant bluestone around the mouth of Stony Creek was initially used as ballast, filling empty ships heading back to the northern hemisphere (Vines, 1993) where, it is mooted, it found another life as paving and building stone for London. In later years, Stony Creek's proximity to abattoirs on the Maribyrnong made it a convenient location for allied fellmonger and tannery operations, offering swift drainage of foaming, acrid industrial wastes into its tidal estuary.

This moon-shaped estuary, or Backwash, as it is unromantically named, sits at the border of Yarraville and Spotswood where Stony Creek meets the larger Maribyrnong and Birrarung rivers beneath the West Gate Bridge. It's part of the *sink*. The estuary has been revegetated, fringed with dark green undulating mangrove crowns, their root systems plunging into the silty mud and seawater. An old wooden pier of red gum sleepers extends out across the Backwash, towards the river. The river banks are lined with bluestone quarried from nearby.

In the early morning, Stony Creek Backwash appears as a mirrored sheet of tidal water against a backdrop of fuel storage tanks and factory chimney stacks. On a still day, smoke rises slowly in an uninterrupted plume as gulls, terns and pelicans fly low. Overhead, the concrete, brutalist West Gate Bridge looms and rumbles, its numerous supporting girders gently curving west to east, like an articulated skeleton stretching across the sky. I am humbled by the scale of industry and transfixed by the eerie serenity of this hazardous and beautiful location. The estuarine past is still here – although severely impacted upon, not obliterated.

The deep past is held here in the monosulphidic black ooze (EPA Victoria, 2009) – the ancient 'soft dark grey silt, clayey silt and silty clay', extending over the Birrarung and Maribyrnong valleys and delta (Thomas, 1967: 49). When I was ten years old, I got stuck in this mud with my sister and friends. We followed the train line between Yarraville and Spotswood to where the tracks crossed Stony Creek, and clambered down the embankment into the waterway. Following the cemented channel, its basalt sides now hollowed out and widened, we reached the Backwash, where Stony Creek meets the river. The tide was out, and the mudflat looked smooth and firm. We dropped over the railing of a pier, keen to reach the opposite side.

This was a mistake. We sank into the thick, grey-black, anaerobic mud. It belched a sulphurous tang. We laughed uneasily, our feet submerged. I dreaded the trouble at home for getting so muddy, hoping there might be a tap somewhere on the way back to wash. One of the little kids started crying, scared. It crossed my mind that the area could be contaminated with industrial waste. The nearby fuel storage tanks I had dismissed as benign, became a portent. We tried to get out, wriggling and squirming, but sank deeper. Over our knees. The mud sucked my shoe away. As the oldest kid in the group, I was accountable for this fix. Lost shoe. Bare foot. Stinking mud. There was no getting help. No one in this industrial wasteland.

That was what drew us there.

The tide started coming in. Fear rose from my gut. If we didn't get out we would drown. Panicking, I lost a sense of boundary with my body and the surrounding materiality, as my skin numbed. The relentless ingress of water. Ebb and flow. Vascular. Fine particulates sucking at my limbs. Sediments filling vesicles. I felt the Backwash claiming us. I started to cry too.

To escape, we had to fall. Relinquish our bodies to the mud. Prostrate, crawling, bellies slick with fine, silty, clayey, oily mud. I slid my way across the flat, to the engineered causeway.

We sat on the pier in the sun, clothes encrusted with reeking sediments. The smell coalesced with sewerage pumping station and petrochemical fumes emanating nearby, creating a distinctive association with place in my mind. It is an emotional memory that vacillates between fear and relief. These sediments: decomposed basalt, untold plants, animals – and what I did not fully comprehend at the time – heavy metals and contaminants, are also an archive (van Wyk, 2013). The State Environmental Protection Authority writes in 2013 of Stony Creek Backwash sediments containing 'very high concentrations of heavy metals (arsenic, cadmium, copper, mercury, nickel, lead and zinc above ISQG-High) and extreme concentrations of TPH (Total Petroleum Hydrocarbons) (over 20,000 mg/kg)', toxicants produced by historical smelting and tanning in the Stony Creek catchment (EPA Victoria, 2013).

Figure 61.2 Stony Creek Song Map, Marita Dyson, 2018.

Sadly, these impositions continue into the present. On 30 August 2018, toxic runoff from a factory fire upstream in West Footscray made its way through the storm water drainage system and into the creek. Contaminants included 'phenol (an industrial chemical and cleaning product), polyaromatic hydrocarbons (fire and soot by-products), lighter petroleum hydrocarbon chemicals called BTEX (benzene, toluene, ethylbenzene and xylene), PFAS, and industrial solvents such as acetone and butanone' (EPA Victoria, 2018). Remediation treatment requires that a section of the creek be diverted, and the removal of more than 670 cubic metres of contaminated silt to landfill (Melbourne Water, 2019). The creek is still fenced off with EPA warning signs (Figure 61.2).

SONG – *STONY CREEK*

Marita Dyson and Stuart Flanagan 2017

When it rains
you flow by

eleven quarry holes
dub the town: Stoneopolis;
of ballast stocks,
interminable.

Blast
Chip
Barge
Ship

Your ancient bed served London well:
her pavement smooth and durable
heels low and high,
as distant feet, hooved, unshod
sink into softer clays and silt
(like mine like mine).

When it rains
you flow high
waste of wool and skin:
the scouring foam
of sulfur, lime and tan,
to hair combed half-moon estuary.
Mangrove fringed
in green and white.
Backwash named
to push away
loss of tea tree
casuarina
and untold lives who passed
(this way this way).

When it rains
you flow high
long cemented sides,
reminder:
Water always finds a way
to this bay home,
to this bay home.

Wales Quarry – Merri Creek

Merri Creek flows north–south and, like other waterways, has been heavily impacted by colonial settlement and industrialisation. Stripped of vegetation. Gouged by bluestone and clay quarries. Filled with waste. Used as a drain and contaminated. By the 1980s, invasive weeds choked its banks: blackberries, hemlock, fennel, thistles, morning glory, onion grass. Neglected, heavily polluted, but still flowing, its path was eyed by authorities as a potential freeway route – a flyover to funnel cars from the north into the city. Public campaigning from locals and environmental advocates saved the Merri from this destructive fate (McGregor, 2019) and, with support from councils and water authorities, began replanting the creek banks with endemic species, being gradually transformed from weedy, rubbish filled drain, to remediated green corridor.

Merri Creek has carved its way over ancient basalt lava flows of between 0.8 and 4.6 million years old. A tributary of the Birrarung, its headwaters form at the northern edge of the Newer Volcanics geological province, flowing through the unceded lands of the Wurundjeri-willem clan of the Wurundjeri (Woi Wurrung) people, who have cared for it over millennia (Elender and Christiansen, 2001). In East Brunswick, the Merri was once bordered by grasslands and lightly wooded with red gums, casuarinas, wattles and tea tree. Murnong daisy and reeds covered its soft flood plains.

The post-settlement history of the Merri Creek in Brunswick, from the 1860s, is one of extraction and exploitation. Its basalt bed and surrounding clay soils provided building materials for the growing settlement of Melbourne. Brunswick bluestone was used for foundations, walls, paving, and crushed for use as an aggregate in roads and cement. Later it was put to use in the expanding railway network as track ballast. Its clays were used in brickmaking and potteries.

An 1866 photograph shows working conditions at the East Brunswick Wales Quarry, situated alongside Merri Creek (Figure 61.3). The quarry site is a desolate moonscape of boulders and blocks, denuded of vegetation. Draught horses are tethered to stone hauling carts. Men pose with picks and crowbars. Trees endemic to the area have been replaced by angular cranes, hooks, stone piles and dust. The undulations of a biodiverse environment,

Figure 61.3 Wales Quarry 1866, State Library of Victoria Collection.

formed over thousands of years through the interactions of water, weather, plant and stone, have been obliterated. The earth is hacked and pitted.

Quarrying of bluestone around the Merri Creek in Brunswick continued until the mid-1950s (Moreland City Council, 2008). Eventually, urban density forced the industry out. Explosives and mechanical means for extracting stone were considered too dangerous to continue near residential areas. Quarries were pushed to Braybrook, Deer Park and Werribee at the city's perimeter.

Located on inclines, adjacent to ephemeral water courses flowing into the Merri Creek valley, the quarry pits accumulated dumped refuse and stagnant waters to become 'bubbling, putrescent [...] offensive' areas (The Age, 1911). They were often inadequately fenced, unlit and unpatrolled, attracting the curiosity of children and causing drownings. As early as 1881, the filling of holes had become a longstanding 'bone of contention' (The Herald, 1881) between local council and State departments, a tension that continued through the twentieth century.

The Wales Quarry hole was purchased in 1965 and used as a tip by demolition firm Whelan the Wrecker. This gaping 50-metre-deep hole, once Melbourne's deepest bluestone quarry, offered 'a-million-and-a-quarter cubic meters of "air space"'. Within ten years, it had 'swallowed the debris of a thousand wreckings or more' (Annear, 2006: 270–271). Robyn Annear highlights the poetic irony of the Wales Quarry site: bluestone extracted from this location ended up back in the same hole within a century, as demolition waste. Today, the 1993 bicycle path cutting along the creek bank reveals Whelan's archaeology: bricks, rags and cut stone embedded as visible strata in compacted rubble. Sometimes shards of patterned china appear in the fill after rain. I collect the unusual ones and display them on my window sill. From kitchen to tip, to kitchen again.

Their instability and potential contamination precludes former quarry sites from being built upon, leaving contemporary Brunswick well-endowed with municipal reserves. Our song *Merri* arrives peripatetically, as we tread the Wales Quarry infill, under replanted eucalypts, wattles and casuarinas. Lyrics emerged from the protruding evidence of the tip, the remnant winch tower, the scent of wattle tannins. *Merri* is a tribute to the creek's resilience despite the violence carried out against it through land clearing, extraction and waste disposal (Figure 61.4).

SONG – *MERRI*

Marita Dyson and Stuart Flanagan 2010

Merri – your water flows into the quarry
Where bluestone
Once lay below
But now the gutters are paved with your heart
And the hole is filled with clay
And fallen homes

Evening, as we ride into the valley
Cool air
And tannin sighs
Allocasuarinas
Whisper the same song

As the water follows home
And so do I

Down past the convent
And under river
And through the headland
Out to sea
Allocasuarinas
Whisper the same song
I will always follow you.

Figure 61.4 Merri Creek Song Map, Marita Dyson, 2018.

Performing *Newer Volcanics*

Our intention is to perform these songs close to the sites they depict. *Newer Volcanics* premiered in November 2018 in the western suburb of Newport in a cavernous former electrical substation, built in 1914 to power the Victorian Railways. Like many of the places we sing about, the *Substation* had experienced its own post-industrial transformation: from abandoned site in 1967, to community arts centre in 2008. It has many connections to the industrial history referenced in our songs: situated downstream from the Birrarung and Maribyrnong confluence, and a few streets away from bluestone quarry site, turned tip, turned nature reserve: Newport Lakes.

The venue has a visible history of industrial use, which we sought to build on by working with eco-artist Aviva Reed as set designer. Endemic plants were installed to evoke the surrounding wetlands. Hard rubbish, hunks of cement, bluestone and introduced weed species exemplified the post-invasion and quarry hole landscapes. Large format prints of our Song Maps covered the walls. Each song was performed to a sequence of visuals comprising historic photographs and contemporary footage by filmmaker Brian Cohen. Through these multi-sensory elements, we expressed the materiality, ambience and history of the waterways and their surrounds. After the performances, audience members – environmentalists, bird watchers, former factory and railway workers, water researchers, activists – shared their personal stories and knowledge with us.

Walking through these watercourses has brought us to psychogeography, to histories and people, environments and industries, and to creative spaces where we try to weave these elements together, acknowledging land on which sovereignty was never ceded, and paying attention to stories of pain and resistance, to the layers that make and remake place.

Acknowledgement

Parts of this text were first published in *Cordite Poetry Review*, February 2018.
Thank you to Tim Edensor for valuable feedback and editing of this chapter.

References

Annear, R. (2006) *A City Lost and Found: Whelan The Wrecker's Melbourne*. Melbourne: Black Ink.

Annear, R. (2009) Blue Lake, In *Mrs Bradley's Melbourne: A Curious Person's Guide to The Sovereign City of the South* (Self Published).

Barnard, J., Butler, G., Gilfedder, F. and Vines, G. (2000) Cumming Smith. *Maribyrnong Heritage Review - Industrial Places*, 3: Appendix 1: 227–229.

Chancellor, J. (11/10/2011) Yarraville Industrial Site for Sale. *Property Observer*, www.propertyobserver.com.au/finding/commercial-investment/industrial/13844-yarraville-industrial-site-for-sale.html, Accessed 21/4/2019.

Cunningham, S. and Wells, D. (2016) *Boundaries*. Melbourne: City of Melbourne.

Debord, G. (1956) Theory of dérive, (Ken Knabb trans.) www.cddc.vt.edu/sionline/si/theory.html, *Situationist International Online*, Accessed 7/4/2019.

Elender, I. and Christiansen, P. (2001) *People of the Merri Merri: The Wurundjeri in Colonial Days*. Melbourne: Merri Creek Management Committee.

EPA Victoria (2009) Acid Sulphate Soil and Rock. Publication 655.1, July 2009 www.epa.vic.gov.au/about-epa/publications/655-1, Accessed 2/05/2019.

EPA Victoria (2013) The Origin, Fate and Dispersion Scientific Report of Toxicants in the Lower Sections of the Yarra River. Environmental Audit Report, Publication number 1529 May 2013, www.epa.vic.gov.au/~/media/Publications/1529.pdf Accessed 27/04/2019.

EPA Victoria (2018) Waterways around the West Footscray Industrial Fire. Publication 1713, 6 September 2018, www.epa.vic.gov.au/~/media/Publications/1713.pdf Accessed 7/3/2019.

Fyfe, M. (22/8/2005) A Great Idea with a Nasty Legacy. *The Age*. www.theage.com.au/national/a-great-idea-with-a-nasty-legacy-20050822-ge0qg9.html Accessed 2/05/2019.

Gadd, N. (15/10/2016) From the Ocean to the Lake. *Melbourne Circle*, https://melbournecircle.net/2016/10/15/from-the-ocean-to-the-lake/, Accessed 26/4/2019.

Gorrie, N. (12/4/2019) The Government wants to Bulldoze my Inheritance: 800-year-old Sacred Trees. *The Guardian Australia*, www.theguardian.com/commentisfree/2019/apr/12/the-government-wants-to-bulldoze-my-inheritance-800-year-old-sacred-trees, Accessed 14/4/2019.

Heritage Council Victoria (2019) 'Commonwealth Fertiliser Wharves and Associated Land', Victorian Heritage Database Report H7822-0528, https://vhd.heritagecouncil.vic.gov.au/places/13793/download-report, Accessed 3/05/2019.

Lack, J. (1991) *A History of Footscray*. Melbourne: Hargreen.

Lang, D. and Vaughan, M. (2013) How does the Pumping Station Work? Museums Victoria, https://museumsvictoria.com.au/website/scienceworks/discoverycentre/pumpingstation/videos/how-does-the-pumping-station-work/index.html, Accessed 21/4/2019.

McGregor, B. (2019), Activism on the Merri Creek, Friends of Merri Creek https://friendsofmerricreek.org.au/activism-history/ Accessed 22/4/2019.

Melbourne Water (2019) Stony Creek Recovery Update, 12 April 2019, www.melbournewater.com.au/what-we-are-doing/works-and-projects-near-me/all-projects/stony-creek-recovery Accessed 27/04/2019.

Moreland City Council (2008) Victorian Heritage Database Report, 59178. Wales Quarry Site (former). https://vhd.heritagecouncil.vic.gov.au/places/59178/download-report Accessed 14/04/2019.

Museums Victoria (circa 1918) Superphosphate – prepared by the Mount Lyell Chemical Works, Yarraville, 1918. Science & Technology Collection, Registration Number ST 10621.

Museums Victoria (2019), Women at Spotswood. Scienceworks Website, https://museumsvictoria.com.au/website/scienceworks/discoverycentre/pumpingstation/open-all-hours/women-at-spotswood/index.html Accessed 02/05/2019.

Scienceworks Museum (2019) Women at Spotswood, https://museumsvictoria.com.au/website/scienceworks/discoverycentre/pumpingstation/open-all-hours/women-at-spotswood/index.html, Accessed 22/4/2019.

Sornig, D. (2018) *Blue Lake*. Melbourne: Scribe.

State Library of Victoria (2015) www.slv.vic.gov.au/about-us/fellowships/creative-fellowships/current-creative-fellows/david-sornig, Accessed 14/4/2019.

The Age (1911) 'Brunswick Death Trap', p. 10.

The Herald (1881) 'Brunswick Quarry Holes', p. 3.

Thomas, D. (1967) Geology of the Melbourne District, Bulletin No. 59, *Geological Survey of Victoria*, Melbourne: Mines Department.

van Wyk, P. (2013) Footbridge at atwater: a chorographic inventory of effects'. In Chen, C., MacLeod, J. and Neimanis, A. (eds) *Thinking with Water*, p. 256. Montreal: McGill.

Vines, G. (1993) *Quarry and Stone: Bluestone Quarrying and Building in Melbourne's West*. Melbourne: Melbourne's Living Museum of the West.

62

Place and music

Composing Concrete Antenna

Rob St. John

Sound is promiscuous. It exists as a network that teaches us how to belong, to find place, as well as how not to belong, to drift.

(LaBelle, 2010: xvii)

Bodies inhabit places; there are places only as lived by bodies.

(Casey, 2005: xviii)

This chapter reflects on the production of Concrete Antenna, a place-based sound installation in Newhaven, Edinburgh, between 2015–2016. It suggests that a relational and dynamic notion of place (as variously articulated by Edward S. Casey, Nigel Thrift, Doreen Massey) can inspire composition techniques which blur sound and music. Following Elizabeth Grosz's (2008) notion of art practice as a (temporary) joining with the chaotic flux of the world, Concrete Antenna represents a temporary immersion in a place that – like any other – is always in-the-making. Place is imagined here as a composition itself: one that has elements which are seemingly permanent and patterned; and others which are fleeting, chaotic or imperceptible. This chapter reflects on how sound and music composition techniques might be arranged to respond to such an 'ecology of place' (Thrift, 1999).

Concrete Antenna was installed at Edinburgh Sculpture Workshop between March 2015 and June 2016 (Figure 62.1). It was an interdisciplinary collaboration between myself (a cultural geographer and artist), Tommy Perman (a musician and graphic designer), Simon Kirby (a language evolution academic). Although this chapter is a single-author piece, the practices it documents are inherently collaborative and indebted to Tommy and Simon. The chapter moves through two registers. In the first, the keywords of the chapter title, 'Place' and 'Music', are considered to offer a perspective on how sound and music composition techniques might engage with a dynamic notion of place. In the second, three key concepts which developed in the making of Concrete Antenna – 'distributed listening', the 'place-

Figure 62.1 The Concrete Antenna tower at Edinburgh Sculpture Workshop.

specific-non-specific' and 'aural topographies of chance' – are outlined to give an impression of the research, practice and presentation of the work.[1]

Place and music

Place

Place is a keyword of modern life, something which, as Nigel Thrift puts it: 'is both so pivotal and so hard to grasp' (1999: 317). As generations of geographers (for example, Tuan, 1977; Agnew, 1987; Cresswell, 2004) have variously teased out, places – however defined – are locations that we as humans somehow attach meaning to, dwell in, and actively shape. Lucy Lippard writes: 'Place is latitudinal and longitudinal within the map of a person's life. It is temporal and spatial, personal and political' (1997: 7). Place, for Lippard, is a situated repository of human history and memory, which forms a proving ground for possible futures. For Edward S. Casey, place can be conceived as an eventful gathering of things, both human and non-human, in ongoing space–time negotiations. Casey writes: 'places gather things in their midst – where "things" connote various animate and inanimate entities. Places also gather experiences and histories, even languages and thoughts [...] What else is capable of this massively diversified holding action?' (1996: 24). Casey's conception of place shares relational and experiential themes with Nigel Thrift's 'ecology of place', which involves: 'conceiving of the world as associational, as an imbroglio of heterogeneous and more or less expansive hybrids performing "not one but many worlds" [...] and weaving all manner of spaces and times as they do so' (1999: 316). As a result, for Thrift: 'places must be seen as dynamic, as taking shape only in their passing [...] places can never be pre-ordained' (1999: 310). Doreen Massey shares Casey and Thrift's notion of place as a dynamic assemblage of human and non-human relations, but expands the scalar boundaries of its 'extroverted' linkages with the wider world, writing: 'What gives place its specificity is not some long internalized history but the fact that it is constructed out of a particular constellation of social relations meeting and weaving together at a particular locus' (1991: 28). Approaching place with Casey, Thrift and Massey, then, encourages us to look for how different social, cultural, political, economic and environmental forces interact to produce a distinctive 'sense' of place, however fleeting and individually experienced.

Music

Musical composers have long sought to channel elements of their surroundings into their work; and there has been growing academic interest in the interconnections of place and music in recent decades, particularly in cultural geography and social anthropology (for example, Feld, 1996; Leyshon et al., 1998; Connell and Gibson, 2003; Revill, 2005, 2014). Concrete Antenna is a work at the blurred edge of musical and soundscape composition. The definitions and boundary-markings of what constitutes 'music' and 'sound' (and their differences) are as complex and contested as those around the notion of place (Licht, 2007; Kim-Cohen, 2009). For John Cage, the sounds of daily life – often rich in dynamic, pattern, timbre and pitch – can be understood to have an inherent musicality. Famously, in his 1952 piece '4′33″', Cage scored a set period of silence (4 minutes and 33 seconds) as a musical composition in which the incidental sounds of the concert hall space formed the emergent work, collapsing any neat distinction between 'sound' and 'music' (Licht, 2007). As Georgina Born writes, subsequent lineages in post-1950s sound and music practices, 'where the ontological distinction between music and sound is disturbed', have the potential to 'foreground the creative possibilities [...] of the mutable boundaries between music, sound and space' (2013: 5).

Canadian composer R. Murray Schafer's work in the 1960s and 1970s – itself influenced by Cage – provides complementary resources for understanding how the boundaries between 'music' and 'sound' are blurred. In his 1977 book, *The Soundscape: Our Sonic Environment and the Tuning of the World*, Schafer outlined a 'notation' schema to 'score' the sonic character of different places. Schafer's urban 'acoustic ecology' was based on three key terms: keynotes, sound signals, and soundmarks. Keynote sounds reference a soundscape's background tonality which shape the character of a place and those who live in it (for example, water, wind, birds, and insects); whilst sound signals are those in the foreground, which are more infrequent and carry various semiotic meanings (for instance, bells, horns, sirens). Soundmarks are akin to the visual notion of the landmark, denoting a place's distinctive soundscape, which: 'make the acoustic life of the community unique' (1994: 10). Schafer's notion of soundscape has been repeatedly critiqued: as ignoring cultural and political complexities and dynamism of place (Thompson, 2004); and as framing the sonic environment as something 'out there' waiting to be tuned into, rather than something inherently ongoing, experiential, and not necessarily descriptive of a specific place (Ingold, 2007). Ingold writes:

> if sound is like the wind, then it will not stay put, nor does it put persons or things in their place [...] To follow sound, that is to *listen*, is to wander the same paths. Attentive listening as opposed to passive hearing, surely entails the very opposite of emplacement.
>
> *(2007: 2, original emphasis)*

However, Schafer, with collaborators including Hildegard Westerkamp and Barry Truax, subsequently developed forms of soundscape composition, a variant of *musique concrète*, in which recordings of both human and non-human things and activities are brought together in a process of 'acoustic design' (Licht, 2007; Truax, 2012). Truax describes soundscape composition as: 'a form of electroacoustic music characterized by the presence of recognizable environmental sounds and contexts. Its purpose is to invoke the listener's associations, memories, and imagination related to the soundscape' (2013, unpaginated). In essence, the original documentary tenets of Schafer's 'acoustic ecology' became a framework for processes of 'acoustic design' championed by Truax and Westerkamp through the 1980s and 1990s at the World Soundscape Project (Truax, 2012). In so doing, the composition of place-based sonic elements gained the potential to become increasingly experimental and recombinant – musical, even – as recently evidenced in Chris Watson's sonic collage of a railway journey on El Tren Fantasma, where soundscape composition evokes a dynamic sense of place (Revill, 2014).

This brief engagement with Cage and Schafer suggests that processes of sonic documentation and musical composition have the potential to significantly overlap and cross-pollinate. Given the characteristics of place outlined above, I suggest that such a hybrid form of composition can productively engage with gatherings of human and non-human life that characterise contemporary places. Sound – thus framed – can help us tease out the nuances and dynamics of place, as David Matless suggests: 'Sonic geographical understanding alerts us to the contested values, the precarious balances, the battles for beauty and peace and excitement, which make up a place. Sounds echo into social debate over what a place has been, is and might be' (2005: 747). Similarly, for George Revill, sound can shape the bodily experience of place:

> Sound is at once medium – the sensuous stuff through which the world is experienced; method – processes of resonance and the practices of embodied and reflexive engagement, hearing and listening which engage the world; and modality – the structure or sensory registers through which the world is engaged, connecting entities and animating experience in its meaningfulness.
>
> *(2016: 6)*

If place is a series of interlinked compositions of human and non-human life – some temporary, some seemingly permanent – then the musical composition approaches used to respond to a place might productively take on a similar form. Perhaps, instead of attempting to create music which somehow evokes or represents a particular place or landscape, we might instead attempt to channel what Anna Tsing (2015) calls the 'polyphonic' assemblage of landscape. Tsing uses the musical metaphor of polyphony to describe 'open-ended gatherings' of humans and non-human things and practices (2015: 28). For Tsing, conceptualising the landscape in this way allows us 'to ask about communal effects without assuming them. They show us potential histories in the making', and to 'pick out separate, simultaneous melodies and to listen for the moments of harmony and dissonance they created together' (2015: 27–28). How, though, might such an approach to composing place and music play out on Edinburgh's coastal fringes? The next section will reflect on how Concrete Antenna, a sound installation attentive to the dynamism of Newhaven, northern Edinburgh, was conceived and created (Figure 62.2). In so doing, it will point towards the possibilities of experimental creative practice as a means of engaging with the 'open-ended gatherings' of life that characterise contemporary places.

Figure 62.2 The Concrete Antenna tower transmitting and receiving beside a disused railway line, now a greenway cycle path.

Score tae the toor: sounding Concrete Antenna

Newhaven is a coastal district in north Edinburgh, previously a separate burgh until 1920 when, like neighbouring Leith, it was incorporated into the city. As a place it is something of a jumble: a cobbled harbour – once a thriving fishing port – fringed by the low-level creep of retail parks, light industry, and lock-up sheds, with waterfront apartments rising from the reclaimed deep-water harbour wall. Between rows of 1880s tenement buildings runs an abandoned railway line, decommissioned in 1956 and now a wooded cycle route, part of a network of 'greenways' which crisscross Edinburgh's urban river, the Water of Leith. Established in 1986, and expanded in 2012, Edinburgh Sculpture Workshop is located on the site of former railway engine sheds and a blacksmith workshop in Newhaven. It provides studio space, workshops, residencies, and exhibition spaces to artists, both from the local community and more widely. In 2015, a 28-metre-tall triangular concrete tower was constructed on the site. The tower is both an exhibition space and a geographical beacon on the local skyline among church spires, tower blocks, lighthouses, factories, and gas towers. Concrete Antenna was the first commissioned work in the tower. Upon entering the in-progress tower for the first time, the rectangular opening at its summit immediately drew the attention of Tommy, Simon, and myself. The tower seemed to act as an antenna for the fluxes of the surrounding landscape, the gusting of what Ingold terms a 'weather world' (2010). What if, as well as receiving the outside soundscape, the tower might also act as a place-particular transmitter of sorts, and so also as a resonating chamber in which these processes of sonic receiving and transmitting could mix and muddle? The process of creating Concrete Antenna took six months of research and recording; composition; and installation. Three themes of place-attentive terms describe this process: *distributed listening*; the *place-specific-non-specific*; and *auditory topographies of chance*.

Distributed listening

The first activity in creating Concrete Antenna was to 'sound out' Newhaven as a place. This unfolded in two ways: through a series of sound walks involving a variety of recording devices; and through the collation of place-specific sound archives – some personal (from art projects in the local area); some institutional (for example, Tobar an Dualchais); some informal (for instance, YouTube videos of foghorn blasts and delivery drivers documenting their daily journeys). This formed an 'active archive' of sounds from the local landscape, both historic and contemporary, as a sonic palette for composition. We might call this a process of *distributed listening* in two ways. First, practically, in utilising recording techniques which stretch at the limits of normal human perception to allow us to 'hear' multiple sonic traces, patterns and rhythms of a place; and second, geographically, in bringing together disparate archives to weave a sonic tapestry which captures something of the character of a place, however fleetingly. This approach to more-than-human[2] composition takes inspiration from composer Pauline Oliveros' notion of 'Deep Listening' – 'learning to expand the perception of sounds to include the whole space/time continuum of sound' (Oliveros, 2005: xxiii) – and the Cagean 'Reduced Listening' of soundscape composer Hildegard Westerkamp, which involves 'listening for the purpose of focusing on the qualities of the sound itself (for example, pitch, timbre) independent of its source or meaning' (in Chion, 1994: 223). These approaches to listening complement what Anna Tsing (2015) terms the 'arts of noticing' – a heightened, patient attunement to the more-than-human patterns and processes of a landscape during fieldwork.

Three primary recording techniques were used in fieldwork: binaural microphones, contact microphones, and hydrophones. Binaural microphones are small, earbud-like objects worn in each ear, which record a soundfield around the recordist's head. They are extremely portable and require minimal, light-weight equipment to operate. This allows for highly mobile recording strategies – in this case largely on foot – where recording can begin as soon as it is sensed that *something is happening*. The process affirms a practice philosophy in which the recordist is *part of* the ongoing landscape being recorded, or, as Ingold puts it, the process of: 'immersion in, and commingling with, the world in which we find ourselves' (2007: 11). Recording can take place unobtrusively – beside the tills at the big Asda, with your head stuck out of a tenement skylight, through the Workshop clatter of builders and artists. Binaurals afford the recordist the ability to document the aural fluxes of a 'soundwalk' (Westerkamp, 2006). In this way, whether following paths, or traversing pre-defined transect lines, the recordist can, as Ingold puts it, document how: '[t]hrough walking [...] landscapes are woven into life, and lives are woven into the landscape', at least for a fleeting moment (2011: 47).

Contact microphones sense vibrations through contact with solid objects. They can be variously clipped (with woodworking clamps), stuck (with electrical tape) or wedged (into cracks in rock or bricks) into material. Contact microphones allow the recordist to access hidden rhythms of the world – invisible water channels running under a manhole cover, the rattle of a long fence wire tuned by local topography, the bell tones of a submerged dock cable chain, the wavering of a sign post fluttering in the wind – thereby opening up spaces of encounter and meaning (St. John, 2018).

Hydrophones are underwater microphones that detect sub-surface vibrations and sound waves. They allow the recordist to hear underwater soundscapes that would otherwise be inaudible. Listening with hydrophones along the Newhaven docks and the Water of Leith produces multiple soundings that are hard to pin down: rumbles, drones, scratches; periodical intensities and silences. Some are artefacts of the hydrophone's movement on the waterbody's bed, often transmitting a sonic reflection of its materiality: the scree-slip sharpness of a gravel bed; the slow gloop of soft, stratified layers of silt. Others are the rhythms of more-than-human life – the pop of pondweed photosynthesising or the burring of aquatic insects stridulating.

Over a series of months Tommy and I explored Newhaven using 'distributed listening' techniques to record short moments of the interplay of its human and non-human lives. Where does a place begin and end? Can any boundaries be realistically drawn between where Newhaven blurs into Granton to the west, Leith to the east, and Edinburgh to the south? Thinking with Casey, we might ask what human and non-human individuals, objects and processes are being 'gathered' in a place at a particular space and time? Alternatively, thinking with Massey, we might ask what 'articulated moments in networks of social relations and understandings' (1991: 28) are produced, however temporarily? Finally, thinking with Thrift, the outlined methods engage with Newhaven as a dynamic place, from which we can only catch sonic traces in its passing. Along with recordings sourced from a number of digital archives, what our hundreds of binaural, contact and hydrophone recordings represent is an 'active archive' of Newhaven: both as heard and experienced by us over six months, and by multiple others. Contained in those recordings are multiple mobilities, rhythms, journeys, and soundings of life of all kinds. This is not a definitive representation of place – whatever can be? – but instead a gathering of threads of experience plucked from it, providing the raw material to be rewoven into the compositional work. The next section outlines how we went about that process.

The place-specific-non-specific

The installation used numerous compositional techniques to rework the sonic palette collected in fieldwork. Three key techniques are outlined here, broadly defined as: the *melodic soundscape, the sonic recreation of an imagined past,* and *granulated architectures.*

The Newhaven soundscape is rich in melodic and rhythmic elements: the peal of church bells, snatched snippets of song through an open car window, chaffinches singing in hawthorn trees, sculptors' chisels ringing, water dripping off disused railway bridges and glooping along the harbour wall, footballs echoing off concrete playgrounds. We took direct inspiration from such 'musical' elements of the local *melodic soundscape* in our compositions. A recording of bells from a local church tower was imported into the Ableton Live production software and converted to a series of MIDI notes – a universal digital music language. In attempting to digitally 'map' the recording, the volume and complexity of sonic information it contained overwhelmed the digitisation process, producing white noise. However, slowly, the 'grid points' of this MIDI 'map' were manually 'chipped away' at by individually removing notes, until emergent chord progressions and melodies began to appear: signals from digital noise. These patterns could be automatically played back on virtual instruments such as the grand piano ('Church Bells'). The resulting compositions can thus be thought of as a collaboration between the algorithmic and human ear: a negotiation between technological and human modes of placed-pattern-creation. A recording of the Hogmanay fireworks made by Tommy through the skylight of his north Edinburgh home proved similarly fertile ground for composition. As Simon writes:

> as the distance from a firework increases, its characteristic sound changes. It is modified by wind and temperature gradients and humidity in the air, as well as reverberation caused by multiple reflections between ground and buildings. When we listen to these recordings of fireworks over Newhaven, what we hear carries with it an imprint of the geography of Edinburgh, its hills, streets and buildings, as well as the neighbouring sea.
> *(2016, unpaginated)*

The stuttering geo-rhythms of the firework pops became the basis of percussion patterns for a minimal electronica track ('Harbour Fireworks'), interplaying digital clicks and drops like a seismograph tracing the pulse of the place.

One aspect of the idea of place might be typified by the ongoing negotiation and reworking of the past. In the work we wanted to include soundings of past lives and activities from Newhaven, without necessarily attempting to recreate a historical soundscape (on this topic, see Coates, 2005). Instead, our archival and cartographic research pointed towards what Schafer would term historical keynote sounds and sound signals from the local area: steam trains, blacksmiths, sail ships and foghorns, to name but a few. By chance, a triad of three YouTube foghorn recordings from locations along the East Coast of the UK layered together produced an almost-pitched chord. This 'fogorgan' – assembled in Ableton Live, and patterned by the lo-fi digital weathering of wind and rain against cheap camera microphones – became the basis for a number of compositions. The triad (triangles recurred in the composition) of foghorns form a slow, melancholy cluster of droning chords, to which the tonal micro-variations and inconsistencies of each individual recording lends a wavering, brass-like character (for example, 'Branch Line', 'Topping Out'). These recordings were shaped again by the contemporary geography of Newhaven using convolution reverb, a process which 'maps' the sonic characteristics of a particular space. Reverberant spaces – in this case including local storage sheds, sewer pipe outflows, tenement stairwells, and the tower itself – were 'mapped' sonically by recording how an 'impulse' (a starting pistol, popped balloon, or a range of white noise frequencies), decays and fades to silence in the space. The resulting 'impulse responses' can then be applied to any other recording, to give an impression of how it might sound in a particular space. As a result, the 'distributed listening' recordings from Newhaven could be remotely replayed through different spaces from the place, potentially even through multiple spaces at once. What emerges in the 'fogorgan' tracks – often mixed with the slow drones of contact mic'd stretched fence wires and harbour chains – is a kind of uncertain, wavering memory, passing from past to present, through the spaces which comprise a dynamic place: the *sonic recreation of an imagined past.*

Concrete is often framed as the archetypal building material of artificial modernity, lacking the 'naturalness' of other materials such as stone and timber. However, as architectural historian Adrian Forty (2012) outlines, this is not entirely accurate, and instead concrete might be thought of as an assemblage of geological materials – cement, sand, aggregate – reformed by human activity and ongoing weathering. With this in mind, we devised composition and production strategies which engaged with the material character of the concrete tower in which the installation was housed. One key technique was granulation, a digital production technique using the Max MSP software, in which audio recordings are 'dissolved' into thousands of tiny particles of sound, and subsequently recombined (or precipitated, to follow this metaphor) in new arrangements in playback (for example, 'Score Tae the Toor', 'Undersea Static'). Small particle sizes give a smooth, almost liquid character to the playback, as the original recordings are dissolved to such an extent that they become abstract micro-second drops of pure sound. Coarser particle sizes create more heterogeneous compositions, in which brief identifiable moments – a seagull call, wind in fence wires, a passing bicycle – of the original recordings can be heard, a little like a shell or shard of glass in a cast concrete wall. Another association is drawn in 'Undersea Static' by using hydrophone recordings from pipe outflows on the harbour docks and the nearby Water of Leith alongside granulated recordings – assembling different forms of dissolved traces of the city in sound. Granulation was used extensively on the recordings gathered in the 'distributed listening' part of the process, in this way dissolving the sonic geographies of Newhaven and producing recombinant compositions, which were at once abstracted from the place itself, whilst largely open and allusive in their sonic character. It was something akin to a radio dial tuning through bands of the sonic geography of the place, producing temporary moments of clarity. In other words, *granulated architecture* as the place-specific-non-specific.

These techniques are inherently exploratory processes, and tended to yield a lot of material which was seemingly noisy, flat or featureless alongside that which seemed useful. This process of sifting and sorting sonic material during composition is intrinsically difficult to describe and justify; often the result of individual affective responses to the emergent soundwork taking shape. There is a productive tension here: in placing uncertainty and chance at the heart of the process, the composer is actively seeking to be surprised by how the work takes shape (and to be regularly disappointed or underwhelmed). The composition process – influenced by Grosz (2008) – is one of channelling the sonic flux and flow of a place to shape work that has a structure, tonality, rhythm and melody which somehow fosters affective responses in us as composers, and perhaps in an audience too. How these audience responses take shape, though, is again out of our control. In other words, in Concrete Antenna, our sensibilities and practices as artists and composers form only one part of a generative network of possible outcomes. All that can be done is to set suitable starting points and let the work find emergent trajectories: *aural topographies of chance.*

Aural topographies of chance

Michael Gallagher (2015) argues that such *in situ* sound installations have the potential to influence visitors' relationships with a place. He suggests that:

> in situ audio can re-make landscapes as well as representing them. It can fold the sounds of a landscape back into it, with sounds returning as revenants that generate, at least for some listeners, uncanny affects of ambiguity, disorientation, haunting and hallucination. Affects of this kind have the potential to influence listeners' relationships to place, opening up new modes of attention and movement.
>
> *(2015: 316)*

The final composition comprised a 30-minute piece of music which looped continuously in the tower across multiple vertical speakers ascending its height (Figure 62.3). The content of this composition, and the way that it interacted with the brief random playback of spoken

Figure 62.3 Inside the tower, looking up past multiple speakers towards the sky.

word sound archives from the local area (triggered by visitor movements detected by a motion sensor) and the flux of the outside soundscape itself, were governed by local environmental conditions. The composition varied in subtle ways depending on the state of the tide at neighbouring Granton ('tide in'/'tide out') and the prevailing weather conditions ('good'/'bad'), which were inputted into the live audio stream through online data feeds. The different parameters caused the compositions to be subtly altered: 'tide out' led to a more minimal arrangement; 'bad weather' to emphasise underlying drones over melodies. As a result, each visit to the tower will have been a unique aural experience, governed not only by the composition itself, but also the tide, the weather, visitor presence and movement within the tower, the chance selection of snippets of local history spoken word audio, and the dynamics of the local soundscape. In this way Concrete Antenna did not only transmit sound and music shaped by Newhaven, but it also acted as a resonating chamber where sounds determined by local environmental conditions could mix and muddle with our work. In other words, a space for place and music to meet in an ongoing, emergent process of (re)composition.

Notes

1 Details of the project (including videos of the tower and surrounding area) can be found here – www.concreteantenna.org – and a book of essays and art from the project was later published (St. John, 2016). A two-channel stereo mix of the installation (subsequently released on 12-inch vinyl LP) can be heard here – https://soundcloud.com/concretentenna/sets/concrete-antenna-tracked – it is recommended that you listen whilst you read this chapter.

2 Defined by Sarah Whatmore (2006, p. 604) as 'modes of enquiry [that] neither presume that socio-material change is an exclusively human achievement nor exclude the "human" from the stuff of fabrication [and] attend closely to the rich array of the senses, dispositions, capabilities and potentialities of all manner of social objects and forces assembled through, and involved in, the co-fabrication of socio-material worlds'.

References

Agnew, J.A. (1987) *Place and Politics: The Geographical Mediation of State and Society*. London: Allen and Unwin.

Born, G. (2013) Introduction – music, sound and space: transformations of public and private experience. In G. Born (ed.) *Music, Sound and Space: Transformations of Public and Private Experience*. Cambridge: Cambridge University Press, pp. 1–70.

Casey, E. (1996) How to get from space to place in a fairly short stretch of time: phenomenological prolegomena. In S. Feld and K.H. Basso (eds.) *Senses of Place*. Santa FE: SAR Press, pp. 13–52.

Casey, E. (2005) *Earth-Mapping: Artists Reshaping Landscape*. Minneapolis, MI: University of Minnesota Press.

Chion, M. (1994) *Audio-Vision: Sound on Screen*. New York: Columbia University Press.

Coates, P.A. (2005) The strange stillness of the past: toward an environmental history of sound and noise. *Environmental History*, 10 (4), 636–665.

Connell, J. and Gibson, C. (2003) *Sound Tracks: Popular Music, Identity and Place*. London: Routledge.

Cresswell, T. (2004) *Place: A Short Introduction*. Oxford: Wiley.

Feld, S. (1996) Waterfalls of song: An acoustemology of place resounding in Bosavi, Papua New Guinea'. In S. Feld and K.H. Basso (eds) *Senses of Place*. Santa FE: SAR Press, pp. 91–135.

Forty, A. (2012) *Concrete and Culture: A Material History*. London: Reaktion Books.

Gallagher, M. (2015) Landscape audio in situ. *Contemporary Music Review*, 34 (4), 316–326.

Grosz, E.A. (2008) *Chaos, Territory, Art: Deleuze and the Framing of the Earth*. New York: Columbia University Press.

Ingold, T. (2007) Against soundscape. In A. Carlyle (ed.) *Autumn Leaves: Sound and the Environment in Artistic Practice*. Paris: Double Entendre, pp. 10–13.

Ingold, T. (2010) Footprints through the weather-world: walking, breathing, knowing. *Journal of the Royal Anthropological Institute*, 16, S121–S139.
Ingold, T. (2011) *Being Alive: Essays on Movement, Knowledge and Description*. Abingdon: Routledge.
Kim-Cohen, S. (2009) *In the Blink of an Ear: Toward a Non-Cochlear Sonic Art*. London: A&C Black.
Kirby, S. (2016) Harbour fireworks. In R. St. John (ed.) *Score Tae the Toor*. Edinburgh: Random Spectacular (unpaginated).
LaBelle, B. (2010) *Acoustic Territories: Sound Culture and Everyday Life*. New York: Bloomsbury.
Leyshon, A., Matless, D. and Revill, G. (eds) (1998) *The Place of Music*. New York: Guilford Press.
Licht, A. (2007) *Sound Art: Beyond Music, Between Categories*. New York: Rizzoli.
Lippard, L.R. (1997) *The Lure of the Local: Senses of Place in a Multicentered Society*. New York: New Press.
Massey, D. (1991) A global sense of place. *Marxism Today*, 38, 24–29.
Matless, D. (2005) Sonic geography in a nature region. *Social & Cultural Geography*, 6 (5), 745–766.
Oliveros, P. (2005) *Deep Listening: A Composer's Sound Practice*. Lincoln: iUniverse.
Revill, G. (2005) Vernacular culture and the place of folk music. *Social & Cultural Geography*, 6 (5), 693–706.
Revill, G. (2014) El tren fantasma: arcs of sound and the acoustic spaces of landscape. *Transactions of the Institute of British Geographers*, 39 (3), 333–344.
Revill, G. (2016) How is space made in sound? Spatial mediation, critical phenomenology and the political agency of sound. *Progress in Human Geography*, 40 (2), 240–256.
Schafer, R.M. (1977/1994) *The Soundscape: Our Sonic Environment and the Tuning of the World*. Rochester and Vermont: Destiny.
St. John, R. (2018) Fluid-sound. In L. Roberts and K. Phillips (eds) *Water, Creativity and Meaning: Multidisciplinary Understandings of Human-Water Relationships*. Abingdon: Routledge, pp. 157–171.
Thompson, E.A. (2004) *The Soundscape of Modernity: Architectural Acoustics and the Culture of Listening in America, 1900–1933*. Cambridge: MIT Press.
Thrift, N. (1999) Steps to an ecology of place. In D. Massey, P. Sarre, J. Allen (eds) *Human Geography Today*. Cambridge: Polity, pp. 295–322.
Truax, B. (2012) "From soundscape documentation to soundscape composition" *Proceedings of the Acoustics 2012 Nantes Conference, 2014–2017*, available at: https://hal.archives-ouvertes.fr/hal-00811391/document. Accessed 01. 06.19
Truax, B. (2013) *Soundscape Composition*, available at: www.sfu.ca/~truax/scomp.html. Accessed 01. 06.19
Tsing, A.L. (2015) *The Mushroom at the End of the World: On the Possibility of Life in Capitalist Ruins*. Princeton: Princeton University Press.
Tuan, Y.F. (1977) *Space and Place: The Perspective of Experience*. Minneapolis, MI: University of Minnesota Press.
Westerkamp, H. (2006) Soundwalking as Ecological Practice. *Proceedings for the International Conference on Acoustic Ecology, November 2006*, available at: www.sfu.ca/~westerka/writings%20page/articles%20pages/soundasecology2.html. Accessed 01.06.19.
Whatmore, S. (2006) Materialist returns: practising cultural geography in and for a more-than-human world. *Cultural Geographies*, 13 (4), 600–609.

63

Of all places ... drama and place

Mike Pearson

Introduction

This chapter traces theatre's spatial and architectural evolution and elaboration. Its model commences with performance making room – through action – for itself; next, delineating and organising areas for its privileged usage; and eventually, formalising and fixing the spatial coupling of performers and spectators in playhouses – particular places of representation and reception, dedicated to manifesting other, fictional places. It examines the scenic and compositional strategies, techniques and technologies developed in conjuring and replicating such locations: how theatre's constructed scenographies, its things, have concrete and symbolic agency; and how its material realities both enhance and impact upon the expressive capacities of performers, theatre's inhabitants. Finally, it considers recent artistic endeavours to unsettle the stage/auditorium divide, and the practical and aesthetic implications of such displacements.

> They are like so many cages, so many small theatres, in which each actor is alone, perfectly individualised and constantly visible. The panoptic mechanism arranges spatial unities that make it possible to see constantly and to recognise immediately.
>
> *(Foucault 1975: 200)*

A place apart

In his reflections on Jeremy Bentham's prison design, Michel Foucault finds his key metaphor in theatre, in that enduring arrangement of stage framed by a proscenium and auditorium. This place of and for drama is configured as two discrete though contiguous spaces: the one often highly illuminated – a scene of noisy 'information rich' expression (Elam 1980: 34) and exhibition, of nervous anticipation, of labour; the other frequently dark, often silent – site of observation and reception, of eager expectation, of diversion. A fixed partitioning of watchers from watched in a see/being seen dyad: the former thrown apart sociofugally in unidirectional face of the display; the latter engaged in virtuosic routines of verbal and physical utterance and interaction, and in creating resonant images that aim to bridge the divide – their 'on-stage' presence enhanced by constructed décor, complementary effects and hidden technologies, and augmented by ancillary 'back-stage' and 'off-stage' extensions.

Theatre as a specialist mechanism of surveillance: a place where – conventionally – audiences are supplied with and/or challenged by fictive portrayals and representations of society: that may pretend to echo reality but that may equally offer caricatured simulacra and fantastical alternatives. This is achieved, above all, spatially: in an alignment of performance and its spectators that is the result of, and the medium for, concrete social practices; that catches performers and spectators in an encounter involving culturally determined styles of exposition and modes of apprehension.

Theatre as both architecture and artistic medium, bracketed off from the everyday yet intimately linked to it: in which an autonomous dramatic world is contrived – coherent around particular compositional principles and *modi operandi* that vary with genre, period and circumstance – that, despite occasional abundances, is fundamentally metonymic and/or synecdochic. It is characterised by selection and by omission, by scenic and kinesic abbreviations that stand in for absences and/or substitute part for whole – as the complexity, density and messiness of the vernacular would appear chaotic corralled on stage.

Here performers engage in vocal and corporeal practices – gestural, proxemic, haptic – that adapt and adjust everyday behaviours and interactive conventions in size, intensity and emphasis: refining, rephrasing, retiming and ultimately stylising and abstracting them in forms of mimetic correspondence. Ever cognisant of the need to be seen and heard, this may lead to aberrant projections of the voice, over-exaggerated signing and feigned emotional excesses. And to the precise though extra-daily distribution and sequencing of activities – planned initially through a rough plotting or 'blocking' of positions on a marked-out floor in the rehearsal room, and thereafter established as a repeatable choreography of movements and occurrences upon the fully elaborated stage.

Here spectators understand that they must assume the things they perceive – everything that enters the frame, however prosaic or absurd, preposterous or extravagant, fragmented or symbolic – to be purposeful; indeed – in their ceaseless inspection – without further direction, they will search for, and generate, meaning in everything they see and hear, including mishaps and chance intrusions.

Although theatre's apparatuses are usually at the service of enacting and illustrating specially formulated literature – play-scripts – in semblances of realism and naturalism, the employment of its devices and staging techniques – alone and in varying formats and concatenations – can itself generate dramatic content, in genres that include text only partially, or that substitute it completely. Or that re-balance, combine and layer its constituents in atypical relationships on stage that may purposefully contradict, obscure and critique Foucault's formulaic requirement 'to see constantly and to recognize immediately' (1975: 200): best exemplified in the deliberate recalibrations of text, action, scenography and soundtrack in post-dramatic manifestations (Lehmann 2006).

Taking place

The history of places of drama is predicated upon a gradual formalisation and institutionalisation of the spatial coupling of performers and spectators in architecturally configured spaces. But as theatre director Peter Brook famously observed: 'I can take any empty space and call it a bare stage. A man walks across this empty space whilst someone else is watching him, and this is all that is needed for an act of theatre to be engaged' (Brook 1972: 11). Whilst no space is surely ever empty – bereft of ambiance, historical resonances, and personal and collective memories – Brook's adage is a model for tracing theatre's elaboration.

First then, imagine a field, and a group of people – standing. As yet no division of those gathered, no set aside space – nothing that resembles a stage, nothing to orientate the gaze.

There may be things to look at – other people, surroundings, sky – but no clues what to watch. The single factor conditioning perception and experience is the scale of the environs and the crush of bodies in this 'locus of multiple exchanges' (Foucault 1975: 201). Suddenly, a fight breaks out. The crowd recoils: steps back to avoid the set-to. Inevitably, they adopt the best orientation for observing: a circle – 'the gathering round of a crowd to look at an incident' (Southern 1962: 57). It's democratic: there are no hierarchies of viewpoint; everyone is equidistant from the centre. There may be a struggle to see better, but the shape can expand to accommodate those who rush to see what's happening; or it can thicken. There is a change in status – an immediate detachment of watchers from watched. Although there may as yet be no direct appeal outwards, the essential bond of theatre materialises. A temporary playing area is created – with an interior and a 'beyond' to the ring – constantly redefined by the rolling brawl of the combatants who remain in three-dimensions, though it is fellow spectators who provide the animate backdrop. The crowd may shout encouragement, pushing in to jostle the participants, encroaching upon the area. Or they may withdraw as the unpredictable contretemps roils and – amoeba-like – distorts the encirclement. Here the complexion of the action is mediated by the qualities of ground surface; by pertaining environmental conditions; by social context; and by the seriousness of intent and degrees of stamina of those involved. Then, just as quickly, the incident ends, the provisional space disappears, and there are no clues as to what to watch.

Alternatively, the protagonist could be a busker, a preacher, a fire-eater who begins to direct and moderate their delivery in acknowledgement of the presence of watchers: situating their address at the centre, or as part of the circumference. Or an English mummers' troupe (Marshfield Mummers 2018): whose mock combat takes place in the middle; who revolve as they declaim; who turn to speak whilst pacing the perimeter. Sometimes, it is the uncompromising progress of the event – blazing tar barrels carried shoulder high through crowded village streets (Ottery St Mary 2018) – that makes room for itself, with little consideration for the safety of onlookers who give way as the onrush presses on relentlessly. If two episodes occur simultaneously, the spectator will need to make a choice – perhaps drawn towards that most ostensibly attractive. If sequentially, she may move to see and hear better; or remaining rooted, witness some exploits in close-up, others at a distance, some half-hidden, and some from behind. Whichever, attention will shift rapidly and repeatedly: without peremptory indications – 'Behold this, now this!' – the observations and interpretations of individuals will vary radically. Happenings may seek to distinguish themselves, temporarily or permanently, through elevation: lifting either the watched above the throng – on a soapbox, in a pulpit, with a pulley system; or the watchers – in serried ranks (the 'emperor's seat' ensuring the best view). Precipitating new directionalities of address and contemplation – now down and up, rather than out.Event is integral to the nascence of theatrical place.

Making place

Theatre begins to settle itself through the reservation, demarcation, delineation and indication of areas intended for its own usage – from a few chalk marks traced on the floor, to the cordoning off of a restricted patch, to the installation of a discernible marker such as a laid carpet (see Heilpern 1977: 65). Or by the indicative placement of seating – chairs set in rows, aisles, squares ....

Insert a wall in the field and it backs the activity, masking visual irrelevance and focussing attention. It has both 'before' and 'behind', and potentially 'on-stage' and 'off-stage': two separate realms – one of exposition, of 'performing' in the public domain, the other of suspension and preparation in the semi-sequestered – with the prospect of narrative advancement in the passage from one to the other through surprising entrances and unexpected exits. And incorporating a flat façade that can be abstractly decorated, or pictorially embellished to depict other times and places or to give the appearance of perspective.

Here, sightlines can be determined, with implications for both presentational approaches employed and their appraisal. Performers are caught half-turned: acting sideways to each other; and acting out towards the spectators. Before the wall, they appear more two-dimensional: though the static ground serves to clarify their deeds, and to foster mounting iconographic friezes and tableaux. Within a rudimentary stage picture, there may be stratifications of activity or information of different intensities and types in both vertical and horizontal dimensions, with activity both close to the wall and some distance in front: with foreground and background; with actions of primary and secondary import; with transits back and forth, and side to side.

Now build several walls – an enclosure with a privileged interior: with thresholds to be crossed, contracts of provision to be made, suspensions of disbelief to be engendered. And, eventually, decorum to be demonstrated in the spectators' ability to sit in the dark – constrained, but given to both spontaneous and accordant demonstrations of appreciation. Within this controllable place, effects both visual and auditory can be generated and orchestrated in a sophisticated balance of concealment and disclosure. As Jean-François Lyotard observes: 'To Hide, to Show: that is theatricality' (see Read 2007: 142). But it requires cultural competence to distinguish what's on show as theatre 'as such' (Elam 1980: 54) – with the risk of dumbfounding or excluding the uninitiated – though it may be aided simply by hanging the word 'THEATRE' above the door. Those who have been before know how to go on and have memories of other occasions here, of other places summoned here in the past.

Richard Southern (1962: 156–158) proposes that this proto-playhouse requires an entrance, a background, a raised stage, a dressing room, a property store and some means to act 'above' (for the *deus ex machina*). And from this basic structure – with, over time, incremental finessing of visual, acoustic and auditory qualities – dispositions develop that indeed 'make it possible to see constantly' (Foucault 1975: 200). A place gradually invested by performers with superstitions (no whistling; no mentioning of Shakespeare's 'Scottish play'); and religiously described from their point of view (stage left/right; upstage/downstage).

Keir Elam suggests that what strikes spectators when they first enter a theatre is

> the physical organization of the playhouse itself: its dimensions, the stage-audience distance, the structure of the auditorium (and thus the spectator's own position in relation to his fellows and to the performers) and the size and form of the awaiting stage – a space "potentially 'fillable' visually and acoustically".
>
> *(1980: 34)*

The essential divide is 'reinforced by symbolic spatial or temporal boundary markers or "brackets"' (Elam 1980: 34) – the dimming of the lights, the raising of the curtain ....

Faking place

In Japanese *noh* theatre, spectators sit asymmetrically on two sides of the roofed, raised stage. In a scenically attenuated genre with origins in the fourteenth century, huts, boats and carts are insubstantial, portable structures, little more than outlines in bamboo unable to withstand weight or force; the design of the *tskurimono* (assembled things) hints only at the form of the actual object, often reduced in size (Komparu 1984: 253). Despite this paucity, severe oscillations, compressions and expansions of place and time are conjured on the *noh* stage, primarily through the words of the principal performers and the commentary of the chorus: a traveller sets out on a journey and almost simultaneously arrives whilst barely moving; a woman steps out of the boathouse into a boat that is the bare stage. Widely in theatre, verbal evocations – or title-cards bearing the words 'The English camp at Agincourt' – have both the capacity to locate, and repeatedly relocate spectators: when informed where they are, they are induced to visualise that place. Uniquely in *noh*, on-stage position is also significant: particular kinds of speech are always delivered at specific locations on the unmarked floor.

In contrast, the eighteenth-century Japanese *kabuki* stage is a complex contraption of hoists, trap doors, flying systems and turntables that conspire sudden transformations and revelations (Brandon et al. 1978). Actors walk from one revolving scene into another; what was ground level becomes a rooftop; a building rolls backwards to reveal another underneath it; whole set-pieces are pushed in on wheels. The wide proscenium accommodates elaborate, colourful panoramic vistas and encourages theatrical legerdemain – as when children replace retreating actors to imply distance travelled.

In the stories it tells and the impressions it makes in the melding of generic style, scenic design and stage technologies, theatre is invariably utopian (and/or dystopian) in aspect – having 'a general relation of direct or inverted analogy with the real space of Society, in a perfected or critical relationship. They [utopias] present society itself in a perfected form, or else society turned upside down' (Foucault 1986: 24). However they remain 'fundamentally unreal spaces' (Foucault 1986: 24): although theatre pretends to replicate reality, it is – in reality – in its fabrications, 'a rickety bricolage of cardboard effects' (Read 2008: 120).

Foucault also includes theatre in his order of heterotopias: counter-sites

> in which the real sites, all the other real sites that can be found within the culture, are simultaneously represented, contested, and inverted. Places of this kind are outside of all places, even though it may be possible to indicate their location in reality.
>
> *(1986: 24)*

In indicating a heterotopia's ability to juxtapose incompatible sites in a single real place, Foucault turns once again to theatre: 'Thus it is that the theater brings onto the rectangle of the stage, one after the other, a whole series of places that are foreign to one another' (1986: 25). Scenes shift, place follows place; but theatre can stage conjunctions and imbrications of era and locality simultaneously, as well as serially.

Distinguishing place

How does theatre 'suspect, neutralize, or invent the set of relations' it happens 'to designate, mirror, or reflect' (Foucault 1986: 24)? Through the distinctive and allegorical world-making manipulation of material constituents including not only props, costumes and scenery, but also architectures and technologies – all that is of the playhouse, its plumbed-in mechanisms; its temporary contrivances; its portable accoutrements. All – though in differing ways, to varying degrees and extents – apparent, germane and effective, for both performers and spectators alike, who themselves constitute substantive media.

Theatre, at base, arising from the co-presence and articulation of bodies and things within abnormal and irregular modulations of time and timing: for extended durations; in instantaneous contractions; in episodic ruptures; in symbolic substitutions of one period for another – as when 'time passes'; in heterochronic displacements – 'a sort of absolute break with their traditional time' (Foucault 1986: 25). This play of materials and time – wrought in aestheticised circumstances, in a place apart – is diagnostic of theatre.

Here – independently and in both planned and chance amalgamations – things can serve representational, decorative, fictive, cognitive and/or functional purposes, that may shift and blur from moment to moment. On stage, diverse things without natural affinities of origin, provenance and type are drawn into and gathered in arrays unique to this bracketed slice of space/time: into unlikely associations; into exceptional convergences and collisions; into disturbing recontextualisations; into contrary narratives that may question and confound everyday ordering. Here, the placement, ratios and combinations of objects are governed by both artistic and operational criteria – by predilection, but also by the need to be discernible.

Scenic formulations can include things drawn out of their originary or familiar setting – found, appropriated; and things specially made – fabricated, interpolated. Here found objects and their replicas can co-exist in illogical arrangements, with excessive repetitions, subversions of taste, and unconventional animations, as when the scenery becomes disconcertingly mobile. Here familiar objects may be gathered in disquieting numbers; or built at startling scale – in different materials and to different dimensions from their quotidian equivalents.

Genre may prescribe a compositional strategy, or act as proscription in setting parameters for inclusion – determining, for instance, the range of constituents and methods of making to be used, and hierarchies of layout – as in the illusionary expansion of a shallow space to bring about *trompe l'oeil* effects. And what gets in can cause astonishment, delight and incredulity at its excision from the everyday: for Italian director Eugenio Barba, witnessing a live horse on stage was a shocking, foundational experience (1995: 81).

Whatever their source, theatre radically transforms all objects and bodies described within it, bestowing upon them a signifying power that they lack – or which at least is less evident – in their normal function. They become freighted. As 'a regime of signs' (Deleuze and Guattari 1988: 504), theatre is unremittingly equivocal, its semiotic fraction profligate – denotative and connotative meanings jostling to inform both creative intent and its mnemonic reading.

And theatre's things are ever unstable, liable to slip their ascription: their identity and their intimations mutating through the applications of performers – another characteristic attribute. The 'poor theatre' of the late Polish director Jerzy Grotowski is reliant upon the fungibility and multivalency of a limited and fixed repertoire of objects that 'must be sufficient to handle any of the play's situations' (1969: 75). Each object in a production has shifting functions – a bathtub becomes an altar and a nuptial bed, whilst 'representing all the bathtubs in which human bodies were processed for the making of soap and leather' (Grotowski 1969: 75). Through recurrent reuse, misuse and abrupt rearrangement, the repertoire carries forward traces of its previous guises, fostering dense, complex and contradictory analogies, anachronisms and ambiguities.

Theatre's things usually help to establish location, social situation, historical period and generic style, conspiring dramatic contexts imitative and/or imaginary. As insignia, personal effects, appurtenances, they aid the identification of characters, their gender, status, class and type. But the principles of arrangement may also include unusual and uncanny elisions, oppositions, conflicts and discontinuities – in procedures of overlapping and simultaneous *hypotaxis* and slammed-together *katachresis* – as much as logical organisation: drawing sundry objects together into extra-daily juxtapositions and contingent taxonomies in which like can stand adjacent to unlike in new, irregular and even inverted correlations. In concocting fabulations: in a place, incidentally, with the tendency to subsume other artistic forms – painting, videography, music – into its on-stage constitution.

And herein are opportunities for the generation of dramatic content (and upset) in productive encounters, syntheses, frictions and amalgamations between bodies and things, and between things and things, as well as between performers and performers. The admixture of actual and constructed can produce dramatic richness – in the ironic, referential appearance of the detritus of past theatre genres in a contemporary context such as cardboard backdrops and pantomime horses (Forced Entertainment 2018) – that, for the spectator, cause both delight and a readjustment of critical faculties in pondering what is typically to be expected or excluded here, and why.

Things also have agency. They make things happen – the characteristic gliding motions of *noh* only result from the contact of *tabi* footwear with polished *hinoki* wood surface. They are inciting: their purposeful activation necessitating and occasioning engagements between body and object that are dramatic in their consequences: requiring extra-daily effort; triggering linkages that contest and invert social norms.

Theatre envisaged – either through prescriptive design or *bricolage* – as a *heterogeneous assemblage*: as a complex of people, things, effects and conditions pertaining – including the immaterial, the intangible, the ephemeral in the shape of environmental phenomena and technical effects and even audiences as fundaments – within the context of exposition; after Jane Bennett, as a living, throbbing confederation (2010: 23). As an entanglement of humans and non-humans, of 'material and technical as well as immaterial, symbolic and conceptual components' (Hodder 2012: 113), tied together in webs or networks of mutual and contingent restraints and reliances – 'the dialectic of dependence and dependency', 'that create potentials, further investments and entrapments' (89). In a dedicated place ....

Reconceiving place

It was the installation of incandescent, focussable electric lighting in the 1880s – with the facility to illuminate the scene and to pinpoint and highlight key moments – that revealed the artifice of the Victorian stage, with its painted backdrops and heavily made-up actors looming melodramatically over the gas footlights. With the stage now in plain sight, designers such as Edward Gordon Craig sought to integrate scenic elements with the work of actors in a total though often non-representational setting – in a unified *mise en scène* (Innes 1998). Italian pioneer Adolphe Appia too conceived a synchronicity of sound, light and movement in stagings that promoted three-dimensional acting styles (Beacham 1993). Both employed lighting suspended above the stage to create mood and indicate location: in concert with the realist dramas of playwrights such as Anton Chekov and Henrik Ibsen, the sense emerged of looking into interiors – both domestic and psychological – through the transparent 'fourth wall' of the proscenium.

Subsequent endeavours to unsettle the stasis of the stage/auditorium divide, and the aesthetic and ideological fixities that the playhouse ostensibly enshrines, have taken three forms. First, by changing the optic: refashioning the picture in situ. Since the early 1970s, American director Robert Wilson has regarded the whole proscenium opening as his canvas, dispersing actions across its vertical and foreshortened horizontal expanse (Wilson 2018); in *The CIVIL warS* (1984) the frame even appears to regard only part of a larger scene – when we see only the enormous legs of the famously tall Abraham Lincoln crossing. Latterly, digital technology has expanded the scope and reach of the confined stage: in marathon productions such as *The Kings of War* (2015), Dutch director Ivo van Hove employs projected pre-recorded video to insert matter from other times and places – and on-stage camerawork to provide close-up, partial and concurrent off-stage views (Toneelgroep Amsterdam 2018).

Second, by conceiving new places: as did both Craig and Appia. Craig:

> Theatre must be an empty space with only a roof, a floor, and wall. Inside this space one must set up for each new type of play a new sort of stage and temporary auditorium. We shall thus discover new theatre, for every type of drama demands a particular type of scenic space.
>
> *(in Wiles 2003: 246)*

Appia: 'No more stage, no circle of seats. Just a bare and empty room [*salle*], in anticipation' (see Wiles 2003: 246).

In 'The Theatre of Cruelty: First Manifesto' of 1938, French visionary Antonin Artaud foresees 'a single, undivided locale without any partitions of any kind and this will become the very scene of the action' (Artaud 2013: 68). With the audience seated at the centre of the unadorned hangar-like building: 'The action will unfold, extending its trajectory from floor to floor, from place to place, with sudden outbursts flaring up in different spots like conflagrations' (69).

The unpartitioned 'black box' studios of the late 1960s – exemplified by Grotowski's Theatr Laboratorium in Wroclaw – were the quintessential implementation of Peter Brook's empty space: 'We have resigned from the stage-and-auditorium plant: for each production, a new space is designed for the actors and spectators. Thus, infinite variation of performer-audience relationships is possible' (Grotowski 1969: 19–20). With designer Jerzy Gurowski, Grotowski conceived scenographies that occupied the whole studio (Grotowski 1969: 157–164; Gurawski 2018) – including a unique synthesis of setting, performers and spectators for each production to concretise its exegesis of the source.

Despite claims to flexibility, however, the closeted black painted room is dogmatic in its demands: disguising scenic mechanisms is impossible; lack of side, rear and overhead appendages cramps activity. Whilst professing neutrality – a putative non-place awaiting places to 'reconstitute themselves in it' (Augé 2009: 78) – the blackness immediately reveals any artifice set against it. Hence Grotowski espoused theatrical poverty – the use of the simplest dramatic means – the accent thrown upon the uninterrupted presence and extended resources of the performer as key dramatic carrier. Here, she is an immaculate subject removed from social and environmental context – 'perfectly individualized and constantly visible' (Foucault 1975: 200) – her motives and actions laid bare; her techniques and expressive capacities open for acute scrutiny.

The fourth wall – distancing act from its apprehension – disappears. Thrown into proximity and intimacy, approaches also develop in which performers – undermining their function as mere 'object of information' (Foucault 1975: 200) – turn and directly address spectators in implicit acknowledgement that all are gathered in the here and now, recognising the limits of simulation and admitting the inherent theatricality.

Third, by occupying other places: with site-specific theatre (Pearson 2010; Birch and Tompkins 2012). Here the specific constitution of the site – its extent, height, ground plan, layout of integral features and distribution of vernacular details – serves to inform and influence themes, dramaturgical structures and scenic deployments. Productions may take account of a site's historical, environmental, architectural, spatial, functional and organisational traits – in their subject matter, in their assumption of pre-existing seating plans, in their co-opting of in-built amenities and plumbed-in services.

Theatre now as three separate but interlocking components – site; constructed scenography; and performance in all its ingredients (text, action, soundtrack): in relationships of reciprocity (complementarity), conflict and indifference with each other; in both fleet and protracted periods of operative predominance (Tschumi 2012: 184–186).

At site – likely visited for the first time in a theatrical guise – there are no prescriptions for what might happen, no moderating influences, no accepted ways of 'going on' for either performers or spectators, who are now cast into – even 'immersed' within (see Machon 2013) – the same place: that resembles a sensorium, not least with the utilisation of prosthetic technologies such as closed audio broadcasting to headsets. The transitory occupation of the 'real' allows the contravention and suspension of the prescribed practices and by-laws of the auditorium. The non-pristine nature and lack of seemliness of the space permits the use of materials and phenomena, unusual, undesirable or dangerous in a playhouse – things not conventionally theatrical but of the site. Equally, the intractability of site may necessitate the substitution of approaches – annexed from what usually goes on here; from, for instance, industrial processes of construction and engineering in order to cope, to embrace and to take advantage of its inherent features. Presaging the substitution of actual task-based activities in performance for their imitation: in a place where change in context now renders the 'real' dramatic – with the inclusion of vehicles, animals, commercial supplies of water and power ....

Imagine a production in a disused factory: Brith Gof's staging of a sixth-century Welsh battle epic in an abandoned twentieth-century car plant (Pearson and Shanks 2001: 102–108; Pearson 2018: 295–296) – with both performers and spectators mobile in a shared space. At scale, it is possible to create a scenic installation – of wrecked cars, trees, sand – as large as the building itself: effectively, to construct another architecture within that existing – employing substances and processes unusual in the playhouse but commonplace at such locations – that are integrated *per se* into the composition, the dramaturgy. It involves the coexistence of two architectonic systems and their associated narratives: that of the extant building, that which is at site – ornamentation, fixtures, history; and that of the built scenography, that which is brought to site. In this superimposition and interpenetration of the found and the emplaced, the latest occupation of a place where other occupations – their traces and residues – are still apparent. Building and production might have unrelated origins, might ignore each other's presence and even appear paradoxical, frictional or anachronistic: they are coterminous but not necessarily congruent, though both are always evident for spectators.

Site and production are inseparable as generators of dramatic meaning: in this one entity, this unitary place. The two architectures are jointly affective: for the performer, their attributes proffer consequential alternations of affordance (Gibson 1997) and the need for ergonomic adjustment. Whilst scenic materials generally facilitate choreographies, their positioning may equally provide obstacles and create barriers to movement, upsetting corporeal equilibrium, occasioning dramatic unbalancing.

Within the built scenography, it may be possible to conspire live climatic effects, the environment becoming active and the conditions – the ecology of surface, climate, illumination and temperature – by turn much better or much worse for performers than in everyday life and requiring improvised responses. But the extant, largely uncontrollable conditions within or of the site – its atmosphere, its temperature, its acoustics – unless somehow ameliorated, will be inescapable, for performers and spectators alike. In engagements and reengagements – body to scene, body to object, body to body – in the clash between energetic activities and both malleable and intractable matters – dramatic effects are forged and vibrant properties accentuated. Expression here is mediated by location, surface, area, volume, climate, light, by hardness, texture and by the levels of flexibility and resistance in substances.

However, there are markers to distinguish it as theatre 'as such'. Despite impediments, the trained performer employs rhetorics – aestheticised ways of going – in the articulation of actions: using more or less time and energy than would be applied in the everyday completion of a similar task, and applying extra-ordinary degrees of tension, exaggeration, repetition, distortion, reversal .... In these modifications, theatre as medium is evidenced.

Imagine the place of theatre here as field once more: as landscape, as a topography where distinct though dissimilar things coexist. A terrain charted by performers as itineraries, as a series of places to be, as 'an ensemble of tasks', as taskscape (Ingold 2000: 195): rehearsal involving close attention to acclimatisation and habituation, to developing ways of going on, to coping, to dwelling in this special world.

Ineluctable place

Yet the notion of theatre as an exclusive building – with attendant practices – persists. In the popular imagination, the faces of the spectators still glow in the reflected light – transported, drawn in in their rapture, as in a Walter Sickert painting; and the performers take their place amongst Foucault's exotic captives – madmen, patients, the condemned ....

In a place for viewing (*theatron*) and listening (auditorium): in a place devoted to manifesting place(s) ....

References

Artaud, A. (2013) *The Theatre and Its Double*, translated by Corti, V. Richmond: Alma Classics.

Augé, M. (2009) *Non-places: An Introduction to Supermodernity*. London: Verso.

Barba, E. (1995) *The Paper Canoe: A Guide to Theatre Anthropology*. translated by Fowler, R. London: Routledge.

Beacham, R. C. (1993) *Adolphe Appia: Texts on Theatre*. London: Routledge.

Bennett, J. (2010) *Vibrant Matter: A Political Ecology of Things*. Durham, NC: Duke University Press.

Birch, A. and Tompkins, J. (2012) *Performing Site-Specific Theatre*. Basingstoke: Palgrave.

Brandon, J., Malm, W. P. and Shively, D. H. (1978) *Studies in Kabuki: Its Acting, Music and Historical Context*. Hawaii: University of Hawaii Press.

Brook, P. (1972) *The Empty Space*. Harmondsworth: Penguin.

Deleuze, G. and Guattari, F. (1988) *A Thousand Plateaus*, translated by Massumi, B. London: Athlone Press.

Elam, K. (1980) *The Semiotics of Theatre and Drama*. London: Routledge.

Forced Entertainment. (2018) Available online at: www.forcedentertainment.com (accessed 15 October 2018).

Foucault, M. (1975) *Discipline and Punish. The Birth of the Prison*, translated by Sheridan, A. New York: Random House.

Foucault, M. (1986) Of other places, translated by Jay Miskowiec *Diacritics*, 16 (1), 22–27.

Gibson, J. J. (1997) *The Ecological Approach to Visual Perception*. Boston: Houghton Mifflin.

Grotowski, J. (1969) *Towards a Poor Theatre*. London: Methuen.

Gurawski, J. (2018) Available online at: www.grotowski.net/en/media/galleries/jerzygurawski-projects-1962-1965 (accessed 15 October 2018).

Heilpern, J. (1977) *Conference of the Birds: The Story of Peter Brook in Africa.* London: Faber and Faber.
Hodder, I. (2012) *Entangled: An Archaeology of the Relationships between Humans and Things.* Chichester: Wiley-Blackwell.
Ingold, T. (2000) *The Perception of the Environment.* London: Routledge.
Innes, C. (1998) *Edward Gordon Craig: A Vision of Theatre.* London: Routledge.
Komparu, K. (1984) *The Noh Theater: Principles and Perspectives.* New York: Weatherhill.
Lehmann, H-T. (2006) *Postdramatic Theatre,* translated by Jürs-Munby, K. London: Routledge.
Machon, J. (2013) *Immersive Theatres.* Basingstoke: Palgrave.
Marshfield Mummers. (2018) Available online at: www.marshfieldmummers.co.uk (accessed 15 October 2018).
Ottery St Mary. (The Tar Barrels of) (2018) Available online at: www.tarbarrels.co.uk (accessed 15 October 2018).
Pearson, M. (2010) *Site-Specific Performance.* Basingstoke: Palgrave Macmillan.
Pearson, M. (2018) Site-specific theatre. In Aronson, A. (ed.) *The Routledge Companion to Scenography.* London: Routledge, pp. 295–301.
Pearson, M. and Shanks, M. (2001) *Theatre/Archaeology.* Abingdon: Routledge.
Read, A. (2007) *Theatre, Intimacy and Engagement.* Basingstoke: Palgrave Macmillan.
Southern, R. (1962) *The Seven Ages of the Theatre.* London: Faber and Faber.
Toneelgroep Amsterdam. (2018) Available online at: https://tga.nl/en (accessed 15 October 2018).
Tschumi, B. (2012) *Architectural Concepts: Red Is Not a Color.* New York: Rizolli.
Wiles, D. (2003) *A Short History of Western Performance Space.* Cambridge: Cambridge University Press.
Wilson, R. (2018) Available online at: www.robertwilson.com (accessed 15 October 2018).

Index

Note: spellings have been standardised with an s.